Discrete Causal Theory

Benjamin F. Dribus

Discrete Causal Theory

Emergent Spacetime and the Causal Metric Hypothesis

Springer

Benjamin F. Dribus
Department of Mathematics
William Carey University
Hattiesburg
USA

ISBN 978-3-319-50081-2 ISBN 978-3-319-50083-6 (eBook)
DOI 10.1007/978-3-319-50083-6

Library of Congress Control Number: 2016958972

Printed on acid-free paper

This Springer imprint is published by Springer Nature
The registered company is Springer International Publishing AG
The registered company address is: Gewerbestrasse 11, 6330 Cham, Switzerland

Preface

Three outstanding problems of modern physics are the following:

1. Description of fundamental spacetime structure.
2. Quantum-theoretic description of gravitation.
3. Unification of physical law.

These problems are deeply intertwined. Since Einstein's discovery of general relativity a century ago, gravity has been understood to be *structural* in nature, rather than a "force," in the usual sense of the word, and it is therefore difficult to imagine the development of a successful approach to quantum gravity without simultaneous acquisition of a deeper structural understanding of spacetime. Quantum theory, meanwhile, though expressed and interpreted in a variety of different ways, is believed by most serious theorists to represent a fundamental aspect of nature; hence, any successful unification of physical law is expected to be quantum-theoretic. It seems, then, that two common ingredients necessary for the solution of these three problems are *suitable structural notions,* and *a suitable approach to quantum theory.* This book attempts to help identify and develop these two ingredients, with Part I devoted principally to structure, and Part II devoted principally to quantum theory. The most important aspects of the approach developed here may be summarized in a few short phrases: cause and effect relationships between pairs of events are taken to be the fundamental building blocks of structure; these events are assumed to form a discrete family rather than a continuum; and quantum theory is formulated as an abstraction and adaptation of Feynman's path summation approach. The resulting theory is called *discrete causal theory.*

At a formal level, discrete causal theory models spacetime via structured sets, called *directed sets,* whose elements represent events; these sets are equipped with special binary relations, whose elements represent causal influences between pairs of events. The term "directed set" is used in this book to mean essentially what the term "directed graph" means in graph theory; it does *not* have the more specific meaning involving upper bounds common in order theory and category theory. Ambitious versions of discrete causal theory attempt to treat "particles and fields" as aspects of the same directed structure, leading to a simple unified picture of nature at the fundamental scale. The theory may also be generalized to admit the possibility of causal relationships between structural components more complicated than individual events. Discrete causal theory has deep historical roots reaching back to the ancient Greeks, and was foreseen to some degree by Riemann, and even by Leibniz. However, Einstein's relativity, the digital information theory of Shannon and others, the path summation approach to quantum theory pioneered by Feynman, and an assortment of modern mathematical tools from algebra, order theory, and graph theory, are all more-or-less necessary to create a natural context for the theory.

The purest version of discrete causal theory presently enjoying an established theoretical niche is *causal set theory,* introduced in 1987 by Bombelli, Lee, Meyer, and Sorkin. Causal set theory is motivated by technical results of Hawking and Malament, called *metric recovery theorems,* which relate the causal, conformal, and metric structures of relativistic spacetime. Rafael Sorkin's well-known phrase, "*order plus number equals geometry,*" summarizes the view that causal structure provides enough information, in the discrete context, to describe emergent spacetime. This phrase represents an early version of what I refer to as the *causal metric hypothesis,* which is the more general idea that the properties of the physical universe, including the metric properties of classical spacetime, arise from causal structure at the fundamental scale. Other versions of discrete causal theory exist; perhaps the best known is *causal dynamical triangulations,* invented by Ambjørn, Jurkiewicz, and Loll. This approach takes for granted certain mysterious properties of nature, such as the local four-dimensional structure of spacetime, in an attempt to shortcut to predictive physics. This is a sensible and worthwhile undertaking, and should be considered part of the same broad program as the more general theory developed in this book. Meanwhile, *category-theoretic* and *topos-theoretic* approaches to physics, pioneered by Isham, Raptis, and others, also incorporate some of the structural ingredients of discrete causal theory, although the interpretations assigned to these structures are generally quite different from the interpretations adopted in this book. In addition, there exist related approaches to certain topics in fundamental physics that are "causal but not discrete," such as approaches based on the mathematical apparatus of *domain theory.*

A major attraction of discrete causal theory is that it admits clean axiomatic presentations. For example, causal set theory may be expressed in terms of six axioms: the *binary axiom*, the *measure axiom*, *countability*, *transitivity*, *interval finiteness*, and *irreflexivity*. The first three of these axioms specify the physical interpretation of a causal set, and restrict its cardinality, while the last three encode its essential mathematical structure. In this book, I develop a new, more general version of discrete causal theory, based on a different choice of axioms, and amplified by the incorporation of new conceptual perspectives and technical methods. These innovations are motivated by the realization that the choice to abstain from continuum-based geometry raises the possibility of qualitatively new types of behavior, such as *irreducibility* and *independence* of influences between pairs of events, which are not completely captured by the conventional order-theoretic machinery of causal set theory. In particular, the transitive binary relations of causal sets are not necessarily optimal for resolving the subtleties of *independent modes of influence,* while the causal set axiom of interval finiteness, sometimes mislabeled as "local finiteness," permits potentially problematic locally infinite behavior. Interval finiteness also imposes unjustified restrictions on the global structure of classical spacetime.

Under the alternative set of axioms proposed in this book, transitivity is no longer taken for granted, and interval finiteness is replaced with a genuinely local finiteness condition called *star finiteness.* The latter condition encodes the physical idea of *causal local finiteness,* which states that each event possesses only a finite number of direct causes and effects. I refer to a binary relation satisfying these new axioms as a star finite *causal relation*; its transitive closure is called the *transitive relation*, and is analogous to the familiar causal order. The resulting models, which I call *star finite directed sets,* generalize both causal sets and Finkelstein's *causal nets*, with the exception of a class of locally infinite objects arising in certain idealized constructions in causal set theory. The latter objects have been recognized as physically unrealistic ever since the theory was founded, and were never intended to serve as actual workhorses for quantum gravity, but they remain a useful source of examples for testing new methods and inferring more general behavior. I also adapt, from mathematical sources, a broader interpretation of directed structure, and more generally, multidirected structure, inspired by modern category theory, and by Grothendieck's scheme-theoretic approach to algebraic geometry. This expansive viewpoint provides an assortment of new methods and insights.

The resulting theory differs significantly from existing versions of discrete causal theory, particularly at the quantum level. A variety of different types of entities more complex than individual spacetime events may be viewed in a unified fashion as elements of "higher-level multidirected sets," in analogy with Isham's topos-theoretic approach to quantum gravity, and Sorkin's *quantum measure theory.* This viewpoint leads to a new background independent quantum theory of spacetime, in which "elements" are classical histories, modeled via directed sets, and "relations" are natural relationships between pairs of classical histories, called *co-relative histories.* For information-theoretic reasons, these natural relationships are subtler than morphisms in the category of directed sets. The resulting multidirected configuration spaces of directed sets, called *kinematic schemes,* provide the basic structural scaffolding for quantum theory, via path summation. Special cases of kinematic schemes appear under different terminology in causal set theory; for example, in Sorkin and Rideout's theory of *sequential growth dynamics.* A striking property of kinematic schemes is that their higher-level structure is of essentially the same type as the structure of their constituent histories, a phenomenon I refer to as *iteration of structure.* Besides providing an attractive conceptual unification of classical and quantum theory as different levels of hierarchy under the same structural paradigm, this property enables parallel use of the same technical methods in both semiclassical approximations and in the full background independent quantum theory.

These perspectives and methods not only flesh out the new version of discrete causal theory developed in this book, but also enable important improvements in existing versions of the theory, even if one ignores most of the axiomatic and conceptual innovations offered here. For example, systematic use of *relation space* circumvents the generic discrete problem of *permeability of maximal antichains,* thereby removing an important obstruction to foliation-type approaches to dynamics. This leads to the derivation of *causal Schrödinger-type equations,* which describe quantum spacetime dynamics in a very broad context. The theory of *generational dynamics* for acyclic directed sets, which generalizes sequential growth dynamics for causal sets, leads to the definition of natural *entropic phase maps,* which supply specific content to these equations. Possible near-term phenomenological applications of the new theory involve existing searches for deviations from Lorentz invariance and spatiotemporal locality, refinement of causal set-theoretic explanations for the size of the "cosmological constant," and non-geometric alternatives to the *inflationary hypothesis* in the cosmology of the early universe.

How to read this book. Faced with a 500-page volume, it is natural to wonder if one can absorb the essentials without reading the entire work. Below, I list several abbreviated reading strategies for various classes of readers. The approximate numbers of pages involved are given below each list. Experienced readers know that it is ill-advised to become bogged down during a first reading, and it is often useful to acquire a broad conceptual picture at the outset by reading through multiple chapters without too much concern about the details. One may always delve deeper on a second reading. Chapters 1 and 2 are mostly motivational, since discrete causal theory remains relatively unfamiliar, both to the scientific mainstream and to the general public. These chapters may be skipped entirely, but they are easy to read, and they make the rest of the book easier to understand. Serious readers will likely spend the most time in Chapters 5–7. It would be unfortunate to put serious effort into understanding Part I without seeing the quantum theory developed in Chapter 6!

1. *Physicists or mathematicians familiar with discrete causal topics such as causal sets or causal dynamical triangulations:* read Sections 3.9, 4.4, 4.5, 4.7, 4.8, 4.10, 5.1, 5.4, 5.7, 5.9, 5.10, 6.2, 6.7, 6.9, 6.10, 7.4, 7.5, 7.8, 8.2–8.4. Total pages: about 155.
2. *Physicists or mathematicians unfamiliar with discrete causal theory:* add to the above list Sections 2.8, 3.1–3.5. Total pages: about 180.
3. *Non-technical readers interested in modern science:* read Chapter 1, then read Sections 2.1, 2.2, 2.10, 3.1, 3.2, 3.9, 4.1, 4.8, 4.10, 5.8, 6.1, 6.5, 7.1–7.3, 8.1, 8.8, and 8.10. Total pages: about 150.

All readers should read the single-page abstracts that open each chapter. Students who have not yet seriously studied general relativity should read the first eight sections of Chapter 2, while keeping handy a good relativity text, such as Wald. Sections 3.6 and 3.7 contain important definitions used throughout the book, but these may be skimmed as needed. The index will be a useful resource for many readers.

Hattiesburg, USA
October 2016

Benjamin F. Dribus

Acknowledgements

Physicists. Rafael Sorkin provided insightful answers to questions about foundational topics in causal set theory, and allowed me to quote personal correspondence. Jorge Pullin provided background on general relativity, and offered valuable opinions on various approaches to quantum gravity. David Finkelstein answered questions about causal nets, and sent me some of his unpublished work. George Ellis made helpful remarks regarding top-down causation. Giacomo Mauro D'Ariano provided useful information from the related viewpoint of quantum automaton theory. Torsten Asselmeyer-Maluga offered general encouragement, and acquainted me with some interesting mathematical topics. David Rideout answered multiple questions about causal sets. An anonymous reviewer pointed out interesting early references on the same topic. Angela Lahee made helpful suggestions about a variety of issues. Eric Abraham, my first physics teacher, was a significant influence on the physical intuition behind the book.

Mathematicians, computer scientists, and others. Jimmie Lawson offered useful advice, and contributed to my understanding of related topics in quantum information theory. James Madden supplied important algebraic and order-theoretic notions. Brendan McKay provided helpful advice on graph-theoretic issues. Johnny Feng offered valuable feedback and pointed out errors in draft versions. Marcel Erné was kind enough to scan and send me copies of his unpublished notes on order theory. Manfred Droste wrote me a very nice response to a question about universal causal sets. Joel David Hamkins and Chip Delzell provided some helpful answers on similar topics. Juan Marin acquainted me with some interesting historical details. Ben Goertzel offered intriguing ideas about related topics in information theory and computer science. Al Vitter, my first algebraic geometry teacher, acquainted me with the philosophy of Grothendieck. J.W. Hoffman, my dissertation advisor, supplied me with hundreds of articles worth of indispensable algebraic background. Thomas Naugle and Stephanie Dribus endured much tedious conversation about the subject, and made helpful conceptual suggestions. Finally, Marian Dribus, Virginia Dribus, Lauren Starnes, Joshua Deaton, Jessica Garriga, Madeline LeBoeuf, Katie Sippel, June Songy, and Jalynn Roberts provided valuable editing assistance.

Contents

Part I Foundations of Discrete Causal Theory

1 Introduction . . . 3
1.1 A Picture and an Equation . . . 4
1.2 The Causal Metric Hypothesis . . . 8
1.3 Emergent Spacetime . . . 12
1.4 Practice with Toy Histories . . . 18
1.5 The Relative Viewpoint . . . 22
1.6 Background Independence . . . 28
1.7 Particles and Fields . . . 30
1.8 Kinematics and Dynamics . . . 36
1.9 Phenomenology . . . 42
1.10 Looking Ahead: Twenty Questions . . . 52
References . . . 61

2 The Causal Metric Hypothesis . . . 65
2.1 General Version of the Hypothesis . . . 66
2.2 Classical Version of the Hypothesis . . . 69
2.3 Structure on Relativistic Spacetime . . . 72
2.4 Metric Structure . . . 78
2.5 Conformal Structure . . . 84
2.6 Causal Structure . . . 91
2.7 Causality Conditions . . . 97
2.8 Metric Recovery . . . 106
2.9 Order Good, Continuum Bad . . . 112
2.10 The Philosopher's Peril . . . 126
References . . . 134

3 Causal Sets and Generalizations . . . 137
3.1 Causal Set Theory: Conceptual Roots . . . 138
3.2 Causal Sets and the Causal Metric Hypothesis . . . 141

3.3 Quantum Causal Set Theory 144
3.4 Causal Set Dynamics and Phenomenology 147
3.5 Axioms of Causal Set Theory 150
3.6 Generalized Causal Sets: Directed Sets and Multidirected Sets 156
3.7 Building Blocks of Structure in Multidirected Sets 161
3.8 The Shadow of Grothendieck 172
3.9 Independent Modes of Influence 179
3.10 The Causal Relation 190
References 193

4 Local Properties 197
4.1 Local Behavior 198
4.2 Topology 203
4.3 Four Topologies for Multidirected Sets 206
4.4 Interval Finiteness and Star Finiteness 212
4.5 Sprinkled Causal Sets; Lorentz Invariance; Spatiotemporal Locality 223
4.6 Relative Multidirected Sets Over a Fixed Base 243
4.7 Seven Arguments Questioning Interval Finiteness 252
4.8 Nine Arguments in Favor of Star Finiteness 256
4.9 Hierarchy of Finiteness Conditions 258
4.10 New Axioms for Discrete Classical Causal Theory 263
References 268

Part II Discrete Quantum Causal Theory

5 Relation Space and Generalizations 273
5.1 Relation Space Over a Multidirected Set 274
5.2 Abstract Element Space 280
5.3 Preservation of Information 285
5.4 Permeability 288
5.5 Analogy Involving Morphism Categories 295
5.6 Power Spaces 298
5.7 Causal Atomic Resolution 305
5.8 Top–Down Causation; Classical Holism 313
5.9 Causal Path Spaces 318
5.10 Causal Path Algebras 326
References 336

6 Quantum Spacetime 339
6.1 Quantum Preliminaries 340
6.2 Path Summation 344
6.3 Iteration of Structure 350
6.4 Co-Relative Histories 359

6.5 Quantum Causal Generalities 365
6.6 Feynman Revisited 369
6.7 Quantum Theory in Relation Space 373
6.8 Schrödinger's Equation via Path Summation 381
6.9 Causal Schrödinger-Type Equations 386
6.10 Generalized Quantum Amplitudes via Causal Path Functionals 391
References 394

7 Kinematics and Dynamics 397
7.1 Frames of Reference 398
7.2 Refinements 401
7.3 Kinematic Accounts 407
7.4 Kinematic Schemes 415
7.5 Positive Sequential Kinematic Scheme 427
7.6 Quantum Causal Metric Hypothesis 433
7.7 Path Summation Over a Kinematic Scheme 436
7.8 Generational Kinematics and Dynamics 442
7.9 Completions of Kinematic Schemes 446
7.10 Hyperquantum Theory 448
References 453

8 Toward Phenomenology 455
8.1 Axioms and Technical Methods Revisited 456
8.2 Phase Theory 463
8.3 Covariance 475
8.4 Emergence of the Poincaré Group 479
8.5 Elementary Graph Dynamical Considerations 487
8.6 Algebraic Structure and Hierarchy 489
8.7 Miscellany Regarding Alternative Formulations 494
8.8 Connections with Other Physical Theories 497
8.9 Connections with Physics-Related Mathematics 507
8.10 Final Remarks 515
References 517

Appendix A: Conventions 519

Index 525

Notation

Preliminary remarks. The following list of notation is organized on a chapter-by-chapter basis. Some duplication of symbols is unavoidable; for example, the letter R denotes a relation space in one context, and the scalar curvature of a pseudo-Riemannian manifold in another. Similarly, lower-case letters such as x and y may stand for elements of relativistic spacetime, events in a directed set, or a variety of other "element-like" entities. To make the list easier to use, some definitions are repeated under the headings for multiple chapters. Further remarks on notation may be found in the Appendix.

Basic Notation

$\mathbb{N}-\{0\}$	Positive integers
$\mathbb{N}$	Natural numbers
$\mathbb{Z}$	Integers
$\mathbb{Q}$	Rational numbers
$\mathbb{R}$	Real numbers
$\mathbb{C}$	Complex numbers
$\mathbb{H}$	Quaternions
$\mathbb{O}$	Octonions
$\mathbb{Z}_n$	Integers modulo n
$\mathbb{Q}_p$	p-adic numbers
$\mathbb{R}^n$	n-dimensional Euclidean space
$[n]$	Finite simplex $\{0,\ldots,n\}$
S_n	nth symmetric group
$S\times S'$	Cartesian product of S and S'
S^n	nth Cartesian power of S
$S^{(n)}$	nth symmetric power of S
$\emptyset$	Empty set
$\binom{n}{k}$	Binomial coefficient

$x \in X$	Means x is an element of X
$U \subset X$	Means U is a subset of X, not necessarily proper
$U \cup V$	Union of U and V
$U \bigsqcup V$	Disjoint union of U and V
$U \cap V$	Intersection of U and V
$f : X \to X'$	Morphism from X to X'; usually a function
$f(U)$	Image in X' of subset $U \subset X$ under $f : X \to X'$
$f^{-1}(U')$	Preimage in X of subset $U' \subset X'$ under $f : X \to X'$
$f \circ g$	Composition of morphisms; $f \circ g(x)$ means $f(g(x))$ for functions

Chapter 1

$\mathbb{S}_{\text{PS}}$	Positive sequential kinematic scheme
$\psi_{R;\theta}^{-}$	Generalized state function
θ	Relation function
R	Space representing co-relative histories
r, r^-	Elements of R
$\prec$	Precursor symbol
i	Imaginary unit $\sqrt{-1}$
$\hbar$	Planck's reduced constant $h/2\pi$
ψ	Ordinary state function
$\mathbf{H}$	Hamiltonian operator
D, D_i, D_t	Directed sets
x, y, x', y'	Elements of a directed set
$R_{\mu\nu}$	Ricci curvature tensor
R	Scalar curvature
$g_{\mu\nu}$	Metric tensor
Λ	"Cosmological constant"
G	Gravitational constant
c	Speed of light
$T_{\mu\nu}$	Stress-energy tensor
ε_{IJKL}	Levi-Civita symbol
e^I	Tetrad 1-form
R^{JK}	Curvature 2-form
T_L	Energy-momentum 3-form
x, y, x', y'	Spacetime events
$\mathbb{R}^4$	Four-dimensional Euclidean space
$\mathcal{R}(D)$	Relation space over D
r_0, r_1, r_2	Elements of $\mathcal{R}(D)$
ϕ	Element field
ϕ_1, ϕ_2, ϕ_3	Values of ϕ
θ	Relation field (same as relation function)

$\theta_1, \theta_2, \theta_3$	Values of θ
T	Target object
ν^+, ν^-	Valence fields
P, p	Probabilities
$D_i \Rightarrow D_t$	D_i evolving to D_t
$\varnothing$	Empty history

Chapter 2

CMH	Causal metric hypothesis
$x \prec y$	Means x directly precedes y
$D \times D$	Cartesian product
$(D, \prec)$	Directed set as an ordered pair
CCMH	Classical causal metric hypothesis
X, X'	Pseudo-Riemannian manifolds
$f : X \to X'$	Morphism of manifolds
$X \cong X'$	Means X is isomorphic to X'
g, g'	Pseudo-Riemannian metrics
T_xX	Tangent space to X at x
v, w	Tangent vectors
g_x	Bilinear map induced by g
$g_{\mu\nu}$	Traditional notation for metric
δ	Euclidean metric on $\mathbb{R}^4$
η	Minkowski metric on $\mathbb{R}^4$
(r, θ, ϕ)	Spherical coordinates
G	Gravitational constant
M	Mass in Schwarzchild metric
$a(t)$	Scale factor in FLRW metric
Λ	"Cosmological constant"
f^*g'	Pullback of g' along f
(p, q)	Metric signature
$(4, 0)$	Euclidean signature
$(3, 1)$	Lorentz signature
$\mathbb{R}^{p+q}$	Pseudo-Euclidean space
$\mathbb{R}^{3+1}$	Minkowski spacetime
γ_1, γ_2	Smooth curves
Ω	Conformal factor
U_x	Open neighborhood containing x
$\prec_{\mathrm{GR}}$	Relativistic causal relation
$J^-(x)$	Causal past of x
$J^+(x)$	Causal future of x
$\prec\prec_{\mathrm{GR}}$	Chronological relation
$I^-(x)$	Chronological past of x

$I^+(x)$	Chronological future of x
$\rightarrow$	Horismos relation
σ	Cauchy surface
ρ, ρ'	Acausal subsets
$\langle\langle x, y\rangle\rangle$	Open interval between x and y
$\mathbb{F}$	(Algebraic) field
$+$	Additive operation on $\mathbb{F}$
$\times$	Multiplicative operation on $\mathbb{F}$
$\preccurlyeq$	(Nonstrict) total order on $\mathbb{F}$
$\prec$	Strict total order on $\mathbb{F}$
$\leq$	Usual total order on $\mathbb{R}$
$<$	Usual strict total order on $\mathbb{R}$
$\mathbb{R}(x)$	Field of rational functions
$d : S \times S \rightarrow \mathbb{R}$	"Metric" on a set S
$(\mathbb{R}, +)$	$\mathbb{R}$ as a totally ordered additive group
L	Linear continuum
S	Subset of L
u	Upper bound
u_{MIN}	Least upper bound
ϕ	Golden ratio
$\mathbb{Q}_p$	Field of p-adic numbers
$\mathbb{H}$	Quaternions
$\mathbb{O}$	Octonions
ψ	State function in Schrödinger's equation

Chapter 3

C, C'	Causal sets
$\prec_{\mathrm{CS}}$	Causal set relation
μ_{CS}	Discrete measure on a causal set
x, y, x', y'	Elements of a causal set
ϕ	Element field on C
ϕ_1, ϕ_2, ϕ_3	Values of ϕ
ψ	Relation field on C
ψ_1, ψ_2, ψ_3	Values of ψ
$\mathbb{S}_{\mathrm{SR}}$	Sorkin–Rideout kinematic scheme
τ	Transition of causal sets
C_i	Initial causal set
C_t	Terminal causal set
$\theta(\tau)$	Phase of transition τ
$\{x_0, x_1, x_2, \ldots\}$	Total labeling of a causal set
M	Measure axiom
TR	Axiom of transitivity
IF	Axiom of interval finiteness

IR	Axiom of irreflexivity
$\langle\langle x, y\rangle\rangle$	Open interval between x and y
$\mathcal{P}(C)$	Power set of C
$\mathbb{R}^+$	Positive real numbers
$(C, \prec_{CS})$	Causal set as an ordered pair
S	Subset of C
$\mathcal{C}$	Category of causal sets
$f : C \to C'$	Morphism of causal sets
$f^{-1}(x')$	Fiber of f over x'
$\preceq_{CS}$	Causal set order
$\prec_{GR}$	Relativistic causal relation
$\prec\prec_{GR}$	Chronological relation
$\rightarrow$	Horismos relation
$\prec$	Causal relation
$\prec_{tr}$	Transitive relation
$\prec_{sk}$	Skeletal relation
D, D'	Directed sets
$(D, \prec)$	Directed set as an ordered pair
S	Subset of D
$\mathcal{D}$	Category of directed sets
$f : D \to D'$	Morphism of directed sets
AC	Acyclicity (definition)
$\mathcal{A}$	Category of acyclic directed sets
M, M'	Multidirected sets
R	Relation set of M
$i : R \to M$	Initial element map
$t : R \to M$	Terminal element map
$x \prec y$	Means there exists $r \in R$ such that $i(r) = x$ and $t(r) = y$
(M, R, i, t)	Multidirected set as an ordered quadruple
$\mathcal{M}$	Category of multidirected sets
$f = (f_{ELT}, f_{REL})$	Morphism of multidirected sets
f_{ELT}	Map of elements
f_{REL}	Map of relations
$\gamma = \ldots \prec x_0 \prec x_1 \prec \ldots$	Chain in M
$\mathrm{Ch}(M)$	Chain set of M
$\mathrm{Ch}_n(M)$	Set of n-chains in M
σ	Antichain in M
$\mathcal{R}(M)$	Relation space over M
Γ, Γ'	Families of chains in M
V	Vector space
S	Linearly independent subset of V
IC	Independence convention
U	Subset of a multidirected set M

$J(x)$	Total domain of influence of x
$J(U)$	Total domain of influence of U
$J^-(x)$	Past of x
$J^-(U)$	Past of U
$J^+(x)$	Future of x
$J^+(U)$	Future of U
$J_0^-(x)$	Direct past of x
$J_0^+(x)$	Direct future of x
$J_{\text{MAX}}^-(x)$	Maximal past of x
$J_{\text{MIN}}^+(x)$	Minimal future of x
∂M	Boundary of M
$\text{Int}(M)$	Interior of M
$\mathcal{A}_{\text{FIN}}$	Class of isomorphism classes of finite acyclic directed sets
$\aleph_0$	Countably infinite cardinal
$S_{\aleph_0}$	Fixed countably infinite set
$2^{\aleph_0}$	Cardinality of $\mathcal{P}(S_{\aleph_0})$
$\mathcal{A}_{\aleph_0}$	Class of isomorphism classes of countable acyclic directed sets
$\mathcal{C}_{\text{SF}}$	Class of isomorphism classes of star finite causal sets
$\mathcal{A}_{\aleph_0,\text{SF}}$	Class of isomorphism classes of countable star finite acyclic directed sets
ORP	Order refinement principle
$\mathcal{C}$	A category; not necessarily causal sets
$\text{Obj}(\mathcal{C})$	Objects of $\mathcal{C}$
$\text{Mor}(\mathcal{C})$	Morphisms of $\mathcal{C}$
$\mathbf{i}$	Initial object map of $\mathcal{C}$
$\mathbf{t}$	Terminal object map of $\mathcal{C}$
γ_X	Identity morphism of object X
RV	Grothendieck's relative viewpoint
HS	Principle of hidden structure
C	Finkelstein's "transitive causal relation"
c	Finkelstein's nontransitive, local, "causal connection relation"
$(D, \prec_{\text{tr}})$	Transitive closure of $(D, \prec)$
$(D, \prec_{\text{sk}})$	Skeleton of $(D, \prec)$
tr	Transitive closure functor
sk	Skeleton operation
$\mathcal{D}_{\text{TR}}$	Category of transitive directed sets

Chapter 4

S	Arbitrary set
X	Real manifold; usually pseudo-Riemannian
g	Pseudo-Riemannian metric on X
$\mathcal{T}, \mathcal{T}'$	Topologies on S
$(S, \mathcal{T})$	Topological space as an ordered pair
$f : S \to S'$	Continuous map
$\overline{S}$	Enlarged set containing S
$\overline{\mathcal{T}}$	Topology on $\overline{S}$
$\mathcal{T}_{\text{DIS}}$	Discrete topology
$\mathcal{T}_{\text{INT}}$	Interval topology
$\langle\langle w, y\rangle\rangle$	Open interval between w and y
$R(x)$	Relation set at x
$M_{\mathbb{R}}$	Continuum model of M
$\mathcal{T}_{\mathbb{R}}$	Continuum topology
$\text{St}_{\rho}(x)$	Open star of radius ρ at x
d	Euclidean metric on a unit interval
U_x	Neighborhood of x in continuum topology
$M_{\star}$	Star model of M
$\text{St}(x)$	Star at x in $M_{\star}$
$\mathcal{T}_{\star}$	Star topology on $M_{\star}$
$v(x)$	Valence of x
v_M	Valence field on M
v_M^-	Past valence field on M
v_M^+	Future valence field on M
SF	Star finiteness (definition)
SFT	Star finiteness in the transitive closure
SFS	Star finiteness in the skeleton
$\mathbb{R}^{3+1}$	Minkowski spacetime
η	Minkowski metric
$\prec_{\text{GR}}$	Relativistic causal relation
$(\rho V)^n e^{-\rho V}/n!$	Probability in Poisson process
$C = (C, \prec_{\text{CS}})$	Sprinkled causal set
A, B	Frames of reference in $\mathbb{R}^{3+1}$
c	Speed of light
β	Boost of frame B with respect to frame A
(t_A, x_A)	Coordinates for frame A
$p, q, s, u, v, w, y, y', z$	Elements of $\mathbb{R}^{3+1}$
$\Omega = (\Omega, \mu, \Sigma)$	Space of sprinklings
H	Hyperboloid of unit future-directed timelike vectors
$\mathbf{u}$	Vector in H
$\Delta : \Omega \to H$	Direction map

L	Lorentz group
L''	Proper orthochronous Lorentz group
Λ	Lorentz transformation
σ	Local constant-time section of $\mathbb{R}^{3+1}$
D, D'	Directed sets
t_0, t_1	Cutoff times
T	Cutoff size
r	Spatial radius
X, Y	Fixed coordinates on $\mathbb{R}^{3+1}$
α	Causal atom in Theorem 4.6.3
$\langle\langle \alpha, y_{k+1} \rangle\rangle$	Generalized open interval in Theorem 4.6.3
P	Poincaré group
$\mathbb{N} \coprod -\mathbb{N}$	Linearly ordered base for Jacob's ladder
EF	Element finiteness
LEF	Local element finiteness
RF	Relation finiteness
PRF	Pairwise relation finiteness
CF	Chain finiteness
AF	Antichain finiteness
B	Binary axiom
M*	Generalized measure axiom
C	Axiom of countability
SF	Axiom of star finiteness
AC	Axiom of acyclicity
IRR	Axiom of irreducibility

Chapter 5

$\mathcal{R}(M)$	Relation space over $M = (M, R, i, t)$
$\prec$	Induced relation on $\mathcal{R}(M)$
r_0, r_1, r_2	Elements of relation space
$\mathcal{R} : \mathcal{M} \to \mathcal{D}$	Relation space functor
$f = (f_{\mathrm{ELT}}, f_{\mathrm{REL}})$	Morphism of multidirected sets
$\mathcal{R}(f)$	Induced morphism of relation spaces
$R = (R, \prec)$	Arbitrary directed set, viewed as a relation set
$\mathcal{E}(R)$	Abstract element space over R
$M^{\pm}$	Abstract sets yielding $\mathcal{E}(R)$ as a quotient
$\widetilde{x_r}, \widetilde{y_r}$	Abstract elements in $\mathcal{E}(R)$ corresponding to $r \in R$
$\mathcal{E} : \mathcal{D} \to \mathcal{M}$	Abstract element space functor
σ	Cauchy surface in spacetime, or maximal antichain in directed or multidirected set
$\mathcal{M}(\mathbb{S})$	Underlying multidirected set of kinematic scheme $\mathbb{S}$
W, X, Y, Z	Objects of a category $\mathcal{C}$

a, b, f, g	Morphisms in $\mathcal{C}$
$(M_\star, R_\star, i_\star, t_\star)$	Star model of M as a multidirected set
M_2	Set of 2-element subsets of M
$\mathcal{P}(M)$	Power set of M
$\mathcal{R}^2(D)$	Second relation space over D
$\prec_n$	nth induced relation
$\overline{\gamma}, \overline{\delta}$	Simplices in M
α, α'	Causal atoms
M^1	Causal atomic decomposition of $M = M^0$
$\mathbf{M} = \{M^0, M^1, M^2, \ldots\}$	Causal atomic resolution of $M = M^0$
M^n	nth level of causal atomic resolution
λ_n	nth growth factor for $\mathbf{M}$
$\Lambda(\mathbf{M})$	Growth sequence for $\mathbf{M}$
T	Twistor space over Minkowski spacetime $\mathbb{R}^{3+1}$
$\mathbb{CP}^3$	Complex projective space
L, L', L''	Linear directed sets
$\alpha, \beta, \gamma, \delta$	Paths in a directed or multidirected set
$\Gamma(M)$	Causal path set of M
$\Gamma_{\mathrm{IRR}}(M)$	Irreducible causal path set of M
ℓ^+, ℓ'^-	Maximal element of L and minimal element of L' in Definitions 5.9.5 and 5.9.7
$L \sqcup L'$	Linear directed set given by joining L, L' via connecting relation $\ell^+ \prec \ell'^-$
$\gamma \sqcup \delta$	Concatenation product of γ and δ
$\prec_\sqcup$	Concatenation relation on $\Gamma(D)$
$(\Gamma(D), \prec_\sqcup)$	Causal concatenation space over $D = (D, \prec)$
$L \vee L'$	Linear directed set given by identifying ℓ^+ with ℓ'^-
$\gamma \vee \delta$	Directed product of γ and δ
$\prec_\vee$	Directed product relation on $\Gamma(M)$
$(\Gamma(M), \prec_\vee)$	Causal directed product space over $M = (M, R, i, t)$
$\mathcal{S}$	Object class of a semicategory
$\Gamma(\mathcal{S})$	Morphism class of a semicategory
$\mathbf{i}, \mathbf{t}$	Initial and terminal object maps of a semicategory
$(\Gamma(D), \sqcup)$	Concatenation semicategory over D
$(\Gamma(M), \vee)$	Directed product semicategory over M
$(\mathrm{Ch}(D), \sqcup)$	Chain concatenation semicategory over D
$(\mathrm{Ch}(M), \vee)$	Chain directed product semicategory over M
$\sim$	Congruence relation on a semicategory
$\mathbb{R}[x]$	Algebra of polynomials with real coefficients
T	A ring, usually commutative, with unit
t_γ	Element of T, appearing as coefficient of path γ in path algebra
$T^\sqcup(D)$	Concatenation algebra over D with coefficients in T

$T^{\vee}(M)$	Directed product algebra over M with coefficients in T
$T^{\sqcup}[D]$	Chain concatenation algebra over D with coefficients in T
$T^{\vee}[M]$	Chain directed product algebra over M with coefficients in T
$P(x)$	Punctual module of x
$\|x\rangle\langle y\|$	Ket-bra notation for Raptis' incidence algebra
δ_{uv}	Kronecker delta function

Chapter 6

$\mathcal{Q}$	Category in Isham's quantization
Q, Q'	Objects of $\mathcal{Q}$
$\mathcal{L}$	Lagrangian for single particle
γ	Particle trajectory in continuum spacetime
$\mathcal{S}$	Classical action for $\mathcal{L}$ over γ
Θ	Phase map
$\hbar$	Planck's reduced constant $h/2\pi$
θ	Relation function
T	Target object of relation function or phase map
$\sum_{\gamma\in\Gamma(M)}\Theta(\gamma)\gamma$	Phase map Θ as a path algebra element
$T^{\Gamma(D)}, T^{\Gamma(M)}$	Mapping spaces of maps $\Gamma(D)\rightarrow T$ and $\Gamma(M)\rightarrow T$
Θ, Ψ	Pair of elements of $T^{\Gamma(D)}$ or $T^{\Gamma(M)}$
$\Phi\vee\Psi, \Phi\sqcup\Psi$	Convolutions; i.e., products in path algebras $T^{\sqcup}(D)$ and $T^{\sqcup}(M)$
$f*g$	Convolution of real-valued functions
$\delta-\gamma$	Path such that $\gamma\sqcup(\delta-\gamma)=\delta$
$T^{\mathrm{Ch}(D)}, T^{\mathrm{Ch}(M)}$	Chain mapping spaces
0_T	Additive identity of T
e_M	Global evaluation map $T^{\Gamma(M)}\rightarrow T$
e_γ	Indicator function for γ
$\Theta(M)$	Shorthand for causal path sum $\sum_{\gamma\in\Gamma(M)}\Theta(\gamma)$
$\mathbb{P}^1(\mathbb{R})$	Real projective line
IS	Principle of iteration of structure
τ	Transition in $\mathcal{D}$
D_i	Initial object, or source, of a transition
D_t	Terminal object, or target, of a transition
$\mathrm{Aut}(D_i)$, $\mathrm{Aut}(D_t)$	Automorphism groups
α, β	Automorphisms of directed sets
F_τ	Group of extensions of symmetries of D_i
ρ_τ	Restriction homomorphism $F_\tau\rightarrow\mathrm{Aut}(D_i)$

K_τ, G_τ	Kernel and cokernel of ρ_τ; also called causal Galois groups
ι	Inclusion $K_\tau \to F_\tau$
$\overline{\rho_\tau}$	Quotient homomorphism $G_\tau \to \mathrm{Aut}(D_i)$
$[F_\tau : G_\tau]$	Subgroup index of G_τ in F_τ
$h : D_i \Rightarrow D_t$	Co-relative history
D_i, D_t	Cobase and target of h
$\mathbf{D}_i$	Cobase family
PSP	Path summation principle
X	Finite subset of Euclidean spacetime $\mathbb{R}^4$
$t = t^\pm$	Spatial sections bounding X
$\partial^\pm$	Upper and lower boundaries of X
$\Gamma = \Gamma_{\mathrm{MAX}}(X)$	Space of maximal directed paths in X
γ	Element of $\Gamma_{\mathrm{MAX}}(X)$
F1, F2	Feynman's "postulates"
$\psi(X; \mathcal{L})$	Feynman's quantum amplitude
$\hbar$	Planck's reduced constant $h/2\pi$
$\Theta(\gamma) = e^{\frac{i}{\hbar}S(\gamma)}$	Feynman's phase map
Δ	Partition of $[t^-, t^+]$
$\mathbf{x}$	Sequence of position values for Δ
γ_i	"Path increment" from (x_i, t_i) to (x_{i+1}, t_{i+1})
$\sigma_i = \sigma_i(\Delta)$	Spatial section $t = t_i$ for $t_i \in \Delta$
$\|\Delta\|$	Norm of partition Δ
$\psi(\Delta; \mathbf{x}; \mathcal{L})$	Amplitude for particular sequence $\mathbf{x}$ of position values
C	Proportionality factor in Feynman's path integral
$M' = (M', R, i', t')$	"Ambient" multidirected set
R	Finite subset of relation space $\mathcal{R}(M')$
$(\mathrm{Ch}(R), \sqcup)$	Chain concatenation semicategory over R
$\Gamma = \mathrm{Ch}_{\mathrm{MAX}}(R)$	Subset of maximal chains in R
γ	Element of $\mathrm{Ch}_{\mathrm{MAX}}(R)$
σ	Maximal antichain in R
$R^- \coprod \sigma \coprod R^+$	Separation of R into past and future regions $R^\pm$ by σ
T	Target object for path sums
Θ	Phase map $\mathrm{Ch}(R) \to T$
θ	Relation function generating Θ
$\Psi(R; \theta)$	Maximal chain functional of R with respect to θ
$\psi(R; \theta)$	Generalized quantum amplitude
ψ^-	Past state function in Schrödinger's equation
$\mathbf{H}$	Hamiltonian operator
V	Potential energy
m	Particle mass
$\langle , \rangle$	Complex inner product
σ'	Spatial section $t = t'$

$X^- \coprod \sigma' \coprod X^+$	Separation of X into past and future regions $X^\pm$ by σ'
Δ^-	Partition of time interval for X^-
$\approx$	Signifies approximation
t''	A time shortly after t'
$\delta\gamma$	Path increment
r	Element of σ
$\Gamma^\pm$	Sets of maximal chains in $R^\pm$, respectively
γ	Element of Γ^-
$\Theta\|_\sigma$	Restriction of phase map Θ to σ
Γ_σ^-	Set of maximal chains in R terminating at σ
Γ_σ^+	Set of maximal chains in R beginning at σ
Γ_r^-	Set of maximal chains in R terminating at r
Γ_r^+	Set of maximal chains in R beginning at r
$\Psi_{R,\theta}^\pm$	Past and future chain functionals
$\psi_{R,\theta}^\pm$	Past and future state functions
r^-	Maximal predecessor of r

Chapter 7

A, B	Objects moving in Minkowski spacetime $\mathbb{R}^{3+1}$ (or corresponding frames)
a, a', b, b'	Events in $\mathbb{R}^{3+1}$
$(t_\mathrm{A}, x_\mathrm{A})$	Coordinates for frame A in $\mathbb{R}^{3+1}$
$(t_\mathrm{B}, x_\mathrm{B})$	Coordinates for frame B in $\mathbb{R}^{3+1}$
$\prec_\mathrm{A}$	Temporal relation with respect to frame A
$\prec_\mathrm{B}$	Temporal relation with respect to frame B
R, R', R''	Refinements of causal relation $\prec$ on $D = (D, \prec)$
(D, R)	D with refined relation R
$\mathbb{S} = (\mathcal{K}, \mathscr{H})$	Kinematic prescheme or scheme
$\mathcal{K}$	Object class of $\mathbb{S}$
$\mathscr{H}$	Class of co-relative histories of $\mathbb{S}$
$\mathcal{D}(\mathbb{S})$	Underlying directed set of $\mathbb{S}$
$\mathcal{M}(\mathbb{S})$	Underlying multidirected set of $\mathbb{S}$
$\mathcal{T}(\mathbb{S})$	Underlying transition structure of $\mathbb{S}$
$x(D)$	Element in $\mathcal{D}(\mathbb{S})$, $\mathcal{M}(\mathbb{S})$, or $\mathcal{T}(\mathbb{S})$
$r(h)$	Relation in $\mathcal{M}(\mathbb{S})$
$r(\tau)$	Relation in $\mathcal{T}(\mathbb{S})$
A	Accessibility
RA	Relative accessibility
FA	Finite accessibility
FRA	Finite relative accessibility
H	Hereditary property

G	Generational property
$\mathscr{H}_1$	Class of co-relative histories adding a single element to a directed set
$\mathbb{S}_{SR} = (\mathcal{C}_{FIN}, \mathscr{H}_1)$	Sorkin and Rideout's kinematic scheme
$\mathbb{S}_{PS} = (\mathcal{A}_{FIN}, \mathscr{H}_1)$	Positive sequential kinematic scheme
QCMH	Quantum causal metric hypothesis
$\theta_1, \theta_2, \theta_3$	Toy relation functions
γ	Co-relative kinematics in $\mathbb{S}$
Γ	Family of co-relative kinematics
$\mathbb{S}_{GEN} = (\mathcal{A}_{FIN}, \mathscr{H}_{GEN})$	Generational kinematic scheme
$\mathscr{H}_{GEN}$	Class of co-relative histories adding a single generation to a directed set
$\mathbb{S}_{FOL} = (\mathcal{A}_{FIN}, \mathscr{H}_{FOL})$	Foliation scheme
$\mathscr{H}_{FOL}$	"Foliation class"
$\widetilde{\mathrm{Ch}_{\mathrm{MAX}}}(\mathbb{S})$	Class of limits of $\mathbb{S}$
$\overline{\mathbb{S}} := (\mathcal{K} \cup \widetilde{\mathrm{Ch}_{\mathrm{MAX}}}(\mathbb{S}), \mathscr{H})$	Chain completion of $\mathbb{S}$
$\mathscr{I} : \mathbb{S} \to \mathbb{S}'$	Inclusion "functor" of kinematic schemes
$(T, \prec)$	Portion of Stern–Brocot tree
$\mathscr{H}(\mathcal{K})$	Class of all co-relative histories between pairs of objects of $\mathcal{K}$
$\mathbb{S}(\mathcal{K}) = (\mathcal{K}, \mathscr{H}(\mathcal{K}))$	Universal kinematic scheme over $\mathcal{K}$
$\mathbf{Kin}(\mathcal{K})$	Kinematic space over $\mathcal{K}$

Chapter 8

$\mathrm{Aut}(h)$	Generational automorphism group
$\overline{\theta}$	Naïve relation function
$\overline{\Theta}$	Naïve phase map
$\log \overline{\theta}$	Boltzmann-type entropy function
$H(D) = -\sum_k p_k \log p_k$	Shannon-type entropy function
ϑ	Relation functor
P	Poincaré group
U(1), SU(2), SU(3)	Gauge groups of the standard model
L	Lorentz group
L'	Orthochronous Lorentz group
L''	Proper orthochronous Lorentz group
G	Zeeman's causality group
$\prec_{GR}$	Relativistic causal relation (here for Minkowski spacetime $\mathbb{R}^{3+1}$)
F	Inertial frame of reference for $\mathbb{R}^{3+1}$
R_F	Refinement of $\prec_{GR}$ corresponding to F
(R, x)	Pointed refinement
$G(D)$	Causality group $\mathrm{Aut}(D)$ of a directed set D

$\mathbf{D} = \{D = D^0, D^1, D^2, \ldots\}$	Causal atomic resolution of $D = (D, \prec)$
$G(D_n)$	Causality groups of $\mathbf{D}$
$\mathbf{P}(D)$	Poincaré semigroup of D
$[(R,x),(R',x')]$	Ordered pair of pointed refinements
$(D, \prec, x)$	Pointed directed set
$\prec^n$	Causal relation on nth level D^n of $\mathbf{D}$
d^n	"Metric" on space of refinements of $\prec^n$
R^n, R'^n	Refinements of $\prec^n$
V	Vertex set of a random graph
(N,p)	Type of random graph
N	Number of vertices in a random graph
p	Probability of an edge between two vertices in a random graph
G_1	Giant component of a random graph
$(T, \prec)$	Binary tree
ℓ	"Local" edge label
g	"Global" edge label
C	Cyclic directed set
$\omega : C \rightarrow M$	Loop in M
$\Omega(M)$	Loop set of M
$\Omega_{\mathrm{SIM}}(M)$	Set of simple loops over M
$\mathrm{Cyc}(M)$	Set of cycles in M (equivalent to $\Omega_{\mathrm{SIM}}(M)$)
$S = (S, R, e_1, \ldots, e_N)$	Generalized multidirected set
e_i	Generalized initial or terminal element map

Part I
Foundations of Discrete Causal Theory

Chapter 1
Introduction

Abstract. This chapter provides a concise qualitative overview of discrete causal theory. Section 1.1 illustrates how discrete causal theory models fundamental spacetime structure at the quantum level, and introduces a "discrete diffusion-type or wave-type equation," called a *causal Schrödinger-type equation,* that serves as an example of the kind of dynamical law that arises in this context. Section 1.2 introduces the *causal metric hypothesis,* which states that the properties of the physical universe, and in particular, the metric structure of spacetime, emerges from causal structure at the fundamental scale. An important version of this hypothesis appears in Sorkin's *causal set theory.* Section 1.3 describes, at a qualitative level, how spacetime geometry can emerge from fundamental causal structure, with motivation from general relativity. Section 1.4 illustrates some basic structural principles, in terms of "toy models" of discrete causal classical histories. Section 1.5 introduces Grothendieck's *relative viewpoint,* which states that "objects should be studied together with their natural relationships, rather than in isolation." An important embodiment of this idea is *relation space,* which is a space whose elements represent causal relationships between pairs of spacetime events. Relation space methods are crucial to the technical apparatus of discrete causal theory. Section 1.6 discusses the principle of *background independence,* which states that "all physical entities invoked by a theory should participate nontrivially in the dynamics of the theory." In particular, as is known from general relativity, spacetime itself is dynamical, rather than static. Section 1.7 gives a brief overview of the subject of *particles and fields* in discrete causal theory. Section 1.8 explains the distinction between *kinematics* and *dynamics,* and describes how these topics are treated in the discrete causal context. Section 1.9 describes some qualitative ideas regarding discrete causal phenomenology, including a possible alternative to the *inflationary hypothesis* in the cosmology of the early universe. Section 1.10 poses a number of obvious questions raised by the foregoing material, and describes how the book attempts to address these questions.

B.F. Dribus, *Discrete Causal Theory*, DOI 10.1007/978-3-319-50083-6_1

1.1 A Picture and an Equation

Quantum spacetime. Discrete causal theory is a relatively new and ambitious approach to fundamental physics, independent from mainstream continuum-based theories, such as string theory and loop quantum gravity, and much less incremental in its structural innovations. Figure 1.1.1, revisited repeatedly throughout the book, provides a tiny glimpse of how discrete causal theory models the quantum universe. The gray highlighting is included for future reference.

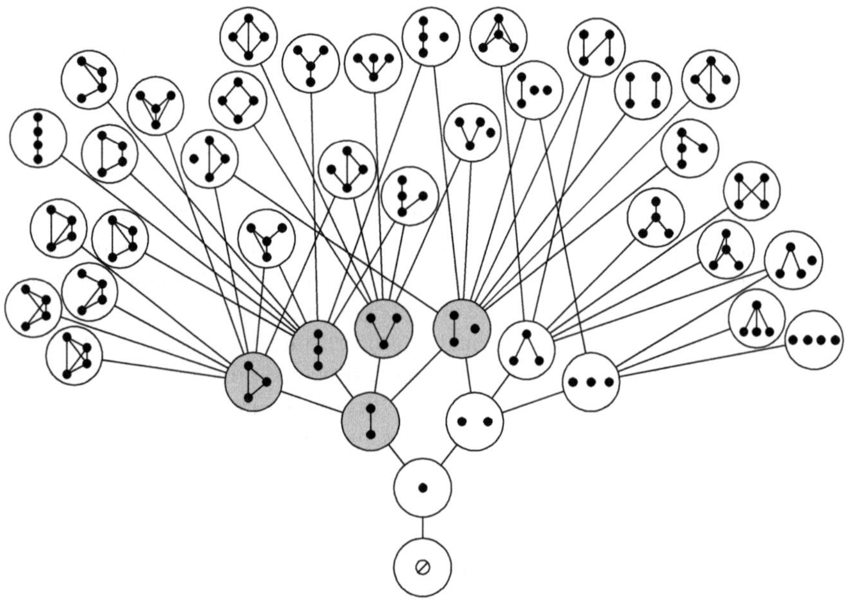

Figure 1.1.1. A glimpse of how discrete causal theory models the quantum universe.

Two different "levels of structural hierarchy" are evident in the figure. The "higher-level structure," represented by the large open nodes and long edges, evokes the world tree of Norse mythology. In fact, it is a small portion of a "quantum multiverse;" not the dubious type of multiverse one encounters in modern string cosmology, but a special type of "structured configuration space," called a *kinematic scheme.* This particular kinematic scheme is called the *positive sequential kinematic scheme,* denoted by $\mathbb{S}_{\text{PS}}$, since it encodes "sequential evolutionary processes." Other equally valid kinematic schemes may be used to study the same theory. Depicted inside each open node is a "lower-level structure," represented by a diagram consisting of small filled nodes and short edges. These diagrams evoke the "ball-and-stick" representations of molecular structure encountered in chemistry. Here, however, they represent extremely simple *classical universes,* or *classical histories,* modeled via mathematical objects called *directed sets.*[1] This extreme simplicity explains why the portion of $\mathbb{S}_{PS}$ appearing in the figure provides only a “tiny glimpse” of the overall structure of the theory, since the classical histories required to model processes in conventional physics, even “local processes” such as “particle interactions,” are expected to be, at the very least, *tens of orders of magnitude “larger” than the histories illustrated here.* Both of the terms “classical universes” and “classical histories” are appropriate for describing these objects. The former term emphasizes that they are regarded as “self-contained,” while the latter term emphasizes that they represent “extended processes,” not merely “snapshots of structure at a particular moment.” I mostly use the term “classical histories” throughout the book, because it is more descriptive of the type of structure involved, and more neutral in its connotations. I sometimes shorten this to just “histories,” when there is no danger of confusion.

The same essential *type* of structure, encoded by nodes and edges, exists at both levels represented in the figure. As explained below, these edges are in fact *directed edges,* with each edge assigned an initial node and a terminal node. At the lower level of structure, nodes and edges represent “cause-and-effect relationships between pairs of events.” At the higher level of structure, nodes and edges represent “stages of evolution” of classical histories. This “layered directed structure” provides the prototypical example of the principle of *iteration of structure* (IS),[2] elaborated in Chapters 6 and 7. A more conventional example of the same general idea is provided by *moduli spaces* in algebraic geometry, in which each element represents a space in its own right, generally possessing formal properties similar to those of the moduli space itself. For now, deferring detailed justification to subsequent sections and chapters, I label these structural levels in the following suggestive way:

lower-level structure: *classical structure*

higher-level structure: *quantum structure*

This designation may seem a priori unusual, for at least two reasons. First, a naïve heuristic is that “quantum” corresponds to “small,” while “classical” corresponds to “large.” For example, it is conventional to say that Einstein’s general theory of relativity, a classical theory, “governs large-scale phenomena,” while quantum theories, such as the standard model of particle theory, “govern small-scale phenomena.” However, among the various established approaches to quantum theory, one approach in particular is distinguished by its explicit practice of building up a quantum-theoretic picture from many classical alternatives, regardless of scale. This is Richard

[1]As noted in the preface, and repeated throughout the book, the term “directed set” does *not* have the same narrow meaning here as in category theory.

[2]The purpose of the acronym (IS) following the term *iteration of structure* is to serve as a clickable hyperlink to the definition of the term in a later chapter. I generally do *not* use acronyms as mere abbreviations, since this would only confuse and annoy the reader.

Feynman's *path summation approach* [FE48], first envisioned by Dirac [DI35], in which "phases" are assigned to each "evolutionary pathway" in a suitable configuration space, and are then summed to yield "quantum amplitudes," which may ultimately be used for explanatory and predictive purposes.[3] This approach generalizes in a straightforward manner to apply to a much broader range of structural paradigms than most other approaches to quantum theory, including those defined in terms of special algebraic structures, such as Hilbert spaces and operator algebras. In particular, the path summation approach provides the most natural and advantageous perspective regarding the "quantum nature" of the kinematic scheme $\mathbb{S}_{PS}$ represented in Figure 1.1.1.

The second reason why the designation of quantum structure as "higher-level structure" may seem unusual is because of the widely-recognized primacy of quantum theory in the description nature, i.e., the belief that "quantum theory ought to be regarded as more fundamental than classical theory." In Figure 1.1.1, the roles seem to be reversed, since "quantum structure" is built from "classical structure." This objection is unlikely to trouble anyone familiar with the path summation approach to quantum theory, but it is nevertheless worthwhile to spell out the potential source of confusion; namely, that two different meanings of the word "classical" are involved. One meaning is technical; it refers to *a specific permissible physical scenario.* This is essentially the meaning I am using in referring to the lower level of structure in the figure as "classical." The word "specific" is important here, since it contrasts with the quantum context, in which *all* permissible scenarios contribute. In more conventional theories, "permissible scenarios" are often identified as "solutions of a classical theory," such as Maxwell's theory of electromagnetism, or Einstein's general theory of relativity. However, in the most interesting versions of discrete causal theory, *one does not begin with classical equations of motion at all;* in particular, one does not distinguish between "on-shell" scenarios that satisfy such equations, and "off-shell" scenarios that do not. Rather, the "permissibility" of histories is determined via the axioms of the theory, and every history satisfying these axioms is treated on the same footing. Histories that do not satisfy the axioms are not included in the theory in any way; i.e., there are no "off-shell" scenarios. Equations of motion, i.e., dynamical laws, arise at the quantum level, as described below.

The other meaning of "classical" is an imprecise and operational meaning, namely, *"whatever can be measured at sufficiently large scales and sufficiently low energies to have been noticed before the early 1900s, when the rudiments of quantum theory were stumbled upon by human physicists."* The fact that these meanings roughly coincide in many conventional scenarios is a nontrivial consequence of the so-called *correspondence principle,* which states, roughly, that "quantum theory leads to approximately the same predictions as classical theory at large scales." In particular, classical

---

[3] An interesting historical detail is that Feynman's paper [FE48] and Shannon's landmark paper on information theory [SH48] both appeared in the year 1948.

equations of motion may be recovered as approximations or limiting cases of quantum dynamical laws. It is true that quantum theory should be regarded as more fundamental than this rough operational meaning of "classical," but it should *not* be regarded as more fundamental than the technical meaning described above. Under the technical meaning, both types of theory are equally fundamental; they merely represent different levels of structure, as represented in the figure. Quantum structure is more *complete* than any individual classical structure, however, since it combines all classical alternatives.

Causal Schrödinger-type equations. In a broad formal sense, *any* structured configuration space may be regarded as a "quantum object," by adopting a sufficiently generalized view regarding path summation. However, in order to build a "physically recognizable quantum theory," one must find a way to *combine contributions from different evolutionary pathways in the configuration space,* and interpret the results. This brings *algebraic* considerations into the picture. In particular, the process of forming actual "sums" requires the availability of algebraic objects equipped with appropriate operations, in which individual contributions may be "added." The same is true for calculating quantum amplitudes, and for performing all the other computational tasks necessary to flesh out an explanatory and predictive quantum-theoretic framework. In the discrete causal context, all or most of the necessary algebraic structure arises naturally from the underlying causal structure itself, and this leads to interesting and physically novel mathematics. Among the results of these developments are dynamical laws for discrete quantum causal theory. A special case of such a dynamical law is the following *causal Schrödinger-type equation:*

$$\psi_{\bar{R};\theta}(r) = \theta(r) \sum_{r^- \prec r} \psi_{\bar{R};\theta}(r^-). \tag{1.1.2}$$

I give here only a brief sketch of the content and meaning of this equation, leaving most of the details for later. First, the equation is analogous, at a deep conceptual level, to the version of Schrödinger's equation encountered in the conventional Hilbert space approach to nonrelativistic quantum theory:

$$i\hbar\frac{\partial \psi}{\partial t} = \mathbf{H}\psi. \tag{1.1.3}$$

Here, $i = \sqrt{-1}$ is the imaginary unit, $\hbar = h/2\pi$ is Planck's reduced constant, ψ is an element of a Hilbert space, representing the state of a quantum system, and $\mathbf{H}$ is the Hamiltonian operator. Schrödinger's equation is related closely to classical *heat* or *diffusion equations*, and more loosely to classical *wave equations.* Its solutions are sometimes called "wave functions;" however, the term *state function* is preferable.

Returning to Equation (1.1.2), $\psi^-_{R;\theta}$ is a *generalized* state function, called the *past state function,* while θ is a *relation function,* analogous to a Lagrangian, rather than a Hamiltonian. R represents a subobject of a special configuration space of relationships between pairs of classical histories, called *co-relative histories,* which is derived from a kinematic scheme such as $\mathbb{S}_{\mathrm{PS}}$. Roughly speaking, sequences of co-relative histories represent *evolutionary pathways* for classical histories. The *relations* r and r^- are elements of R representing specific co-relative histories. The *precursor symbol* $\prec$ in the expression $r \prec r^-$ indicates that r precedes r^- in a "generalized causal sense."[4] These structural analogies are the reason why I refer to Equation (1.1.2) as a causal Schrödinger-type equation. In particular, it may be viewed as a *discrete heat or diffusion-type equation* from a mathematical perspective. The technical background and derivation of this equation, together with its relationship to mainstream physics, is one of the main topics of Chapter 6.

The ultimate purpose of deriving equations such as Equation (1.1.2) is to help solve the central problems of twenty-first century physics; namely, the problems of describing spacetime structure at the fundamental scale, providing a workable quantum-theoretic description of gravitation, and unifying the standard model of particle theory with general relativity via a deeper theory subsuming both. In particular, given the proper dynamical input, and assuming that the underlying hypotheses of discrete causal theory are valid, Equation (1.1.2) *ought* to contain the necessary information to enable recovery of the formulae of conventional quantum field theory, together with Einstein's field equation for general relativity, as "smooth approximations at low energies," at least in special cases.[5] However, there remain substantial issues yet to be resolved in demonstrating that this approach actually works. These issues are elaborated and analyzed throughout the book.

1.2 The Causal Metric Hypothesis

Elements and relations in a small family of classical histories. I now return to the diagram in Figure 1.1.1 above, and focus on the portion highlighted in gray. A "zoomed-in version" is reproduced in Figure 1.2.1:

---

[4]In the present "quantum-level" context, the expression $r^- \prec r$ represents a *relationship between a pair of co-relative histories,* i.e., a "relationship between a pair of evolutionary relationships," in which the evolutionary relationship represented by r is a possible sequel to the evolutionary relationship represented by r^-.

[5]For example, the specific derivation of Equation (1.1.2) in Chapter 6 assumes that spacetime is *acyclic* (AC), although the overall development of the theory is much more general. It is worth noting that Einstein's field equation, which appears as Equation (1.3.1) below, involves only a small part of the emergent structure that should arise from equations such as (1.1.2), since it takes for granted so much geometric information; for example, the *integer dimensionality* of spacetime. This topic is revisited in Section 8.2.

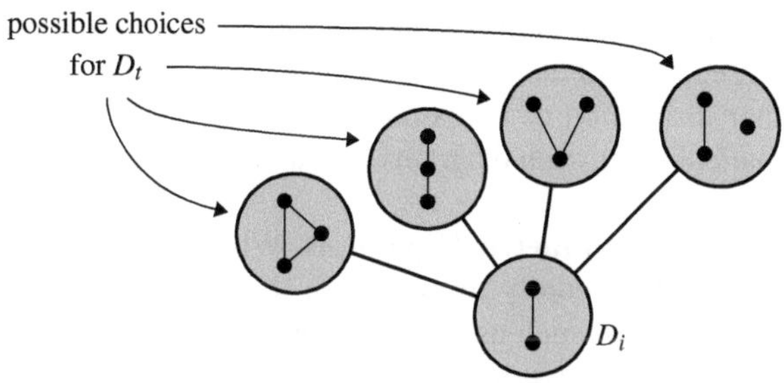

Figure 1.2.1. A "parent history" D_i, and four "child histories".

The simplest way to describe the meaning of this figure is to say that it represents four different ways in which an "initial classical history" D_i, which has two elements and one causal relation connecting them, may evolve into a "terminal classical history" D_t, which has three elements, and anywhere from one to three relations. In this special case, D_i may be suggestively called the "parent history," while the four possible choices for D_t may called "child histories." This terminology reflects the fact that the evolutionary relationships involved here are *direct,* i.e., without intervening stages. The letter "D" is used because discrete causal theory models classical histories via *directed sets,*[6] which are sets equipped with special binary relations. The physical interpretation of these sets, already mentioned above, is that their elements represent "spacetime events," while their relations represent "cause-and-effect relationships between pairs of events." The word "directed" refers to the direction from cause to effect; for example, the existence of a relation $x \prec y$ between a pair of elements x and y indicates the flow of causal influence from x to y.[7] It is worth repeating that the term "directed set," as used in this book, does not carry the more restrictive meaning involving common "successors" or "upper bounds" for arbitrary pairs of elements, which is conventional in order theory and category theory. Here, the "directions" are purely *local* directions between specific pairs of elements. Elements are represented in the figure by small black nodes, and relations are represented by line segments joining pairs of these nodes.

Informally, the entire diagram may be viewed as a "family tree," directed upward. The "parent history" D_i appears at the bottom of the diagram as the "root," while the four "child histories" are the "branches." As indicated above, the rationale for calling these histories "child histories" is that they are *direct descendants* of the parent history, since they are each constructed by adding one new element to the parent. The reason for taking the relation of "family descent" to flow *up* the page, contrary to the ordinary English meaning of the word "descent," is because this is the direction in which *time* is conventionally taken to flow in Minkowski diagrams

[6]The fact that the word "discrete" begins with the same letter is a happy coincidence, but many of the results of this book also apply to non-discrete directed sets.

[7]The appearance of the same precursor symbol $\prec$ in the causal Schrödinger-type equation (1.1.2) is a reflection of *iteration of structure* (IS).

in relativity. This *connection between time and order-theoretic structure,* encoded by relations between pairs of elements in a directed set, is of great significance, but I postpone further discussion of temporal issues until Sections 1.3 and 1.4. At an abstract mathematical level, such diagrams are called *generalized Hasse diagrams.*[8]

First glimpse of the causal metric hypothesis. What *are* the "elements" of these "classical histories?" At a purely formal level, they may be taken to represent a variety of different entities, such as steps in a process, or computers in a network. However, in the specific context of fundamental spacetime structure, they are taken to encode the most localized possible information about "time and place." This is the same conceptual role played by "events" in relativistic spacetime, which explains the foregoing statements that "elements represent events." In particular, these elements are taken to possess *no internal structure.*[9] However, it is reasonable to ask how they relate to more tangible entities, such as "elementary particles." For now, it suffices to understand that they are, in all probability, vastly "smaller" than any particle observable in modern accelerators, such as the Large Hadron Collider, insofar as such statements about "size" have meaning. Ultimately, however, the logic goes the other way around; in discrete causal theory, the attributes of physical objects, including their sizes, are defined *in terms of* elements and relations. A more succinct and definitive statement of this idea is the following:

Causal metric hypothesis (CMH). *The properties of the physical universe are manifestations of causal structure.*

The causal metric hypothesis was first stated explicitly in general form in my papers [DR15, DR13].[10] The statement of the hypothesis appearing in the present chapter is purely qualitative in nature; quantitative versions are introduced beginning in Chapter 2. In general terms, the word "metric," refers to *measuring* things, i.e., assigning "sizes" or "scales" to physical objects. In a more specific context, it refers to a particular way of describing geometric structure, which plays a familiar role in general relativity and related areas of modern physics. The causal metric hypothesis is intended to involve *both* meanings of "metric." Under the general meaning, it says that "whatever can be measured is ultimately attributable to causal structure." Under the more specific meaning, it says that "what appears to be geometric structure, in the context of what is called spacetime, is really just causal structure." At a technical level, the latter statement is based on a radical and far-reaching interpretation of the *metric recovery theorems* of Stephen Hawking [HA76] and David Malament

[8]The qualifier "generalized" is necessary because "true" or "strict" Hasse diagrams encode partially ordered sets, which are less general than directed sets under the conventions used in this book. See Section 2.2 or Chapter 3 for more details.

[9]That is, in the most straightforward versions of discrete causal theory. Section 8.7 includes brief speculation about "subclassical structure."

[10]The reason why these references seem to appear in reverse chronological order is because the paper [DR15] was first presented in 2012; the reference dated 2015 includes it as an individual chapter. The link to the 2012 version appears in the list of references along with the bibliographical information for the 2015 version.

[MA77], which relate topological, conformal, and metric structure for certain types of *pseudo-Riemannian manifolds,* the mathematical objects used to model classical spacetime in general relativity. These results have since been further amplified in various contexts, for example, in *domain theory* [MP06]. The subject of metric recovery is discussed in much greater detail in Chapter 2.

Roots of the hypothesis. The historical and philosophical roots of the causal metric hypothesis stretch back to the ancient Greeks; for example, *Zeno's paradoxes* play a role. More modern influences include Leibniz, Gauss, Riemann, Einstein, Kaluza and Klein, Weyl, Dirac, Feynman, Wheeler, Finkelstein, Grothendieck, Hawking, Malament, Sorkin, and many others, although I will not attempt to trace all these contributions here. It is a very ambitious idea, in the sense that it attempts to explain much in terms of little. Of course, many physical hypotheses outside the scientific mainstream are ambitious in this sense, and most are not worthy of serious consideration. What distinguishes the causal metric hypothesis from "Time Cube," and other amusing "theories" of this ilk, is the existence of rigorous technical results about relativistic spacetime structure, namely, the metric recovery theorems mentioned above, which strongly suggest that the hypothesis is likely to be, at the very least, "almost true." The great ambition of the hypothesis lies in two radical conclusions it draws regarding the physical implications of these results: first, that whatever aspects of geometric structure prevent "almost true" from being "true" in this context do not actually exist, except in an emergent sense; second, that the scope of description of causal structure may be extended beyond the mere "fabric of spacetime," to include *all physical interactions.* From a historical viewpoint, the first conclusion might be compared to Einstein's rejection of the luminiferous aether in special relativity, while the second follows the spirit of other "unification" efforts in modern physics. Further contextual and historical information regarding the causal metric hypothesis appears in Chapters 2 and 3.

Versions and interpretations of the hypothesis. The foregoing remarks apply principally to the most literal and far-reaching interpretation of the causal metric hypothesis (CMH), called the *strong interpretation,* discussed further in Chapter 2, which seeks to ascribe *all of physics* to causal structure at the fundamental scale. Despite this lofty aspiration, the "approximate truth" of the causal metric hypothesis, established by the metric recovery theorems, renders it unnecessary to approach discrete causal theory from an "all-or-nothing" perspective. The theory remains quite promising even under much weaker versions of the causal metric hypothesis. For example, perhaps only "spacetime" may legitimately be attributed to causal structure, while "particles and fields" may involve some auxiliary data, expressed in the form of algebraic structure "attached to" or "interacting with" causal structure. This, in fact, is how such "material entities" are typically treated in the most popular existing versions of discrete causal theory, such as causal set theory. Adopting this relatively conservative approach, one may choose to temporarily set aside ambitions of unification, and pursue discrete causal theory as an approach to quantum gravity. However, there are some striking reasons to hope that a stronger interpretation of the hypothesis may ultimately hold true. In particular, *nontransitive* directed sets already contain more information than is necessary for "emergent geometry," in the sense that transitivity is taken for granted in the metric recovery theorems. These details are discussed in Chapter 3.

1.3 Emergent Spacetime

Classical spacetime, beginning with general relativity. In Section 1.1, I remarked that the classical histories represented in Figure 1.1.1 are vastly smaller than their common counterparts in conventional physics; for example, the classical histories used to model typical "particle interactions." Figure 1.3.2 illustrates a discrete causal description of a slightly larger region of classical spacetime. The left-hand diagram in the figure depicts a region of spacetime in the neighborhood of an event x, viewed in the usual relativistic fashion. The vertical and horizontal curves represent a particular choice of coordinates centered at x. As usual in such schematic diagrams, only two dimensions are shown: the "vertical dimension" for "time," and the "horizontal dimension" for "space." By convention, the time coordinate increases from top to bottom along the vertical curves, while the horizontal curves represent "spatial sections." The fact that these curves are not straight lines indicates that the region exhibits nontrivial curvature. According to general relativity, this curvature interacts dynamically with particles and fields, as described by Einstein's field equation,

$$R_{\mu\nu} - \frac{1}{2}Rg_{\mu\nu} + \Lambda g_{\mu\nu} = \frac{8\pi G}{c^4}T_{\mu\nu}, \tag{1.3.1}$$

where $R_{\mu\nu}$ is the Ricci curvature tensor, R is the scalar curvature,[11] $g_{\mu\nu}$ is the metric tensor, Λ is the "cosmological constant,"[12] G is Newton's gravitational constant, c is the speed of light, and $T_{\mu\nu}$ is the stress-energy tensor.[13] The spacetime region

[11]The scalar curvature has nothing to do with the relation space referred to in Equation (1.1.2) above, also denoted by R. The justification for using the same symbol twice is that the scalar curvature plays little direct role in this book, and the notation is standard.

[12]The reason for the quotation marks here is that the term "cosmological constant" itself imposes a very specific and conventional physical interpretation on certain observed data.

[13]This is a "classical" description of Einstein's equation and its constituents. Mathematically, it is very old-fashioned, and even in conventional physics there exist more modern descriptions enjoying certain advantages. For example, in the *tetrad formalism,* Einstein's equation may be written as

$$\varepsilon_{IJKL}\left(e^I \wedge R^{JK} - \frac{2}{3}\Lambda e^I \wedge e^J \wedge e^K\right) = 2\pi G T_L,$$

where ε_{IJKL} is the totally antisymmetric Levi-Civita symbol, e is a 1-form called the tetrad field, R is the curvature 2-form, and T is the energy-momentum 3-form. One advantage of this formalism is that it can incorporate *fermionic fields.* Rovelli gives a useful description of different formalisms for general relativity in the second chapter of his book [RO04].

illustrated here is assumed to be "small" by ordinary standards, but much "larger" than the discrete causal classical histories illustrated in Figure 1.1.1. The pair of "warped dark gray triangles" represent the set of events that "could be causally related" to x. This region is the natural generalization of the *light cone,* or *null cone,* of an event in Minkowski spacetime in special relativity. The bottom triangle represents the *causal past* of x; it consists of all events that "could conceivably" influence x. The top triangle represents the *causal future* of x; it consists of all events that x "could conceivably" influence. Below, I explain how discrete causal theory removes certain awkward counterfactual speculation in relativity, expressed here by the phrase "could conceivably."

Discrete microstructure and its consequences. The middle diagram in Figure 1.3.2 shows a magnified view of the spacetime region near x. According to discrete causal theory, "granularity" appears at sufficiently small scales. If this is true, then Einstein's field equation (1.3.1) begins to lose precision at this level, and some deeper and more sophisticated law describes the details of the microstructure. This particular diagram depicts an "intermediate scale," at which the structure is still "approximately geometric," despite noticeable deviations. How these deviations might actually manifest themselves experimentally is a question of great subtlety and interest, since the scales involved are likely many orders of magnitude smaller than the scales directly accessible to present-day technology. One possible way might be via systematic violations of *Lorentz invariance,* meaning, roughly, evidence that "not all inertial frames of reference are created equal."[14] Here, "Lorentz" refers to the *Lorentz group* L, which is a subgroup of the *Poincaré group* P, the group of symmetries[15] of the flat Minkowski spacetime $\mathbb{R}^{3+1}$ of special relativity. Of course, curved spacetime in general relativity is itself not Lorentz symmetric at a *global* level; rather, Lorentz invariance is exhibited as a local property of relativistic spacetime in a limiting

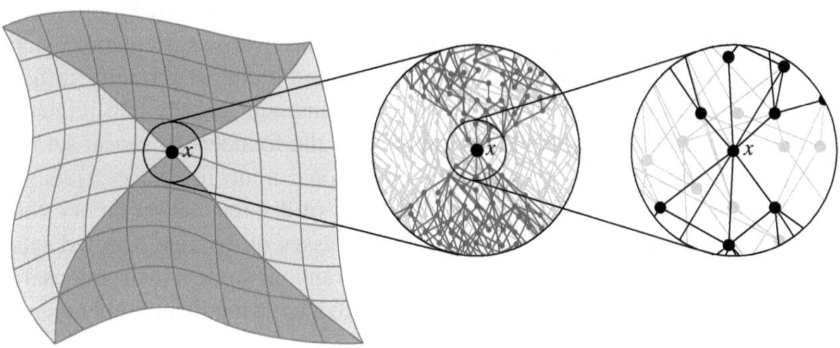

Figure 1.3.2. Classical spacetime according to discrete causal theory.

---

[14]The notion of "systematic" Lorentz invariance is described precisely in Chapter 4.

[15]In general, a *symmetry* of a mathematical object means an *automorphism* of the object in the *category* to which it belongs. In this case, symmetries are *isometries,* i.e., self-maps preserving the *Minkowski metric*. See Chapter 2 for more details.

sense.[16] However, the subtle topic of local structure in discrete causal theory is not seriously examined until Chapter 4, so I use the term "Lorentz invariance" informally in the present introductory context. The Lorentz group is a Lie group, which means that it possesses both a group structure and a "compatible" geometric structure, with the latter "abstract geometry" essentially attributable to the "physical geometry" of Minkowski spacetime. It would be absurd to expect discrete models of spacetime to respect such a continuous structure precisely, and Lorentz invariance violation is a generic feature of discrete causal theory. In particular, the directed set depicting discrete microstructure illustrated in Figure 1.3.2 violates Lorentz invariance.[17] However, it is well-known that discrete microstructure need not produce *systematic* violations large enough to be noticeable at large scales.

Some physicists reject out of hand any theory that violates Lorentz invariance "nontrivally," i.e., that violates the local, general relativistic version of Lorentz invariance, described above, in a systematic way. The search for such violations is the subject of numerous ongoing experiments and observations, whose results have yet to reveal any robust evidence of them. This body of empirical data imposes strong constraints on the possible type and scope of deviations from Lorentz invariance. While many of the mechanisms examined in the context of such searches remain essentially in the continuum-based realm, the consequences of discrete microstructure have enjoyed a reasonable degree of scrutiny as well; for example, Sabine Hossenfelder [HO13] provides an interesting discussion of constraints on violations of Lorentz invariance arising from *"local defects in underlying, non-geometric theories,"* an informal description for a broad class of non-continuum-based approaches that includes discrete causal theory. Certain types of discrete directed sets defined a posteriori, via "Poisson sprinkling" of elements into relativistic spacetime, may be constructed in such a way as to avoid such violations entirely; for example, the paper [BHS09] explains how this may be accomplished in the context of causal set theory. However, as described in Chapter 4, there are steep structural prices to pay for working exclusively with such objects, as noted, for example, by Moore [MO88]. Fortunately, better-behaved directed sets may be defined whose systematic deviations from Lorentz invariance remain arbitrarily small, and it almost certainly suffices to restrict attention to such objects when studying quantum gravity and related topics in the discrete causal context, as explained by the founders of causal set theory [BLMS88]. This means that experimental bounds on Lorentz invariance violation do not present a serious obstacle to the development of discrete causal theory, though they do offer important clues regarding which classes of discrete directed sets are most physically relevant. In particular, they favor random models such as causal sets, in which a given element may be directly related to a large number of other elements,

---

[16] That is, the tangent spaces T_xX at each point $x \in X$ of a relativistic spacetime manifold are isomorphic to $\mathbb{R}^{3+1}$. This ignores singularities, such as those resulting from gravitational collapse, which are not considered to be part of the actual spacetime.

[17] In particular, referring to the discussion below, this directed set looks nothing like a "sprinkled" causal set, in which most of the irreducible relations are "almost null."

and disfavor more regular structures, such as lattices. These topics are explored in more detail in Chapter 4.

It is worth noting that aesthetic motivations surely play a role in the common reluctance to sacrifice Lorentz invariance. The entire development of theoretical physics over the last century and a half has tended toward greater and greater use of symmetry principles, especially those expressed via the representation theory of Lie groups. The choice to challenge the systematic validity of one of the most sacred of these principles produces an unmistakable impression of theoretical recidivism from a naïve point of view. From the perspective of discrete causal theory, however, Lorentz invariance must be regarded as an emergent phenomenon; it applies remarkably well to a wide range of physical scenarios, but cannot be expected to govern the most extreme regimes. Should the theory prove successful, a similar fate likely awaits many other Lie groups prominent in modern physics, even those whose structure does not derive entirely from spacetime geometry. These include, in particular, the gauge groups of the standard model of particle theory, which encode "internal symmetries." The specific topic of the emergence of the Poincaré group P is discussed in Section 8.4.

Spacetime at the fundamental scale. The right-hand diagram in Figure 1.3.2 shows the hypothesized granularity of classical spacetime "up close;" at this scale, the structure looks nothing like smooth geometry, and Einstein's field equation (1.3.1) is essentially irrelevant for describing its local properties. The black nodes in the diagram, with the obvious exception of the node labeled x itself, represent events causally related to x, with events "below x" exerting influence on x, and events "above x" influenced by x, either directly or indirectly. It is worth noting that this diagram is still highly idealized and simplified compared to what one might actually expect to see at the fundamental scale. In particular, it would not be very surprising if x were directly related to a very large number of other elements, say 10^{10}, or 10^{20}, or even 10^{50} or more. The light gray nodes in the diagram represent events causally unrelated to x, i.e., events that neither influence x nor are influenced by x. No awkward counterfactual statements appear in this description, in contrast to its relativistic analogue. An incidental point illustrated by these diagrams is that the idea of "temporal structure" carries over much more naturally to discrete causal theory than the idea of "spatial structure." For this reason, discrete causal theory is sometimes referred to as a "time-first theory," in contrast to "space-first theories," such as Julian Barbour's *shape dynamics.* In particular, the role of the vertical "time curves" appearing in the left-hand diagram in Figure 1.3.2 is subsumed by sequences of relations running up the page, but corresponding analogues of the horizontal "spatial sections" in the diagram cannot be constructed, in any straightforward way, from families of relations. In fact, a necessary condition for a pair of elements to belong to the same "spatial section" is that there is *not* a sequence of relations between them, since this condition is the natural discrete causal analogue of *spacelike separation* in relativity. However, spatial notions involving larger families of elements, such as discrete causal analogues of *Cauchy surfaces,* are much subtler. This important topic is revisited repeatedly throughout the book, particularly in Chapter 5.

Preliminary plausibility considerations. The preeminence of continuum-based theories in conventional physics may encourage a skeptical reaction to the discrete causal viewpoint regarding fundamental spacetime structure. This is true despite the general belief that "the spacetime continuum breaks down somehow in the context of quantum gravity." What is radical about discrete causal theory in this context is that it *completely ignores the continuum from the outset,* rather than "arriving at discreteness" via some "quantization procedure" beginning with continuum-based geometry. Of course, causal set theory has been advocating an a priori discrete approach for the last thirty years, but has been dwarfed in its influence by continuum-based alternatives, especially string theory and loop quantum gravity. Further, even if one is prepared to consider abstaining from the continuum entirely, it is by no means obvious that the structural role of the real numbers, and of real manifolds more generally, should be replaced by discrete directed sets. In particular, there exist many possible structural paradigms that are neither continuous nor discrete, and many others than involve a mixture of "spatial" and "causal" structure. Given these factors, one might be inclined to conclude that, all other things being equal, a continuous, smooth structure for classical spacetime is simpler and more natural than the "chaotic-looking discrete jumble" illustrated in Figure 1.3.2.

However, all other things are *not* equal. To begin with, forbidding technical problems, such as divergences, plague most, if not all, well-explored continuum-based approaches to fundamental physics, and discrete causal theory is automatically immune to many of these problems. Also, there is nothing simple or natural about a choice of smooth structure, especially in four real dimensions. Any doubts about the veracity of this statement may be settled by studying "exotic $\mathbb{R}^4$," or by reading about four-dimensional versions of *geometrization* and the *Poincaré conjecture.*[18] Further, as mentioned above in the context of the causal metric hypothesis (CMH), the metric recovery theorems provide surprising mathematical justification for the idea that causal structure supplies "most" of the information content of classical spacetime, even without bringing discrete considerations into the picture. Following the proof of these theorems in the late 1970s, the development of "pure causal" approaches to spacetime microstructure was virtually inevitable; in particular, the basic principles of causal set theory emerged from multiple independent sources within a few years. The slow subsequent development of discrete causal theory is perhaps due more to the long-standing particle/string orientation of the physics community than to the challenges of the theory itself.

One might still argue, however, that spacetime *appears* to be continuous and smooth from an empirical standpoint, and that continuum-based geometry should

[18] Torsten Asselmeyer-Maluga has built an interesting research program in fundamental physics by focusing on the special properties of four-dimensional real manifolds in the context of differential topology. In particular, four-dimensional manifolds may possess an infinite, or even uncountable, number of inequivalent *smoothness structures,* and it is natural to wonder if it is a coincidence that four is the "physical dimension." Asselmeyer-Maluga [AS15] goes as far as to suggest that *"the plethora of exotic smoothness structures in dimension four could be the cornerstone of quantum gravity."* This line of thought goes in a completely different direction from discrete causal theory, arguing that fundamental spacetime discreteness is simply the wrong idea!

therefore remain the default paradigm until experimental evidence refutes it. However, there is nothing about the available experimental data itself that singles out either continuity or smoothness from among an infinite number of other structural possibilities. What singles out these properties is their mathematical *convenience*, which is very different from physical plausibility. Physicists as recently as a few generations ago lacked suitable technical alternatives to the continuum-based machinery of calculus and differential equations, and natural outgrowths of these methods, such as differential geometry and Lie theory. It is therefore not surprising that the physical theories developed during this time modeled basic physical structure in terms of such tools. A useful metaphor is provided by the popular fable about searching for lost keys only near streetlights after dark. There may be no logical reason to believe that the keys are in fact near a streetlight, but it is only sensible to search where there is some hope of achieving success. However, it would be unreasonable to limit one's search to the same vicinities after daybreak. In a similar manner, it would be imprudent for the physics community to continue to restrict most of its attention to unconsummated continuum-based approaches to fundamental physics, now that there exist viable alternatives to the real numbers as sources of basic structure.

For the benefit of any reader still unwilling to consider sacrificing the continuum, I will point out that even if spacetime *were* continuous and smooth, or even if "granularity" *were* properly understood only via some quantization procedure, the ideas developed in this book would remain useful, and indeed unavoidable, in many physical contexts. The ultimate reason for this conclusion arises from the very nature of modern science, in which the existence of directed relationships between such entities as causes and effects, apparatus and results, and data and conclusions, demands systematic study of directed sets, and related structures, in a physical context. At a more concrete level, interesting applications of the same ideas, independent of any discreteness hypothesis for classical spacetime, arise in the contexts of quantum information theory and domain theory. In the former context, quantum circuits may be viewed as superpositions of discrete "classical histories," in a manner similar to, though less general than, the one hinted at in Figure 1.1.1. In the latter setting, one finds natural "continuous analogues" of many of the results developed in this book. In a broad sense, therefore, the long-term relevance of discrete causal theory seems inevitable. The same cannot be said for specific continuum-based approaches, which tend to lose most of their physical interest once they prove unsuited to achieve their original goals. A good example is the a priori compelling Georgi-Glashow SU(5) grand unification theory, proposed soon after the completion of the standard model, which predicted too high a rate of proton decay. The entire compendium of continuum-based theories involving *supersymmetry* is presently at risk of a similar fate, although these theories have made interesting contributions to mathematics.

1.4 Practice with Toy Histories

Nuts and bolts of elements and relations. I now consider in more detail the two-element "parent history" D_i appearing in Figure 1.2.1 above. What exactly is the meaning of the diagram illustrating this classical history? Figure 1.4.1 expands the picture, and labels the structural constituents of D_i, which possesses two elements and one relation. Mathematically, D_i may be regarded as a *directed graph* with two vertices and one edge, but I will usually use set-theoretic terminology, and refer to such objects as *directed sets.* I repeat again that the term "directed set," by itself, does not carry any additional structural connotations in this context, such as the common upper bound property typically associated with the term in order theory and category theory. The physical interpretation of this history is that it represents a "classical universe with two events;" the first event causes the second event, and then "the show is over." One may imagine a clock that ticks just once; the single tick, which may be suggestively referred to as a *classical chronon,*[19] stretches from the beginning of time to the end of time in this universe. The element representing the first event is called the *initial element* of the relation, and the element representing the second event is called the *terminal element* of the relation. Note that this constitutes a "lower-level analogue" of the notion of initial and terminal *histories* discussed in Section 1.2. Unrealistically simple histories such as D_i may be referred to as **toy histories,** since they provide convenient examples for examining basic structural ideas at a minimum level of complexity.

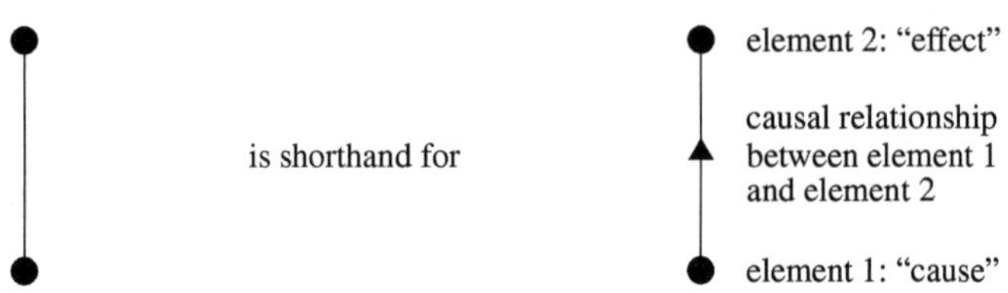

Figure 1.4.1. Meaning of the "parent history" D_i from Figure 1.2.1.

As this example illustrates, it is relatively simple to think about "time" in discrete causal theory: cause precedes effect, and a single classical chronon, i.e., a single fundamental unit of local time, is associated with each relation. As mentioned in Section 1.3, however, the idea of "space" does not carry over so easily from conventional physics to discrete causal theory. Rather, spatial structure must be *inferred* from causal structure in a nontrivial way. For the classical history D_i, "space" is not very interesting; it consists of a single point. More precisely, the two elements of D_i may be viewed as "the same point in space at different times." It is instructive, therefore, to examine a classical history possessing "more space." One example

[19]This terminology can sometimes be useful for rendering "concrete" the physical interpretation of the mathematical structures involved, but the more neutral terminology of "elements and relations" is usually preferable.

is the third "child history" from the left in Figure 1.2.1. I reproduce this history in Figure 1.4.2.

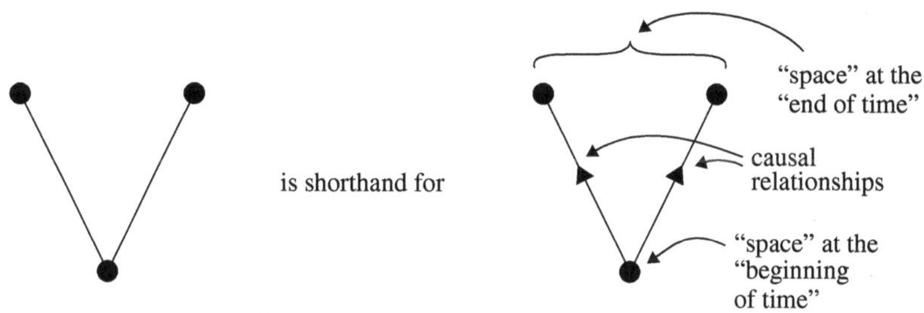

Figure 1.4.2. The simplest "expanding universe" in discrete causal theory.

This classical history represents the simplest possible example of an "expanding universe" in discrete causal theory. Once again, one may imagine a clock that ticks just once. At the "beginning of time," "space" consists of one event; this event causes two more independent events, and then "the show is over." These two terminal events constitute "space" at the "end of time." For future reference, I will remark that these two events may be considered "spatially adjacent," since they are not themselves related, but share a common direct cause. To understand why this makes sense, one may recall the particle-theory analogy of "pair production," in which two particles emerge, heading in different directions, from a single event. After a very short time interval; in the present case, a single classical chronon, the particles are "side-by-side" in space, and this provides a reasonable intuitive picture of spatial adjacency. It is worth noting that one may define two other "frames of reference" for this simple "expanding universe," in which one of the two terminal events is taken to precede the other. In this context, the original "frame," in which the two terminal events are regarded as simultaneous, is roughly analogous to the frame of the cosmic microwave background, i.e., a "universal rest frame." Chapters 4 and 7 contain much more information regarding discrete causal notions analogous to relativistic frames of reference.

These concrete descriptions highlight the extreme simplicity of the classical histories illustrated in Figure 1.1.1. These histories, together with the natural relationships among them, make up the only local region of the corresponding "quantum universe" that is reasonably easy to represent graphically. In Figure 1.4.3, I illustrate a slightly more complicated classical history, with 60 elements. This history is still "tiny" by conventional standards, yet attempting to draw its immediate neighborhood in the positive sequential kinematic scheme $\mathbb{S}_{\rm PS}$ of Figure 1.1.1 would be a fool's errand, since it possesses roughly 2^{60} child histories in $\mathbb{S}_{\rm PS}$. I say "roughly," because the number of child histories is reduced whenever the parent history possesses non-trivial symmetries, and the existence of such symmetries is generally difficult to determine. This problem is related to a famous problem in graph theory, called the *graph isomorphism problem.*[20] According to certain asymptotic results in random graph theory [BR88], "most" finite classical histories in discrete causal theory possess no nontrivial symmetries, and this *rigidity* property has profound implications. In particular, it suggests that symmetry is much less central than generally believed to the notion of *covariance,* i.e., the principle that the laws of physics do not depend on one's "point of view." This realization further undermines the primacy of group representation theory in theoretical physics, already mentioned in Section 1.3 in the context of Lorentz invariance. The topic of discrete causal covariance is revisited in Section 8.3.

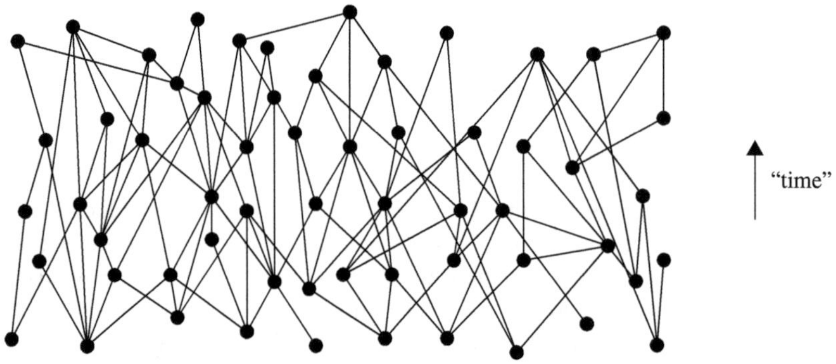

Figure 1.4.3. A slightly more complicated classical history.

Evolution of classical histories. The foregoing discussion provides a reasonable first glance at discrete causal classical histories, viewed as "self-contained universes." However, it is often useful to view a classical history, such as the "parent history" D_i detailed in Figure 1.4.1, as an "intermediate stage" in the evolution of a more complicated classical history, such as one of its "child histories." In this context, one is faced with the general question of what the "next stage of evolution" of D_i might look like. Many possible choices exist, but if one restricts attention to the simple *sequential growth* model represented by the positive sequential kinematic scheme $\mathbb{S}_{\mathrm{PS}}$, then the only choices are the four child histories illustrated in Figure 1.2.1. In this context, a step in the evolutionary process is given by adding a single "new element," which may be related to any subset of elements of D_i. In this case, $4 = 2^2$ is the number of subsets of D_i, so the fact that D_i also possesses four "essentially distinct," i.e., nonisomorphic, child histories, indicates that it possesses no nontrivial symmetries. This is because child histories whose "new elements" are related to subsets of D_i that are interchanged by symmetries of D_i are themselves isomorphic. This rather subtle topic is elaborated in Chapter 6. It is useful to examine in greater detail the

[20]It is worth noting that significant and possibly decisive progress on this problem has recently been claimed by László Babai [BA15], a mathematician and computer scientist at the University of Chicago. As of early 2016, this result remained under review by other experts.

child history on the far left in Figure 1.2.1. I label this history D_t, where the subscript "t" stands for "terminal." This is the "most complicated" of the child histories of D_i, in the sense that it possesses the largest number of relations. I reproduce this history in Figure 1.4.4.

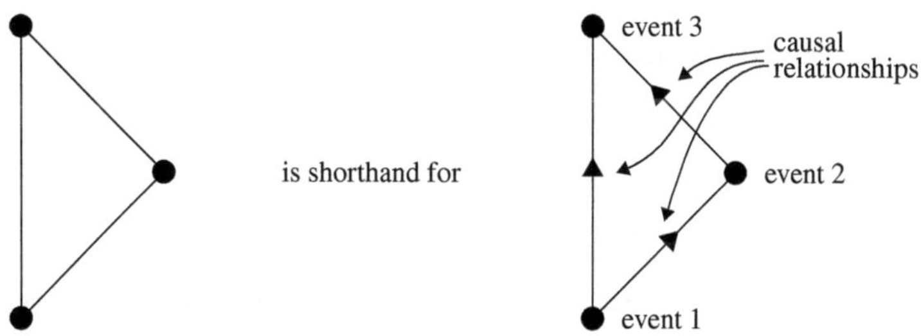

Figure 1.4.4. Leftmost "child history" D_t from Figure 1.2.1.

Perhaps the first observation to make about this child history D_t is that it "really does descend from the parent history D_i," in the sense that it may be constructed by "adding to D_i structure compatible with its existing structure." Of course, this is a qualitative description; one requirement for making it precise is that it should be possible to identify a "copy of the parent history in the child history," and this relationship extends in an obvious way to apply to initial and terminal histories separated by many evolutionary stages. Figure 1.4.5 identifies such a copy in the present case. This figure also illustrates an important convention that is probably self-evident from the foregoing discussion, but which is worth spelling out nonetheless: the exact size and shape of a generalized Hasse diagram of a directed set is immaterial; only the cause-and-effect information represented by the diagram is significant. In particular, the fact that the relation between the two elements in D_i is "drawn vertically," on the left-hand side of the figure, while the corresponding relation in D_t is "drawn diagonally," is of no consequence whatsoever; all that matters is that each relation represents the flow of causal influence "up the page," from the "lower" element to the "higher" element.

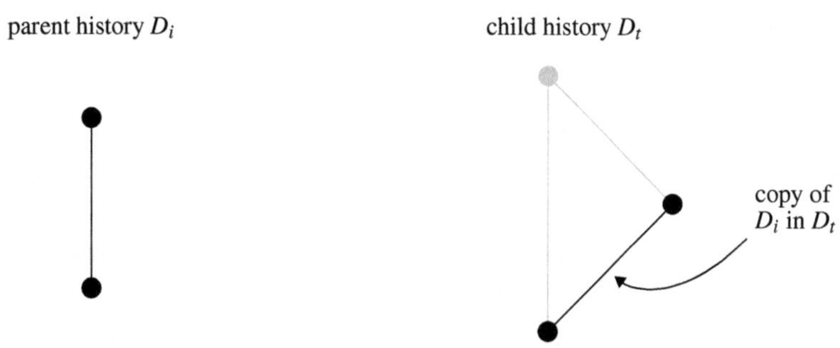

Figure 1.4.5. Identifying a copy of the parent history D_i in the child history D_t.

The arrows on the edges representing the relations in the right-hand diagrams of Figures 1.4.1 and 1.4.4 may be viewed as local versions of the "arrow of time," corresponding to the increasing time coordinate in the relativistic spacetime diagram in Figure 1.3.2. Up to this point in the discussion, I have restricted attention to a special class of classical histories, namely, *acyclic* classical histories, in which each history possesses a "global time direction." Informally, acyclic classical histories represent "universes in which communication with the past is forbidden," i.e., universes in which there exists no causally-directed path that both begins and ends at the same event. Many physicists believe that physical spacetime is acyclic in this sense, at least on large scales, but general relativity is ambivalent on the subject. This latter fact carries considerable weight, since general relativity is the only experimentally established theory that abstains from "fudging in the answers" regarding causal structure. This issue is discussed further in Section 2.7, and again in Section 8.7.

1.5 The Relative Viewpoint

Grothendieck. Figure 1.4.5 provides a simple example of a very important and general viewpoint, called the *relative viewpoint* (RV), championed beginning in the 1960s by the late mathematical titan Alexander Grothendieck, primarily in the context of algebraic geometry. According to the relative viewpoint, mathematical *objects* should be studied together with their *natural relationships,* rather than in isolation. In general, this viewpoint remains insufficiently appreciated and ineffectively applied in mainstream modern physics. In discrete causal theory, however, the relative viewpoint applies naturally in a variety of important ways. At the lowest level of structure, "objects" are merely elements of a directed set, and "natural relationships" are relations between pairs of elements. In the context of Section 1.4, however, "objects" are directed sets, viewed as classical histories, and "natural relationships" are *evolutionary* relationships, such as the relationship between the parent history D_i and the child history D_t illustrated in Figure 1.4.5. Following the relative viewpoint, one should resist the temptation to consider these histories separately; instead, *the relationship between them should be considered as a single coherent entity.* Numerical or algebraic quantities measuring the "likelihoods" or "favorabilities" of such relationships are roughly analogous to entries of the *S-matrix* in conventional quantum field theory,[21] which encodes the "likelihoods" of transitions between specified pairs of "initial" and "terminal" states. Quantities associated with arbitrary *evolutionary pathways* between pairs of classical histories, which may involve multiple "intermediate stages," are analogous to the *phases* associated with particle trajectories in

[21]One reason why the analogy is merely "rough" is because multiple distinct relationships may exist between a given pair of classical histories, even a pair differing by a single element. This subtlety is explained in Section 6.4. Also, the S-matrix was introduced to treat *asymptotic* behavior, "ignoring the details of the interaction," while the relationships of principal interest in the discrete causal context are often "direct" or "immediate" relationships. Such relationships generally do not exist in the continuum-based setting, due to the *interpolative property* of the real numbers $\mathbb{R}$.

Feynman's *path summation approach* to quantum theory. These quantum-theoretic notions are studied in detail in Part II of the book, particularly in Chapter 6.

Category theory; morphisms. A standard mathematical framework for studying "natural relationships between pairs of objects" is provided by *category theory.* Informally, a *category* is a collection of *objects* of a particular type, together with a collection of special relationships between pairs of objects, called *morphisms.* In most of the "concrete" examples encountered in elementary algebra or geometry, these morphisms are in fact *maps,* i.e., functions, which "respect the structures" of the corresponding objects in an appropriate way. For example, maps such as *diffeomorphisms* and *conformal isometries* are special types of morphisms between pairs of spacetime manifolds in general relativity; these morphisms are revisited in Chapter 2 in the context of metric recovery. In the present case, the relationship between the parent history D_i and the child history D_t illustrated in Figure 1.4.5 may be understood in terms of a *morphism of directed sets,* i.e., a special map from D_i to D_t that "preserves the direction of the relation in D_i," as explained below. This morphism is illustrated in Figure 1.5.1.

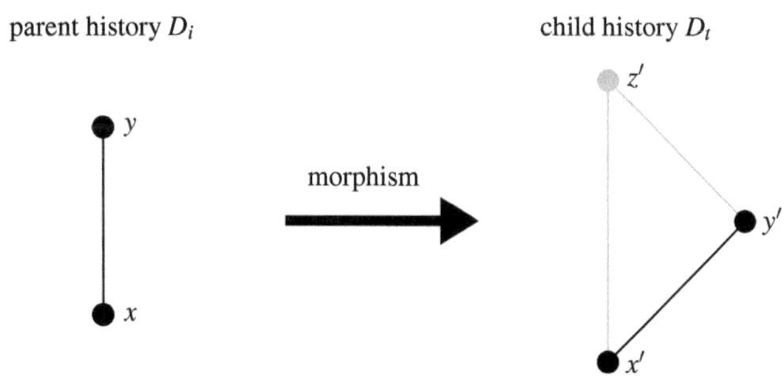

Figure 1.5.1. Relationship between a parent history and child history in terms of morphisms. Here, the morphism sends x to x' and y to y'.

For convenience, I have assigned labels to the elements of the two histories D_i and D_t in the figure, namely x, y, x', y', and z'. The morphism maps the elements x and y of the parent history D_i to the elements x' and y' of the child history D_t, respectively. It "preserves structure," in the sense that the relation between the elements x and y in D_i is reflected in the relation between their images x' and y' in D_t. Using the imprecise analogy of family descent, one might refer to this as a "family resemblance:" the "child looks like the parent" in certain ways. In a more rigorous sense, this morphism renders precise the statement in Section 1.4 that the child history "contains a copy" of the parent history.

Co-relative histories. For future reference, I should warn the reader that, in general, category theory is *not* ideally suited for describing natural relationships between pairs of classical histories in discrete causal theory. In particular, the family of natural relationships between two such histories, defined so as to encode information of actual physical significance, generally does *not* identify, in any straightforward manner, with a family of category-theoretic morphisms between the corresponding directed sets. There are at least two reasons for this. First, not all such morphisms encode "essentially distinct" information about relationships. For example, some morphisms differ only due to properties specific to one of the directed sets involved, in particular, its symmetries. Second, the sense in which morphisms of directed sets are "natural" is generally not "natural enough" for the desired applications; it is generally preferable to restrict attention to relationships that faithfully represent some type of physical process, such as the evolution of one classical history into another. Hence, a natural relationship between two classical histories, represented by directed sets D_i and D_t, is defined to be a subtler entity, namely, an equivalence class h of special morphisms from D_i to D_t, usually denoted in this book by $h : D_i \Rightarrow D_t$. I refer to such a relationship as a *co-relative history.* Here, D_i is called the *initial history,* or *source history,* and D_t is called the *terminal history,* or *target history.*

The long edges connecting the large open nodes in the illustration of the positive sequential kinematic scheme $\mathbb{S}_{PS}$ in Figure 1.1.1 represent co-relative histories. The history "lower on the page" in each co-relative history is the source history, while the history "higher on the page" is the target history. Similar relationships appear in the context of causal set theory; in particular, the *transitions*[22] in Sorkin and Rideout's theory of *sequential growth dynamics* [SR99] *represent* particular types of co-relative histories, in a sense made precise in Chapter 6. Families of such transitions, organized into diagrams very similar to the one appearing in Figure 1.1.1, may be found in the literature. Co-relative histories play a similar role to that of transitions, but this role is both more general and more precise; essentially, transitions are examples of the "special morphisms" representing co-relative histories mentioned above, but some pairs of transitions are physically equivalent. Further, the theory of co-relative histories takes into account "higher-level multidirected structure" that seems to have been overlooked in existing versions of discrete causal theory. Chapters 6 and 7 explain these ideas in detail.

Deeper analysis is necessary to actually justify the assertion that category theory is not ideally suited for describing co-relative histories. For example, one might suggest simply defining a new category, in which the morphisms *are* co-relative histories. This naïve "fix" fails, however; co-relative histories do not compose in the same way that category-theoretic morphisms do. What this deeper analysis reveals is that category theory is simply not the optimal language for describing discrete causal theory in general, although it is "close to optimal," due to its emphasis on structural hierarchy and directed relationships. From a historical perspective, this lack of a perfect match between the two theories is not very surprising; category theory was invented by Eilenberg and MacLane in the 1940s to describe certain special relationships, namely, *natural transformations of functors,* which they had encountered in the context of algebraic topology. It would be a miracle if the same formalism were precisely tuned to describe theoretical physics. Nevertheless, category theory has become extremely

[22]These transitions have nothing to do with the *graph-dynamical phase transitions* discussed in Section 1.9.

popular in physics over the last generation, and some of the ideas arising out of this viewpoint are both important in their own right, and relevant to discrete causal theory. Foremost among these ideas is Christopher Isham's general *topos-theoretic approach to physics,* revisited in Chapter 6.

Before proceeding further, I should briefly explain the reasoning behind the choice of the term "co-relative history." The word "history" means, of course, "classical history." The necessity for considering multiple classical histories simultaneously arises in the quantum-theoretic context, where families of such histories are studied collectively under a suitable adaptation of the path summation approach to quantum theory. The word "relative" indicates that the construction involves a relationship between two classical histories. The prefix "co" is borrowed from category theory, and indicates that the "direction of the relationship" in a co-relative history is the opposite of the corresponding direction in familiar "relative constructions," such as the construction of "relative schemes over a fixed base" in algebraic geometry. Again, these ideas are discussed in more detail in Chapter 6.

Relation space. If one takes the relative viewpoint seriously, it is natural to try to apply the same philosophy *inside* each classical history, i.e., to describe such a history "first and foremost in terms of its relations," rather than in terms of its elements. Carrying this approach to the extreme, one might even choose to regard isolated elements as oxymoronic, rather like the "sound of one hand clapping," and might treat elements in general as mere "intersections of relations," with no independent ontological role. This juggling of viewpoints turns out to be much more significant than a mere philosophical exercise. In particular, an important and vexing technical problem central to the dynamics of discrete systems; namely, the *permeability problem* discussed in Chapter 5, disappears when the relative viewpoint is applied in an appropriate way. What does it mean, precisely, to view a classical history "in terms of its relations?" Here, I merely present an example, involving the "child history" D_t illustrated in Figure 1.4.4. I streamline the notation in this example, denoting this history by just D, since its relationship to the "parent history" D_i is immaterial to the present consideration of its internal structure. The first step in translating to the "relation only" perspective is to identify all the relations in D. In this case there are three relations, which I label r_0, r_1, and r_2.[23] The next step is to construct a *new* classical history $\mathcal{R}(D)$, called the *relation space over* D, whose *elements* correspond to the relations of D. The choice to study the structure of D in terms of its relation space represents a specific realization of the relative viewpoint in the context of classical spacetime, which I sometimes refer to as the "relation space viewpoint."[24] For comparative purposes, I sometimes refer to the "original," or "naïve," or "default" viewpoint regarding D, as the "element space viewpoint." Construction of the rela-

---

[23]The letters r, s, and t would be a bad choice, because t is used in Chapter 5 to denote a "terminal element map" in the same context. Even here, t would clash with the subscript in D_t.

[24]However, as described in Chapter 5, the mathematical construction of relation space may be applied to structures with physical interpretations different than models of classical spacetime. In fact, one may construct the relation space over any multidirected set.

tion space $\mathcal{R}(D)$ is illustrated in Figure 1.5.2. Square nodes are used to represent elements of $\mathcal{R}(D)$.

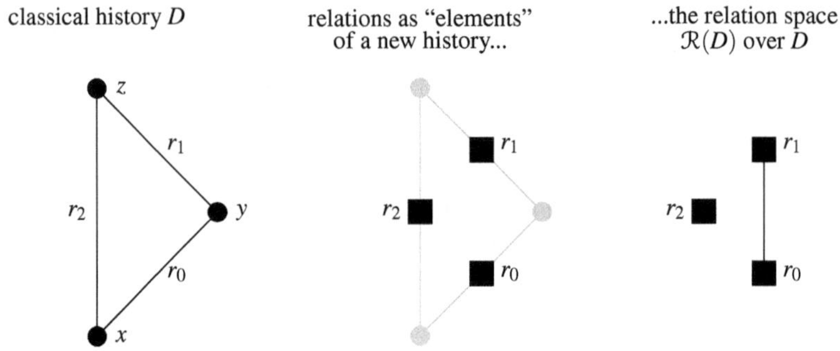

Figure 1.5.2. Relation space viewpoint.

It is simple enough to understand how the elements of $\mathcal{R}(D)$ are derived from the structure of D, but the reader will notice that $\mathcal{R}(D)$ also possesses a relation, as shown in the right-hand diagram in Figure 1.5.2. Where does this "induced relation" come from? It turns out that the "correct" convention is that r_0 and r_1 are related in $\mathcal{R}(D)$ if and only if the terminal element of r_0 in D coincides with the initial element of r_1 in D. Since the terms "element" and "relation" have two possible meanings in this context, the easiest way to avoid confusion is to rely on the concrete example illustrated in the figure. Here, the element y in D is both the terminal element of r_0 and the initial element of r_1, so there is a relation between r_0 and r_1 in $\mathcal{R}(D)$. I remark that this is one of several important areas in which naïve adaptation of category-theoretic concepts to the discrete causal context leads in the wrong direction. In particular, a relation space is "something like a morphism category," but the most popular definition of "morphisms between morphisms" leads to the wrong concept of induced relations in this context. The analogy between relation spaces and morphism categories is further elaborated in Chapter 5.

Recovery of a classical history from its relation space; boundary issues. Experimenting with the relation space construction, one quickly discovers that it is not quite possible to uniquely reconstruct a classical history from its relation space. For example, the three classical histories shown in Figure 1.5.3 all share essentially the same relation space as the history D appearing in Figure 1.5.2. By "essentially the same," I mean that their relation spaces are isomorphic; i.e., they share the same abstract structure as directed sets.

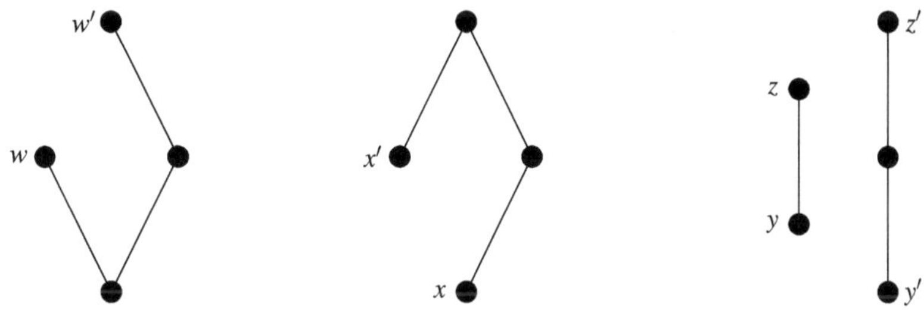

Figure 1.5.3. Three classical histories with isomorphic relation spaces.

This failure of "perfect information preservation" upon passage to relation space might at first appear to be an indication that the relation space viewpoint is seriously deficient. After all, what sense is there in describing a classical history "in terms of its relations," if such a description does not uniquely specify the history? It turns out that this problem is not nearly as serious as it appears in these examples; the "tiny" sizes of the histories involved produces a misleading impression. In fact, the relation space $\mathcal{R}(D)$ over a classical history D *does* uniquely specify D up to "information on the causal boundary," i.e., information about what happens "at the beginning and end of time." The missing information involves questions about whether or not certain relations "intersect" at their initial and/or terminal elements. For example, if one "joins up" the pairs of elements w and w', x and x', y and y', and z and z' in Figure 1.5.3, one recovers three copies of the familiar history D from Figure 1.5.2.

Figure 1.5.4 illustrates the relation space over a somewhat more complicated classical history, which I again denote by D to avoid proliferating symbols. The observant reader may notice that D is a "piece hacked out of" the classical history appearing in Figure 1.4.3. There is nothing particularly special about D, but because it is both "large enough to be interesting" and "small enough to be tractable," I use it repeatedly in illustrations throughout the remainder of the book. It is interesting to reflect that D and $\mathcal{R}(D)$, despite their very different appearances, contain essentially the same information, up to some ambiguity about what happens on the causal boundary.

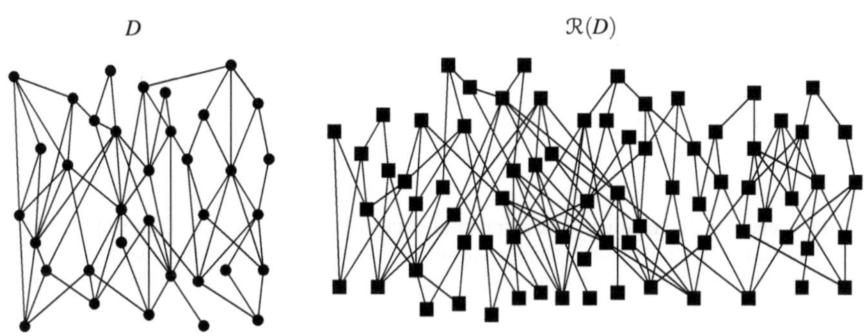

Figure 1.5.4. Relation space over a more complicated classical history.

Anyone who has studied partial differential equations will recognize that the problem of recovering a classical history from its relation space looks very much like a *boundary value problem.* This is not an accident; in both cases, two types of data are involved: "information about local relationships in the interior of a region," and "information living on the boundary of the region." In the context of partial differential equations, the "information about local relationships" is expressed in terms of partial derivatives, while in the discrete causal context, it is encoded in the cause-and-effect relationships that collectively make up relation space. Since most of mainstream physics involves partial differential equations, structural similarities of this type are of obvious interest. It turns out that many of the conceptual tools familiar from the study of partial differential equations carry over to discrete causal theory with surprisingly little modification. For example, discrete causal analogues of *Cauchy surfaces* play a central role in some of the major technical developments in this book, due to their importance in the solution of discrete initial value problems.

1.6 Background Independence

Dynamical character of spacetime. In many areas of conventional physics, such as the versions of quantum field theory involved in the standard model of particle theory, spacetime is treated as an "arena in which particles and fields interact." This viewpoint survives, in an operational sense, even though the consensus of the modern theoretical physics community is that it is mistaken. Since the advent of general relativity, it has been understood that spacetime should be treated as an active part of the dynamical picture, not merely as a static background. The general principle that *all the constituents of a theory should play an active and coherent role in the dynamics of the theory* is called **background independence.**[25] In particular, background independence implies that spacetime is "part of what happens," not merely "a place where things happen." The most obvious reason why background *dependent* theories continue to enjoy currency is simply because it has proven very difficult to construct workable background independent quantum theories; in particular, quantum theories of gravity. The reason why gravity is exceptional in this context is precisely *because* it is understood as an aspect of spacetime structure, rather than as an interaction, i.e., a "force" in the usual sense.[26] One of the principal criticisms of string theory is that it is usually formulated and studied in a background dependent context.

Perfect background independence. Even general relativity is not "perfectly background independent," because a background of sorts still exists in the theory. The

[25]This is one possible qualitative description of background independence. Others describe the principle quite differently; for example, Thiemann [TH07] equates it with *general covariance* in the relativistic context. In discrete causal theory, the terms "background independence" and "covariance" are assigned very different meanings.

[26]Of course, the strong interpretation of the causal metric hypothesis (CMH) assumes "interactions," or "forces," to be ultimately structural in nature as well.

"spacetime arena containing particles and fields" interacts with its material content according to Einstein's field equation (1.3.1), but one may still distinguish between the two. The same essential distinction between "spacetime" and "nongravitational matter" appears in causal set theory, as discussed in Chapter 3. A colorful metaphor is supplied by the "dynamical arenas" of the ancient gladiators, which interacted with the chief participants via trapdoors, large felines, and other interesting features. The precise status of the "spacetime background" in general relativity, and similar theories, calls for careful consideration. Rovelli [RO04], for example, prefers to regard general relativity as a theory of *"fields on fields,"* and argues that there is "no meaning in talking about" physical spacetime elements. Following this perspective, he expresses the theory in terms of a *tetrad field,* which he calls the "gravitational field," rather than in the classical manner via the metric tensor, as already described in one of the footnotes in Section 1.3. This "geometry-deprecating" viewpoint is supported by Einstein's famous "hole argument" regarding *general covariance,* whereby he concluded that only "spacetime coincidences of material points," not individual elements of a pseudo-Riemannian manifold, are physically meaningful. Rovelli asserts that:

> *What disappears... is precisely the background spacetime that Newton believed to be able to detect...* ([RO04], p. 71)

While it is certainly true that general relativity assigns no physical meaning to specific, individual elements of a pseudo-Riemannian manifold, there remains a great deal of topological and geometric background structure that *is* taken for granted in the theory; specifically, the smooth, four-dimensional real manifold on which the gravitational field is defined. In particular, the absence of a fixed spacetime *metric* as a *"distinguished, externally prescribed, background structure"* on this manifold,[27] only incrementally decreases the amount of structure taken for granted. Regardless of how one chooses to view this structure, it can be distinguished from the "material entities" of the theory, and this leaves a lingering potential for inconsistencies. This is the reason for including the word "coherent" in the above description of background independence. The "inconsistencies" I am referring to here are the "causality paradoxes," and "time-travel paradoxes," that have wasted so much effort, and so much ink, since the development of general relativity. These are briefly revisited in Section 2.7. The degree of background independence exhibited by a theory provides one way of gauging whether, and to what extent, it improves on the present state of structural understanding in physics. From this perspective, certain mainstream modern approaches to physics beyond the scope of relativity and the standard model are actually *less* satisfactory than general relativity itself. Under the strong interpretation of the causal metric hypothesis (CMH), one may aspire to build a theory that is *perfectly* background independent, describing "spacetime, particles, and fields" as aspects of something more fundamental, namely, causal structure.

Background independence and universal properties. The observant reader may have already noted that the version of discrete causal theory developed in this book

[27] As expressed by Thiemann [TH07], p. 9.

makes use of "fixed backgrounds" at the *quantum* level, namely, structured configuration spaces of classical histories, such as the *positive sequential kinematic scheme* $\mathbb{S}_{\mathrm{PS}}$, illustrated in Figure 1.1.1. One might argue, on this basis, that the theory merely "hides" background dependence by moving it to a different level of structure. There are at least two answers to this objection. First, although $\mathbb{S}_{\mathrm{PS}}$ is the only kinematic scheme illustrated in the present chapter, there exist other equally valid kinematic schemes constructed from the same class of classical histories, and describing the same evolutionary processes. These include the *generational kinematic scheme* $\mathbb{S}_{\mathrm{GEN}}$, and the *foliation scheme* $\mathbb{S}_{\mathrm{FOL}}$, introduced in Chapter 7. Hence, the theory does not depend on a specific choice of "quantum background."

Second, and more fundamentally, it is important to recognize that the most objectionable attribute of background dependent theories is not their description in terms of specific objects among a class of objects, but the fact that these objects are *arbitrary;* i.e., there is no "God-given" reason why they *must* contribute to the theory, and no mechanism within the theory itself to explain how or why they *do* contribute. In Newtonian physics, for example, one may immediately ask why the spacetime background has the structure of $\mathbb{R}^4$, rather than S^4, or $\mathbb{R}^5$, or any number of other specific choices. General relativity is better in this regard, because it provides at least a partial explanation of *how spacetime structure arises physically.* Still better, however, is to incorporate *all* structures of a given "physically reasonable" class into a theory, thereby eliminating any questions regarding why one specific structure should contribute rather than another. This is one of the desirable features resulting from the use of a "quantum background" such as $\mathbb{S}_{\mathrm{PS}}$.[28] One may express this situation by saying that $\mathbb{S}_{\mathrm{PS}}$ possesses a *universal property,* which renders it "natural," rather than "arbitrary." In this case, the universal property may be described informally by the statement that *"there exists at least one evolutionary pathway in $\mathbb{S}_{\mathrm{PS}}$ to describe the evolution of every permissible classical history."*

1.7 Particles and Fields

Ranking theories via background independence. The idea of comparing theories of fundamental physics according to their degrees of background independence provides a convenient way of describing how particles and fields may be modeled in different versions of discrete causal theory. These range from versions that are "less background independent than general relativity," but possibly useful for investigative or computational purposes, to highly ambitious versions, exhibiting perfect

[28] The unfortunate modern prominence of the *string theory landscape* and the *anthropic principle* places a burden on new theories to offer assurances against the use of similarly non-explanatory mechanisms, under which "anything goes, somewhere." In the present case, the incorporation of all "physically reasonable" structures means essentially the same thing as in Feynman's path summation approach to ordinary quantum theory: structures favored by the dynamics of the theory are reinforced, while disfavored structures are damped out via destructive interference.

background independence. At this stage of the discussion, I present only illustrative glimpses of these various approaches.

Fields in static discrete spacetime. The "least background independent way" to model particles and fields in discrete causal theory is to simply *fix a particular classical history* and build a field theory on it. As one might expect, this approach is also generally the easiest with which to perform computations. From a long-term perspective, such theories might be considered "underwhelming," since they obviously do not provide a truly fundamental description of physics. However, they do supply "topologically random" generalizations of *lattice field theories,* and discrete analogues of *quantum field theories in curved spacetime,* both of which play useful roles in a variety of physical contexts. From a more practical viewpoint, such theories should be considered "interesting" if they provide any chance of describing or predicting behavior beyond the scope of relativity and the standard model. As it happens, there are a number of existing research projects devoted to approaches of this nature. Some of these are briefly discussed in Chapter 3.

Figure 1.7.1 illustrates a very simple example of fields on a "toy history" D. Here, ϕ is a field on the set of elements of D; it has just three values, ϕ_1, ϕ_2, and ϕ_3, corresponding to the three elements of D. Similarly, θ is a field on the set of relations of D, with values θ_1, θ_2, and θ_3. Either or both kinds of fields may play a role in this type of theory. Here, of course, I am using the word "field" in the physicist's sense, not in the sense of abstract algebra; for example, the "element field" ϕ may be expressed as a map $\phi : D \rightarrow T$, for some "target object" T. To draw an analogy with conventional continuum-based physics, D might be compared to a manifold, T to a bundle of some type over D, and ϕ to a section of T. In the present context, however, T could be practically any object imaginable; a particularly simple choice would be $T = \mathbb{Z}$, in which case ϕ would be an "integer-valued classical scalar field." In a quantum-theoretic setting, T might be an algebra of operators on a Hilbert space, with ϕ assigning a different operator to each element of D. Or, ϕ might assign an entire Hilbert space to each element, or a Hilbert space together with a distinguished algebra of operators, or any number of other possibilities.

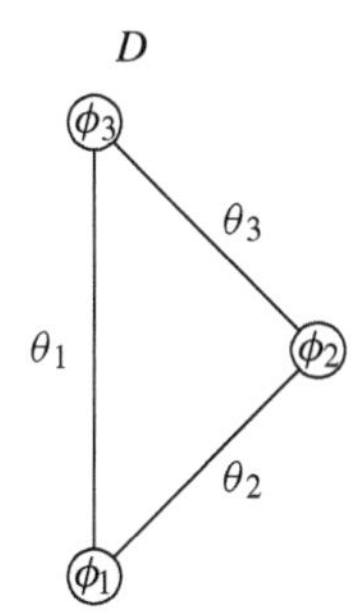

Figure 1.7.1. Fields on a toy history D.

"Relation fields" such as θ, which I often refer to as *relation functions,* are entities of great interest in this book, particularly in the quantum-theoretic setting. The relations of D themselves resemble *timelike tangent vectors,* discussed in Section 2.6, so θ is roughly analogous to a function on the *tangent bundle* over a spacetime manifold. A conventional example of such a function is the *Lagrangian,* which measures the difference between a particle's kinetic and potential energies in the context of elementary classical mechanics. The Lagrangian determines the *classical action* along any "permissible" path in spacetime, viewed as a particle trajectory, which in turn determines the value of the *phase map* associated with this path in Feynman's path summation approach to quantum theory. All of these relationships are explained in detail in Chapter 6. Relation functions play an analogous, and equally crucial, role in discrete causal theory. They have been studied extensively in causal set theory; for example, by Sorkin and Rideout [SR99]. It is no coincidence that the same symbol θ is used for the "relation field" illustrated in Figure 1.7.1 and the relation function appearing in the causal Schrödinger-type equation (1.1.2). Of course, as explained in Section 1.1, the relations appearing in Equation (1.1.2) actually represent co-relative histories, so in this context θ is a "higher-level relation function" for a background independent theory. However, similar "higher-level functions" may be associated with relations even in the simple background *dependent* case illustrated in Figure 1.7.1. In particular, if algebraic objects are assigned to elements of D, as described above, then it is natural to assign *morphisms* to the relations of D. In this case, ϕ and θ might represent the "object map" and "morphism map" of a single *functor,* i.e., a category-theoretic version of a "higher-level function." The theory of *quantum causal histories,* developed by Fotini Markopoulou and her collaborators [MA00, HMS03], represents an important example of this type of construction. Christopher Isham's *quantization on a category* [IS05], revisited in Sections 3.8 and 6.1, provides a compelling general context for such approaches. In the present toy scenario, however, ϕ and θ are just functions between finite sets, whose targets are left unspecified.

It should be clear that only a weak interpretation of the causal metric hypothesis is possible under a theory like the one illustrated in Figure 1.7.1. Although the underlying classical history D in the figure is defined in terms of "causal structure," the actual meaning of this designation, as it pertains to any reasonable understanding of cause and effect, is highly questionable. Most of the information content of the theory is contained in the fields, so "what happens" in the theory is largely dependent on the algebraic structure of ϕ and θ, not the directed structure of D. The latter structure is referred to as "causal structure," but it plays only a limited role in the *actual process of cause and effect;* in fact, its only obvious contribution is to impose certain constraints on the fields. For example, the value of ϕ_3 "should depend on both the values of ϕ_1 and ϕ_2," since there are relations between the corresponding elements. Further, there remains the problem of explaining where the algebraic data encoded by ϕ and θ comes from, and why this data should take on certain values rather than others. All this is very awkward.

Fields in dynamical discrete spacetime. A much more interesting and serious type of discrete causal field theory is one that applies to *all* classical histories of a particular class, and to all fields of a particular type on these histories, and that involves constraints relating field values to causal structure, and vice versa, in a nontrivial way. In such a theory, *the causal structure affects the fields, and the fields affect the causal structure.* This type of theory is background independent in the same sense that general relativity is background independent; each classical history still serves as a background of sorts, but not a static background. The most obvious way to build a quantum theory in this context is via a "sum over evolutionary pathways for decorated classical histories," where "decorated" means that each history is equipped with appropriate fields. Causal set theorists such as Sorkin [SO12] have done notable work along these lines, although the "classical histories" of causal set theory do not overlap exactly with those emphasized in this book. Isham's category-theoretic methods [IS05] apply equally well in this setting.

Emergent field theories. The most interesting and ambitious versions of discrete causal theory are *perfectly background independent.* They are based on the strong interpretation of the causal metric hypothesis (CMH), which treats "particles and fields," not only "spacetime," as emergent manifestations of discrete causal structure. How could such an approach possibly account for all the variety of modern particle theory? The first point to understand in attempting to answer this question is that there is a great deal of natural "particle-like" and "field-like" information available "for free" in discrete causal structure itself, and that this information is "new" in the sense that it is generally absent in continuum-based theories. The simplest such information exists at the classical level, and may be expressed in the form of *integer-valued scalar fields* that "count combinatorial aspects of local structure." To give a concrete example, it is useful to borrow some graph-theoretic terminology: the **valence** $v(x)$ of an element x in a directed set D, viewed as a classical history, is the number of relations r in D such that x is either the initial or terminal element of r.[29] The function v defines a natural number-valued scalar field on D, which I refer to as the *valence field.* An example of the valence field on a classical history D is illustrated in the left-hand diagram in Figure 1.7.2.

The valence field v may be expressed as a sum of two more-primitive fields v^+ and v^-, which count the number of "outgoing" and "incoming" relations at each element, respectively. The difference $v^+ - v^-$ of these two fields is illustrated in the right-hand diagram of Figure 1.7.2. Elements with positive values of $v^+ - v^-$ are "net sources," in an obvious sense, while elements with negative values of $v^+ - v^-$ are "net sinks." These are illustrated by white nodes and black nodes in the diagram, respectively. Gray nodes indicate elements with a "balance" of incoming and outgoing relations, i.e., nodes where $v^+ - v^-$ vanishes. It is worth noting, for future reference, that the valence field v is most interesting under an appropriate notion of *local finiteness* for directed sets; otherwise its values can be infinite, and one faces the same types of

[29] As explained in Section 4.3, *reflexive relations* $x \prec x$ are counted twice in determining $v(x)$, because such relations both begin and terminate at x. The examples considered in this section involve only *acyclic* directed sets, which do not exhibit such reflexive relations.

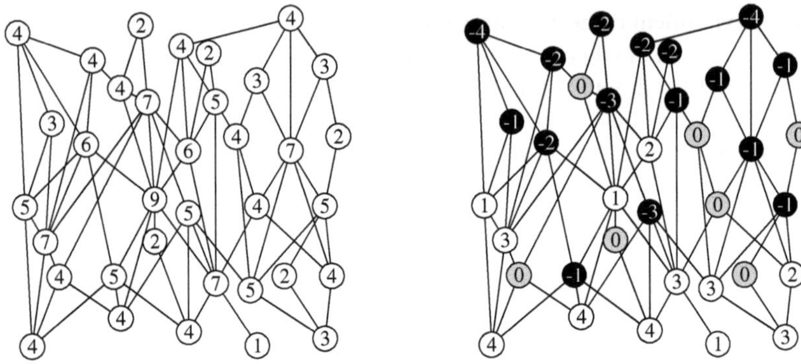

Figure 1.7.2. Simple integer-valued scalar fields on a classical history D.

divergence issues that often arise in continuum-based theories. Considerations of this nature are important in the study of *local properties* of directed sets, undertaken in Chapter 4. However, as already hinted in Section 1.3, the values of v might be very large for "realistic" classical histories.

The valence field v, and the related fields v^+ and v^-, are only the simplest of many natural fields on D, where "natural" in this context means "inherent in the causal structure." It would be unreasonable to expect the familiar fields of conventional physics to emerge easily from such a picture; indeed, the present focus is to understand how more basic properties such as spacetime dimension emerge dynamically. The objective here is merely to provide a glimpse of the vast quantity of natural arithmetic and algebraic data lurking in the causal structure of a typical discrete directed set. This suggests that discrete causal theory can draw heavily from existing order theory, number theory, and combinatorics in modeling physical interactions. As indicated above, the most ambitious and optimistic approaches to the theory *restrict* attention to this natural content, resisting the temptation to import extrinsic objects such Hilbert spaces and operator algebras, although such objects may emerge from intrinsic data. It is worth noting that the available information content is greatly amplified in discrete *quantum* causal theory, where entire families of classical histories are available to work with.

Emergent particles. I now risk illustrating a drastically oversimplified view of how particles might emerge from discrete causal structure without the necessity of importing auxiliary mathematical content such as Hilbert spaces. Again, the classical context suffices for illustrative purposes. Figure 1.7.3 shows a "particle trajectory in a classical history," viewed at several different scales. This trajectory is represented by a black curve in the left-hand diagram, and by black nodes and edges in the middle and right-hand diagrams. At the moment, I am relying on a purely intuitive notion of what the word "particle" means in an emergent sense; namely, a "spatially localized family of events that retains similar internal structure over a time interval." Ultimately, one must be mathematically precise about such notions; for example, conventional background dependent quantum field theory describes particles in terms of the representation theory of Lie groups. Of course, "space" and "time" are themselves taken to be emergent in discrete causal theory, but it should be reasonably obvious that the black nodes in the right-hand diagram represent a family of "spatially localized events" in a rough sense, and that the nodes near the bottom of the diagram, which exist at an "earlier time," are interrelated in approximately the same way as the nodes near the top of the diagram, which exist at a "later time." At the end of Section 8.2, I offer a speculative idea for how such emergent particles might be related to the theory of *finite simple groups.*

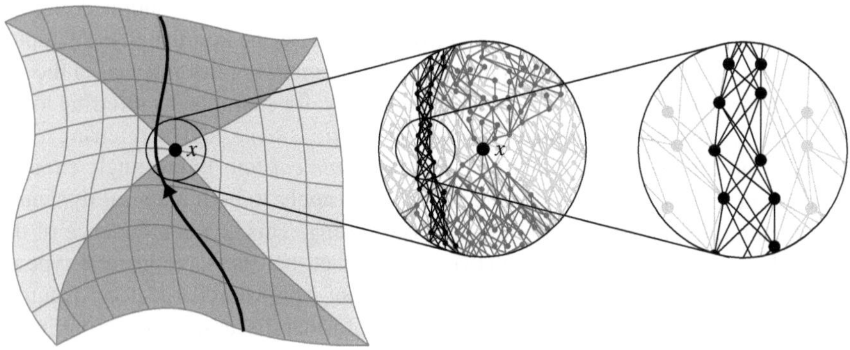

Figure 1.7.3. A "particle trajectory" in a classical history.

The emergent treatment of particles has potentially important consequences, even in relatively conventional settings. For example, Dreyer [DRE15] and others have advocated the view that a consistent treatment of particles as *emergent aspects of spacetime,* rather than as "separate entities" existing *on* spacetime, can lead to possible insights into famous problems such as the magnitude of the "cosmological constant," and the nature of "dark matter,"[30] using arguments otherwise based on a relatively conventional paradigm involving four-dimensional spacetime geometry.[31] The reader may already be familiar with historical attempts to model particles as geometric or topological "ripples" or "defects" in a real manifold; Wheeler's *geons* [WH98] are one example. The discrete causal analogue is similar, but the type of basic structure involved is much more flexible, and hence much friendlier to such constructions. Generally speaking, one must do drastic things to a continuum-based model of spacetime in order to obtain particle-like entities. This is because such models are, by definition, very uniform at the local level, and particles necessarily break up this uniformity in some way. In discrete causal theory, however, the structure is uniform only at "sufficiently large scales," and in an emergent sense, and this

[30] As in the case of the "cosmological constant," the use of quotation marks is intended to highlight the fact that the term "dark matter" itself imposes a rather specific and conventional, though well-supported, interpretation on certain observed phenomena.

[31] See the recent preprint of Saravani and Aslanbeigi [SA15] for an interesting discussion about the possible origin of dark matter due to nonlocal effects in discrete gravity.

tendency toward *local irregularity* is an advantage for modeling a physical universe that is manifestly locally irregular. In a metaphorical sense, it is easier for the material participants of physical interactions to blend into the "jungle" of discrete causal theory than the "desert" of conventional continuum spacetime.

1.8 Kinematics and Dynamics

Distinguishing between kinematics and dynamics. I have now supplied enough conceptual background to enable a brief overview of how the familiar physical topics of kinematics and dynamics are approached in the context of discrete causal theory. So far, I have touched on these topics only indirectly; for example, by referring to the structure $\mathbb{S}_{\mathrm{PS}}$ appearing in Figure 1.1.1 as a *kinematic scheme,* and by referring to the causal Schrödinger-type equation (1.1.2) as a *dynamical law.* Interestingly enough, it turns out that many of the same technical tools apply to both classical dynamics and quantum dynamics in the discrete causal setting. This is a consequence of the principle of *iteration of structure* (IS), i.e., the fact that the same general type of abstract directed or multidirected structure is present at both the classical and quantum levels of the theory. This means that even the informal discussion appearing in the present section captures many of the essential concepts of the much more rigorous discrete quantum causal dynamics developed in Part II of the book. Following the relative viewpoint (RV), I describe these concepts in terms of relationships between pairs of classical histories, i.e., co-relative histories.

I begin by spelling out the essential distinction between kinematics and dynamics in general terms. It is perhaps easier to perceive the information-theoretic source of this distinction in the discrete causal context than in any other field of modern physics. Some variation in the intended meaning of these terms may be found in the literature, but the description given here is relatively conventional from a sufficiently broad viewpoint. The *kinematics* of a theory describes *permissible* behavior, without specifying what *actual* behavior is determined, or favored, under any given set of conditions. This latter role is left to the *dynamics* of the theory. For example, the kinematics of special relativity permits any future-directed timelike trajectory for a massive particle, but to determine the trajectory of a *specific particle under specific conditions* requires dynamical information; in this case, an appropriate version of Newton's second law. Similarly, the kinematics of general relativity permits any spacetime metric among a particular class of metrics, but to determine how the metric *actually interacts with matter and energy* requires dynamical information; in this case, Einstein's field equation (1.3.1). In discrete causal theory, the strong interpretation of the causal metric hypothesis (CMH) "unifies kinematics and dynamics," in the sense that it requires the existence of a unique "correct" dynamics corresponding to a given choice of kinematics.

Kinematic schemes. I have already informally introduced the "structured configuration space" $\mathbb{S}_{\mathrm{PS}}$, illustrated in Figure 1.1.1, as the *positive sequential kinematic* *scheme.* It is now possible to give a preliminary explanation of the notion of kinematic schemes in general. The overall role of a kinematic scheme $\mathbb{S}$ is to specify families of permissible "evolutionary pathways," whereby "simpler" classical histories may evolve into "more complicated" classical histories. The "higher-level structure" of $\mathbb{S}$, which governs the corresponding quantum theory, and which is generally multi-directed in nature, is built from a suitable class of co-relative histories, viewed as individual "stages of evolution" for the corresponding class of classical histories. In many cases, it is useful to isolate this "higher-level structure," by ignoring the "internal structure" of the individual classical histories contributing to $\mathbb{S}$. This yields a *multidirected set* $\mathcal{M}(\mathbb{S})$, called the *underlying multidirected set of* $\mathbb{S}$, whose elements represent classical histories, and whose relations represent co-relative histories. The structure of $\mathcal{M}(\mathbb{S})$ alone does *not* determine which types of evolution are actually favored; rather, $\mathcal{M}(\mathbb{S})$ provides "kinematic scaffolding" for a particular version of discrete causal theory. To "flesh out" the theory and "bring it to life," one must supply dynamical information.[32] Under the strong interpretation of the causal metric hypothesis, the "correct" dynamical information for $\mathbb{S}$ is *ultimately determined by the directed structures of its individual classical histories.* Different kinematic schemes built from the same class of classical histories are assigned "different but equivalent" dynamical information, which describes the same physics under a suitable notion of *covariance.* Hence, discrete causal theory provides both a remarkably clear picture of the distinction between kinematics and dynamics, and an obvious method of unifying the two. These ideas are elaborated in Chapter 7.

Toy dynamics. A naïve way to supply dynamical information for a kinematic scheme $\mathbb{S}$ is to assign *probabilities* to each of its co-relative histories. The values of these probabilities define a "relation function on $\mathbb{S}$," by abuse of terminology, since they correspond to relations in the underlying multidirected set $\mathcal{M}(\mathbb{S})$ of $\mathbb{S}$. Conceptually, this function specifies the "likelihoods" that each source history will evolve into each target history. Such probabilities may be viewed as rough classical analogues of entries of the *S-matrix* in conventional quantum field theory, which are complex amplitudes measuring the "likelihoods of evolution" from specified "initial states" to specified "terminal states." However, a kinematic scheme may include multiple co-relative histories between a given pair of classical histories, and it is a priori necessary to assign separate "likelihoods" to each of these separate relationships. More generally, one must consider "likelihoods" associated with arbitrary *evolutionary pathways* between pairs of classical histories, which may involve multiple "intermediate stages of evolution." These "generalized likelihoods" are analogous to the *phases* of paths in the *path summation approach* to quantum theory, as explained below. Figure 1.8.1 illustrates a choice of probabilities for the co-relative histories of a small finite "subscheme" $\mathbb{S}$ of the positive sequential kinematic scheme $\mathbb{S}_{\text{PS}}$, including only classical histories with at most three elements. For future reference, I include labels on some of the individual classical histories appearing in the figure.

[32]This basic idea is not restricted to discrete causal theory; in particular, one may define kinematic schemes involving "classical histories" represented by other objects; for example, spin foams.

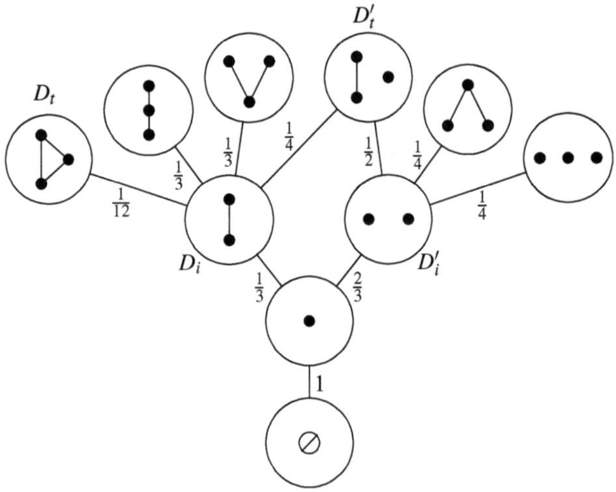

Figure 1.8.1. A choice of classical stochastic dynamics on a kinematic scheme $\mathbb{S}$.

The standard technical name for what I have just specified is a choice of *classical stochastic dynamics* for $\mathbb{S}$. This general class of dynamics includes, as a special case, the *sequential growth dynamics* introduced by Sorkin and Rideout in their 1999 paper *Classical sequential growth dynamics for causal sets* [SR99]. To avoid confusion, it is important to point out that the word "classical" in the phrase "classical stochastic dynamics" refers *not* to the individual classical histories, but to the fact that *real* probabilities are assigned to each co-relative history, rather than complex "quantum" amplitudes. The word "stochastic" means "probabilistic," or "nondeterministic." While there is nothing particularly special about the specific choice of dynamics appearing in the present example, it does illustrate some features of general interest. For example, the probabilities assigned to each family of co-relative histories sharing a particular *source* history sum to 1. This reflects the assumption that each source history will continue to develop in *some* fashion. In particular, the co-relative history whose source is the empty history $\emptyset$, and whose target is the history with one element, is assigned a probability of 1. The physical interpretation of this assignment is simply that "something exists;" i.e., "at least one element appears." At a more mathematical, and less philosophical, level, these assignments reflect a type of unidirectional "conservation law" or "current law," analogous to the current laws familiar from elementary electrodynamics. Ultimately, the direction of the "quantum arrow of time" may be inferred from such laws. This is related to the fact that the properties of $\mathbb{S}$ distinguish the "past to future" temporal direction over the "future to past" direction, even though the individual classical histories belonging to $\mathbb{S}$ do not. This *essentially quantum theoretic attribute of time* is revisited in Sections 6.5 and 7.5.

A familiar fact from elementary probability theory is that "independent probabilities multiply;" for example, the probability of obtaining the "child history"

D_t of Figure 1.4.4, which appears at the top left of the diagram in Figure 1.8.1, is $(1)\left(\frac{1}{3}\right)\left(\frac{1}{12}\right) = \frac{1}{36}$, under this particular choice of dynamics. Hence, in the present case, the "likelihoods" assigned to evolutionary pathways involving multiple stages are obtained by simply multiplying the "likelihoods" assigned to the separate stages. Altogether, there are six possible "final states" in $\mathbb{S}$, and seven possible evolutionary pathways to reach one of these final states, indicated by the gray arrows in the Figure 1.8.2. The fact that "independent probabilities multiply," together with the "current law" mentioned above, ensures that the sum of the probabilities over these seven evolutionary pathways is 1; these probabilities are $\frac{1}{36}, \frac{1}{9}, \frac{1}{9}, \frac{1}{12}, \frac{1}{3}, \frac{1}{6}$, and $\frac{1}{6}$, as shown in the figure. Two of these probabilities, namely, $\frac{1}{12}$ and $\frac{1}{3}$, represent the likelihoods of the two *different* pathways leading to the history labeled D_t'. It is important to point out that the two different *kinematic accounts* of the evolution of D_t', supplied by these two pathways, do not involve *physical distinctions,* any more than different choices of inertial frames of reference in special relativity involve physical distinctions. The physical information contained in D_t' merely indicates that "one event causes another, and there is an additional isolated event." The kinematic account supplied by the evolutionary pathway assigned probability $\frac{1}{12}$ adds the fictional detail that "the isolated event came after the two related events," while the pathway assigned probability $\frac{1}{3}$ argues, to the contrary, that "two isolated events appeared first, then one of these events caused a third event." This superficial descriptive conflict represents a discrete causal analogue of the *relativity of simultaneity,* under which the evolutionary pathways involved represent two different, equally valid, "frames of reference," or "points of view." This analogy is developed in much greater detail in Chapter 7.[33]

The "fictional time" involved in these statements, which involves arbitrary choices about "what happens first," is sometimes called *kinematic time.* The physically insignificant, or *extraphysical,* information involved in discriminating between different kinematic accounts, is sometimes called *gauge-like information.* This is because passage from one kinematic account to another may be viewed as analogous to a *gauge transformation* in conventional physics, which does not alter the dynamical laws of the theory. *Covariance* is a related term, which I use more generally to describe the systematic study of extraphysical information in a physical theory.[34] I give a brief description of discrete causal covariance in Section 8.3.

Toy analogue of causal Schrödinger-type equation. Using this very simple example of a classical stochastic dynamics, I now attempt to convey at least a rudimentary idea of how the causal Schrödinger-type equation (1.1.2) arises. I focus on the classical history D_t' in Figure 1.8.2. In particular, I express the probability $P(D_t')$ that

[33] A closer analogy exists between frames of reference and *refinements* of the causal relation on D_t', as explained in Sections 7.1 and 7.2.

[34] For historical reasons, the terms *invariance, covariance,* and *gauge* are all used in the study of extraphysical information, and these terms are imbued with many different shades of meaning in different contexts. I choose the term "covariance" to describe such considerations in the discrete causal setting. One reason for this choice is that the formation of the word "covariance" suggests the intended meaning that *"varying one's point of view demands a corresponding systematic variation in associated quantities."* In particular, the choice of phase map varies *along with* the choice of kinematic scheme in discrete causal dynamics.

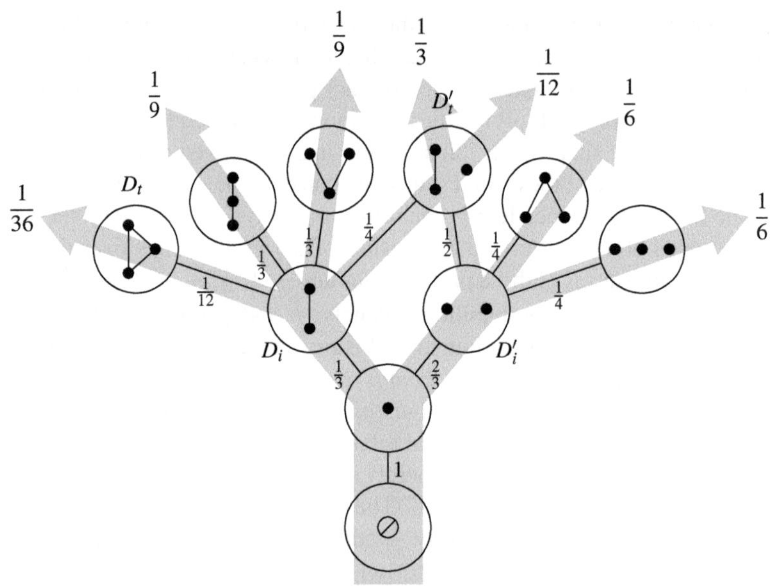

Figure 1.8.2. Evolutionary pathways, or kinematic accounts, in a kinematic scheme $\mathbb{S}$.

D'_t is the "final state," under this choice of dynamics, in terms of certain related probabilities. These are the "penultimate probabilities" $P(D_i)$ and $P(D'_i)$, associated with the two classical histories D_i and D'_i "immediately preceding" D'_t in the kinematic scheme $\mathbb{S}$, together with the "evolutionary probabilities" $p(D_i \Rightarrow D'_t)$ and $p(D'_i \Rightarrow D'_t)$, which encode the "likelihoods that D_i, and respectively D'_i, will evolve into D'_t." The reason for using lower-case p for these latter probabilities is to distinguish the fact that they correspond to *particular* co-relative histories,[35] while upper-case P represents probabilities over *all evolutionary pathways* terminating at the specified history. In the present case, it is easy to read off what these probabilities are:

$$P(D_i) = \left(1\right)\left(\frac{1}{3}\right) = \frac{1}{3}, \quad P(D'_i) = \left(1\right)\left(\frac{2}{3}\right) = \frac{2}{3},$$

$$p(D_i \Rightarrow D'_t) = \frac{1}{4}, \quad p(D'_i \Rightarrow D'_t) = \frac{1}{2}.$$

The desired probability $P(D'_t)$ may therefore be expressed as follows:

$$P(D'_t) = p(D_i \Rightarrow D'_t)P(D_i) + p(D'_i \Rightarrow D'_t)P(D'_i)$$

[35]In general, more than one co-relative history may exist between a given source and target, even in the case of sequential growth, but such details are postponed until Chapter 6.

$$= \left(\frac{1}{4}\right)\left(\frac{1}{3}\right) + \left(\frac{1}{2}\right)\left(\frac{2}{3}\right) = \frac{1}{12} + \frac{1}{3} = \frac{5}{12}.$$

This simple example already incorporates many of the necessary conceptual features of a quantum-theoretic *path sum* in the discrete causal context, i.e., a *sum over kinematic accounts of the evolution of classical histories.* With subsequent developments in mind, I introduce a suggestive use of the precursor symbol $\prec$, by writing $D_i \prec D_t'$ and $D_i' \prec D_t'$ to indicate that D_i and D_i' "directly precede" D_t' in $\mathbb{S}$. This is analogous to writing $0 < 1$ to indicate that 0 directly precedes 1 in the ordered set of integers $\mathbb{Z}$. Note, however, that *only* direct precedence is indicated by the symbol $\prec$; corresponding analogues of "indirect" integer relations, such as $0 < 7$, are omitted.[36] With this notational adjustment, $P(D_t')$ may be expressed as a sum:

$$P(D_t') = \sum_{D \prec D_t'} p(D \Rightarrow D_t')P(D), \tag{1.8.3}$$

where in this case D runs over the two-element set $\{D_i, D_i'\}$. This equation is clearly amenable to broad generalization.[37]

One may now compare Equation (1.8.3) to the causal Schrödinger-type equation (1.1.2):

$$\psi_{R;\theta}^{-}(r) = \theta(r) \sum_{r^- \prec r} \psi_{R;\theta}^{-}(r^-).$$

There are obvious similarities between these two equations. First, $\psi_{R;\theta}^{-}(r)$ corresponds roughly to $P(D_t')$, and is therefore "something like the likelihood of obtaining a particular terminal history in $\mathbb{S}$." The relations r and r^- correspond to D_t' and D, respectively, and are therefore "something like histories." In fact, they are elements of the relation space $\mathcal{R}(\mathcal{M}(\mathbb{S}))$ over the underlying multidirected set $\mathcal{M}(\mathbb{S})$ of $\mathbb{S}$, and therefore represent *co-relative histories. R* is simply a portion of this relation space; its role is to accommodate the practical consideration that one does not always wish to "look back to the beginning of time," i.e., to the empty history $\emptyset$, in comput-

[36] Chapter 3 explains in detail the distinctions among *direct, indirect, maximal,* and *minimal* relationships in this context. In general, the version of discrete causal theory developed in this book is described principally in terms of *direct* relationships, whether or not these relationships are maximal or minimal.

[37] In Equation (1.8.3), one considers the family of all co-relative histories with a fixed *target* D_t'. It can also be interesting to consider such a family with a fixed *source*. For example, in the present classical stochastic context, the *Shannon entropy* associated with the family of co-relative histories $\{D_i \Rightarrow D'\}$, for a fixed source set D_i, is given by the formula

$$H = -\sum_{D_i \prec D'} p(D_i \Rightarrow D') \log p(D_i \Rightarrow D').$$

See Section 6 of Shannon's paper [SH48] for an analogous discussion involving abstract probabilities. The subject of entropy returns in Section 8.2 in the context of *entropic phase maps*.

ing quantities relevant to the present state of the universe. Finally, the value $\theta(r)$ of the relation function θ corresponds roughly to the "evolutionary probability" $p(D \Rightarrow D_t')$, and is therefore "something like the probability that a particular source history will evolve into a particular target history." Of course, $\theta(r)$ appears *outside* the sum, while $p(D \Rightarrow D_t')$ appears *inside* the sum, but this is merely a consequence of working in relation space, rather than element space.

At a grand conceptual level, Equations (1.8.3) and (1.1.2) share similar content: both signify the eminently reasonable assertion that *"information about the present stage of evolution may be reconstructed from knowledge about the previous stage, together with knowledge about the relationships between the two stages."* The same reasoning applies to the ordinary Schrödinger equation of conventional nonrelativistic quantum theory, although in this case one must take limits in order to obtain the equation. This is, in fact, the essence of Feynman's approach in "re-deriving" the Schrödinger equation in his famous 1948 paper [FE48]. In Chapter 6, I review Feynman's arguments, and explain in detail how they may be abstracted and adapted to provide a dynamical framework for discrete causal theory. Interestingly enough, the vast apparatus of technical methods associated with path integration in modern quantum field theory provides no significant advantage over Feynman's original conceptions in this context!

1.9 Phenomenology

Competing with unproven conventional theories. With a few possible exceptions, discrete causal theory is not yet sufficiently developed to make definitive, testable predictions about physics beyond the scope of general relativity and the standard model of particle theory. The same is generally true, of course, of more mainstream modern approaches to fundamental physics, including string theory, which has been under consideration for more than 45 years, and loop quantum gravity, which is about 20 years younger. However, these theories are "at least relevant to conventional physics," because they *are* largely conventional from a historical and conceptual standpoint. This is true despite the novel consequences of these theories. String theory was developed as an extension of familiar "point-particle technology" to higher-dimensional objects, not as an expression of any radical new idea of the same magnitude as relativity or quantum theory when they were first introduced. Innovations such as *compactified extra dimensions, supersymmetry,* and the *string theory landscape* entered the picture only later, via efforts to "rescue" the theory. Loop quantum gravity, meanwhile, is even more conservative in its conception, and this is arguably one of its greatest strengths; it seeks simply to quantize the familiar spacetime of general relativity, without trying to solve all the problems of modern physics at once. Novelties enter the picture via the technical methods used in this quantization effort, and in the resulting quantum-theoretic objects. Given this background, it is not surprising that these theories exhibit close and potentially fruitful connections with established physics. What remains less clear is whether or not the *new* ideas brought to the table by these theories are truly pivotal. For example, is supersymmetry a physically important idea, or not? No one knows.

The principal reason why conventional physics remains conventional is "because it works," i.e., because it explains most of what is known experimentally, even if these explanations may seem unsatisfying or incomplete. Due to their close connections with this working body of knowledge, theories such as string theory and loop quantum gravity involve ideas that most physicists would consider to be "worth trying," even if they ultimately prove to be unsuccessful. Discrete causal theory, however, takes much less of the apparatus of conventional physics for granted, and therefore enjoys much less automatic affinity for the current state of the art. The metric recovery theorems, discussed in Chapter 2, constitute a crucial exception to this statement, but the significance of these theorems remains surprisingly under-appreciated in the modern physics community. In any case, these theorems themselves provide no obvious clue regarding how directed sets reproducing the observed properties of spacetime might actually arise out of a vast spectrum of other possibilities. For these reasons, discrete causal theory faces a much larger gap to fill than conventional continuum-based approaches in convincing the typical theorist that it is sufficiently relevant to justify serious study. This is true despite the fact that the new ideas involved in the theory; in particular, the assertions that discreteness and directed structure are themselves fundamental, *are* physically compelling in a way that ideas such as supersymmetry are not.

The best way for discrete causal theory to fill this gap would be to achieve some striking phenomenological success that would be impossible to ignore. Causal set theory has already offered a simple mechanism that approximately reproduces the observed value of the "cosmological constant," which arguably provides a better explanation for this value than conventional alternatives. This result is well-explained in the literature; see, for example, the paper *Everpresent Lambda* [ADGS04], by Ahmed, Dodelson, Greene, and Sorkin. Essentially the same type of argument carries over to the version of discrete causal theory developed in this book, but this modest success is unlikely to move the needle much, without a great deal of other supporting evidence. For the most part, discrete causal phenomenology is at the stage where it must focus on promising general possibilities for where breakthroughs might occur, rather than examining precise quantitative details of specific phenomena. Perhaps the most obvious such possibilities involve deviations from *local Lorentz invariance* and *spatiotemporal locality.* These topics are revisited in later chapters, particularly in Section 4.5. In the remainder of the present section, however, I focus on just one such general idea; namely, the prospect of discrete causal alternatives to the *inflationary hypothesis* in conventional modern cosmology. The idea presented here is sufficiently novel to be considered "speculative," but it does enjoy some degree of context and support from the literature. The general mechanism proposed is purely classical, and is likely too simple in its own right, but it provides a useful illustration of how discrete causal theory brings to the table physically suggestive mathematical "phenomena" unavailable in the continuum-based context.

Cosmology; inflation. The inflationary hypothesis was invented by Alan Guth [GU81] and Andrei Linde [LI82] about 35 years ago to explain certain large-scale cosmological phenomena, such as the overall properties of the *cosmic microwave background.* To be brief, the **inflationary hypothesis** holds that *spacetime expanded at an extraordinary rate over a very short time interval in the early universe,* growing by a factor of perhaps 10^{30} or more in less than 10^{-32} s. A field called the *inflaton field* is invoked to explain this expansion. After this extremely short, explosive interval, spacetime "decayed" to a state of more gradual expansion, continuing to the present era. This thumbnail description omits many details and variations of the inflationary idea. For example, in one popular class of scenarios, called *eternal inflation* [VI83, LI86], the inflating region of spacetime grows so rapidly that inflation forever dominates the overall behavior of the universe, even though any particular region quickly decays to a non-inflating state. In these scenarios, the "observable universe" is taken to be just one of a multitude of "bubbles" of non-inflating spacetime in an incomparably vaster region of inflating spacetime. The nature and origin of the hypothetical inflaton field has been studied and elaborated under various models.[38] A concise modern description of inflation may be found in the standard relativistic cosmology text of Ellis, Maartens, and MacCallum [EMM12]. A leisurely and informal description appears in Guth's popular book [GU97].

To anyone unfamiliar with the history of modern cosmology, inflation may appear ridiculous on its face. However, it has strong empirical arguments in its favor; it successfully accounts for certain phenomena that seem difficult to explain by other conventional means. For the sake of brevity, I mention here only the most common and general argument in favor of inflation, which involves the large-scale homogeneity of the observable universe. This argument begins reasonably enough, by taking this homogeneity to indicate "mixing" of some kind, in which regions of spacetime that appear widely separated exchanged information in the distant past. A more sophisticated way to express the same idea is to say that these regions were once in *causal contact;* i.e., they possess "common ancestors." At this point, however, a conundrum arises: based on actual observations, there does not appear to have been any opportunity for this "mixing" to occur, since the regions involved appear to have always been beyond each other's *causal horizons,* i.e., the boundaries separating those events that "may have influenced" a given region from those that could not have influenced it, under the usual relativistic prescription forbidding "superluminal communication." This issue is sometimes called the *horizon problem.* As expressed by Guth in his inaugural paper on inflation [GU81]:

> *The standard model of hot big-bang cosmology requires initial conditions which are problematic... the early universe is assumed to be highly homogeneous, in spite of the fact that separated regions were causally disconnected (horizon problem)...*

In the context of the horizon problem, the inflationary hypothesis seems much less ridiculous. Qualitatively, it merely expresses the idea that *"regions of spacetime that were once close enough together to exchange information, and thereby approach*

[38]For a string theory viewpoint, see [BBS07], Section 10.7.

equilibrium, abruptly became widely separated, which is why they appear causally disjoint from our present viewpoint."

Discrete causal alternatives to inflation. Provided that one insists on modeling classical spacetime as a real manifold, the inflationary hypothesis may indeed offer the most reasonable explanation for large-scale spacetime homogeneity yet proposed.[39] However, discrete causal theory provides another type of explanation, which is much more natural from a basic structural perspective, namely, that *"the causal structure of spacetime abruptly became much sparser in the early universe."* The reason why this explanation is "more natural" than inflation is because discrete structures often exhibit abrupt qualitative changes due to simple combinatorial considerations, without the need to invoke hypothetical entities such as the "inflaton field." In particular, the mathematical field of *random graph dynamics* [ER59, BO01], which is intimately related to discrete causal theory, supplies many examples of sudden, qualitative changes in structure, called *graph-dynamical phase transitions.* Such phase transitions enable natural modeling of phenomena such as abrupt narrowing of causal horizons, as demonstrated in the example below. Phase transitions from "pre-geometric" to "geometric" states in the early universe have been considered recently in the context of discrete theories that are not "purely causal," including *causal dynamical triangulations* [AJJR12], and *quantum graphity* [KMS06, KMS08]. A relevant general discussion of discrete causal structure in the cosmological context appears in the recent paper *Network Cosmology* [KKSR12], by Krioukov and five coauthors, which discusses connections between cosmological models involving "sprinkled" causal sets, and the behavior of large networks, such as those arising in neuroscience and social science. For a different viewpoint regarding inflation in causal set theory, see the recent preprints by Gudder [GU15, GU16].

Toy phase transition: setup. Most serious models of the early universe involve "phase transitions" of some sort, so the idea of invoking a phase transition in the context of inflation may seem perfectly conventional. However, the familiar phase transitions arising in continuum-based quantum field theories are much different than the graph-dynamical phase transitions of interest in the present context. The phase transition I actually illustrate here involves a toy model, a drastic simplification of the type of model one would need to consider to actually construct a serious quantitative alternative to the inflationary hypothesis. It involves the toy history D shown at the bottom of Figure 1.9.1, which has 58 events and 673 relations. At a conceptual level, one imagines that this history "duplicates itself, but with fewer and fewer relations," and thereby "runs down to an approximately geometric structure," similar to the history D' shown at the top of the figure, which has 58 events and 92 relations. Though it is impossible to discern by merely looking at the figure, D contains an isomorphic copy of D'. Informally, one imagines that the events in D' are "later versions" of events in D, but are "less interconnected," because D' possesses fewer

[39]Of course, inflation solves more than just the horizon problem, and a serious discrete causal alternative would have to address its other successes in a suitable way. It is worth noting, however, that inflation has some very credible critics whose objections have nothing to do with discreteness hypotheses; these include Roger Penrose [PE10].

relations. Succeeding figures illustrate this "run-down process" in more detail. Before proceeding, however, I will briefly discuss the physical reasons for considering such a process.

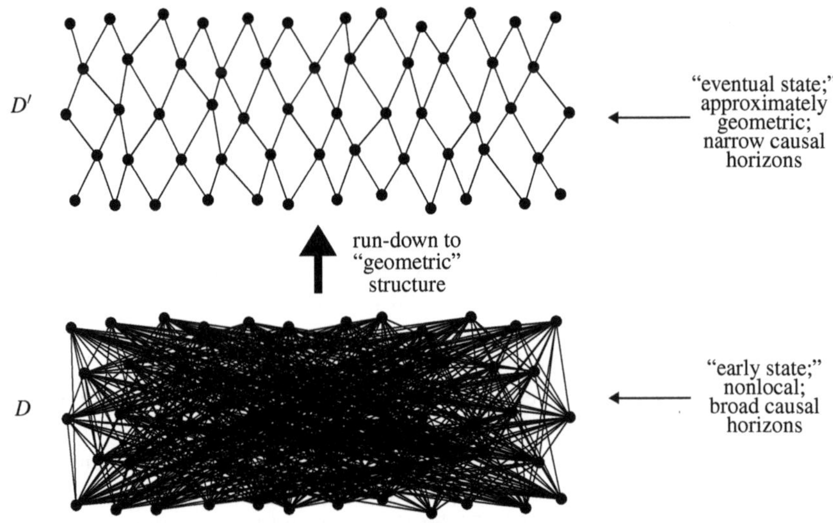

Figure 1.9.1. Run-down of a classical history to an approximately geometric state.

The reader will surely notice that the arrangement of the nodes representing events in the classical histories D and D' does not appear to be completely random. Instead, there are "five rows," each containing 11 or 12 nodes. This arrangement is chosen in anticipation of the "eventual state" D', in which the relations encode an "approximately geometric" structure, in contrast to the relatively random structure of D. The reason for constructing the example in this way is that actual observations of physical spacetime *seem* to reveal a structure that is "geometric" in certain important ways. In particular, physical spacetime has a well-defined dimension at ordinary scales; if not four, then at least very close to four. It also seems to exhibit a strong degree of *spatiotemporal locality*[40]; i.e., it is possible to speak unambiguously about whether events are "close together" or "far apart" in a given frame of reference. In this particular case, the structure of D' actually looks approximately like a *lattice,* but this is merely for convenience of illustration. True lattices can be problematic in discrete causal theory; for example, they tend to produce uncomfortably large systematic violations of Lorentz invariance. However, it is reasonably obvious that the qualitative behavior illustrated here does not depend on specific "lattice-like" structure, but only on more general properties, such as approximate spatiotemporal locality.

[40]The qualifier "seems to" is added here in anticipation of Section 4.5, which discusses the tension between spatiotemporal locality and "approximate Lorentz invariance" in discrete causal theory.

Without considering possible constraints imposed by dynamical considerations, a "typical" classical history in discrete causal theory appears "non-geometric" in the extreme. This follows from simple counting; there exist many more histories with relations scattered "randomly" than histories sufficiently "well-arranged" to exhibit approximate geometry. In particular, for a history D with n events, there exist $\binom{n}{2}$ "possible" relations, one for each pair of elements of D. Hence, for any positive integer $k \leq \binom{n}{2}$, there exist $\binom{\binom{n}{2}}{k}$ such histories with exactly k relations. This number is, of course, largest when k is equal to about half of $\binom{n}{2}$, which means that any given pair of events has a roughly even chance of being related. These figures are altered somewhat, for practical purposes, if one imposes extra structural conditions, such as acyclicity.[41] For example, the relatively "random" acyclic history D illustrated in Figure 1.9.1 possesses considerably fewer than $\frac{1}{2}\binom{58}{2} \approx 827$ relations, meaning that a typical "random" history is even more "cluttered" than this. In any case, there is no reasonable way to infer any structure resembling low-dimensional geometry from such random histories. An important reason for this is that they are highly nonlocal in a spatiotemporal sense; no matter how one may attempt to define distance, events in one "spatial section" will always be related to events "on the other side of the universe" in succeeding "spatial sections." This, incidentally, is exactly the kind of property that allows for the sort of large-scale "mixing" believed to have occurred in the early universe prior to inflation, according to conventional physics.[42]

Perhaps, then, "what really happened" in the early universe is that spacetime "ran down" from a relatively random causal structure, with a large number of relations, like the classical history D, to a sparser but more regular structure, like the classical history D'. This does not mean that the universe did not *also* expand, in the sense of possessing "more events in later spatial sections than in earlier spatial sections." Indeed, naïve entropic considerations encourage such expansion for discrete causal classical histories, as discussed in Section 8.2. However, for the sake of simplicity, I choose to leave the number of events in each section roughly constant in this particular illustration. The next task is to try to argue that this type of behavior could reproduce the drastic alteration of basic structure necessary to replace the role of the inflationary hypothesis.

Toy phase transition: computation. Of course, what these classical histories are really intended to illustrate are "thickened horizontal slices" of a larger classical history, whose causal structure "grows sparser from bottom to top." To simplify the picture, however, I "simulate" such a larger history by beginning with D and removing relations one by one, until only the underlying copy of D' remains. The process is carried out randomly, except for the fact that relations in D' are immune

---

[41] Also, this naïve enumeration ignores isomorphisms between pairs of "physically equivalent" histories, but this does not alter the conclusions.

[42] A "tamer" type of spatiotemporal nonlocality occurs in causal sets constructed via global sprinkling into Minkowski spacetime $\mathbb{R}^{3+1}$, as discussed in Section 4.5. In this context, a given element is directly related to elements that are distant in *both* a spatial and temporal sense in a given frame of reference; the separation between such elements is therefore small under the Minkowski metric.

to removal. The feature of interest is *how the sizes of the causal horizons of events change during this process.* To simplify further, I focus only on the sizes of the "initial cross-sections" of the causal pasts of terminal events in the histories D and D', and in the "intermediate histories" involved in the process. Conceptually, these cross-sections measure "how much of the beginning of time may be observed from a given event at the end of time." For each classical history illustrated in the figure, the events in the initial cross-section are labeled x_1 through x_{12}, and the events in the terminal cross-section are labeled y_1 through y_{12}. Each history has a preferred frame of reference, consisting of five successive layers of events, beginning with the "x-layer" and ending with the "y-layer." Hence, each history is "four chronons deep." I focus here on the simple question of "how much of the x-layer may be observed from each event in the y-layer." Figure 1.9.2 illustrates some interesting stages in this process, for a particular event in the y-layer, labeled y_4. It is convenient to re-label the histories involved as D_0, D_1, D_2, and D_3, so as to keep track of the order of the process. Under this labeling, D_0 is the original "random" history D, and D_3 is the final "geometric" history D'.

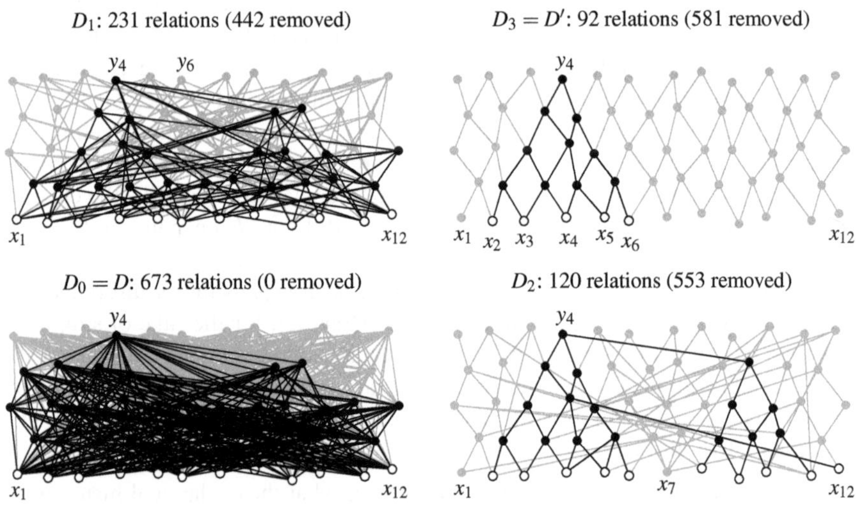

Figure 1.9.2. Abrupt narrowing of initial cross-sections of causal pasts as relations are removed.

The crux of the whole exercise is that the sizes of initial cross-sections change *abruptly* during a specific, and relatively brief, interval in the removal process. The bottom-left diagram shows the original history $D_0 = D$, along with the causal past of the event y_4 in this history. Due to the large number of relations, this causal past contains most of the events in D_0, and forms a complicated tangle. Events in the initial cross-section of this causal past are indicated by white nodes. The top-left diagram shows the history D_1, which results from removing 442 relations, or about 2/3 of the relations in D_0. Although the causal past of y_4 is "thinned out" considerably by this process, the difference cannot be detected by looking at its initial cross-section, which still contains the entire bottom layer of events in D_1. In fact, it is still possible to observe the entire bottom layer of D_1 from *any* event in the top layer. The next relation removed in this particular random process decreases the size of *one* of the initial cross-sections from 12 to 11; namely, the initial cross-section of the causal past of the event labeled y_6. However, no other initial cross-section changes size in the process until an additional 39 relations are removed.

After this, the sizes of initial cross-sections begin to decrease rapidly. Removing an additional 71 relations, about 1/7 of the total already removed, yields the history D_2, shown in the bottom-right diagram. In D_2, *none* of the events in the top layer can "see" the entire bottom layer. The largest initial cross-sections have size 10; for example, the events labeled x_1 and x_7 are absent from the initial cross-section of the causal past of y_4. Removing an additional 28 relations yields the "geometric" history $D_3 = D'$, reproduced in the top-right diagram. D' differs drastically from D_1 in the sizes of initial cross-sections. In particular, the initial cross-section of the causal past of y_4 in D_3 contains only the five events labeled x_2 through x_6. About half of the total decrease in sizes of initial cross-sections occurs during the last five percent of the removal process, between the histories D_2 and D_3.

Toy phase transition: summary. Figure 1.9.3 gives an overall view of this process. The black curve illustrates how the average size of initial cross-sections changes as relations are removed, until the "geometric" structure of the history $D_3 = D'$ is reached. The locations of $D_0 = D$, D_1, D_2, and $D_3 = D'$ in the process are marked by square white nodes on the plot. The "process time" is plotted on the horizontal axis. In each unit of process time, a single relation is removed. The process time has nothing to do with "physical time" in the individual histories, viewed as models of classical spacetime. However, in a slightly more serious model, in which these histories represent "thickened horizontal slices of a larger classical history, whose

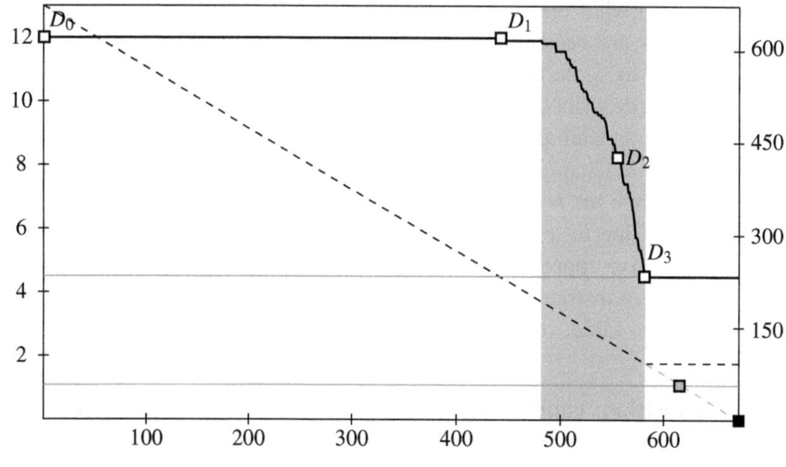

Figure 1.9.3. Plot illustrating the phase transition. Nearly 2/3 of the relations are removed before the sizes of initial cross-sections begin to shrink. After this point, the behavior is abrupt.

causal structure grows sparser from bottom to top," the process time is closely related to physical time. In this type of scenario, the plot gives a good qualitative idea of how the causal structure of the larger history changes in physical time in a particular frame of reference.

The feature of principal interest in the plot is the abrupt change of behavior highlighted by the gray band. This is the phase transition. The black dashed line indicates the number of relations remaining after each step, measured along the right vertical axis. The gray dashed line near the bottom right shows what would happen if the process were allowed to continue, with the "geometric" history $D_3 = D'$ decaying first to a "frozen history," indicated by the gray node, in which events merely perpetuate themselves without any "cross-relations," then to a "timeless history," indicated by the black node, i.e., a family of isolated events. Neither of these outcomes is physically realistic. More reasonable is the scenario actually considered here, in which the decay process terminates at a "geometric" history; as indicated by the sudden flattening out of the black curve. However, this sudden termination of the process is admittedly artificial; one could construct more-detailed and more-natural processes, designed to "gradually peter out" in some way. In such processes, the qualitative behavior is smoothed out to some degree, but generally not enough to hide the phase transition. Qualitatively, it appears that "run-down of a classical history from a random state to a geometric state features a very large and abrupt narrowing of causal horizons." This is the general type of behavior necessary to construct plausible alternatives to the inflationary hypothesis.

It is important to emphasize that the example presented here is merely suggestive, and there are a number of basic questions to be answered before one should begin to take this particular idea seriously. For example, *why* would causal structure grow sparser? Why would it stabilize in a "geometric" structure? Why one particular geometric structure rather than another? How does the "process time" relate to "physical time?" If relations are removed at, say, an exponentially decreasing rate, rather than a constant rate, to what extent does this "cancel out" the abrupt behavior? Some of these questions are easier to answer than others. Entropic considerations offer potential answers to some of them; indeed, interesting variants of discrete causal dynamics may be described in terms of a version of the second law of thermodynamics, involving special algebraic structures associated with directed sets. These ideas are revisited in Chapter 8. In any case, while this example, by itself, cannot be expected to convince the reader that the large-scale homogeneity of the observable universe *really is* due to a graph-dynamical phase transition, it does illustrate an interesting alternative approach to the problem of explaining this phenomenon. At a broader level, it demonstrates how discrete causal theory can bring completely novel mechanisms to the table, mechanisms which are unavailable in continuum-based theories.

Toy phase transition: concluding remarks. A few points regarding this example are worth elaborating. First, as mentioned above, the term "phase transition" is used routinely in cosmology in a more conventional sense, both in the inflationary context and elsewhere. For example, *symmetry breaking* in gauge theories, such as the electroweak theory, is commonly described in terms of phase transitions. The graph-dynamical phase transitions discussed here are qualitatively different. They were originally discovered in a purely mathematical setting, and were so-named by *analogy* to familiar physical processes, such as the melting of ice. Hence, their potential application to fundamental physics is ironic in a historical sense, since it renders the target of the analogy more basic than its source. Second, students of general relativity, causal set theory, and/or twistor theory, may recall studying relationships between *null rays* and *causal links,* and may draw the conclusion that one may "travel faster" in the random universe D than in the "geometric" universe D'. From this viewpoint, it is reasonable to wonder how the discrete causal alternative to the inflationary hypothesis sketched here relates to widely discredited proposals involving drastic alteration of the speed of light in the early universe. I will not go into the details, but the main arguments against these proposals involve specific structural assumptions that, as far as I know, have no bearing on the general applicability of the ideas outlined here.

Third, one might complain that the phase transition in this example is not nearly sharp enough to match the extreme characteristics of inflation. While this is undoubtedly true, it is important to recognize that this particular phase transition involves only very small classical histories. "Realistic" phase transitions, involving histories with perhaps 10^{100} or more elements, present a much different picture. A basic lesson of thermodynamics is that statistical properties tend to become very sharp, in a relative sense, when large numbers of elements are involved. A familiar illustration of this phenomenon may be described in terms of flipping coins: if one flips a fair coin a hundred times, there is a reasonable chance that the number of heads will differ by an appreciable percentage, say five percent, from the expected value of fifty. If one flips the coin a billion times, however, the chance of missing the expected value by five percent is vanishingly small. The strategy of invoking such thermodynamic mechanisms to explain"fundamental" properties of the universe, such as the existence of gravitation [VE11], or the dimensionality of spacetime [CGA16], has a long and distinguished history, both in conventional and unconventional settings. It is self-evident that this general approach is most relevant in situations involving very large but finite[43] numbers of particles, degrees of freedom, or, in the present case, "spacetime elements." In particular, such considerations are ubiquitous in discrete causal theory.

Finally, the particular graph-dynamical recipe described in this section completely ignores another "magical mechanism" for converting "fuzzy, pathological behavior" into "sharply-defined, reasonable behavior" at appropriate scales; namely, the path summation approach to quantum theory. Whatever the correct explanation for the observations underlying the inflationary hypothesis, it would be naïve to expect a purely classical explanation to cover the essence of the problem. At a qualitative level, classical thermodynamic mechanisms and quantum-theoretic mechanisms may be expected to play complementary roles in this context, with thermodynamics ensuring that "reasonable" or "desirable" scenarios are not completely overwhelmed by

[43]Or at least, *locally* finite.

"pathological" scenarios, and quantum theory damping out the influence of the latter scenarios, via "destructive interference."

1.10 Looking Ahead: Twenty Questions

Preliminary questions. The material presented in this chapter provides a quick overview of the content, philosophy, and goals of discrete causal theory. However, the informal level of the discussion up to this point raises many questions, and provides only imprecise and incomplete answers. Below, I list some of these questions, and briefly describe if, where, and how this book attempts to address them more rigorously. I begin with some preliminary conceptual questions about the overall viewpoint of discrete causal theory, its historical background, and its current state of development.

1. *What is the overall scientific and philosophical viewpoint underlying discrete causal theory?* Despite its radical structural content, discrete causal theory is ultimately intended to fit into the usual scientific tradition of offering specific explanations and predictions regarding observable physical phenomena, which may be verified or falsified by experiment and observation. If this goal cannot be achieved after a reasonable amount of effort, then the theory makes no claim to any special exception for further consideration. Of course, I have already explained in Section 1.3 how the fundamental concept of *directed relationships* is innate to the practice of modern science, and for this reason alone, the abstract content of discrete causal theory is almost certain to remain relevant in some capacity. Its focus, however, could conceivably be redirected away from fundamental spacetime structure, should the causal metric hypothesis (CMH) be decisively discredited. More specific aspects of discrete causal theory, including spacetime discreteness itself, the local properties ascribed to directed sets, the chosen approach to quantum theory, and so on, are all subject to being modified or discarded. In this sense, discrete causal theory is rather orthodox. In particular, it has no role to play in modern debates about "changing the way science is done" to accommodate the weaknesses of popular mainstream theories.

   In other ways, however, discrete causal theory is unorthodox in the extreme. The choice to abstain completely from continuum-based geometry in the construction of the theory is already unconventional, but the causal metric hypothesis itself, which is the real cornerstone of the theory, is arguably its most radical feature. The general qualitative statement of this hypothesis appearing in Section 1.2; namely, that *"the properties of the physical universe are manifestations of causal structure,"* may seem, a priori, to express something obvious, and not very interesting. In particular, this statement might be viewed as an expression of the conventional assumption that events have causes, rather than "appearing out of nowhere, for no reason." However, the ultimate goal of the theory is to use *only* very simple and specific mathematical representations of causal structure to explain *all* of physics. This goal is clearly implied by the decision to emphasize the strong interpretation of the hypothesis, together with the choices to model classical histories in terms of discrete directed sets, and quantum theory in terms of configuration spaces of such sets. Section 2.10 revisits the overall philosophical viewpoint underlying discrete causal theory.

2. *What problems is discrete causal theory intended to help solve?* Discrete causal theory is aimed principally at addressing the three outstanding problems of modern physics mentioned in the preface; namely, *(1) describing fundamental spacetime structure, (2) modeling gravitation at the quantum level, and (3) unifying physical law.* It is therefore a very ambitious project. It is safest to interpret the word "fundamental" here in an operational sense, as "something more fundamental than what is already known," since future scientists may always posit still-deeper structure. Similarly, near-term progress in quantum gravity and unification is more likely to offer new viewpoints and opportunities for further improvements than to provide final answers. Nevertheless, it is desirable to consider what might be "truly fundamental," insofar as the present development of science enables the researcher to even ask the right questions. Discrete causal theory makes a much deeper attempt in this direction than most other existing approaches. Solution of the three problems mentioned above would be monumental,[44] and the odds are steep against rapid and complete success. However, any noticeable progress toward the solution of any of the three would render the program worthwhile. Hungarian mathematician Paul Erdös, co-founder of random graph dynamics, described the typical difficulty of "interesting" problems by remarking that *"problems worthy of attack prove their worth by fighting back."* These three problems have certainly run up the score against the physics community over the last century. However, Erdös' aphorism is not literally accurate; problems remain "stationary," and any progress toward their solution is permanent, barring global cataclysm. These problems are kept in view throughout the book, although unification receives the least attention of the three, since recovering the details of the standard model is beyond the current state of development of the theory. Part I of the book focuses mostly on fundamental spacetime structure, while Part II focuses mostly on quantum gravity.

3. *What justification is there for believing that discrete causal theory can contribute to solving these problems?* There are many good reasons for optimism about the theory. These include (1) strong theoretical support from the metric recovery theorems; (2) modest success in recovering or explaining certain aspects of known physics, such as the value of the "cosmological constant;" (3) use of simple structural building blocks of undeniable importance to modern science; (4) absence of certain notorious technical problems that encumber more conventional approaches; (5) natural unification of classical theory and quantum theory as different levels of structural hierarchy; and (6) automatic incorporation of certain properties believed to be necessary for a fundamental theory, such

[44] Comparable to finding the Holy Grail and the two Arks in a single archaeological expedition.

as background independence. To these I would *personally* add plausibility, elegance, and good physical motivation, but the latter factors could be considered a matter of opinion. These topics are revisited multiple times throughout the book, with the first two chapters devoted almost entirely to qualitative motivation for the theory. The single greatest justification for optimism remains the metric recovery theorems, discussed in Chapter 2.

4. *What is the historical background of discrete causal theory?* The study of cause and effect has an ancient history, but the inception of discrete causal theory as a field of modern physics may be localized around the late 1970s, when Hawking [HA76] and Malament [MA77] proved the metric recovery theorems. *Causal set theory* [BLMS87], championed principally by Rafael Sorkin, emerged within a few years of these developments. It is important to recognize that Finkelstein [FI69], Myrheim [MR78], and t'Hooft [TH78] all anticipated certain important aspects of causal set theory. *Causal dynamical triangulations* [ADJL02], which might be considered the most pragmatic existing version of the theory for the purpose of short-term progress, followed about 15 years later. Modern *category-theoretic approaches* to fundamental physics, developed by Isham [IS05] and others, share considerable overlap with discrete causal theory. Recent work on *quantum automatons* by D'Ariano [DP14] and others is also closely related, as are *tensor networks.* The version of the theory developed in this book was conceived in 2008-2009, introduced publicly in 2012 [DR15], and elaborated in 2013 [DR13]. Finkelstein and a few others had previously considered formally similar structures such as *causal nets* [FI88], although with different interpretations. Chapter 3, though mostly technical, describes some of this history, focusing on the case of causal sets.

5. *Which active approaches to fundamental physics are most closely related to discrete causal theory as developed in this book?* These overlap substantially with the approaches mentioned in the answer to the previous question. As noted in the preface, *causal set theory* is the purest version of discrete causal theory enjoying an established theoretical niche, and is the version most similar to the one developed in this book, despite crucial differences. Related approaches, such as *causal dynamical triangulations,* tend to take a larger proportion of conventional apparatus for granted. *Category-theoretic approaches, quantum automata, tensor networks,* and a number of other such approaches share some features of the theory, but also tend to import a great deal of structure from special mathematical constructions. *Domain theory* [MP06] is very closely related at an abstract level, although it is nondiscrete. Many of the ideas central to discrete causal theory may also be applied in mainstream *quantum information theory,* and there is at least a limited converse relationship. *Loop quantum gravity* [RO04, TH07], also shares interesting similarities with discrete causal theory, despite being a continuum-based theory. A few other continuum-based approaches, including *twistor theory* [PE67] and *shape dynamics* [BA12], are related to discrete causal theory in at least a limited way. This book discusses causal set theory in some detail, particularly in Chapter 3. Other related approaches are mentioned only obliquely. Section 8.8 briefly outlines some of these connections.

6. *What is the current state of development of these approaches?* Although I make a reasonable effort to point the reader in the right direction, this book does not attempt to supply an adequate treatment of this question. The closest I come to addressing this topic is in Chapter 3, where I briefly discuss causal set dynamics and phenomenology. It is worth noting that several existing versions of discrete causal theory have progressed far enough to study specific problems in fundamental physics and cosmology. For example, causal dynamical triangulations has reached the stage of carrying out significant computations and simulations involving the evolution of discrete classical histories [AJL05].

Technical questions. Next, I provide some preliminary guidance on technical aspects of discrete causal theory. To understand the details, it is necessary to work through the remainder of the book, but this brief series of questions and answers provides a concise preview of the most important aspects. Probably the most useful feature of this list is simply that it explains where in the book the various ideas are discussed, in a more thorough manner than the table of contents.

7. *What are the details of how discrete causal theory models classical histories?* Mathematically, the classical histories of discrete causal theory are modeled via *directed graphs* of a particular type, according to conventional terminology. However, the viewpoint and language of graph theory is idiosyncratic and awkward in this context, so I refer to these models as *directed sets* instead. This choice of terminology is still not ideal, partly because the term "directed set" is usually assigned a more specific meaning in conventional order theory and category theory. However, it seems to be the best choice available. Terminology aside, these classical histories are usually taken to be *countable* and *acyclic.* Further properties may be assigned in a variety of different ways, depending on the specific version of the theory. For example, causal sets are assumed to be *interval finite* and *transitive,* while causal dynamical triangulations are assumed to be "locally Lorentzian." The directed sets studied in this book are usually assumed to be *star finite,* i.e., *locally finite* in an appropriate sense. They are *not* assumed to be transitive or interval finite, and these are important distinctions when comparing this version of the theory to causal set theory. The book also includes limited analysis of classical histories containing cycles, since general relativity does not rule out closed causal curves. Multidirected sets arise inevitably at the quantum level, and are more complicated. Mathematically, they may be viewed as *directed multigraphs* or *quivers* of a special type. Directed sets and multidirected sets are defined and described in Section 3.6. Section 4.10 discusses the specific classes of directed sets most useful as models of classical spacetime from the perspective of this book.

8. *What are the justifications for these choices?* The manner in which I choose to model classical histories in this book is based on serious and detailed consideration of the physical meaning of causal structure, together with the novel possibilities associated with discreteness. A few prominent physicists who have thought deeply about these topics, such as Finkelstein, have agreed with many

of these choices, while some others have disagreed, and still others have apparently never seriously considered how the change to the discrete paradigm alters the picture. I have tried very carefully to avoid two particular pitfalls in making these choices. The first is the common tendency to automatically borrow concepts and properties from continuum-based theories, which are more familiar, and are therefore dangerous sources of unconscious assumptions. The second is the temptation to import convenient constructions from mathematics. Seminal *ideas* originating in mathematics, such Grothendieck's *relative viewpoint* (RV), are fair game in this regard, but one must be very careful in appropriating entire *systems,* such as category theory, which were invented to serve specific purposes far removed from the physical applications of present interest. Chapters 2, 3, 4, and 5 elaborate the reasoning behind these choices. In particular, Sections 3.9, 4.4, 4.5, 4.7, and 4.8 carefully explain some of the differences between the version of discrete causal theory developed in this book and causal set theory. Section 4.10 presents a "suggested list" of axioms for classical histories in discrete causal theory.

9. *What are the most important resulting properties of these histories?* Configuration spaces of classical histories arising in the version of discrete causal theory developed in this book are generally better-behaved than the corresponding configuration spaces appearing in previously-developed approaches. In particular, a relatively large proportion of the classical histories studied in this book possess "good properties." For example, many countable, star finite, acyclic directed sets exhibit a reasonable balance between "spatial" and "causal" extent, while almost all causal sets are "squashed" into just a few generations. Of course, since "good behavior" is not taken for granted in discrete causal theory, and since the definitions are so general, it remains true that *most* of the "permissible histories" considered in this book look nothing like physical spacetime. The same is true of most other modern theories of physics beyond the scope of relativity and the standard model. For example, the physics community is now quite familiar with the philosophy surrounding the *string theory landscape,* in which "anything goes, somewhere." However, discrete causal theory *does* exhibit a reasonable supply of classical histories whose properties closely match what is actually observed in nature, rather than insisting on the wrong dimension for spacetime, or requiring the existence of huge families of particles that no one has ever observed. The properties of discrete causal classical histories are developed in detail in Chapters 3 and 4.

10. *What is the technical definition of relation space?* From a conventional mathematical perspective, the relation space $\mathcal{R}(D)$ over a classical history D in discrete causal theory is merely the *line digraph* over the corresponding directed graph. From the alternative set-theoretic viewpoint adopted in this book, $\mathcal{R}(D)$ is the directed set whose elements are relations in D, and whose relations are ordered pairs of relations in D, such that the terminal element of the first relation coincides with the initial element of the second relation. These notions are explained and illustrated in Section 5.1.

11. *What are the properties of relation space?* Relation space is special, in the sense that not every directed set of a given type is the relation space over another directed set of the same type. For example, it is easy to show that the "child history" in Figure 1.4.4 is not isomorphic to the relation space over any acyclic directed set. This "specialness" turns out to be very important from a technical perspective, since certain crucial properties that fail to hold for directed sets in general, *do* hold for relation space. Prominent among these properties is the *impermeability of maximal antichains,* which is important for dynamical reasons. At a more abstract level, the assignment $D \mapsto \mathcal{R}(D)$ is a *functor from the category of multidirected sets to the category of directed sets.* In certain interesting special cases, this functor preserves all the information in a classical history, except possibly certain information on the *causal boundary.* These details are explained in Chapter 5, particularly Sections 5.1, 5.2, and 5.3.

12. *What technical advantages result from working in relation space?* One of the most important advantages, already mentioned above, is that passage to relation space solves the *permeability problem,* which interferes with "foliation-type" approaches to dynamics in previously-developed versions of discrete causal theory. Briefly, this problem involves the possibility of information "leaking through spatial sections" of directed or multidirected sets, which inhibits satisfactory treatment of certain *discrete boundary value problems.* This problem disappears upon passage to relation space, since *maximal antichains in relation space are impermeable.* The technical advantages of relation space are discussed in Chapter 5, particularly Section 5.4, and are exploited in Chapters 6 and 7. In particular, the more general theory of *path semicategories* and *path algebras* developed and applied in these chapters is based on the relation space construction.

13. *Why is the relative viewpoint important?* The simplest *physical* explanation of the importance of the relative viewpoint (RV), in the context of discrete causal theory, is that it enables a *precise description of evolution,* which is necessary for any useful theory of physics. The type of evolution of principal interest here is the evolution of an "initial" classical history into a more complicated "terminal" classical history, encoded by a *co-relative history* between the two. However, the conceptual origin of the relative viewpoint has nothing to do with physics. Instead, it arose in a purely mathematical context, in the philosophy of Alexander Grothendieck, who made seminal use of it in algebraic geometry, category theory, and related areas of abstract algebra. In this context, the idea is indispensable, because it is almost always ill-advised to study algebraic objects in isolation. For example, even if one is principally interested in a particular object, such as the ring of integers $\mathbb{Z}$, it is often very illuminating to study an entire family of related objects, such as *number fields,* or *rings* in general. It is ironic to note the odd way in which this algebraic connection to the study of physical evolution in discrete causal theory evokes Hamilton's temporary, embarrassing desire to view algebra as the "science of pure time." The relative viewpoint is formally introduced in Section 3.8, and is put to serious use in Chapters 4, 5, 6, and 7.

14. *How is the idea of relationships between pairs of classical histories, i.e., co-relative histories, made precise?* Co-relative histories are subtle. They are *represented* by morphisms of directed sets, but are not themselves morphisms. The origins of this subtlety may be understood by carefully keeping track of the distinction between "physical" and "extraphysical" information in discrete causal theory. Technically, the co-relative histories of greatest interest are *equivalence classes of transitions,* which are special morphisms in which the target history may be viewed as a "later stage of evolution" of the source history. I have borrowed the term "transition" from causal set theory, which uses it in a more restricted sense. Two transitions are regarded as equivalent if there exist automorphisms of the target history "interchanging" their images. These details are discussed in Chapter 6, particularly Sections 6.3 and 6.4.

15. *What are the technical properties of co-relative histories?* Co-relative histories have some unusual properties from a conventional perspective. They generally do not compose in the same way that category-theoretic morphisms do, and there generally exist multiple distinct co-relative histories between a given pair of classical histories, even a pair "differing by the addition of a single element." The former property provides one demonstration of why category theory is not the optimal language for describing discrete causal theory, while the latter property explains why multidirected sets, rather than merely directed sets, play a central role in the theory at the quantum level. Chapter 6 describes these properties, particularly Section 6.4.

16. *What is the best way to approach quantum theory in the discrete causal context?* The path summation approach to quantum theory, envisioned by Dirac and developed by Feynman, may be abstracted and generalized to yield a promising quantum version of discrete causal theory. Other approaches are possible, but no other approach of which I am aware is as natural and adaptable as the path summation approach. One reason why this approach provides such an advantage is because it does not rely on convenient, "prepackaged" mathematical constructions, such as Hilbert spaces. In fact, all it requires in order to make sense is the existence of a map from the space of directed paths over a multidirected set into a set with an operation. Of course, additional properties are useful for obtaining "physically recognizable" versions of quantum theory. In particular, it is preferable if the "set with an operation" is in fact an algebra over a commutative ring, whose multiplication is compatible with the operation of *concatenation* on the path space. In conventional quantum theory, the object playing this role is the familiar field of complex numbers $\mathbb{C}$, but in discrete causal theory, many other options are available. These topics are the main focus of Chapter 6, but they are mentioned repeatedly throughout Part II of the book.

17. *What are the similarities and differences between classical theory and quantum theory in the discrete causal context?* One of the striking features of discrete causal theory is the principle of *iteration of structure* (IS), whereby "classical" and "quantum" versions of the theory fit together harmoniously as different structural levels in a comprehensive unifying hierarchy. The *positive sequen-*

tial kinematic scheme $\mathbb{S}_{PS}$, first illustrated in Figure 1.1.1, provides a glimpse of this feature. "Causally suggestive" mathematical notions, such as *partial orders, directed paths,* and *path algebras,* may be leveraged at both the classical and quantum levels. In fact, the classical theory leads naturally to the quantum theory in a manner that is both interesting and convenient. The study of classical histories points inevitably to questions of evolution, and thence to *co-relative histories,* which fit together into higher-level structures; namely, *kinematic preschemes* and *kinematic schemes,* whose abstract similarities to their individual constituent histories are impossible to ignore. One may even imagine an intelligent alien race discovering the conceptual essentials of relativity, such as covariance and background independence, in a discrete setting, long before acquiring the technology necessary to conduct "particle theory experiments," and thereby stumbling upon quantum theory from a theoretical direction, without any recognizable experimental impetus. It is difficult to imagine this occurring via continuum-based approaches. Of course, there are important differences between the two levels of structure; for example, *directed* sets at the classical level lead to *multidirected* sets at the quantum level, and appropriate algebraic data is needed to perform summation over evolutionary pathways for classical histories. Chapters 5 and 6 describe the natural buildup from "classical" to "quantum" in discrete causal theory.

18. *What are the technical details of discrete causal kinematics and dynamics?* Discrete causal theory provides an extraordinarily clear picture of the distinction between kinematics and dynamics. Kinematic data may be expressed in terms of structured configuration spaces of classical histories, called *kinematic schemes,* which organize possible evolutionary relationships among families of such histories, while dynamical input may be supplied by means of *relation functions,* or more generally, *phase maps,* which assign weights to evolutionary pathways in a kinematic scheme. At an abstract level, kinematic schemes may be viewed, via *iteration of structure* (IS), as "higher-level multidirected sets," whose "relations" are *co-relative histories.* The prototypical example of a kinematic scheme is the *positive sequential kinematic scheme* $\mathbb{S}_{PS}$, illustrated in Figure 1.1.1. In $\mathbb{S}_{PS}$, each "relation" represents a "single step in the evolution of a classical history," defined by the "addition of a single new element." A similar kinematic scheme $\mathbb{S}_{SR}$ appears in Sorkin and Rideout's theory of *sequential growth dynamics* for causal sets [SR99]. However, there are important differences between these two kinematic schemes: the classical histories in $\mathbb{S}_{SR}$ are causal sets, and its higher-level structure is usually treated in the literature as a partially ordered set, rather than as a multidirected set. Kinematic schemes and phase maps provide the underlying structural ingredients for the *path summation approach to quantum theory* in the discrete causal context. In particular, phases assigned to evolutionary pathways in a region of the "relation space over a kinematic scheme" may be summed to yield *generalized quantum amplitudes.* This leads naturally to the derivation of dynamical laws, such as the *causal Schrödinger-type equation* (1.1.2), first mentioned near the beginning of the present chapter. Chapters 6 and 7 describe these details.

Practical questions. The final two questions involve preliminary information about some practical issues involving short and long-term applications of discrete causal theory.

19. *How might discrete causal theory make contact with experimental physics?* As mentioned earlier in the chapter, causal set theory has already produced a "reasonable explanation" for the apparent value of the "cosmological constant." This observed value, as it is usually interpreted, differs immensely and notoriously from the predictions of conventional continuum-based theories. It is a *very small but nonzero value,* around 10^{-122} in "natural units." This is the type of value that could easily arise as a "statistical fluctuation away from zero" in a universe which is large and finite, or at least locally finite. More generally, since discrete causal theory takes a host of "exact" properties from conventional physics to be merely emergent, other such small deviations from "sensible continuum-theory values" may be anticipated across a range of phenomena. This provides a new and compelling way of thinking about otherwise-problematic issues involving *hierarchy* and *naturalness,* and enables qualitatively different kinds of explanations for present and future observational anomalies than are available in conventional settings. For example, if the average dimension of spacetime over a region of the observable universe were discovered to be approximately 3.9999999994387, instead of exactly 4, then discrete causal theory could easily account for this phenomenon, whereas most continuum-based approaches would be forced to interpret the same data in a baroque manner, in terms of otherwise unidentified "particles" or "fields." Due to the generic expectation of Lorentz invariance violation in discrete causal theory, an obvious place to search for evidence of the theory is in the context of highly boosted frames of reference; for example, in the study of ultra high energy cosmic rays. As elaborated in Section 4.5, there is a "tension," or "balance," between local Lorentz invariance and spatiotemporal locality in discrete causal theory, so searches for nonlocal effects are also of interest. Some researchers, such as Sorkin [SO09A], have suggested that nonlocal effects might be within the range of current experimental capabilities, or at least very close to this range. Another promising avenue of investigation, already discussed in Section 1.9, involves discrete causal alternatives to the *inflationary hypothesis* in cosmology. Chapter 8 includes some further material about issues related to phenomenology, including the *emergence of the Poincaré group.*
20. *What are the long-term prospects of discrete causal theory as a viable predictive theory of fundamental physics?* From a neutral perspective, it is too early to answer this question. The overall amount of research time, effort, and resources devoted by the scientific community to discrete causal theory over the last generation has been minuscule, in comparison to the attention lavished on mainstream continuum-based approaches, such as string theory and loop quantum gravity. For example, there are specific and relatively-untouched mathematical topics related to discrete causal theory that are probably no more difficult to understand than some of the string-related topics that have led to numerous famous results by some of the world's strongest mathematicians since the 1980s. The

most obvious problem, namely, how to obtain emergent geometry in a natural dynamical fashion from discrete causal structure, has received a limited degree of serious attention, but nothing resembling the mania surrounding, say, the *Maldacena conjecture.* A few other mathematical topics especially relevant to discrete causal theory have also seen development, but rarely in the most helpful contexts. For example, as far as I know, many of the random graph-dynamical scenarios arising most naturally in discrete causal theory have not yet been seriously examined, even though the field of random graph theory itself has seen substantial development since Erdös and Renyi invented it in the 1960s.

However, there are good reasons for optimism about the future of discrete causal theory. Astrophysical and cosmological observations will likely enable certain aspects of the theory to be tested within the next generation or so, even if no special effort is expended in this direction by experimentalists or funding agencies. It is also likely that the sociological environment in theoretical physics will become more favorable to discrete causal theory within a reasonably short period of time. Factors encouraging such a change include the continued failure of existing continuum-based approaches, the increasing age of powerful academics committed to these approaches, the impending experimental crisis in particle theory precipitated by the difficulty and cost of reaching higher energies, and the increasing prominence of quantum information theory. These factors may soon free up sufficient resources to provide a much clearer evaluation of the long-term potential of discrete causal theory. The overall outlook for theory is touched on periodically throughout the book.

References

[FE48] Richard Feynman. *Space-Time Approach to Non-Relativistic Quantum Mechanics.* Reviews of Modern Physics, **20**, 2, pp. 367–387, 1948.

[DI35] P. A. M. Dirac. *The Principles of Quantum Mechanics, Second Edition.* Oxford University Press. 1935.

[SH48] Claude Shannon. *Space-Time Approach to Non-Relativistic Quantum Mechanics.* The Bell System Technical Journal, **27**, 3, pp. 379–423, 1948.

[DR15] Benjamin F. Dribus. *On the Foundational Assumptions of Modern Physics.* Questioning the Foundations, The Frontiers Collection, Springer, pp. 45–60, 2015. Early version: http://www.fqxi.org/community/essay/winners/2012.1#dribus.

[DR13] Benjamin F. Dribus. *On the Axioms of Causal Set Theory.* Preprint. arXiv link: http://arxiv-web3.library.cornell.edu/pdf/1311.2148v3.pdf.

[HA76] S. W. Hawking, A. R. King, and P. J. McCarthy. *A new topology for curved spacetime which incorporates the causal, differential, and conformal structures.* Journal of Mathematical Physics, **17**, 2, pp. 174–181, 1976.

[MA77] David B. Malament. *The class of continuous timelike curves determines the topology of spacetime.* Journal of Mathematical Physics, **18**, 7, pp. 1399–1404, 1977.

[MP06] Keye Martin and Prakash Panangaden. *A Domain of Spacetime Intervals in General Relativity.* Communications in Mathematical Physics, **267**, 3, pp. 563–586, 2006.

[RO04] Carlo Rovelli. *Quantum Gravity.* Cambridge Monographs on Mathematical Physics. Cambridge University Press, 2004.

[HO13] Sabine Hossenfelder. *Phenomenology of Space-time Imperfection II: Local Defects*. Physical Review D, **88**, 12, 124031, 2013. arXiv preprint: http://arxiv.org/pdf/1309.0314v2.pdf.

[BHS09] Luca Bombelli, Joe Henson, Rafael Sorkin. *Discreteness without symmetry breaking: a theorem*. Modern Physics Letters A, **24**, 32, pp. 2579–2587, 2009. arXiv preprint: http://arxiv.org/pdf/gr-qc/0605006v1.pdf

[MO88] Christopher Moore. *Comment on "Space-Time as a Causal Set."* Physical Review Letters, **60**, 7, pp. 655, 1988.

[BLMS88] Luca Bombelli, Joohan Lee, David Meyer, and Rafael Sorkin. *Bombelliet al.Reply to Comment on "Space-Time as a Causal Set."* Physical Review Letters, **60**, 7, pp. 656, 1988.

[AS15] Torsten Asselmeyer-Maluga. *A Chicken-and-Egg Problem: Which Came First, the Quantum State or Spacetime?* Questioning the Foundations of Physics, The Frontiers Collection, Springer, pp. 205–217, 2015.

[BA15] László Babai. *Graph Isomorphism in Quasipolynomial Time*. arXiv preprint: https://arxiv.org/pdf/1512.03547v2.pdf

[BR88] Edward A. Bender and Robert W. Robinson. *The Asymptotic Number of Acyclic Digraphs II*. Journal of Combinatorial Theory, Series B, **44**, pp. 363–369, 1988.

[SR99] David Rideout and Rafael Sorkin. *Classical sequential growth dynamics for causal sets*. Physical Review D, **61**, 2, 024002, 2000. arXiv preprint: http://arxiv.org/pdf/gr-qc/9904062v3.pdf.

[TH07] Thomas Thiemann. *Modern Canonical Quantum General Relativity*. Cambridge Monographs on Mathematical Physics. Cambridge University Press, 2007.

[MA00] Fotini Markopoulou. *Quantum Causal Histories*. Classical and Quantum Gravity, **17**, 10, pp. 2059, 2000. arXiv preprint: https://arxiv.org/pdf/hep-th/9904009v5.pdf.

[HMS03] Eli Hawkins, Fotini Markopoulou, and Hanno Sahlmann. *Evolution in Quantum Causal Histories*. Classical and Quantum Gravity, **20**, 16, pp. 3839, 2003. arXiv preprint: http://arxiv.org/pdf/hep-th/0302111.pdf.

[IS05] Christopher Isham. *Quantising on a Category*. Foundations of Physics, **35**, 2, pp. 271–297, 2005. arXiv preprint: http://arxiv.org/pdf/quant-ph/0401175v1.pdf.

[SO12] Rafael Sorkin. *Toward a Fundamental Theorem of Quantal Measure Theory*. Mathematical Structures in Computer Science, **22**, 05 (special issue), pp. 816–852, 2012. arXiv preprint: http://arxiv.org/pdf/1104.0997v2.pdf.

[DRE15] Olaf Dreyer. *Not on, but of*. Questioning the Foundations, The Frontiers Collection, Springer, pp. 131–138, 2015.

[SA15] Mehdi Saravani and Siavash Aslanbeigi. *Dark Matter from Spacetime Nonlocality*. arXiv preprint: http://arxiv.org/pdf/1502.01655v2.pdf.

[WH98] John Archibald Wheeler and Kenneth W. Ford. *Geons, Black Holes, and Quantum Foam: A Life in Physics*. W. W. Norton and Company, New York, 1998.

[ADGS04] Maqbool Ahmed, Scott Dodelson, Patrick B. Greene, and Rafael Sorkin. *Everpresent* Λ. Physical Review D, **69**, 10, 103523, 2004. arXiv preprint: http://arxiv.org/pdf/astro-ph/0209274v1.pdf.

[GU81] Alan Guth. *Inflationary universe: A possible solution to the horizon and flatness problems*. Physical Review D, **23**, 2, pp. 347–356, 1981.

[LI82] A. D. Linde. *A new inflationary universe scenario: A possible solution of the horizon, flatness, homogeneity, isotropy, and primordial monopole problems*. Physics Letters B, **108**, 6, pp. 389–393, 1982.

[VI83] Alexander Vilenkin. *The birth of inflationary universes*. Physical Review D, **27**, 12, pp. 2848–2855, 1983.

[LI86] A. D. Linde. *Eternal chaotic inflation*. Modern Physics Letters A, **1**, 2, 1986.

[BBS07] Katrin Becker, Melanie Becker, and John H. Schwarz. *String Theory and M-Theory*. Cambridge University Press, 2007.

[EMM12] George F. R. Ellis, Roy Maartens, and Malcolm A. H. MacCallum. *Relativistic Cosmology*. Cambridge University Press, 2012.

[GU97] Alan H. Guth. *The Inflationary Universe*. Basic Books, 1997.

[PE10] Roger Penrose. *Cycles of Time*. Vintage Books, New York, 2010.

[ER59] Paul Erdös and Alfred Rényi. *On random graphs I*. Publ. Math. Debrecen, **6**, pp. 290–297, 1959.

[BO01] Béla Bollobás. *Random Graphs*. Cambridge Studies in Advanced Mathematics, **73**, Second Edition. Cambridge University Press, 2001.

[AJJR12] Jan Ambjørn, S. Jordan, J. Jurkiewica, and R. Loll. *Second- and first-order phase transitions in causal dynamical triangulations*. Physical Review D, **85**, 12, 124044, 2012. arXiv preprint: http://arxiv.org/pdf/1205.1229.pdf.

[KMS06] Tomasz Konopka, Fotini Markopoulou and Lee Smolin. *Quantum Graphity*. Preprint, 2006. arXiv preprint: http://arxiv.org/pdf/hep-th/0611197v1.pdf.

[KMS08] Tomasz Konopka, Fotini Markopoulou, and Simone Severini. *Quantum graphity: A model of emergent locality*. Physical Review D, **77**, 10, 104029, 2008.

[KKSR12] Dmitri Krioukov, Maksim Kitsak, Robert S. Sinkovits, David Rideout, David Meyer, and Marián Boguñá. *Network Cosmology*. Nature Scientific Reports **2**, 793, 2012.

[GU15] Stan Gudder. *Inflation and Dirac in the Causal Set Approach to Discrete Quantum Gravity*. Preprint, 2015. arXiv preprint: http://arxiv.org/pdf/1507.01281v1.pdf.

[GU16] Stan Gudder. *Discrete Quantum Gravity and Quantum Field Theory*. Preprint, 2016. arXiv preprint: http://arxiv.org/pdf/1603.03471v1.pdf.

[VE11] Erik Verlinde. *On the origin of gravity and the laws of Newton*. Journal of High Energy Physics, **04**, 29, 2011.

[CGA16] Julian Gonzalez-Ayala, Rubén Cordero, and F. Angulo-Brown. *Is the* $(3+1)-$ d *nature of the universe a thermodynamic necessity?* Europhysics Letters, **113**, 4, 2016.

[BLMS87] Luca Bombelli, Joohan Lee, David Meyer, and Rafael Sorkin. *Space-Time as a Causal Set*. Physical Review Letters, **59**, 5, 1987.

[FI69] David Finkelstein. *Space-Time Code*. Physical Review, **184**, 5, pp. 1261–1271, 1969.

[MR78] J. Myrheim. *Statistical Geometry*. CERN preprint, 1978.

[TH78] Gerard 't Hooft. *Quantum Gravity: A Fundamental Problem and some Radical Ideas*. Recent Developments in Gravitation (Proceedings of the 1978 Cargese Summer Institute), edited by M. Levy and S. Deser.

[ADJL02] J. Ambjorn, A. Dasgupta, J. Jurkiewicz, R. Loll. *A Lorentzian cure for Euclidean troubles*. Nuclear Physics B, Proceedings Supplements, **106-107**, pp. 977–979, 2002. arXiv preprint: http://arxiv.org/pdf/hep-th/0201104v1.pdf.

[DP14] Giacomo Mauro D'Ariano and Paolo Perinotti. *Derivation of the Dirac Equation from Principles of Information Processing*. Physical Review A, **90**, 6, 062106, 2014. arXiv preprint: http://arxiv.org/pdf/1306.1934v1.pdf.

[FI88] David Finkelstein. *"Superconducting" Causal Nets*. International Journal of Theoretical Physics, **27**, 4, pp. 473–519, 1988.

[PE67] Roger Penrose. *Twistor Algebra*. Journal of Mathematical Physics, **8**, 2, 345–366, 1967.

[BA12] Julian Barbour. *Shape Dynamics: an Introduction*. Quantum Field Theory and Gravity: Conceptual and Mathematical Advances in the Search for a Unified Framework, Edited by Felix Finster, Olaf Müller, Marc Hardmann, Jürgen Tolksdorf, and Eberhard Zeidler. arXiv preprint: http://arxiv.org/pdf/1105.0183v1.pdf.

[AJL05] Jan Ambjørn, J. Jurkiewicz, and R. Loll. *Spectral Dimension of the Universe*. Physical Review Letters, **95**, 171301, 2005. arXiv preprint: https://arxiv.org/pdf/hep-th/0505113v2.pdf.

[SO09A] Rafael Sorkin. *Does locality fail at intermediate length scales?* Approaches to Quantum Gravity, edited by Daniele Oriti. Cambridge University Press, 2009.

Chapter 2
The Causal Metric Hypothesis

Abstract. This chapter formally introduces the *causal metric hypothesis,* and describes in detail its motivations and justifications. Foremost among these are the *metric recovery theorems* of Hawking and Malament, which state, roughly, that *"the causal structure of relativistic spacetime determines its metric structure up to scale."* As understood in causal set theory, the novel assumption that spacetime is discrete provides a natural notion of scale, given by the "sizes of fundamental elements and relations." This suggests that causal structure alone can account for emergent geometry in the discrete context. Section 2.1 introduces a general version of the causal metric hypothesis, which states, very broadly, that *"the properties of the physical universe are manifestations of causal structure."* This basic idea may be modified and/or interpreted in various ways; in particular, the *strong interpretation* of the causal metric hypothesis ascribes *all* of physics, including "nongravitational matter," to causal structure at the fundamental scale. Section 2.2 introduces a *classical version* of the causal metric hypothesis, which states that classical spacetime may be modeled in terms of mathematical objects called *directed sets,* or, more conventionally, *directed graphs.* The term "directed set" has a different conventional meaning, but I prefer to re-purpose the term than to use awkward graph-theoretic terminology. Section 2.3 begins the study of metric recovery by introducing five types of structure on relativistic spacetime; namely, *metric, conformal, causal, smooth,* and *topological* structure, in decreasing order of detail. Section 2.4 discusses metric structure in the relativistic context, i.e. *pseudo-Riemannian geometry.* Section 2.5 covers conformal structure, which defines "geometry up to scale." Section 2.6 discusses causal structure, which generalizes the "null cone structure" on Minkowski spacetime. The metric recovery theorems state that *"causal structure determines conformal structure"* under suitable assumptions. Section 2.7 introduces *causality conditions* on relativistic spacetime, which play a technical role in the metric recovery theorems. Section 2.8 gives a formal statement of metric recovery, sketches its proof, and describes how it motivates the causal metric hypothesis in the discrete setting. Section 2.9 explains why continuum-based theories are inherently awkward for modeling fundamental physics. Section 2.10 outlines some of the basic principles underlying the technical developments of subsequent chapters.

B.F. Dribus, *Discrete Causal Theory*, DOI 10.1007/978-3-319-50083-6_2

2.1 General Version of the Hypothesis

Background and context. Discrete causal theory is founded on a single motivating idea, which I refer to as the *causal metric hypothesis.* While an informal sketch of this hypothesis appears in Section 1.2, a much deeper analysis of its meaning and implications is required to support the developments in the remainder of the book. The choice of terminology for the causal metric hypothesis is my own, but variants of the same basic idea may be distilled from a number of previous sources, stretching back to at least the 1960s. The clearest and most explicit of these is Rafael Sorkin's structural *ansatz* for causal set theory, *"order plus number equals geometry."* This statement reflects the conviction that the metric recovery theorems of Hawking [HA76] and Malament [MA77], proven in the late 1970s, suggest subtle new aspects of spacetime structure *beyond the scope of their native relativistic paradigm,* just as the Lorentz invariance of Maxwell's equations suggested special relativity, despite the fact that these equations were formulated in the Newtonian context. In the case of metric recovery, the suggested new structure is *discrete directed structure.* Causal set theory, which emerged in the early 1980s, was the first sustained theoretical program attempting to describe spacetime in such terms, although a few abortive individual efforts along similar lines enjoy chronological priority. I postpone further discussion of such historical details until Chapter 3, which provides a concise overview of the origins of discrete causal theory before launching into the technical apparatus of directed sets and multidirected sets. The present chapter involves a modest amount of standard technical material, but focuses as much as possible on basic conceptual topics.

The causal metric hypothesis itself does not require any discreteness assumptions, but the metric recovery theorems described in Section 2.8 provide stronger motivation for the hypothesis in the discrete context than in the continuum-based setting.[1] Sorkin's version of the hypothesis, which applies at the classical level, is *explicitly* discrete, since it invokes *counting,* and therefore requires a notion of local finiteness in order to make sense.[2] This book devotes the most attention to versions of the hypothesis that are very similar to Sorkin's. However, it is possible to imagine continuous versions of the causal metric hypothesis, or versions that are neither discrete nor continuous. Further, due to the ubiquity of directed relationships in modern science, there exist many "non-fundamental" settings in which the causal metric hypothesis serves as a useful source of *analogy,* without being taken literally. For example, one may compare certain abstract architectures in computer science to spacetime, and may define "frames of reference," and other related notions for these structures, without insisting on any exact correspondence between the two, or sug-

[1] Indeed, in the latter setting, the metric recovery theorems essentially say that causal structure is *not quite sufficient* to recover geometry, at least under relativistic assumptions.

[2] The "local finiteness" condition used in causal set theory, which I refer to more descriptively as *interval finiteness* (IF), is not necessarily ideal for this purpose, as explained in Chapter 4. However, causal sets actually appearing in physically realistic scenarios in the literature generally *do* satisfy a suitable notion of local finiteness, even in the context of cosmology.

gesting that spacetime *is* a computer in some sense. In *quantum information theory*, such considerations are more than just an analogy, but this subject is not explored in this book.

Restatement of the general version of the hypothesis. The philosophical content of the causal metric hypothesis is that *the observed properties of the physical universe arise from causal relationships between pairs of events,* or more generally, from causal relationships *among families* of events. The latter generalization is included to allow for the possibility of *classical holism,* although I focus almost exclusively on classically reductionist models in this book. The following statement of the causal metric hypothesis, repeated from Section 1.2, is sufficiently general to use as a starting point:

Definition 2.1.1. Causal metric hypothesis (CMH). *The properties of the physical universe are manifestations of causal structure.*

The causal metric hypothesis may be regarded as an expression of the longstanding idea, examined explicitly by Leibniz, Gauss, Riemann, Einstein, Kaluza and Klein, Weyl, Wheeler, and many others, that physics is *essentially structural* in nature. The hypothesis takes the familiar relationship between cause and effect to be the fundamental building block of this structure.

Scope of the hypothesis; strong interpretation. The proper scope of the causal metric hypothesis is debatable. A conservative approach is to soften the statement in Definition 2.1.1 by replacing the words "physical universe" with the word "spacetime." This approach abandons any attempt to explain the "material content of spacetime" by means of causal structure. As noted in Section 1.7, this alternative leads to theories that possess, at best, a limited degree of background independence. These include "discrete quantum field theories," and "theories of gravity," but not "unified theories" in the deepest sense. At the opposite extreme, one may choose to take the statement in Definition 2.1.1 at face value, and interpret the opening phrase "the properties," as "*all* the properties." This is the **strong interpretation** of the causal metric hypothesis. Its radical nature was already highlighted in Section 1.10. The strong interpretation leads to a version of discrete causal theory that is ambitious and optimistic, but also quite pleasing at a structural and aesthetic level. In particular, it enables *perfect background independence,* by removing any possibility of tension between "material bodies" and "spacetime." The advantages of this approach are elaborated in Section 2.7 in the context of "causality paradoxes," and are revisited periodically throughout the book. Definition 2.1.1 is deliberately phrased in such a way as to *suggest* the strong interpretation of the causal metric hypothesis, but weaker interpretations are possible, and most of the methods and results of the book do not require the strong interpretation.

Prescription versus description. An important philosophical distinction between the relativistic viewpoint and the causal metric hypothesis, particularly its strong interpretation, involves the choice between *prescription of possible behavior* and *description of actual behavior.* Relativity employs spacetime geometry to *prescribe* which events *may* influence a given event, while the causal metric hypothesis interprets the same structure as merely an approximate way of *describing* which events *actually do* influence others. This distinction enables discrete causal theory to eliminate "awkward counterfactual speculation" regarding causality in relativity, as described in Section 1.3. In particular, the discrete causal rejoinder to Wheeler's famous statement that *"spacetime tells matter how to move; matter tells spacetime how to curve"* [WH98], is that *"things happen; "spacetime" and "matter" are ways of describing them."* A general preference for description over prescription in theoretical physics constitutes one of the philosophical principles informing the overall development of discrete causal theory, as described in Section 2.10. This does not mean that the theory seeks to avoid the necessary criteria of *explaining* and *predicting* physical behavior, but merely that "causal structure" should mean neither more nor less than the aggregate of *actual* causes and effects. This viewpoint is closely linked to the notion of perfect background independence, because it removes the distinction between a "spacetime" that prescribes behavior, and "material participants" in this behavior.

Technical implementations. A "technical implementation" of the causal metric hypothesis specifies what the words *causal structure* in Definition 2.1.1 are taken to mean in mathematical terms. Many different such implementations are possible, both at the classical level and the quantum level. In Section 2.2, I introduce a *classical version* of the causal metric hypothesis (CCMH), which identifies *directed sets* as the chosen mathematical models of classical spacetime. This version is more specific than the version appearing in Definition 2.1.1, but is still quite general, since its purpose is to accommodate any "reasonable" variant of the theory. Hence, additional conditions must be imposed in order to obtain a specific theory capable of precise quantitative description of nature. This is accomplished by specifying a set of *axioms* that restrict attention to a "desirable" class of directed sets, together with a "plausible" physical interpretation of these sets. Physical intuition plays an unavoidable role in this process, but data from experimentally-established physics should be used for guidance whenever possible. Examples of the types of considerations that might be involved in selecting such a set of axioms are the questions of how to implement the idea of *discreteness,* what type of *local behavior* the chosen class of directed sets should exhibit, and whether or not directed sets containing *causal cycles* should be included. Chapters 3 and 4 are largely devoted to identifying a suitable set of axioms for discrete classical causal theory, based both on experimental evidence and on basic structural considerations. A "suggested list" of axioms is offered in Section 4.10, along with "conservative" and "radical" alternatives. A specific choice of axioms leads, via iteration of structure (IS), to a specific version of quantum theory, as described in Part II of the book.

Quantum causal metric hypothesis. A detailed treatment of the *quantum causal metric hypothesis* (QCMH) is postponed until Chapter 7, since much conceptual and technical ground must be covered before it can be adequately explained. In general terms, the role of classical causal structure is superseded in discrete quantum causal theory by the "higher-level multidirected structures" of *kinematic schemes,* introduced informally in Chapter 1 as "structured configuration spaces" of classical histories, whose "relations" are *co-relative histories.* Given a kinematic scheme, the general strategy is to "superpose" evolutionary processes for its constituent classical histories, thereby building the "quantum universe." Feynman's *path summation approach to quantum theory* provides the basic conceptual ingredients of this viewpoint. In a more modern context, this approach shares important features with Isham's *quantization on a category* [IS05], and Sorkin's *quantum measure theory* [SO12]. A formal statement of the quantum causal metric hypothesis appears in Section 7.6.

2.2 Classical Version of the Hypothesis

Review of basic building blocks of causal structure. The content of the causal metric hypothesis (CMH) must be expressed mathematically before its physical consequences may be explored in a precise quantitative fashion. In the classical context, this may be accomplished by means of the "classical histories" introduced informally in Chapter 1. In the present chapter, it is necessary to consider the mathematical properties of these histories in slightly more detail, although formal definitions and analysis are postponed until Chapter 3. The basic unit of mathematical structure used to model a *particular* instance of cause and effect between two events is an *ordered pair of abstract elements,* with the first element representing the cause, and the second element representing the effect. In Figure 1.4.1, I introduced a convenient way to represent this structure via diagrams, called *generalized Hasse diagrams,* in which elements are represented by nodes, and relations are represented by directed line segments connecting pairs of nodes, with directions inferred by using an "up the page" convention. For convenience, I reproduce this picture in Figure 2.2.1.

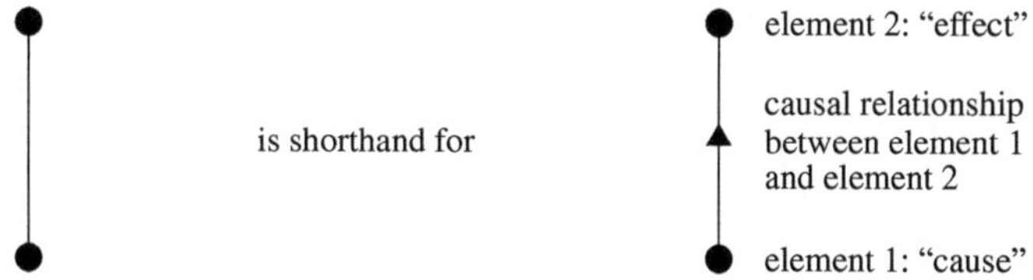

Figure 2.2.1. Abstract representation of a single instance of cause and effect.

Directed sets. Elements 1 and 2 in Figure 2.2.1 are naturally ordered, with element 1, representing the "cause," preceding element 2, representing the "effect." It is convenient to name these elements x and y; one may then write $x \prec y$ to represent this order. As explained in Section 1.8, the *precursor symbol* $\prec$ is analogous to the familiar "less than" symbol $<$ in integer arithmetic, although $\prec$ is more specific, since $x \prec y$ means that x *directly* precedes y. In discrete classical causal theory, the local "arrow of time" is defined by this primitive order, with a single fundamental unit of local time, i.e., a single "classical chronon," separating x and y. For a larger family D of elements, say $D = \{w, x, y, z\}$, one must consider a corresponding family of relations, encoding each individual instance of cause and effect. It is standard to denote this entire family, say $\{w \prec y, x \prec y, z \prec w\}$, by the single symbol $\prec$. In technical terms, this means that $\prec$ is a **binary relation** on D, i.e., a subset of the Cartesian product $D \times D$. Using this definition, the statement $x \prec y$ means that the ordered pair (x, y) in $D \times D$ is an element of $\prec$. In this book, the pair $(D, \prec)$ is called a *directed set.* Equivalent structures are called *directed graphs* in conventional mathematical settings, with the term "directed set" usually assigned a more specific meaning; however, in this book, a "directed set" means simply a set equipped with a binary relation. An individual relation between a pair of elements x and y in D is almost always represented by the notation $x \prec y$, rather than the alternative notation $(x, y) \in \prec$.

The "up the page" convention for inferring the directions of relations may be used for any directed set D in which "causal influence always flows one way, never looping back." Technically, this means that D is *acyclic,* as noted in Section 1.4. All of the classical histories used for illustrative purposes in Chapter 1 are modeled via acyclic directed sets, but it is sometimes interesting to adopt a broader viewpoint, and allow directed sets containing cycles. In spite of their counterintuitive properties, such sets can be physically interesting, partly because general relativity does not rule out the existence of "closed causal curves." Generalized Hasse diagrams cannot be used to represent such sets; rather, arrows must be included in the diagrams to explicitly indicate the direction of each relation. The vast majority of directed sets considered in this book, however, are acyclic. In Figure 2.2.2, I reproduce the generalized Hasse diagram of the "slightly more complicated classical history" of Figure 1.4.3, now viewed abstractly as an acyclic directed set.

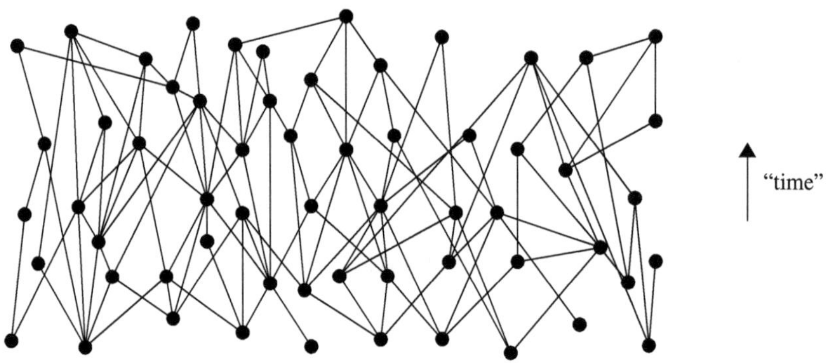

Figure 2.2.2. An acyclic directed set $(D, \prec)$.

Classical causal metric hypothesis in terms of directed sets. In the classical context, a general mathematical version of the causal metric hypothesis may be formulated in terms of directed sets, as follows:

Definition 2.2.3. Classical causal metric hypothesis (CCMH). *Classical histories may be modeled via directed sets.*

Though more specific than Definition 2.1.1, this is still a very general statement. For example, it includes Sorkin's version of the causal metric hypothesis, since causal sets are a special case of directed sets, but it also includes continuous *domain-theoretic* versions, since domains are directed sets as well. As noted in Section 2.1, practical application of discrete causal theory requires narrowing the focus to a *specific* class of "physically relevant directed sets," and this involves nontrivial choices. In this book, *the most substantial such choices are to work in the discrete context, to focus mainly on the acyclic case, and to work primarily in terms of direct, i.e., immediate, relationships.* In technical terms, discreteness is expressed via an appropriate *local finiteness condition;* in this case, *star finiteness* (SF). *Acyclicity* (AC) simply means that no sequence of relations may both begin and end at the same element. The choice to focus on direct relationships is implemented by modeling classical causal structure in terms of generally *nontransitive* binary relations, and interpreting each individual relation to encode independent influence. This means abstaining from the axiom of *transitivity* (TR), and adopting the *independence convention* (IC). These choices are briefly explained below, and are examined and justified more systematically in Chapters 3 and 4.

Discreteness. The single most significant restriction on the types of directed sets studied in this book as models of classical spacetime is that they are *discrete.* The term "discrete" has different meanings in topology, measure theory, and order theory, so further explanation is necessary to render the precise meaning of this choice sufficiently clear. In causal set theory, the axiom of *interval finiteness* (IF) ensures that causal sets are discrete in an order-theoretic sense; i.e., every nonextremal element in a causal set has at least one *maximal predecessor* and *minimal successor.* In addition, the use of a *discrete measure* that "counts fundamental volume units" ensures that causal sets are discrete in a measure-theoretic sense.[3] This lends plausibility to the appearance of "number" in Sorkin's version of the classical causal metric hypothesis, without any need to "quantize spacetime," as is typically attempted in continuum-based approaches to quantum gravity. The causal set version of discreteness *may* take too literally the idea of "volume" at the fundamental scale, since "familiar geometric notions" are generally expected to emerge only at relatively large scales in discrete causal theory.[4] For the purposes of this book, the order-theoretic meaning of discreteness, suitably generalized, is the most important. The measure-theoretic meaning, meanwhile, is "relevant, but less precise," while the topological meaning is almost completely irrelevant. The desired order-theoretic and measure-theoretic

---

[3]To be precise, the causal set measure allows "statistical fluctuations" in the assignment of volume. This topic is revisited in Sections 3.2, 3.5, and 4.5.

[4]One such familiar notion that turns out to be very complicated is the emergent notion of spacetime *dimension.* In particular, a variety of approaches to discrete spacetime structure suggest possible *scale-dependence* of dimension. See the recent paper of Carlip [CA15] for an interesting discussion and list of references.

properties may be achieved by imposing the local finiteness condition of *star finiteness* (SF), together with the *generalized measure axiom* (M*), both discussed in Chapter 4. Some of the general motivations for working in the discrete context are discussed further in Sections 2.9 and 2.10, as well as in later chapters.

Acyclicity. Another attractive "physical relevance criterion" for directed sets is acyclicity (AC), which justifies the "up the page" convention for the generalized Hasse diagrams appearing in this book.[5] Acyclicity is a natural abstraction of the apparently *unidirectional* nature of causality, i.e., the fact that events do not seem to contribute to their own causes, either directly or indirectly. The acyclic binary relations studied in this book are generally *not* assumed to be transitive (TR), since transitive relations cannot distinguish naturally between direct and indirect relationships. By abstaining from transitivity, one is free to use the independence convention (IC), which specifies that each individual relation encodes direct influence, independent of other modes of influence between its initial and terminal elements. A binary relation $\prec$ on a set D generates a *transitive relation* $\prec_{\rm tr}$ under the operation of *transitive closure*, as described in Chapter 3. In the acyclic case, $\prec_{\rm tr}$ is a *partial order*, which is essentially why partial orders play such a prominent role in causal set theory. In particular, this lends plausibility to the appearance of "order" in Sorkin's version of the classical causal metric hypothesis. However, as indicated above, passage from $\prec$ to $\prec_{\rm tr}$ destroys information about direct relationships; i.e., one cannot recover $\prec$ from $\prec_{\rm tr}$. For this reason, I work almost exclusively with the nontransitive binary relation $\prec$ itself, which I call the *causal relation*. These details are discussed in Chapter 3, particularly in Sections 3.9 and 3.10.

2.3 Structure on Relativistic Spacetime

"Geometric intuition" versus physical geometry. Given the "geometric" picture of directed sets afforded by graphical representations such as generalized Hasse diagrams, it is natural to view these sets as "spaces" in a *mathematical* sense, quite apart from any specific physical interpretation. Of course, I have already explained the discrete causal interpretation of these sets at an informal level, but this preliminary description falls far short of providing a precise quantitative description of how discrete causal theory models fundamental spacetime structure, or enabling a meaningful comparison of the theory with more conventional approaches to fundamental physics. This situation calls for a deeper examination of the physical role of directed structure, followed by a careful explanation of its consequences in the discrete setting. The best understanding of such structure available within the scope of experimentally-established physics comes from general relativity. Hence, much of the present chapter focuses on reviewing a few important aspects of relativistic

[5] For diagrams of directed sets that are *not* assumed to be acyclic, arrows are added to the edges to explicitly indicate the direction of causal influence. Examples appear in Figures 2.7.2, 3.6.5, 5.4.4, and 8.7.2. It is occasionally convenient to add arrows even in the acyclic case.

spacetime structure, from a perspective that may be readily adapted to the discrete causal context. Central to this picture are the *metric recovery theorems.*

The pedestrian view of directed sets as "spaces in a mathematical sense" represents an instance of the common and useful practice of applying "geometric intuition" to help analyze the properties of mathematical objects, whether or not these objects are "geometric" in a traditional sense. While the original source of such intuition is often partly physical in nature, the practice itself has no necessary connection to physics at all; indeed, fields of pure mathematics such as functional analysis and algebraic geometry abound with such "geometric" methods. In functional analysis, for example, one studies "spaces of functions," which are typically infinite-dimensional vector spaces. The elements of such spaces are functions on some other space, such as the real line $\mathbb{R}$. Notions originating in geometry, such as projections and orthogonality, play a central role. In algebraic geometry, meanwhile, one studies "spaces" called *algebraic schemes*, whose elements are prime ideals in commutative rings; for example, the ideal of all polynomials $f(x, y)$ in the polynomial ring $\mathbb{R}[x, y]$ vanishing on an irreducible algebraic curve[6] such as $\{(x, y) \in \mathbb{R}^2 | y = x^2\}$. In this case, familiar "geometric" concepts are applied in ways unimagined by mathematicians over the first two millennia of studying such objects; for example, individual points may possess nonzero dimension. Generalization of these ideas to the noncommutative setting leads back in a curious manner to topics in fundamental physics, via Connes' *noncommutative geometry,* revisited briefly in Chapter 8.

The purpose of rehearsing this bit of pure mathematics is to emphasize that the vague "geometric" character of arbitrary directed sets does not, by itself, constitute evidence that physical spacetime, from which humans have acquired much of their geometric intuition, may actually *be* a directed set, or a structure "built from directed sets." Indeed, "geometric data" in some form may be squeezed out of almost any type of mathematical object one might choose to work with. For example, when studying any suitable class[7] of structured sets, one may always pass to a *category* of such sets, then ignore the "internal structure" of the sets themselves, regarding them as merely "higher-level elements," just as directed sets are viewed as "elements" of a kinematic scheme. This yields an abstract *multidirected set*, whose elements represent the original structured sets, and whose relations represent morphisms between pairs of structured sets. This multidirected structure, in turn, provides natural "geometric" notions of "directions," "neighbors," "paths," distances," and so on. Of course, neither multidirected sets nor categories enter the picture in any serious manner until Chapter 3, aside from a brief explanation in the present section regarding the role of category theory in organizing different types of structure on relativistic spacetime. However, the details of this particular example involving categories of structured sets are immaterial at present. Its role is merely to illustrate why the causal metric

[6]In this context, the word "curve" means "locus of points," *not* "map from a real interval into a manifold," as it does later in the chapter.

[7]The reason for the qualifier "suitable" here is that the class must be "small enough" so that the resulting multidirected structure will actually be a set, rather than a *proper class*.

hypothesis (CMH) requires much stronger and more specific justification than I have demonstrated thus far.

Fortunately, such justification exists, in the form of the metric recovery theorems of Hawking and Malament. The basic idea of metric recovery is that *almost all of the apparent geometric structure of relativistic spacetime is encoded in its causal structure.* The only information missing is "scale data;" or, more precisely, a *conformal factor* in the metric. If spacetime is actually discrete, however, then the combinatorial details of discrete microstructure can supply scale data "for free." Hence, one may construct discrete models whose *only* structure is causal structure, yet which "look just like relativistic spacetime at ordinary scales." This suggests that, under the limitations of present observations, discrete causal models of classical spacetime are "just as good" as the geometric models used in relativity. In fact, they turn out to be much better in a number of significant ways. The principal reason why such models are not yet ready to replace relativistic spacetime root and branch is because general relativity explains *how* specific geometry arises dynamically, while discrete causal dynamics is still in its infancy.

Pseudo-Riemannian manifolds; diffeomorphism invariance. To properly understand the subject of metric recovery, it is necessary to examine a few of the geometric ingredients of general relativity. In this context, the "spaces" of interest are special types of *real manifolds,* called *pseudo-Riemannian manifolds,* viewed as models of classical spacetime. To be precise, I should point out that even in general relativity, such manifolds are not properly regarded as "physical" in their own right. Einstein himself understood that individual elements of a pseudo-Riemannian manifold do not possess intrinsic physical meaning, as already mentioned in Section 1.6. To understand, at an informal level, why this is true, one might imagine "painting" certain physical information on the surface of a sphere, then mapping each element of the sphere to another element by means of a rotation. In this context, the physical information is "re-associated" with different abstract points on the sphere, but there is no intrinsic physical distinction between the two associations.[8] More formally, general relativity is *diffeomorphism invariant;* i.e., Einstein's equation (1.3.1) does not change its form under "smooth transformations" of relativistic spacetime. With this understanding, I often lapse into the common habit of treating elements of a pseudo-Riemannian manifold as "spacetime events" in the relativistic context, even though these elements really only *represent* spacetime events.

This distinction between mathematical elements and physical events actually turns out to be important when comparing general relativity to discrete causal theory, due to the relative *rigidity* of discrete directed sets, mentioned periodically throughout Part I of the book, and revisited more thoroughly in Section 6.3. I will not go into the details in the present chapter, but one of the basic physical implications of this rigidity is that physical spacetime events may be associated *much more directly* with elements of a discrete directed set than with elements of a pseudo-Riemannian manifold. On a historical note, the "non-physicality" of elements and coordinate systems presented Einstein with such severe difficulties that it contributed to several

---

[8]This familiar thought experiment paraphrases part of an analogous discussion in Rovelli [RO04].

years of delay in the publishing of his first papers on general relativity, even after most of the mathematical and physical essentials were in place. Had Einstein been working with discrete directed sets instead, this particular conceptual issue would likely have posed far less of a obstacle.[9]

Five types of structure. Pseudo-Riemannian manifolds are endowed with a number of different types of structure, with varying types and degrees of physical significance. Two manifolds that are "the same" with respect to one type of structure may be "different" with respect to another. Five important types of structure on a pseudo-Riemannian manifold X are *topological, smooth, causal, conformal,* and *metric* structure. These are listed in a suggestive way in Figure 2.3.1. For future reference, the left-hand side of the figure, which looks like part of a curved two-dimensional surface, really represents part of a pseudo-Riemannian manifold, usually assumed to be four-dimensional, connected, and without boundary. In particular, the "edges of the surface" do not represent actual boundary points of the manifold, but merely delimit the portion being illustrated. Auxiliary structure represented by graphical features that intersect the edges, such as "curves drawn on the surface," should be assumed to "keep on going," rather than "stopping at the edge."

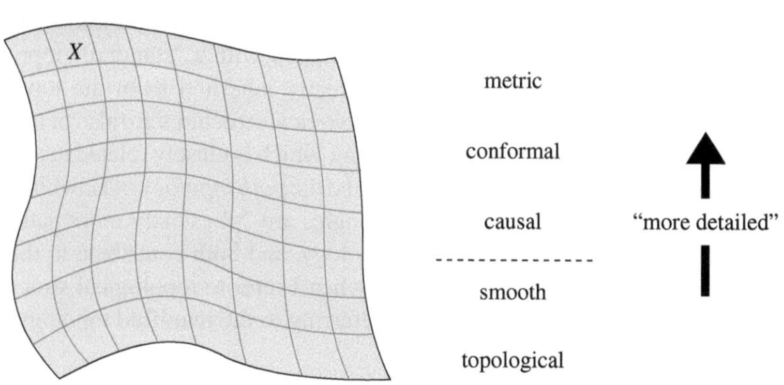

Figure 2.3.1. Informal view of types of structure on a relativistic spacetime manifold X.

"Relativistic spacetime geometry" is an informal term for metric structure in relativity, which is the "most detailed" of the five types of structure listed in the figure. A question of central importance, both in relativity and in related theories, is how much of this metric structure one may "recover" from knowledge of one or more of the other four types of structure. This is a generalized version of the metric recovery problem. In the present context, of course, the focus is directed toward the more specific question considered by Malament; namely, how much of the *metric* structure may be recovered from knowledge of the corresponding *causal* structure. However, these five types of structure are intimately bound together, and it

[9]Rovelli [RO04] gives a splendid explanation of this topic in his chapter 2.

is necessary to consider all of them to some degree in order to understand this more specific problem.

Ranking of structures. The list appearing on the right-hand side of Figure 2.3.1 gives an informal "ranking" of these five types of structure in the specific context of relativistic spacetime. The qualitative idea of this ranking is that it is "easy" to recover structures lower on the list from structures higher on the list, but difficult or impossible to do the opposite. For example, metric structure is specified by a particular choice of metric, as described below, while conformal structure is specified by an equivalence class of metrics, related to each other by "scaling functions" called *conformal factors.* Knowledge of metric structure, i.e., of a specific metric, immediately yields knowledge of the corresponding equivalence class of metrics, i.e., of the conformal structure. However, knowledge of an equivalence class of metrics, by itself, does not yield a method of choosing a unique metric from among them. Hence, it is generally impossible to recover metric structure from conformal structure alone.

The reason for the dashed line in the figure is that it is possible to talk about the top three types of structure in "non-smooth" contexts; in particular, *discrete* causal structure is the main subject of Part I of this book. Hence, the "ranking" only applies to relativistic spacetime, where an underlying smooth manifold structure is taken for granted. Such a manifold comes equipped with a "standard" topological structure, called the *manifold topology,* indirectly inherited from the topological structure of the real line $\mathbb{R}$. However, other topological structures are also of interest in relativity; for example, the *Alexandrov topology,* which is closely related to the axiom of interval finiteness (IF) in causal set theory, and the *path topology* of Hawking, King, and McCarthy [HA76]. The latter two topologies are "physically more natural," in certain specific ways, than the manifold topology, and both contribute to the proof of the metric recovery theorems. However, when I refer to topological structure on relativistic spacetime in this chapter, I am referring to the manifold topology unless stated otherwise.

For the sake of relevance and brevity, I focus primarily on the top three types of structure listed in the figure, examining them in reverse order of detail, beginning with metric structure and ending with causal structure. The choice to present the material in this order is motivated by the fact that conformal structure and causal structure in relativity are usually expressed *in terms of* metric structure. I discuss smooth structure and topological structure only briefly, and in purely auxiliary ways. For example, smooth structure enters the picture in the discussion of metric structure, since the bilinear maps on tangent spaces defined by the metric are taken to vary smoothly. It also plays a role in the discussion of causal structure, since the distinction between *causal isomorphisms* and *enhanced causal isomorphisms* in Definition 2.6.3 is described in terms of the distinction between smooth causal curves and causal curves that are merely continuous. I discuss topological structure last of all, and only briefly, after the description of metric recovery in Section 2.8. The purpose of this brief topological detour is partly to aid the reader in understanding the literature on metric recovery, which makes use of all three topologies mentioned above, and partly to facilitate the discussion in Chapter 4 regarding how the Alexandrov topology relates to the interval topology and the causal set axiom of interval finiteness (IF).

The qualitative claim of the classical causal metric hypothesis (CCMH) is that "metric structure" is merely *a way of describing certain aspects of causal structure.* This assertion seems to turn on its head the ranking of structures appearing in Figure 2.3.1, but the causal metric hypothesis is not intended to apply, at a precise level, in the relativistic setting. Indeed, if classical spacetime *really is* precisely represented by a pseudo-Riemannian manifold in the usual relativistic sense, then the classical causal metric hypothesis is *wrong;* at least, under the standard definitions and assumptions specifying how causal structure should be modeled in this context. In this case, the metric recovery theorems tell exactly what is missing; namely, scale data, and this is where conformal structure enters the picture. Hence, the plausibility of the causal metric hypothesis depends on the conviction that what is conventionally viewed as relativistic spacetime is *really some alternative structure that merely mimics a pseudo-Riemannian manifold at large scales.* Although the metric recovery theorems say nothing about discreteness explicitly, they do demonstrate indirectly that "relativistic spacetime looks suspiciously like a discrete directed set," since discrete directed structure carries natural scale data. These theorems may therefore be interpreted to suggest that discrete directed sets represent a particularly promising candidate for such alternative structure. It is therefore reasonable to consider the possibility that general relativity is merely a "smooth approximation" of discrete causal theory, just as special relativity is a "flat approximation" of general relativity, and Newtonian mechanics is a "low-velocity approximation" of special relativity.

Category theory as an organizing principle. Category theory provides a useful, though incomplete, method of organizing the five types of structure listed in Figure 2.3.1. It is useful because it treats these structures in a unified and coherent manner, but it is incomplete because it emphasizes only the *active viewpoint,* in which structural relationships are studied via *morphisms,* in this case, by actively mapping each element of spacetime to another element. The complementary *passive viewpoint,* in which one compares different instances of the same type of structure without any active mapping procedure, is implicitly deprecated in this context. In conventional continuum-based physics, these two viewpoints often, though not always, involve essentially equivalent treatment of physically relevant information. In discrete causal theory, however, they lead in profoundly different directions. I return to this subject later in the chapter, and again in Section 8.4. At present, the distinction introduces no serious difficulties.

If X and X' are "spaces," in a general and unspecified sense, with each space possessing different instances of a particular type of structure, then X and X' may be viewed as objects in a category characterized by this common structural type, ignoring for the moment any other type of structure they might possess. For example, if X and X' are "relativistic spacetimes," then one may choose to view them as simply topological spaces, i.e., as objects in an appropriate *topological category.* For the sake of clarity, it is useful to note that there is a different category for each type of topology; for example, the *manifold, Alexandrov,* and *path topologies.*

Alternatively, one may choose to view X and X' as smooth real manifolds, i.e., as objects in an appropriate *smooth category*. Smooth real manifolds are the underlying structures on which geometry is built in general relativity. Once a particular type of structure has been chosen to study, it is natural to turn attention to the class of maps $f : X \rightarrow X'$ preserving this structure. These are the *morphisms* between X and X' in the chosen category. For more abstract categories, morphisms are not necessarily maps, but in the context of relativity, the categories involved are very "concrete," and maps suffice. If the morphism $f : X \rightarrow X'$ possesses an inverse that is also a morphism, then f is called an *isomorphism*, and X and X' are called *isomorphic*. This is sometimes denoted more succinctly by the expression $X \cong X'$, which expresses the information that there exists *at least one* isomorphism between X and X', without specifying a *particular* isomorphism. Isomorphic objects X and X' are considered to be "essentially the same" with respect to whatever type of structure is being studied. "Self-morphisms" $f : X \rightarrow X$ are called *endomorphisms*, and "self-isomorphisms" are called *automorphisms*. It is easy to see why this approach emphasizes only the active viewpoint, since different instances of a particular type of structure on X are compared by *actively transforming* X.

Traditionally, different names are assigned to morphisms in different categories. For example, a topological isomorphism is called a *homeomorphism*, and one must specify which type of topological structure is being considered for this notion to be well-defined. A smooth isomorphism is called a *diffeomorphism*, a metric isomorphism is called an *isometry*, and a conformal isomorphism is called a *conformal isometry*. As far as I know, causal morphisms do not possess separate traditional names, probably because they were not seriously studied until after category theory became the standard structural paradigm in abstract algebra. In particular, Zeeman [ZE64] seems to have coined the term *causal automorphism* in his 1964 paper on causality and the Lorentz group.[10] In the topological category for the manifold topology, relativistic spacetime is assumed to be locally homeomorphic to $\mathbb{R}^4$; i.e., it is a four-dimensional real manifold. In the more detailed smooth category, it is assumed to be locally diffeomorphic to $\mathbb{R}^4$ with its usual smoothness structure; i.e., it is a *smooth* four-dimensional real manifold. This supplies enough underlying structure to facilitate the specification of spacetime geometry.

2.4 Metric Structure

Pseudo-Riemannian metrics. The type of geometry of principal interest in general relativity is a special kind of *pseudo-Riemannian geometry*, sometimes called *Lorentzian geometry*. It is the tool Einstein finally settled on for modeling classical spacetime structure after several years of painful self-education in then-relatively-

[10] In fact, both Zeeman [ZE64] and Malament [MA77] define causal morphisms in terms of *timelike* rather than *causal* relationships. The latter may be *either* timelike or null. Discrete causal theory generally does not make such distinctions. See Sections 2.6 and 2.8 for more details.

modern mathematics. A significant proportion of readers will probably be grateful for the inclusion of a few extra pages recalling some of the rudiments of this particular type of geometry. However, I do assume that the reader is familiar with the basic definitions of *real manifolds, tangent* and *cotangent spaces, smooth maps,* and a few related ideas from elementary differential geometry.

Definition 2.4.1. *Let X be a smooth real manifold. A* **pseudo-Riemannian metric** *g on X is a smoothly-varying family of real-valued, non-degenerate, symmetric, bilinear maps on the tangent spaces of X. A* **pseudo-Riemannian manifold** *is a smooth real manifold together with a choice of pseudo-Riemannian metric.*

To spell this out in more detail, the definition means that for each $x \in X$, g assigns a real value $g_x(v, w)$ to each pair of tangent vectors v and w in the tangent space T_xX at x. For any fixed tangent vector v in T_xX, one may define a map $g_x(v, -) : T_xX \to \mathbb{R}$, sending each tangent vector w to $g_x(v, w)$. A similar map $g_x(-, w) : T_xX \to \mathbb{R}$ may be defined by fixing the second argument in g_x. The "non-degenerate" property of g means that the map $g_x(v, -)$ is identically zero if and only if v is itself the zero vector, and similarly for $g_x(-, w)$. The "symmetric" property means, of course, that $g_x(v, w) = g_x(w, v)$ for every choice of v and w. The "bilinear" property means that g_x is linear in each of the variables v and w. For the first variable v, this means that $g_x(a_1v_1 + a_2v_2, w) = a_1g_x(v_1, w) + a_2g_x(v_2, w)$, for any tangent vectors v_1, v_2, and w at x, and any scalars a_1 and a_2. An alternative way to say this is that the maps $g_x(v, -)$ and $g_x(-, w)$ are linear; i.e., they are *dual vectors* or *cotangent vectors* at x.

The metric g is a *tensor,* which is a general term denoting a family of multilinear maps of an appropriate type, whose arguments are tangent vectors and/or cotangent vectors on X. The assignment $(v, w) \mapsto g_x(v, w)$ may be viewed as a "generalized inner product" on the vector space T_xX. In the special case of a *Riemannian* manifold, dropping the prefix "pseudo," this assignment really *is* an inner product, i.e., a symmetric, bilinear, positive-definite map, where "positive" means that $g_x(v, v) \geq 0$, and "definite" means that $g_x(v, v) = 0$ if and only if $v = 0$. More generally, however, it is possible that $g_x(v, v) \leq 0$ even when the vector v is nonzero; in this case, g does not define true inner products on the tangent spaces of X. This occurs, in particular, in the relativistic case, where the sign of $g_x(v, v)$ determines whether v is *timelike, null,* or *spacelike.* These designations, along with their physical interpretations, are discussed further below. For notational clarity, I remark that it is sometimes convenient to denote a pseudo-Riemannian manifold by a pair (X, g), when one wishes to make the choice of metric explicit.

In the context of relativity, one is interested in the specific case of four-dimensional spacetime. In this case, the metric g may be represented at each point $x \in X$ by a 4×4 symmetric matrix:

$$(g_{\mu\nu}) = \begin{pmatrix} g_{00} & g_{01} & g_{02} & g_{03} \\ g_{10} & g_{11} & g_{12} & g_{13} \\ g_{20} & g_{21} & g_{22} & g_{23} \\ g_{30} & g_{31} & g_{32} & g_{33} \end{pmatrix}. \tag{2.4.2}$$

The numerical values of the entries $g_{\mu\nu}$ in this expression depend on a choice of basis for T_xX, which is often derived from a choice of local coordinates on X near x. However, the metric g itself, as opposed to a particular matrix representation of g at a particular point, is often written as $g_{\mu\nu}$, for historical reasons. For example, this notation appears in the usual expression of Einstein's equation (1.3.1). Mathematically, this is a bit awkward, and constitutes one of the reasons why much of the literature on general relativity, and also on quantum field theory, is difficult for many mathematicians to read. A reasonable compromise between traditional and modern conventions is Penrose's *abstract index notation,* in which the indices appearing in the expression for a tensor have nothing to do with bases or coordinates, but merely indicate the type and order of its arguments, i.e., the number and arrangement of tangent vectors and cotangent vectors on which it operates. However, such notational details play essentially no role in this book beyond the present chapter.

Examples of metrics. The typical student of relativity usually encounters two specific, and particularly simple, pseudo-Riemannian metrics, before studying the properties of metrics in general. The first of these metrics is the *Euclidean metric* δ on $\mathbb{R}^4$, for which the diagonal entries in the above matrix representation (2.4.2) are $1, 1, 1, 1$, and the off-diagonal entries are 0, for every point $x \in \mathbb{R}^4$, under a standard choice of basis. The second is the *Minkowski metric* η on $\mathbb{R}^4$, for which the corresponding diagonal entries are $-1, 1, 1, 1$, and the off-diagonal entries are 0. Since the generalized inner products defined by these metrics do not vary across spacetime, they are called *constant metrics.* In both cases, the entire pseudo-Riemannian manifold involved is isomorphic to any of its tangent spaces. This means that properties that generally apply only locally, or in a limiting sense, such as *Lorentz invariance* in the case of relativistic spacetime, actually hold at a global level in these special cases. A simple example of a non-constant metric is the *Schwarzchild metric* on $\mathbb{R}^4$, which may be represented by the matrix

$$(g_{\mu\nu}) = \begin{pmatrix} -\left(1-\frac{2GM}{c^2r}\right) & 0 & 0 & 0 \\ 0 & \left(1-\frac{2GM}{c^2r}\right)^{-1} & 0 & 0 \\ 0 & 0 & r^2 & 0 \\ 0 & 0 & 0 & r^2\sin^2\theta \end{pmatrix},$$

using coordinates (ct, r, θ, ϕ), where c is the speed of light, G is Newton's gravitational constant, t is the time coordinate, and the spatial coordinates (r, θ, ϕ) are the usual spherical coordinates on $\mathbb{R}^3$.[11] In the limit of a vanishing "cosmological constant," the Schwarzchild metric describes relativistic spacetime near an appropriate spherically-symmetric body of mass M, such as an ideal, non-charged, non-rotating black hole.

---

[11]It is common to "choose units" in such a way that the numerical values of constants such as c and G are set to 1. For example, Carroll [CA04], p. 193, omits explicit inclusion of c in his expression of the Schwarzchild metric, and Hawking and Ellis [HE73], p. 149, omit both c and G. I include these factors so that the "units work out" in a naïve sense.

A general class of metrics of particular interest in conventional relativistic cosmology is the class of *Friedman-Lemaître-Robertson-Walker* (FLRW) *metrics,* which are special solutions to Einstein's equation (1.3.1), describing homogeneous, isotropic spacetimes with time-varying "scale factors." Under a suitable choice of coordinates (ct, r, θ, ϕ), a FLRW metric may be represented by the matrix

$$(g_{\mu\nu}) = \begin{pmatrix} -1 & 0 & 0 & 0 \\ 0 & a^2(t) & 0 & 0 \\ 0 & 0 & a^2(t)f^2(r) & 0 \\ 0 & 0 & 0 & a^2(t)f^2(r)\sin^2\theta \end{pmatrix},$$

where $a(t)$ is the scale factor, and where $f(r)$ takes on one of a few simple values, depending on the curvature. Two specific "maximally symmetric" examples of particular prominence are the positive-curvature *de Sitter metric,* which describes an idealized cosmology dominated by a positive "cosmological constant," and the negative-curvature *anti-de Sitter metric,* whose higher-dimensional analogues are central to Maldacena's *AdS/CFT correspondence* [MA99] in string theory. The most popular "realistic" models in modern cosmology involve so-called "almost FLRW metrics," which are perturbed away from an exact FLRW form by inhomogeneities on relatively small scales, in an effort to model the manifest local irregularity of matter content in the observable universe. In particular, the *ΛCDM model,* named to reflect the fact that it involves a nonzero "cosmological constant" Λ and "cold dark matter" (CDM), uses such metrics. A good standard reference on the subject is [EMM12].[12] FLRW models make a brief reappearance in Chapter 4, where I quote an assertion by the founders of causal set theory [BLMS88] that star finite causal sets suffice for studying their discrete causal analogues.

Pulling back a metric; metric morphisms. Given a smooth morphism[13] $f : X \to X'$, between a pair of smooth real manifolds X and X', together with a choice of pseudo-Riemannian metric g' on X', one may define a smoothly-varying family of symmetric, bilinear maps f^*g' on the tangent spaces of X, called the *pullback* of g' along f. This family is defined by "pushing forward" tangent vectors from X to X', then applying g' to these tangent vectors. If the morphism f possesses "suitable properties;" in particular, if it is a diffeomorphism, or more generally, an *immersion,* then f^*g' is nondegenerate, and therefore constitutes a pseudo-Riemannian metric on X. The reason for such a brief and informal description of this construction is that the pullback plays a very limited and specific role in this chapter. Most of the technical details are unnecessary for this purpose, but may be found in any appropriate reference on differential geometry or general relativity, such as Wald [WA84]. A *metric morphism* between a pair of pseudo-Riemannian manifolds (X, g) and (X', g') is a map $f : X \to X'$ that preserves metric structure, in the sense that the pulled-back metric f^*g' on X coincides with the chosen metric g on X. In particular, if $X = X'$,

[12] See [EMM12], p. 205. Most texts based on general relativity use the "line element" notation to describe such metrics; in this case, $ds^2 = -dt^2 + a^2(t)\big[dr^2 + f^2(r)(d\theta^2 + \sin^2\theta d\phi^2)\big]$.

[13] This means a morphism in the smooth category.

then such a morphism f is necessarily bijective, and is therefore a self-isometry, i.e., an automorphism of X in the category of pseudo-Riemannian manifolds.

Metric signature. The *signature* of a pseudo-Riemannian metric g on a smooth n-dimensional real manifold X is an ordered pair of natural numbers (p, q), with sum n, where p and q are the "numbers of plus 1's and minus 1's along the diagonal," when g is represented in matrix form, via appropriate choices of bases for the tangent spaces T_xX. Again, an informal description suffices for the present purposes. The signature of the Euclidean metric δ on $\mathbb{R}^4$ is (4, 0), since the corresponding diagonal entries are $1, 1, 1, 1$ in this case, while the signature of the Minkowski metric η on $\mathbb{R}^4$ is (3, 1), since the diagonal entries are $-1, 1, 1, 1$.[14] More generally, signatures of the form $(p, 1)$, i.e., signatures with "exactly one minus sign," are so important in theoretical physics, regardless of the dimension of the underlying manifold, that they are collectively referred to by the single name **Lorentz signature.** Other metric signatures may also be physically relevant; for example, the signature $(2, 2)$ plays a role in Penrose's *twistor theory.* The fact that the signature of g is independent of the choice of point $x \in X$ and the choice of basis for T_xX, is due to *Sylvester's law of inertia,* which says that *"the number of positive and negative coefficients in a diagonalized quadratic form does not depend on the choice of diagonalization,"* together with the "smoothly varying" hypothesis on g.

Riemannian geometry is the study of real manifolds of metric signature $(p, 0)$, for some p, called *pure signatures.* In this sense, Riemannian geometry is a generalization of Euclidean geometry, since the spaces involved "look like Euclidean spaces locally," in a limiting sense, even though their underlying smooth manifold structures may be very complicated. **Pseudo-Riemannian geometry** expands this picture further, to allow *mixed signatures,* i.e., signatures for which both p and q are nonzero; the case $q = 1$ is the Lorentzian case discussed above. Altogether, there exist $n + 1$ possible signatures (p, q) for a pseudo-Riemannian metric on a smooth real manifold of dimension n, running from $(n, 0)$ to $(0, n)$. In particular, in the special case where the manifold under consideration is merely $\mathbb{R}^n$, there exist $n + 1$ different *pseudo-Euclidean spaces* $\mathbb{R}^{p+q}$, whose metrics are the constant metrics of signatures (p, q). This is why Minkowski spacetime is often denoted by $\mathbb{R}^{3+1}$. Physically, this choice of notation emphasizes the fact that one of the dimensions, viewed here as the "temporal dimension," is distinguished from the others, due to its association with the single minus sign in the metric signature. More generally, it is easy to understand why Lorentz signature is so important from the perspective of causal structure: the single minus sign in the signature corresponds to the "local direction from cause to effect," regardless of the dimension of the underlying manifold. For a more general pseudo-Riemannian manifold X with signature (p, q), the tangent spaces T_xX at each point $x \in X$ are naturally isomorphic to $\mathbb{R}^{p+q}$ as pseudo-Euclidean spaces, so these manifolds are "locally pseudo-Euclidean of type (p, q)," in a limiting sense. In

[14]The inconsistency of the order of plus and minus signs between the list of entries $-1, 1, 1, 1$ and the abbreviation (3, 1) for the metric signature is an annoying historical artifact. One *ought* to either flip the signs to $1, -1, -1, -1$ and denote the signature by $(1, 3)$, as is done in certain texts on quantum field theory, or else flip the entries to $1, 1, 1, -1$, and denote the signature by (3, 1).

particular, the tangent spaces of pseudo-Riemannian manifolds of Lorentz signature (3, 1) are isomorphic to $\mathbb{R}^{3+1}$, so relativistic spacetime is "locally approximated by Minkowski spacetime."

Recovering "physically relevant metric structure." Among all pseudo-Riemannian manifolds of Lorentz signature (3, 1), those actually arising in "physically relevant scenarios" in general relativity are sometimes given special names. For example, *vacuum* solutions to Einstein's equation, in which the stress-energy tensor vanishes, are called *Einstein manifolds.* Minkowski spacetime is the prototypical example. The class of Einstein manifolds is too restrictive for the study of metric recovery in the relativistic setting, since the matter-energy content of the observable universe is non-negligible. On the other hand, certain classes of solutions to Einstein's equation involve types of "exotic matter" whose existence is doubtful, or configurations of matter and energy that may be difficult or impossible to achieve dynamically, even if they are theoretically possible. Hence, it can sometimes be desirable to restrict attention to the recovery of a smaller class of manifolds than the class of *all* solutions to Einstein's equation for *all* possible configurations of matter and energy. It is convenient to refer to such "physically relevant" manifolds as *generalized Einstein manifolds.* The following definition is deliberately vague, simply because the actual results necessary to motivate the developments in this book are much more general.

Definition 2.4.3. *A pseudo-Riemannian manifold of Lorentz signature* (3, 1), *satisfying Einstein's equation* (1.3.1) *for a "physically reasonable" choice of stress-energy tensor, is called a* **generalized Einstein manifold**.

General relativity is very successful experimentally, and any theoretical effort to improve upon it must eventually reproduce its empirical success. This is a very demanding task, involving detailed quantitative behavior across scales from the everyday scale up to at least the stellar scale, and quite possibly to the scales of "dark matter," "dark energy,"[15] and beyond. For this reason, the most promising new theories of fundamental spacetime structure are those that naturally approximate the content of general relativity in a comprehensive manner, rather than attempting to reproduce a host of experimental results in a purely coincidental way. This is one reason why "theories" such as *modified Newtonian dynamics* (MOND) are problematic; such approaches may explain a limited range of phenomena quite well, but the scope of explanation of general relativity is so great that it is difficult to imagine a successful replacement for the theory that is not intimately connected to it at a deep structural level. Despite the crucial common thread of causal structure, discrete directed sets might a priori be expected to prove absurdly inadequate for this purpose, since smoothness is one of the basic properties on which pseudo-Riemannian geometry is built. However, there is "plenty of room" for such models to converge with relativity somewhere between the hypothesized fundamental scale and the scales

[15] As in the case of the "cosmological constant" and "dark matter," the quotation marks serve to warn the reader that the term "dark energy" itself suggests a conventional interpretation of certain observed phenomena.

accessible to present-day experiment. The crucial requirement, then, is not that the fundamental constituents of a new theory must match those of relativity in every respect, but that the new theory must adequately approximate relativistic spacetime structure in a uniform and natural way across a suitable range of scales. These considerations narrow the general problem of describing physical spacetime in terms of causal structure to the following much more specific problem:

> **Relativistic metric recovery problem:** *Can directed sets, and preferably discrete directed sets, adequately approximate a suitable class of generalized Einstein manifolds at sufficiently large scales?*

If the answer to this metric recovery question were *negative,* then the classical causal metric hypothesis (CCMH) would be in serious jeopardy. Indeed, general relativity would have to be basically wrong across a broad range of observable scales for the hypothesis to be true. This is not out of the realm of possibility; for example, many physicists have questioned whether "dark matter" and/or "dark energy" might actually represent MOND-like dynamical deviations from Einstein's equation. At present, however, it is unnecessary to explore these issues further, because the metric recovery theorems solve a much more general problem:

> **Solution:** *Discrete directed sets can adequately approximate the entire class of four-dimensional*[16] *pseudo-Riemannian manifolds of Lorentz signature at sufficiently large scales.*

The next four sections of the chapter explain the meaning and significance of this solution.

2.5 Conformal Structure

Spacetime "angles" and "scales." Before stating an appropriate version of the metric recovery theorems, I must supply some preliminary information about conformal structure and causal structure in general relativity. To avoid the nuisance of copying lists of technical properties, and becoming bogged down in mostly irrelevant discussions about which specific properties apply in which cases, I will sometimes refer to the manifolds involved as merely "relativistic spacetime manifolds," or even just "spacetimes," in what follows, even though they may not be actual solutions to Einstein's equation. A more descriptive term might be "relativistic classical histories," but this term is not quite accurate, due to the imperfect background independence of general relativity, i.e., the fact that typical relativistic scenarios include *material content* distinct from pure geometry.

---

[16]The same is true, in fact, for any dimension at least three, as discussed in Section 2.8.

The metric g on a relativistic spacetime manifold X encodes information that enables measurement of several different types of geometric quantities in X, such as "angles" and "scales." The reason for the quotation marks here is that a mixed signature, in particular, the Lorentz signature $(3, 1)$, alters the naïve Euclidean picture of such quantities. For example, the angle between two smooth intersecting curves in a Euclidean space, and by extension, in a Riemannian manifold X, is defined, reasonably enough, to be the angle between their tangent vectors at their point x of intersection. This angle, in turn, is defined in terms of the inner product on the tangent space T_xX, which is supplied by the metric. However, if X is a pseudo-Riemannian manifold of mixed signature, then the metric g does not define true inner products on the tangent spaces of X, and this makes the picture subtler. In particular, this is what leads to consideration of *analogues* of Euclidean angles in relativity, such as the *hyperbolic angles* measuring the "rapidity of reference frames in relative motion." An iconic feature of mixed signature, which distinguishes it from the Riemannian case, is the existence of *null vectors* in the tangent spaces T_xX, i.e., nonzero vectors v that are "orthogonal to themselves," in the sense that $g_x(v, v) = 0$. In the special case of Minkowski spacetime $\mathbb{R}^{3+1}$, these vectors represent the trajectories of light rays, and define the "light cone," or **null cone**, of x in $\mathbb{R}^{3+1}$.[17] Despite such distinctions, the intuition associated with Euclidean "angles" and "scales" remains valuable, and I make informal use of these concepts in some of the examples and illustrations below. The purpose of these examples is merely to *motivate* the notions of "conformal equivalence of metrics" and "conformal maps between spacetimes." Hence, little precision is needed.

Separating "angle data" and "scale data." The information encoded in the metric g on a relativistic spacetime manifold X may be partitioned in various ways; in particular, one may study "angle data" and "scale data" separately. To understand how these two types of information may be distinguished, it is instructive to consider a pair of smooth curves γ_1 and γ_2 in X that "intersect with angle θ" at a point $x \in X$, as illustrated in the left-hand diagram in Figure 2.5.1. There are several choices for how to make this scenario more precise, if one wishes to do so, and any of these choices serve adequately for the purposes of illustration. For example, one may choose to take X to be a Riemannian manifold in this particular example, and view the "angle" between the curves at x as an actual angle; or one may ignore the "up the page rule," and think of the figure as representing a "spacelike section" of X, which possesses a natural "induced Riemannian structure." Finally, one may take the tangent vectors of the curves to be timelike, in the sense described below, and view this "angle" as a hyperbolic angle.

To illustrate the distinction between "angle data" and "scale data" on X, one may replace the metric g with a new metric $g' = \Omega^2 g$, for some positive real

[17] As explained in Section 4.5, the "near-zero" Minkowski spacetime intervals between an event and other events near its null cone in a given frame of reference translates to extreme *spatiotemporal nonlocality* in certain idealized types of causal sets induced by global sprinklings into $\mathbb{R}^{3+1}$. This leads to interesting general considerations regarding *local structure* in discrete causal theory.

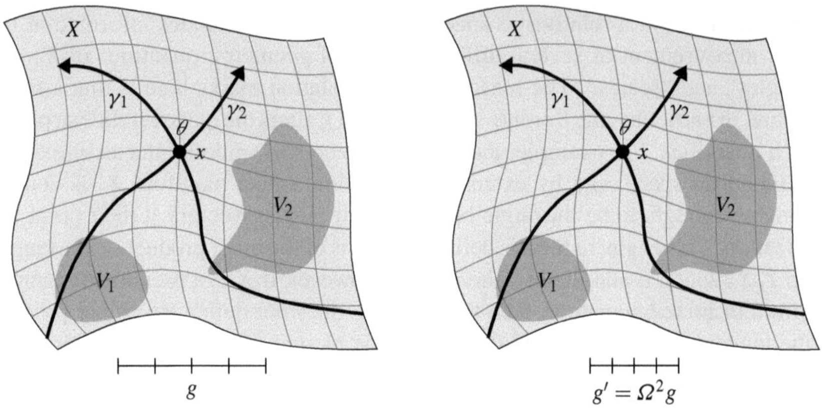

Figure 2.5.1. "Angle data" and "scale data" in relativistic spacetime; multiplying the metric by a constant factor changes "scales," but does not affect "angles".

number Ω. To be precise, g' is shorthand for the metric defined by the formula $g'_x(v, w) = \Omega^2 g_x(v, w)$ at each point x in X. In this particular example, Ω is chosen to be greater than 1; later, it will be allowed to assume any positive magnitude, and will also be allowed to vary over X. In the present case, the new metric g' will yield larger "scale" measurements for regions in X than g does; this is illustrated heuristically in the figure, by "drawing the tick marks on the measuring stick representing g' as closer together than the tick marks on the measuring stick representing g." However, both metrics yield the *same* measurement for the "angle" between the curves γ_1 and γ_2. Hence, "angle data" and "scale data" are separate, in the sense that two metrics may encode exactly the same "angle data," even if they encode totally different "scale data."

Conformal equivalence. The two metrics g and g' illustrated in Figure 2.5.1 provide a simple example of *conformally equivalent* metrics. The factor Ω, which distinguishes the two metrics, is called a constant *conformal factor.* The qualitative meaning of "conformal" is "same shape," just as one would expect on etymological grounds. Since the conformal factor Ω is constant in this particular example, it is clear that g and g' measure any subset of X to be the "same shape, but different sizes;" examples of such subsets include the images of the curves γ_1 and γ_2, and the shaded regions labeled V_1 and V_2. The general concept of conformal equivalence is less restrictive: two pseudo-Riemannian metrics g and g' on a smooth manifold X are called *conformally equivalent* if they measure the "same shapes" in an *infinitesimal* sense, i.e., if "sufficiently small regions are arbitrarily close to being the same shape" with respect to the two metrics. This notion is made precise in Definition 2.5.2 below.

In Riemannian geometry, the idea that two metrics g and g' on a manifold X measure the same shapes in an infinitesimal sense may be elegantly re-expressed by the simple statement that g and g' measure the same *angles* between pairs of tangent vectors in the tangent spaces T_xX of X. For this reason, "angle preservation" is often the central motivating concept offered when conformal structure is introduced in elementary settings; for example, in single-variable complex analysis. In *pseudo-*Riemannian geometry, the same intuition remains useful, but true angles do not play the same role in the case of mixed signature, since in this case a metric on X does not define true inner products on its tangent spaces. In this context, it is simpler to describe conformal equivalence of metrics in terms of *how such metrics may differ;* namely, with regard to scale measurements. This is accomplished by simply allowing the constant conformal factor Ω in the previous example to vary smoothly over X.

Definition 2.5.2. *Two pseudo-Riemannian metrics g and g' on a smooth manifold X are called* **conformally equivalent** *if there exists a smooth positive function*

$$\Omega : X \to \mathbb{R},$$

called the **conformal factor,** *such that $g'_x(v, w) = \Omega(x)^2 g_x(v, w)$ for every point $x \in X$, and every pair of tangent vectors $v, w \in T_x X$.*

A **conformal geometry** on a smooth manifold X is an equivalence class of conformally equivalent pseudo-Riemannian metrics on X. As a branch of mathematics, conformal geometry is concerned with "scale-independent" properties of manifolds. Conformal geometry is important in many physics-related contexts besides general relativity. Perhaps the most famous of these, already mentioned in Section 2.4, is Maldacena's AdS/CFT correspondence in string theory, where "AdS" stands for "anti-de Sitter," and "CFT" stands for "conformal field theory." A less conventional application is Julian Barbour's *shape dynamics* [BA12], which studies the "evolution of three-dimensional conformal spatial geometries." More recently, Penrose's *conformal cyclic cosmology* [PE10] examines models of the universe in which the contribution of scale data is "transient" in a limiting sense, allowing cosmological epochs to be "stitched together," despite the "initial smallness of Big-Bang type scenarios," and the "terminal largeness of expanding spacetimes."

"Scaling the manifold instead of the metric." A familiar scientific fable features an observer who wakes up one morning to find that everything, except for all the measuring sticks, has increased in size during the night. The question then becomes whether "the world has really grown," or whether "the measuring sticks have shrunk." Silly as this scenario may seem, reputable physicists have actually worked on very similar ideas; for example, the question of whether or not there is any important distinction between the conventional wisdom that *spacetime is expanding,* and the alternative hypothesis that *its material content shrinking.* The answers to such questions depend on a number of factors, but one of the most obvious of these is the actual *nature* of fundamental spacetime structure. It is useful to consider this question in the specific context of conformal structure, since changes of scale are "allowed" in this setting. In trading the metric g on X for the conformally equivalent metric $g' = \Omega^2 g$, as illustrated in Figure 2.5.1, one is not "doing anything" to the underlying manifold X, so in this case it is clear that "the measuring sticks have shrunk." Intuitively, it is easy to imagine the alternative scenario, in which "X grows;" this scenario is illustrated in Figure 2.5.3. In attempting to make this idea precise, however, an obvious problem arises: how can one "scale the manifold instead of the metric," since a metric is required to determine "sizes in X" in the first place?

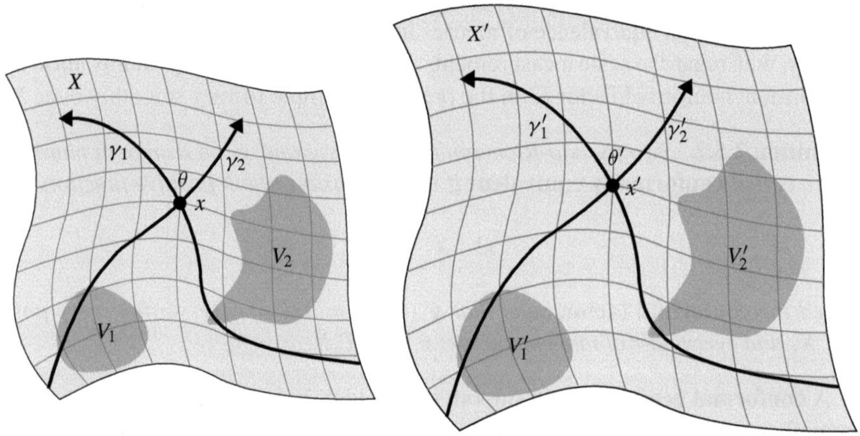

Figure 2.5.3. "Scaling the manifold instead of the metric" makes sense only if scale data is available from some other source.

The only possible solution to this problem is that *one must obtain scale data about X from some other source,* and this is exactly what I have done in the figure. Here, for once, it is instructive to think for a moment about the figure itself, temporarily forgetting about what it is supposed to represent abstractly. The illustration of X *looks* like a manifold; or, more precisely, like "part of a manifold," *embedded in a higher-dimensional space, which is itself equipped with a metric.* This metric provides a "natural way to measure X." Given this setup, it is easy to "copy," and "dilate" X to yield another manifold, called X', embedded in the same ambient space. The new manifold X' may then be measured using the same metric, and it is clear that in this case "the world has grown," while "the measuring sticks have stayed the same size." Of course, in pseudo-Riemannian geometry, there is generally no ambient space to appeal to, and therefore no "natural way to measure X;" instead, one must *choose* a metric. In the discrete causal context, however, there *is* a natural source of scale data for classical histories; namely, discrete causal structure itself. This is one of the principal reasons why the metric recovery theorems provide "enough evidence" to motivate the classical causal metric hypothesis (CCMH) in the discrete setting.

Smooth conformal isometries. Returning to Figure 2.5.3, what is represented is a "pair of spacetimes" X and X' embedded in an ambient space, whose metrics g and g' are induced by a choice of metric on this space. In this context, there exist natural diffeomorphisms between X and X', which may be called *dilation* and *contraction maps.* In particular, the dilation map sends x to x', and the contraction map sends x' to x. These maps are simple examples of *smooth conformal isometries,* i.e., isomorphisms between X and X' in the smooth conformal category. It is worth emphasizing that referring to a map f as a "conformal isometry" does *not* mean that f is "an isometry that is conformal," since isometries are *always* conformal by definition. Rather, it means that f is an "isometry up to conformal equivalence." This is made precise by the following definition:

Definition 2.5.4. *A* **smooth conformal isometry** $f : (X, g) \to (X', g')$, *between two pseudo-Riemannian manifolds* (X, g) *and* (X', g'), *is a diffeomorphism* $f : X \to X'$, *such that the pulled-back metric* f^*g' *on* X *is conformally equivalent to the metric* g *on* X.

The reason for the qualifier "smooth" in the phrase "smooth conformal isometry" is that it is possible to study conformal properties of objects in more general categories, as already mentioned in Section 2.3. A *smooth conformal morphism* $f : (X, g) \to (X', g')$ is defined by demoting "diffeomorphism" to "smooth map" in Definition 2.5.4. Since conformal equivalence of metrics "preserves shapes" only infinitesimally, smooth conformal morphisms and smooth conformal isometries generally do *not* "preserve shapes" at a finite level, and they obviously do not preserve volumes, lengths, and other related measurements. Figure 2.5.5 illustrates two spacetimes related by a smooth conformal isometry, which are *not* "the same shape" macroscopically. The checkerboard pattern in the figure is included to show how corresponding regions "approach the same shape as they shrink in size;" in particular, the individual black and gray regions in the right-hand diagram are "much closer to being squares" than the region itself.

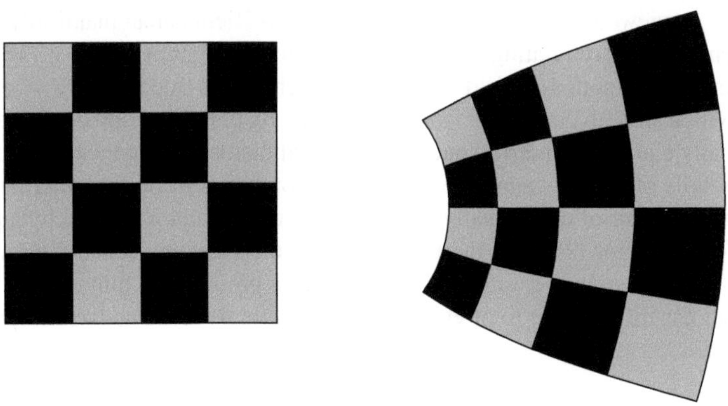

Figure 2.5.5. Spacetimes related by a smooth conformal isometry.

Passive viewpoint versus active viewpoint. The specific distinction between *conformal equivalence of metrics* and *conformal morphisms between manifolds* provides an excellent illustration of the more general difference between the passive and active viewpoints regarding structural comparisons, already mentioned in Section 2.3

in the context of category theory. Exchanging one metric for another involves the passive viewpoint, since one begins with an underlying structure, which is "left alone," while alternative additional structures are superimposed on it. Such an exchange of metrics is mathematically analogous to changing "coordinate systems" or "frames of reference," but is much more drastic, at least in the relativistic context. This is because different metrics are generally assumed to encode different physics,[18] while different frames of reference merely represent different points of view regarding the same physics. Conformal morphisms, by contrast, involve the active viewpoint, since they map each element of one manifold to an image element in another manifold. Of course, nothing *physically* passive or active is involved in either case; but the choice of terminology offers a convenient way of describing the difference in viewpoint.

Subtle relationships exist between the active and passive viewpoints in the context of conformal structure. For example, if g and $g' = \Omega^2 g$ are conformally equivalent metrics on an underlying smooth manifold X, then the identity map $\mathrm{Id} : X \to X$ may be viewed as a map between two *different* pseudo-Riemannian manifolds (X, g) and (X, g'). From this viewpoint, the map "Id" is no longer the identity if $g \neq g'$; i.e., it is not the identity morphism in the pseudo-Riemannian category, because it is a map between two different objects in the category. In fact, it is not a morphism in this category at all, because it fails to preserve metric structure. However, it *is* a conformal isometry relating the two metrics g and g', because $\mathrm{Id}^* g' = g' = \Omega^2 g$. Given such a pair of conformally equivalent metrics on X, there generally exist many different self-diffeomorphisms $f : X \to X$ relating g and g' via the same conformal factor; for example, the underlying diffeomorphism of any smooth self-isometry f of (X, g) relates g to itself in the trivial way: $f^* g = g = 1^2 g$.

One may also ask the question of when a nontrivial self-diffeomorphism $f : X \to X$, viewed as a self-map on a *fixed* pseudo-Riemannian manifold (X, g), is a conformal isometry relating g to *another* previously-chosen metric $g' = \Omega^2 g$ on the underlying smooth manifold X, i.e., for which f it is true that $f^* g = g'$. If one begins with a smooth conformal isometry f, then by definition one obtains a pair of conformally equivalent metrics, but the conformal factor relating them is determined by the details of f, and cannot be specified beforehand. If, on the other hand, one first *chooses* a pair of conformally equivalent metrics g and g' related by a particular conformal factor Ω^2, then there may or may not exist any self-diffeomorphism $f : X \to X$ such that $f^* g = g'$.[19] This provides a preliminary hint that the passive viewpoint has information to offer that is unavailable from the active viewpoint.

[18] Here I am referring to the many attempts, from Weyl to CFT's to conformal cyclic cosmology, to extend the relativity principle to include *conformal invariance*. In such a theory, an appropriate change of conformally equivalent metrics would *not* affect the physics. Note that under the passive viewpoint, the question of whether or not "different metrics encode different physics" does *not* involve the active idea of "re-associating physical data with different points in a manifold," discussed in Section 2.3 in the context of diffeomorphism invariance. Here, the question is merely, "if the metric near x is changed, does the physics near the event represented by x change?".

[19] See Wald [WA84], Appendix D.

2.6 Causal Structure

Relativistic causal structure and the classical causal metric hypothesis. What the metric recovery theorems demonstrate is that *the causal structure of relativistic spacetime determines the corresponding metric structure up to smooth conformal isometry.* Informally, this means that information about cause and effect reveals everything that there is to know about relativistic spacetime geometry, *except* for scale data, i.e., except for the choice of conformal factor. As mentioned in Section 2.3, if classical spacetime *really is* precisely represented by a pseudo-Riemannian manifold, then the metric recovery theorems are not quite sufficient to support the classical causal metric hypothesis (CCMH), since there is no way to obtain the "missing conformal factor" without turning to some auxiliary, "non-causal" source. Hence, if general relativity is *absolutely correct* in how it models classical spacetime, then the classical causal metric hypothesis is *wrong,* and the metric recovery theorems explain exactly why, and to what extent. This does not necessarily mean that the physical existence of continuum-based structure *in general* would doom the causal metric hypothesis, since one could conceive of continuum-based spacetime models other than pseudo-Riemannian manifolds. For example, *domain theory* provides a context in which "continuous, pure causal" structure makes sense. On the other hand, if classical spacetime is discrete, and merely *looks* like a pseudo-Riemannian manifold at large scales, then a natural scale comes for free: the scale of the fundamental elements and relations.[20] This is what the founders of causal set theory realized around 1980.

Directions of curves in relativistic spacetime. What, precisely, is meant by *"the causal structure of relativistic spacetime?"* The reader is no doubt well-aware that relativity "forbids superluminal communication," and this means that influence may travel only along certain curves[21] in spacetime, called *causal curves.* These curves are determined by the metric g on a relativistic spacetime manifold X, which supplies information about "what direction a differentiable curve is pointing" at each point along the curve. At a fixed point x in X, these various "directions" may be partitioned into three classes, usually called *timelike, null,* and *spacelike.* Timelike and null directions may be further subdivided into classes of *past* and *future* directions. From a physical standpoint, future timelike or null directions are directions in which "causal influence may propagate," while past timelike or null directions are directions from which "causal influence may arrive." Hence, these directions have absolute physical significance, and their identification does not depend on a choice of reference frame. For spacelike directions, however, the distinction between "past" or "future" *does* depend on the frame of reference, as elaborated below. If X admits a consistent, continuously-varying designation of "past" and "future" for timelike and null directions, then it is called *time-orientable,* and most references on general

---

[20] As explained in Section 2.1, there may be more than one reasonable way to define "emergent volume" in terms of this fundamental scale data.

[21] Here I am really referring to the *images* of such curves; the distinction between a curve and its image is discussed in more detail below.

relativity eliminate non-time-orientable spacetime manifolds from consideration at the outset, on basic physical grounds. The possibility of non-time-orientability is one of many "continuum-related pathologies" that discrete causal theory avoids entirely, since there is no need for anything to vary continuously.

The "directions" at a point x in a relativistic spacetime manifold X may be represented precisely by tangent vectors to X at x. A tangent vector v at x in X is called *timelike* if $g_x(v, v) < 0$, *null* if $g_x(v, v) = 0$, and *spacelike* if $g_x(v, v) > 0$. These designations make sense for any pseudo-Riemannian manifold of Lorentz signature. In the special case in which X is Minkowski spacetime $\mathbb{R}^{3+1}$, X is isomorphic to T_xX, and each nonzero tangent vector v at x "actually points to another element y of X," in an obvious sense. In this case, g is the Minkowski metric η, and the Minkowski *spacetime interval,* or more precisely, "squared interval," between x and y, is just $\eta_x(v, v)$. The events x and y are called *timelike separated* if $\eta_x(v, v) < 0$, *null separated* if $\eta_x(v, v) = 0$, and *spacelike separated* if $\eta_x(v, v) > 0$. In the general case of a curved spacetime manifold X, the corresponding relationships between pairs of events x and y are described in terms of curves from x to y, as explained below.

In relativistic kinematics, a future timelike direction is a "permissible direction" for the motion of a massive material object, such as an electron. A future null direction is a "permissible direction" for the propagation of electromagnetic radiation, or any other form of energy involving massless particles. In particular, future null directions at a point x in a relativistic spacetime manifold X may be viewed as the possible directions of light rays emanating from x, and these determine the "future light cone," or *future null cone,* of x in the tangent space T_xX. Spacelike directions are "forbidden directions" for all forms of influence. Due to the relativity of simultaneity, a spacelike direction that points toward the future in some frames of reference will point toward the past in other frames. Timelike and null directions do not suffer from this ambiguity. The left-hand diagram in Figure 2.6.1 illustrates timelike, null, and spacelike future directions at a point x in X, with respect to a particular frame of reference, represented by the curved "coordinate lines." Of course, the timelike and null directions illustrated here retain their future orientation in any frame of reference.

Smooth causal curves. The "directional information" associated with tangent vectors in a relativistic spacetime manifold X may be used to classify special families of curves in X. As in the case of tangent vectors themselves, the resulting definitions make sense for any pseudo-Riemannian manifold of Lorentz signature. A smooth curve passing through $x \in X$ is called *timelike at x* if its tangent vector at x is timelike. It is called *globally timelike,* or just *timelike,* if it is timelike at each of its points. A pair of distinct events x and y in X are called *timelike separated* if they are connected by a smooth timelike curve. *Null* and *spacelike* smooth curves, and *null* and *spacelike separation* of pairs of distinct events, are defined in an analogous manner. A smooth curve is called **causal** if it is either timelike or null at each of its points. The right-hand diagram in Figure 2.6.1 illustrates timelike, null, and spacelike future-directed smooth curves at a point x in X, with respect to the same frame

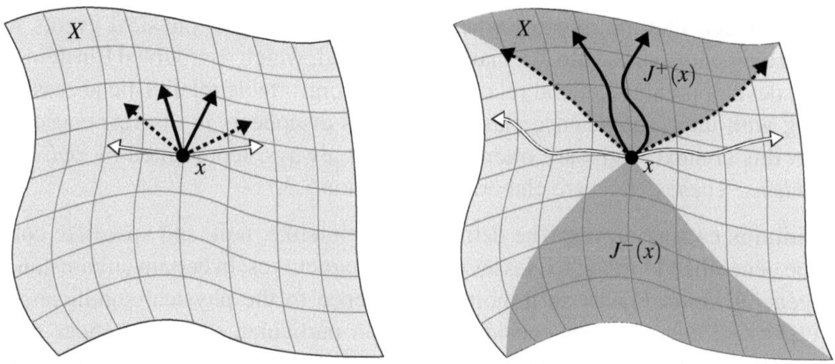

Figure 2.6.1. Timelike (*black*), null (*dashed*), and spacelike (*white*) future directions in a given frame of reference; corresponding timelike, null, and spacelike smooth curves.

of reference illustrated in the left-hand diagram. The warped, dark gray "triangular" regions labeled $J^-(x)$ and $J^+(x)$ in the figure represent the *causal past* and *causal future* of x, respectively; these are discussed more thoroughly below. The definitions of timelike, null, and spacelike curves given here do not require actual smoothness in order to make sense; these curves need only possess unique tangent vectors at each of their points.

A few mathematical details are worth mentioning here for the purpose of clarity. In the context of general relativity, a *curve* in a spacetime manifold X is usually defined to be a *map* from an interval in $\mathbb{R}$ into X, often with additional conditions imposed, such as smoothness, or non-vanishing of the tangent vector along the curve. However, in the present context, one is principally interested in the properties of the *image* of such a map, rather than the details of the map itself. For example, "reparameterization" of such a map generally does not change its physical interpretation. In particular, the elementary-calculus heuristic of "moving at different speeds along the same curve" is irrelevant for a map into spacetime, whose image encodes all physical motion-related quantities. This means that very large classes of physically equivalent curves lurk beneath every curve-related notion in relativity. In particular, different curves sharing the same image in X will generally have different tangent vectors at a given point in X, so one might a priori worry about whether or not the notions of timelike, null, and spacelike curves, defined in terms of these tangent vectors, are actually physically meaningful. The reason why these particular notions *are* meaningful is because only the *sign* of the tangent vector, and not its magnitude, is involved in defining them. It is therefore common, and often innocuous, to refer to such curves and their images interchangeably, and many instances of such language appear throughout the remainder of the book. However, the presence of such equivalence classes must be kept in mind in more general settings. For example, the *paths* involved in Feynman's path summation approach to ordinary quantum theory are large equivalence classes of curves in a spacetime manifold, sharing a common image.[22] Much like the issue of whether or not to associate physical events with specific elements in a pseudo-Riemannian manifold, which so troubled Einstein during the development of general relativity, the uncomfortable relativistic necessity of dealing with large equivalence classes of curves disappears in discrete causal theory. In this context, paths are often, though not always, represented by *individual morphisms.* These details are elaborated in Section 5.9.

Continuous causal curves. The definitions of timelike, null, and spacelike curves may be generalized further, to classes of *continuous* curves. In certain circumstances, this generalization makes an important difference to the physical significance of the classes of curves under consideration.[23] In particular, strictly stronger metric recovery results may be proven if one chooses to describe causal structure in terms of continuous causal curves, rather than restricting attention to smooth causal curves. This is the reason for the distinction between *causal morphisms* and *enhanced causal morphisms* of relativistic spacetime manifolds, appearing in Definition 2.6.3 below. The definitions of timelike, null, and spacelike curves in the continuous context are slightly subtle. For example, a continuous curve is called *timelike* if each of its points x possesses a *convex normal open neighborhood* U_x, such that any two points w and y on the curve in U_x are connected by a smooth timelike curve "in the proper order."[24] *Null* and *spacelike* continuous curves are defined in an analogous manner. A continuous curve is called **causal** if it is everywhere timelike or null. The left-hand diagram in Figure 2.6.2 illustrates a continuous causal curve, with two points w and y in a suitable neighborhood U_x of a point x on the curve connected by a smooth causal curve.

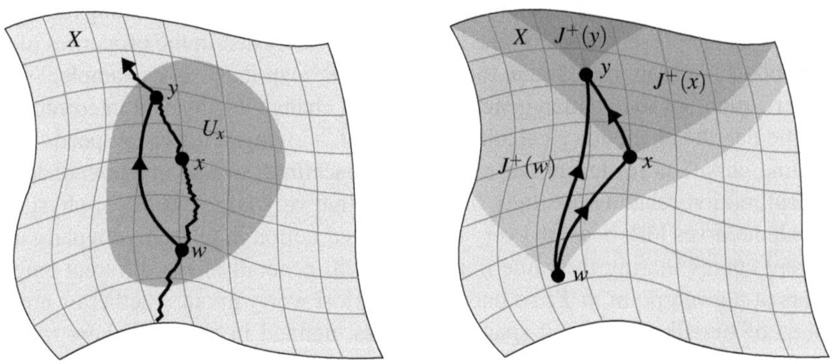

Figure 2.6.2. A continuous causal curve; transitivity of the relativistic causal relation.

---

[22]For examples of the distinction between "curves" and "paths" in general relativity, see Rovelli [RO04], pp. xxii–xxiii, Thiemann [TH07], p. 164, Hawking and Ellis [HE73], p. 15, or Wald [WA84], p. 17.

[23]See Malament, p. 1400.

[24]See Wald [WA84], p. 193, for details.

Relativistic causal relation; causal pasts and futures. It is very useful to define a binary relation $\prec_{\mathrm{GR}}$ on a relativistic spacetime manifold X, called the **relativistic causal relation**, where $x \prec_{\mathrm{GR}} y$ if and only if there exists a nontrivial[25] smooth causal curve from x to y in X. In other words, $x \prec_{\mathrm{GR}} y$ if and only if x and y are timelike or null separated. The subscript "GR" in the expression $\prec_{\mathrm{GR}}$ stands for "general relativity," and is included in order to avoid confusion with other binary relations appearing in this book, particularly the *causal relation* $\prec$ on a directed set in discrete causal theory, introduced in Section 3.10. As usual, however, the definition of $\prec_{\mathrm{GR}}$ makes sense for any pseudo-Riemannian manifold of Lorentz signature. The set $J^-(x) := \{w \in X | w \prec_{\mathrm{GR}} x\}$ is called the **causal past** of x, while the set $J^+(x) := \{y \in X | x \prec_{\mathrm{GR}} y\}$ is called the **causal future** of x. These sets may be viewed as relativistic prototypes of the more general *domains of influence* introduced in Section 3.7, which are subsets or subobjects of a directed set D, defined to encode information about influences between pairs of events represented by elements of D. In the relativistic context, $J^-(x)$ is interpreted as the set of all events in X which "could conceivably" influence x, and $J^+(x)$ is interpreted as the set of all events in X which "could conceivably" be influenced by x. The union $J(x) = J^-(x) \cup J^+(x)$ is called the **total domain of influence** of x.

The causal past $J^-(x)$ and causal future $J^+(x)$ of an element x in a relativistic spacetime manifold X were illustrated as the warped, dark gray "triangular" regions appearing in the right-hand diagram in Figure 2.6.1 above. In the right-hand diagram in Figure 2.6.2, the progressively darker-shaded regions represent the causal futures of the events labeled w, x, and y. The boundaries of these regions generalize the future "light cones," or *null cones*, of events in Minkowski spacetime $\mathbb{R}^{3+1}$. Two rather trivial points are worth mentioning here for the sake of clarity. First, the reason why these regions appear triangular, rather than cone-shaped, in these particular illustrations, is because only two of the dimensions involved are actually shown in the diagrams. Second, the term "light cone" is popular in the relativistic context, but is not ideal in *general* relativity. In this setting, the actual *cones* involved exist only in the tangent spaces T_xX of a relativistic spacetime manifold X, and not in X itself, while rays of light follow curved geodesics in X, determined by Einstein's equation (1.3.1). Hence, one must choose between speaking about a "light cone" in T_xX, which does not really describe the propagation of light, or a "light cone" in X, which is not really a cone. Penrose offers the reasonable suggestion to simply use the more precise term *null cone* to describe the desired object in T_xX; one may then describe the more general corresponding structures in X itself as *pasts* and *futures*, eliminating the term "light cone" entirely. I mostly follow Penrose's convention in this book.

The relativistic causal relation $\prec_{\mathrm{GR}}$ is *transitive*, which means that if $x \prec_{\mathrm{GR}} y$ and $y \prec_{\mathrm{GR}} z$, then $x \prec_{\mathrm{GR}} z$. For relativistic spacetime manifolds satisfying the *causal condition*, defined in Section 2.7, $\prec_{\mathrm{GR}}$ is also *irreflexive*, which means that $x \not\prec_{\mathrm{GR}} x$. At a conceptual level, transitivity of $\prec_{\mathrm{GR}}$ encodes the "common sense" that "if x influences y, and y influences z, then x influences z." The choice to abstain from

[25]"Nontrivial" means "nonconstant;" see the discussion of the *causal condition* in Section 2.7 for more details.

imposing transitivity on the more general *causal relation* introduced in Section 3.10 does not contradict this common sense, but merely recognizes the fact that discrete structure demands a more fundamental notion of *direct* causation, encoded by $\prec$, while "possibly indirect" causation is encoded by a transitive relation $\prec_{\mathrm{tr}}$, *generated* by $\prec$. Irreflexivity of $\prec_{\mathrm{GR}}$ means that "x does not influence itself;" however, if there exists a nontrivial *closed causal curve* in X, beginning and terminating at x, then transitivity implies that x *does* influence itself. The causal condition rules out the existence of such curves, which is why irreflexivity of $\prec_{\mathrm{GR}}$ holds only for spacetime manifolds satisfying this condition. Conveniently, it turns out that the actual identity of $J^+(x)$ and $J^-(x)$, as subsets of X, is unaffected if "smooth causal curve" is replaced by "continuous causal curve" in the definition. This follows from the transitivity of the relativistic causal relation, together with an easy compactness argument. By replacing causal curves with timelike curves in the definition of the relativistic causal relation, one may define a corresponding relativistic *chronological relation,* which I denote by $\prec\prec_{\mathrm{GR}}$. In particular, the set $I^-(x) := \{w \in X | w \prec\prec_{\mathrm{GR}} x\}$ is called the *chronological past* of x, while the set $I^+(x) := \{y \in X | x \prec\prec_{\mathrm{GR}} y\}$ is called the *chronological future* of x. Finally, one may define a *horismos relation,* sometimes denoted by $x \rightarrow y$, by using only null curves in the same setting.[26]

Causal morphisms and "enhanced causal morphisms." The relativistic causal relation $\prec_{\mathrm{GR}}$ on a relativistic spacetime manifold X is what is usually meant by the *causal structure of relativistic spacetime.* It is curious, however, that incorporating information about *continuous* causal curves, as opposed to merely *smooth* causal curves, yields stronger metric recovery results. In discrete causal theory, there are no such distinctions, and hence no ambiguities about how structure-preserving maps, i.e., morphisms, should be defined. To deal with this particular continuum-induced complication, I give *two* definitions of causal morphisms: the first following the usual relativistic conventions, and the second included to account for the difference between smooth curves and continuous curves in this context.

Definition 2.6.3. *Let $f : X \rightarrow X'$ be a map between two pseudo-Riemannian manifolds of Lorentz signature, with relativistic causal relations $\prec_{\mathrm{GR}}$ and $\prec'_{\mathrm{GR}}$, respectively.*

1. *The map f is called a* **causal morphism** *if it preserves relativistic causal relations, i.e., if $f(x) \prec'_{\mathrm{GR}} f(y)$ whenever $x \prec_{\mathrm{GR}} y$. It is called a* **causal isomorphism** *if it possesses an inverse that is also a causal morphism.*
2. *The map f is called a* **enhanced causal morphism** *if it preserves future-directed continuous causal curves. It is called an* **enhanced causal isomorphism** *if it possesses an inverse that is also an enhanced causal morphism.*

The term "enhanced causal morphism" is awkward; the alternative term "strong causal morphism" is a priori more attractive. However, the former term is chosen here to avoid confusion with the *strongly causal condition* on relativistic spacetime manifolds, discussed in Section 2.7, which is ubiquitous in the early-modern relativity

[26]See Malament p. 1400 for details.

literature. The definitions given here are stronger than necessary to prove the version of metric recovery stated in Theorem 2.8.1; in particular, Malament [MA77] works mostly in terms of maps f and f^{-1} that are assumed only to preserve certain classes of *timelike* curves. The necessary properties for causal curves in general are then *proven,* using the assumptions regarding timelike curves, together with the topological structures of the manifolds involved. However, these details are not essential to the present discussion. Causal morphisms generalize in a natural way to the context of directed sets, but enhanced causal morphisms do not, because there is generally no concept of continuity in this setting. Note that this distinction cannot be eliminated by defining a new binary relation on a relativistic spacetime manifold X in terms of *continuous* causal curves, since one merely recovers the usual relativistic causal relation $\prec_{\text{GR}}$ in this manner. This, of course, is just a pointwise statement of the fact that the causal pasts and futures $J^+(x)$ and $J^-(x)$ of an event x in X do not depend on whether one uses smooth curves or continuous curves to define them. This situation foreshadows the conclusion that *something more than a simple binary relation is necessary to specify geometry in the absence of natural scale data.*

2.7 Causality Conditions

Avoiding "causality violations." The degree to which the metric structure of a relativistic spacetime manifold X may be recovered from its causal structure depends, informally, on "how close X comes to violating causality." In this context, "violating causality" means that X includes events that influence their own causes, and hence, that indirectly influence themselves. *Causality conditions* are technical conditions that give precise meaning to the words "how close," in this qualitative description. As discussed below, the potential consistency issues posed by "causality-violating" relativistic spacetime manifolds may be completely avoided in discrete causal theory, by working in a perfectly background independent setting. However, causality conditions remain important in this context, due to their role in metric recovery.

Overview of causality conditions. Seven causality conditions that feature prominently in general relativity are the *chronological, causal, past or future distinguishing, past and future distinguishing, strongly causal, stably causal,* and *globally hyperbolic* conditions. Figure 2.7.1, which is an elaboration of one of Malament's [MA77] diagrams, lists these conditions in ascending order of restrictiveness. For example, every strongly causal spacetime is past and future distinguishing, but the converse is false. Conditions possessing natural analogues in discrete causal theory appear in bold font in the figure. From a modern viewpoint, the past and future distinguishing condition is the condition that determines whether *enhanced* causal isomorphisms, or merely causal isomorphisms, are necessary in the hypotheses for metric recovery. This is the reason for the dashed line in the figure, which separates conditions "restrictive enough" to render enhanced causal isomorphisms unnecessary from conditions which are "too weak." One of Malament's achievements was to eliminate the use of the "strongly causal" hypothesis from Hawking's earlier proof. The remainder of this section examines these seven conditions. Due to the relationships among them, it is convenient to discuss them in a somewhat different order: the chronological and causal conditions are discussed first, followed by the stably causal condition, the past and/or future distinguishing conditions, the strongly causal condition, and finally, the globally hyperbolic condition.

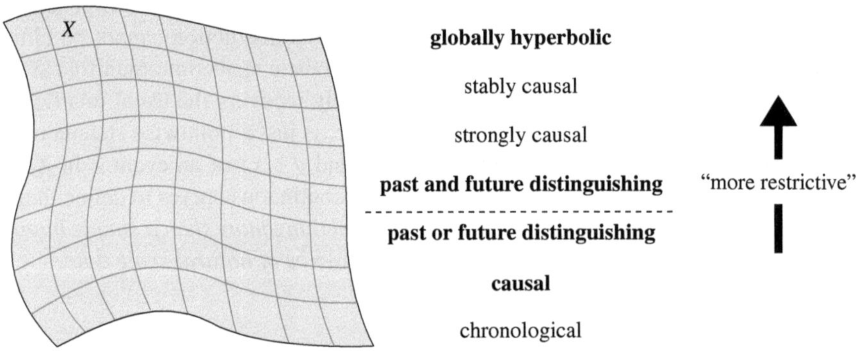

Figure 2.7.1. Seven causality conditions on a relativistic spacetime manifold X.

Closed timelike curves; chronological condition. The "worst" possible type of behavior, from the viewpoint of "causality violation," is the existence of *closed timelike curves,* like the one illustrated in the left-hand diagram in Figure 2.7.2. A relativistic spacetime manifold sufficiently "well-behaved" not to contain such curves is called *chronological.* Since relativity permits the transport of a material body in any future timelike direction, an observer could theoretically travel to his or her own past along such a curve. From a naïve viewpoint, this possibility raises serious consistency issues. In particular, familiar "causality paradoxes," such as the *grandfather paradox,* originate from this scenario. It is likely a waste of time to devote serious consideration to such "paradoxes" in their own right, since their actual significance seems to be mostly restricted to illustrating potential issues arising from a lack of perfect background independence in general relativity. Indeed, the common source of such paradoxes is a potential clash between two different types of structure: the relativistic spacetime manifold on which a material body could conceivably "move around and return to where it started," and the material body itself. The strong interpretation of the causal metric hypothesis (CMH) eliminates such issues once and for all, by admitting *only one type of basic structure;* namely, causal structure, of which "spacetime" and "material bodies" are viewed as different manifestations. It does not rule out the possibility of discrete causal analogues of closed timelike curves, i.e., *cycles* such as the one illustrated in the right-hand diagram in Figure 2.7.2, but it does guarantee their physical consistency, should they exist. In particular, it avoids the potential conflicts involved in traversing a cycle, by simply disallowing any extrinsic entity that *could* traverse it. Of course, most versions of discrete causal theory do not distinguish precisely between analogues of timelike and null curves, so it is generally not obvious which cycles in a directed set should be regarded as "timelike." For this reason, the chronological condition has no simple analogue in discrete causal theory.

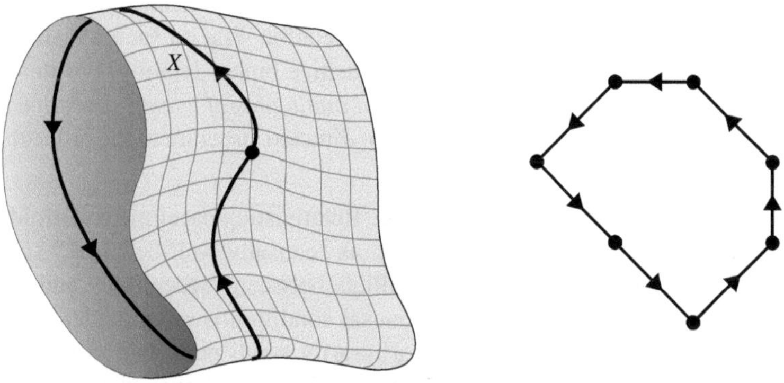

Figure 2.7.2. A closed timelike curve in relativistic spacetime; a cycle in a directed set.

Closed causal curves; causal condition. A slightly more restrictive causality condition than the chronological condition is the *causal condition,* which rules out not just closed timelike curves, but *all* nontrivial closed causal curves, including null curves. Any pseudo-Riemannian manifold X admits trivial "constant curves," mapping an entire interval in $\mathbb{R}$ to a single point x in X, and these curves are technically "null," because their tangent vectors vanish. Such curves are *not* interpreted as encoding "self-causation." Throughout the remainder of the book, null curves are assumed to be nontrivial unless stated otherwise. The principal qualitative difference between a closed timelike curve and a closed null curve is that relativity permits the latter to convey only *information,* and not material bodies. In the absence of perfect background independence, however, closed null curves are "nearly as bad" as closed timelike curves from a consistency standpoint. For example, one may imagine sending instructions "back in time" along a closed null curve, detailing how to construct the very device one is using to send the instructions. This creates obvious problems if one distinguishes "spacetime" from "information," since once again two types of structure are pitted against each other.

By contrast, discrete causal analogues of non-causal spacetimes, viewed according to the strong interpretation of the causal metric hypothesis (CMH), are immune to consistency issues involving cycles, since they possess only one type of structure. If such cycles *do* exist, they merely represent "part of what the causal structure is like," even if the resulting behavior is alien to ordinary experience. However, it is worth noting that the class of relativistic spacetime manifolds admitting the "best" metric recovery results, i.e., results that do *not* require the use of enhanced causal isomorphisms, is a subclass of the class of causal spacetimes; namely, the class of *past and future distinguishing* spacetimes. In other words, it is technically more difficult to recover metric structure from causal structure for spacetimes that fail to satisfy the causal condition. It is possible to interpret this fact as a hint that acyclicity *should* be taken as an axiom of discrete causal theory. In my own opinion, however, the relativistic evidence is not overwhelming one way or the other; for instance, closed causal curves appear in rather generic relativistic situations, such as the *Kerr black hole.* An important characteristic of relativistic spacetime manifolds satisfying the causal condition is that the relativistic causal relation $\prec_{\mathrm{GR}}$ on such a manifold is in fact a *strict partial order,* i.e., an acyclic transitive binary relation. For this reason, *causal sets may be viewed as discrete causal analogues of causal spacetimes,* since causal set theory restricts its consideration of directed structure to the order-theoretic paradigm.

Stability issues; stably causal condition. Relativistic spacetime manifolds satisfying the causal condition may be viewed as "causally well-behaved," since they avoid, by definition, causality-based conflicts between "spacetime" and auxiliary "matter-energy content." However, the causal condition is often inadequate from a practical perspective. In particular, an arbitrarily small perturbation of the metric in a causal spacetime can produce closed causal curves. Issues of this nature arise in almost any conceivable continuum-based theory, as part of the cost to be paid for the availability of convenient interpolation and limiting procedures. To prove certain desirable results, it is sometimes necessary to impose more restrictive conditions that "bound systems away from bad behavior," instead of merely ruling out the bad behavior itself. The *stably causal* condition, appearing second on the list in Figure 2.7.1, accomplishes this by explicitly requiring that a particular type of finite perturbation of the metric preserves the chronological condition.[27] In the context of metric recovery, however, the stably causal condition is unwieldy. Moreover, the condition does not possess a natural discrete causal analogue.

Past and/or future distinguishing conditions. Less-restrictive conditions, called past and/or future distinguishing conditions, are more useful in this setting. Informally, these conditions govern the extent to which a relativistic spacetime manifold X "separates pasts and futures of individual events." In particular, X is called *past distinguishing* if distinct elements possess distinct causal pasts, and is called *future distinguishing* if distinct elements possess distinct causal futures. Conditions of this nature are easy to conceptualize by considering relationships among small finite sets of elements, and this reflects the fact that these conditions possess natural analogues in discrete causal theory. The directed sets depicted in the left-hand diagram in Figure 2.7.3 illustrate these conditions. The "diamond-shaped" set D is neither past nor future distinguishing, since the elements x_1 and x_2 both possess the same past and the same future. The set D' is future distinguishing but not past distinguishing, because the elements y_1 and y_2 both possess the same past. The set D'' is past and future distinguishing. For simplicity, I am ignoring the "sameness" of empty pasts and/or futures in these examples. It is worth noting that a "large random directed set" may easily fail to satisfy either or both of these conditions, due solely to small, local

[27] See Wald [WA84] p. 198 for details.

"defects," such as the existence of "diamond-shaped" subsets. Hence, one should be very reluctant to impose such conditions as *axioms* for discrete causal theory.

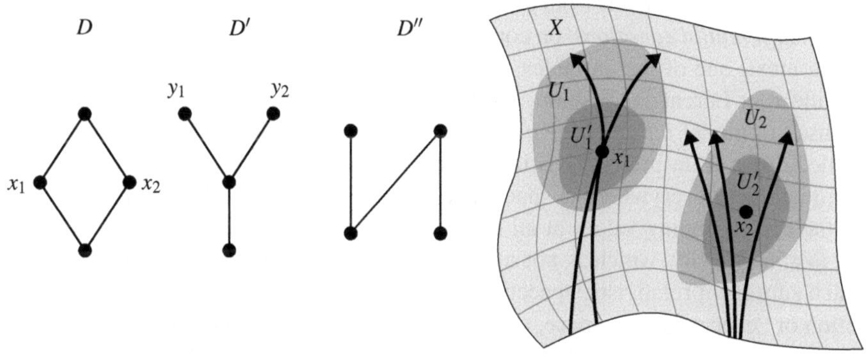

Figure 2.7.3. Directed sets illustrating past and/or future distinguishing conditions; relativistic versions of the future distinguishing condition and the strongly causal condition.

Returning to the relativistic context, it is convenient to reformulate past and/or future distinguishing conditions in *topological* terms. In particular, it turns out that a relativistic spacetime manifold X is future distinguishing if and only if for every event x in X, and every open set U containing x, there exists a "smaller" open set U' in U, containing x, such that no future-directed smooth timelike curve through x that leaves U' ever returns to it.[28] This situation is illustrated for the event x_1 in the right-hand diagram in Figure 2.7.3. The past distinguishing condition may be reformulated in an analogous manner, by replacing the word "future" with "past" in the reformulation of the future distinguishing condition.

Strongly causal condition. The topological reformulation of the past and/or future distinguishing conditions may be slightly modified to yield a stronger condition, called the strongly causal condition. A relativistic spacetime manifold X is called *strongly causal* if for every element x in X, and every open set U containing x, there exists a "smaller" open set U' in U, containing x, such that no future-directed smooth timelike curve leaving U', *whether or not it passes through* x, ever returns to U'. Heuristically, this condition provides "slightly more room between pasts and futures" than the past and/or future distinguishing conditions. The strongly causal condition is illustrated for the event x_2 in the right-hand diagram in Figure 2.7.3. There is generally no natural analogue of this condition in discrete causal theory, and it plays a less-significant role even in relativity than it did a few generations ago. As mentioned above, Malament succeeded in removing this condition from the hypotheses of his version of metric recovery. More recently, Bernal and Sanchez [BS07] have expunged it from topological reformulations of the globally hyperbolic condition, discussed below.

---

[28] See Malament p. 1400 for details.

Cauchy surfaces; globally hyperbolic condition. The globally hyperbolic condition is the most restrictive causality condition of the seven listed in Figure 2.7.1, and requires a bit of preliminary explanation before it may be properly introduced. First, the terminology originates from the study of *wave equations,* i.e., special *hyperbolic partial differential equations,* on continuum-based models of spacetime. In this general context, one of the variables is interpreted as representing time, whether in a Euclidean or Lorentzian fashion, and the differential equations under consideration are used to model *initial value problems.* The informal idea behind this approach, handed down from pre-relativistic physics, is that if one knows the values of certain quantities "everywhere in space, at a given instant in time," then one can solve for the corresponding values at all later times. Since this description refers implicitly to *simultaneity,* which is not absolute in relativity, one must be precise about which subsets of relativistic spacetime manifolds are suitable representatives of the notion of "everywhere in space, at a given instant in time." Such subsets are called *Cauchy surfaces.* The corresponding initial value problems are classified as special types of *Cauchy problems,* i.e., problems involving the solution of partial differential equations satisfying specified conditions on hypersurfaces. Since discrete causal analogues of Cauchy surfaces play a crucial role in this book, it is worthwhile to give a formal definition:

Definition 2.7.4. *A* **Cauchy surface** *in a relativistic spacetime manifold X is a subset σ of X such that every inextensible causal curve in X intersects σ exactly once.*

The term "inextensible causal curve" in the definition means what the terminology suggests: a causal curve that cannot be extended to yield a "longer" causal curve. Several different types of inextensible causal curves exist. One obvious type is *closed* causal curves, such as the closed causal curve illustrated in Figure 2.7.2. A few other types of inextensible causal curves are illustrated in the left-hand diagram in Figure 2.7.5. The white region represents a "hole" in the relativistic spacetime manifold W, with its boundary "stripped away." Of course, it is not a hole in a physical sense, but a topological feature of W. Causal curves in W that "approach this hole," such as the curves γ_2, γ_3, and γ_4 illustrated in the figure, cannot be extended. A "simpler" type of inextensible causal curve is one that "runs on forever in both directions," such as the curve γ_1.

A Cauchy surface in a relativistic spacetime manifold X is a special case of an *acausal subset* of X, which is defined to be a subset intersected *at most once* by any causal curve. In particular, no pair of events belonging to such a subset are causally related. The right-hand diagram in Figure 2.7.5 illustrates two acausal subsets of X, represented by thick horizontal curves. The lower, "broken" curve, labeled ρ, is *permeable,* in the sense that it has "gaps," through which causal curves may pass without intersecting it. The upper curve, labeled σ, is a Cauchy surface, which is by definition *impermeable;* no causal curve may pass from its past to its future without intersecting it. From a modern information-theoretic viewpoint, a Cauchy surface "samples," or "filters," data flowing from its past to its future. A permeable acausal subset is "faulty" in this sense, because information may flow from its past to its future without being "sampled," by following causal curves permeating the subset.

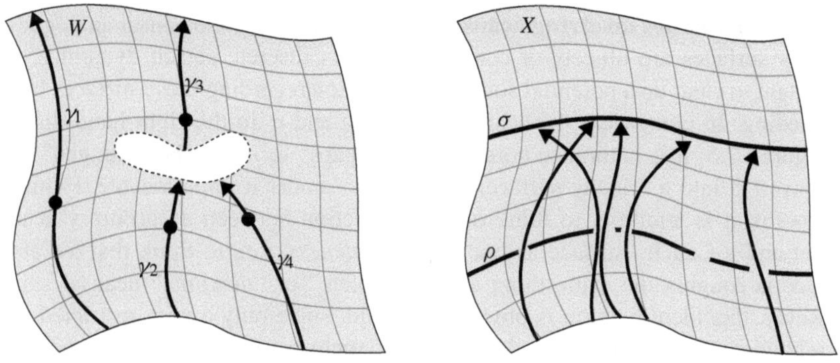

Figure 2.7.5. Inextensible causal curves; a permeable acausal subset ρ and a Cauchy surface σ.

This creates "predictability issues," since data associated with a permeable acausal subset is insufficient to determine "what will happen" in the future of the subset. This is sometimes expressed by saying that the initial value problem for such a subset is not "well-posed." I refer to this general deficiency of arbitrary acausal subsets as the **permeability problem.** As explained below, the discrete causal version of the permeability problem is of great significance in discrete causal dynamics.

A relativistic spacetime manifold is called **globally hyperbolic** if it contains a Cauchy surface. The spacetime W illustrated in the left-hand diagram in Figure 2.7.5 is *not* globally hyperbolic; no acausal subset of W can intersect all inextensible causal curves in W, due to the presence of the "hole" in W. It may be easy to construct acausal subsets intersecting *particular* families of causal curves; for example, the events represented by the four nodes in the figure form a finite acausal subset of W intersecting the four curves γ_1 to γ_4. However, if one attempts to extend such a subset, one is forced to make choices; for example, the extended subset may intersect certain curves "below the hole" *or* other curves "above the hole," but not both. The spacetime X illustrated in the right-hand diagram of the figure is globally hyperbolic by definition, since σ is a Cauchy surface. The "global" part of the term "globally hyperbolic" may seem inappropriate, since a particular Cauchy surface represents just one "slice" of spacetime. However, the existence of *even one Cauchy surface* implies much more in the relativistic setting. In particular, a globally hyperbolic relativistic spacetime manifold may be *foliated* by Cauchy surfaces. Informally, this means that the entire spacetime may be viewed as a "stack of Cauchy surfaces;" the individual surfaces are called the *leaves* of the foliation. Each leaf represents a "moment in time" in a particular frame of reference. More precisely, a suitable choice of transition functions defining X as a manifold separates the local temporal variable from the corresponding spatial variables, and the resulting spatial "plaques" patch together across coordinate charts to form the leaves of the foliation. From a dynamical perspective, all the information flowing from past to future may be sampled at any given leaf. This, together with an appropriate dynamical law, enables a "global solution" for the physical behavior modeled by this dynamics; i.e., it allows one to predict "what happens anywhere in spacetime."

Permeability issues involving acausal subsets. Since discrete causal analogues of Cauchy surfaces are objects of central interest in discrete causal dynamics, it is important to clear up a potential source of confusion regarding these surfaces before proceeding. In comparing the acausal subsets ρ and σ in the right-hand diagram in Figure 2.7.5, it is natural to notice that the "gaps" in ρ may be "plugged" so as to convert ρ into a Cauchy surface, i.e., so as to render it impermeable. From this viewpoint, it is tempting to think of the distinction between an arbitrary acausal subset and a Cauchy surface as one of *completeness,* i.e., to think that a Cauchy surface is roughly the same thing as a "complete" or "maximal" acausal subset. However, this identification is obviously invalid, since only a very restricted class of relativistic spacetime manifolds possess Cauchy surfaces at all. The left-hand diagram in Figure 2.7.6 illustrates, at an informal level, how a maximal acausal subset of a non-globally hyperbolic spacetime may fail to qualify as a Cauchy surface. The acausal subset illustrated here is the union $\rho \cup \rho'$ of two smaller acausal subsets ρ and ρ' of the spacetime manifold W first illustrated in Figure 2.7.5. Here, ρ is represented by the thick black curve "below the hole" in W, which has a "missing point" at x_1, while ρ' is represented by the shorter thick black curve "above the hole," which includes its left endpoint x_2, but not its right endpoint x_3. The latter subset ρ' is "shielded from ρ by the shadow of the hole," in the sense that the "hole" prevents causal curves passing through ρ from reaching ρ'. The "shadow of the hole" is represented by the dark gray region.

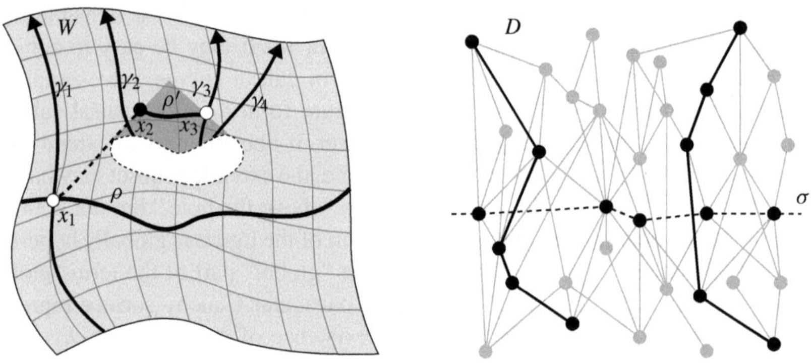

Figure 2.7.6. Permeable maximal acausal subset in relativistic spacetime; a typical maximal antichain in a directed set is highly permeable.

Note that the extension "back in time" of the boundary of this "shadow" intersects the "missing point" x_1; i.e., there is a null curve from x_1 to x_2, represented by the dashed curve in the figure. Hence, the "gap" in $\rho \cup \rho'$ at x_1 cannot be "plugged by adding x_1;" the resulting subset $\rho \cup \{x_1\} \cup \rho'$ of W would not be acausal. If the "hole" were absent, then $\rho \cup \{x_1\}$ *would* be a Cauchy surface. As it is, however, the curves γ_1, γ_2, γ_3, and γ_4 are examples of inextensible causal curves that do not intersect the maximal acausal subset $\rho \cup \rho'$. The curve γ_1 "permeates" $\rho \cup \rho'$ in an obvious way, since it "goes through the missing point x_1;" this curve behaves much like the curves permeating the acausal subset labeled ρ in Figure 2.7.5. The other curves, γ_2, γ_3, and γ_4, "avoid" $\rho \cup \rho'$ in subtler ways, whose details are unimportant at present. The purpose of the illustration is merely to demonstrate how a maximal acausal subset of a relativistic spacetime manifold may fail to be a Cauchy surface.

In discrete causal theory, the permeability problem is a priori "much worse" than in relativity. It occurs more naturally, and is less-strongly tied to global structure. In particular, after defining appropriate discrete causal analogues of Cauchy surfaces in a directed set, one finds that the existence of one such "Cauchy surface" does *not* guarantee that the set may be "foliated by such surfaces." Further, "most" maximal acausal subsets, called *maximal antichains* in this context, are not even close to being "Cauchy surfaces." Indeed, they are "riddled with permeations," rather than merely "missing a few points." These details are worth remembering when consulting the literature; for example, Bleybel and Zaiour [BZ15] prove a "foliation theorem" for causal sets in a recent paper; however, the "leaves" of the "foliations" involved are *not* close analogues of Cauchy surfaces, since they are generally permeable. Similarly, the *foliation scheme* $\mathbb{S}_{\mathrm{FOL}}$ introduced in Chapter 7 describes a special type of "generational growth" via generally permeable antichains.

The right-hand diagram in Figure 2.7.6 illustrates a permeable maximal antichain σ in a directed set D, represented by the roughly-horizontal row of nodes connected by dashed lines. These dashed lines are merely a visual aid, included to suggest the "surface-like" characteristics of σ; they are not part of the actual structure of D. At a mathematical level, the statement that σ is an antichain means that no pair of its elements, distinct or otherwise, is connected by a chain, i.e., a sequence of relations. Chains, which are natural discrete causal analogues of causal curves,[29] are studied formally in Chapter 3. The antichain σ is maximal because every other element of D is connected to at least one element of σ by a chain[30]; hence, it is impossible to add any more elements to σ while preserving its properties as an antichain. The remaining black nodes in the diagram, and the edges connecting them, represent two chains in D permeating σ. The "flow of information" represented by these chains is "invisible to σ," and this means that the corresponding initial value problem is not well-posed. The resulting complications for the dynamics of directed sets have already been recognized in the literature, particularly in the special case of causal sets [MRS06]. As described in Chapter 5, passage to relation space provides a pleasing solution to this problem. At least in the acyclic case, the relation space $\mathcal{R}(D)$ over a directed set D is roughly analogous to a "globally hyperbolic spacetime" hidden behind the seemingly intractable structure of D itself. This enables associated dynamical problems to be treated in a much more satisfactory manner.

---

[29]It would be more precise to say "analogues of causal *paths;*" see Section 5.9 for details.

[30]Johnny Feng pointed out to me the a posteriori obvious fact that maximality of an antichain in a multidirected set containing cycles *cannot* be characterized by the condition that "every other element is connected to it by a chain." For example, according to Definition 3.7.1, the empty subset of a cyclic multidirected set is a maximal antichain, since no element of such a set can belong to an antichain.

2.8 Metric Recovery

Historical context. The previous five sections of this chapter provide enough background information on relativistic spacetime structure and causality conditions to enable the statement of an appropriate version of metric recovery. No formal proof is included, since the details are mostly tangential to the discussion, and may be easily found in the literature. Historically, the theorems of Hawking [HA76] and Malament [MA77], published in the late 1970s, supplied the "critical pieces of the puzzle," in the sense that they led almost immediately to the birth of discrete causal theory as a serious approach to fundamental spacetime structure. These results have since been amplified in various ways. For example, Hawking and Malament consider only the four-dimensional case, since relativistic spacetime is modeled via four-dimensional manifolds, but the same basic results have since been shown to apply to any dimension at least three. Further, Hawking and Malament assume knowledge of the causal structure of the entire manifold under consideration in the hypotheses of their theorems, but subsequent results have demonstrated that knowledge of the causal structure of a *countable dense subset* suffices. While newer results of this nature are important for establishing the precise details of how causal structure relates to geometry, the original metric recovery theorems are themselves sufficient to motivate the classical causal metric hypothesis (CCMH) in the discrete context.

Statement of the theorem. For the sake of simplicity, I state here a version of metric recovery that may be extracted from Malament's paper [MA77] alone, although Malament does make use of Hawking's theorem. The result here is expressed in a somewhat different way than Malament's main theorem; in particular, it includes two statements, one for enhanced causal isomorphisms of arbitrary relativistic spacetime manifolds, and one for causal isomorphisms of past and future distinguishing relativistic spacetime manifolds. As stated at the end of Section 2.6, Malament works mostly in terms of *timelike* curves, which means that the properties of causal isomorphisms and enhanced causal isomorphisms are actually somewhat stronger than necessary to prove the theorem.

Theorem 2.8.1. Metric recovery. *Let X and X' be smooth four-dimensional real manifolds without boundary, and let g and g' be smooth pseudo-Riemannian metrics of Lorentz signature on X and X', respectively.*

1. *If $f : X \to X'$ is an enhanced causal isomorphism, then f is a smooth conformal isometry.*
2. *If X and X' are past and future distinguishing, and $f : X \to X'$ is a causal isomorphism, then f is a smooth conformal isometry.*

In particular, in either case, knowledge of g enables recovery of g' up to conformal equivalence, and vice versa.

Sketch of Proof. Let X and X' be as described in the statement of the theorem. Informally, the proof involves combining the consequences of the following two statements:

Hawking: *"Topological structure determines conformal structure."*

Malament: *"Causal structure determines topological structure."*

Here, I outline the proof as it appears in Malament's paper [MA77]. Due to Malament's choice to work mostly in terms of *timelike* curves, certain aspects of the proof are more detailed than necessary to establish Theorem 2.8.1 as I have stated it. However, the additional detail is not too cumbersome, and may help the interested reader to follow the proof in its original context.

The first step in the proof is simply to note Hawking's result that if $f : X \to X'$ is a homeomorphism with respect to the manifold topologies on X and X', and if both f and f^{-1} preserve future-directed continuous null geodesics, then f is a smooth conformal isometry.[31] Malament refers to Hawking, King, and McCarthy for the proof, while noting that the theorem is described there in a slightly different way.[32] The essential argument is described as an "unpublished result of Hawking," which justifies Malament's attribution of the theorem to Hawking specifically.

The second step is to extend Hawking's theorem by means of an easy lemma[33] demonstrating that if $f : X \to X'$ is a homeomorphism with respect to the manifold topologies on X and X', and if f and f^{-1} preserve future-directed continuous timelike curves, then f and f^{-1} also preserve future-directed continuous null geodesics.

The third step is to prove that if $f : X \to X'$ is a bijection, and if f and f^{-1} preserve future-directed continuous timelike curves, then f is a homeomorphism with respect to the manifold topologies on X and X', and hence, by Hawking's theorem, a conformal isometry. This is the lengthiest part of the proof; it comprises the entire fifth section of Malament's paper. This is more than enough to establish the first statement in Theorem 2.8.1. Indeed, if f is an enhanced causal isomorphism, then by Definition 2.6.3, f is a bijection, and f and f^{-1} preserve future-directed continuous *causal* curves, which include future-directed continuous *timelike* curves.

The fourth and final step is to establish that if X and X' are past and future distinguishing, and if $f : X \to X'$ is a bijection such that f and f^{-1} preserve relativistic *chronological* relations, then f and f^{-1} preserve future-directed continuous timelike curves. Hence, by the previous steps in the proof, f is a conformal isometry. This is more than enough to establish the second statement in Theorem 2.8.1. Indeed, if f is a causal isomorphism, then by Definition 2.6.3, f is a bijection such that f and f^{-1} preserve relativistic *causal* relations, and hence, relativistic chronological relations. □

Topological details. A few technical details are worth clarifying before resuming a more qualitative examination of how metric recovery motivates the classical causal metric hypothesis (CCMH). First, I explain some topological details relating to the proof of Theorem 2.8.1. The reason for emphasizing that the maps involved in the proof are *homeomorphisms with respect to the manifold topologies* on X and X' is

[31] See Malament [MA77], p. 1400.

[32] See Hawking, King, and McCarthy [HA76], p. 174.

[33] See Malament [MA77], Lemma 1, p. 1400.

to avoid potential confusion involving two other topologies that play a role in the papers of Hawking, King, and McCarthy [HA76], and Malament [MA77]; namely, the *Alexandrov topology* and the *path topology.* It is convenient to give here a brief account of these topologies, thereby completing the discussion of the five types of structure on relativistic spacetime, listed in "reverse order of detail" in Figure 2.3.1. Besides assisting the reader in deciphering the literature on metric recovery, this information serves the additional purpose of preparing the ground for later examination of topologies and local properties for directed sets and multidirected sets.

A topology defines which subsets of a set X are "open," and by complementation, which subsets are "closed." Most of the details regarding topologies are postponed until Chapter 4. In the present section, I take the informal viewpoint that a topology on X is a collection of subsets of X, called *open sets,* satisfying certain properties abstracted from the properties of intervals on the real line $\mathbb{R}$. Topologies are particularly useful for describing "local properties of spaces." Generally, a property is considered to be local near a point x in X if it may be detected by examining any open set containing x, usually called an *open neighborhood* of x. The heuristic that open sets in a topology are analogues of open intervals in $\mathbb{R}$ plays a role both in relativity and in discrete causal theory. In the relativistic case, the Alexandrov topology is a type of *order topology,* or *interval topology,* with respect to the partial order on a relativistic spacetime manifold X satisfying the causal condition. The path topology on X, meanwhile, is defined in terms of *maps* from open intervals in $\mathbb{R}$ into X. In the discrete causal context, however, the naïve idea that "intervals measure local properties," abstracted from the relativistic setting, leads to serious conceptual and technical issues, as explained in Chapter 4.

The manifold topology on a relativistic spacetime manifold X is inherited from the "usual topology" on $\mathbb{R}^4$, via the coordinate charts defining X as a real manifold. This latter topology is the *metric topology* for the usual Euclidean metric on $\mathbb{R}^4$. It is defined by taking a subset U of $\mathbb{R}^4$ to be open if and only if for every element $x \in U$, there exists a positive number ε, such that every element of $\mathbb{R}^4$ within distance ε of x with respect to the Euclidean metric is also in U. The resulting structure is then transported to X via its coordinate charts in the obvious way. The dark gray "circular" regions illustrated in Figure 2.8.2 represent typical open sets in the manifold topology on a relativistic spacetime manifold X. For $\mathbb{R}$ itself, which may be viewed as one-dimensional Euclidean space $\mathbb{R}^1$, the metric topology coincides with the order topology, and hence involves intervals. However, there is no obvious, unique, physically significant partial order on a higher-dimensional Euclidean space, so it is not surprising that the manifold topology on X fails, in certain important ways, to mesh naturally with the physical attributes of X arising from its pseudo-Riemannian structure.

A topology is called "coarse" if it has "few open sets," and is called "fine" if it has "many open sets." The **Alexandrov topology** on a relativistic spacetime manifold X is defined to be *the coarsest topology such that the chronological past $I^-(x)$ and the chronological future $I^+(x)$ of each event x in X are open sets.* In particular, every open set in the Alexandrov topology is automatically open in the manifold topology. The "basic open sets" in the Alexandrov topology are the "diamond-shaped" subsets

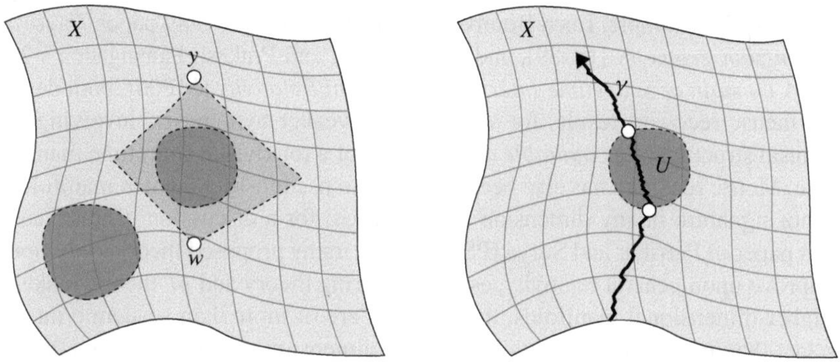

Figure 2.8.2. Manifold topology and Alexandrov topology; defining the path topology.

$I^+(w) \cap I^-(y)$; one of these sets is illustrated in the left-hand diagram in Figure 2.8.2. By the definition of the chronological relation $\prec\prec_{\mathrm{GR}}$ on X, these basic open sets are *intervals* of the form $\langle\langle w, y \rangle\rangle := \{x | w \prec\prec_{\mathrm{GR}} x \prec\prec_{\mathrm{GR}} y\}$. From a physical standpoint, the basic open set $\langle\langle w, y \rangle\rangle$ consists of all events that may be reached from the event w by the motion of a massive body, and may also reach the event y by the motion of a massive body. This description demonstrates the obvious physical significance of the Alexandrov topology in the relativistic setting.

A topology on a set X *induces* topologies on each of its subsets S, by defining a subset V of S to be open if and only if $V = U \cap S$ for some open subset U of X. The **path topology** on a relativistic spacetime manifold X is defined to be *the finest topology that induces, on the images of all continuous timelike curves in X, the same topology induced by the manifold topology.* In particular, every open set in the manifold topology is automatically open in the path topology. The right-hand diagram in Figure 2.8.2 illustrates how an open subset U of X in the manifold topology, represented by the dark "circular" region, defines an open subset $V = U \cap \gamma$ of the image of a continuous timelike curve γ in X, represented by the "part of γ inside U," which has its "endpoints missing." The physical significance of the path topology is slightly less direct than that of the Alexandrov topology; continuous timelike curves are obviously significant in their own right, but the finest choice of "open sets" in X that reproduces the usual manifold-induced topology on the images of such curves is a bit of a nuisance to describe. The real advantage of the path topology is that its homeomorphisms are precisely the smooth conformal isometries of X. To paraphrase Malament,[34] the path topology *"simultaneously encodes information about the manifold structure, the smooth structure, and the conformal structure of X."*

Modern improvements on metric recovery. A second technical point, already mentioned in passing above, is that the strength and scope of metric recovery results has been significantly improved since the original metric recovery theorems of the

[34] Malament [MA77], p. 1399.

late 1970s. For example, Luca Bombelli and David Meyer's 1989 paper *The origin of Lorentzian geometry* [BO89], and Keye Martin and Prakash Panangaden's 2006 paper *A Domain of Spacetime Intervals in General Relativity* [MP06], both demonstrate metric recovery results for which much weaker hypotheses, involving only the causal structure of a *countable dense subset* of a relativistic spacetime manifold, suffice. Metric recovery has also been extended to pseudo-Riemannian manifolds of Lorentz signature in any dimension at least three; for a discussion of this, see the recent paper of Parrikar and Surya [PS11]. Since many proposed theories attempting to improve upon general relativity, especially string theory and M-theory, make use of higher-dimensional manifolds, these results are of more than academic interest. The fact that analogous results fail to hold in dimension two is also interesting; for example, because of simulations suggesting *dimension reduction* in certain fields of quantum gravity, and because of the prominence of conformal field theories on Riemann surfaces in other areas of theoretical physics.

Motivation for the causal metric hypothesis. The simplest way to express the meaning of metric recovery in the context of general relativity is to say that *causal structure determines metric structure up to scale.* As discussed in Section 2.3, this statement approaches, but does not quite reach, the elegant and tempting conclusion represented by the classical causal metric hypothesis (CCMH); namely, that *metric structure is merely an approximate way of describing causal structure.* The qualifier "up to scale" obstructs such a conclusion in the relativistic case, but the idea is sufficiently compelling that it is natural to ask if one can somehow justify it by shifting attention to "causal structures possessing a natural scale." As mentioned in Section 1.3, and again in Section 2.6, the founders of causal set theory perceived the most obvious way to accomplish this, almost immediately after the original metric recovery theorems were established: to work with discrete models, and to assign volume to subsets by counting their elements. This is the strategy that Sorkin later encapsulated in his phrase, *"order plus number equals geometry,"* where "order" stands for causal structure, and "number" stands for the counting procedure. Perhaps the most straightforward way to realize this idea is by means of causal sets constructed via "sprinklings" into pseudo-Riemannian manifolds, such as Minkowski spacetime $\mathbb{R}^{3+1}$. Such causal sets are discussed briefly in Section 3.2, and more thoroughly in Section 4.5.

Options for realizing the hypothesis. However, the causal set approach is only one of many possible ways to realize the classical causal metric hypothesis, even if one restricts attention to the discrete context. "Order" is a very restrictive, and likely inadequate, proxy for causal structure, even in the relativistic case. In particular, the relativistic causal relation $\prec_{\text{GR}}$ defines a partial order only for relativistic spacetime manifolds satisfying the causal condition. "Number," meanwhile, is a very specific, and possibly oversimplified, proxy for scale. A general hazard to be avoided when converting a compelling conceptual motif; in this case, the classical causal metric hypothesis, into a specific technical approach; for example, causal set theory, is the risk of ignoring equally viable approaches that may ultimately reach further. Hence, it is crucial not to narrow down the possibilities prematurely. In particular, even if one restricts attention to spacetimes satisfying the causal condition, the transition to the discrete context is nontrivial, and automatically transferring over all the axioms of partially ordered sets is unjustified. Similarly, it is by no means obvious that each element in a discrete directed set should contribute equally, or even approximately equally, to the "volume" of the set. After all, one of the most obvious ways in which a typical such a set differs from a manifold is that it is *locally irregular;* i.e., its local structure is generally *not* the same near each element.

The only truly essential feature of causal structure, beginning from first principles, is the directed relationship between cause and effect. A partial order takes this *local* building block of structure, and adds additional *nonlocal* properties that are not necessarily appropriate; for example, transitivity (TR). Similarly, the only truly essential requirement regarding the introduction of scale data in the context of the classical causal metric hypothesis is that this data should arise *naturally* from the causal structure itself. Following these lines of thought, it is useful to juxtapose Sorkin's version of the classical causal metric hypothesis with a more general statement that, while less succinct, avoids the risk of placing the subsequent technical development in a structural straitjacket:

Sorkin: *"Order plus number equals geometry."*

Generalization: *"Directed structure plus natural scale equals geometry."*

Figure 2.8.3 illustrates three different ways of assigning "volume data" to a discrete directed set. The left-hand diagram illustrates Sorkin's original prescription, taken literally; every element is assigned exactly the same volume. The middle diagram illustrates the incorporation of "statistical fluctuations" in the computation of volume. Such fluctuations are invoked in the causal set literature for technical reasons; in particular, to avoid systematic violations of Lorentz invariance for causal sets constructed via global "sprinkling" into Minkowski spacetime $\mathbb{R}^{3+1}$. The right-hand diagram illustrates a much different method of assigning volume, in which not only the number of elements, but also the *local causal structure,* plays a role. In this particular case, the "volume" of each element is determined by its *valence,* i.e., by the number of relations for which it is the initial or terminal element.[35] In this context, the valence field illustrated in Figure 1.7.2 of Chapter 1 serves as a "volume field," i.e., a "discrete conformal factor."

From the *relation space viewpoint,* introduced briefly in Section 1.5, and developed in detail in Chapter 5, the latter method of assigning scale data is perhaps the most attractive of the three rather naïve methods illustrated in the figure. This is because relations, rather than elements, are considered to be fundamental in this context, and the valence field essentially "counts relations." However, discrete directed sets are rich in combinatorial structure, and there are many other possible methods of deriving scale data from this structure. In a way, this is a disadvantage, because

[35] As explained in Section 4.3, *reflexive relations* $x \prec x$ are counted twice in enumerating $v(x)$, because such relations both begin and terminate at x. Of course, such relations do not occur in *acyclic* directed sets.

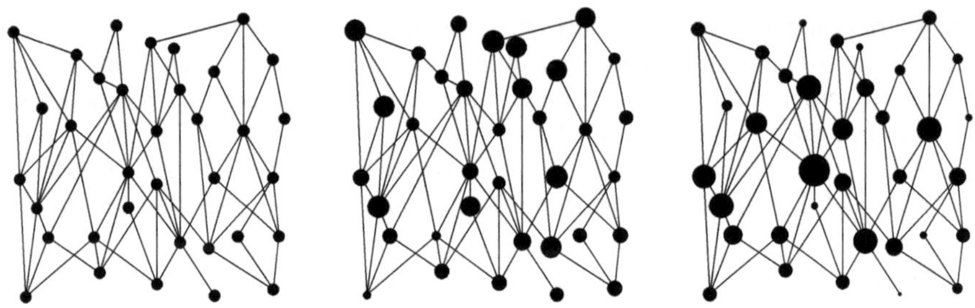

Figure 2.8.3. Alternative methods of assigning volume to a discrete causal structure: constant; incorporating statistical fluctuations; valence-based.

it introduces a risk that the selection of one particular method among these will be unconvincing. It would be preferable if one of the simplest methods, involving a straightforward counting of elements or relations, could be proven to be the "right one." However, there are other quantities just as basic as scale from a geometric viewpoint, such as dimension, whose emergence from discrete causal structure almost certainly does not arise from a method as direct as a simple counting procedure. Perhaps the most reasonable approach to this situation, while discrete causal theory remains in a relatively early stage of development, is to focus on the simplest models, while keeping in mind the possibility that subtler considerations may ultimately be incorporated as "corrections" in more mature versions of the theory. This strategy is supported by the plausible reflection that a variety of different methods of assigning scale data may very well "converge" within just a few orders of magnitude of the fundamental scale. In particular, the "variable volume" method illustrated in the right-hand diagram in Figure 2.8.3 could, in many cases, be replaced by a simpler causal set-type method, using the "average valence" of the elements, leading to only negligible differences in the resulting computations above the very smallest scales. In fact, if the typical valences of elements in a physically realistic directed set turn out to be very large, then the differences between the volumes assigned to elements under a valence-based approach to volume could be negligibly small in proportion even at the fundamental scale itself. Finally, Bombelli, Henson, and Sorkin [BHS09] suggest the possibility of adding "distance information" to the relations of a causal set, but conjecture that causal sets represent *"in some sense, the minimal* [systematically] *Lorentz invariant discrete* [structures] *from which* [Minkowski spacetime] *can be reconstructed at macroscopic scales."*

2.9 Order Good, Continuum Bad

Criticizing continuum-based theories. The foregoing sections of this chapter provide preliminary evidence suggesting that discrete causal theory is a reasonable candidate to serve as an alternative structural paradigm for modeling classical spacetime. However, it is worthwhile to consider the basic question of *why such alternatives are needed in the first place;* in particular, why real manifolds are ultimately not ideal for this role, despite their distinguished history in theoretical physics. In this section, I present a perspective on this subject that might be considered "underrepresented in the literature;" namely, that *the prominence of continuum-based theories is partly just a historical accident,* resulting from broad shortcomings in human knowledge, and human computational capabilities, throughout the early development of modern science. From this point of view, there is nothing natural or inevitable about the role of the real numbers to recommend them as the default building block of basic physical structure to any sufficiently advanced scientific community. In particular, certain well-known obstructions to progress in standard continuum-based theories, including divergence issues and problems with renormalizability, are regarded, from this perspective, less as "great problems to be solved," than as evidence that the wrong questions are being asked, based on the wrong presuppositions. Discrete causal theory aims, as far as possible, to completely circumvent such obstructions.

A nonspecific hypothesis of fundamental discreteness no longer seems *avant-garde* in its own right, since it is now widely expected that "the spacetime continuum breaks down at the fundamental scale" in the context of quantum gravity. Indeed, a large number of different *non*-continuum-based approaches to fundamental physics, and more specifically, discrete approaches, may be found in the literature.[36] However, even a cursory examination of leading publications suggests that the vast majority of mainstream modern theoretical physicists still devote their efforts almost exclusively to theories that use the real numbers as a structural "starting point" in one form or another. In particular, string theory, which remains, by a huge margin, the dominant approach, both in terms of its number of researchers and its quantity of resources, is entirely continuum-based. Its main competitor, loop quantum gravity, "arrives" at a form of fundamental discreteness in a circuitous manner, via a novel quantization procedure, beginning in the continuum-based context of general relativity.[37] Hence, even though it is generally believed that "quantum spacetime is not a real manifold," the actual practice of modern theoretical physics mostly fails to reflect this belief.

Many of the specific technical problems arising in continuum-based theories are completely irrelevant in the discrete causal context. For example, *renormalization* is a standard device in quantum field theory for "curing" certain divergence issues that arise, ultimately, from properties of the real numbers. This approach succeeds only in special cases; theories for which it fails are called *nonrenormalizable.* One of the major historical obstacles in formulating successful theories of quantum gravity has been the fact that "standard" approaches to "quantizing general relativity" lead to nonrenormalizable theories. In discrete causal theory, however, the original sources of the divergences eliminated by renormalization are generally absent, and

---

[36]An interesting list of such approaches appears in the recent paper on *tensor networks* by Chen, Sasakura, and Sato [CSS16]. The majority of these approaches are less relevant to this book than causal set theory, since they rely to a large degree on auxiliary, "non-causal" structure.

[37]Of course, as discussed in Section 8.8, much of this reliance on the continuum may be stripped away a posteriori.

this renders the whole subject of such devices a priori irrelevant, at least in its original context.[38] Of course, objects such as renormalization groups, or analogues thereof, may reappear in interesting mathematical roles, possibly intersecting with discrete causal theory. As originally conceived, however, such methods may be regarded as "chemotherapy for continuum-based theories," i.e., as tortuous cures for problems that one would prefer to avoid entirely.

Similarly, the deeply-entrenched and tiresome controversy over the *string theory landscape*, and its implications regarding the *anthropic principle,* arises ultimately from the properties of certain families of manifolds, including the iconic *Calabi–Yau manifolds,* which are likely of greater mathematical than physical interest. These manifolds have been "imported" into string theory from algebraic geometry, in order to "cure" the inconvenient fact that string theory requires the wrong dimension for spacetime, as indicated by all available observational evidence. A serious description of such problems would contribute nothing to the subject of this book, so I choose instead to focus on more basic and foundational objections to the entire corpus of continuum-based theories, in particular, objections arising from generic structural properties of real manifolds that are almost certainly physically irrelevant. Many of the technical struggles of modern theoretical physics, including those mentioned above, may be regarded as mere symptoms of these deeper problems.

The real numbers $\mathbb{R}$; order and completeness. Every physicist is familiar with the real number system $\mathbb{R}$, which supplies the structural scaffolding for the continuum-based theories that dominate conventional modern physics. $\mathbb{R}$ is a *linear continuum,* in a sense made precise below; for the moment, it suffices to remark that the word "linear" refers to a purely order-theoretic property of $\mathbb{R}$, while the word "continuum" refers to a "completeness property," which is also *essentially* order-theoretic, but which may be generalized to apply to non-ordered sets. In this book, the term *continuum-based theory* refers to a physical theory described in terms of real manifolds. Such manifolds inherit a completeness property from $\mathbb{R}$, but generally possess no natural order. Hence, in passing from $\mathbb{R}$ itself to manifolds defined over $\mathbb{R}$, one loses most of the original order-theoretic structure. To construct continuum-based theories that yield even an approximate description of nature, one must "add back in" order-theoretic structure artificially, by means of a metric. In the discrete causal context, order-theoretic properties are essential, because the directed sets used to model discrete causal classical histories derive their local structure from individual ordered relationships between pairs of elements. Hence, even though order theory is not quite general enough to model classical causal structure in a global sense, it remains part of the conceptual core of the theory.

Despite the universal familiarity of the real numbers, it is instructive to re-examine $\mathbb{R}$ in detail as a *mathematical object,* but with a view toward physical applications. This enables a better understanding of some of the basic objections to continuum-based theories of fundamental physics. These objections strongly suggest a need for alternative approaches, of which discrete causal theory is an obvious candidate. It

[38] See, however, the discussion in Section 4.5 regarding "sprinkled" causal sets, which may exhibit locally infinite behavior.

turns out that most of the order-theoretic structure of $\mathbb{R}$ may be regarded as "good" in this context, with totally ordered and partially ordered sets retaining a prominent role in discrete causal theory. However, the completeness property of $\mathbb{R}$, which is the property that "makes it a continuum," leads to deep trouble. In fact, even the weaker *interpolative property* of $\mathbb{R}$, which says that "one can always find a real number strictly between any two distinct real numbers," is problematic. In particular, even the "incomplete" field of rational numbers $\mathbb{Q}$ is already "bad" from the discrete causal perspective, since it shares the interpolative property with $\mathbb{R}$.

$\mathbb{R}$ in terms of universal properties. Standard definitions of $\mathbb{R}$, though given in almost every university analysis course, appear surprisingly obtuse when approached from first principles. $\mathbb{R}$ may be defined, in terms of universal properties, as *the unique isomorphism class of Archimedean complete totally ordered fields.* To review a bit of algebra, a field is an "optimally behaved number system," in the sense that it possesses additive and multiplicative operations that satisfy familiar properties, and that "cooperate" with each other in familiar ways. To be precise, a **field** is a set $\mathbb{F}$, together with two operations $+$ and $\times$, called *addition* and *multiplication,* respectively, such that $(\mathbb{F}, +)$ is an abelian group with identity 0, $(\mathbb{F} - \{0\}, \times)$ is an abelian group with identity $1 \neq 0$, and multiplication distributes over addition.

Total order. A *total order,* or synonymously, a *linear order,* on a field $\mathbb{F}$, is a *transitive, antisymmetric, total* binary relation $\preceq$ on $\mathbb{F}$. Letting x, y, and z be elements of $\mathbb{F}$, "transitive" means that if $x \preceq y$ and $y \preceq z$, then $x \preceq z$, "antisymmetric" means that if $x \preceq y$ and $y \preceq x$, then $x = y$, and "total" means that either $x \preceq y$, or $y \preceq x$ for every choice of x and y, including $x = y$. In particular, $\preceq$ is *reflexive;* i.e., $x \preceq x$ for every $x \in \mathbb{F}$. Omitting the "total" property, but retaining reflexivity, yields a *partial order.* The set of real numbers $\mathbb{R}$ is a totally ordered field under the familiar "less than or equal to" relation $\leq$. Given a total order $\preceq$ on $\mathbb{F}$, one may define a unique irreflexive binary operation $\prec$ on $\mathbb{F}$, sometimes called a *strict total order,* by setting $x \prec y$ if and only if $x \preceq y$ and $x \neq y$. The familiar "less than" relation $<$ on $\mathbb{R}$ is the strict total order corresponding to $\leq$. Conversely, given a strict total order $\prec$ on $\mathbb{F}$, one may define a unique non-strict total order $\preceq$ on $\mathbb{F}$ in the obvious way, by setting $x \preceq y$ if and only if $x \prec y$ or $x = y$. It is convenient here to dispose of some nuisances of terminology. First, strict orders such as $<$ are more useful in this book than nonstrict orders such as $\leq$, even though the latter are often more popular in mathematical settings. Second, partial orders play a larger role in this book than total orders. Hence, most encounters with "order theory" in this book actually involve *strict partial orders,* whether or not the words "strict partial" appear explicitly. By contrast, the discussion of ordered fields in the present section follows the usual mathematical conventions; in particular, the "order" on $\mathbb{R}$ is taken to be the usual non-strict total order $\leq$.

It is important to note that the total order $\leq$ on $\mathbb{R}$ is a natural aspect of the structure of $\mathbb{R}$, not an arbitrary auxiliary structure added a posteriori. Following the order refinement principle (ORP), discussed in Section 3.8, *any* set, and hence any field, may be endowed with a total order. For example, the field of complex numbers $\mathbb{C}$, whose elements are of the form $a + bi$, where a and b are real numbers and

i means the imaginary unit $\sqrt{-1}$, may be endowed with the total order borrowed from the familiar *lexicographic order* on $\mathbb{R}^2$, i.e., the order $\preceq$ defined by setting $a + bi \preceq c + di$ if and only if $a < c$, or $a = c$ and $b \leq d$, under the usual order on $\mathbb{R}$. The distinction between this sort of ad hoc total order imposed on a field, and a natural total order, such as the usual order $\leq$ on $\mathbb{R}$, is that the latter "respects the field structure." For example, given positive elements w, x, y, and z of $\mathbb{R}$, it is true that if $w \leq x$ and $y \leq z$, then $wy \leq xz$. The analogous property fails to hold for the total order on $\mathbb{C}$ defined above; for example, let $w = y = 1$ and $x = z = 1 + i$; then $wy \npreceq xz$, since the real part of wy is 1 and the real part of xz is 0. Generally, when one speaks of an "ordered" algebraic object, the order is assumed to respect the algebraic structure, unless stated otherwise.

Archimedean property. The order $\preceq$ on a totally ordered field $\mathbb{F}$ provides a way of comparing any pair of elements x and y of $\mathbb{F}$. In particular, any nonzero element x satisfies either $x \prec 0$ or $0 \prec x$ under the corresponding strict total order; in the first case, x is called *positive,* and in the second case, x is called *negative.* The Archimedean property says, informally, that "given any pair of positive elements x and y in $\mathbb{F}$, either element may be re-scaled to become larger than the other." More precisely, a positive element x is called *infinitesimal* with respect to a positive element y if every natural-number multiple of x is less than y; the Archimedean property says that $\mathbb{F}$ has no pairs x and y such that x is infinitesimal with respect to y. To readers without much background in abstract algebra, the Archimedean property may seem "obvious," but there exist familiar and important examples of non-Archimedean total ordered fields. For example, the field $\mathbb{R}(x)$ of rational functions in one variable x, with real coefficients, possesses a natural total order, defined in terms of the leading coefficients of numerator polynomials. However, it possesses infinitesimal elements; for example, $1/x$ is infinitesimal with respect to 1.

Completeness. The remaining property of $\mathbb{R}$ cited above is *completeness,* and it is this property that distinguishes $\mathbb{R}$ as the only continuum, up to isomorphism, among the class of Archimedean totally ordered fields. The set of rational numbers $\mathbb{Q}$, for example, is an Archimedean totally ordered field, but it is not complete, since it "leaves out" certain "limiting values," such as the algebraic number $\sqrt{2}$, and the transcendental numbers π and e. Completeness is defined in terms of *Cauchy sequences* in a totally ordered set, which are sequences whose elements "eventually become arbitrarily close to each other," in a manner familiar from elementary calculus. However, a potential problem of self-reference arises in this context, because Cauchy sequences are usually defined *in terms of* $\mathbb{R}$, which, of course, is the object of present scrutiny. In more detail, Cauchy sequences in a set S are usually defined in terms of a generalized distance function d, called a "metric," which quantifies the "closeness" of pairs of elements of S. To be clear, such a "metric" d does *not* represent the same type of structure as a metric g on a smooth real manifold; e.g. a pseudo-Riemannian metric, although a metric of Euclidean signature *induces* a "metric" d in the present sense. Rather, d is a map $S \times S \to \mathbb{R}$, which is *positive-definite, symmetric,* and satisfies the *triangle inequality.* Defining "closeness" with respect to such a "metric" d is circular when discussing the real numbers themselves, since the target of d is $\mathbb{R}$.

A more general notion is needed in this context, and this is supplied by the totally ordered group structure of $(\mathbb{R}, +)$, which defines what is called a *uniform structure.* Generalized Cauchy sequences may be defined with respect to any uniform structure, and this method may be used to characterize the completeness of $\mathbb{R}$. In general, a set S endowed with a uniform structure is called *complete* if every Cauchy sequence defined in terms of this structure converges to an element of S. The real numbers are complete with respect to the uniform structure defined by $(\mathbb{R}, +)$.

$\mathbb{R}$ as a linear continuum. Technically, a *linear continuum* L is an *interpolative totally ordered set satisfying the least-upper bound property.* Here, letting x, y, and z be elements of L, and denoting the strict total order on L by $\prec$, "interpolative" means that for any pair of elements x and z in L with $x \prec z$, there exists a third element y "between the two," i.e., such that $x \prec y \prec z$. The least upper bound property says, as one would expect, that any subset S of L that is bounded above has a least upper bound in L. More precisely, a subset S of L is bounded above if there exists an element $u \in L - S$ such that $s \leq u$ for every $s \in S$. A least upper bound for S is an element $u_{\text{MIN}} \in L$ that is, first of all, an upper bound of S, and secondly, is less than any other upper bound of S. By antisymmetry, u_{MIN} is necessarily unique if it exists. The rational numbers $\mathbb{Q}$ fail to satisfy the least upper bound property; for example, the subset of rational numbers $\{\frac{1}{1}, \frac{3}{2}, \frac{8}{5}, \frac{13}{8}, \ldots\}$, defined in terms of the Fibonacci numbers, is bounded above by the rational number 2, by an easy induction argument, but does not possess a rational least upper bound. Its least upper bound in $\mathbb{R}$, of course, is the golden ratio $\phi = \frac{1+\sqrt{5}}{2}$.

In the special case of Archimedean totally ordered fields, completeness is enough to guarantee the interpolative property and the least upper bound property defining a linear continuum, but in more general contexts, completeness does not imply continuum structure. For example, the set of integers $\mathbb{Z}$ is complete, since every Cauchy sequence is "eventually constant" at a specific integer n, to which it therefore converges. However, $\mathbb{Z}$ is not a continuum, because it does not satisfy the interpolative property. Even for the rational numbers, there exist different, non-Archimedean, completions, which are not continua; namely, the p-adic fields $\mathbb{Q}_p$, in which "distance" is quantified in terms of divisibility properties. It is interesting to note that there has recently emerged an entire field of fundamental physics devoted to non-Archimedean versions of quantum theory and related topics, sometimes called *p-adic quantum mechanics.*

Real manifolds as "unordered continua." As suggested above, the word "continuum," without the qualifier "linear," is often used in a general manner to refer to certain spaces that possess an appropriate "completeness property," but which generally do not possess natural order-theoretic structure. For example, a *topological continuum* is defined to be a compact, connected topological space equipped with a "metric," i.e., a generalized distance function like the "metric" d discussed above. In this context, the necessary completeness property is embodied by the compactness condition, which may be expressed in terms of the convergence properties of generalized sequences. However, in the context of theoretical physics, the word "continuum" is often used as a synonym for "real manifold."

Real manifolds are generally neither linear continua nor compact connected metric spaces. In particular, essentially the only "ordinary" real manifold possessing the natural structure of a linear continuum is $\mathbb{R}$ itself,[39] while an n-dimensional real manifold is, by definition, locally isomorphic to $\mathbb{R}^n$, which itself possesses no natural order-theoretic structure for $n \geq 2$. However, real manifolds are "locally complete," in the sense that every point in a real manifold X possesses a neighborhood which "contains all limit points of sequences in the neighborhood." This is illustrated in Figure 2.9.1, which shows a sequence $\{x_n\}_{n\in\mathbb{N}}$ of points converging to a point x in X. Order theory plays only an indirect role in this notion of completeness; the points of the sequence are ordered, but this order is borrowed from the natural numbers $\mathbb{N}$, and does not reflect any essential structural aspect of X itself. At a formal level, the sequence $\{x_n\}_{n\in\mathbb{N}}$ may be viewed as a map from $\mathbb{N}$ into X, and this is how it is represented in the figure.[40] A *path* in a set X, in a generalized order-theoretic sense, is an equivalence class of maps from a "linear directed set" into X, so this sequence may be regarded as representing a "discrete path" in X. However, it is not a "discrete directed path," i.e., an equivalence class of morphisms of directed sets from $\mathbb{N}$ into X, since X is not assumed to be a directed set in this context. Indeed, the only structure on X taken for granted here is its real manifold structure.

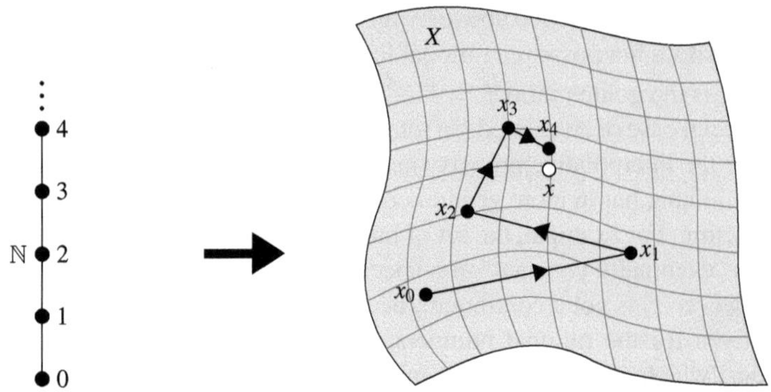

Figure 2.9.1. A sequence $\{x_n\}$ converging to a point x in a real manifold X, viewed as a map $\mathbb{N} \rightarrow X$.

It is both unfortunate and ironic that the ubiquitous use of the word "continuum" to mean "real manifold" in theoretical physics *emphasizes only the completeness property* of the linear continuum $\mathbb{R}$, the object from which every real manifold derives most of its structure, and *suppresses the role of the other order-theoretic properties*

[39]The informal qualifier "ordinary" is included to rule out "exotic examples" such as the *long line*.

[40]The directed set illustrated in the left-hand diagram in Figure 2.9.1 is really the *skeleton* sk($\mathbb{N}$) of $\mathbb{N}$, not $\mathbb{N}$ itself, since reducible relations are not included in the figure. Skeletons are introduced in Chapter 3.

of $\mathbb{R}$. This completeness property, despite its mathematical convenience in the context of calculus and differential equations, is probably the least relevant and most problematic feature of $\mathbb{R}$ from a physical standpoint. From the perspective of the causal metric hypothesis (CMH), it is the *linear order* of $\mathbb{R}$ that stands out as its principal redeeming characteristic, since this order enables subsets of real intervals to parameterize sequences of events.

Constructive view of $\mathbb{R}$. Of course, mainstream modern physics devotes little attention to the basic properties of $\mathbb{R}$, whether order-theoretic, topological, analytic, or algebraic. Instead $\mathbb{R}$ is usually treated as a collection of numbers which can serve, at least in principle, as the values of measurements, and which is moreover *large enough and well-enough behaved* to enable convenient methods of mathematical analysis; in particular, calculus and differential equations. Here, the viewpoint represented by the phrase "large enough and well-enough behaved" may be understood in terms of the historical emergence of $\mathbb{R}$ as a "number system," which encourages the unfortunate impression that "progress" involves a constructive process of adding more and more structure until "enough structure is present to describe nature." In fact, such a process risks, at each step, the inclusion of mathematically convenient but physically irrelevant structure, which may ultimately lead the researcher, or the entire scientific community, in the wrong direction.[41]

The history of the constructive process leading to $\mathbb{R}$ might be told in the following naïve way: the positive integers, which had been used since paleolithic times to count quantities such as the number of bear skins harvested in a given month, required augmentation by a zero element, to facilitate such innovations as the place value system; this led to the natural numbers $\mathbb{N}$. Similarly, negative integers and fractions were added to the picture to describe such concepts as directions on the number line, and equal partitions of objects; this yielded the integers $\mathbb{Z}$ and the rational numbers $\mathbb{Q}$. Early geometry, and later algebra, raised awareness of the existence of pairs of idealized quantities not in integer ratios to each other; for example, the Greeks noticed that the diagonal of a square in the Euclidean plane is incommensurable with its edges. This eventually motivated the recognition of irrational numbers such as $\sqrt{2}$, and transcendental numbers, such as π and e. The table in Figure 2.9.2 shows some of the "number systems" appearing in this process, along with their basic algebraic and order-theoretic properties. As indicated by the three bottom rows of the table, the process may be continued "beyond $\mathbb{R}$."

[41] It is prudent to balance this warning with the *principle of hidden structure* (HS), introduced in Chapter 3, which emphasizes the utility of *enlarging* the structural picture, if necessary, with new features that *are* relevant! The moral, ultimately, is that one must always keep in mind the physical motivation, or lack thereof, of structural features in physical theories.

symbol	name	algebraic structure	order structure
$\mathbb{N}-\{0\}$	positive integers	semigroup under $+$, monoid under $\times$	discrete total order
$\mathbb{N}$	natural numbers	monoid under $+$, monoid under $\times$	discrete total order
$\mathbb{Z}$	integers	ring (under $+,\times$)	discrete total order
$\mathbb{Q}$	rational numbers	field	interpolative total order
$\mathbb{R}$	real numbers	field	linear continuum
$\mathbb{C}$	complex numbers	algebraically closed field	no natural order
$\mathbb{H}$	quaternions	non-commutative division algebra	no natural order
$\mathbb{O}$	octonions	non-associative division algebra	no natural order

Figure 2.9.2. Comparison of "number systems," increasing in "size" from *top* to *bottom.*

From this constructive viewpoint, $\mathbb{R}$ may ultimately be defined as *the set of equivalence classes of Cauchy sequences of rational numbers,* under an appropriate equivalence relation. For example, the sequence $\{\frac{1}{1},\frac{3}{2},\frac{7}{5},\frac{17}{12},...\}$ is a representative of the equivalence class identified with $\sqrt{2}$.[42] Now suppose that there really is a minimum physical length; for example, the Planck length, and suppose that one wishes to "describe the diagonal of a macroscopic square-shaped region in flat space in terms of its edge." The naïve description involves the irrational number $\sqrt{2}$, but the precise physical description must be rational; for instance, there might be exactly 1.414213562373095048801688724209698 units of length along the diagonal for each unit of length along the edge. This example illustrates one reason why hypotheses involving discrete microstructure might generate little sympathy from a naïve point of view, since it is often a waste of time, practically speaking, to worry about so many decimal places. It might seem that "nothing is lost" by enlarging the set of possible values to admit quantities such as $\sqrt{2}$, even if they may not be precisely physical.

Probably no serious modern theorist actually holds a view so naïve as this, but the example nonetheless provides a reasonably accurate illustration of some of the conceptual pitfalls surrounding the physical role of $\mathbb{R}$, which *do* seem to exert a profound influence on the way mainstream theoretical physics is done. In particular, the impression that "nothing is lost" by enlarging, or completing, a number system, is egregiously wrong in general. In the context of classical causal structure, the nat-

[42]The method of *Dedekind cuts* is a different way of constructing $\mathbb{R}$ in terms of subsets of $\mathbb{Q}$.

ural scale is lost thereby, and with it, metric recovery. In this setting, it is striking to consider the fact that smooth manifolds, by themselves, are *manifestly* inadequate for modeling classical spacetime; even in relativity, one must import a metric as auxiliary structure. Discrete directed sets, by contrast, can *at least* approximate classical spacetime to a high degree of accuracy, even if spacetime is ultimately nondiscrete. In a broader setting, an unfortunate tendency exists to automatically place objections to the *physical* status of $\mathbb{R}$ in the same category as ancient Greek qualms about irrational numbers, or the constructivist arguments of Kronecker, Brouwer, and others, made more than a century ago. In fact, the objections to $\mathbb{R}$ most relevant to the motivations for discrete causal theory have nothing to do with the ontology of number systems, or with mathematical constructivism more generally.

Larger related number systems. Before enumerating some of these specific objections, I briefly outline the remaining content of the table in Figure 2.9.2, including the "larger number systems" $\mathbb{C}$, $\mathbb{H}$, and $\mathbb{O}$. First, I must explain the algebraic terminology appearing in the table. A *semigroup* is a set equipped with an associative binary operation, but generally without an identity or inverses. The prototypical example is the set $\mathbb{N} - \{0\}$ of positive integers[43] under addition, since, informally speaking, the "obvious candidates for an additive identity and additive inverses," namely, the zero element and the negative integers, respectively, are "left out." A *monoid* is a semigroup with identity; the prototypical examples are the positive integers under multiplication, with identity 1, and the natural numbers $\mathbb{N}$ under addition, with identity 0. A *ring* is "like a field," in the sense that it possesses "additive" and "multiplicative" operations, but it is more general; in particular, nonzero elements of a ring may not possess multiplicative inverses, multiplication may not be commutative, and so on. The prototypical example of a ring is the set of integers $\mathbb{Z}$, with the usual multiplication and addition. The construction of $\mathbb{Q}$ from $\mathbb{Z}$ is algebraic in nature; it is an example of what is called *localization.* The terminology arises from the fact that the same construction is used to examine local properties of algebraic schemes in algebraic geometry. The construction of $\mathbb{R}$ from $\mathbb{Q}$ is order-theoretic, as described above.

The complex numbers form an algebraically closed field, which means that the roots of any polynomial with coefficients in $\mathbb{C}$ are also in $\mathbb{C}$. Hence, the construction of $\mathbb{C}$ from $\mathbb{R}$ is algebraic. The complex numbers play a central role in ordinary quantum theory; in particular, the "state function" ψ appearing in Schrödinger's equation (1.1.3) takes on complex values, and the phases associated with spacetime paths in Feynman's path summation approach to quantum theory are also complex-valued. The apparent physical importance of progressively larger number systems, up to and including $\mathbb{C}$, has provoked natural interest in the even larger systems of *quaternions* $\mathbb{H}$ and *octonions* $\mathbb{O}$, the latter of which is the "largest normed division algebra over the real numbers." For example, since passage from $\mathbb{R}$ to $\mathbb{C}$ "produces quantum theory from classical theory," it is natural to ask if there might exist "hyperquantum

---

[43]This book follows the convention that the natural numbers $\mathbb{N}$ include zero; hence, the set of positive integers is given by removing zero from $\mathbb{N}$.

theories," based on $\mathbb{H}$ and $\mathbb{O}$.[44] Examination of these larger number systems has borne only modest fruit in physical settings; both $\mathbb{H}$ and $\mathbb{O}$ play a limited role in quantum information theory, and $\mathbb{H}$ makes appearances in spin dynamics and a few other contexts. However, these contributions remain miniscule compared to those of $\mathbb{R}$ and $\mathbb{C}$.

The table of "number systems" in Figure 2.9.2 may be placed into a larger context, in which "simple objects," such as $\mathbb{N}$, and even "simpler objects," such as finite sets, may be generalized and/or extended in a variety of different ways. For example, finite sets may be converted into "number systems," by endowing them with group or ring structure. Another example is given by rings of *adeles,* which generalize the completion $\mathbb{Q}_p$ of $\mathbb{Q}$. Examples more directly relevant to discrete causal theory are *ordinal numbers* and *cardinal numbers,* which generalize naïve "counting" of finite sets in different ways. It is sometimes useful to regard directed sets as generalizations of ordinal numbers; for example, this viewpoint is useful in the theory of *relative directed sets over a fixed base,* introduced in Section 4.6. From this perspective, Sorkin's phrase *"order plus number equals geometry"* "almost" says that *"ordinal plus cardinal equals geometry,"* since the partial orders involved in causal set theory generalize the total orders defining ordinal numbers, while the "quantity," or "measure," or "scale," represented by Sorkin's use of the word "number," refines the "pure size" meaning associated with cardinal numbers.

Quantum-theoretic significance of $\mathbb{C}$. A common heuristic is to associate the real numbers $\mathbb{R}$ with "classical physics," and the complex numbers $\mathbb{C}$ with "quantum physics." For example, the reader may recall that the use of *real* probabilities in the "toy dynamics" discussed in Section 1.8 is what distinguishes this choice of dynamics as a *classical stochastic theory.* A natural, and indeed inevitable, question arising in this context is *why,* at a fundamental level, the transition from "classical" to "quantum" is accompanied by a transition from $\mathbb{R}$ to $\mathbb{C}$. Most attempts to address this question involve significant structural assumptions that narrow the focus to the specific properties of particular "state spaces." For example, a 2001 paper of Caves, Fuchs, and Schack [CF02] explains why complex Hilbert spaces, but not real or quaternionic Hilbert spaces, support "reasonable behavior" for density operators in quantum theory.[45] A common "continuity" argument in favor of the use of complex numbers in quantum theory is that the unitary operations representing time evolution in this context should possess square roots, enabling the "time interval" under consideration to be subdivided.[46] Obviously, arguments of this type are generally irrelevant in the discrete causal setting, which does not involve the interpolative property.

---

[44] Discrete causal theory raises the possibility of much more natural types of "hyperquantum theories," defined by adding additional levels of hierarchy "above the level of kinematic schemes." This topic is revisited in Section 7.10.

[45] Another virtue of the paper [CF02] is that it supplies a spectrum of useful references on the subject.

[46] For example, this argument is raised in the course of an illuminating general discussion of the subject on Scott Aaronson's blog on quantum information theory.

However, discrete causal theory itself offers interesting insight into the quantum-theoretic role of $\mathbb{C}$, and also suggests that $\mathbb{H}$ and $\mathbb{O}$ remain worth considering in similar roles. The basic argument, which applies most naturally to the path summation approach to quantum theory in the background independent setting, is that one should not "artificially discriminate" among "evolutionary pathways" for classical histories, by assigning them weights of different *magnitudes* in path sums. This immediately focuses attention on *spheres,* or close analogues of spheres, as the natural target objects for the "phases" of these pathways, since spheres are the prototypical examples of "spaces whose elements all possess the same magnitude." The desire to preserve the algebraic structure associated with "splicing together evolutionary pathways" then leads to the consideration of "multiplicative structures on sphere-like objects," while the procedure of path summation requires "additive structures" on larger objects containing these "spheres." Without attempting to be precise, these very general reflections lead immediately to the consideration of the 1-sphere S^1, viewed as a subobject of $\mathbb{C}$, the 3-sphere S^3, viewed as a subobject of $\mathbb{H}$, and the 7-sphere S^7, viewed as a subobject of $\mathbb{O}$. Coincidentally or not, S^1 *is* the target of Feynman's phase map in his original description of the path summation approach [FE48]. In the discrete setting, of course, it is more natural to consider subobjects of these spaces that are not real manifolds, or possibly abstract analogues of such spaces. These rather speculative ideas are revisited in Section 6.7, in the context of adapting the path summation approach to the discrete causal setting. Note that *none of this reasoning has anything to do with the "completeness" of such number systems, or suggests that they should serve as basic structural scaffolding for spacetime.*

The right-hand column of the table in Figure 2.9.2, which lists the order-theoretic structure of the number systems appearing in the table, reveals how the buildup from $\mathbb{N} - \{0\}$ to $\mathbb{R}$, and beyond, first obscures, then destroys, the order-theoretic significance of the term "number." The notion of "quantity" suffers a similar fate, at least from a physical perspective, since cardinality provides much too coarse a description to support a useful notion of scale in continuum-based settings. Given the ongoing troubles of continuum-based theories, as well as the relatively barren physical role of the quaternions and octonions to date, it is worth considering the possibility that this progression of "more-and-more-complete" number systems is *simply not well-suited to describing fundamental physics.* In particular, as noted above, the general quantum-theoretic virtues of the field of complex numbers $\mathbb{C}$ do not seem to depend on its "completeness." For the simplest number systems, order-theoretic information, *together with* sufficiently-refined notions of quantity, are embodied in the same objects; for example, in the set of natural numbers $\mathbb{N}$, and in subsets of $\mathbb{N}$. Historically, this is one reason why the distinctions between ordinals and cardinals were not clearly recognized until the late nineteenth century. In the present context, $\mathbb{N}$ stands in stark contrast to larger and more complicated number systems such as $\mathbb{C}$, which fail to preserve suitable notions of order and quantity, at least in physical settings. Fortunately, the discrete directed sets central to discrete causal theory, while they are much more complicated than the natural numbers $\mathbb{N}$, at least share with $\mathbb{N}$ the important feature of supporting useful versions of *both* concepts. Among other important consequences, this is why metric recovery "works" in the discrete setting.

Physical objections to continuum-based theories. It is useful to briefly focus specific attention on a few of the many objections to the use of $\mathbb{R}$ as a basic source of structure in fundamental physics. Some of these objections involve technical issues, while others are more conceptual in nature, or arise from the theory and practice of experimental science. The persistence of these problems suggests that the mainstream modern physics community should seriously consider devoting a more equitable proportion of its efforts to alternative approaches, which can potentially avoid these problems entirely.

1. *Divergence issues.* Many of the theoretical problems of continuum-based theories of fundamental physics involve divergences that arise precisely because certain quantities are permitted to become arbitrarily small. This is true, in particular, of general relativity and quantum field theory. It is especially difficult to construct continuum-based background independent quantum theories that avoid such divergences. These problems are deeply-rooted, and remain mostly intractable, despite several generations of intensive effort by the world's foremost physicists. There is no convincing evidence that they can be surmounted or circumvented without a significant change of structural paradigm. Similar problems are to be expected in almost any theory in which $\mathbb{R}$ plays a substantial role.
2. *Lack of natural scale.* Real manifolds do not possess natural scale data. In particular, different coordinate charts on a real manifold X yield different "sizes" for a given subset. The only natural way to measure the "quantity" of a subset in this context is by its cardinality, which is much too coarse to provide a meaningful notion of scale in physical settings. For example, $\mathbb{R}^n$ has the same cardinality as $\mathbb{R}$ for any positive integer n. This means that scale data must be *supplied artificially* by the addition of auxiliary information, such as a metric. In the context of general relativity, this implies that metric recovery from causal structure is possible only up to a conformal factor, as established by Malament's theorem.
3. *Experimental discreteness.* The choice to discard continuum-based assumptions, even without prior theoretical justification, has proven strikingly successful historically. The prototypical example is Planck's solution of the blackbody radiation problem, which illustrates how divergence issues, arising from an irrelevant assumption of underlying continuum-based structure, can be cured by changing to a discrete paradigm. More generally, quantum theory has already replaced continua with discrete sets in a host of physical situations. Given this precedent, it seems imprudent to automatically retain continuum-based assumptions everywhere that experimental evidence has not already rooted them out.
4. *Discreteness arising from continuum-based assumptions.* Even continuum-based theories tend to predict important instances of discreteness in quantum-theoretic contexts. For example, in conventional quantum theory, certain operators on Hilbert spaces of functions over a real manifold may happen to possess discrete spectra of eigenvalues, leading to *derived* discreteness for the values of the associated observable quantities. At a more basic level, continuum-based approaches to quantum gravity and fundamental spacetime structure tend to *arrive at a form of fundamental discreteness,* via "quantization of spacetime." Most notably, loop

quantum gravity features "area" and "volume" operators that measure the "smallest meaningful units" of these observables. Thus, fundamental discreteness forces its way into the picture regardless of continuum-based assumptions.

5. *Discreteness via the philosophy of measurement.* It is impossible to directly establish, via measurement, the existence of a continuum of values of any observable quantity. In particular, one may always posit discrete structure at smaller scales, and such structure may be experimentally detectable, directly or indirectly. This realization, by itself, should not rule out continuum-based theories, but it should be considered as a mark against them, since it is awkward to invoke structure whose existence can never be established, even in principle. This view is closely related to a philosophical preference for background independence, since it treats as undesirable the practice of carrying along a "continuum background" in which discrete families of measurement values are taken to be "embedded." As the theory of metric recovery illustrates, assuming the existence of such a "background" can lead to difficulties that are more than merely aesthetic, such as the loss of natural scale.

Historical and sociological objections to continuum-based theories. After demonstrating the existence of serious experimental, mathematical, and/or logical problems associated with an existing approach to fundamental physics, or a class of such approaches, it can be instructive to consider possible historical and sociological factors that might have contributed to these shortcomings. This is a risky exercise, due to the hazard of historical bias. In particular, it is tempting to view the development of scientific thought in a *teleological* sense, as "leading up to" the present state of scientific understanding, and this type of presumption often generates serious misconceptions. For example, popular "explanations" of Zeno's paradoxes, which fortunately are ignored by serious physicists, mathematicians, and philosophers, focus on the largely irrelevant development of the rational and real number systems, boasting that "it is now known that an infinite number of terms can sum to a finite answer." In fact, the original statements of these paradoxes are *physical* in nature, and essentially question whether or not spacetime possesses the interpolative property. As another example, Riemann seems to have been as ready to consider "discrete manifolds" in the 1850's as he was to consider continua, but many physicists regard his work as "leading up to" general relativity. Despite these risks, it can still be useful to think about the development of science itself in terms of cause and effect. In this spirit, I mention the following factors as possibly contributing to the historical preeminence of continuum-based theories of fundamental physics.

1. *Pragmatism of continuum-based theories.* Until recently, continuum-based theories have been remarkably successful in providing solutions to the physical problems of greatest immediate interest, particularly in the context of applied science and engineering. In other words, these theories have prospered not because they seem likely to be *true,* but because they have been *useful.* A modern analogy is illuminating: during the construction of the standard model of particle theory, physicists were well-aware that the tools involved; in particular, background-dependent quantum field theories on Minkowski spacetime $\mathbb{R}^{3+1}$, were very

unlikely to address fundamental problems such as the unification of relativity and quantum theory.[47] However, these tools facilitated short-term progress, so they achieved temporary ascendancy. Failure to address deeper issues has largely squelched further progress over the last generation, and the lengthy run of success of continuum-based methods in fundamental physics since the time of Newton may well have ended with the completion of the standard model.

2. *Early ignorance of experimental discreteness.* Reasonable judgment regarding the adequacy of a structural paradigm is strongly influenced by the results of experiment and observation. In this regard, the last century has left continuum-based approaches to fundamental physics on far shakier ground than previously, due to recognition of numerous discrete phenomena via advances in microtechnology, which spurred the development of ordinary quantum theory. By this measure, early preeminence of continuum-based theories may be partly attributable to early ignorance of such discrete behavior at small scales.
3. *Early lack of structural alternatives to the continuum.* The continuum-based focus of early-modern physics was likely influenced by an absence of recognized mathematical alternatives to the real number system as a source of basic structure. Many structurally promising and physically suggestive alternatives arising in information theory, order theory, graph theory, and category theory were nonexistent or unrecognizable. Of course, the formal properties of $\mathbb{R}$ were not explicitly understood during this period either, but this technical imprecision seems to have had little effect on the conceptual development of physics.
4. *Early lack of computational alternatives to continuum-based techniques.* The predominance of continuum-based methods may be partly attributed to a lack of computational alternatives to techniques from calculus, differential equations, and other areas of mathematics involving real analysis. Computational science was in its infancy when modern physics arose. Discrete models involving more than a few elements would have seemed computationally intractable even if they had been considered conceptually promising. For example, Riemann might have had difficulty actually *studying* examples of "discrete manifolds," whatever his estimation of their fundamental merits. By contrast, real analysis is remarkably congenial to the computational limits of the unaided human brain, regardless of its ultimate physical relevance.

2.10 The Philosopher's Peril

Scientific philosophy. A significant task of Chapters 3 and 4 of this book is to analyze the *physical plausibility* of various mathematical models for encoding discrete causal structure, and of axiomatic systems governing these models. I would prefer to regard much of this analysis as "appeals to the self-evident," i.e., as common sense, but

[47] Of course, there are important connections between Yang–Mills theory and general relativity in the context of loop quantum gravity.

objectively speaking, it belongs to the field of "scientific philosophy." This field, of course, is one of the principal avenues whereby scientists and philosophers alike have made fools out of themselves since antiquity. Aristotle, for example, still receives what is probably an undeserved degree of ridicule for his unfortunate non-empirical conclusions regarding falling bodies. However, there is no shortage of other examples, both ancient and modern. Perhaps partly as a response to this, many modern physicists have taken the view that attempting to formulate novel approaches to fundamental physics based on "physical intuition" and "general principles" is a waste of time, and instead favor the more conservative approach of trying to first match experimental data before drawing any philosophical conclusions.[48] At an individual level, this strategy is pragmatic, since such efforts frequently achieve modest success, while more ambitious approaches usually fail. However, these "successful" models are often merely incremental updates of previous models, and suffer from the same obvious foundational issues as their progenitors. They also tend to exhibit what might be referred to as the "more-and-more disease," in which a good idea from an established theory is extended beyond its scope of applicability. Historically, this has led to spurious consideration of more and more epicycles, particles, symmetries, dimensions, and so on, often accompanied by less and less progress in basic understanding.

Of course, *ignoring* experimental data is even worse; doing so pits the philosopher against the Almighty, a losing proposition.[49] Perhaps the sensible middle ground is to avoid choosing sides between science and scientific philosophy at all. Einstein certainly harbored a healthy respect for scientific philosophy, remarking that the deep physical principles underlying the natural world cannot be logically deduced, but must be reached by *"intuition, resting on sympathetic understanding of experience"* [EI34]. While mostly above reproach in his own right, Einstein is sometimes cited as a misleading exception to the "rule" that intuition is relatively useless compared to close grappling with experimental evidence. This rule seems valid, however, only if intuition is allowed to operate unconstrained by what is actually known. If the "sympathetic understanding of experience" is taken into consideration, then exceptions accumulate rapidly. For example, the Lagrangian and Hamiltonian formulations of mechanics and the path summation approach to quantum theory are all based on intuition involving deep general principles. Deep general ideas about spontaneous symmetry breaking, including the Higgs mechanism, were formulated years before acquiring a specific use in the Glashow–Weinberg–Salam electroweak theory. Non-abelian gauge theories were developed on basic structural grounds long before they were used to describe the electroweak and strong interactions. These examples sug-

[48]Hawking has gone so far as to remark that "philosophy is dead." While I disagree with this statement at face value, Hawking was primarily expressing a low opinion of the scientific competence of the philosophical community, *not* suggesting that philosophical issues themselves are vacuous. Ironically, Malament works as a philosopher.

[49]Erdös remarked facetiously that life is a game against the "Supreme Fascist," his fanciful conception of God. The human contestant can never score any points, but can keep the S.F.'s score as low as possible, by avoiding error.

gest that even intuition has its merits.[50] In any case, it may be healthy to balance the philosophical skepticism of the modern physics community, which has not enjoyed the privilege of seeing any revolutionary advances, with the outlook of a more fortunate generation. For example, Hermann Weyl expressed the following viewpoint in the introduction to his classic text *Space Time Matter* [WE52]:

> *And now, in our time, there has been unloosed a cataclysm which has swept away space, time, and matter, hitherto regarded as the firmest pillars of natural science, but only to make place for a view of things of wider scope, and entailing a deeper vision. This revolution was promoted essentially by the thought of one man, Albert Einstein... Philosophy, mathematics, and physics have each a share in the problems presented here... I shall only touch lightly on the philosophical implications, for the simple reason that in this direction nothing final has yet been reached, and that for my own part I am not in a position to give such answers to the epistemological questions involved as my conscience would allow me to uphold... As things stand today... the separate sciences... should follow in good faith the paths along which they are led by reasonable motives proper to their own peculiar methods and special limitations. The task of shedding philosophic light onto these questions is nonetheless an important one... This is the point at which the philosopher must exercise his discretion. If he keeps in view the boundary lines determined by the difficulties inherent in these problems, he may direct, but must not impede, the advance of sciences whose field of inquiry is confined to the domain of concrete objects. Nevertheless, I shall begin with a few reflections of a philosophical character...* (pp. 2–3)

Weyl's carefully balanced perspective respects the necessity for caution in the practice of philosophy in the physical sciences, but never questions the legitimacy or importance of the discipline itself. Similar acknowledgment of the role of scientific philosophy has remained relatively mainstream throughout most of the history of modern science, despite current prejudices. This is not very surprising, because the subject involves questions of great scope and depth, whose difficulties remain as formidable, and whose consequences remain as important, as ever. Ultimately, the peril of squarely facing such difficulties must be accepted if these questions are ever to be adequately addressed.

Experimental challenges. In the last generation or so, an unfortunate practical limitation has grown increasingly prominent in fundamental physics: new experimental results in the realm of high energy particle theory have become almost prohibitively difficult and costly to obtain, thereby *forcing* philosophical substitutes to play a greater role in the way physics is done. For example, I have already mentioned the controversies surrounding the anthropic principle and the multiverse, in the context of string theory. Over much of the previous century, experiment decided the fate of many theoretical approaches within a few years of their inception. Today this is less-often true; a large proportion of theories that can be readily dismissed on experimental grounds are manifestly unworthy of attention in the first place, while most of the "interesting" theories are so difficult to test definitively that even decades-long research projects, and vast, internationally-funded engineering operations, cannot reliably decide their viability. This remains true, in particular, of string theory and

[50]The mathematical reader will recognize here a slightly facetious reference to Gordan's reluctant endorsement of Hilbert's "theology."

loop quantum gravity. Under these conditions, every possible source of insight into the conceptual integrity and technical viability of a theory is useful, including knowledge of where the theorist stands on philosophical grounds. Such information is far from definitive, since some great physicists have been suspect philosophers, while many outstanding philosophers have been terrible physicists. In this book, however, I wish to provide the reader with a viewpoint as conceptually comprehensive as possible. Hence, I devote the remainder of this section to the risky task of outlining some crucial conceptual and philosophical underpinnings of discrete causal theory, as I choose to approach it. The reader should be aware that important elements of this philosophy differ significantly from certain viewpoints associated with previously-existing versions of discrete causal theory.

Six basic principles. The following basic principles of scientific philosophy help to frame the overall viewpoint underlying the version of discrete causal theory developed in this book. In stating these principles, I make no attempt to be original; most of them fall well within the scientific mainstream. I also make no attempt to be definitive; it is possible to add or subtract a statement or two from this list without substantially altering the overall viewpoint. The remarks accompanying these statements provide illustrative examples, and indicate some of the places in the book where these principles are employed.

1. *Physics should seek not to prescribe what may be, but to describe and explain what is.* As explained in Section 2.1, the distinction between prescription and description is well-illustrated by comparing certain aspects of general relativity and discrete causal theory. General relativity treats the mathematical structure used to model classical spacetime as *prescribing* which pairs of events *may be* causally related. As noted in Section 1.3, this leads to "awkward counterfactual speculation" about actual events. Discrete causal theory, on the other hand, treats such structure as *describing* which pairs of events *are* causally related. Some might argue, with a degree of justification, that general relativity is merely *misinterpreted* along these lines,[51] but this is indisputably the mainstream interpretation. Of course, some descriptions are more satisfying than others. An ideal description should *explain,* or "render intelligible," the associated formalism, by showing it to faithfully represent clear fundamental principles.[52] As outlined in the discussion of causality conditions in Section 2.7, the descriptive philosophy avoids meaningless inconsistencies, such as "time-travel paradoxes." It also abstains from unjustified assumptions, such as the presumed steady state of the universe before Hubble's observations. Looking ahead to Chapters 3 and 4, the axioms of *transitivity* (TR) and *interval finiteness* (IF), which feature prominently

[51] Here, I am thinking about influential scientists such as Rovelli [RO04], who offers a viewpoint about certain aspects of relativity that may give too much credit to the theory itself. I agree with much of the physical content of Rovelli's viewpoint, without necessarily agreeing that Einstein's theory itself suffices to adequately embody it.

[52] Much has been written about whether or not physics should be *expected* to be intelligible, whatever the meaning of the term. In any case, it would be difficult to formulate a clearer fundamental principle than the causal metric hypothesis (CMH).

in the existing literature, are worrisomely prescriptive. For this reason and others, neither of these two axioms plays a role in the version of discrete causal theory developed in this book. This illustrates the fact that shifting to the discrete causal paradigm does not automatically cure prescriptive issues; the specific choice of axioms is also important. In a broader context, prescriptive issues are often related to a lack of perfect background independence, and more generally to the use of structures that are arbitrary, rather than structures possessing a universal property, as discussed in Section 1.6.

2. *Mathematics and physics are distinct; each informs the other.* Mathematical structures in physics should be chosen for their conceptual merits, not for their familiarity or convenience. This principle, which may be shortened to the phrase *"concept over convenience,"* recalls Gauss' *"notions over notations."* Historically, this practice has led not only to good physics, but also to interesting mathematics, while the mathematical community has returned the favor by introducing concepts and methods whose physical significance has only been appreciated much later. Care must be taken to distinguish between mathematical and physical properties. An unfortunate byproduct of the long-standing success of continuum-based theories in physics has been the automatic and unjustified attribution of certain *mathematical* properties of the continuum, such as the interpolative property and the completeness property discussed in Section 2.9, to physical spacetime. An important example of distinguishing between mathematical and physical properties in discrete causal theory is offered by the *independence convention* (IC), introduced in Section 3.7. In this case, the two properties to be distinguished are mathematical *irreducibility* and physical *independence* of relations between pairs of elements in a directed or multidirected set.

3. *Basic structural concepts are crucial.* As mentioned in Section 2.9, it is instructive to consider the poverty of structural alternatives to continuum-based geometry available to physicists during the early 20th century, when relativity and quantum theory were developed. Many physically suggestive ideas from fields such as order theory, graph theory, information theory, computer science, category theory, algebra, and algebraic geometry, were not yet known. Even group theory faced a difficult reception, as evidenced by Wigner's description of Schrödinger's *"gruppenpest."* The twenty-first century scientific community is much better equipped, at least on paper, to follow up Einstein's intuition that physics is essentially structural in nature, brilliantly vindicated by general relativity, but largely unconsummated thereafter. Many of the structural ideas appearing in this book have roots in modern algebra, particularly in the work of Alexander Grothendieck. Especially important is Grothendieck's *relative viewpoint* (RV), formally introduced in Section 3.8, and applied in Chapters 4–7.

4. *Local and global properties must be properly distinguished.* The history of physics is littered with errors resulting from specious local-to-global reasoning and dubious extrapolation across scales. Often, such errors arise from failure to recognize the limitations of "obvious" observations, such as the apparent motionlessness of the earth, or the apparent flatness of spacetime in the vicinity of the earth, or the apparent possibility of assigning definite values of position and momentum simul-

taneously to macroscopic material bodies. Particularly troublesome are "local" conditions adopted without recognition of their global consequences. Looking ahead to Chapter 4, the axiom of interval finiteness (IF), sometimes mislabeled as "local finiteness" in the literature, prescribes *global* structure to an uncomfortable degree. This provides another illustration of why the specific set of axioms chosen for discrete causal theory is of crucial importance.

5. *The nature of experimentation has theoretical significance.* A specific instance of this principle, involving the uncomfortable status of continuum-based theories with regard to the philosophy of measurement, was mentioned in Section 2.9. More generally, besides attempting to explain specific experimental results, theorists should consider *general demands and prohibitions* associated with the experimental method. For example, in Section 1.3, I noted the unavoidable scientific role of directed relationships between experimental conditions and results. The nature of experimentation also favors axiomatizing local rather than global properties, since the latter may be experimentally inaccessible. For example, in the context of relativity, it is reasonable to assume that classical spacetime is four-dimensional, since dimension is defined locally, but it is unreasonable to assume a specific global topology. Of course, conclusions about large-scale topology could conceivably be *derived* on dynamical grounds, or inferred from unlikely observational scenarios, such as "circles in the sky."

6. *Censor the fatal, not the merely unexpected.* A reasonable facet of theory-building is to impose conditions "censoring" properties that are so qualitatively contrary to observation that any theory exhibiting them is immediately discredited. More succinctly, it is reasonable to "ignore the irrelevant." This is the rationale behind the discussion of narrowing the focus of the classical causal metric hypothesis (CCMH) to "physically relevant directed sets," in Section 2.2. More generally, this principle defines the boundary of the prohibition against "prescription" mentioned above, by permitting the *proscription* of "fatal phenomena." However, this idea must be applied with great care, due to the limitations of human judgment and imagination. Planck's approach to black-body radiation, eliminating all but a discrete set of emission frequencies, to avoid the fatal ultraviolet catastrophe, is an example of justified censorship. However, Einstein's fixing of the "cosmological constant," to achieve his *expectation* of a steady-state universe, is not. In the context of discrete causal theory, the axioms of *transitivity* (TR) and *interval finiteness* (IF), already mentioned above, censor *nonfatal* phenomena in problematic ways; the former by ignoring distinctions among certain modes of influence between pairs of events, and the latter by drastically constraining the global structure of classical spacetime.

Ten qualitative assumptions of discrete causal theory. The six principles listed above are very general, and provide only a partial overview of a broad approach to doing science. They *inform* the developments described in this book in important ways, but do not come close to determining them. Hence, it is useful to gather together some more-specific assumptions underlying the version of discrete causal theory developed here. Many of these assumptions have already been mentioned and applied, at least implicitly, in previous sections, but they have not yet appeared explicitly in one place. Among other advantages, this listing provides the opportunity to revisit some of the conceptual topics discussed in the questions and answers at the end of Chapter 1, with the benefit of additional information from the present chapter.

1. *There is no physical continuum.* In Section 2.9, I discussed some of the general shortcomings of physical models that rely on the real number continuum $\mathbb{R}$ as a source of basic structure. Discrete causal theory treats these objections as sufficient to completely rule out the use of $\mathbb{R}$ in this role.

2. *The physical universe is basically discrete.* Having dispensed with $\mathbb{R}$, along with derivative structures such as real manifolds, it is necessary to offer an alternative structural paradigm for fundamental physics. Discrete structures seem to furnish more-natural and more-promising physical models. It is important to emphasize that discreteness is *not* "the exclusive physical alternative" to the continuum; most of the structural paradigms one could choose to explore are neither continuous nor discrete. However, there are good reasons why most modern approaches to fundamental physics rely heavily on one or both of these extremes: they possess special properties that are either physically suggestive, or amenable to mathematical analysis, or both. From the viewpoint of discrete causal theory, the conceptual advantages of discrete models far outweigh the mathematical convenience of continuum-based models.

3. *Physics is about cause and effect.* This statement is, of course, a paraphrase of the causal metric hypothesis (CMH), which is the subject of the present chapter, and one of the main themes of the book. The first two chapters have presented a number of different versions, paraphrases, and shades of meaning of this hypothesis. A crucial part of the picture, not yet discussed in detail, is the *quantum causal metric hypothesis* (QCMH), which requires additional theoretical background before it can be stated in a precise manner. The necessary developments are carried out in Chapters 5–7.

4. *Classical spacetime may be modeled in terms of directed sets.* This statement is a paraphrase of the classical causal metric hypothesis (CCMH). As explained in Section 2.2, discrete classical causal structure may be modeled, at the local level, in terms of directed relationships between pairs of elements, and a directed set is merely a collection of such related pairs, considered as a single object. One advantage of describing causal structure in this way is that it avoids prescribing dubious global properties, such as transitivity (TR). More-detailed discussion of axiomatic systems for directed sets appears in Chapters 3 and 4.

5. *Quantum spacetime may be modeled in terms of multidirected sets.* This is the essence of the quantum causal metric hypothesis, although the details are postponed until Chapter 7. Multidirected sets are natural generalizations of directed sets, in which a given pair of elements may have multiple relations between them in either or both directions. The elements in discrete quantum causal theory represent classical histories, and the corresponding relations represent relationships between classical histories, i.e., co-relative histories. Similarly, the multidirected

sets arising in this context represent kinematic schemes. The reason why multi-directed structure, and not merely directed structure, is necessary in this context, is because of a subtle technical property of directed sets that permits the existence of multiple distinct co-relative histories between a given pair of classical histories.

6. *Classical histories are generally nontransitive.* The conceptual basis of this assertion is merely that direct and indirect relationships are physically different; the details are explained in Chapter 3, particularly in Section 3.9. Using a nontransitive binary relation, direct causation may be modeled by an individual relation between a pair of events, while indirect causation may be modeled by chains of relations. In general relativity, influence is taken to flow along causal curves, which implies that *every* instance of causation is indirect, due to the interpolative property of $\mathbb{R}$. However, in the discrete setting, both direct and indirect influences are possible, and it is necessary to use models that enable distinction between the two. Following this reasoning, the possibility of direct causation is an *essentially new feature,* introduced by exchanging continuum-based models for discrete ones.

7. *Classical histories are locally finite, i.e., star finite* (SF), *but not necessarily interval finite* (IF). The motivation for imposing a local finiteness condition on discrete causal structure is, roughly speaking, that one expects individual elements to possess a "finite size" in the discrete context. Hence, one faces convergence issues if an infinite number of elements are permitted to coexist in a "local region." The subject of local behavior in discrete causal theory is examined in much more detail in Chapter 4, but I briefly elaborate on the need for a local finiteness condition here. The metric recovery results discussed earlier in the present chapter imply that if natural scale data can somehow be derived from causal structure, then one can recover the apparent geometric properties of relativistic spacetime at ordinary scales. The obvious way to proceed in the discrete context is by "using local combinatorial data to determine scale," generalizing the causal set prescription. To avoid physical pathologies, such as the instantaneous expansion of a minimal region of spacetime into an infinite volume, it is natural to impose the condition that "each element is directly related to only a finite number of other elements." This local finiteness condition, called *star finiteness,* is formally introduced in Section 4.4. Unfortunately, the term "local finiteness" is sometimes used in the literature to denote the very different condition of *interval finiteness*, which is not a local condition at all, and which permits the very type of physical pathologies described above.

8. *The relative viewpoint is indispensable.* As explained in Section 1.5, Grothendieck's relative viewpoint (RV) embodies the philosophy that objects should not be analyzed in isolation, but should be studied along with their natural relationships. At the classical level, this viewpoint leads to the definition of the *relation space* $\mathcal{R}(D)$ over a directed set D, viewed as a discrete causal classical history. Relation space was briefly introduced in Section 1.5, and is studied in detail in Chapter 5. At the quantum level, the relative viewpoint leads to the theory of *co-relative histories* and *kinematic schemes,* developed in Chapters 6 and 7.

9. *"Classical" and "quantum" may be understood as levels of structural hierarchy.* As described in the first few pages of the book, discrete causal theory exhibits an attractive self-similarity, called *iteration of structure* (IS), in which quantum structure naturally occupies a level of mathematical hierarchy above that of classical structure. A hint of this relationship is evident even in the ordinary Hilbert-space approach to quantum theory, which shifts the focus from individual states, and relationships among them, to spaces of states, and operators on these spaces. Category theory provides a more general analogy, elaborated in a striking manner by Christopher Isham [IS05]: "quantization" corresponds roughly to passage from "elements and relations" to "objects and morphisms." What is special, and perhaps unique, about discrete causal theory, in this context, is that its "higher-level quantum objects" possess essentially the same type of structure as its "lower-level classical objects." A concise way to paraphrase this viewpoint is to say that *"classical physics is about relationships between pairs of events; quantum physics is about relationships between pairs of histories."*
10. *Quantum dynamics arises from generalized path summation.* The import of this assertion is that one particular approach to quantum theory; namely, Feynman's *path summation approach,* is sufficiently general to apply to discrete causal theory, for which other popular approaches are inadequate. This is because most approaches to quantum theory take for granted a great deal of structure arising from the properties of the real and complex numbers, which is generally unavailable in the discrete causal setting. For example, ordinary quantum theory and quantum field theory begin with Hilbert spaces of complex-valued functions over real manifolds, which depend on constructions that discrete causal theory treats as emergent.[53] By contrast, the path summation approach may be abstracted to apply to any situation involving families of directed relationships between pairs of classical histories.

References

[HA76] S. W. Hawking, A. R. King, and P. J. McCarthy. *A new topology for curved space-time which incorporates the causal, differential, and conformal structures*. Journal of Mathematical Physics, **17**, 2, pp. 174–181, 1976.

[MA77] David B. Malament. *The class of continuous timelike curves determines the topology of spacetime*. Journal of Mathematical Physics, **18**, 7, pp. 1399–1404, 1977.

[WH98] John Archibald Wheeler and Kenneth W. Ford. *Geons, Black Holes, and Quantum Foam: A Life in Physics*. W. W. Norton and Company, New York, 1998.

[IS05] Christopher Isham. *Quantising on a Category*. Foundations of Physics, **35**, 2, pp. 271–297, 2005. arXiv preprint: http://arxiv.org/pdf/quant-ph/0401175v1.pdf.

[SO12] Rafael Sorkin. *Toward a Fundamental Theorem of Quantal Measure Theory*. Mathematical Structures in Computer Science, **22**, 05 (special issue), pp. 816–852, 2012. arXiv preprint: http://arxiv.org/pdf/1104.0997v2.pdf.

[53] An alternative, of course, is to abstract, elevate and generalize this algebraic structure; this is what is done in Connes' *noncommutative geometry.*

[CA15] S. Carlip. *Dimensional reduction in causal set gravity*. Classical and Quantum Gravity, **32**, 23,232001, 2015. arXiv preprint: http://arxiv.org/pdf/1506.08775v3.pdf.

[RO04] Carlo Rovelli. *Quantum Gravity.* Cambridge Monographs on Mathematical Physics. Cambridge University Press, 2004.

[ZE64] E. C. Zeeman. *Causality Implies the Lorentz Group*. Journal of Mathematical Physics, **5**, 4, pp. 490–493, 1964.

[CA04] Sean Carroll. *Spacetime and Geometry: An Introduction to General Relativity* Addison Wesley, 2004.

[HE73] S. W. Hawking and G. F. R. Ellis. *The large scale structure of space-time.* Cambridge Monographs on Mathematical Physics. Cambridge University Press, 1973.

[MA99] Juan Maldacena. *The Large N Limit of Superconformal Field Theories and Supergravity*. International Journal of Theoretical Physics, **38**, 4, pp. 1113–1133, 1999.

[EMM12] George F. R. Ellis, Roy Maartens, and Malcolm A. H. MacCallum. *Relativistic Cosmology.* Cambridge University Press, 2012.

[BLMS88] Luca Bombelli, Joohan Lee, David Meyer, and Rafael Sorkin. *Bombelli et al. Reply to Comment on "Space-Time as a Causal Set."* Physical Review Letters, **60**, 7, pp. 656, 1988.

[WA84] Robert M. Wald *General Relativity.* University of Chicago Press, 1984.

[BA12] Julian Barbour. *Shape Dynamics: an Introduction.* Quantum Field Theory and Gravity: Conceptual and Mathematical Advances in the Search for a Unified Framework, Edited by Felix Finster, Olaf Müller, Marc Hardmann, Jürgen Tolksdorf, and Eberhard Zeidler. arXiv preprint: http://arxiv.org/pdf/1105.0183v1.pdf.

[PE10] Roger Penrose. *Cycles of Time.* Vintage Books, New York, 2010.

[TH07] Thomas Thiemann. *Modern Canonical Quantum General Relativity.* Cambridge Monographs on Mathematical Physics. Cambridge University Press, 2007.

[BS07] Antonio N. Bernal and Miguel Sanchez. *Globally hyperbolic spacetimes can be defined as "causal" instead of "strongly causal"*. Classical and Quantum Gravity, **24**, 3, 745, 2007. arXiv preprint: http://arxiv.org/pdf/gr-qc/0611138v1.pdf.

[BZ15] Ali Bleybel and Abdallah Zaiour. *A general theorem on temporal foliation of causal sets*. Preprint, 2015. arXiv preprint: http://arxiv.org/pdf/1508.01052v1.pdf.

[MRS06] Seth A. Major, David Rideout, and Sumati Surya. *Spatial Hypersurfaces in Causal Set Cosmology.* Classical and Quantum Gravity, **23**, 14, pp. 4743–4751, 2006. arXiv preprint: http://arxiv.org/pdf/gr-qc/0506133v2.pdf.

[BO89] Luca Bombelli and David Meyer. *Origin of Lorentzian geometry*. Physics Letters A, **141**, 5-6, pp. 226–228, 1989.

[MP06] Keye Martin and Prakash Panangaden. *A Domain of Spacetime Intervals in General Relativity*. Communications in Mathematical Physics, **267**, 3, pp. 563–586, 2006.

[PS11] Onkar Parrikar and Sumati Surya. *Causal topology in future and past distinguishing spacetimes*. Classical and Quantum Gravity, **28**, 15, 155020, 2011.

[BHS09] Luca Bombelli, Joe Henson, Rafael Sorkin. *Discreteness without symmetry breaking: a theorem*. Modern Physics Letters A, **24**, 32, pp. 2579–2587, 2009. arXiv preprint: http://arxiv.org/pdf/gr-qc/0605006v1.pdf

[CSS16] Hua Chen, Naoki Sasakura, and Yuki Sato. *Emergent classical geometries on boundaries of randomly connected tensor networks*. Preprint, 2016. arXiv preprint: http://arxiv.org/abs/1601.04232

[CF02] Carlton M. Caves, Christopher A. Fuchs, and Rüdiger Schack. *Unknown Quantum States: The Quantum de Finetti Representation.* Journal of Mathematical Physics, **43**, 9, pp. 4537–4559, 2002. arXiv preprint: http://arxiv.org/pdf/quant-ph/0104088v1.pdf

[FE48] Richard Feynman. *Space-Time Approach to Non-Relativistic Quantum Mechanics*. Reviews of Modern Physics, **20**, 2, pp. 367–387, 1948.

[EI34] Albert Einstein. *Essays in Science*. Philosophical Library, New York, 1934.

[WE52] Hermann Weyl. *Space Time Matter*. Dover, 1952.

Chapter 3
Causal Sets and Generalizations

Abstract. This chapter describes properties of *directed sets* and *multidirected sets,* which supply the basic structural content of discrete causal theory. Mathematically, these objects are usually called *directed graphs* and *directed multigraphs,* respectively, but set-theoretic terminology seems preferable for physical applications, despite the fact that the term "directed set" has a more specific conventional meaning. Directed sets and multidirected sets may be viewed as generalizations of *causal sets,* which have been studied as discrete models of spacetime since the 1980s. One of the main tasks of this chapter is to begin to describe some of the ways in which the version of discrete causal theory developed in this book *differs* from causal set theory, and to explain the reasons for these differences. Section 3.1 describes the preliminaries of causal set theory, along with some historical details. Section 3.2 explains how Sorkin's *ansatz* for causal sets, *"order plus number equals geometry,"* may be viewed as an early version of the classical causal metric hypothesis. Section 3.3 briefly outlines quantum causal set theory, which serves as a point of comparison for the version of discrete quantum causal theory developed in Part II. Section 3.4 describes some topics in causal set dynamics and phenomenology, with a focus on Sorkin and Rideout's theory of *sequential growth dynamics.* Section 3.5 presents a formal axiomatic description of causal set theory. Section 3.6 introduces the more general classes of directed sets and multidirected sets necessary to describe the version of discrete causal theory developed in this book. Directed sets serve as models of classical spacetime, while multidirected sets supply the "higher-level" structure of configuration spaces of classical histories arising in the corresponding quantum theory. Section 3.7 describes basic structural notions for multidirected sets, such as *chains, antichains,* and *domains of influence.* Section 3.8 outlines useful analogies involving order theory and category theory, with emphasis on the contributions of Grothendieck. Section 3.9 explains why transitive binary relations, such as those used in causal set theory, are likely inadequate to model fundamental causal structure in the discrete setting. A few prominent physicists, such as David Finkelstein, have previously recognized this problem. Section 3.10 introduces the *causal relation,* which is a generally nontransitive binary relation that generates the transitive relation usually associated with causal structure in conventional contexts.

B.F. Dribus, *Discrete Causal Theory*, DOI 10.1007/978-3-319-50083-6_3

3.1 Causal Set Theory: Conceptual Roots

Causal sets as "discrete relativistic spacetimes." Causal set theory is an important version of discrete causal theory that emerged in the early 1980s in direct response to the metric recovery theorems of Hawking [HA76] and Malament [MA77]. Currently, it enjoys the attention of a small but highly active research community. Informally, causal sets are "as much like relativistic spacetime manifolds as possible," with the crucial exception of being discrete. Ultimately, discreteness itself introduces the possibility of new structural subtleties, absent in relativity, which the apparatus of causal set theory does not seem to completely capture. This leads to potential limitations in the theory, both at the conceptual and technical levels. Despite these issues, causal set theory does incorporate many, if not most, of the features necessary for an effective modern version of discrete causal theory. Hence, an overview of causal set theory serves as an excellent introduction for a more general study of the subject. This chapter begins with a concise description of causal set theory, then expands the picture to include the basic objects of discrete causal theory in general; namely, directed sets and multidirected sets.

Structural preview. Causal sets are discrete order-theoretic models of classical spacetime. The following definition of a causal set is a bit unwieldy, and requires substantial explanation, which is not completed until Section 3.5. However, I present it here at the beginning of the chapter, for easy reference.

Definition 3.1.1. *A* **causal set** *is a countable set C, equipped with a transitive, interval finite, irreflexive binary relation $\prec_{\text{CS}}$. Elements x, y of C represent spacetime events, and relations $x \prec_{\text{CS}} y$ in C represent causal relationships between pairs of spacetime events. C is also equipped with a discrete measure μ_{CS}, which assigns to each subset of C a volume equal to its number of elements in fundamental units, up to Poisson-type fluctuations.*

I refer to the binary relation $\prec_{\text{CS}}$ on C as the **causal set relation.** It plays a role analogous to that of the relativistic causal relation $\prec_{\text{GR}}$ appearing in Chapter 2. The subscript "CS" in the expression $\prec_{\text{CS}}$ stands for "causal set;" it is included to avoid confusion with other binary relations appearing in this book, which generally possess different properties. Analogous remarks apply to the discrete measure μ_{CS}, which I refer to as the **causal set measure.** Definition 3.1.1 makes implicit reference to six axioms for causal sets, which I call the *binary axiom,* the *measure axiom, countability, transitivity, interval finiteness,* and *irreflexivity.* Formal definition and description of these axioms is deferred until Section 3.5; in the present section, I merely provide some brief contextual remarks. The literature does not describe causal sets in exactly this way, but the information content is identical. The reason why causal sets may be classified as "order-theoretic models" is because the causal set relation $\prec_{\text{CS}}$ determines an *interval finite partial order* $\preceq_{\text{CS}}$ on C, in a sense made precise in Section 3.5. In particular, this means that causal sets are analogous to relativistic spacetime manifolds satisfying the *causal condition* described in Section 2.7. The causal set relation $\prec_{CS}$ itself is a *strict* partial order, analogous to the familiar "less than" relation $<$ on the integers.

The statement that the causal set measure μ_{CS} on C is *discrete* means, in this context, that singleton sets are assigned positive measure. Many different kinds of discrete measures exist, but the ones of most obvious relevance in the context of metric recovery are those that assign a finite volume to a subset of classical spacetime if and only if it contains a finite number of fundamental elements.[1] Rafael Sorkin's now-familiar phrase, *"order plus number equals geometry,"* invokes the simplest possible such measure; namely, the **counting measure**, which assigns one fundamental unit of volume to each element. In practice, Poisson-type fluctuations are permitted, in order to avoid systematic violations of Lorentz invariance. This topic is revisited briefly in Sections 3.2 and 3.5, and much more thoroughly in Section 4.5. In more sophisticated versions of discrete causal theory, the emergent volume assigned to a subspace of classical spacetime may be taken to depend in a nontrivial way on the details of its causal structure, as discussed in Section 2.8.

One of the principal technical limitations of causal set theory is that it ignores a large proportion of potentially relevant directed sets that are *not* causal sets; in particular, nontransitive acyclic directed sets. Figure 3.1.2 illustrates this deficit, using the *positive sequential kinematic scheme* $\mathbb{S}_{PS}$, introduced informally in Figure 1.1.1 of Chapter 1, as an example. The directed sets appearing in the large white nodes are causal sets, while those appearing in the large gray nodes are nontransitive acyclic directed sets. "Almost all" finite acyclic directed sets are nontransitive in an asymptotic sense, so causal set theory "ignores most of the information in $\mathbb{S}_{PS}$." Further, the subscheme of $\mathbb{S}_{PS}$ consisting of finite causal sets is made up almost entirely of physically irrelevant objects possessing only a "few generations between the beginning and end of time," as demonstrated by a graph-theoretic result of Kleitman and Rothschild [KR75]. This issue, which I refer to as the *Kleitman–Rothschild pathology,* is of limited concern in the classical context, where one is interested in particular "manifold-like" causal sets. However, it presents an a priori larger obstacle in quantum theory, where such irrelevant causal sets threaten to dominate path sums. Efforts to remedy this problem are not discussed in any detail in this book, since the version of discrete causal theory developed here is much less vulnerable to such issues. However, causal set theorists are aware of this problem, and have offered some possible strategies for addressing it.

A brief history of causal set theory. As explained Chapters 1 and 2, the basic physical questions that causal set theory attempts to address have roots long predating any formal mathematical notion of discreteness or order. In the generations leading up to the emergence of relativity and quantum theory, however, the overwhelming success of classical continuum-based methods in producing short-term progress in physics relegated serious examination of fundamental structural questions to the periphery of scientific thought. A few notable figures, such as Leibniz and Riemann, considered such ideas, but the scientific community as a whole followed more pragmatic,

[1] An example of a discrete measure that does *not* satisfy this condition is the measure on the set $\{2^{-n}\}_{n\in\mathbb{N}}$ defined by taking the volume of each singleton $\{2^{-n}\}$ to be 2^{-n}.

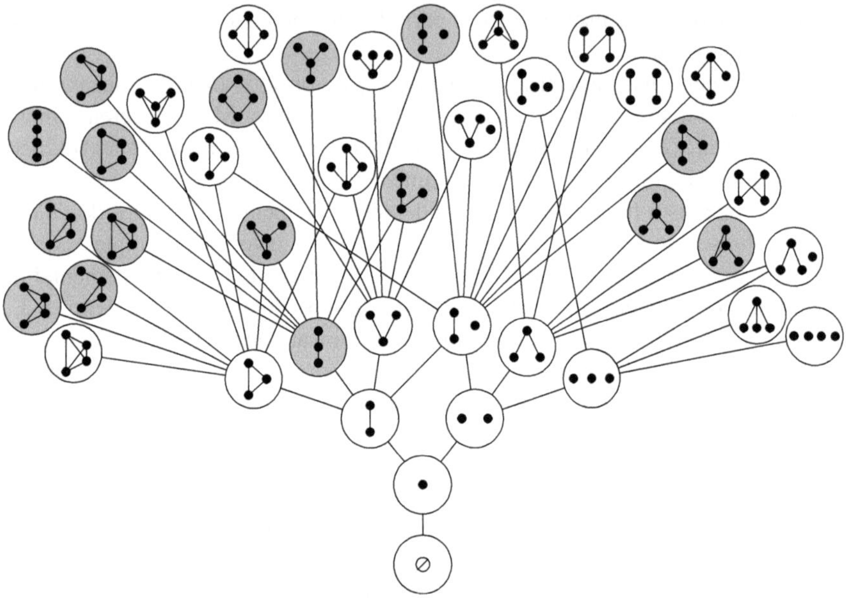

Figure 3.1.2. Positive sequential kinematic scheme $\mathbb{S}_{PS}$, including causal sets (*white*) and acyclic directed sets that are not causal sets (*gray*).

though less fundamental, lines. Relativity brought fresh scrutiny to basic structural issues, gradually elevating *causal structure* to its present central role. For historical background on the relativistic roots of causal set theory, it is difficult to improve upon the efforts of Sorkin. For example, in his 2005 article *Causal Sets: Discrete Gravity* [SO05], Sorkin details the isolation and elevation of causal structure in the relativity literature as early as the 1930s, and even quotes Einstein's misgivings about continuum-based geometry, and Riemann's early remarks about *discrete manifolds.* David Finkelstein [FI88] also provides useful historical context reaching back before 1950. Finkelstein's 1969 paper *Space-Time Code* [FI69] is perhaps the earliest "modern" paper on discrete causal theory, described by Sorkin [SO05] as having "adumbrated" the causal sets program.

The metric recovery theorems of Hawking [HA76] and Malament [MA77], discussed in Chapter 2, provided a striking new perspective regarding the scope of causal structure in general relativity. Almost immediately following the proof of these theorems, ideas very similar to causal set theory appeared in Myrheim's 1978 CERN preprint *Statistical Geometry* [MR78], and't Hooft's 1978 notes *Quantum Gravity: A Fundamental Problem and some Radical Ideas* [TH78]. Besides the strong motivation provided by the metric recovery theorems themselves, these efforts were spurred by the continued failure of continuum-based methods to provide a viable avenue toward a theory of quantum gravity, and the growing expectation that fundamental spacetime structure must involve discreteness at some level. Finally, "causal set theory proper" was introduced by Bombelli, Lee, Meyer, and Sorkin, in their

1987 paper *Space-Time as a Causal Set* [BLMS87]. The identification of this particular paper, rather than the earlier efforts cited above, as the "official beginning" of causal set theory, reflects the persistence of the authors in the *isolation and pursuit of a specific seminal idea,* and effective advocacy of this idea, rather than conceptual priority.[2] Indeed, Bombelli, Lee, Meyer, and Sorkin credit Finkelstein, Myrheim, and't Hooft's efforts with the statement that *"The picture of space-time as a causal set is by no means new,"* but go on to explain that Myrheim and't Hooft's proposals were *"undeveloped,"* and that Finkelstein's *"led to formulations in which the issues we would like to address here were not dealt with."* In any case, there is no doubt that Sorkin and his collaborators have done more to advocate and develop causal set theory than anyone else.

3.2 Causal Sets and the Causal Metric Hypothesis

Causal sets and metric recovery. It is convenient here to briefly revisit the metric recovery results described in Section 2.8, in order to explain their specific contribution to the causal sets program. The result of principal importance is the 1977 theorem of David Malament [MA77], relating the causal and conformal structures of classical spacetime in the context of general relativity. Hawking [HA76] had already established that *topological structure determines conformal structure* under suitable hypotheses; Malament showed, in turn, that *causal structure determines topological structure.* Combining these two results, Malament concluded that the only obstruction to the recovery of relativistic spacetime geometry from causal structure is a *lack of appropriate data about scale,* i.e., a missing conformal factor. In the continuum-based relativistic setting, the necessary "scale data" must be added to the picture artificially, since real manifolds possess no natural scale. However, natural scale data may be ascribed to *discrete* causal structure by combinatorial means. Of course, such structure does not *precisely* recover continuum-based geometry; but it can mimic continuum-based geometry on sufficiently large scales. Turning this idea around, discrete causal structure can provide a viable explanation for the observed properties of spacetime, *if* the fundamental scale is sufficiently small. Causal set theory adopts the "most straightforward" approach to replacing relativistic spacetime geometry with discrete causal structure. Of course, as explained in Section 3.1, this approach might be considered a little *too* straightforward, in the sense that it does not incorporate certain intrinsic structural subtleties arising from discreteness. However, it does capture the basic idea of a very fine "approximate geometry" emerging from discrete causal structure.

"Order plus number equals geometry." One of the advantages of discrete causal theory in general is that it admits a very succinct basic description. This conceptual

---

[2] By way of comparison, Lorentz, Poincaré, and others knew much of the content of special relativity before Einstein's 1905 paper, but it was Einstein who first grasped, isolated, and advocated the essential aspects of the theory in an effective manner.

elegance is fully realized in causal set theory, epitomized by Sorkin's phrase, *"order plus number equals geometry."* This phrase encapsulates the physical conclusions drawn by the early causal set theorists in response to the metric recovery theorems. The word "order" stands for the causal set relation $\prec_{\rm CS}$ on a causal set C, which determines a partial order $\preceq_{\rm CS}$, while the word "number" represents the scale data supplied by the causal set measure $\mu_{\rm CS}$ on C, which is analogous to Malament's "missing conformal factor." In particular, since this scale data may be ascribed to C itself, Sorkin's phrase expresses the view that the properties of classical spacetime are *essentially attributable to causal structure alone.* This is the reason why I have cited this phrase as an early instance of the classical causal metric hypothesis (CCMH).

The idea of using the natural "counting measure" to supply scale data appeared in the literature almost immediately after Malament's paper [MA77]. For example, in his 1978 preprint [MR78], Myrheim writes,

> *If spacetime is assumed to be discrete, then the counting measure is the natural measure, and the causal ordering is the only structure needed. Coordinates and metric may be derived as secondary, statistical concepts.* (page 1)

Later, in their 1987 paper [BLMS87], Bombelli, Lee, Meyer, and Sorkin express the conceptual content of the classical causal metric hypothesis very clearly, writing:

> *In this view volume is number, and macroscopic causality reflects a deeper notion of order in terms of which all the "geometrical" structures of space-time must find their ultimate expression.* (page 522)

These statements, which implicitly *identify* the causal set measure $\mu_{\rm CS}$ with the counting measure, historically precede the appearance of the caveat *"up to Poisson-type fluctuations"* in the later literature. Section 3.5 contains a more detailed discussion of this subtlety.

Causal set "hauptvermutung." As explained above, the discrete data encoded in a causal set can only *approximately* supply the causal and conformal ingredients of pseudo-Riemannian geometry involved in the metric recovery theorems. Hence, causal set theory hypothesizes a very fine "approximate geometry" emerging from discrete causal structure, not a precise geometry. However, it is important to emphasize that, from a basic ontological perspective, *causal set theory treats "less-fundamental" relativistic spacetime as an approximation of a "more-fundamental" causal set, not vice versa.* In other words, continuum-based geometry is regarded as merely a *large-scale smoothing* of the structure encoded by C and $\prec_{\rm CS}$. The same viewpoint applies to discrete causal theory in general. This distinction is especially relevant in studying the properties of "sprinkled" causal sets, which are induced by pre-existing pseudo-Riemannian structure, as described below. In particular, while such objects provide an invaluable supply of causal sets exhibiting "physically reasonable" properties for study purposes, one should not attribute undue significance to *any* class of discrete models whose structures are completely derived from the very continuum-based objects which they are ultimately intended to replace.

The extent to which a "manifold-like" causal set *uniquely* determines large-scale manifold structure remains, as far as I know, an outstanding issue.[3] The conjecture that an appropriate uniqueness result exists in this context is called the causal set *hauptvermutung*, or *fundamental conjecture*, a historical reference to an analogous conjecture in geometric topology.[4] For the interested reader, Sorkin [SO05] provides a useful discussion of the causal set *hauptvermutung*. While the status of this conjecture would certainly be worth knowing, it does not necessarily have a critical bearing on the physical viability of causal set theory, or of discrete causal theory in general. For example, if the conjecture turns out to be false, this may merely reflect the existence of "physically irrelevant degrees of freedom in continuum-based geometry." This would not be very surprising, since continuum-based theories typically contain vast quantities of manifestly irrelevant information.

Working backwards: sprinkling. As mentioned in Section 3.1, "most" causal sets bear no resemblance whatsoever to relativistic spacetime manifolds, and are therefore not directly relevant to classical spacetime structure.[5] Here, the quantifier "most" is made precise by the asymptotic enumeration of Kleitman and Rothschild [KR75], discussed in more detail in Section 3.9. An "effective realization" of the classical causal metric hypothesis (CCMH) for causal set theory, whereby one may *derive* the emergence of approximate geometry from just the axioms of the theory and suitable dynamical laws, has not yet been found, and the possible limitations of the theory, due to certain technical issues, render its pursuit a questionable enterprise. However, a useful intermediate step is to study causal sets *induced* by "sprinkling" elements discretely on a pseudo-Riemannian manifold, according to a *Poisson process.* The approximate "spacetime volume" represented by each element varies statistically under this method, since some regions of the underlying manifold happen to be "more densely sprinkled" than others. The resulting causal sets automatically exhibit certain "desirable properties," such as systematic Lorentz invariance. One may thereby obtain physically relevant causal sets for investigative purposes, without the necessity of knowing how these sets might arise dynamically. The method of sprinkling first appeared in the inaugural paper on causal set theory [BLMS87]. The role of sprinkled causal sets in the study of local properties in discrete causal theory is described in much more detail in Section 4.5.

Sprinkling is illustrated schematically in Figure 3.2.1. In the left-hand diagram in the figure, a region of a relativistic spacetime manifold X is shown, together with a distinguished event x in X, and a portion of its causal future. In the middle diagram, a family C of elements is sprinkled on X. In the right-hand diagram, X is removed from the picture, and only the abstract causal structure of C, induced by the geometric

---

[3]For example, Martin and Panangaden's recent metric recovery result [MP06] involves a *dense* subset.

[4]Namely, that any two triangulations of a triangulable space admit a common refinement. John Milnor proved in 1961 that this conjecture is false.

[5]Of course, this statement does not a priori apply to the corresponding quantum theory, since the path summation approach incorporates *all* histories in an appropriate class.

structure of X, remains. The latter diagram is a *generalized Hasse diagram,*[6] like those appearing in the previous two chapters; in particular, each edge represents a relation between its initial and terminal elements, with the direction of relations inferred by the "up the page" convention.

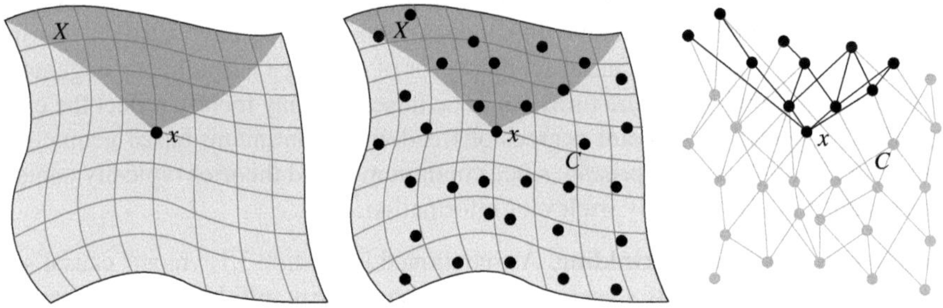

Figure 3.2.1. Relativistic spacetime manifold X; family C of elements sprinkled on X; abstract causal structure of C, induced by X.

3.3 Quantum Causal Set Theory

Background dependent theories. Following the reasoning introduced in Sections 1.6 and 1.7, approaches to the quantum theory of causal sets may be filtered according to their degree of *background independence.* Background dependent approaches involve the study of auxiliary structures, such as particles and fields, existing *on* a fixed causal set, viewed as a "discrete background spacetime," while background independent approaches involve the quantum nature of spacetime itself. From a technical perspective, background dependent methods are much easier to study, and a variety of such formulations have been described in the literature.[7] Figure 3.3.1, which reprises the diagram in Figure 1.7.1 of Chapter 1, illustrates fields on a simple background causal set. Two basic types of fields exist in this context; these may be called *element fields,* and *relation fields,* respectively. An element field assigns algebraic data to each element in a causal set. In the figure, the triple $\{\phi_1, \phi_2, \phi_3\}$ defines an element field ϕ on the causal set C. In the context of quantum theory, each ϕ_i might be an operator on a Hilbert space. Similar remarks apply to the triple $\{\theta_1, \theta_2, \theta_3\}$, which defines a relation field θ on C. Among many others, Sorkin and Rideout [SR99] have studied relation fields arising in causal set dynamics; in

---

[6] This particular diagram is actually a "true" or "strict" Hasse diagram, since it depicts the *transitive reduction* of a partially ordered set, i.e., the structure given by removing reducible relations from the partial order, while preserving its *accessibility* properties. The transitive reduction is generally different from the *skeleton,* introduced in Section 3.10, which does *not* necessarily preserve accessibility.

[7] Recent work of Fritz; e.g., [FR16], goes beyond causal set theory to examine general features of quantum theory on general acyclic directed sets.

particular, fields of *"Ising spins living on the relations of the causal set."*[8] A topic of recent interest in the causal sets community is the study of *generalized d'Alembertian operators,* which describe propagation of fields on causal set backgrounds. Interesting considerations involving deviations from *spatiotemporal locality* arise in this context; these issues are described in more detail in Section 4.5. A few contributors to this approach are Sorkin [SMS14], Dowker [BBD15], Belenchia [BE16], Glaser [GL14], and their coauthors.

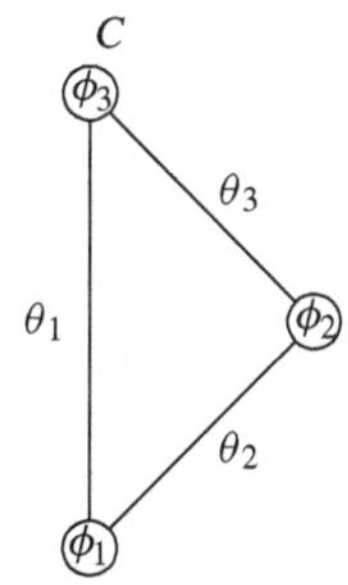

Figure 3.3.1. Fields on a causal set C.

While interesting in their own right, background dependent approaches to quantum causal set theory generally differ too much in emphasis from the version of discrete quantum causal theory developed in this book to provide ideal sources for study. An exception is the work of Ioannis Raptis, beginning with his 2000 paper *Algebraic Quantization of Causal Sets* [RA00], which provides useful context for the theory of *causal path algebras* introduced in Section 5.9, and the *quantum theory in relation space* developed in Section 6.7.

Dynamical causal set field theories. Versions of quantum causal set theory involving some degree of background independence are much more interesting and difficult than their background dependent counterparts. As in the case of discrete causal theory in general, Feynman's path summation approach to quantum theory [FE48] probably represents the most promising method of constructing such theories, and this has certainly been the approach favored by the causal sets community. Different degrees of background independence are possible in this context, depending on how much, if any, "extra structure" is added to the basic causal structure encoded by the causal sets themselves. For example, paraphrasing the discussion of Section 1.6, one may use causal sets to construct discrete quantum field theories that are "as background independent as general relativity," by working with *all* causal sets of a particular class, and *all* fields of a particular type on these causal sets, and invoking

[8] The approach of this particular paper [SR99] is actually "mostly background independent," because the authors study *configuration spaces* of causal sets equipped with such relation fields, rather than fixing a particular background causal set. Hence, this approach is "as background independent as general relativity," in the sense of Section 1.7.

constraints relating field values to causal structure, and vice versa, in a nontrivial way. In such theories, causal structure "interacts with" fields; each causal set still serves as a background of sorts, but not a static background. The most obvious way to build a quantum theory in this context is via "sums over evolutionary pathways for decorated causal sets," such as causal sets equipped with relation fields. The classical stochastic theory of *sequential growth dynamics*, developed by Sorkin and Rideout [SR99], serves as a prototype for such approaches; this theory is briefly discussed in Section 3.4, and generalized in Chapter 7.

Perfectly background independent theories. Alternatively, one might try to construct quantum causal set theories exhibiting perfect background independence, by attributing "matter and energy," rather than merely "spacetime," to the unadorned structure of causal sets. Such a theory would represent an effort to realize the strong interpretation of the causal metric hypothesis (CMH), as described in Section 2.1. However, this approach is not currently popular in the causal sets community. A quote from Sorkin and Rideout [SR99] is illustrative:

> *A question commonly asked of the causal set program is, "how could non-gravitational matter arise from only a partial order?" One obvious answer is that matter can emerge as a higher-level construct via the Kaluza-Klein mechanism, but this possibility has nothing to do with causal sets as such. The theory developed herein suggests... that the dynamics* [arises] *from... a field of Ising spins residing on the relations of the causal set.* (page 1)

Within the confines of causal set theory itself, this point of view seems perfectly reasonable, due to the genesis of the theory as a natural response to the metric recovery theorems. In particular, there is nothing in these theorems that directly suggests the emergence of anything *in addition* to pseudo-Riemannian geometry from discrete causal structure. However, in a broader context, the nature of discrete directed sets themselves hints that more than merely emergent geometry may be available. For example, at a technical level, the choice to drop the questionable causal set axiom of transitivity (TR) immediately furnishes additional information, in the form of natural data about which relations represent *direct* influence.[9] Recalling the "jungle-desert" metaphor of Section 1.7, the *local irregularity* of a typical directed set, which contrasts so sharply with the "local regularity" inherent in the definition of a manifold, is naturally amenable to modeling matter and energy, which are manifestly locally irregular.

Influence from beyond causal set theory. It is worth noting that some of the most useful existing references relating to quantum causal set theory do not belong to the causal set literature proper, but rather to closely-related research fields, often pioneered by some of the same authors who have contributed the most to the causal sets program. In particular, Christopher Isham's category-theoretic treatment of histories, introduced in his 2005 paper *Quantising on a Category* [IS05], is very illuminating.

---

[9] In fact, an important class of directed sets of this type may be viewed as *decorated causal sets*, where the "decoration" is a $\mathbb{Z}_2$-valued relation field encoding direct and indirect influences. Sorkin mentioned this viewpoint in response to my inquiry regarding his current views on the subject; see Section 3.9 for more details.

Isham's work provides part of the motivation for the theory of *co-relative histories* and *kinematic schemes,* developed in Chapters 6 and 7 of this book. Also relevant is Sorkin's work on the general topic of *quantum measure theory,* including his 2012 paper *Toward a Fundamental Theorem of Quantal Measure Theory* [SO12], which discusses quantum-theoretic measures on configuration spaces of histories, including both causal sets and discrete lattices. These ideas foreshadow the theory of *co-relative kinematics,* and the theory of *completions of kinematic schemes,* both discussed in Chapter 7.

3.4 Causal Set Dynamics and Phenomenology

Sequential growth dynamics. There is presently no consensus in causal set theory regarding exactly what the dynamical laws of the theory should be, either at the classical or quantum level. Nevertheless, important and suggestive "exploratory models" have been examined. The most prominent of these is a classical stochastic model called *sequential growth dynamics*, introduced by Sorkin and his student David Rideout in their 1999 paper *Classical sequential growth dynamics for causal sets* [SR99]. The toy dynamics introduced in Section 1.8 is of essentially the same type as those considered by Sorkin and Rideout; the only basic structural difference is that nontransitive directed sets are included in the former. In sequential growth dynamics, a causal set is "built up one element at a time," with each "evolutionary step" represented by a special causal set morphism, called a *transition.*[10] As mentioned in Section 1.5, transitions are not quite the optimal device for modeling natural relationships between pairs of classical histories in discrete causal theory; co-relative histories are more suitable for this purpose. However, transitions suffice for the present discussion of causal set dynamics.

Transitions may be used to organize classes of causal sets, considered up to isomorphism,[11] into a higher-level structure $\mathbb{S}_{\mathrm{SR}}$, which is a causal set analogue of the *positive sequential kinematic scheme* $\mathbb{S}_{\mathrm{PS}}$ first appearing in Figure 1.1.1, and reproduced in Figure 3.1.2. Here, the subscript "SR" stands for "Sorkin–Rideout." The general theory and terminology of kinematic schemes is developed in Chapter 7. The kinematic scheme $\mathbb{S}_{\mathrm{SR}}$ possesses the abstract structure of a *multidirected set;* this terminology is explained in Section 3.6. To be clear, it is better *not* to view $\mathbb{S}_{\mathrm{SR}}$ as merely a partially ordered set, since there exist "inequivalent" transitions in $\mathbb{S}_{\mathrm{SR}}$ sharing the same source and target.[12] $\mathbb{S}_{\mathrm{SR}}$ is analogous to a mathematical category, with its member causal sets corresponding to objects, and transitions between pairs of

[10] As mentioned in one of the footnotes to Section 1.5, these transitions have nothing to do with the *graph dynamical phase transitions* discussed in Section 1.9.

[11] This follows the convention used throughout most of this book for structured sets in general, since isomorphic sets generally share the same physical interpretation.

[12] See Section 7.4 for an example of this, due to Brendan McKay. This distinction does not seem to be recognized in the causal set literature.

these causal sets corresponding to morphisms. Of course, transitions *are* morphisms in the category of causal sets, but the point of emphasis in the present context is that they *serve a similar structural role in* $\mathbb{S}_{\mathrm{SR}}$*, which is not a category.* At a more precise level, transitions should be taken to merely *represent* the "morphisms" of $\mathbb{S}_{\mathrm{SR}}$, which should be recognized as co-relative histories. These details are explained in Chapter 6. The general analogy between category theory and kinematics has been developed in a striking way by Christopher Isham and his collaborators, although they use different terminology. In Chapter 7, I offer a somewhat different perspective regarding such constructions.

The left-hand diagram in Figure 3.4.1 illustrates a portion of the kinematic scheme $\mathbb{S}_{\mathrm{SR}}$ underlying Sorkin and Rideout's theory of sequential growth dynamics. Comparing this figure to Figure 1.8.1 of Chapter 1, the only obvious distinction is that one of the directed sets in Figure 1.8.1 is omitted here. The reason for this omission is that this particular directed set is *nontransitive,* and hence is not a causal set. Returning to Figure 3.4.1, individual causal sets appear in the large open nodes, and transitions are represented by edges between pairs of such nodes. To each transition τ is associated an *initial causal set* C_i, and a *terminal causal set* C_t, given by adding a single element to C_i, along with a family of relations terminating at the "new element." The right-hand diagram shows the abstract structure of the same portion of $\mathbb{S}_{\mathrm{SR}}$, viewed as a "higher-level multidirected set." Here, only *directed* structure is visible, since multiple inequivalent transitions between pairs of causal sets do not enter the picture until the sets involved become somewhat larger.

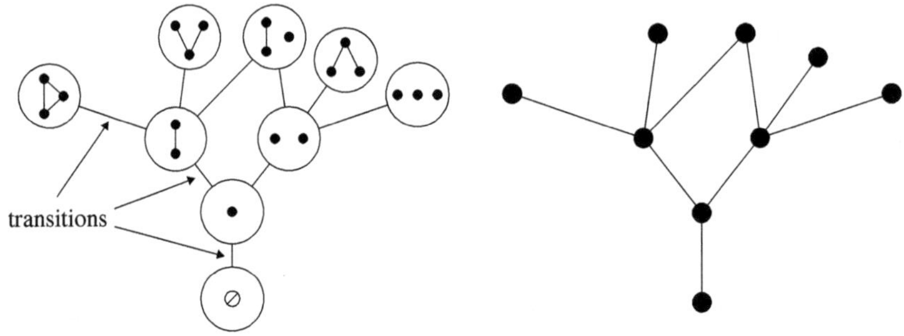

Figure 3.4.1. Portion of Sorkin and Rideout's kinematic scheme $\mathbb{S}_{\mathrm{SR}}$ describing sequential growth of causal sets; abstract structure of this portion of $\mathbb{S}_{\mathrm{SR}}$.

The Sorkin–Rideout kinematic scheme $\mathbb{S}_{\mathrm{SR}}$ only describes *possible* evolutionary behavior of causal sets. The actual *dynamics* of sequential growth is supplied by assigning a *weight* or *phase* $\theta(\tau)$ to each transition τ in $\mathbb{S}_{\mathrm{SR}}$, encoding the *"likelihood that* C_i *transitions to* C_t*."* This deliberately vague description is intended to cover a range of possible choices for the values of θ, from real probabilities, to complex amplitudes, to more exotic possibilities. The "likelihood" of evolution between any pair of causal sets C and C' in $\mathbb{S}_{\mathrm{SR}}$ may then be expressed in terms of the weights of all transitions along all upward-directed sequences of transitions from C to C'

in $\mathbb{S}_{SR}$. Each such sequence is viewed as an "evolutionary pathway" connecting C and C'. In the specific context of classical sequential growth dynamics, Sorkin and Rideout take these weights to be real probabilities, which is why their model is classified as a classical stochastic model. The same basic recipe is followed in the toy dynamics illustrated in Figure 1.8.1 of Chapter 1. However, the same kinematic scheme $\mathbb{S}_{SR}$ may also be approached in a quantum-theoretic context, as in Sorkin's paper [SO12] on quantum measure theory. Brightwell and Georgiou [BG10] have recently proved that sequential growth models for causal sets of the type studied in [SR99] generically produce objects that look nothing like Minkowski spacetime, but this is not very surprising, given the exploratory nature of such models, and the broader structural questions about causal set theory.[13] The *conceptual* content of such models remains of great interest, however.

Given any pair of causal sets C and C' in $\mathbb{S}_{SR}$, each upward-directed sequence of transitions from C to C' in $\mathbb{S}_{SR}$ generates a *sequential labeling* of the elements of $C' - C$. If C is taken to be the empty set $\emptyset$, then such a sequence defines a *total labeling* $\{x_0, x_1, x_2, \ldots\}$ of C', which is equivalent to a *total refinement* of the causal set relation $\prec_{CS}$ on C', or alternatively, to a *bijective morphism* from C' into a totally ordered subset of the natural numbers.[14] Ultimately, the definition of such a labeling is ambiguous unless $\mathbb{S}_{SR}$ is viewed as a multidirected set, which is one of the technical reasons why the common view of $\mathbb{S}_{SR}$ as a partially ordered set is inadequate.[15] Ignoring these details for the moment, such a sequence of transitions is roughly analogous to a relativistic *frame of reference,* which supplements the causal structure of its terminal causal set with arbitrary "extraphysical information," just as relativistic coordinates assign arbitrary temporal orders to pairs of spacelike-separated events.[16] In Chapter 7, I generalize and formalize such sequences as choices of *co-relative kinematics.*

Phenomenology. Like other modern approaches to fundamental spacetime structure, causal set theory cannot yet boast a robust record of phenomenological achievement. However, given the general difficulty of distilling definitive and experimentally accessible predictions from a relatively young theory of fundamental physics, it is reasonable to exercise patience in evaluating the ongoing viability of causal set theory. The theory is well-motivated, clean, and plausible, and its possible technical limitations are not sufficiently worrisome to rule out the possibility that it could make further important contributions to describing nature. Probably the most well-known phenomenological success of causal set theory to date is an approximate prediction of the size of *fluctuations* of the "cosmological constant," as discussed in [ADGS04]. Assuming that such fluctuations are centered about zero, this prediction falls within

---

[13]See [BL15] for an updated discussion of this result.

[14]Here, "totally ordered" means *strictly* totally ordered.

[15]At a technical level, a total labeling of an "initial" finite causal set C_i, together with the isomorphism class of the "next" causal set C_t, generally does *not* determine a class of labelings of C_t related by automorphisms of C_t. This is explained in more detail in Chapter 6.

[16]A closer analogy exists between frames of reference and *refinements* of the causal set relation $\prec_{CS}$, as explained in Sections 7.1 and 7.2.

the range suggested by observation, in contrast to the embarrassingly large predictions of conventional continuum-based theories. As a note of caution, this result relies on modest assumptions about the interpretation of scale, the emergent dimensionality of spacetime, and the large-scale structure of the universe, that one might a priori prefer to avoid in the most general context. For a fine overview of this effort, and other notable work relevant to causal set phenomenology, I refer the reader to Sumati Surya's review article *Directions in Causal Set Quantum Gravity* [SU11]. Surya discusses a wide variety of recent developments, including Dowker, Henson, and Sorkin's work on *diffusion in a causal set background* [DHS04], examination of *causal set curvature* and analogues of the *Einstein–Hilbert action* by Dowker and her collaborators [BD10], and the construction of a *scalar field Feynman propagator* by Johnston [JF09].

Causal set phenomenology has thus far been directed more toward the preliminary question of *how* causal set theory might make contact with the experimental realm than toward making specific quantitative predictions that can actually be tested. Many of the active research topics cited above involve background dependent versions of the theory, and are therefore of only limited relevance to the principal focus of this book. In particular, as discussed in Section 3.3, a common phenomenological question is how "nongravitational matter" might arise in causal set theory. Under the strong interpretation of the causal metric hypothesis (CMH), the answer to this question is that *"there is no such thing as nongravitational matter,"* in the sense that "matter and energy" emerge from the same fundamental directed structure to which gravitation is attributed. However, this viewpoint is alien to causal set theory proper. As noted in Section 3.3, Sorkin and Rideout [SR99] suggest *"a kind of "effective action" for a field of Ising spins living on the relations of the causal set"* to explain the origin of matter. Belenchia [BE15] discusses the topic of phenomenological tests of causal set theory, emphasizing the possibility of deviations from *spatiotemporal locality* involving scalar fields on a causal set background. Sorkin [SO09A], Glaser [GL14], and a few others have investigated similar questions. Sorkin goes so far as to hazard a guess of 10^{-14} meters as a possible "intermediate length scale" at which nonlocal effects might begin to manifest themselves. The specific topic of nonlocality in causal set theory is revisited in a broader context in Section 4.5.

3.5 Axioms of Causal Set Theory

Two formulations. The causal set literature contains two "almost equivalent" formulations of causal set theory, which I refer to as the *irreflexive formulation,* and the *partial order formulation.* The irreflexive formulation is the one more commonly used. In this section, I introduce and discuss both formulations in some detail. In later sections, the irreflexive formulation is used by default, although I occasionally borrow convenient terminology from the partial order formulation.

Irreflexive formulation. The "standard" formulation of causal set theory is expressed in terms of the *causal set relation* $\prec_{\text{CS}}$, which is an irreflexive binary relation, analogous to the familiar "less than" relation $<$ on the integers. It is typically denoted by just $\prec$, but this symbol is reserved for more general directed sets in this book. The properties of $\prec_{\text{CS}}$ may be described in terms of the six axioms first cited implicitly in Definition 3.1.1, which I now explain in detail. The reader should be aware that many of these same axioms, or modified versions of them, also play a role in the more general version of discrete causal theory developed in this book. For this reason, they reappear in various forms in Sections 3.7, 4.10, and 8.1.[17] In particular, the binary axiom and countability are carried over essentially unchanged, while the measure axiom is replaced by the generalized measure axiom (M*), and irreflexivity is usually subsumed by acyclicity (AC). Transitivity and interval finiteness play little role beyond Chapter 4.

Definition 3.5.1. *The irreflexive formulation of causal set theory is defined by the following six axioms:*

1. **Binary axiom**[18]: *Classical spacetime may be modeled via a set C, called a causal set, whose elements represent spacetime events, together with a binary relation $\prec_{\text{CS}}$ on C, called the causal set relation, whose elements represent causal relationships between individual pairs of spacetime events.*
2. **Measure axiom** (M): *C is equipped with a discrete measure μ_{CS}, called the causal set measure, which assigns to each subset of C a volume equal to its number of elements in fundamental units, up to Poisson-type fluctuations.*
3. **Countability**: *C is countable.*
4. **Transitivity** (TR): *Given three elements x, y, and z in C, if $x \prec_{\text{CS}} y \prec_{\text{CS}} z$, then $x \prec_{\text{CS}} z$.*
5. **Interval finiteness** (IF): *For every pair of elements x and z in C, the open interval*[19]

$$\langle\!\langle x, z \rangle\!\rangle := \{y \in C \mid x \prec_{\text{CS}} y \prec_{\text{CS}} z\}$$

   has finite cardinality.
6. **Irreflexivity** (IR): *Elements of C are not self-related with respect to $\prec_{\text{CS}}$; i.e., $x \not\prec_{\text{CS}} x$.*

The discrete measure μ_{CS} is a special map $\mathcal{P}(C) \to \mathbb{R} \cup \{\infty\}$, where $\mathcal{P}(C)$ is the *power set* of C, i.e., the set of all subsets of C. Use of the real numbers here is purely

---

[17] In the electronic version of the book, the hyperlinks for the binary axiom (B) and countability (C) take the reader to definitions in later sections, which is why the statements of these axioms given in Definition 3.5.1 do not include the usual "tags." However, the hyperlinks for the measure axiom (M), transitivity (TR), interval finiteness (IF), and irreflexivity (IR) link to Definition 3.5.1.

[18] Here, the binary axiom is stated specifically in the language of causal set theory. However, the information content is the same as in later statements, since no other properties of $(C, \prec_{\text{CS}})$ appear in the axiom itself.

[19] In the case of nontransitive directed or multidirected sets, it is necessary to redefine open intervals in terms of *chains*. This is done in Definition 3.7.3. I thank Johnny Feng for the suggestion to warn the reader here about this redefinition.

for convenience; any "sufficiently large" extension of the positive integers suffices. In particular, no essential properties of the continuum are necessary. Transitivity and irreflexivity together imply the condition of *acyclicity*, which rules out discrete causal analogues of closed causal curves. Interval finiteness is called "local finiteness" in the causal set literature, but it is not a local condition in any suitable sense. A genuinely local alternative condition, called *star finiteness* (SF), is introduced in Section 4.4. Cardinality conditions rarely appear explicitly in the causal set literature, but countability is usually implicit. For example, Sorkin and Rideout's theory of sequential growth dynamics involves specific enumerations of causal sets. More generally, interval finite but uncountable analogues of causal sets are "physically pathological," as explained in Section 4.6.

It is often useful to make explicit reference to the causal set relation $\prec_{\text{CS}}$ on a causal set C. In such cases, the set may be introduced as a pair $(C, \prec_{\text{CS}})$. Alternatively, it may be denoted by the single symbol C, with the understanding that this is an abbreviation. A subset S of a causal set $C = (C, \prec_{\text{CS}})$ inherits from C a binary relation, called the *subset relation*, also denoted by $\prec_{\text{CS}}$, where $x \prec_{\text{CS}} y$ in S if and only if $x \prec_{\text{CS}} y$ in C. It is often useful to view the class of causal sets, together with the class of morphisms between pairs of causal sets, as a category $\mathcal{C}$, called the **category of causal sets**. To be explicit, a morphism between causal sets $C = (C, \prec_{\text{CS}})$ and $C' = (C', \prec_{\text{CS}}')$ is a set map $f : C \to C'$ that "preserves directed structure," in the sense that $f(x) \prec_{\text{CS}}' f(y)$ in C' whenever $x \prec_{\text{CS}} y$ in C. Here, C is sometimes called the *source* of f, and C' is called the *target* of f. The *index* of f is the supremum of the cardinalities of its fibers, where the *fiber* $f^{-1}(x')$ of f over x' in C' is the set of elements of C that map to x' under f. An *isomorphism* is an invertible morphism. In almost all cases, only the isomorphism class of a causal set is significant. A bijective morphism is generally *not* an isomorphism, since the target may possess "extra relations," extending the binary relation on the source. Such bijective morphisms play an important role in Sections 3.8 and 4.6. A *monomorphism* is an injective morphism. I sometimes refer to a monomorphism $f : C \to C'$ as a morphism "embedding its source as a subobject of its target."

Partial order formulation. A less-common alternative formulation of causal set theory employs an interval finite partial order $\preceq_{\text{CS}}$, which I refer to as the *causal set order*, in place of the irreflexive causal set relation $\prec_{\text{CS}}$ described above. A *partial order* on a set C is a *reflexive, antisymmetric, transitive* binary relation on C. The symbol $\preceq$ is usually used to denote a partial order; in the present context, I add the subscript "CS" for "causal set." "Reflexive" means that $x \preceq_{\text{CS}} x$ for every x in C, and "antisymmetric" means that if $x \preceq_{\text{CS}} y$ and $y \preceq_{\text{CS}} x$ for two elements x and y in C, then $x = y$. Transitivity and antisymmetry together imply that C has no cycles *except for* the reflexive cycles $x \preceq_{\text{CS}} x$. Hence, in the partial order formulation of causal set theory, the data represented by the phrase "interval finite partial order" replaces axioms 1, 4, 5, and 6 from the irreflexive formulation, while axioms 2 and 3 are carried over unchanged. As the notation suggests, the causal set order $\preceq_{\text{CS}}$ is analogous to the familiar "less than or equal to" relation $\leq$ on the integers. Interval finiteness is an extra condition, not part of the definition of a partial order. Interval finite partial orders are special cases of *discrete orders,* which are partial orders in which every nonextremal element has at least one *maximal predecessor* and *minimal successor.*[20] However, the reader should also be aware that the term "discrete order" is sometimes assigned other, mutually contradictory meanings, such as a discrete *total* order, or a "minimal" order involving only reflexive relations. *Predecessors* and *successors* are discussed in more detail in Section 3.7.

Comparing the two formulations. The irreflexive and partial order formulations of causal set theory are "equivalent at the level of objects," in the following sense: a causal set $(C, \prec_{\mathrm{CS}})$ may be viewed as a countable, interval finite, partially ordered set, whose causal set order $\preceq_{\mathrm{CS}}$ is defined by taking $x \preceq_{\mathrm{CS}} y$ if $x \prec_{\mathrm{CS}} y$ or $x = y$; conversely, a countable, interval finite, partially ordered set $(C, \preceq_{\mathrm{CS}})$ may be viewed as a causal set, whose causal set relation $\prec_{\mathrm{CS}}$ is defined by taking $x \prec_{\mathrm{CS}} y$ if $x \preceq_{\mathrm{CS}} y$ and $x \neq y$. This correspondence generalizes the correspondence between strict total orders, such as the "less than" relation $<$ on $\mathbb{Z}$ or $\mathbb{R}$, and non-strict total orders, such as the corresponding "less than or equal to" relation $\leq$, discussed in Section 2.9. However, as explained below, the two formulations are *not* "equivalent at the level of categories," because their morphisms do not correspond in the same way as their objects correspond.

The partial order formulation of causal set theory has the superficial advantage of general familiarity, since partial orders are more popular than irreflexive relations in most mathematical settings. In the context of general relativity, binary relations encoding chronological or causal structure on a relativistic spacetime manifold X are *sometimes* taken to be reflexive, in which case they define partial orders, provided that X satisfies the appropriate chronological or causal condition. This is an unfortunate choice of convention from the standpoint of transitioning to the discrete causal paradigm, because reflexive relations in the latter context are naturally interpreted as indicating "self-causation" of events, which is generally not the intended meaning in relativity. Hence, the relativistic chronological relation $\prec\prec_{\mathrm{GR}}$ and the relativistic causal relation $\prec_{\mathrm{GR}}$ are defined in this book to be *irreflexive* for chronological and causal spacetime manifolds, respectively. More generally, special care is required in order to avoid confusion when comparing binary relations encoding causal structure in relativity, causal set theory, and discrete causal theory as presented in this book. The table in Figure 3.5.2 clarifies the notation and meaning associated with various binary relations appearing in this book, although some of the notation and terminology is not explained until later sections. The reader should take note that *neither* of the irreflexive binary relations $\prec_{\mathrm{CS}}$ or $\prec$, nor the partial order $\preceq_{\mathrm{CS}}$, corresponds to the relativistic chronological relation $\prec\prec_{\mathrm{GR}}$, which "excludes the entire boundaries of causal pasts and futures."

---

[20] The quote from Sorkin in Section 3.9 provides an example of the use of this terminology.

general relativity		causal set theory		discrete causal theory	
name and symbol	meaning	name and symbol	meaning	name and symbol	meaning
relativistic causal relation $\prec_{\mathrm{GR}}$	$x \prec_{\mathrm{GR}} y$ means there exists a nontrivial smooth causal curve from x to y	causal set relation $\prec_{\mathrm{CS}}$	$x \prec_{\mathrm{CS}} y$ means x influences y (self-influence ruled out)	causal relation $\prec$	$x \prec y$ means x *directly* influences y ($x \prec x$ allowed)
chronological relation $\prec\!\prec_{\mathrm{GR}}$	$x \prec\!\prec_{\mathrm{GR}} y$ means there exists a nontrivial smooth timelike curve from x to y	causal set order $\preceq_{\mathrm{CS}}$	$x \preceq_{\mathrm{CS}} y$ means $x \prec_{\mathrm{CS}} y$ or $x = y$	transitive relation $\prec_{\mathrm{tr}}$	$x \prec_{\mathrm{tr}} y$ means x influences y (generally *not* a partial order!)
horismos relation $\rightarrow$	$x \rightarrow y$ means $x \prec_{\mathrm{GR}} y$ but $x \not\prec\!\prec_{\mathrm{GR}} y$			skeletal relation $\prec_{\mathrm{sk}}$	$x \prec_{\mathrm{sk}} y$ means $x \prec y$ and $\nexists\, w$ such that $x \prec w \prec y$

Figure 3.5.2. Comparison of binary relations in general relativity, causal set theory, and the version of discrete causal theory developed in this book.

From a structural perspective, the partial order formulation of causal set theory is inconvenient, and the irreflexive formulation is clearly preferable. Fortunately, the latter formulation is predominant in the causal set literature. Partial orders are technically not acyclic, since they include reflexive relations, i.e., "self-relations" $x \preceq_{\mathrm{CS}} x$.[21] Hence, the *categories* of causal sets and interval finite partially ordered sets are not equivalent, since the latter category admits "structure-destroying morphisms," that wrap "meaningful" relations in the source around "non-meaningful" reflexive relations in the target, under the intended "essentially acyclic" interpretation of the structures involved. These inconveniences do not reflect any essential difference in physically relevant *information content* between the two formulations, however. Hence, although I use the irreflexive formulation by default throughout the remainder of this book, I employ order-theoretic notions and terminology whenever convenient.

Treatment in the literature. The binary axiom, the measure axiom, and countability do not appear as explicit axioms in the causal set literature. However, the *content* of the binary axiom and measure axiom appear explicitly, and countability is implicit in the focus and methods. Treatment of the remaining axioms depends on whether the irreflexive formulation or the partial order formulation is being used. Here, I reproduce a few representative quotations to illustrate this treatment. In their inaugural

[21] The term "reflexive relation" is potentially confusing, because it is a priori unclear whether the entire binary relation $\preceq_{\mathrm{CS}}$ is intended to satisfy the reflexive property, or if one is merely referring to a particular element which is "self related," for example, $x \preceq_{\mathrm{CS}} x$. I usually make an effort to spell out the difference, except in the case of multidirected sets, where the second meaning is automatic, since no binary relation is involved.

paper *Space-Time as a Causal Set* [BLMS87], Bombelli, Lee, Meyer, and Sorkin use the partial order formulation of causal set theory, with the exceptions that an explicit statement of countability, and the nuance "up to Poisson-type fluctuations" in the measure axiom, do not appear:

> *...when we measure the volume of a region of space-time, we are merely indirectly counting the number of "point events" it contains... ...volume is number, and macroscopic causality reflects a deeper notion of order in terms of which all the "geometrical" structures of spacetime must find their ultimate expression... ...Before proceeding any further, let us put the notion of a causal set into mathematically precise language. A partially ordered set... ...is a set... ...provided with an order relation, which is transitive... ...noncircular... ...*[and] *reflexive. A partial ordering is locally finite if every "Alexandroff set"... ...contains a finite number of elements... ...a causal set is then by definition a locally finite, partially ordered set.* (page 522)

Here, *"noncircular"* means "acyclic, except for reflexive cycles." An *"Alexandroff set"* is just an open interval $\langle\langle x, z\rangle\rangle := \{y \in C \mid x \prec_{\mathrm{CS}} y \prec_{\mathrm{CS}} z\}$; hence, *"local finiteness"* really means interval finiteness (IF) in this context. The name "Alexandroff" is an alternative spelling of the name "Alexandrov," appearing in Chapter 2 in the context of the *Alexandrov topology,* which is defined in terms of the obvious continuum-based analogues of such intervals, encoded via the relativistic causal relation $\prec_{\mathrm{GR}}$. The irreflexive causal set relation $\prec_{\mathrm{CS}}$ may be viewed here as a derivative construct of the causal set order $\preceq_{\mathrm{CS}}$, in the manner described above.

Later papers on causal set theory are split between the irreflexive formulation and the partial order formulation. For example, in their influential paper *Classical sequential growth dynamics for causal sets* [SR99], Sorkin and Rideout use the irreflexive formulation.[22] Later papers also amend the measure axiom to allow for Poisson-type fluctuations. For example, in *Everpresent Λ* [ADGS04], Ahmed, Dodelson, Greene, and Sorkin write,

> *In order to do justice to local Lorentz invariance, the correspondence between number and volume... ...must be subject to Poisson-type fluctuations...* (page 2)

This particular subtlety is discussed in detail in Section 4.5. Absence of explicit cardinality conditions in the causal set literature may be attributed to the fact that countability is automatic from the viewpoint of recovering a relativistic spacetime manifold, up to a specified degree of precision, from a discrete subset. More fundamentally, allowing uncountable analogues of causal sets greatly exacerbates the existing imbalance of the category of causal sets toward objects of "large spatial and small causal size," as indicated in the finite case by the *Kleitman–Rothschild pathology.* This pathology is explained in Section 3.9, and revisited in Section 4.6. It arises, ultimately, because of the questionable axioms of transitivity (TR) and interval finiteness (IF).

[22] See [SR99], p. 2.

3.6 Generalized Causal Sets: Directed Sets and Multidirected Sets

Generalizing causal sets. Having examined causal sets in some detail, I now turn attention to the more general classes of structured sets that supply the objects of principal interest in this book. I refer to these sets as *directed sets* and *multidirected sets*, respectively. *Acyclic* directed sets are a special case of directed sets, which represent a "blank canvas for conservative discrete causal models of classical spacetime," i.e., discrete causal analogues of relativistic spacetime manifolds satisfying the *causal condition,* described in Section 2.7. More general directed sets and multidirected sets, while *possibly* unnecessary for modeling classical spacetime structure, arise unavoidably in the theory of kinematic schemes, and hence, in discrete quantum causal theory.

Directed sets. Directed sets have already been introduced in Section 2.2 and used informally throughout the first two chapters. The formal definition is simple and straightforward:

Definition 3.6.1. *A* **directed set** *is a set D equipped with a binary relation $\prec$.*

As explained in Section 2.2, a binary relation $\prec$ on a set D is a subset of the Cartesian product $D \times D$, and the notation $x \prec y$ is used to indicate that the ordered pair (x, y) of elements of D belongs to $\prec$. A directed set may be introduced explicitly as a pair $(D, \prec)$, or it may be denoted in abbreviated fashion by a single symbol D. A subset S of a directed set $(D, \prec)$ inherits a natural *subset relation,* given by restricting $\prec$ to S. The **category of directed sets** is the category $\mathcal{D}$ whose objects are directed sets, and whose morphisms $(D, \prec) \rightarrow (D', \prec')$ are set maps $f : D \rightarrow D'$, such that $f(x) \prec' f(y)$ in D' whenever $x \prec y$ in D. Related notions, such as sources, targets, fibers, indices, isomorphisms, and monomorphisms, generalize from the case of causal sets in the obvious manner. At an intuitive level, the term *directed set* captures the desired conceptual meaning very clearly: it is simply a set with "directions added between certain pairs of elements." For clarity, however, I reiterate that this term is usually assigned a more specific meaning in conventional settings, particularly in order theory and category theory.[23]

Directed sets may also be understood in graph-theoretic terms, as *small, simple, directed graphs,* under the correspondence sending elements to vertices and relations to directed edges. Here "small" means that the vertex class of the graph under consideration is a set, rather than a *proper class,* and "simple" means that for any pair of vertices x and y, there is at most one edge from x to y. In this case, the equivalence extends to the level of categories. Both sides of this equivalence are useful: the set-theoretic side for its convenient terminology, and the graph-theoretic side for the visual perspective afforded by generalized Hasse diagrams. Note that the definition of a simple directed graph allows for the coexistence of "reciprocal edges" $x \prec y$

[23]Namely, that any pair of elements share a common "upper bound," or "successor." Predecessors and successors are defined in Section 3.7.

and $y \prec x$, but prohibits multiple edges between x and y in the same direction. This corresponds to the fact that a binary relation may include both ordered pairs (x, y) and (y, x), but cannot include "multiple copies of a given ordered pair." Besides their use as models of classical spacetime, directed sets contribute naturally to discrete causal theory in more abstract roles, in which their elements represent entities more general than structureless "spacetime events." Examples include the *relation spaces* and *causal path spaces* studied in Chapter 5.

Acyclic directed sets: a special case. An acyclic directed set is merely a directed set whose binary relation is acyclic. The meaning of acyclicity has already been discussed, both in the relativistic context, and in the context of causal set theory. However, it is convenient, for future reference, to spell out the notion here.

Definition 3.6.2. *A* **cycle** *in a directed set* $(D, \prec)$ *is a nontrivial*[24] *sequence of relations with the same initial and terminal element:* $x = x_0 \prec x_1 \prec \cdots \prec x_{n-1} \prec x_n = x$. *The binary relation* $\prec$ *is* **acyclic** (AC) *if it possesses no cycles.*[25]

Looking ahead to Section 3.7, a general sequence of the form $x_0 \prec x_1 \prec \cdots \prec x_{n-1} \prec x_n$ is called a *chain* in D; in this context, a cycle is a nontrivial "closed chain." The simplest example of a cycle in a directed set is a *reflexive relation* $x \prec x$. It is also useful to include here a formal definition of acyclic directed sets:

Definition 3.6.3. *An* **acyclic directed set** *is a directed set* D *whose binary relation* $\prec$ *is acyclic.*

Irreflexivity alone does not imply acyclicity; for example, cycles of the form $x \prec y \prec x$ are allowed under irreflexivity. However, irreflexivity and transitivity together *do* imply acyclicity. This distinction is important in this book, since most of the binary relations considered here are nontransitive. Transitive acyclic directed sets are equivalent *as objects* to partially ordered sets. This extends the correspondence between causal sets and countable, interval finite, partially ordered sets, introduced in Section 3.5, to a broader correspondence permitting arbitrary cardinalities and arbitrary sizes of intervals. Acyclic directed sets form a category $\mathcal{A}$, which embeds into the category $\mathcal{D}$ of directed sets as a full subcategory. Conceptually, this means that pairs of acyclic directed sets share the same relationships, whether viewed as objects of $\mathcal{A}$ or $\mathcal{D}$. The category $\mathcal{C}$ of causal sets, in turn, embeds into $\mathcal{A}$ as a full subcategory, provided that one chooses to use the irreflexive formulation of causal set theory. Hence, causal sets may be viewed as objects of $\mathcal{C}$, $\mathcal{A}$, or $\mathcal{D}$, without ambiguity concerning their morphism classes.

---

[24]The word "nontrivial" means that at least one relation is involved. In Definition 3.7.1, cycles are redefined as special types of *chains* in the more general context of multidirected sets. In this setting, it is convenient to include individual elements as 0-chains. The condition of nontriviality is replaced by the statement that a cycle must have "nonzero length." This is necessary to exclude individual elements from being regarded as cycles.

[25]In the electronic version of the book, hyperlinks labeled (AC) do not link to this definition, but to the statement of acyclicity among the "suggested list" of axioms in Section 4.10.

Acyclic directed sets are equivalent to *small, simple, acyclic directed graphs,* under the correspondence sending elements to vertices and relations to directed edges. As discussed in Chapter 2, general relativity admits solutions involving closed causal curves, whose discrete causal analogues are the cycles of Definition 3.6.2. In fact, closed causal curves appear in rather generic situations in relativity, such as the *Kerr black hole.* Acyclic directed sets are clearly inadequate to model discrete causal analogues of such relativistic spacetimes. This is one of the principal reasons why more general directed sets are worthy of study, although quantum theory forces their consideration in any case. It is tempting to invent suggestive terminology for distinguished classes of directed sets that seem to possess special physical significance. However, the best such term is "causal sets," which has already been claimed.[26] Hence, I choose to retain the general terminology of directed sets.

Multidirected sets. The most general class of structured sets enjoying broad use in this book is the class of *multidirected sets,* which plays a central role in the study of configuration spaces of directed sets. The most important such configuration spaces are *kinematic schemes,* which supply the basic structural scaffolding underlying the path summation approach to discrete quantum causal theory, as described in Chapter 7. In fact, the multidirected sets of principal interest in this book are defined by "decategorifying" kinematic schemes, i.e., demoting them to a lower level of algebraic hierarchy by "forgetting" the internal structure of their member sets. Possible applications of multidirected sets in classical physics are much less obvious, but are mentioned very briefly in Section 8.7.

Definition 3.6.4. *A* **multidirected set** *consists of a set of elements M, a set of relations R, and* **initial** *and* **terminal element maps** $i : R \to M$ *and* $t : R \to M$*, assigning to each relation initial and terminal elements.*

A multidirected set for which each ordered pair (x, y) of elements has at most one relation r satisfying the conditions that $i(r) = x$ and $t(r) = y$, may be viewed as a directed set, whose binary relation $\prec$ is defined by setting $x \prec y$ whenever there exists a relation r such that $i(r) = x$ and $t(r) = y$. However, the structure of a general multidirected set *cannot* be expressed in terms of a single binary relation, since an ordered pair (x, y) of elements of M generally does not *uniquely identify* a relation from x to y. In this context, the expression $x \prec y$ may be used to refer to a *particular* relation with initial element x and terminal element y, or it may merely indicate that at least one such relation exists. A multidirected set may be introduced as a quadruple (M, R, i, t), or it may be denoted in abbreviated fashion by a single symbol M. It is often convenient to refer to the set R of relations in a multidirected set $M = (M, R, i, t)$ as the **relation set** of M. A *subobject* of $M = (M, R, i, t)$ consists of a subset of M, together with a subset of R, chosen in such a way that the initial and terminal element maps i and t map the latter subset into the former. Subsets of causal

[26]Finkelstein [FI88] uses the term *causal nets.* Benincasa and Dowker [BD10], following Riemann, use the term *discrete manifolds* in a general context, including discrete causal models as a special case. Other possible choices are *causal graphs* and *discrete chronofolds,* but "chronofold" is the name of an Apple smartphone application.

sets, acyclic directed sets, and directed sets, together with their subset relations, are all special cases of subobjects of multidirected sets.

Abstractly, multidirected sets are structurally analogous to *small categories,* i.e., categories whose object classes are sets. As in the case of directed sets, the elements of a multidirected set are often taken to represent entities more complex than individual events. From a graph-theoretic perspective, multidirected sets are equivalent to *small directed multigraphs,* or equivalently, *small quivers.*[27] Use of the term "quiver" often carries algebraic connotations; for example, the *concatenation algebra* over a multidirected set, introduced in Section 5.9, is a special type of *quiver algebra.* The use of multidirected sets as "algebraic substrata" goes back at least to *Gabriel's theorem* [GA72] in 1972, which classifies connected quivers of finite type, along with their representations, in terms of *Dynkin diagrams* and *root systems.* Under Definition 3.6.4, a multidirected set is precisely what Abrams and Pino [AP05] call a "directed graph." Despite these varied choices of viewpoint and terminology, the term "multidirected set" is the most convenient choice for this book, for at least five reasons: (1) it is descriptive; (2) it is congenial to the causal set viewpoint; (3) it is relatively neutral in its connotations; (4) it is amenable to generalized order-theoretic notions and terminology; and (5) it eliminates potential confusion regarding proper classes.

The **category of multidirected sets** is the category $\mathcal{M}$ whose objects are multidirected sets, and whose morphisms are pairs of maps of elements and relations that respect initial and terminal element maps. More precisely, a morphism f between two multidirected sets (M, R, i, t) and (M', R', i', t') consists of *two* maps: a **map of elements** $f_{\text{ELT}} : M \to M'$, and a **map of relations** $f_{\text{REL}} : R \to R'$, satisfying the conditions that $f_{\text{ELT}}(i(r)) = i'(f_{\text{REL}}(r))$ and $f_{\text{ELT}}(t(r)) = t'(f_{\text{REL}}(r))$, for each relation r in R. Students of algebraic geometry will recognize an interesting analogy here: a morphism of algebraic schemes *also* consists of two maps: a map of topological spaces, and a map of structure sheaves. Of course, the latter map is *contravariant,* i.e., "in the opposite direction." Nevertheless, one may recognize a loose analogy between the set of relations R of a multidirected set and the *tangent sheaf* of a manifold in geometry. This analogy was implicitly invoked in Section 1.7, where relations were compared to timelike tangent vectors. Morphisms of multidirected sets play a prominent role in Section 5.1, where I use them to analyze the properties of the *relation space functor* and the *abstract element space functor.* Related notions, such as sources, targets, fibers, indices, isomorphisms, and monomorphisms, may be defined by applying familiar notions to *both* maps f_{ELT} and f_{REL}.

Figure 3.6.5 illustrates the structural differences among totally ordered sets, causal sets, acyclic directed sets, directed sets, acyclic multidirected sets, and multidirected sets containing cycles.[28] The top left diagram in the figure illustrates a totally ordered, or linearly ordered, set. This description may seem puzzling, from a naïve perspective,

[27] This rather quaint term deliberately evokes the "quiver of arrows" carried by an archer.

[28] Strictly speaking, I have not yet defined the meaning of a cycle in a *multidirected* set, but the generalization from directed sets is relatively straightforward. A formal statement appears in Definition 3.7.3.

because the elements in this set do not "look as though they are arranged linearly." However, this is a trivial artifact of the way in which the generalized Hasse diagram depicting this set is drawn; what is important is that the order of the elements inferred by the "up the page" convention is linear: $w \prec x \prec y \prec z$.[29] The top middle diagram illustrates a causal set. Note, in particular, the transitivity of the binary relation in this case. The top right diagram illustrates a nontransitive acyclic directed set; more precisely, only *some* of the relations "implied by transitivity" are present. For example, there is a relation $w \prec z$ corresponding to the pair of relations $w \prec x \prec z$, but no relation $w \prec y$ corresponding to the pair of relations $w \prec x \prec y$.

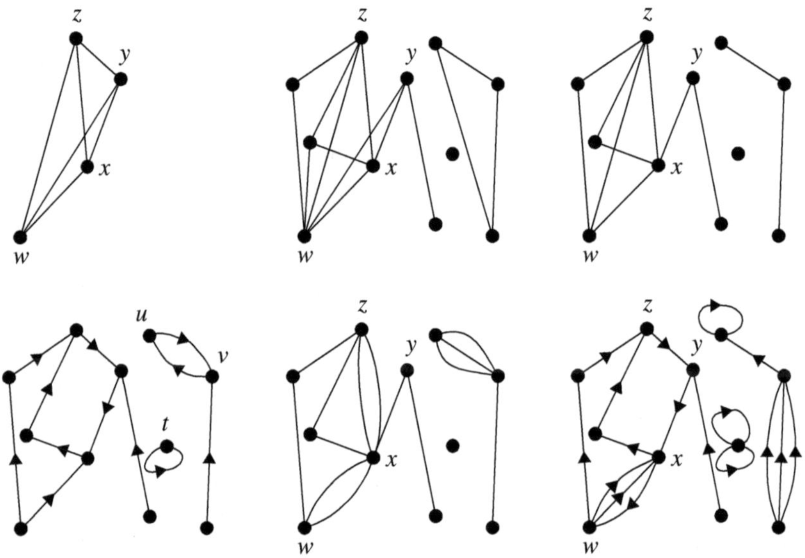

Figure 3.6.5. *Top:* totally ordered set; causal set; acyclic directed set. *Bottom:* directed set; acyclic multidirected set; multidirected set with cycles.

The bottom left diagram illustrates a directed set containing cycles. Note, in particular, that reflexive "self-relations," such as $t \prec t$, and "reciprocal relations," such as $u \prec v$ and $v \prec u$, are present in this case. The bottom middle diagram illustrates an acyclic multidirected set. While I have so far defined acyclicity only for binary relations, generalization to multidirected sets is straightforward; this is spelled out in Section 3.7, along with "multidirected generalizations" of other basic properties of binary relations. Informally, in an acyclic multidirected set, multiple independent relations between a given pair of elements are allowed, but "only in one direction." In this case, for example, there are two independent relations from w to x and from x to z. The bottom right diagram illustrates a general multidirected set, containing cycles.

[29]This clarification is repeated in Section 5.9, where more general *linear directed sets* are introduced as "sources" of *causal paths*.

The bottom left and bottom right diagrams highlight the fact that cycles cannot be represented by generalized Hasse diagrams; the "up the page" convention does not suffice in this case. Hence, whenever cycles are present, the directions of relations are indicated explicitly by arrows.

3.7 Building Blocks of Structure in Multidirected Sets

Chains and antichains. Many important properties of multidirected sets may be described in terms of sequences of consecutive elements and relations, called *chains.* For example, a multidirected set M is *acyclic* if and only if no chain of nonzero length in M has the same element as its initial and terminal element. Also important are subsets of multidirected sets with no chains of nonzero length between any pair of elements, distinct or otherwise; these are called *antichains.* Both chains and antichains made informal appearances in Section 2.7, in the context of *Cauchy surfaces* and the *permeability problem.* They are formally introduced in Definition 3.7.1. Chains are then used in Definition 3.7.3 to generalize familiar mathematical conditions on directed sets, such as transitivity (TR) and interval finiteness (IF), to the case of multidirected sets, where such conditions can no longer be defined in terms of binary relations.

Definition 3.7.1. *Let $M = (M, R, i, t)$ be a multidirected set.*

1. *A* **chain** *γ in M is a sequence of elements and relations of the form $... \prec x_0 \prec x_1 \prec ...$ in M, where the notation $x_n \prec x_{n+1}$ refers to a particular relation r in R such that $x_n = i(r)$ and $x_{n+1} = t(r)$. The* **chain set** $\text{Ch}(M)$ *of M is the set of all chains in M.*
2. *A* **chain of length** *n, or n-***chain***, between x and y in M, is a chain γ of the form $x = x_0 \prec x_1 \prec \cdots \prec x_{n-1} \prec x_n = y$. The element x is called the* **initial element** *of γ, and the element y is called the* **terminal element** *of γ. The* **set of n-chains** $\text{Ch}_n(M)$ *in M is the subset of* $\text{Ch}(M)$ *consisting of all chains of length n. A* **complex chain** *is a chain of length at least two.*
3. *A* **cycle** *in M is a chain $x_0 \prec x_1 \prec \cdots \prec x_{n-1} \prec x_n$ of nonzero length*[30] *such that $x_0 = x_n$; i.e., its initial element coincides with its terminal element.*
4. *A relation r in R is called* **reducible** *if there exists a complex chain from its initial element to its terminal element. Such a chain is called a* **reducing chain** *for r. If r is not reducible, it is called* **irreducible**. *M itself is called* **irreducible** *if all its relations are irreducible.*

[30]In particular, an individual element, viewed as a 0-chain, is *not* regarded as a cycle, even though it is its own initial and terminal element. Including 0-chains as "trivial cycles" would necessitate the use of the word "nontrivial" preceding almost every instance of the word "cycle" throughout the remainder of the book.

5. *An* **antichain** σ *in* M *is a subset of* M *admitting no chain of nonzero length in* M *between any pair of elements* x *and* y *in* σ, *distinct or otherwise.*[31]

A 0-chain in M is just an element of M, and a 1-chain in M is just an element of the relation set R of M. The relation set R is the underlying set of the *relation space* $\mathcal{R}(M)$ over M, introduced briefly for directed sets in Section 1.5, and studied in detail in Chapter 5. More generally, it is useful to define *chain spaces* over M, whose underlying sets are the chain sets $\text{Ch}_n(M)$ and $\text{Ch}(M)$; such chain spaces are introduced in Section 5.9 as special cases of *causal path spaces*. The notions of initial and terminal elements generalize in the obvious way to the case of infinite chains that are "bounded in only one direction." For example, a chain $x_0 \prec x_1 \prec x_2 \prec \ldots$ with elements indexed by the natural numbers $\mathbb{N}$ has initial element x_0, but has no terminal element. A chain γ in a multidirected set M may include a given element or relation *more than once;* whenever this occurs, γ contains a cycle as a "subchain."[32] The possibility of repeated elements or relations in a chain reflects the fact that chains are defined as *sequences* of elements and relations, not as *sets* of elements and relations. However, it is common, and usually innocuous, to refer to "elements of a chain," or "relations in a chain," as if it were a set. Any relation r in M belonging to a cycle is reducible, since one may identify reducing chains "traversing the cycle." No element of M lying on a cycle in M can belong to an antichain in M, because such an element is connected to itself by a chain of nonzero length along the cycle. The reason for defining chains in terms of elements *and* relations, with the somewhat awkward use of the expression $x_n \prec x_{n+1}$ to denote a particular relation, is to incorporate individual elements in the definition, as 0-chains. For other purposes, it suffices to specify a chain purely in terms of its relations, since the intervening elements may be recovered via the initial and terminal element maps. An irreducible relation $x \prec y$ is called a *link* in the context of causal set theory.

Chains may be viewed as special cases of *causal paths,* which are discussed in more detail in Section 5.9. The distinction is that causal paths are defined in terms of morphisms from *linear directed sets*[33] into M, and the source objects of these morphisms may be more complicated than mere sequences of elements and relations. For example, they may include multiple chains between a given pair of elements. If a chain γ in a multidirected set M containing cycles possesses a terminal element z, then z need *not* be maximal among the set of elements of M belonging to γ, in the sense that there may exist another chain in M with initial element z and terminal element belonging to γ. Similarly, if γ has an initial element w, then w need not be minimal among the elements of M belonging to γ. In particular, a chain beginning on

---

[31] I thank Johnny Feng for suggesting explicit clarification prohibiting such chains in the case where $x = y$. Admitting such chains would render every individual element an antichain, and the *impermeability theorem* 5.4.3 would be false under this definition.

[32] Since chains are special cases of causal paths, which are properly defined in terms of morphisms, one should really speak in terms of *restrictions,* not "subobjects." However, the meaning is reasonably clear in the present context.

[33] A *linear directed set* is a directed set whose transitive closure is totally ordered. See Section 5.9 for details.

a cycle has no minimal elements in this sense, and a chain terminating on a cycle has no maximal elements, since no element of a cycle can be either minimal or maximal. These notions are clarified and elaborated in Section 5.9 in the more general context of causal paths.

A chain is roughly analogous to a causal curve in the context of relativity,[34] and may be regarded, from an intuitive perspective, as a "pathway along which influence or information may flow from past to future." In fact, the future of an element x in a multidirected set M is *defined* to be the set of all elements of M that may be reached from x by a chain in M, and the past of x is defined to be the set of all elements of M from which x may be reached via a chain. Pasts and futures are formally introduced in Definition 3.7.7, in the more general context of *domains of influence.* In a similar manner, an antichain is analogous to an acausal subset of relativistic spacetime, and a *maximal* antichain is roughly analogous to a Cauchy surface. In particular, the past and future of an antichain σ in M are necessarily disjoint, since given in an element x in their intersection, one could identify chains in M from σ to x and from x to σ, contradicting the hypothesis that σ is an antichain. In this sense, an antichain "separates past and future," much like a Cauchy surface. However, there are important conceptual distinctions between relativity and discrete causal theory in regard to such structural notions. For example, as explained in Section 2.6, if γ is a differentiable causal curve in a relativistic spacetime manifold X with tangent vector $v = v(x)$ at each point x on γ, then one may unambiguously distinguish between the case where γ is *timelike;* i.e., $g_x(v, v) < 0$ for every x, and the case where γ is *null;* i.e., $g_x(v, v) = 0$ for every x. In discrete causal theory, there generally is no such distinction. Returning to the subject of acausal subsets, discrete causal theory faces the generic *permeability problem* for maximal antichains, also mentioned in Section 2.7, and ultimately resolved in Chapter 5 via passage to relation space.

A simple but important property of antichains is that one may "add new elements" to an antichain σ as long as these elements do not belong to the past or future of σ, or to a cycle. Like much of the preceding discussion, this observation anticipates the formal introduction of pasts and futures in Definition 3.7.7 below. However, it is convenient to include it here as a lemma, in close proximity to the definition of antichains in Definition 3.7.1.

Lemma 3.7.2. *Let $M = (M, R, i, t)$ be a multidirected set, and let σ be an antichain in M. Suppose that x is an element of M belonging neither to σ itself, nor to the past or future of σ, nor to a cycle in M. Then the subset $\sigma' = \sigma \cup \{x\}$ of M is an antichain in M.*

Proof. Since σ is an antichain, there is no chain of nonzero length in M between any pair of elements of σ, distinct or otherwise. Since x is not in the future σ, there is no chain in M from any element of σ to x. Since x is not in the past of σ, there

[34]Again, the more general notion of *causal paths* enters the picture here; a causal path in a relativistic spacetime manifold is a special equivalence class of causal curves sharing the same image. Rovelli [RO04] and Thiemann [TH07] describe the distinction between curves and paths, using slightly different conventions.

is no chain in M from x to any element of σ. Finally, since x does not belong to a cycle in M, there is no chain of nonzero length in M from x to itself. Hence, there is no chain of nonzero length in M between any pair of elements of σ', distinct or otherwise, so σ' is an antichain in M. □

Mathematical conditions on multidirected sets. Special mathematical conditions on directed sets, such as transitivity, interval finiteness, irreflexivity, and acyclicity, defined in Sections 3.5 and 3.6 in terms of binary relations, have useful generalizations in the broader context of multidirected sets. Since binary relations are no longer part of the definition in this setting, these conditions must be expressed in terms of relation sets and initial and terminal element maps. Chains are particularly convenient for this purpose.

Definition 3.7.3. *Let $M = (M, R, i, t)$ be a multidirected set.*

1. *M is* **transitive** *if every complex chain in M is a reducing chain for a relation r in R.*
2. *Let x and z be two elements of M, not necessarily distinct. The* **open interval** *$\langle\langle x, z\rangle\rangle$ in M is the subset of M consisting of all elements y admitting chains of nonzero length from x to y and from y to z. M is* **interval finite** *if every open interval in M is finite.*
3. *A relation r in R is called a* **reflexive relation** *at x in M if $i(r) = t(r) = x$. M is* **irreflexive** *if R includes no reflexive relations.*
4. *M is* **acyclic** *if it contains no cycles.*

Note that the definition of an open interval $\langle\langle x, z\rangle\rangle$ in a multidirected set is formally different than the definition of an open interval in a causal set, introduced in Section 3.5 in the causal set axiom of interval finiteness (IF). However, both definitions capture the "correct" meaning of all elements lying "strictly between the two endpoints" x and z of the interval, under the relevant directed or multidirected structure. In particular, the two definitions coincide in the special case of transitive directed sets, because in this case, the existence of a chain of nonzero length between two elements implies the existence of a relation between them. Since causal sets are transitive directed sets, no actual conflict of definition arises. Other important conditions on multidirected sets, such as *star finiteness* (SF), are examined in Chapter 4.

Distinguishing between mathematical and physical properties. Whenever a multidirected set M is assigned a physical interpretation, for example, as a model of information flow or causal structure, it is vital to distinguish between its mathematical properties, which are "objective" or "absolute," and the physical characteristics ascribed to it, which are "subjective" or "relative." An important example of such a distinction exists between the mathematical notion of *irreducibility* of a relation between a pair of elements x and y in M, and the physical notion of *independence* of the information encoded by this relation, with respect to the information encoded by other chains between x and y. In particular, the above Definitions 3.7.1 and 3.7.3 are expressed in terms of *purely mathematical* notions, such as reducibility and irreducibility, making no reference to physical interpretations. However, Definition 3.7.4

below involves the physical property of *independence of families of chains between pairs of elements,* with individual relations included as an important special case. This definition, and the accompanying discussion, provide the necessary background for the *independence convention* (IC), introduced in Definition 3.7.5, which is crucial to the physical interpretation of directed sets and multidirected sets throughout the remainder of this book. The role of the independence convention is to *enforce objectivity* in discrete causal theory, by ensuring that isomorphic objects encountered in equivalent structural contexts are always interpreted in the same way. This means that once one has specified the intended role of a given directed or multidirected set; for example, as a model of classical spacetime, or as the underlying multidirected set of a kinematic scheme, then all the information necessary to interpret its physical attributes is encoded in its mathematical structure alone. This is necessary for the strong interpretation of the causal metric hypothesis (CMH) to make sense, because the hypothesis says that all physical interactions may be described in terms of such structure. Of course, this convention is intended to apply only in the context of fundamental physics; other applications of the same objects as information-theoretic models in fields such as computer science may require different conventions.

Independence: a physical property. Independence of a family of chains in a multidirected set is a *physical notion,* subjectively associated with such a family. This subjectivity may either be deliberately retained; i.e., isomorphic multidirected sets arising in similar contexts may be deliberately ascribed different meanings, depending on some type of auxiliary information, or it may be removed from the picture by fixing a specific convention, to be followed unless explicitly stated otherwise. I choose to follow the second option, since the overwhelming majority of the book deals with applications of such sets for which one particular convention is appropriate. This convention, called the *independence convention* (IC), is stated in Definition 3.7.5 below. First, it is necessary to spell out the meaning of independence in this context.

Definition 3.7.4. *Let $M = (M, R, i, t)$ be a multidirected set, interpreted as a model of information flow or causal structure, and let x and y be elements of M.*

1. *A family Γ of chains between x and y in M is called* **dependent** *if there exists another such family Γ', not containing Γ, encoding all information or causal influence encoded by Γ.*
2. *In particular, a chain γ from x to y in M is called* **dependent** *if there exists a family Γ' of chains from x to y, not including γ, encoding all information or causal influence encoded by γ.*
3. *If a chain or family of chains is not dependent, it is called* **independent**.

The mathematically inclined reader should be careful to observe that the notion of independence introduced in Definition 3.7.4 is much different than mathematical notions of independence appearing in fields such as linear algebra and matroid theory, and that this difference is not merely one of distinction between physical and mathematical properties. In these particular mathematical contexts, independence is, informally speaking, an absolute, "internal" property of the independent set under consideration, signifying *absence of information-theoretic redundancy.* In the present physical context, independence is a *relative* property, signifying *uniqueness of information content.* To understand the difference, consider a linearly independent subset S of a vector space V. The linear independence of S does not depend on its complement $V - S$; for example, one cannot "spoil" the linear independence of S by enlarging the vector space V to include additional dimensions. Further, S has no "monopoly" on the part of V it encodes; any basis for the span of S in V encodes, by definition, the same subspace of V.[35] By contrast, independence of a family Γ of chains in a multidirected set M means that Γ encodes, possibly in a redundant fashion, *at least some information* that cannot be recovered from any subobject of M not containing Γ. Here, one *can* "spoil" the independence of Γ by simply adding additional chains to M that carry the same information. Of course, *dependence* of a family Γ of chains in M *guarantees* a degree of redundancy in the information encoded by M as a whole, since it ensures, by definition, the existence of another such family Γ' containing the same information as Γ.

Dependence and independence may easily be generalized to the case of families of chains between pairs of *subsets* of M. These notions may also be refined by considering the distinctions between "partial overlap" of information between different families of chains, and *complete independence* of information content. The entire picture may also be broadened by replacing chains with more general *causal paths,* introduced in Section 5.9. However, Definition 3.7.4 suffices for the purposes of this book. Independence is constrained, but not determined, by the mathematical structure of M. Of principal interest is the dependence or independence of individual relations, i.e., 1-chains. For example, irreducible relations in directed sets are necessarily independent, since no other chains exist to "carry information" between their initial and terminal elements. Reducible relations, on the other hand, may be interpreted as either dependent or independent. For multidirected sets, multiple relations may share the same initial and terminal elements, so even irreducible relations may be interpreted as dependent in this context. These ambiguities illustrate the desirability of fixing a convention governing which relations are to be considered dependent or independent. Causal set theory implicitly treats reducible relations as *dependent,* as reflected in the practice of representing causal sets by "true" or "strict" Hasse diagrams. However, this convention leads to information-theoretic limitations in causal set theory, as explained in Section 3.9. The directed sets and multidirected sets of principal interest in this book satisfy a local finiteness condition called *star finiteness* (SF), which renders dependent relations information-theoretically superfluous. I therefore adopt the following convention:

Definition 3.7.5. Independence convention (IC). *Every relation in a multidirected set is interpreted as independent unless stated otherwise.*

[35] Of course, different bases encode different information in a *set-theoretic* sense, but this is not the type of information one is concerned with in the context of linear dependence and independence.

It is reasonable to ask why one would even *consider* including dependent relations in a physical model. In the presence of a suitable local finiteness condition, dependent relations only add useless redundancy, but many areas of theoretical physics call for locally infinite directed or multidirected sets in which *every* relation is both reducible and dependent. The principal example is, of course, relativistic spacetime, which exhibits this behavior due to the *interpolative property* inherited from the real numbers $\mathbb{R}$, whereby every individual relation $x \prec_{GR} z$ in spacetime admits an "interpolating event" y such that $x \prec_{GR} y \prec_{GR} z$. The interpolative property has been studied since antiquity; for example, it forms the crux of Zeno's *dichotomy paradox.* In a more modern context, the interpolative property is closely related to the use of Cauchy surfaces to separate "past" and "future" regions of spacetime in general relativity. This connection is illustrated in Figure 4.5.5 of Section 4.5. *Domain theory* is an interesting branch of "nondiscrete causal theory" involving interpolative causal structures, which therefore requires a different viewpoint than the one represented by the independence convention.[36] A useful reference for domain theory in the context of classical spacetime structure is the paper of Martin and Panangaden [MP06], already mentioned in preceding chapters.

The independence convention makes physical sense only if one asserts the freedom to use relations between pairs of elements to encode *actual* influence or information flow. The causal set axiom of transitivity (TR) removes this freedom by *prescribing* the existence of a relation between two elements in a causal set whenever there is a chain between them. This eliminates the option of representing *dependent* influence solely by means of complex chains, thereby limiting the variety of causal structures that may be modeled. This is the root of the Kleitman–Rothschild configuration space pathology [KR75], elaborated in Section 3.9. However, an even greater concern is the manner in which dependent relations cloud the overall structural picture in causal set theory, hampering the recognition and development of useful perspectives and methods. In particular, the theory of *relation space,* developed in Chapter 5, which is crucial to the derivation of dynamical laws for discrete causal theory, is severely muddled in the transitive context.

Domains of influence. Causal pasts and futures of subsets of relativistic spacetime manifolds were featured prominently in Chapter 2, particularly in the context of the *causality conditions* examined in Section 2.7. These pasts and futures are defined in terms of the potential influence of certain families of events on others, and may therefore be viewed as special cases of *domains of influence,* a general term used in this book to describe any subset or subobject of a "causal structure" admitting such a definition. A necessary step in the development of an effective version of discrete causal theory is to define and describe important domains of influence for more general directed sets and multidirected sets. Since membership of an element in a given domain of influence depends on whether or not this element precedes or succeeds certain other elements with respect to the relation structure of the set under consideration, the concepts of *predecessors* and *successors* are crucial to defining

[36]The domains studied in domain theory are entirely distinct from the *domains of influence* introduced in the present section.

domains of influence. These notions, along with important refinements specific to the discrete context, such as *direct* predecessors and successors, are made precise in Definitions 3.7.7 and 3.7.8 below.

Before stating these formal definitions, I briefly review some of the conceptual background for domains of influence. The prototypical example of such a domain in conventional physics is the *total domain of influence* $J(x)$ of an event x in Minkowski spacetime $\mathbb{R}^{3+1}$, which is the region consisting of all events separated from x by a timelike or null interval, bounded by the *null cone* of x. $J(x)$ may be written as the union $J^-(x) \cup J^+(x)$, where, qualitatively, $J^-(x)$ is set of all events that "could conceivably" influence x, and $J^+(x)$ is the set of all events that "could conceivably" be influenced by x. These regions are illustrated in the left-hand diagram in Figure 3.7.6. As explained in Section 2.6, $J^-(x)$ and $J^+(x)$ are called the *causal past* and *causal future* of x, respectively, even though there is no guarantee that every event in these regions is *actually* causally related to x. In general relativity, the same qualitative definitions of $J^-(x)$ and $J^+(x)$ in terms of "potential influence" apply. The geometry is more complicated, however, so timelike and null intervals are superseded by causal curves in the corresponding mathematical definitions. The middle diagram in Figure 3.7.6 illustrates the general relativistic situation in the case of a spacetime manifold satisfying the causal condition. For manifolds violating this condition, the causal past and future of a given event are not necessarily disjoint.

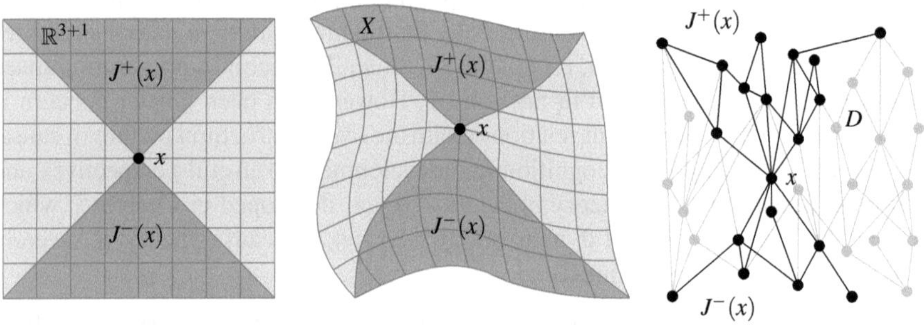

Figure 3.7.6. Total domain of influence of an event in Minkowski spacetime $\mathbb{R}^{3+1}$; corresponding general relativistic picture; total domain of influence of an element in a discrete directed set.

As mentioned at several points in Chapters 1 and 2, the uneasy distinction between which events "could conceivably" influence or be influenced by an event x, and which events *actually do* exert or experience such influence, arises from the *prescriptive* nature of spacetime geometry in general relativity. This distinction disappears in discrete causal theory. In particular, for a discrete directed set $D = (D, \prec)$, viewed as a model of classical spacetime under the classical causal metric hypothesis (CCMH), the causal past $J^-(x)$ of x consists of all elements representing events that actually *do* influence the event represented by x. Similar remarks apply to the causal future $J^+(x)$ of x, and the total domain of influence $J(x)$ of x. The right-hand diagram in Figure 3.7.6 illustrates these domains of influence. The terms "past" and "future" almost always mean "causal past" and "causal future" in the discrete causal context, so I generally drop the qualifier "causal" in what follows. The same physically suggestive notions remain useful for multidirected sets in general, although in this context "pasts and futures of elements" often possess "higher-level" physical meanings, ultimately understood in terms of *previous and potential stages of evolution* of classical histories in a kinematic scheme. Notions such as "past" and "future" may also be generalized to apply to arbitrary subsets of a multidirected set, rather than merely single elements, just as they apply to extended objects such as Cauchy surfaces in relativistic spacetime. These generalized definitions are useful, for example, in the theory of *causal atoms,* used in the proof of Theorem 4.6.3, and discussed more thoroughly in Section 5.7.

Precise definitions of predecessors, successors, pasts, futures, and other domains of influence in a multidirected set are expressed in terms of its relation structure.

Definition 3.7.7. *Let $M = (M, R, i, t)$ be a multidirected set, and let w, x, and y be elements of M. Let U be a subset of M.*

1. *If there exists a chain of nonzero length from w to x in M, then w is called a* **predecessor** *of x. The set $J^-(x)$ of all predecessors of x is called the* **past** *of x. More generally, the set $J^-(U) := \bigcup_{x\in U} J^-(x)$ is called the* **past** *of U.*
2. *If there exists a chain of nonzero length from x to y in M, then y is called a* **successor** *of x. The set $J^+(x)$ of all successors of x is called the* **future** *of x. More generally, the set $J^+(U) := \bigcup_{x\in U} J^+(x)$ is called the* **future** *of U.*
3. *The union $J(x) := J^-(x) \cup J^+(x)$ of the past and future of x is called the* **total domain of influence** *of x. Similarly, the union $J(U) := J^-(U) \cup J^+(U)$ is called the* **total domain of influence** *of U.*
4. *A* **domain of influence** *of x is a subset of $J(x)$. A* **domain of influence** *of U is a subset of $J(U)$.*

The identification of arbitrary subsets of $J(x)$ and $J(U)$ as "domains of influence" of x and U, respectively, is convenient due to its simplicity. However, domains of influence of actual physical interest are almost always defined in much more specific ways. For example, in the relativistic setting, the *chronological past* $I^-(x)$ and *chronological future* $I^+(x)$ of an event x, which made brief appearances in Chapter 2, are defined in terms of timelike curves, which are special cases of causal curves. Hence, although $I^-(x)$ and $I^+(x)$ *are* subsets of $J(x)$, they are far from being arbitrary subsets. In a similar manner, interesting domains of influence in discrete causal theory are usually defined in terms of conditions *at least as restrictive* as the existence of families of chains between certain families of events. In this sense, consideration of $I^-(x)$ and $I^+(x)$ provides a good conceptual point of reference for the study of domains of influence in more general settings, even though precise analogues of the relativistic chronological relation are generally absent in discrete causal theory.

Despite the conceptual tension between potential and actual influence in relativity, the mathematical classification of physically relevant domains of influence is comparatively simple in this context. This is because relations between pairs of events in relativistic spacetime are automatically both reducible and dependent, due to the interpolative property. For more general directed sets and multidirected sets, the possibilities of irreducibility and independence introduce qualitatively new features, giving rise to more nuanced domains of influence. These include *direct pasts and futures, maximal pasts,* and *minimal futures,* introduced in Definition 3.7.8 below. Even more complicated domains of influence may be defined in terms of *filtrations* of the pasts and futures of general subsets of a multidirected set. While these details are ultimately relevant to a more systematic study of the *entropic phase maps* introduced in Section 8.2, they are not pursued further in this book.

Definition 3.7.8. *Let $M = (M, R, i, t)$ be a multidirected set, and let x be an element of M. Let w be a predecessor of x, and let y be a successor of x.*

1. *If there exists a relation r in R such that $i(r) = w$ and $t(r) = x$, then w is called a* **direct predecessor** *of x. If r is irreducible, then w is called a* **maximal predecessor** *of x. The sets of direct predecessors and maximal predecessors of x are denoted by $J_0^-(x)$* and *$J_{\text{MAX}}^-(x)$, and are called the* **direct past** *and the* **maximal past** *of x, respectively.*
2. *If there exists a relation r in R such that $i(r) = x$ and $t(r) = y$, then y is called a* **direct successor** *of x. If r is irreducible, then y is called a* **minimal successor** *of x. The sets of direct successors and minimal successors of x are denoted by $J_0^+(x)$ and $J_{\text{MIN}}^+(x)$, and are called the* **direct future** *and the* **minimal future** *of x, respectively.*

It follows immediately from Definition 3.7.8 that $J_{\text{MAX}}^-(x) \subset J_0^-(x) \subset J^-(x)$, and that $J_{\text{MIN}}^+(x) \subset J_0^+(x) \subset J^+(x)$. In particular, the maximal past $J_{\text{MAX}}^-(x)$ of x is *not* maximal in terms of its "size;" the word "maximal" is inherited from the properties of the predecessors of x belonging to $J_{\text{MAX}}^-(x)$. The direct past $J_0^-(x)$ of x and the past $J^-(x)$ of x coincide in the case of transitive acyclic multidirected sets, by the definition of transitivity, as do the direct future $J_0^+(x)$ of x and the future $J^+(x)$ of x. In the case of interpolative multidirected sets, maximal pasts and minimal futures are always empty. For multidirected sets including cycles, pasts and futures are not necessarily disjoint. For example, if M includes a reflexive relation $x \prec x$, then the element x belongs to its own direct past and future, though not its maximal past or minimal future. Unambiguous physical interpretation of direct pasts and futures requires the independence convention (IC), or a suitable alternative, since without such a convention, the physical significance of reducible relations is subjective. Figure 3.7.9 illustrates the future, direct future, and minimal future of an element x in a multidirected set.

Boundary and interior; full subobjects. An element x in a multidirected set $M = (M, R, i, t)$ is called an *extremal element* if either its past or its future is empty. From a naïve physical perspective, a "Big Bang-like" or "Big Crunch-like" event might be modeled as an extremal element. "Free agency" might also be represented in this way. In the context of classical spacetime structure, the set of all extremal elements serves as a *causal boundary* of spacetime. The notion of a *spatial* boundary, by contrast, is much less straightforward, and similar statements apply to most

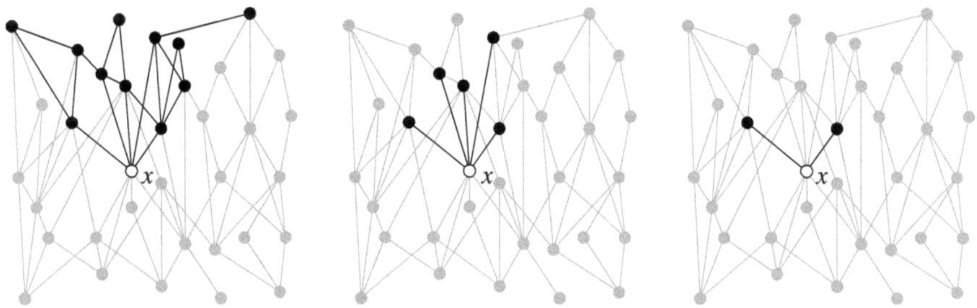

Figure 3.7.9. Future $J^+(x)$ of an element x in a multidirected set; direct future $J_0^+(x)$ of x; minimal future $J_{\text{MIN}}^+(x)$ of x.

spatial notions in discrete causal theory. Hence, the term "boundary" means "causal boundary," unless stated otherwise. The *interior* of a multidirected set is the complement of its boundary. In the context of classical spacetime structure, the interior is where "ordinary physics" occurs: every event in the interior has causes, and produces effects. These concepts are made precise in the following definition:

Definition 3.7.10. *Let $M = (M, R, i, t)$ be a multidirected set.*

1. *An element x in M is called* **maximal** *if $J^+(x) = \emptyset$, and is called* **minimal** *if $J^-(x) = \emptyset$. It is called* **extremal** *if it is either maximal or minimal.*
2. *The* **boundary** *∂M of M is the subset of all extremal elements of M, i.e., elements with either empty past or empty future.*
3. *The* **interior** *$\text{Int}(M)$ of M is the subset of all nonextremal elements of M, i.e., the complement of the boundary of M.*

The left-hand diagram in Figure 3.7.11 illustrates the boundary ∂M of a multidirected set M, represented by white nodes, and the interior $\text{Int}(M)$ of M, represented by black nodes. Note that Definition 3.7.10 recognizes only *elements* of M as possible constituents of ∂M or $\text{Int}(M)$; relations between interior elements are not included in the definition. However, it is sometimes useful to regard ∂M and $\text{Int}(M)$ as *subobjects* of M, rather than merely structureless subsets of M. The same is true for the domains of influence appearing in Definitions 3.7.7 and 3.7.8. Elevating a subset to a subobject requires specifying sets of relations to complement the sets of elements specified in these definitions. Unless stated otherwise, *all* relations between pairs of elements in the specified subset are included in the corresponding subobject; the resulting subobjects are called **full subobjects.** For example, the interior $\text{Int}(M)$ of $M = (M, R, i, t)$, viewed as a full subobject of M, consists of all nonextremal elements of M, *together with* all relations in M between pairs of nonextremal elements. By contrast, the only possible relations in the boundary ∂M of M are relations from minimal to maximal elements. The convention of working with full subobjects is important for interpreting some of the results of later sections, such as Theorem 5.3.1, which relates the interiors of certain multidirected sets. The right-hand diagram in Figure 3.7.11 illustrates the interior of a multidirected set M, viewed as a full subobject of M.

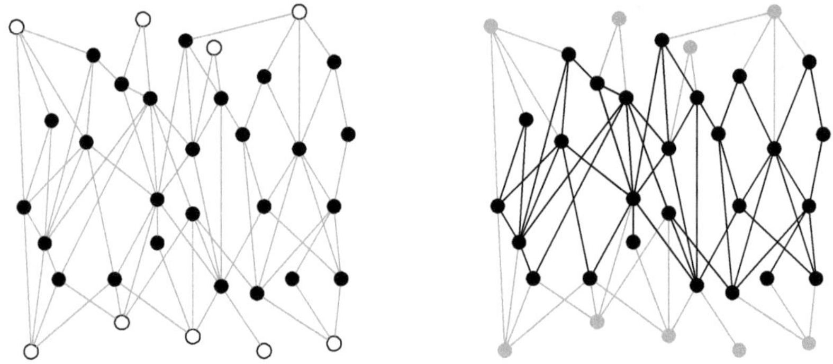

Figure 3.7.11. Boundary ∂M (*white*) and interior $\mathrm{Int}(M)$ (*black*) of a multidirected set M; $\mathrm{Int}(M)$ as a full subobject of M.

3.8 The Shadow of Grothendieck

Structural paradigms. Two broad structural paradigms that contribute significantly to this book are *order theory* and *category theory*. It turns out that neither paradigm is precisely suited for describing discrete causal theory; hence, it is actually *generalized or modified versions* of these paradigms that are most important here. The main shortcoming of standard order theory in this context is that it is simply not general enough. However, all the structured sets considered in this book, including totally ordered sets, causal sets, partially ordered sets, acyclic directed sets, directed sets, and multidirected sets, may be considered as *generalized* order-theoretic objects. Category theory, meanwhile, plays a dual role in this book, both as a general mathematical framework, and as a source of novel structural analogies for discrete causal theory. The version of discrete causal theory developed here returns the favor, by suggesting structural ideas similar to, but distinct from, those appearing in category theory. These "modified category-theoretic notions" describe *enriched versions* of the generalized order-theoretic structures mentioned above, in which elements are replaced by more complicated objects, such as directed sets, and relations are replaced by more complicated relationships, such as *co-relative histories.*

In this section, I gather together some basic ideas and results related to order theory and category theory, which will be useful throughout the remainder of the book. On the order-theoretic side, these include computation of the cardinalities of certain isomorphism classes of acyclic directed sets, as well as some standard properties of *order refinements* and *total orders.* These results are relevant to the theory of *relative multidirected sets over a fixed base,* introduced in Section 4.6, and the theory of kinematic schemes, detailed in Chapter 7. Turning to category theory, I give a preliminary glimpse of Christopher Isham's *topos-theoretic approach* to quantum gravity, revisited in Section 6.1, which offers an important example of the interplay between category theory and discrete causal theory.[37] I also discuss the philosophical distinctions between the popular modern pursuit of seeking precise category-theoretic descriptions of fundamental physics, and the alternative approach of merely exploiting whatever aspects of structure from category theory seem the most physically relevant. In my view, the latter approach is preferable. The mathematical work of Alexander Grothendieck is one of the principal sources of influence behind the perspective presented here. In particular, Grothendieck is the source of the *relative viewpoint* (RV), invoked repeatedly throughout the book, and is largely responsible for the idea of *hidden structure* (HS), which plays a special role in Chapter 4.

Counting classes of finite and countable acyclic directed sets. For future reference, it is useful to establish the cardinalities of a few distinguished classes of acyclic directed sets, considered up to isomorphism. These classes serve as the "object classes" of the "smallest physically relevant kinematic schemes," introduced and studied in Chapter 7. This type of computation may be easily generalized to determine the cardinalities of other isomorphism classes of multidirected sets with specified cardinality bounds on their sets of elements and relations, as well as the cardinalities of isomorphism classes of related *power spaces,* discussed in Section 5.6. First, since a finite set admits only a finite number of distinct binary relations, and since a countable union of countable sets is countable, the class $\mathcal{A}_{\mathrm{FIN}}$ of finite acyclic directed sets, considered up to isomorphism, is a set of cardinality $\aleph_0$, where $\aleph_0$ is the unique countably infinite cardinal. The class $\mathcal{A}_{\aleph_0}$ of countable acyclic directed sets, considered up to isomorphism, is a set of cardinality $2^{\aleph_0}$, i.e., the cardinality of the set of real numbers $\mathbb{R}$.[38] To see why this is true, first note that the unit interval $[0, 1]$ in $\mathbb{R}$, and hence $\mathbb{R}$ itself, may be placed in bijective correspondence with a subclass of $\mathcal{A}_{\aleph_0}$. One way of doing this, by associating each decimal expansion with a countable acyclic directed set, is illustrated in Figure 3.8.1. Anticipating the introduction of the local finiteness condition called *star finiteness* (SF), in Section 4.4, the construction illustrated in Figure 3.8.1 involves only the class $\mathcal{C}_{\mathrm{SF}}$ of isomorphism classes of star finite causal sets. The larger classes $\mathcal{A}_{\aleph_0}$, $\mathcal{C}$, and $\mathcal{A}_{\aleph_0,\mathrm{SF}}$ of isomorphism classes of countable acyclic directed sets, causal sets, and countable star finite acyclic directed sets, respectively, are therefore *at least* as large as $\mathbb{R}$, in terms of cardinality. It remains to show the converse. Let $S_{\aleph_0}$ be a fixed countably infinite set. The class of binary relations on $S_{\aleph_0}$ includes a representative of every infinite member of $\mathcal{A}_{\aleph_0}$; the finite members may be ignored since they are countable in number. Hence, $\mathcal{A}_{\aleph_0}$ corresponds bijectively with a subset of the power set of $S_{\aleph_0} \times S_{\aleph_0}$, which has cardinality $2^{\aleph_0}$, the cardinality of $\mathbb{R}$. Therefore, the four classes $\mathcal{C}_{\mathrm{SF}}$, $\mathcal{A}_{\aleph_0,\mathrm{SF}}$, $\mathcal{C}$, and $\mathcal{A}_{\aleph_0}$ also have cardinality *at most* as large as the cardinality of $\mathbb{R}$.

---

[37] Throughout the book, I offer repeated "comparisons" between category theory and discrete causal theory. These are obviously not "apples-to-apples" comparisons, since category theory is merely a way of organizing mathematical data, which cannot be "right" or "wrong," while discrete causal theory represents an attempt to describe the physical universe, which may or may not prove successful. The point of these comparisons is merely to evaluate the conditions under which category-theoretic notions are physically useful from the discrete causal viewpoint.

[38] Under the *continuum hypothesis,* $2^{\aleph_0}$ is equal to the smallest uncountable cardinal $\aleph_1$.

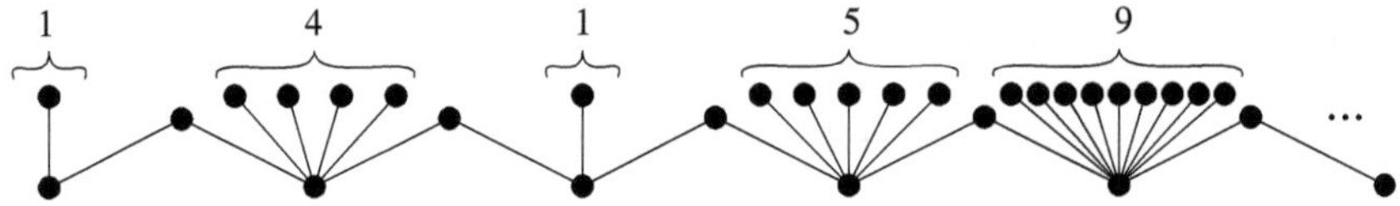

Figure 3.8.1. Portion of a locally finite causal set encoding the decimal $\pi - 3 = 0.14159...$

Total orders; order refinements. *Total orders,* also called *linear orders,* are, in a sense, the simplest "interesting" orders. For this reason, it is often useful to "compare" multidirected sets, and particularly acyclic directed sets, to totally ordered sets. The theory of *order refinements* provides a means of formalizing such comparisons, by extending an acyclic binary relation to a total order. Here, I mean a *strict* total order, such as the "less than" relation $<$ on $\mathbb{Z}$ or $\mathbb{R}$, but I often omit the qualifier "strict," since non-strict orders such as the "less than or equal to" relation $\leq$ do not play a significant role in the remainder of the book. Construction of such an order refinement defines a morphism of directed sets from the original acyclic directed set to a totally ordered set. Sorkin and Rideout's theory of *sequential growth dynamics* for causal sets [SR99], already discussed briefly in Section 3.4, illustrates the importance of such refinements, and their corresponding morphisms, in discrete causal theory. In particular, the *total labeling* $\{x_0, x_1, x_2, ...\}$ of a causal set $C = (C, \prec_{\text{CS}})$ associated with a particular sequence of transitions describing the sequential growth of C, is equivalent to a bijective morphism from C into a totally ordered subset of $\mathbb{N}$. In this book, it is necessary to compare acyclic directed sets to more general totally ordered sets, particularly those possessing *discrete total orders.* Such sets are "locally isomorphic to the integers," except possibly at extremal elements. This topic is discussed in more detail in Section 4.6.

The basic idea of order refinements may be expressed in the following way. Let $(D, \prec)$ be an acyclic directed set. Regarding the binary relation $\prec$ on D as a subset of the Cartesian product $D \times D$, a *refinement* of $\prec$ is a subset $\prec'$ of $D \times D$ containing $\prec$. In terms of individual relations between pairs of elements of D, $\prec'$ is a refinement of $\prec$ if and only if each relation $x \prec y$ implies the relation $x \prec' y$. A *strict total order* on D is a transitive acyclic binary relation $\prec''$ on D, satisfying the property that either $x \prec'' y$ or $y \prec'' x$ for every pair of *distinct* elements x and y in D. A refinement of the acyclic binary relation $\prec$ on D to yield a strict total order $\prec''$ on D is equivalent to a bijective morphism of directed sets $(D, \prec) \rightarrow (D, \prec'')$. The existence of total refinements of acyclic binary relations is surprisingly subtle. A conventionally-worded version of the following *order refinement principle,* often called the *order extension principle,* is sometimes used as an *axiom* in modern set theory.

Definition 3.8.2. Order refinement principle (ORP). *Every acyclic binary relation $\prec$ on a set D may be extended to a strict total order on D.*

In Zermelo–Fraenkel set theory, the order refinement principle may be *proven,* using the *power set axiom* and *Zorn's lemma.* The power set axiom states that the power set $\mathcal{P}(D)$ of a set D is itself a set, rather than a *proper class,* i.e., a class "too large" to be a set.[39] Zorn's lemma is actually an axiom, closely related to the *axiom of choice;* it states that any partially ordered set satisfying the property that every chain is bounded above possesses at least one maximal element. Zorn's lemma may be applied to the class of refinements of an acyclic binary relation $\prec$ on D, ordered by inclusion, which is itself a set, by an elementary application of the power set axiom. The union of the elements of any chain of refinements of $\prec$ serves as an upper bound for the chain, and the maximal element guaranteed by Zorn's lemma is a linear order on D, generally *not* unique.

Category theory revisited. In Chapter 2, I introduced category theory informally, as a way of organizing different types of structure on relativistic spacetime manifolds. In Sections 3.5 and 3.6, I introduced the specific categories of causal sets, acyclic directed sets, directed sets, and multidirected sets. Here, I revisit the subject of category theory in slightly more detail, and outline its subsequent use in this book. The same subject is revisited a third time in Section 5.10, in the context of *causal path semicategories* and *causal path algebras.*

Definition 3.8.3. *A* **category** $\mathcal{C}$ *consists of a class of objects* $\text{Obj}(\mathcal{C})$*, and a class of morphisms* $\text{Mor}(\mathcal{C})$*, such that every morphism* γ *possesses an initial object* $\mathbf{i}(\gamma)$ *and a terminal object* $\mathbf{t}(\gamma)$*, every object* X *possesses an identity morphism* γ_X *with initial and terminal object* X*, and morphisms compose associatively.*

A morphism γ with initial object X and terminal object X' is often denoted by the "mapping notation" $\gamma : X \to X'$, but in abstract settings, a morphism need not involve a map between sets. The distinguishing property of the identity morphism γ_X is that it is *absorbed under composition;* i.e., for any morphism $\gamma : X \to X'$, $\gamma_{X'} \circ \gamma = \gamma$ and $\gamma \circ \gamma_X = \gamma$. For many categories, objects possess "internal structure," and morphisms are chosen to "preserve structure" in some way. This is the case, for example, for the applications of category theory in general relativity appearing in Chapter 2, in which the morphisms under consideration are maps between relativistic spacetime manifolds, chosen to preserve topological, smooth, causal, conformal, or metric structure. In the present chapter, morphisms of multidirected sets preserve initial and terminal elements of relations. However, there also exist important settings in which the objects of a category do *not* possess internal structure, and the entire structure of the category resides in its morphisms. An example is the category-theoretic approach to *groups,* in which group elements are represented by endomorphisms of a single structureless object. A generalization of this viewpoint is useful in the context of *generalized semicategories,* which feature prominently in Section 5.10.

Category theory plays two distinct roles in this book; the first conventional, and the second relatively novel. First, as in Chapter 2, it serves as a *general structural paradigm,* i.e., a convenient method of organizing information about mathematical

[39] As von Neumann and others have pointed out, "size" is actually a rather poor way to understand the distinction between sets and proper classes, but no great sophistication is needed in the present context.

objects, and relationships between pairs of such objects. Throughout modern mathematics, this use of category theory is not only conventional, but ubiquitous; for example, category theory plays a role in many, if not most, mathematical papers involving modern algebra. Second, as illustrated throughout the book, mathematical structures of special interest in discrete causal theory, including directed sets, multidirected sets, and kinematic schemes, *closely resemble* categories, without necessarily satisfying all of the category-theoretic axioms. This leads to interesting interplay between discrete causal theory and category theory. In one direction, category theory serves as a useful source of analogy for discrete causal theory; for example, relations "resemble morphisms" in an obvious sense. In the other direction, discrete causal theory suggests natural structural viewpoints that play essentially the same role as category theory, but differ from it in important ways.

Isham's program. Christopher Isham and his collaborators have carried the analogy between category theory and (generally nondiscrete) causal theory very far in a particular direction, organizing families of causal sets, as well as many other types of physical models, into special categories called *topoi,* in an effort to build up a general *topos-theoretic* framework for physics. Many of these models are not based on discrete structure or causal structure at all, but an overarching connection to causal theory remains evident, due to the underlying directed structure of categories. Though Isham's work is one of the major motivations for the version of discrete causal theory developed in this book, my own perspective regarding the proper role of category theory in physics is somewhat different. To understand why, it is helpful to consider the following two questions:

> **Question 1:** *How can existing structural notions in mathematics, such as category theory, be applied to theoretical physics?*
>
> **Question 2:** *Does conventional mathematics supply suitable structural notions for theoretical physics, and if not, what more can physics teach us about structural notions?*

Isham's program seems to focus mainly on the first question, with the principal "existing structural notion" being topos theory. To be sure, Isham considers the second question as well; for example, in his recent paper *Topos Methods in the Foundations of Physics* [IS11], he points out the dangers of taking for granted the real numbers $\mathbb{R}$ and complex numbers $\mathbb{C}$ in the context of quantum gravity. In addition, he proposes replacing the notion of probabilities with *"generalized truth values,"* and suggests that *"conventional quantum formalism is inadequate to the task of quantum gravity."* All these points are well taken. However, Isham advocates topos-theoretic solutions to these issues in this particular paper, rather than new structural paradigms motivated a priori by physical considerations.

Physical limitations of category theory. My own view is that category theory, including topos theory, is not ideally suited to describe fundamental physics, and that causal theory, discrete or otherwise, motivates *essentially different* structural notions that are worth studying in their own right. Category theory and topos theory arose from the fields of homological algebra and algebraic geometry between

1940 and 1960, in response to specific mathematical issues only distantly related to fundamental spacetime structure. These issues included the necessity of formalizing the essential uniqueness of certain cohomology theories,[40] and assigning useful algebraic structure to generalized topological spaces.[41] While sufficiently rich and general to *accommodate* any conceivable physical theory, categories and topoi may not always be *optimal* for this purpose, and the choice to assign undue weight to the properties of these theories themselves may lead in the wrong direction from a physical perspective. By contrast, the clear physical relevance of structures essentially different from categories, but playing similar roles, such as the *kinematic schemes* arising in discrete causal theory, suggests that these alternative structural notions may be more appropriate in many physical contexts.

Of course, structures very similar to kinematic schemes already appear, in a variety of mathematical settings, as "enriched" or "categorified" *graphs* or *quivers,* but these are seldom accorded broad structural significance in their own right. However, the legacy of interaction between mathematics and physics suggests than any concept of fundamental physical significance should be taken seriously, even from a purely mathematical perspective. An excellent example of the continued mathematical potency of physically-motivated methods, setting aside the question of their actual physical relevance, is the interplay between string theory and algebraic geometry and topology, for example, in *Gromov–Witten theory, "Monstrous Moonshine," mirror symmetry,* and so on. This book contains many examples of "category-like" structures, arising naturally in discrete causal theory, for which naïve category-theoretic intuition leads in the wrong direction. Examples include "objects without identity morphisms," and "morphisms which do not compose," which appear repeatedly in the theory of multidirected sets, *relation spaces,* and *power spaces;* "morphisms without initial and terminal objects," which appear in the theory of *causal path semicategories;* and "morphisms whose compositions are families of morphisms, rather than individual morphisms," which appear in the theory of *co-relative histories.* These examples provide ample evidence that the analysis of basic structural questions is still very interesting and open-ended in theoretical physics, not merely an exercise in choosing and applying conventional mathematical tools.

Influence of Grothendieck. Many of the structural ideas in this book, both conventional and novel, have *algebraic* roots in the work of Alexander Grothendieck, the father of modern algebraic geometry. In light of the foregoing discussion regarding category theory, it is interesting to note that it was Grothendieck, along with his student Verdier, who introduced topos theory around 1960. Most of the topics to which Grothendieck devoted his attention play little direct role in discrete causal theory, although *noncommutative* geometry provides methods for studying the *causal path algebras* introduced in Section 5.10. However, noncommutative geometry took its modern shape beginning around 1980, after Grothendieck's most productive period. Hence, it is the *broad structural concepts* championed by Grothendieck, not his specific technical achievements, that are of principal interest here. Two such concepts are

[40]That is, the theory of *natural transformations* of functors.

[41]That is, the theory of *sheaves, sites,* etc.

the *relative viewpoint* and the principle of *hidden structure.* The relative viewpoint is well-known as such, and may be ascribed explicitly to Grothendieck. "Hidden structure" is my own terminology, and refers more to a general lesson that may be drawn from Grothendieck's modern version of algebraic geometry than to a specific idea deliberately advocated by Grothendieck.

Definition 3.8.4. Relative viewpoint (RV). *Objects should be studied together with their natural relationships, rather than in isolation.*

Grothendieck applied the relative viewpoint systematically to study the properties of *commutative algebraic varieties* and *schemes,* which are geometric objects defined locally by the vanishing of polynomials. In this context, one fixes a "base scheme" Z, and considers the *category of schemes over* Z, i.e., the category whose objects are morphisms of schemes $X \rightarrow Z$. The signature application of this idea was Grothendieck's proof of the *Grothendieck–Riemann–Roch theorem*[42] in 1957, and similar methods have since become ubiquitous in modern algebra. In the context of discrete causal theory, the relative viewpoint plays a particularly striking role. I have already foreshadowed its use in the brief introduction to *relation space* in Section 1.5. The first formal application, however, appears in Section 4.6, where I use it to study *relative multidirected sets over a fixed base,* in the context of local and global conditions on multidirected sets. For example, causal sets may be viewed as relative directed sets over the integers. In Chapter 5, the theory of relation space is developed in detail, along with a number of generalizations and applications. These lead to many of the most important technical methods of discrete quantum causal theory. In particular, the relative viewpoint serves as the conceptual foundation of the theory of *co-relative histories* and *kinematic schemes,* developed in Chapters 6 and 7, which formalizes the idea of "spacetime evolution" in discrete causal theory.

The relative viewpoint is generally distinct from the long-standing physical and metaphysical philosophy of *relationism,* which advocates the idea that the properties of objects are meaningful only relative to those of other objects. However, interesting connections may be recognized between the two ideas in the context of discrete causal theory. For example, the theory of relation space enables the physical universe to be regarded, roughly speaking, as "built of nothing but relationships." Here, the qualifier "roughly speaking" is included for two reasons. First, the validity of the statement assumes the strong interpretation of the causal metric hypothesis (CMH); without this assumption, auxiliary data must be added to relation space in order to recover certain physical interactions. Second, an arbitrary directed set can only be "approximately" recovered from its relation space, since some ambiguity exists regarding information on the causal boundary. This topic is discussed in detail in Chapter 5.

Definition 3.8.5. Hidden structure (HS). *Objects presented as structured sets may often be viewed as enlarged sets with "extra elements," naturally induced by their native structures.*

---

[42]This theorem may be viewed as a special case of the *Atiyah–Singer index theorem.* It has become so prominent in string theory that at least one string theorist has proposed renaming it the "Atiyah–Singer string index theorem."

More general statements of this principle, omitting any mention of sets or elements, may be formulated, but the above statement suffices for the purposes of this book. Grothendieck invoked hidden structure in his theory of algebraic schemes, by adding new elements of positive dimension to the sets of zero-dimensional points of algebraic varieties. Since algebraic schemes are more natural and better-behaved than algebraic varieties in many respects, the "promotion of varieties to schemes" may be viewed as a belated recognition of hidden structure latent in varieties, unnoticed for more than two millennia. In Chapter 4, I treat the *relations* in a multidirected set $M = (M, R, i, t)$ as elements of a topological space called the *star model* $M_\star$ of M, which plays an important role in the study of *causal locality.* As a set, $M_\star$ is the union of M and its relation set R. Hence, "promotion of M to $M_\star$" is roughly analogous to "promotion of varieties to schemes."[43] The local finiteness condition called *star finiteness* (SF) derives its name from the star model. In Chapter 5, I elevate the relation set R to the *relation space* $\mathcal{R}(M)$ over M, already introduced informally in Section 1.5, and briefly revisited above in the context of the relative viewpoint. Hidden structure is further exploited in the context of *power spaces, causal path spaces,* and *kinematic schemes,* studied in Chapters 5, 6, and 7.

Much of Grothendieck's mathematical output came during a period of relative estrangement between fundamental physics and abstract mathematics, which continued until the string theory revolutions, Alain Connes' work on noncommutative geometry, and other developments during the 1980s and 1990s. With a few exceptions, such as Penrose's *twistor theory,* physicists relied during this period on mathematical notions conceived in the century from Riemann to Weyl. Even in the modern physics community, the structural ideas of Grothendieck and his contemporaries are not as broadly appreciated as one might wish; for example, Isham's paper [IS05] implicitly recognizes this as late as 2005 by offering an "apology" for using methods as abstract as category theory! This unfortunate state of affairs may be attributed partly to a legacy of physically unilluminating exposition emanating from certain mathematical sources,[44] and partly to "pragmatism" in the physics community during the heyday of experimental particle physics.

3.9 Independent Modes of Influence

Transitivity. In this section, I examine in detail one of the major technical limitations of causal set theory; namely, its reliance on *transitive* binary relations to encode causal structure at the fundamental scale. Like many other technical difficulties throughout

[43] One could attempt to draw closer analogies involving "larger" objects arising in discrete causal theory; for example, *causal path spaces* are analogous to "two-dimensional schemes," in the sense that they possess elements corresponding to "zero-dimensional" and "one-dimensional" aspects of structure.

[44] An example is the *Bourbaki* group, to which Grothendieck briefly contributed. One of the principal scribes for *Bourbaki* was Jean Dieudonné, who also penned Grothendieck's monumental *Éléments de Géométrie Algébrique* (EGA).

mathematics and theoretical physics, this particular problem is closely related to a specific *conceptual* issue; namely, the risk of assigning undue emphasis to the long-studied mathematical properties of the relativistic causal relation $\prec_{GR}$ on a relativistic spacetime manifold X. An imperfect analogy may be found in linguistics: when translating a passage of text from one language to another, it is rarely the case that word-for-word transcription using a dictionary captures the essential meaning in the most effective way. In a similar fashion, transporting the abstract mathematical properties of the relation $\prec_{GR}$ directly to the discrete context, without modification, leads to information-theoretic limitations in the resulting theory. This topic requires careful explanation, since potential for confusion arises from at least three different sources. First, the concept of causality itself is subtle and controversial; there exist competing points of view regarding what "causal structure" *should mean.* Second, since the study of cause and effect reaches back to antiquity, every term associated with the subject is already charged with various meanings, whether or not these meanings are "correct" or "useful;" hence, it is easy to become confused about what a given expression is *intended to mean.* Third, there is the problem of *capturing the desired meaning* in the form of mathematically precise definitions.

Transitivity (TR) is deeply ingrained in conventional thought as part of the *definition* of classical causal structure. For example, the relationship between a grandmother and grandson, though typically indirect, is recognized, via transitivity, as "causal:" the grandmother A is directly related to her child B, who in turn is directly related to the grandson C; hence, the grandmother is related to the grandson. Challenging this convention would merely incite terminological controversy, to no good purpose. However, to classify both "relationships of direct descent," such as the parent-child relationship, *and* "relationships of indirect descent," such as the grandparent-grandchild relationship, as "causal," does *not* mean that such relationships are indistinguishable, or that their distinctions are physically unimportant. This may seem "obvious" in hindsight, and it may be that no serious reader will dispute the conclusions of this section after reading it carefully. However, the historical timing of this particular work renders careful treatment of this issue a necessity, since the problem addressed here is prevalent in the recent literature. Authors a generation from now may enjoy the luxury of treating this topic in a much more streamlined fashion.

In continuum-based theories such as general relativity, *every causal relationship is indirect* due to the interpolative property, i.e., due to the fact that there always exists a third event between any pair of distinct events on a causal curve. In discrete causal theory, however, direct relationships are possible, and must be accounted for. Hence, I choose to model "relationships of direct descent" in terms of a more fundamental, generally *nontransitive* binary relation $\prec$, which I refer to as the *causal relation,* and which is interpreted to encode only *independent* influences between pairs of events, following the independence convention (IC). A transitive binary relation $\prec_{tr}$, called the *transitive relation,* may be recovered from the causal relation $\prec$, by applying the *transitive closure functor,* since every "relationship of indirect descent" is composed of "relationships of direct descent." This is the reason for the subscript "tr" in the expression $\prec_{\text{tr}}$.[45] In the acyclic case, $\prec_{\text{tr}}$ is a strict partial order, which plays essentially the same role as the causal set relation $\prec_{\text{CS}}$ on a causal set. In this context, it is tempting to refer to the causal relation $\prec$ as the "causal preorder," and the transitive relation $\prec_{\text{tr}}$ as the "causal order," since $\prec$ serves as a "precursor" to $\prec_{\text{tr}}$. However, too much confusion results if one attempts to extend this choice of terminology to the case of directed sets containing cycles, for which $\prec_{\text{tr}}$ is not an actual partial order.[46]

In the remainder of the present section, I analyze the basic deficiency of transitive binary relations for modeling classical spacetime structure: they fail to distinguish among scenarios involving *multiple independent modes of influence* between pairs of events. I discuss some of the rare treatments of this issue appearing in the literature, due to David Finkelstein, Ioannis Raptis, Fotini Markopoulou, and a few others, and also record Rafael Sorkin's current view of the subject, expressed via private communication. I then present six arguments against using transitivity as an axiom in discrete causal theory. I focus on the case of directed sets, but the ideas involved generalize easily to the multidirected context.[47] The principal conclusion of this section is that *transitive binary relations are information-theoretically inadequate to model classical spacetime structure in discrete causal theory.*

Basic deficiency of transitive "causal relations." The need to invoke a nontransitive causal relation, more fundamental than a partial order even in the acyclic case, arises from the possibility of multiple independent ways in which one event may influence another, called *modes of influence.* A simple example of this is illustrated in Figure 3.9.1. Consider three events, labeled x, y, and z. If x influences z, then this influence may be either exclusively independent of y, as illustrated in the left-hand diagram, exclusively mediated by y, as illustrated in the middle diagram, or it may involve *both* independent influence *and* influence mediated by y, as illustrated in the right-hand diagram. In accordance with the independence convention (IC), the relations indicated by solid lines in the figure all represent independent influences; the dashed line in the middle diagram indicates the *hypothetical* relation $x \prec z$ prescribed under the axiom of transitivity (TR). Certain popular approaches to discrete causal structure, such as causal set theory, which do *not* employ the independence convention, and which *do* impose transitivity, *prescribe* the presence of this relation, and therefore cannot distinguish between the physical scenarios represented by the middle and right-hand diagrams. This is a serious information-theoretic deficiency, which extends in an obvious way to apply to any relation in a multidirected set admitting a reducing chain.

---

[45] Most subscripts appearing in this book use capital letters; the reason for the use of lowercase letters in this case is that it is more convenient to denote the transitive closure of a directed set D by $\text{tr}(D)$ than by $\text{TR}(D)$.

[46] I learned this the hard way by using the term "causal preorder" for $\prec$ in the paper [DR13].

[47] It is safe to omit multidirected sets from the present discussion, since the existence of multiple independent "modes of influence" is usually obvious in the settings in which such sets arise; for example, in the "decategorification" of a kinematic scheme.

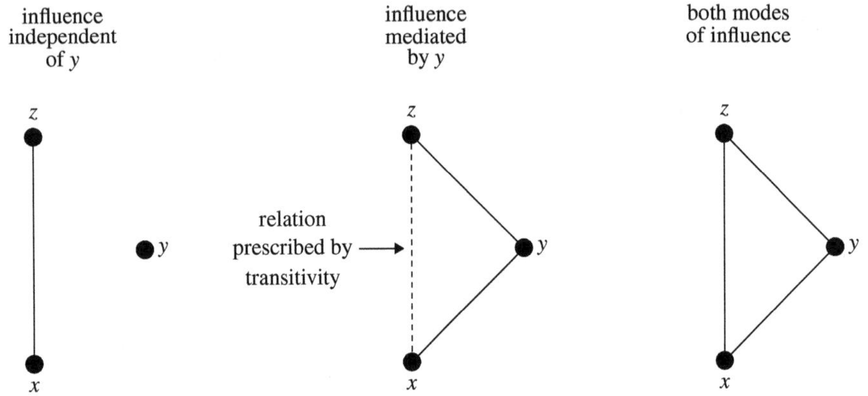

Figure 3.9.1. Possible modes of influence between related elements in three-element directed sets.

Naïve analogies from information theory. A cryptographic analogy is illustrative of this deficiency. Consider the problem of sending a secret message from x to a remote recipient z, either independently, as in the left-hand diagram in Figure 3.9.1, or via an intermediary y, as in the middle diagram. The intermediary can guarantee delivery without third-party tampering, but may not be trustworthy. One possible strategy is to send part of the message independently, and the other part via the intermediary, as in the right-hand diagram, in such a way that its meaning can only be deciphered by combining both parts. Transitive relations are clearly deficient for describing these three scenarios, since they fail to distinguish between the last two. Transitive relations can encode *which* pairs of events communicate, but cannot encode *how* they communicate. This problem only becomes worse for larger families of events, in which there may be many independent modes of influence between a given pair of events. The consequences of ignoring such independence may be illustrated by imagining an academic conference, at which the chair announces that only *one* member need return the following year, since the information brought to the second meeting by all the other members is sure to be redundant!

Finkelstein's causal nets. The foregoing ideas are so simple, and the reasonable alternatives so few, that the same basic conceptual steps have surely been followed many times previously. However, it is surprisingly difficult to find explicit treatments of this issue in the literature, at least in the context of classical spacetime structure. Here, I briefly detail a few instances that come as close as any I could find to addressing this problem. David Finkelstein, in his 1988 paper *"Superconducting" Causal Nets* [FI88], writes,

> *...the* [transitive] *causal relation* $x\mathbf{C}y$ *is not local, but may hold for events as far apart as the birth and death of the universe. Since we have committed ourselves to local variables, we abandon* **C** *for a local causal relation* **c**... (page 476)

and later,

> *We call* **c** *the (causal) connection relation, understanding that this connection is... ...immediate (x***c***y and y***c***z do not necessarily imply x***c***z). Events in a continuum theory have no immediate causal relations...* (page 477)

and again,

> *Every causal set* **C** *may also be regarded as a causal net* **c**, *with the connection x***c***y defined to hold if and only if* [the relation $x\mathbf{C}y$ holds] *and no event z exists with x***C***z***C***y... ... but* [causal nets] *are more general than causal sets.* (page 477)

Finkelstein's "local causal connections" $x\mathbf{c}y$ are *irreducible* relations, i.e., what causal set theorists refer to as *links* in a less-general context. These remarks suggest that Finkelstein's principal objections to transitive binary relations were[48] based on issues of *locality* rather than independence, although the greater generality afforded by dropping the axiom of transitivity (TR), noted in the final quote above, is information-theoretically meaningful only if reducible relations can encode independent influence. Finkelstein's locality-based critique of transitivity is valid, and in my opinion, unanswerable, though the worrisome structural issues involving locality appearing in approaches such as causal set theory do not arise from transitivity alone. This topic is elaborated in Chapter 4.[49] Due to Finkelstein's use of a local causal relation, his *causal nets* share a greater abstract mathematical similarity with the directed sets of principal interest in this book than do causal sets, at least in the specific context of modeling classical spacetime structure. However, the overall approach developed here is much closer to causal set theory than to the theory of causal nets, particularly in the quantum-theoretic context.

Recent literature on transitivity. A few relatively recent papers examining the use of nontransitive causal relations in the study of fundamental spacetime structure may also be found in the literature. In particular, Ioannis Raptis, in his 2000 paper *Algebraic Quantization of Causal Sets* [RA00], deprecates transitive binary relations as unsuitable for modeling *quantum* spacetime, writing,

> *...the physical causal connection between events in the quantum deep should be one connecting nearest-neighboring events.* (page 1238)

Raptis cites Finkelstein's paper [FI88] as the source of this idea. However, Raptis uses *irreducible* relations in his "quantum causal sets," thereby sidestepping the issue of independence. Also, the objections to transitivity outlined above apply not only in the *"quantum deep,"* but also at the classical level. Very recently, Stoica [ST15] has defined *horismotic sets* via a nontransitive relation similar to the *skeletal relation* appearing in the table in Figure 3.5.2.

Fotini Markopoulou seriously considers the implications of transitivity in her inaugural paper [MA00] on *quantum causal histories,* which are essentially causal sets enhanced by the addition of algebraic structure. This construction is very similar

---

[48] I say "were," rather than "are," because Finkelstein suggested to me, via private communication, that his views on the topic have changed considerably since 1988.

[49] For an interesting recent view on the subject, one may consider Finster's theory of *causal fermion systems;* see, for example, [FI16].

to Isham's *quantization on a category* [IS05], in the sense that objects of a category, rather than individual events, are associated with elements of a directed set.[50] In this case, the objects are Hilbert spaces. Markopoulou recognizes the crucial difference between the continuum-based and discrete settings in regard to transitivity, writing,

> *It is, therefore, an open question whether it is sensible to impose... transitivity... on the ordering of the Hilbert spaces.* (page 10)

Markopoulou credits Isham with some of the background on the question of transitivity, including the papers [IS94, IS97], as well as private correspondence. She demonstrates that one simplification resulting from the choice to impose transitivity is that it ensures "good behavior" under *directed coarse-graining*[51] of quantum causal histories, which essentially involves "telescoping" chains of relations to obtain simpler directed structure. At an algebraic level, it is interesting to compare this construction to the operations in Raptis' *incidence algebras,* mentioned at the end of Section 5.10, and Isham's *monoids of arrow fields,* briefly discussed in Section 6.1. From the perspective of this book, this simplification does not provide a convincing *physical* motivation for restricting attention to transitive relations, any more than the convenience of working with commuting families of operators justifies ignoring the uncertainty principle. Markopoulou is very explicit about the possible need to consider nontransitive relations, particularly under more general assumptions involving "non-point-like" elements of directed structure.

Sorkin's current view. An interesting *historical* point, concerning the alternative between transitive and nontransitive causal relations in the study of fundamental spacetime structure, is that Finkelstein [FI88] *credits Sorkin* with suggesting the idea of the nontransitive local "causal connection relation" **c** appearing in Finkelstein's paper, via private communication in 1987! This indicates that Sorkin has considered, and to some extent encouraged, the development of *both* approaches. Sorkin recently expressed to me his current view of the matter in the following remarks, reproduced here with his permission:

> *I suppose the main reason* [not to express causal structure in terms of generally nontransitive binary relations] *is that* [doing so appears] *to add nothing to the theory, insofar as it tries to explain how a Lorentzian spacetime can emerge from an underlying discrete order. In the continuum one has well defined causal relations like* [the relativistic causal relation], *and they are automatically transitive, whence suited to correspond to a discrete transitive relation of precedence. But what in the continuum would correspond to a more general... ...not-necessarily transitive relation...?*

These remarks seem to suggest the view that the choice of whether or not to impose transitivity is largely one of focus and ambition; if one seeks merely to *recover* relativistic spacetime, at a suitable level of approximation, from a binary relation on a "discrete set," then it may indeed *"add nothing to the theory"* to consider the possible

[50] See Section 8.6 for a brief discussion of the more general topic of *categorification.*

[51] In the context of "faithful approximation of directed structure," directed coarse-graining may be compared to *causal atomic resolution,* discussed in Section 5.7.

independence of reducible relations.[52] There is perhaps something humorous about associating the adverb "merely" with a research objective which would represent a monumental advance in theoretical physics, should it succeed; namely, the recovery of relativistic spacetime geometry via causal set dynamics. However, if one *does* aspire to go further, and to recover "matter and energy" along with spacetime, under the strong interpretation of the causal metric hypothesis (CMH), then the possible role of independence becomes clearer. Sorkin goes on to suggest this very idea, while expressing open-minded reservations about its viability:

> [the question is] *whether such additional structure could play a useful kinematical or dynamical role in the theory, perhaps as a kind of "matter" living on the* [causal set] *(a rudimentary scalar field). At present that looks unlikely to me, but of course I'd never say it was impossible.*

My own view is that, given the extreme simplicity of binary relations, any approach to fundamental physics based on such relations should a priori attempt to leverage every bit of natural structure available. This is particularly true if one can *obtain more structure by assuming less;* in this case, obtaining a "scalar field" by replacing the axioms of transitivity (TR) and irreflexivity (IR) with the weaker axiom of acyclicity, (AC) or omitting these axioms altogether. I do not know the details of Sorkin's reasons for doubting that such a generalization can play a *"useful kinematical or dynamical role"* in discrete causal theory, but subsequent chapters of this book present, in my view, strong reasons to believe that it can. Hence, I regard the omission of nontransitive binary relations in causal set theory as an unfortunate example of "censoring the merely unexpected," in the sense of Section 2.10.[53]

Independence at the fundamental scale. Regardless of the opinions of the experts, one might argue that the naïve information-theoretic analogies presented above make the deficiency of transitive binary relations in discrete causal theory seem more obvious than it actually is. I do not personally believe that the issue admits any serious doubt, but it is prudent to "play the devil's advocate," and exert all reasonable efforts to pick holes in such analogies, so as to accurately ascertain if any room remains for exclusively transitive relations in modeling fundamental spacetime structure. In this spirit, one might argue that the intuition of "breaking a message into multiple parts," in the above cryptographic analogy, or of "different agents bringing different information to a meeting," in the above "academic conference" analogy, is dubious in a truly fundamental setting. However, the notion of independent modes of influence does not depend on such intuition; in particular, the issue is not *what* information is transmitted, but *how* it is transmitted, and the significance of different modes

[52] However, if one relaxes the measure axiom (M) to admit volume dependence on local causal structure, as discussed at the end of Section 4.4, then independence becomes geometrically relevant.

[53] It is worth noting that some recent papers involving causal sets quietly remove transitivity from the table; for example, the paper *Network Cosmology* [KKSR12] describes a causal set as a *"network in which nodes are spacetime quanta, and links* [i.e., *irreducible* relations] *are causal relationships between them."* This viewpoint is necessary because the valences of elements, called *degrees* in [KKSR12], which are "large and meaningless" in the transitive paradigm, play an important role in the argument.

of transmission seems indubitable. One might attempt to construct *positive* arguments for transitivity, either by citing structural conveniences such as those noted by Markopoulou [MA00], or by insisting on a closer qualitative analogy between abstract causal structure and relativistic spacetime geometry. However, adopting any such argument seems to endanger both the correctness and the explanatory power of the resulting theory. Indeed, the safer position is to abstain from transitivity, since this leads to a more general theory, which includes transitive relations as a special case. Altogether, it seems that the only possible defense of transitivity in this setting is to argue that some special property of the fundamental scale negates information-theoretic distinctions such as those illustrated in Figure 3.9.1. In particular, any such defense must establish that *only one of the two scenarios* illustrated in the middle and right-hand diagrams of the figure is physically relevant at the fundamental scale. Otherwise, the need to distinguish between them seems unavoidable.

Hence, to rescue transitivity, it seems that one must argue either that independent modes of influence along a relation and a reducing chain *never exist* at the fundamental scale, or, alternatively, that they *always exist.* The latter argument is manifestly absurd, since it leads to the conclusion that, whenever one event influences another, however long the chain of intervening influences, *independent direct influence* must also propagate between the two events; e.g., between the *"birth and death of the universe,"* as Finkelstein puts it.[54] In particular, the independence convention (IC) makes no sense in the transitive paradigm, as already noted in Section 3.7. Eliminating this choice leaves two possibilities: either independent modes of influence along a relation and a reducing chain never exist at the fundamental scale, or transitive binary relations are deficient for modeling classical spacetime structure in discrete causal theory.

Six arguments against transitivity. I now present six arguments that the scenario illustrated in the right-hand diagram in Figure 3.9.1, in which independent modes of influence exist along a relation and a reducing chain between two events, *should not be ruled out at any scale* in classical physics. This leads immediately to the conclusion that transitive binary relations are deficient for modeling classical spacetime structure in discrete causal theory. Some of the arguments presented below are direct, while others merely make note of suggestive structural analogies. These arguments are reinforced and supplemented by the general principles of scientific philosophy offered in Section 2.10.

1. *Multiple independent modes of influence between pairs of events are ubiquitous in conventional physics.* An important example of this is illustrated by the widespread use of Cauchy surfaces in continuum-based theories, both in the classical and quantum settings. As stated in Definition 2.7.4, a Cauchy surface in a relativistic spacetime manifold X is a subset σ of X such that every inextensible

[54] Certain types of idealized causal sets, constructed via global sprinklings into Minkowski spacetime $\mathbb{R}^{3+1}$, are interpreted to exhibit extreme nonlocality in a spatiotemporal sense, as discussed in Section 4.5. However, even these physically unrealistic objects do not represent direct influence along every individual relation.

causal curve in X intersects σ exactly once. The prominence of Cauchy surfaces is based on the assumption that different causal curves from past to future can carry independent information. Extension of this assumption to the fundamental scale allows, as a special case, the scenario illustrated in the right-hand diagram in Figure 3.9.1. The left-hand diagram in Figure 3.9.2 illustrates "Cauchy surfaces" intersected by the two "independent paths" from x to z depicted in the former diagram. These "Cauchy surfaces" may be viewed as primitive examples of *maximal antichains of relations,* which play a major role in the theory of relation space developed in Chapter 5, and in the discrete quantum causal theory developed in subsequent chapters.

2. *Independence of influences exerted by an event should not be constrained by details of its future.* This is particularly true in the acyclic case, in which such constraints may be legitimately viewed as pathological, rather than merely dubious. In particular, given events x, y, and z, together with the knowledge that x influences both y and z, the question of whether or not x *directly* influences z, independently of y, should not depend on the existence of a relation $y \prec z$ lying in the *future* of x. For example, the right-hand diagram in Figure 3.9.2 illustrates an acyclic directed set in which the existence of a reducing chain for the relation $x \prec z$ depends on the existence of a relation $y \prec z$ in the "distant future" of x. It seems very implausible that the presence or absence of this relation should impose a "retroactive" constraint on influences emanating from x.

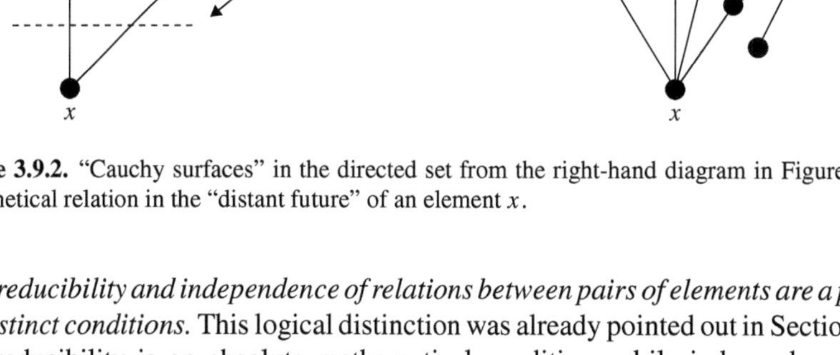

Figure 3.9.2. "Cauchy surfaces" in the directed set from the right-hand diagram in Figure 3.9.1; hypothetical relation in the "distant future" of an element x.

3. *Irreducibility and independence of relations between pairs of elements are a priori distinct conditions.* This logical distinction was already pointed out in Section 3.7. Irreducibility is an absolute mathematical condition, while independence is a subjective physical condition. In a directed set, irreducible relations between pairs of elements are necessarily independent, but the converse is false. The question then becomes whether or not there are compelling physical reasons to *equate* the two conditions in the context of discrete causal theory. In my view, the answer to this question is negative. If this is correct, then it is natural to encode discrete

causal structure by means of independent relations between pairs of elements, following the independence convention (IC).

4. *Configuration spaces of transitive binary relations are pathological, particularly from a physical perspective.* A classic example of such a pathology occurs in the acyclic directed case, in which the distribution of finite partial orders is asymptotically dominated by objects manifestly unsuitable for modeling classical spacetime. This has been known since before the advent of causal set theory, due to Kleitman and Rothschild's 1975 paper *Asymptotic Enumeration of Partial Orders on a Finite Set* [KR75]. Kleitman and Rothschild show that a "generic large finite" partially ordered set of cardinality n has just three generations, of cardinalities roughly $n/4$, $n/2$, and $n/4$, respectively, with the number of irreducible relations beginning at a typical nonmaximal element depending linearly on n. A Kleitman–Rothschild-type partial order is illustrated in the left-hand diagram in Figure 3.9.3. In the context of classical spacetime structure, such a partial order represents a universe of large "spatial" size, and negligible "causal" or "temporal" size, with an unreasonably large "luminal velocity." A priori, this raises the concern that the path summation approach to discrete quantum causal theory might be dominated by such objects under the transitive paradigm. This is sometimes referred to in the literature as an *entropy problem*[55] for the path sum; in this book, I refer to it as the **Kleitman–Rothschild pathology.**

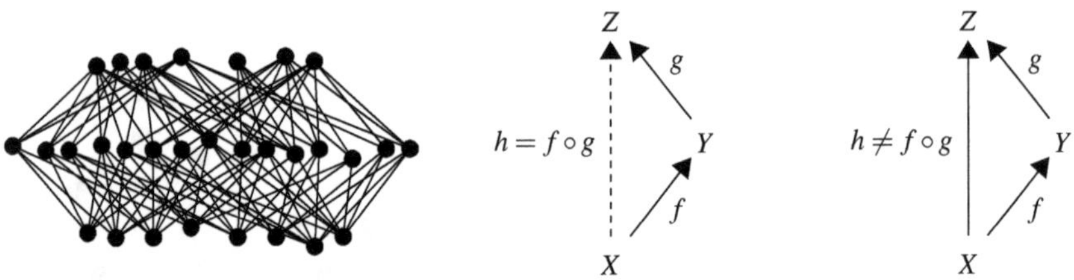

Figure 3.9.3. Kleitman–Rothschild-type partial order (*irreducible relations shown*); category-theoretic analogues of the middle and right-hand diagrams in Figure 3.9.1.

Avoiding "enumerative problems" like the Kleitman–Rothschild pathology was one of the principal motivations for Sorkin and Rideout's theory of sequential growth dynamics [SR99] of causal sets. In particular, Sorkin and Rideout write,

> *Maybe this is not so different from the situation in ordinary quantum mechanics, where the smooth paths, which form a set of measure zero in the space of all paths, are the ones which dominate the sum over histories in the classical limit.* (page 4)

Here, the term "sum over histories" refers to Feynman's path summation approach to quantum theory. In Feynman's original background dependent setup, different

[55] This terminology comes from statistical thermodynamics, in which the large-scale behavior of a system is governed by the multiplicities of microscopic states sharing the same macroscopic properties.

paths always *do* represent different histories. However, this is no longer the case in sequential growth dynamics, or in background independent discrete quantum causal theory in general, where different paths sharing initial and terminal elements represent different *kinematic accounts* of the same evolutionary processes, as explained in Chapter 7. In any case, if one chooses to endorse this tentative statement of Sorkin and Rideout, then perhaps the most reasonable explanation is that *both* conventional quantum theory and causal set theory are obscured, to a degree, by physically irrelevant structure; in the first case, structure from the continuum; and in the second case, structure from dependent relations prescribed by transitivity. While it may be possible to circumvent such problems at a dynamical level, it seems more promising to begin with a class of models that does not present these troubling features at all. Often, pathological generic behavior of this type indicates some form of *degeneracy,* or *lack of naturality,* in the class of structures under consideration. By contrast, the asymptotic behavior of (generally nontransitive) acyclic directed sets seems to be much more reasonable, as indicated by Brendan McKay's 1989 paper *On the shape of a random acyclic digraph* [MC89], and Stephan Wagner's 2013 paper *Asymptotic Enumeration of Extensional Acyclic Digraphs* [WA13]. In particular, the configuration space of finite acyclic directed sets is *not* asymptotically dominated by *either* "near antichains," such as the Kleitman–Rothschild partial orders, *or* by "near-chains," i.e., directed sets of "large causal and negligible spatial size."

5. *Structural notions from mathematics motivate the existence of independent modes of influence.* The most obvious examples of such structural notions, of course, come from category theory, in which morphisms between pairs of objects correspond to relations between pairs of elements in a directed or multidirected set. This analogy is illustrated in the middle and right-hand diagrams of Figure 3.9.3 above, where X, Y, and Z are objects of a category, and f, g, and h are morphisms. These particular diagrams are analogous to the two physical scenarios illustrated in the middle and right-hand diagrams of Figure 3.9.1. The distinction between such diagrams is expressed, in category-theoretic parlance, by saying that the middle diagram "commutes;" i.e., $h = g \circ f$, while the right-hand diagram does not commute; i.e., $h \neq g \circ f$. In information-theoretic terms, the morphism h in the middle diagram is "merely a consequence of f and g," while in the right-hand diagram, h encodes information *independent* of f and g. In this sense, the choice to allow only transitive binary relations in modeling fundamental spacetime structure, as in causal set theory, is analogous to studying only categories in which *every diagram commutes*, a very restrictive condition. Readers familiar only with "concrete categories," in which objects are structured sets of some type, may argue in *favor* of such "commutativity" in the case of directed sets, on the grounds that the unique map of singleton sets from $\{x\}$ to $\{z\}$ is equal to the composition of unique maps from $\{x\}$ to $\{y\}$ and from $\{y\}$ to $\{z\}$. This observation is true, but irrelevant, since the appropriate analogue of a "morphism" in this case is not a set map. Indeed, there exists a unique set map between *any* pair of singletons in *any* set, but this does not mean that there exists a unique *relation* between each pair of singletons in a directed or multidirected set. Further, there exist important

examples of “abstract categories” whose objects possess no internal structure, but which nonetheless admit multiple independent morphisms between pairs of objects, as noted in Section 3.8.

6. *Recognition of nontransitive relations leads naturally to other improvements in discrete causal theory.* As pointed out by Finkelstein, nontransitive relations enable an improved treatment of *causal locality.* This notion is discussed in detail in Chapter 4. The same basic reasoning leads to superior notions of local conditions for multidirected sets in general. Of particular importance is *star finiteness* (SF), introduced in Section 4.4, which replaces the questionable causal set axiom of interval finiteness (IF). Building on these improvements is the theory of *relation space,* introduced in Chapter 5, which is almost unrecognizable, and relatively useless, in the transitive setting. In particular, the use of maximal antichains of relations as discrete causal analogues of Cauchy surfaces, which circumvents the *permeability problem* mentioned in Chapter 2, depends crucially on the independence convention (IC), and therefore makes sense only in the nontransitive context. Similarly, much of the algebraic machinery associated with the theory of path summation over a multidirected set, developed in Chapters 5 and 6, is seriously hampered if dependent relations are forced into the picture. Ultimately, the discrete causal analogues of Feynman’s path integral and Schrödinger’s equation, presented in Sections 6.7 and 6.9, respectively, find full expression only in the nontransitive setting.

3.10 The Causal Relation

Definition of the causal relation. To remedy the shortcomings of transitive binary relations in modeling classical spacetime structure in discrete causal theory, I propose that these relations should be viewed as *derivative constructs,* generated by more fundamental, generally nontransitive, binary relations, which I refer to as *causal relations.* The causal relation on a directed set corresponds to Finkelstein’s [FI88] *“causal connection relation”* **c**, while the transitive relation generated by this causal relation corresponds to Finkelstein’s *“transitive causal relation”* **C**. It is important to reiterate that causal relations in this book may contain cycles.

Definition 3.10.1. *Let* $D = (D, \prec)$ *be a directed set, viewed as a model of causal structure under the independence convention* (IC). *In this context, the binary relation* $\prec$ *on* D *is called the* **causal relation** *on* D.

The independence convention endows each individual relation $x \prec y$ defined by the causal relation $\prec$ with an unambiguous physical meaning: *every such relation encodes direct, independent influence, not represented by any other chain, or family of chains, between* x *and* y. Indirect, dependent influence, meanwhile, is encoded by complex chains of relations. This interpretation meshes with the notions of direct predecessors and successors given in Definition 3.7.8. For example, the causal relation illustrated in the middle diagram in Figure 3.9.1 does *not* include the relation $x \prec z$, since x does not influence z independently of y. The causal relation illustrated in the right-hand diagram in Figure 3.9.1 *does* include the relation $x \prec z$. The transitive relations generated by these two causal relations are the same, since transitive relations encode only *which* influences occur, not *how* they occur.

Transitive closure; skeleton. Two useful operations on directed sets are the *transitive closure* functor and the *skeleton* operation. The transitive closure functor is particularly important because it "generates" transitive relations from causal relations. These two operations realize the opposite extremes of transitivity and irreducibility for binary relations. The transitive closure is defined by "adding all possible reducible relations between pairs of elements connected by chains," while the skeleton is defined by deleting all such reducible relations. Several different generalizations of these operations to the case of multidirected sets are possible, but these details are not needed in this book.

Definition 3.10.2. *Let $D = (D, \prec)$ be a directed set.*

1. *The* **transitive closure** *of D is the directed set* $\mathrm{tr}(D) := (D, \prec_{\mathrm{tr}})$ *whose binary relation $\prec_{\mathrm{tr}}$ is defined by setting $x \prec_{\mathrm{tr}} y$ if and only if there exists a chain of nonzero length between x and y in D. The binary relation $\prec_{\mathrm{tr}}$ is called the* **transitive relation** *on D.*
2. *The* **skeleton** *of D is the acyclic directed set* $\mathrm{sk}(D) := (D, \prec_{\mathrm{sk}})$ *whose binary relation $\prec_{\mathrm{sk}}$ is defined by setting $x \prec_{\mathrm{sk}} y$ if and only if $x \prec y$ is an irreducible relation in D. The binary relation $\prec_{\mathrm{sk}}$ is called the* **skeletal relation** *on D.*

The skeleton $\mathrm{sk}(D)$ of a directed set D is automatically acyclic, since any relation belonging to a cycle admits a reducing chain "traversing the cycle." The skeleton of a directed set is generally not the same as its *transitive reduction,* familiar from graph theory, which preserves the accessibility properties of the set.[56] However, the two coincide for finite acyclic directed sets. Since every relation is itself a chain, the transitive closure of a directed set D possesses *at least* as many relations as D, with equality obtained if and only if D is transitive. Similarly, since every irreducible relation is a relation, the skeleton of D possesses *at most* as many relations as D, with equality obtained if and only if D is irreducible. The transitive closure of a directed set is closely related to its *chain space,* discussed in a more general context in Chapter 5. This relationship arises from the fact that every relation in $\mathrm{tr}(D)$ corresponds to a chain in D. This suggests one way to generalize the transitive closure to multidirected sets: by *adding a relation corresponding to each complex chain.* Figure 3.10.3 illustrates the transitive closure and the skeleton of a directed set.

[56] A closely-related accessibility condition plays a role in the definition of *kinematic schemes* in Section 7.4.

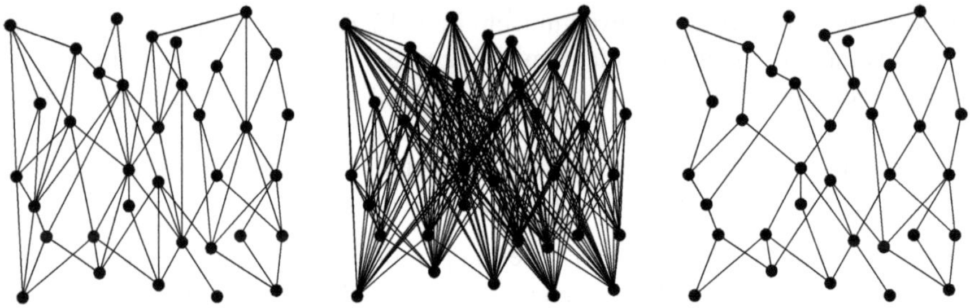

Figure 3.10.3. Directed set D; transitive closure $\mathrm{tr}(D)$ of D; skeleton $\mathrm{sk}(D)$ of D.

Degeneracy. In general, many different directed sets share the same transitive closure and skeleton, up to isomorphism. For example, the transitive closure and skeleton of the directed set D illustrated in the left-hand diagram in Figure 3.10.3 differ by about 200 reducible relations. Adding distinct families of these relations to $\mathrm{sk}(D)$ produces distinct directed sets; hence, there exist roughly 2^{200} *information-theoretically distinct* directed sets sharing $\mathrm{tr}(D)$ and $\mathrm{sk}(D)$ as their transitive closures and skeletons, respectively. Working, as usual, in terms of isomorphism classes, this number may be reduced somewhat by the existence of symmetries, but directed sets tend to be surprisingly *rigid,* as explained in Section 6.3. The **transitive degeneracy class** of a directed set D is the set of isomorphism classes of directed sets with transitive closure isomorphic to $\mathrm{tr}(D)$, and the **skeletal degeneracy class** of D is the set of isomorphism classes of directed sets with skeleton isomorphic to $\mathrm{sk}(D)$. The two classes coincide for finite acyclic directed sets, since the transitive closure of such a set may be recovered from its skeleton, and vice versa. The **transitive** and **skeletal degeneracies** of D are the cardinalities of the corresponding degeneracy classes.

Degeneracy plays an important role in the structure of configuration spaces of directed sets. In particular, degeneracy distinguishes configuration spaces of acyclic directed sets from the corresponding configuration spaces of partial orders, whose elements correspond to transitive degeneracy classes of acyclic directed sets. This partly explains the source of the *Kleitman–Rothschild pathology* for finite partial orders, discussed in Section 3.9, in which "near-antichains" dominate the configuration space asymptotically, since enumeration of partial orders counts only *one* member of each degeneracy class. Referring again to the directed set illustrated in the left-hand diagram in Figure 3.10.3, a Kleitman–Rothschild-type order of the same cardinality has roughly 80 reducible relations, and hence a transitive degeneracy of roughly 2^{80}, about 120 powers of 2 less than the "more physically realistic" directed set in the figure. This illustrates how enumeration of partial orders *relatively overcounts* "near-antichains" by many of orders of magnitude, even for small cardinalities.

Functorial properties. The transitive closure and the skeleton are *idempotent operations* on the category $\mathcal{D}$ of directed sets. This means that repeated application of these operations produces nothing new: $\mathrm{tr}\big(\mathrm{tr}(D)\big) = \mathrm{tr}(D)$ and $\mathrm{sk}\big(\mathrm{sk}(D)\big) = \mathrm{sk}(D)$. In the finite acyclic directed case, the two operations are "roughly inverse" to each other, but this is not true in general; for example, the skeleton of the rational numbers $\mathbb{Q}$, with its usual strict total order $<$, possesses *no relations*, since every relation in $\mathbb{Q}$ is reducible. Similarly, the skeleton of a directed set consisting of a single element with a reflexive relation, is a single element with *no* relations, since the reflexive relation is reduced by any chain "traversing it multiple times." Taking morphisms into account, the transitive closure operation extends to a *functor* from $\mathcal{D}$ to its subcategory $\mathcal{D}_{\mathrm{TR}}$ of transitive directed sets, essentially because chains map to chains under morphisms. There exists an inclusion functor $\mathcal{D}_{\mathrm{TR}} \to \mathcal{D}$ in the opposite direction, and the two functors are *adjoint*. This means that for any directed set $(D, \prec)$, and any transitive directed set $(D', \prec')$, there is a natural identification of morphism sets

$$\mathrm{Mor}_{\mathcal{D}}\big(D, D'\big) \cong \mathrm{Mor}_{\mathcal{D}_{\mathrm{TR}}}\big(\mathrm{tr}(D), D'\big), \tag{3.10.4}$$

given by identifying morphisms sharing the same underlying set maps. The skeleton operation does *not* define a functor, since irreducible relations do not always map to irreducible relations under morphisms. For example, the inclusion $\mathbb{Z} \to \mathbb{Q}$, with the usual orders, is a morphism in $\mathcal{D}$, but does not induce a morphism $\mathrm{sk}(\mathbb{Z}) \to \mathrm{sk}(\mathbb{Q})$.

References

[HA76] S. W. Hawking, A. R. King, and P. J. McCarthy. *A new topology for curved space-time which incorporates the causal, differential, and conformal structures*. Journal of Mathematical Physics, **17**, 2, pp. 174–181, 1976.

[MA77] David B. Malament. *The class of continuous timelike curves determines the topology of spacetime*. Journal of Mathematical Physics, **18**, 7, pp. 1399–1404, 1977.

[KR75] D. J. Kleitman and B. L. Rothschild. *Asymptotic Enumeration of Partial Orders on a Finite Set*. Transactions of the American Mathematical Society, **205**, pp. 205–220, 1975.

[SO05] Rafael Sorkin. *Causal Sets: Discrete Gravity*. Lectures on Quantum Gravity, edited by Andrés Gomberoff and Donald Marolf. Springer, 2005. arXiv preprint: http://arxiv.org/pdf/gr-qc/0309009v1.pdf.

[FI88] David Finkelstein. *"Superconducting" Causal Nets*. International Journal of Theoretical Physics, **27**, 4, pp. 473–519, 1988.

[FI69] David Finkelstein. *Space-Time Code*. Physical Review, **184**, 5, pp. 1261–1271, 1969.

[MR78] J. Myrheim. *Statistical Geometry*. CERN preprint, 1978.

[TH78] Gerard 't Hooft. *Quantum Gravity: A Fundamental Problem and some Radical Ideas*. Recent Developments in Gravitation (Proceedings of the 1978 Cargese Summer Institute), edited by M. Levy and S. Deser.

[BLMS87] Luca Bombelli, Joohan Lee, David Meyer, and Rafael Sorkin. *Space-Time as a Causal Set*. Physical Review Letters, **59**, 5, pp. 521–524, 1987.

[MP06] Keye Martin and Prakash Panangaden. *A Domain of Spacetime Intervals in General Relativity*. Communications in Mathematical Physics, **267**, 3, pp. 563–586, 2006.

[FR16] Tobias Fritz. *Beyond Bell's Theorem II: Scenarios with Arbitrary Causal Structure*. Communications in Mathematical Physics, **341**, 2, pp. 391–434, 2016. arXiv preprint: https://arxiv.org/pdf/1404.4812v2.pdf.

[SR99] David Rideout and Rafael Sorkin. *Classical sequential growth dynamics for causal sets*. Physical Review D, **61**, 2, 024002, 2000. arXiv preprint: http://arxiv.org/pdf/gr-qc/9904062v3.pdf.

[SMS14] Siavash Aslanbeigi, Mehdi Saravani, and Rafael Sorkin. *Generalized Causal Set d'Alembertians*. arXiv preprint: http://arxiv.org/abs/1403.1622.

[BBD15] Alessio Belenchia, Dionigi Benincasa, and Fay Dowker. *The continuum limit of a 4-dimensional causal set scalar d'Alembertian*. arXiv preprint: http://arxiv.org/pdf/1510.04656.pdf.

[BE16] Alessio Belenchia. *Universal behavior of generalized causal set d'Alembertians in curved spacetime*. arXiv preprint: http://arxiv.org/pdf/1510.04665.pdf.

[GL14] Lisa Glaser. *A closed form expression for the causal set D'Alembertian.* Classical and Quantum Gravity, **31**, 9, pp. 5007, 2014. arXiv preprint: http://arxiv.org/pdf/1311.1701v3.pdf.

[RA00] Ioannis Raptis. *Algebraic Quantization of Causal Sets*. International Journal of Theoretical Physics, **39**, 5, pp. 1233–1240, 2000. arXiv preprint: http://arxiv.org/pdf/gr-qc/9906103v1.pdf.

[FE48] Richard Feynman. *Space-Time Approach to Non-Relativistic Quantum Mechanics*. Reviews of Modern Physics, **20**, 2, pp. 367–387, 1948.

[IS05] Christopher Isham. *Quantising on a Category*. Foundations of Physics, **35**, 2, pp. 271–297, 2005. arXiv preprint: http://arxiv.org/pdf/quant-ph/0401175v1.pdf.

[SO12] Rafael Sorkin. *Toward a Fundamental Theorem of Quantal Measure Theory*. Mathematical Structures in Computer Science, **22**, 05 (special issue), pp. 816–852, 2012. arXiv preprint: http://arxiv.org/pdf/1104.0997v2.pdf.

[BG10] Graham Brightwell and Nicholas Georgiou. *Continuum limits for classical sequential growth models*. Random Structures and Algorithms, **36**, 2, pp. 218–250, 2010.

[BL15] Graham Brightwell and Malwina Luczak. *The Mathematics of Causal Sets*. arXiv preprint: http://arxiv.org/pdf/1510.05612v1.pdf.

[ADGS04] Maqbool Ahmed, Scott Dodelson, Patrick B. Greene, and Rafael Sorkin. *Everpresent Λ*. Physical Review D, **69**, 10, 103523, 2004. arXiv preprint: http://arxiv.org/pdf/astro-ph/0209274v1.pdf.

[SU11] Sumati Surya. *Directions in Causal Set Quantum Gravity*. In Recent Research in Quantum Gravity, edited by Arundhati Dasgupta. Nova Science Publishing Incorporated, 2012. arXiv preprint: http://arxiv.org/pdf/1103.6272v1.pdf.

[DHS04] Fay Dowker, Joe Henson, and Rafael Sorkin. *Quantum Gravity Phenomenology, Lorentz Invariance and Discreteness*. Modern Physics Letters A, **19**, 24, pp. 1829–1840, 2004. World Scientific Publishing Company. arXiv preprint: http://arxiv.org/pdf/gr-qc/0311055.pdf.

[BD10] Dionigi M. T. Benincasa and Fay Dowker. *Scalar Curvature of a Causal Set*. Physical Review Letters, **104**, 18, 181301, 2010. arXiv preprint: http://arxiv.org/pdf/1001.2725v4.pdf.

[JF09] Steven Johnston. *Feynman propagator for a free scalar field on a causal set*. Physical Review Letters, **103**, 18, 180401, 2009. arXiv preprint: http://arxiv.org/pdf/0909.0944v2.pdf.

[BE15] Alessio Belenchia. *Causal Set theory, non-locality and phenomenology*. Preprint, 2015. arXiv preprint: http://arxiv.org/pdf/1512.08485v1.pdf.

[SO09A] Rafael Sorkin. *Does locality fail at intermediate length scales?* Approaches to Quantum Gravity, edited by Daniele Oriti. Cambridge University Press, 2009.

[GA72] Pierre Gabriel. *Unzerlegbare Darstellungen. I.* Manuscripta Mathematica, **6**, pp. 71–103, 1972.

[AP05] Gene Abrams and Gonzalo Aranda Pino. *The Leavitt path algebra of a graph*. Journal of Algebra, **293**, 2, pp. 319–334, 2005. arXiv preprint: http://arxiv.org/pdf/math/0509494v1.pdf.

[RO04] Carlo Rovelli. *Quantum Gravity.* Cambridge Monographs on Mathematical Physics. Cambridge University Press, 2004.

[TH07] Thomas Thiemann. *Modern Canonical Quantum General Relativity.* Cambridge Monographs on Mathematical Physics. Cambridge University Press, 2007.

[IS11] Christopher Isham. *Topos Methods in the Foundations of Physics.* In Deep Beauty: Understanding the Quantum World through Mathematical Innovation, edited by Hans Halvorson. Cambridge University Press, 2011. arXiv preprint: http://arxiv.org/pdf/1004.3564v1.pdf.

[DR13] Benjamin F. Dribus. *On the Axioms of Causal Set Theory.* Preprint. arXiv link: http://arxiv-web3.library.cornell.edu/pdf/1311.2148v3.pdf.

[FI16] Felix Finster. *Causal Fermion Systems: An Overview.* Quantum Mathematical Physics, Springer, 2016.

[ST15] O. C. Stoica. *From Horismos to Relativistic Spacetimes.* arXiv preprint: http://arxiv.org/pdf/1504.03265.pdf.

[MA00] Fotini Markopoulou. *Quantum Causal Histories.* Classical and Quantum Gravity, **17**, 10, pp. 2059, 2000. arXiv preprint: https://arxiv.org/pdf/hep-th/9904009v5.pdf.

[IS94] Christopher Isham. *Quantum Logic and the Histories Approach to Quantum Theory.* Journal of Mathematical Physics, **35**, 2157, 1994. arXiv preprint: http://arxiv.org/pdf/gr-qc/9308006v1.pdf.

[IS97] Christopher Isham. *Topos Theory and Consistent Histories: The Internal Logic of the Set of all Consistent Sets.* International Journal of Theoretical Physics, **36**, 785, 1997. arXiv preprint: http://arxiv.org/pdf/gr-qc/9607069v1.pdf.

[KKSR12] Dmitri Krioukov, Maksim Kitsak, Robert S. Sinkovits, David Rideout, David Meyer, and Marián Boguñá. *Network Cosmology.* Nature Scientific Reports **2**, 793, 2012.

[MC89] Brendan McKay. *On the shape of a random acyclic digraph.* Mathematical Proceedings of the Cambridge Philosophical Society, **106**, 03, pp. 459–465, 1989.

[WA13] Stephan Wagner. *Asymptotic Enumeration of Extensional Acyclic Digraphs.* Algorithmica, **66**, 4, pp. 829–847, 2013.

Chapter 4
Local Properties

Abstract. This chapter examines *local properties* of discrete causal spacetime, and culminates with a suggested list of axioms for a new version of discrete classical causal theory, which underlies the discrete quantum causal theory developed in Part II of the book. Section 4.1 discusses the topic of *local behavior* in general terms, and introduces the notion of *causal locality,* whereby pairs of events are regarded as being in local proximity if one event directly influences the other. Section 4.2 reviews some of the basics of *topology,* which is the standard mathematical tool for describing local structure. An important justification for including such elementary material here is that what is sometimes called "local" in causal set theory is *not* local in a topological sense! Section 4.3 introduces four specific topologies for multidirected sets. Of these, the *star topology* is the most useful for the purposes of this book, since it effectively captures the physical notion of causal locality. Section 4.4 discusses *finiteness conditions* in discrete causal theory. An appropriate version of *local finiteness* is necessary for discrete spacetime structure to make physical sense. For this role, causal set theory employs the questionable notion of *interval finiteness,* which is not a local condition, and which permits problematic locally infinite behavior. An alternative choice for a local finiteness condition is *star finiteness,* which meshes well with causal locality. Section 4.5 describes "justifiably" locally infinite behavior in certain types of idealized causal sets, and explains some of the motivations for studying such objects, such as *systematic Lorentz invariance.* Section 4.6 introduces the theory of *relative multidirected sets over a fixed base,* which enables streamlining of certain axiomatic descriptions of directed sets. In particular, causal sets may be viewed as relative directed sets over the integers. Section 4.7 presents seven arguments questioning the use of interval finiteness as a local finiteness condition in discrete causal theory. Section 4.8 offers nine arguments in favor of star finiteness as an alternative condition. Section 4.9 presents two theorems describing relationships among different finiteness conditions for directed sets and multidirected sets. Section 4.10 introduces a suggested list of axioms for discrete causal theory, which serves as an alternative to the axioms of causal set theory.

B.F. Dribus, *Discrete Causal Theory*, DOI 10.1007/978-3-319-50083-6_4

4.1 Local Behavior

Local behavior as a key to new physics. The principal goal of discrete *classical* causal theory is to describe the building blocks of *fundamental spacetime structure.* Progress on the related problems of *quantum gravity* and *unification* may be expected to follow from this improved structural understanding. The word "fundamental," in this particular context, usually refers to *ultra-local behavior,* where the precise meaning of "local" is explained later in this section, and elaborated throughout the chapter. In conventional physics, this local emphasis is often tied to a number of assumptions, which are not always related in a necessary way to the logical meaning of "fundamental." One such assumption is *classical reductionism,* which may be roughly paraphrased by the statement that "what is fundamental in classical physics is local," i.e., that one may acquire progressively more precise descriptions of a system by "breaking it into smaller and smaller pieces," and studying these pieces individually. This need not be true from a logical perspective; it is entirely possible for certain types of data to be distributed in an essentially nonlocal fashion. However, classical reductionism has proven to be very useful in the development of modern physics. General relativity and conventional quantum field theory lead to more specific assumptions, including educated guesses about "how close one must look in order to see truly fundamental processes." Although the smooth manifolds involved in these theories possess no *natural* scale data, factors such as the values of certain physical constants, the limitations imposed by the uncertainty principle, and the gravitational instability of very small measuring apparatus, suggest the existence of a "smallest meaningful scale," often identified as the *Planck scale.*

Regardless of the theoretical justification for equating the concepts of "fundamental" and "local," the historical success of the local viewpoint is difficult to dispute. In fact, one of the best ways to measure the progress of physics over the last few centuries is to track its success in supplying increasingly precise local descriptions of nature. This is true because "zooming in" on interactions has proven not only to enable better descriptions of previously-established physics, but also to reveal *new physics,* which human physicists have often been insufficiently clever to infer on their own. The physical phenomena known to conventional physics as the *strong* and *weak interactions, electromagnetism, gravity, dark matter,* and *dark energy,* each dominate across different ranges of scales, and there is every reason to expect that additional phenomena will manifest themselves at other scales yet to be examined. This *qualitative scale dependence* of physical phenomena may be illustrated in an evocative, though down-to-earth, manner: humans "live near the boundary between electromagnetism and gravity," in terms of which phenomena dominate at everyday scales, while insects are noticeably less sensitive to gravitation, and elephants are noticeably more so. Speculation about a figurative "desert" in particle theory, in which no new phenomena manifest themselves across many orders of magnitude in scale, still seems quite dubious from a historical perspective, despite the recent period of relatively barren results in "high energy" physics.

Ballpark figures regarding the fundamental scale. In particle theory, experimental accessibility of small scales is closely related to the available beam energy of particle accelerators, which, in turn, depends on their size and cost. Present handwringing over the possible unavailability of "even-more-local" data, once the Large Hadron Collider has exhausted its potentialities, illustrates this difficulty.[1] Just how small of scales must be probed in order to capture "interesting" local behavior depends on the context, and the requirements have become progressively more demanding as theory and experiment have advanced. The experiments of Geiger and Marsden in 1908–1910, which led to Rutherford's recognition of the atomic nucleus, revealed local behavior on scales several orders of magnitude smaller than the previously-recognized size of the atom. The entire succeeding century, though favored by the discovery of many interesting new phenomena, witnessed only a modest increase in the actual *scope* of description, measured in terms of scales. Roughly speaking, atoms measure about 10^{-10} m, nuclei about 10^{-15} m, and the smallest currently-accessible scale about 10^{-18} m. Meanwhile, the *Planck length,* conjectured in conventional settings to represent the "fundamental scale," measures about 10^{-35} m. For simplicity, I have been describing *spatial* scales, but *temporal* scales are more natural from the perspective of discrete causal theory. Certain "elementary particles" known to conventional physics possess lifetimes of around 10^{-24} s, although the smallest individually-measured unit of time at present is several orders of magnitude larger than this. The *Planck time,* meanwhile, measures about 10^{-43} s.

According to discrete causal theory, a fundamental scale *does* exist, defined by the "sizes" of elementary elements and relations. Under the causal metric hypothesis (CMH), this scale determines conventional scales, not vice versa. For example, it might make sense to venture the rough guess that "about 10^{43} fundamental relations make up a second in a given frame of reference," whether or not this guess is actually true. However, the arguments of Chapter 3 suggest that "some relations may be longer than others," and not merely in a "frame-dependent" sense. Figure 2.8.3, meanwhile, illustrates the possibility that "some elements may be larger than others." Hence, it may be necessary to "zoom out" to fairly large subsets of discrete causal spacetime before the structure looks "smooth enough" to make meaningful comparisons regarding conventional distance and time. Keeping this in mind, one may still carry out such rough numerical comparisons, for heuristic purposes. From this perspective, *the fundamental scale of discrete causal theory is almost certainly much smaller than the current "experimental scales" of* 10^{-18} m *and* 10^{-24} s, *respectively, but may be either smaller or larger than the Planck scale.*

Importance of local finiteness in discrete causal theory. Discrete causal theory treats the fundamental scale defined by elements and relations in a discrete directed set as a "ground floor" for local behavior, since such elements and relations are

[1] In December 2015, theorists began flooding the arXiv with frantic speculation about a potential "non-standard-model" signal at the LHC, involving photon pairs near an energy of 750 gigaelectronvolts. A similar frenzy followed initial reports of "superluminal neutrinos" several years ago. By early summer 2016, the excitement had mostly deflated, and by mid summer, the disappearance of the signal was confirmed via reliable leaks from CERN.

ascribed no internal structure.[2] In Section 1.10, I explained the risk of speculating about what is "truly fundamental," and concluded that it is a worthwhile enterprise, regardless of the potential for future embarrassment. Hence, in this book, when classical spacetime is modeled via a directed set D, following the classical causal metric hypothesis (CCMH), the local behavior of D is regarded as local *in an absolute sense,* i.e., "as local as possible." Informally, local behavior in this context means "whatever happens near individual elements or relations." Examples include "how many relations begin or terminate at a particular element," or "which relations share the terminal element of a given relation as their initial element," or "which elements are connected to a given element by irreducible relations." *Local conditions* imposed on families of directed sets or multidirected sets constrain such local behavior, with the purpose of restricting attention to "physically reasonable" models.

The most important such conditions in discrete classical causal theory are *local finiteness conditions,* since these conditions help to encode what "discreteness" *means* at a technical level. In the specific context of causal set theory, Rafael Sorkin [SO05] describes the role of local finiteness as a *"a formal way of saying that a causal set is discrete."*[3] The necessity of imposing some type of local finiteness condition is evident from the role of the causal set measure μ_{CS}, under Sorkin's version of the classical causal metric hypothesis, since infinite "number" corresponds to infinite "volume" under this measure. Indeed, permitting an infinite number of elements to be directly related to a given element allows for manifestly nonphysical scenarios in this setting, such as the instantaneous emergence of an infinite volume of spacetime from a single event. As explained in Sections 4.4 and 4.5, certain types of locally infinite behavior in discrete directed sets may be assigned innocuous interpretations, but the risk of extreme pathologies remains worrisome. The same reasoning generalizes to any version of the classical causal metric hypothesis for which infinite sets are assigned infinite measure. In particular, the version of "local finiteness" used as an axiom in causal set theory, which I rename more descriptively as *interval finiteness* (IF), permits such locally infinite behavior. Hence, as an alternative, I propose a different local finiteness axiom, called *star finiteness* (SF), which seems to enjoy both physical and mathematical advantages.

Basic structural meaning of "local". A property of a physical system or mathematical structure is called *local* if and only if it may be established by examining "arbitrarily small neighborhoods" of events or elements. It is easy to invent everyday examples of such properties. For instance, if a zoo consists of several enclosures, one may establish that the zoo has the property of "no giraffe" by checking that there is no giraffe in any *individual* enclosure. Hence, "no giraffe" is a local property of zoos. It is useful to emphasize the "if and only if" relationship here; to establish the property, it is both *necessary and sufficient* to check each enclosure. Turning to a

---

[2] Of course, as mentioned in Section 1.2, one may always speculate about the existence of "subclassical structure." Section 8.7 briefly describes how "small-scale causal cycles" could play a role in this context.

[3] See [SO05], page 309. Sorkin really means *interval finiteness* (IF) in this context, but the spirit of this description involves local finiteness.

more precise context, the "real manifold" property is a local property of a *topological space,* because a real manifold is *defined* to be a topological space for which each element possesses a neighborhood isomorphic to $\mathbb{R}^n$ for some fixed n.[4] More specifically, the dimension of a real manifold is a local property, and this property takes on physical connotations in general relativity and other continuum-based theories.

In general mathematical settings, the notion of a small neighborhood is usually made precise in terms of a *topology* on a set, although alternatives and generalizations are possible.[5] The discussion of metric recovery in Section 2.8 contains informal descriptions of a few specific topologies on relativistic spacetime manifolds, but the present chapter is more systematic, albeit in a completely different context. In classical continuum-based theories, including general relativity, neighborhoods may be defined in *metric* terms; informally, in terms of "distances" in space or spacetime. Under the causal metric hypothesis (CMH), the roles are reversed: local conditions are more fundamental than metric notions in this context. Definition 4.1.1 below provides a physical description of what "local" means under the causal metric hypothesis, expressed in terms of direct causes and effects. The choice to model classical spacetime in terms of special classes of directed sets, motivated by the material developed in Chapters 2 and 3, then leads to precise mathematical definitions of local behavior, which formalize the above notion of "whatever happens near individual elements or relations." It is useful here to refer again to the "jungle-desert" metaphor mentioned previously in Sections 1.7 and 3.3, which emphasizes the distinction between *local regularity* in general relativity (the "desert") and *local irregularity* in discrete causal theory (the "jungle"). Real manifolds, by definition, are locally regular; indeed, as noted above, each element in a real manifold possesses a neighborhood isomorphic to $\mathbb{R}^n$. A discrete directed set or multidirected set, however, *may look very different near one element or relation than it does near another.* Since the physical universe is manifestly locally irregular in its distribution of "matter" and "energy," discrete directed sets carry a natural advantage over real manifolds in modeling what is conventionally viewed as "nongravitational matter." In this way, at least, the discrete causal theorist seems to have an a priori easier task than Kaluza and Klein, Wheeler, or the modern string theorist.

Review of local notions from relativity. Before proceeding, I briefly revisit some of the material from Chapter 2 involving local structure in general relativity. First, as explained in Section 2.9, the word "metric" has a different meaning in the abstract mathematical setting of a *metric space* than it does in differential geometry, and more concretely, in general relativity. In the former setting, a "metric" is just a "generalized distance function" on a set S, i.e., a positive-definite, symmetric function $S \times S \rightarrow \mathbb{R}$, satisfying the triangle inequality. In the latter context, a metric is an

---

[4] Here, I suppress some details for convenience. First, there is no need to specify whether the smooth category, or merely the topological category, is involved. Second, the *Hausdorff condition* and the condition of *second-countability* are usually added, to rule out "pathologies" such as the *doubled origin* and the *long line.*

[5] For example, the theory of *frames* and *locales* abstracts and generalizes some of the order-theoretic properties of topologies.

entire family of bilinear maps, assigning to the tangent spaces of a smooth manifold X smoothly varying "generalized inner products," as described in Section 2.4. In the prototypical case of Euclidean space $\mathbb{R}^n$, the two notions "essentially coincide," since $\mathbb{R}^n$ is isomorphic to each of its tangent spaces, and since the Euclidean metric signature is pure. Any "metric" on a set S, under the simple "metric space" meaning, induces a *metric topology* on S. The Euclidean metric, for example, which generalizes the distance formula arising from the Pythagorean theorem, induces the Euclidean metric topology on $\mathbb{R}^n$. This topology, in turn, induces a natural *manifold topology* on any n-dimensional real manifold X, via its coordinate charts. In the physical context of general relativity, X is a four-dimensional pseudo-Riemannian manifold of Lorentz signature, and the manifold topology on X inherited from the Euclidean metric topology on $\mathbb{R}^4$ is the same manifold topology repeatedly invoked in Section 2.8 while discussing the proof of metric recovery. However, as described in Section 2.9, X does *not* inherit any preferred notion of conformal structure or natural scale from $\mathbb{R}^4$. The pseudo-Riemannian metric g on X is a much-more-detailed special additional structure that endows X with specific conformal and scale data. One *may* express properties of the manifold topology of X in terms of g, but it is generally *not* possible to express metric or conformal structure in terms of the manifold topology. Other useful topologies, besides the manifold topology, may also be defined on relativistic spacetime manifolds. These include the *path topology* and the *Alexandrov topology*, illustrated in Figure 2.8.2. The Alexandrov topology is closely related to the *interval topology* on a multidirected set, discussed in Section 4.2.

Causal locality. Under the causal metric hypothesis (CMH), no metric topology on spacetime, and certainly no pseudo-Riemannian metric, may be taken for granted at the fundamental level. Hence, local notions must be treated more abstractly. The choice to formulate the classical causal metric hypothesis (CCMH) in terms of directed sets provides a mathematical setting in which this may be done, but many different topologies may be defined on such sets, and not all of these topologies incorporate the necessary physical considerations. At an even more basic level, it should not be taken for granted that the conventional apparatus of topology is the best way to formulate local notions in this setting in the first place. It is therefore prudent to draw guidance from *physical* notions of what it means for a condition to be local. For the purposes of discrete causal theory, it is most natural to describe locality in terms of cause and effect.

Definition 4.1.1. *A condition on a physical system is called* **causally local** *if it may be checked by examining all independent causes and effects associated with each event.*

This definition accords with Finkelstein's [FI88] and Raptis' [RA00] notions of locality, briefly described in Section 3.9. Sorkin expresses essentially the same idea in his recent paper [SO09A] describing deviations from *spatiotemporal locality* for *generalized D'Alembertians* governing background dependent scalar field theories in the causal set context, writing:

> *Locality in the discrete context, if it meant anything at all, would imply that the action of* [a generalized D'Alembertian operator] *would be built up in terms of "nearest neighbor couplings..."* ([SO09A], page 27)

It is worth pointing out that causal locality automatically dispenses with the troublesome notion of "action at a distance," just as the strong interpretation of the causal metric hypothesis dispenses with "violations of causality." In particular, if there *is* an "action" of one event on another, which in this context means a *direct* action, independent of any intervening events, then by definition the two events are *not* "at a distance;" rather, they are local. It is easy to distill from Definition 4.1.1 a precise mathematical statement applying to the specific context of directed sets, but it is convenient to defer this statement until Definition 4.4.3 of Section 4.4, since this allows it to be expressed in terms of the local machinery developed in intervening sections. The underlying intuition of causal locality remains useful in the more general context of multidirected sets, although the physical interpretation is generally different in this case, since the "relations" involved usually represent complicated relationships, such as *co-relative histories* between pairs of classical histories, rather than causal relationships between pairs of events in classical spacetime. The choice to take Definition 4.1.1 as the starting point for the study of local structure in discrete causal theory reflects the philosophy that *the physical notion of causal locality should precede any specific choice of mathematical model.* However, it is worth noting that this definition may be applied effectively only to models for which the independence convention (IC) is physically reasonable. A natural class of such models, of course, is the class of discrete directed sets, interpreted according to the classical causal metric hypothesis (CCMH). However, the notion of causal locality applies to the entire class of *discrete manifolds,* as described by Benincasa and Dowker [BD10], following Riemann. In particular, it applies to models such as *spin foams* and *causal dynamical triangulations,* which incorporate "spatial" as well as causal structure at the fundamental scale. Interpolative models of causal structure, such as *domains,* require modified definitions.

4.2 Topology

General topology. In mathematics, local conditions are usually formalized via the theory of *topology,* or modern generalizations thereof. A **topology** $\mathcal{T}$ on a set S, with or without any additional structure, is a special family of subsets of S, which may be broadly viewed as "generalizations of open intervals in $\mathbb{R}$." Members of $\mathcal{T}$ are called *open sets* of S. An open set containing a given element x of S is called a *neighborhood* of x.[6] To qualify as a topology, $\mathcal{T}$ must satisfy the axioms that the "total set" S and the empty set $\emptyset$ are open, that finite intersections of open sets are open, and that arbitrary unions of open sets are open. A *basis* for $\mathcal{T}$ is a family of

[6]The word "neighborhood" is sometimes used in a more general sense, to mean "a subset of S containing an open set containing x." To avoid ambiguity, the term *open neighborhood* is sometimes

open sets such that every open set of S with respect to $\mathcal{T}$ is a union of sets in the basis. A *subbasis* for $\mathcal{T}$ is a *generating family* of open sets with respect to $\mathcal{T}$, meaning that any topology on S containing the subbasis contains all of $\mathcal{T}$. More concretely, this means that every open set of S with respect to $\mathcal{T}$, with the possible exceptions of S and $\emptyset$, is a union of finite intersections of sets in the subbasis. In this context, the basic structural meaning of locality described in Section 4.1, namely, that a property of S is local if and only if it may be established by examining "arbitrarily small neighborhoods," is made precise by saying that *a property is local if and only if it may be established by examining any basis for* $\mathcal{T}$.

A **topological space** is a set S together with a topology $\mathcal{T}$ on S. It is sometimes convenient to express a topological space as an ordered pair $(S, \mathcal{T})$; this is analogous to expressing a directed set as an ordered pair $(D, \prec)$. A morphism in the category of topological spaces is called a *continuous map*; an isomorphism is called a *homeomorphism.* A continuous map $f : S \rightarrow S'$ is defined to be structure-preserving, in the sense that the preimage $f^{-1}(U')$ of any open set U' in the target set S' is open in the source set S. It is worth pointing out that an alternative notion, called *Scott continuity,* is often used in order theory, and particularly in domain theory. A set S may be endowed with multiple different topologies, which may be compared in the following way: a topology $\mathcal{T}$ on a set S is called *finer* than an alternative topology $\mathcal{T}'$ on S if every element of $\mathcal{T}'$ contains an element of $\mathcal{T}$. If the converse is true, then $\mathcal{T}$ is called *coarser* than $\mathcal{T}'$. If neither condition is true, then $\mathcal{T}$ and $\mathcal{T}'$ are called *incomparable*. Note that a given topology $\mathcal{T}$ is both finer and coarser than itself under these definitions.

Since many different topologies may be defined on a given set, the *mathematical* notion of "local properties" is contextual, and is usually chosen so as to feature the specific aspects of structure one is most interested in studying. In general, the topologies most suitable for encoding useful local conditions are those that incorporate a careful balance between "fineness" and "coarseness," with special respect for any auxiliary structure on S, such as a binary relation. If a topology is too fine, then "arbitrarily small neighborhoods" of an element x are generally too small to capture all the behavior one would like to classify as occurring "near" x. On the other hand, if a topology is too coarse, then the family of neighborhoods of x generally does not adequately *resolve* structure near x; i.e., each neighborhood is so large that it contains behavior "irrelevant to x" in some sense. Among the topologies associated with multidirected sets to be examined in Section 4.3, the *discrete topology* is too fine, and the *interval topology* is too coarse, to facilitate effective study of the behavior one wishes to regard as occurring "near" a given element on physical grounds, i.e., in terms of *causal* locality.

Topological hidden structure. An interesting and subtle feature of discrete causal theory is that not every useful topological space that may be associated with a multidirected set $M = (M, R, i, t)$ involves a topology on the underlying set M itself.

(Footnote 6 continued)
used to specify that the set under consideration is itself open. In this book, neighborhoods are assumed to be open unless stated otherwise.

Rather, additional topologies of interest may be defined on "enlarged sets" containing M, whose "extra elements" facilitate the analysis of M. The *continuum model* $M_{\mathbb{R}}$ of M, and the *star model* $M_\star$ of M, both defined in Section 4.3, are examples of such enlarged sets. The continuum model $M_{\mathbb{R}}$ contains an uncountable number of extra elements, drawn from an *extrinsic* source; namely, the unit interval $[0, 1]$ in $\mathbb{R}$. This limits the relevance of $M_{\mathbb{R}}$ to some degree, since the resulting topological space is not actually induced by the multidirected structure of M. By contrast, the extra elements appearing in the star model $M_\star$ correspond to the elements of the relation set R of M, and therefore represent *intrinsic,* though "hidden," topological structure associated with M. The construction of $M_\star$ is therefore an application of the *principle of hidden structure* (HS).

As explained in Section 3.8, the principle of hidden structure is inspired by Alexander Grothendieck's approach to commutative algebraic geometry, in which an algebraic variety is "augmented" by the addition of "extra elements" corresponding to its closed subsets, thereby producing an *algebraic scheme.* The star model $M_\star$ of a multidirected set $M = (M, R, i, t)$ represents a primitive example of an analogous, and equally useful, type of construction in discrete causal theory, in which the underlying set of M is augmented by the addition of "extra elements" corresponding to nontrivial structural components of M; in this case, relations. More elaborate constructions may be defined, by adding elements corresponding to more complicated aspects of structure, such as *causal paths,* introduced in Section 5.9.[7] Going in the opposite direction, it is sometimes useful to adjoin to M "extra elements smaller than ordinary elements." This situation arises, in particular, in the theory of *abstract element space,* introduced in Section 5.2. Similar ideas appear in the theory of *causal atoms,* introduced in Section 5.6, which may be viewed as "generalized elements" of a multidirected set M. If M contains cycles, then "ordinary elements" of M may be "smaller than any causal atom."

Topological local finiteness. I now turn to the specific topic of local *finiteness* conditions for topological spaces. In an abstract context, each topology on a set S is equipped with its own notion of local finiteness. As described above, it is often necessary in discrete causal theory to consider the case of topologies on enlarged sets containing a multidirected set, so I formalize topological local finiteness in the following generalized way:

Definition 4.2.1. *Let S and $\overline{S}$ be sets, with S a subset of $\overline{S}$, and let $\overline{\mathcal{T}}$ be a topology on $\overline{S}$. Then S is called* **topologically locally finite** *with respect to the pair $(\overline{S}, \overline{\mathcal{T}})$ if every element of S is contained in an element of $\overline{\mathcal{T}}$ of finite cardinality.*

Topological local finiteness may be paraphrased informally by the statement that "every element has a finite neighborhood." However, the definition of "neighborhood" depends on the choice of topology, and neighborhoods are generally subsets of the enlarged set $\overline{S}$, rather than subsets of S itself.

---

[7] Such "extra elements" do not necessarily correspond to *subobjects* of M; in particular, causal paths are not subobjects, but equivalence classes of morphisms into M.

4.3 Four Topologies for Multidirected Sets

Four examples out of many. A large number of different topological spaces may be associated with a multidirected set $M = (M, R, i, t)$. Some of these spaces are constructed by imposing topologies on the underlying set M itself, while others involve an "enlarged set" containing M, following the ideas described in Section 4.2. In the present section, I discuss just four illustrative examples of such spaces. Their topologies are called the *discrete*, *interval*, *continuum*, and *star topologies*, respectively. The discrete topology and the interval topology are topologies on the underlying set of M, while the continuum topology and the star topology are topologies on enlarged sets $M_{\mathbb{R}}$ and $M_\star$ containing M, called the *continuum model* and the *star model* of M, respectively. The interval and continuum topologies are each analogous to the Euclidean metric topology on $\mathbb{R}$, but in different ways. The star topology is distinct from the identically-named notion appearing in the study of computer networks, although the underlying idea is similar.

In topological terms, the goals of this section are modest, and even primitive, compared to the sophisticated and comprehensive analyses appearing in the order-theoretic and graph-theoretic literature. The four topological spaces discussed here are chosen specifically to facilitate and clarify the study of local properties, and specifically local finiteness, in discrete causal theory. The star topology on the star model $M_\star$ of a multidirected set M is the most useful of these four topologies for this purpose, since it is the most natural choice for mathematically representing the physical idea of causal locality in the context of discrete classical spacetime structure. In particular, the "mathematization of causal locality for multidirected sets," embodied in Definition 4.4.3 below, coincides with topological locality in the star topology. More specifically, Lemma 4.4.2 demonstrates that topological local finiteness for the star topology coincides with *star finiteness* (SF), which is the most important local finiteness condition for the version of discrete causal theory developed in this book. It is also useful to compare the viewpoint adopted here with some of the more conventional literature. In the discrete context, particularly in the special case of causal sets, the sources cited in Sumati Surya's review article [SU11] provide an excellent starting point for exploring topological issues. For a broader perspective reaching outside the discrete setting, it is instructive to study the various topologies arising in domain theory, such as those discussed by Martin and Panangaden in [MP06].

Figures 4.3.1 and 4.3.2 illustrate small neighborhoods of an element x in a multidirected set M, for each of the four topologies listed above. The unusual appearance of the diagrams in Figure 4.3.2 is due to the fact that these diagrams represent the enlarged sets $M_{\mathbb{R}}$ and $M_\star$; the precise meaning of these diagrams is explained below. As usual, the diagrams in these figures depict only acyclic directed sets, but the associated definitions apply to multidirected sets in general. However, the reader should keep in mind that "strange things can happen" when cycles are admitted; for example, an open interval of the form $\langle\langle x, x\rangle\rangle$ may be nonempty in this context.

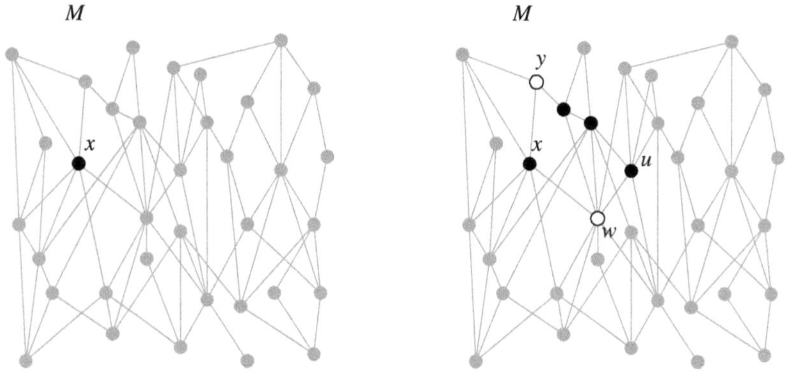

Figure 4.3.1. Small neighborhood of an element x in the discrete topology on a multidirected set M; small neighborhood of x in the interval topology on M.

Discrete topology. The discrete topology $\mathcal{T}_{\mathrm{DIS}}$ on a multidirected set M takes the individual elements of M to be open sets. Since arbitrary unions of open sets are open, *every* subset of M is open in the discrete topology. This topology is too fine to be useful in discrete causal theory. Indeed, as illustrated in the left-hand diagram in Figure 4.3.1, every element x of M defines a singleton neighborhood of itself. Hence, "arbitrarily small neighborhoods" of x in the discrete topology may contain *no information whatsoever* about "what happens near x" from a causal perspective. More precisely, if M is a directed set, viewed as a model of classical spacetime under the classical causal metric hypothesis (CCMH), then one can decipher nothing about local causal structure near the event represented by x, by examining only x itself. Similarly, if M is the underlying multidirected set of a kinematic scheme $\mathbb{S}$, then one can decipher nothing about the family of co-relative histories in $\mathbb{S}$ involving the classical history represented by x, by examining only x itself.

Interval topology. The interval topology $\mathcal{T}_{\mathrm{INT}}$ on a multidirected set M is perhaps the most "obvious" topology to study in the special case when M is a causal set, due to the causal set axiom of interval finiteness (IF). A convenient subbasis for the interval topology on M is the family of pasts $J^-(x)$ and futures $J^+(x)$ of elements x of M. An open interval $\langle\!\langle w, y\rangle\!\rangle$, for two elements w and y in M, may be expressed as the intersection $J^+(w) \cap J^-(y)$ of subbasis elements, which explains the terminology "interval topology." One reason for defining the interval topology in terms of this subbasis, instead of using open intervals directly, is that this definition automatically incorporates "half-open intervals" involving extremal elements, which must otherwise be explicitly, and awkwardly, added. As noted above, "strange properties" arise in the case where M contains cycles; for example, the open interval $\langle\!\langle x, x\rangle\!\rangle$ is nonempty whenever x lies on a cycle.

The reader should note that the term "interval topology" has many alternative meanings besides the one used here, even in the specific context of partially ordered sets.[8] As I have defined it, the interval topology is one possible discrete causal analogue of the *Alexandrov topology* on a relativistic spacetime manifold X, which has subbasis consisting of the chronological pasts $I^-(x)$ and chronological futures $I^+(x)$ of events x in X, as described in Section 2.8. This is the source of the term *"Alexandroff set,"* appearing in the quote of Bombelli, Lee, Meyer, and Sorkin [BLMS87] in Section 3.5. The term "Alexandrov topology" has a somewhat different, and more general, meaning in modern mathematics, but the above usage is too deeply rooted to amend, at least in the physical context of classical spacetime structure; for example, it appears in standard relativity texts such as Wald [WA84], and Hawking and Ellis [HE73]. As noted in Section 2.8, the Alexandrov topology is generally coarser than the manifold topology, but the two coincide for strongly causal spacetimes,[9] which is one reason why the interval topology is a priori interesting in causal set theory. However, nontrivial choices are involved in abstracting the Alexandrov topology to the discrete causal context. For example, the chronological past $I^-(x)$ and chronological future $I^+(x)$ of an element x in X exclude the boundaries of the corresponding causal past $J^-(x)$ and causal future $J^+(x)$ in the relativistic setting. One naïve way to abstract this condition to apply to discrete directed sets would be to exclude y from $J^+(x)$ whenever $x \prec y$ is an *irreducible* relation, and similarly for $J^-(x)$, but these details are peripheral to the present discussion.[10]

Although the interval topology respects multidirected structure in an obvious sense, it is nonetheless a poor choice for defining local properties of multidirected sets. In particular, for a directed set, viewed as a model of classical spacetime under the classical causal metric hypothesis (CCMH), the interval topology generally fails to either *capture* or *isolate* local causal structure near a given element. This is partly because it is too coarse, but it also has other deficiencies. These are elaborated in Section 4.4, and illustrated in the examples appearing in Section 4.7. The right-hand diagram in Figure 4.3.1 illustrates a small neighborhood of an element x in the interval topology on M, consisting of the four elements represented by the dark nodes. In this case, the chosen neighborhood is itself an open interval. Observe that this neighborhood contains elements, such as u, that are *causally unrelated* to x. This is not very surprising, given the relativistic analogue involving the Alexandrov topology: a typical Alexandrov set $I^+(w) \cap I^-(y)$ contains an entire local spacelike section passing through any element x such that $w \prec_{\text{GR}} x \prec_{\text{GR}} y$. However, this property is inconvenient in the pure causal setting, and reflects a generic problem in using the interval topology to describe local causal structure. For future reference, note that *interval finiteness is not the same as topological local finiteness for the interval topology,* since interval finiteness requires that *every* open interval must be finite, not merely that every element is contained in a finite member of $\mathcal{T}_{\text{INT}}$. Also note that a member of $\mathcal{T}_{\text{INT}}$ does not always contain an open interval; i.e., open

[8]Consider, for example, the "interval topologies" discussed by Marcel Erné in [ER80].

[9]In fact, they coincide *if and only if* the spacetime is strongly causal. As noted in Section 2.7, the strongly causal condition does not possess an obvious discrete causal analogue.

[10]This idea is motivated by the fact that irreducible relations are typically "almost null" in sprinkled causal sets, as explained in Section 4.5.

intervals do not form a basis for the interval topology. By contrast, the Alexandrov sets $I^+(w) \cap I^-(y)$ *do* form a basis for the Alexandrov topology in the relativistic context.

Continuum topology. A concrete and useful, though cumbersome, enlarged set containing a multidirected set $M = (M, R, i, t)$, may be constructed by identifying each element of the relation set R of M with a line segment, i.e., a closed interval in $\mathbb{R}$, then gluing these segments together at their endpoints in an appropriate way. This approach leads to a useful topology on the resulting quotient space. More precisely, the **relation set** $R(x)$ at an element x in M is defined to be the subset of the total relation set R of M consisting of all relations r such that either $i(r) = x$, or $t(r) = x$, or both. One may identify every element of R with a copy of the closed unit interval $[0, 1]$ in $\mathbb{R}$, equipped with its Euclidean metric topology. One may then label the initial and terminal points of each interval with the initial and terminal elements of the corresponding relation. Finally, one may define the **continuum model** $M_{\mathbb{R}}$ of M to be the *quotient space* given by identifying all endpoints of intervals with matching labels. In particular, the set of intervals with at least one endpoint labeled by x corresponds to $R(x)$. The quotient topology[11] $\mathcal{T}_{\mathbb{R}}$ on $M_{\mathbb{R}}$ is called the **continuum topology**.

The continuum topology on $M_{\mathbb{R}}$ involves a great deal of structure extrinsic to the multidirected set (M, R, i, t) itself; namely, the uncountable family of points belonging to each of the line segments corresponding to the elements of the relation set R. However, the continuum topology has the advantage of providing natural families of neighborhoods that both capture and isolate local multidirected structure near each element x of M, in a manner that is useful for the desired physical applications. In particular, for every real number ρ such that $0 < \rho < 1$, the **(open) star of radius ρ at** x is defined to be the "star-shaped subset" $\mathrm{St}_\rho(x)$ of $M_{\mathbb{R}}$ with "limbs" of length ρ glued at x, i.e., the subset

$$\mathrm{St}_\rho(x) := \{t \in M_{\mathbb{R}} \mid d(t, x) < \rho\},$$

where d is the distance function on $M_{\mathbb{R}}$ induced by the Euclidean metrics on each unit interval. The terminology of "limbs" is intended to evoke an image similar to the legs of a starfish; this imagery may be understood by examining the left-hand diagram in Figure 4.3.2. Any neighborhood U_x of x in the continuum topology on $M_{\mathbb{R}}$ contains $\mathrm{St}_\rho(x)$ for sufficiently small values of ρ. Hence, given such a neighborhood U_x, one may ascertain the cardinality of the relation set $R(x)$ at x, and hence the local multidirected structure near x, by examining an appropriate star $\mathrm{St}_\rho(x)$ contained in U_x. Note that the term "star-shaped" is merely a suggestive description, carried over from the acyclic case; $\mathrm{St}_\rho(x)$ may not look "star-shaped" in general. For example, if r is a reflexive relation $x \prec x$, and if $\rho \geq 1/2$, then $\mathrm{St}_\rho(x)$ contains the *entire interval*

[11] In general, if $(S, \mathcal{T})$ is a topological space, and $\sim$ is an equivalence relation on S, then the *quotient topology* on the set $S/\!\sim$ of equivalence classes under $\sim$ is defined by taking a subset of $S/\sim$ to be open if and only if it has an open preimage under the natural *quotient map* $S \to S/\sim$ taking each element to its equivalence class.

corresponding to r. Figure 4.3.2 illustrates the star $\mathrm{St}_{1/2}(x)$ at an element x in the continuum topology on $M_{\mathbb{R}}$, with limbs represented by the dark segments. No attempt is made to draw the diagram to scale, i.e., with equal-length line segments, which would clutter up the drawing with curved lines. However, the diagram *is* intended to suggest a "one-dimensional subspace of three-dimensional Euclidean space," in order to emphasize the difference between M and $M_{\mathbb{R}}$, particularly the fact that $M_{\mathbb{R}}$ contains a large number of "extra elements."

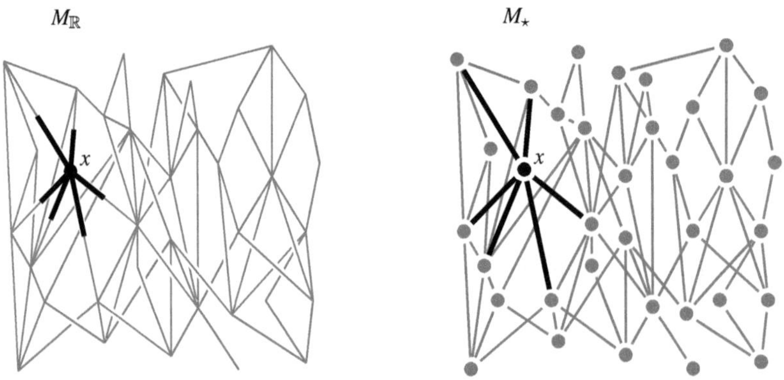

Figure 4.3.2. Star $\mathrm{St}_{1/2}(x)$ of an element x in the continuum topology on the continuum model $M_{\mathbb{R}}$ of a multidirected set M; star $\mathrm{St}(x)$ of x in the star topology on the star model $M_\star$ of M.

Star topology. The most useful features of the continuum topology for a multidirected set $M = (M, R, i, t)$ may be retained, without the burden of the extrinsic structure contained in the uncountable real line segments used to build the continuum model $M_{\mathbb{R}}$ of M. Informally, this may be accomplished by assigning only *one* "extra element" to each relation in M, rather than an uncountable number of extra elements. More precisely, the **star model** $M_\star$ of M is defined, as a set, to be the union $M_\star := M \cup R$ of M with its relation set R. For each element x in M, the **star at** x **in** $M_\star$ is defined to be the subset

$$\mathrm{St}(x) := \{x\} \cup R(x),$$

of $M_\star$. It is worth pointing out that in the *Bass–Serre theory* of group actions on trees, the star of a vertex x in a graph is defined to be the set of all edges terminating at x, which is essentially the same notion as the one used here. The **star topology** $\mathcal{T}_\star$ on $M_\star$ is defined to be the topology with subbasis $\{\mathrm{St}(x) \mid x \in M\}$. It is useful to compare $\mathrm{St}(x)$ with the corresponding open stars $\mathrm{St}_\rho(x)$, defined above in the context of the continuum topology. An important distinction between the two cases is that $\mathrm{St}(x)$ is unique, while there exist uncountably many open stars $\mathrm{St}_\rho(x)$ of different radii at x in $M_{\mathbb{R}}$. At a global level, the extra elements R of $M_\star$ are an intrinsic part of the structure of $M = (M, R, i, t)$, unlike the extra elements of $M_{\mathbb{R}}$. These characteristics confer on $M_\star$ a clear advantage over $M_\mathbb{R}$ in modeling local structure in the discrete causal setting. Variations of the star topology may be defined in terms of *irreducible stars, past stars, future stars,* and so on.

The right-hand diagram in Figure 4.3.2 illustrates the star $\mathrm{St}(x)$ at an element x in the star topology on $M_\star$. As in the case of the continuum topology on $M_\mathbb{R}$, the appearance of this diagram warrants some additional explanation. It is a "hybrid diagram," intended to reflect the fact that the elements of the relation set R of $M = (M, R, i, t)$ now play a *dual role:* they serve both as *relations* between pairs of elements of M in the usual sense, and as *elements* of $M_\star$ in their own right. I have schematically highlighted this difference by drawing the edges in the diagram slightly thicker than in previous diagrams, and by separating the edges from the nodes representing the elements of M. Usually, however, it is more convenient to represent the dual role of R via *pairs* of diagrams, one showing the "relation" role, and the other showing the "element" role. Simple examples of this method have appeared in Figures 1.5.2 and 1.5.4 of Chapter 1, in which *relations* in discrete causal classical histories are also viewed as *elements* of the corresponding *relation spaces.* This viewpoint is greatly expanded in Chapter 5.

Cardinal-valued scalar fields and the star topology. It is interesting to consider the question of how much information about a multidirected set $M = (M, R, i, t)$ may be recovered from the star topology on $M_\star$. The star topology detects the *valences* $v(x)$ at each element x of M, i.e., the numbers of relations incident at each x, which may a priori be finite or infinite. Reflexive relations $x \prec x$ are counted twice in enumerating $v(x)$, because such relations both begin and terminate at x. The valences $v(x)$ form a *cardinal-valued scalar field* v_M on M, called the **valence field** of M, already introduced informally in Section 1.7 in the special case of directed sets. In this book, the cardinal numbers defining the valence field are almost always natural numbers, because the most interesting versions of discrete causal theory, from a physical perspective, involve directed sets satisfying the condition of *star finiteness* (SF). The meaning of this condition is now obvious from the foregoing discussion, but its formal introduction is deferred until Definition 4.4.1. However, it is not a mere exercise in empty generalization to refer to "cardinal-valued fields," rather than natural number-valued fields. For example, the value of the valence field corresponding to the relativistic causal relation $\prec_{\mathrm{GR}}$ at each element x in a relativistic spacetime manifold X is $2^{\aleph_0}$, since x is related, in one direction or the other, to every element of its total domain of influence $J(x)$. Of course, a constant scalar field with an uncountable cardinal value seems, at least to me, to be completely useless for physical purposes. This illustrates, from yet another viewpoint, how discrete causal theory makes available aspects of structure that play no sensible role in continuum-based theories. In particular, it suggests that discrete causal theory has something more to offer, at a local level, than the "mere geometry" suggested by the metric recovery theorems. *Countably* infinite values of the valence field arise in certain idealized constructions in causal set theory, as explained in Sections 4.4 and 4.5.

An arbitrary cardinal-valued scalar field on an arbitrary set S is generally *not* a valence field for any multidirected structure on S. For example, there exists no multidirected structure on a singleton $S = \{x\}$ such that $v_S(x)$ is an odd number, since any relation must begin and terminate at x in this context. More generally, the "inverse problem" of "recovering a multidirected set from an arbitrary valence field" suggests consideration of *generalized multidirected sets,* in which the initial and terminal element maps are only partially defined. For example, one may consider relations that "go off to infinity," instead of "terminating at an element." Again, this is more than just an empty generalization, due partly to an analogy involving the *causal path semicategories* studied in Chapter 5, which include paths with no initial or terminal element. Similar constructions also appear in the literature in the context of quantum information theory; for example, Blute, Panangaden, and Straßburger's *pre-causal graphs* [BPS08] include "half-edges" as part of their structure.

As mentioned in Section 1.7, the valence field v_M of a multidirected set M may be decomposed into a component v_M^-, called the **past valence field** of M, which measures the cardinality of the set of relations *terminating* at each element in M, and a component v_M^+, called the **future valence field** of M, which measures the cardinality of the set of relations *beginning* at each element in M. Unlike the valence field v_M itself, the past and future valence fields $v_M^\pm$ generally cannot be detected by the star topology. Valence fields also induce interesting *relation fields* on a multidirected set M. In particular, one may consider relation fields whose values are integer linear combinations of $v_M\big(i(r)\big)$ and $v_M\big(t(r)\big)$, or of $v_M^\pm\big(i(r)\big)$ and $v_M^\pm\big(t(r)\big)$, for each relation r in the relation set R of M. As noted in Section 1.7, relation fields in general are analogous to *timelike vector fields;* this is because such fields convey *directional information,* rather than merely numerical quantities. For example, the values $v_M\big(t(r)\big) - v_M\big(i(r)\big)$ collectively measure the "change in the valence field in the causal direction," along each relation. In the star finite case, all such fields take on integer values, and their arithmetic properties are of great interest. Figure 4.3.3 illustrates the past and future valence fields on a multidirected set. In "physically realistic" scenarios involving classical spacetime structure, it would not be surprising to encounter much larger valences, as discussed in Section 4.5.

4.4 Interval Finiteness and Star Finiteness

Initial discussion of interval finiteness. It is possible that treatments of discrete causal theory written a generation from now will devote little attention to interval finiteness (IF) in the specific role of a "local finiteness axiom" for directed sets, since it is not a local condition mathematically, and since it does not imply local finiteness in a causal sense. However, as in the case of transitivity (TR), examined in Section 3.9, interval finiteness is commonly used as an axiom in the recent literature, so a careful treatment is necessary. Such an examination also affords an opportunity to elucidate some important physical, structural, and historical issues, and to explore a "menagerie" of directed sets that are interesting in their own right. Of course, many physically interesting types of directed sets *do* satisfy interval finiteness in an incidental manner; for example, many such sets are *globally* finite, rendering the

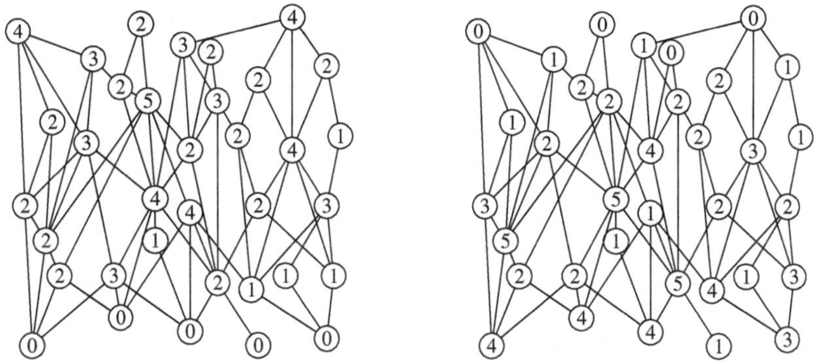

Figure 4.3.3. Past and future valence fields on a multidirected set.

choice between different local finiteness conditions irrelevant. The moral is not that there is anything wrong with directed sets that *happen to be* interval finite, but rather that the condition itself is not ideal for restricting attention to a class of directed sets suitable for modeling classical spacetime in discrete causal theory.

As explained below, the causal sets community has been aware of some of the structural issues associated with the condition of interval finiteness since the birth of the theory in the 1980s. An important example of such issues arises in the context of certain idealized causal sets globally induced, a posteriori, via "sprinkling" into pseudo-Riemannian manifolds, such as Minkowski spacetime $\mathbb{R}^{3+1}$. Such causal sets can exhibit locally infinite behavior at every element, in the sense that each element possesses an infinite number of maximal predecessors and minimal successors. Of course, no one views such objects as physically realistic, any more than $\mathbb{R}^{3+1}$ itself is viewed as a physically realistic global model of classical spacetime in the context of general relativity. It is clear from the literature that the founders of causal set theory never intended to use such locally infinite objects as actual workhorses for quantum gravity. However, the ongoing prominence of such objects, due partly to their automatic similarity to relativistic spacetime, has perhaps contributed to preventing the systematic incorporation of causal local finiteness into discrete causal theory at the axiomatic level.

In a broader context, interval finiteness permits the very types of physically pathological behavior that any appropriate local finiteness condition should rule out, such as the instantaneous emergence of an infinite volume of spacetime from a single event. The local infinities arising in sprinkled causal sets are not of this type, but this is not very reassuring, since the structure of these objects is completely determined by the "well-behaved" manifolds by which they are induced. Locally infinite structure is tenable in these special cases only because the maximal predecessors and minimal successors of a given element x stretch an infinite distance into the past and future of x, respectively, in any given frame of reference, rather than accumulating locally, in a temporal sense, where they could "cause trouble." Taken seriously, this suggests that what happened "long ago, in a galaxy far, far away" *directly* influences present events on earth. This may seem dubious enough in its own right, but it is a natural consequence of a naïve marriage between discreteness and special relativity, as described in Section 4.5. It is also worth pointing out that locally infinite behavior in interval finite *transitive* models of classical spacetime, including causal sets, may sometimes be regarded as an innocuous artifact of transitivity itself, since "most" of the relations involving a given element do not represent direct influence in this setting. This is analogous to the situation in the set of integers $\mathbb{Z}$, where there exist an infinite number of relations of the form $0 < n$, all of which are indirect except for the relation $0 < 1$.

Interval finiteness also leads to serious technical inconveniences that render local causal structure difficult to study, regardless of its physical ramifications. As mentioned above, it is not a causally local condition, in the sense of Definition 4.1.1, since the direct causes and effects associated with a given event generally cannot be isolated in an open interval containing the element representing the event. Mathematically, it is not a local condition for the interval topology, since it involves arbitrarily large neighborhoods, nor is it a consequence of the topological local finiteness condition induced by the interval topology. Further, open intervals do not even *capture* local multidirected structure in any suitable sense, let alone *isolate* such structure. All of these problems suggest that interval finiteness is simply not a natural condition in the discrete causal context. These issues, and others, are examined in detail over the next few sections.

Star finiteness. Sorkin's version of the classical causal metric hypothesis (CCMH) strongly suggests the need for *some* type of local finiteness condition, and the same is true of any version of the hypothesis assigning infinite "spacetime volumes" to infinite subsets of a directed set. As noted above, locally infinite structure in this context permits absurd behavior of the worst sort from a physical perspective. The mere possibility of such behavior poses a serious threat to the plausibility of any theory permitting it, even if special types of local infinities may be assigned reasonable interpretations. This is particularly true in the context of quantum theory, in which *all* models permitted by the axioms are potentially relevant. In causal set theory, for example, one already faces worrisome configuration space issues even in the finite case, as demonstrated by the asymptotic enumeration of Kleitman and Rothschild [KR75]. The possibility of locally infinite behavior renders these problems incomparably worse in the countably infinite setting. One could attempt to formulate different versions of the classical causal metric hypothesis in an effort to ameliorate these issues, without imposing a local finiteness condition at all. For example, one might consider altering the measure axiom (M) to admit discrete measures assigning finite "volumes" to certain *infinite* subsets of a directed set. However, this strategy would lead far away from the basic insight that discrete combinatorial data can serve as a proxy for scale in the context of metric recovery. In any case, axiomatic changes of such significance should be made for physical reasons, not for the purpose of rescuing other axioms.[12]

[12]Of course, as discussed in Section 2.8, there are excellent arguments for considering less-drastic modifications of the causal set version of the measure axiom (M), allowing volume to depend on

How, then, *should* local finiteness be axiomatized in discrete causal theory? The crucial requirement is that local finiteness should adequately encode, at a mathematical level, the physical notion of *causal* local finiteness, in the sense of Definition 4.1.1. In the specific context of classical spacetime, modeled as a directed set $D = (D, \prec)$ under the classical causal metric hypothesis (CCMH), and interpreted according to the independence convention (IC), causal local finiteness means that *the relation set* $R(x)$ *at each element* x *of* D *is finite.* Equivalently, the star $\mathrm{St}(x)$ at each element x in D is finite. Since the same condition is useful for general multidirected sets, I state the axiom in this broader setting:

Definition 4.4.1. Star finiteness (SF). *A multidirected set* $M = (M, R, i, t)$ *is called* **star finite** *if for every element* x *of* M, *the star* $\mathrm{St}(x)$ *of* x *is of finite cardinality.*

In the special case of acyclic directed sets, the cardinality of $\mathrm{St}(x)$ is one greater than the number of direct predecessors and successors of x. Hence, star finiteness may be stated in terms of *elements* in this case. For directed sets including cycles, however, this correspondence is altered by the possibility of reflexive relations, and relations in both directions, between a given pair of elements. Finally, for multidirected sets, one may imagine scenarios such as the existence of an infinite number of reflexive relations at a singleton, for which star finiteness cannot be determined by counting predecessors and successors. In fact, this scenario is not very strange from a basic structural perspective, since it is a causal analogue of the category-theoretic representation of an infinite group via endomorphisms of a single structureless object. For these reasons, it seems best, in the general case, to express star finiteness in terms of relation sets or stars. Star finiteness may be detected by either the continuum topology or the star topology. Indeed, as mentioned informally in Section 4.3, star finiteness is the same condition as topological local finiteness for a multidirected set M with respect to the star topology on the star model $M_\star$ of M. This is the content of the following lemma[13]:

Lemma 4.4.2. *Star finiteness for a multidirected set* $M = (M, R, i, t)$ *coincides with topological local finiteness of* M *with respect to the pair* $(M_\star, \mathcal{T}_\star)$.

Proof. Star finiteness means that the star $\mathrm{St}(x)$ at x is finite for each x in M. Since $\mathrm{St}(x)$ is a neighborhood of x in $M_\star$ with respect to $\mathcal{T}_\star$, this immediately implies topological local finiteness of M with respect to $(M_\star, \mathcal{T}_\star)$. Conversely, the intersection of two stars $\mathrm{St}(x) \cap \mathrm{St}(y)$ is either a family of relations between x and y, or empty. Hence, the smallest neighborhood of x in the star topology on $M_\star$ is $\mathrm{St}(x)$. Therefore, topological local finiteness of M with respect to $(M_\star, \mathcal{T}_\star)$, implies that $\mathrm{St}(x)$ is finite for every x in M. □

(Footnote 12 continued)
local causal structure in a nontrivial way. However, this very notion requires a satisfactory treatment of *causally* local properties, which open intervals cannot supply.

[13]Another equivalent condition is that the valence field v_M of M takes values in $\mathbb{N}$.

Formalizing local conditions. It is easy to generalize from the specific case of star finiteness to formalize what it means for an arbitrary condition on a multidirected set $M = (M, R, i, t)$ to be a local condition. Star finiteness of $M = (M, R, i, t)$ may be determined by merely *counting* the star $\mathrm{St}(x)$ at x for each element $x \in M$, but it is possible to imagine other conditions involving more subtle or intricate data encoded in these stars. Hence, the appropriate notion of what it means for a condition to be local in general is given by simply stipulating that the stars $\mathrm{St}(x)$, considered separately, *contain sufficient information* to decide whether or not the condition holds.[14]

Definition 4.4.3. *Let $M = (M, R, i, t)$ be a multidirected set, and x an element of M.*

1. *A condition on M is called* **local at** x *if it may be established by examining the star* $\mathrm{St}(x)$ *at x.*
2. *A condition on M is called* **local** *if it may be established by examining the disjoint union* $\bigsqcup_{x\in M} \mathrm{St}(x)$.

The significance of the disjoint union in the second part of Definition 4.4.3 is that a local condition must be detectable in terms data *native to the individual stars* $\mathrm{St}(x)$. In particular, this does not allow for consideration of the relationships in M among elements or relations belonging to different stars, or relationships in M among these stars themselves. For the purposes of this book, this is the "default" version of locality, since it corresponds in an suitable way to the desired physical interpretation of the structured sets of principal interest. However, alternative notions of locality are possible even if one remains in the star finite context; for example, one may restrict attention to irreducible relations. A few such variations on star finiteness are outlined below; treatment of other local conditions may be carried out in an analogous manner.

Variations on star finiteness for directed sets. The axiom of star finiteness (SF) may be modified in a number of different ways to obtain related conditions that emphasize or isolate different aspects of structure. For example, multidirected sets permit multiple relations of the form $x \prec y$ between a given pair of elements x and y, but one is sometimes interested only in how many different elements a given element is related to, not in "how many different ways" two specific elements are related. In this context, one may first replace all relations of the form $x \prec y$, for each pair of elements x and y in M, with a single relation, before testing for star finiteness. This leads to the condition of *local element finiteness* (LEF), which is briefly examined, along with a number of other variations on star finiteness, in Section 4.9. At present, however, I restrict attention to two specific variations on star finiteness for *directed* sets, called *star finiteness in the transitive closure* (SFT), and *star finiteness in the skeleton* (SFS). These conditions differ from star finiteness, and from each other, in terms of how they treat reducible relations. In fact, they represent two different extremes in this regard; the first condition "adds all relations implied by transitivity" before testing for star finiteness, while the second condition "removes all reducible

[14] Of course, different notions are necessary in the interpolative context.

relations." Although these conditions are formulated here only for directed sets, generalizations to the case of multidirected sets are possible. The reason for introducing these conditions at this point in the discussion is because they are useful for some of the analysis in the sections preceding Section 4.9. The rationale for restricting attention to the case of directed sets is that the principal intended application of the material introduced in these sections is the study of *classical* spacetime structure.

Definition 4.4.4. *Let* $D = (D, \prec)$ *be a directed set.*

1. *D is called* **star finite in the transitive closure** (SFT), *if* $\mathrm{tr}(D)$ *is star finite.*
2. *D is called* **star finite in the skeleton** (SFS), *if* $\mathrm{sk}(D)$ *is star finite.*

Neither of these conditions offers a plausible alternative to star finiteness as an actual axiom for directed sets in the context of classical spacetime structure, but both conditions are useful for comparative purposes. Star finiteness in the transitive closure is not a causally local condition, since it requires the entire past and future of each element of D to be finite. This follows from Definition 3.10.2, because any element y in the future of an element x in a directed D is related to x under the transitive relation $\prec_{\mathrm{tr}}$ defining the transitive closure $\mathrm{tr}(D)$. Star finiteness in the transitive closure does *not* imply that D itself has finite "causal size," however; for example, a disjoint union of chains of each natural number length is star finite in the transitive closure. However, this condition is still much too restrictive of global structure to serve as a physically useful axiom. Star finiteness in the skeleton, meanwhile, is far too weak a condition to effectively restrict attention to physically plausible directed sets under any version of the classical causal metric hypothesis (CCMH) that assigns infinite "spacetime volumes" to infinite subsets of a directed set. This, of course, includes Sorkin's version of the hypothesis. In particular, star finiteness in the skeleton permits any interpolative directed set, including $\mathbb{R}$, as well as relativistic spacetime manifolds, domains, and so on, since the skeleta of such directed sets have no relations. It also permits "pathologically complex cycles in otherwise simple directed sets," since the skeleton operation destroys cycles. However, star finiteness in the skeleton can be interesting when combined with other axioms. For example, if one insists on retaining interval finiteness (IF) as an axiom, star finiteness in the skeleton cures some the associated structural issues.

Incomparability of interval finiteness and star finiteness. Interval finiteness (IF) and star finiteness (SF) are *incomparable* conditions, meaning that neither condition implies the other. This is true even in the special case of acyclic directed sets. More comprehensive results about the relationships among various finiteness conditions, both local and nonlocal, for both acyclic directed sets and multidirected sets, are given in Section 4.9. In the present section, I merely exhibit a few illustrative examples; or rather, *counterexamples,* since they demonstrate the incomparability result cited above. At a heuristic level, this incomparability means that interval finiteness and star finiteness do *not* merely represent "more or less conservative versions" of the same general type of condition; rather, they govern essentially different types of behavior. The most basic reason for this is that star finiteness is a genuinely local condition, while interval finiteness is not.

To see that interval finiteness does not imply star finiteness, consider the causal set with underlying element set $\{w\} \cup \{x_i \mid i \in \mathbb{Z}\}$, and relations $w \prec x_i$, for all $i \in \mathbb{Z}$, illustrated in the left-hand diagram in Figure 4.4.5. For future reference, I refer to this causal set as the **infinite bouquet**. The metaphor here is obvious: the relations $w \prec x_i$ are the "stems" of the bouquet, and the elements x_i are the "flowers." The infinite bouquet possesses no nonempty open intervals, since its longest chain has length one; it is therefore interval finite. However, it is not star finite, or even star finite in the skeleton, because the element w has an infinite number of minimal successors. This example illustrates how interval finiteness permits physically absurd locally infinite behavior under any version of the classical causal metric hypothesis (CCMH) for which infinite sets are assigned infinite measure. In particular, if one considers the infinite bouquet in isolation, rather than as a subset of a larger directed set, its behavior appears "as bad as possible," since in the "obvious frame of reference" in which the elements x_i are mutually simultaneous, an infinite volume of spacetime instantaneously appears from a single event. Dismissing implausible speculation about the Big Bang, the Big Crunch, eternal inflation, vacuum collapse, and other universal cataclysms, such scenarios have no reasonable place in discrete causal theory.

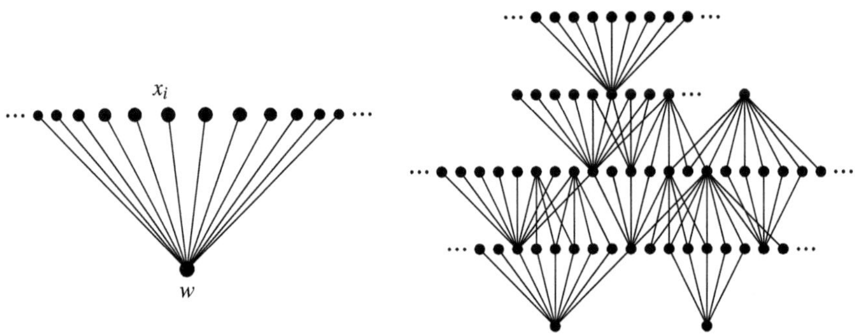

Figure 4.4.5. The infinite bouquet is interval finite but not star finite, or even star finite in the skeleton; absurd interval-finite behavior.

Interestingly enough, however, causal sets induced by global sprinkling into Minkowski spacetime $\mathbb{R}^{3+1}$ typically contain subsets isomorphic to the infinite bouquet; in fact, the irreducible future of almost every element in such a set possesses this structure! As mentioned above, manifestly nonphysical behavior is avoided in this context because the global structure of such causal sets "conspires" to stretch the infinite set $\{x_i \mid i \in \mathbb{Z}\}$ into the infinite temporal future of the element w in any given frame of reference. This strange behavior ultimately originates from the loss of the interpolative property upon passage from the continuum-based paradigm to the discrete paradigm, which confronts the theorist with the completely new structural feature of irreducible relations. If one attempts to merely *copy* structure from pseudo-Riemannian geometry in this context, one encounters a priori unexpected issues of this sort. The alternative is to hypothesize that *spacetime exhibits structural features that do not come from pseudo-Riemannian geometry at all,* which is what one expects, in any case, in the context of quantum gravity. Star infinite sprinkled causal sets are studied in much more detail in Section 4.5.

One of the most serious practical issues arising from the possibility of locally infinite behavior in interval finite directed sets is that it presents tremendous difficulties in attempting to formulate dynamical principles. This is true even for examples such as sprinkled causal sets, in which such behavior, by construction, may be assigned a reasonable physical interpretation. As noted above, the way in which locally infinite behavior is "tamed" in this context is by "stretching it out to infinity," but this results in an infinitely nonlocal theory in a spatiotemporal sense, which a priori renders experimental physics difficult to describe or predict. In particular, it is important in experimental settings to be able to assume that the direct causes and effects of an event can be captured within the laboratory. For example, if some of the direct products of a scattering experiment were to appear a billion years in the future, and nearly a billion light years away, with respect to the laboratory frame, it would be difficult to formulate sensible conservation laws governing this experiment! All available experience suggests that nature does not behave in this way. Also dubious is the manner in which interval finiteness requires local infinities located in different parts of a directed set to conspire to avoid "complementing each other" in such a way as to create infinite intervals. For example, the right-hand diagram in Figure 4.4.5 illustrates part of the skeleton of a physically absurd causal set with multiple local infinities. If it is "allowed" for one element x to possess an infinite direct future and another element y to possess an infinite direct past, how should classical spacetime "know" that it is "not allowed" for the two to coincide? This is just one way of expressing the implausibility of attempting to prescribe global structure.

To see that star finiteness does not imply interval finiteness, consider the acyclic directed set with elements $\{x_0, x_1, ...\} \cup \{..., y_{-1}, y_0\}$, and relations

$$x_i \prec x_{i+1}, \qquad x_i \prec y_{-i}, \qquad y_{-j} \prec y_{-j+1},$$

illustrated in Figure 4.4.6, which I refer to as **Jacob's ladder**.[15] Each element x_i, with the exception of x_0, possesses exactly one direct predecessor and two direct successors, while each element y_{-j}, with the exception of y_0, possesses exactly two direct predecessors and one direct successor. The elements x_0 and y_0 are unique global minimal and maximal elements of Jacob's ladder, respectively, with two direct successors and two direct predecessors, respectively. Hence, Jacob's ladder is star finite. However, for every choice of natural numbers i and j, there exist arbitrarily long chains of the form

$$x_i \prec x_{i+1} \prec \ldots \prec x_{i+n} \prec y_{-i-n} \prec y_{-i-n+1} \prec \ldots \prec y_{-j},$$

[15] This is a reference to Genesis 28:12, in which the patriarch Jacob dreams of a ladder reaching from earth to heaven.

such as the chain illustrated by the dark nodes and edges in the figure. Thus, every open interval of the form $\langle\langle x_i, y_{-j}\rangle\rangle$ in Jacob's ladder is infinite.

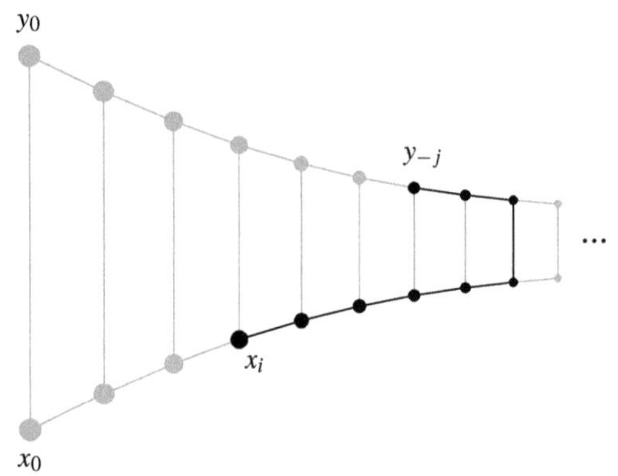

Figure 4.4.6. Jacob's ladder is star finite but not interval finite; dark nodes and edges illustrate a "long chain" from x_i to y_{-j}.

Local finiteness and transitivity. The foregoing criticisms of interval finiteness (IF) might lead the reader to question why this condition was ever axiomatized as a local finiteness condition at all. A very plausible reason, involving sprinkled causal sets and systematic Lorentz invariance, is discussed in detail in Section 4.5. Another possible explanation is that interval finiteness is partly an artifact of the transitive paradigm, inherited from early continuum-based theories, perhaps bolstered by the prominence of Alexandrov subsets in general relativity, and the mathematical popularity of partial orders. In an atmosphere dominated by these notions, the idea of star finiteness (SF) might seem unworkable or irrelevant, leaving interval finiteness as the "default candidate for a local finiteness condition." One reason why the virtues of star finiteness would likely be overlooked in this context is because the direct past $J_0^-(x)$ and direct future $J_0^+(x)$ of an element x in a transitive directed set D coincide with its total past $J^-(x)$ and total future $J^+(x)$, respectively. This is illustrated in the left-hand diagram in Figure 4.4.7. In this setting, the independence convention (IC) makes no sense from a physical perspective, because "most" relations terminating at x come from its "distant past," and "most" relations beginning at x reach to its "distant future." As observed by Finkelstein [FI88], it is absurd to factor in the effects of *"events as far apart as the birth and death of the universe,"* in the study of local structure.[16] However, the relation set $R(x)$ at x in a transitive directed set is dominated by information that is both nonlocal and redundant, and therefore relatively useless.

[16]Some researchers might possibly view causal sets induced by global sprinklings, or similar structures, as *realistic* models of classical spacetime, and might therefore disagree with this statement. This does not seem to include the founders or main proponents of causal set theory, however.

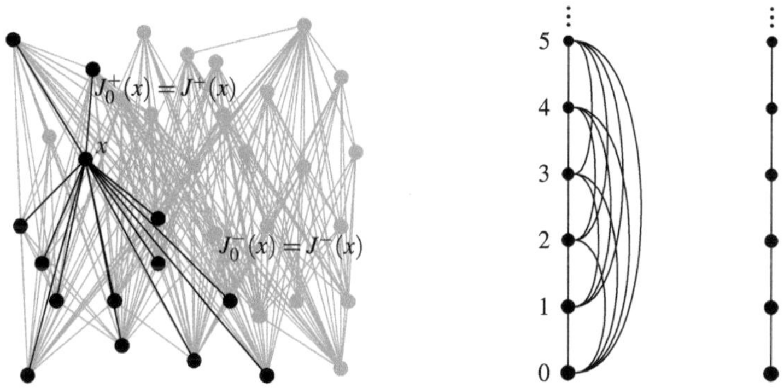

Figure 4.4.7. Transitivity obscures the advantages of star finiteness; $\mathbb{N}$ is not star finite, but *is* star finite in the skeleton.

Viewing this same situation from a different perspective, star finiteness (SF) is generally *too restrictive* to serve as a suitable local finiteness condition under the transitive paradigm, since transitivity "generates innocuous locally infinite behavior." For example, the set $\mathbb{N}$ of natural numbers is star infinite at each element, but is star finite in the skeleton, as illustrated in the right-hand diagram in Figure 4.4.7. Since $\mathbb{N}$ may be recovered from its skeleton by applying the transitive closure functor, this locally infinite behavior is really "no worse," in an information-theoretic sense, than the superficially "simpler" structure of sk($\mathbb{N}$). More generally, any transitive acyclic directed set containing an unbounded chain is star infinite for essentially the same reason, regardless of its actual information-theoretic complexity. Such sets may be used to model discrete causal analogues of certain cosmological models involving "repeated expansions and contractions of classical spacetime." Some such models, such as Penrose's *conformal cyclic cosmology* [PE10], may be physically interesting. Hence, star finiteness and transitivity do not mesh in a promising way.

Even in the transitive case, however, arguably better local finiteness conditions than interval finiteness are available. For example, restricting attention to transitive directed sets that may be recovered from their skeleta,[17] star finiteness in the skeleton (SFS) might be considered a better condition than interval finiteness, since it is a causally local condition, and since it rules out potential pathologies such as the infinite bouquet. The left-hand diagram in Figure 4.4.8 illustrates this condition, showing the set $J^-_{\mathrm{MAX}}(x)$ of maximal predecessors and the set $J^+_{\mathrm{MIN}}(x)$ of minimal successors of an element x in a directed set D. The cardinalities of such sets, as x ranges over D, determine whether or not D is star finite in the skeleton. Among other advantages, star finiteness in the skeleton avoids the particular interval-theoretic nuisance of including irrelevant structure in arbitrarily small neighborhoods of x, such as the element u in the interval $\langle\langle w, y\rangle\rangle$ in D, illustrated in the right-hand diagram in the figure.

[17]This condition rules out objects such as $\mathbb{Q}$, which is star finite in the skeleton because of the interpolative property. It also rules out cycles, but may be easily amended to permit them.

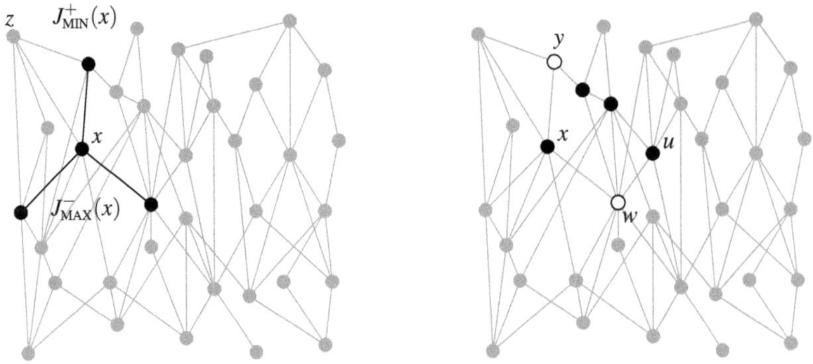

Figure 4.4.8. Star finiteness in the skeleton is arguably a better condition than interval finiteness in many cases.

Of course, the requirement that a directed set should be recoverable from its skeleton is very restrictive. The reason why this requirement is necessary for star finiteness in the skeleton to be a suitable local condition is because otherwise "small neighborhoods" of an element x fail to capture all of the direct causes and effects associated with x under the independence convention (IC). For example, the element z in the left-hand diagram in Figure 4.4.8 is related to x via a reducible relation, and therefore disappears from the star of x after performing the skeleton operation on D. The most obvious way to ensure that a directed set *is* recoverable from its skeleton is simply to impose irreducibility from the outset, i.e., to *begin* with "skeletal objects." Interestingly, this approach "makes information-theoretic sense" for causal sets, since reducible relations are already regarded as redundant in this context. Some prominent papers involving causal set theory, such as [KKSR12], take this approach implicitly, by focusing attention only on *links,* i.e., irreducible relations in a causal set. However, imposing irreducibility reduces the condition of star finiteness in the skeleton to star finiteness (SF), and the independence convention (IC) is automatic in this case. Hence, one recovers the same basic conditions and interpretations regarding directed sets suggested in this book, merely restricted to the class of irreducible star finite directed sets. This class may well be *too* restrictive from a physical point of view, due to the arguments regarding transitivity presented in Chapter 3. However, it is still much different, and actually richer in some ways, than the class of causal sets.

Local finiteness and the measure axiom. The clear picture of local structure on a star finite multidirected set supplied by the star topology suggests reconsideration of the specific version of the measure axiom (M) used in causal set theory. Under this axiom, the causal set measure μ_{CS} on a causal set C assigns approximately one fundamental unit of "spacetime volume" to each element x of C, *regardless of the details of local causal structure near x*. The qualifier "approximately" refers to the modern statistical caveat "up to Poisson-type fluctuations," added to the measure axiom to avoid systematic violations of Lorentz invariance. However, causal set theory does not incorporate the possibility of *systematic* differences in the measures assigned to different elements, even though the regions near these elements may exhibit completely different structure. It would be both simple and attractive if "order plus number" really *did* "equal geometry" in an approximate sense. However, the fact that this simple notion does not incorporate the essentially new feature of *local irregularity,* encountered upon moving from the continuum-based paradigm to the discrete paradigm, raises heuristic doubts about its adequacy. Indeed, the real lesson of the metric recovery theorems is more general; namely, that *"directed structure plus natural scale equals geometry,"* as discussed in Section 2.8. In this context, it seems very likely that local causal structure *should* be factored into the interpretation of volume. This topic represents just one facet of the important and difficult general problem of precisely quantifying emergent geometric properties of directed sets, but I make no further attempt here to examine its details. As noted in Section 2.8, causal set theorists have considered similar modifications [BHS09].

4.5 Sprinkled Causal Sets; Lorentz Invariance; Spatiotemporal Locality

Star infinite behavior in sprinkled causal sets. Before proceeding, it is instructive to examine some idealized examples of causal sets that are star *infinite,* and to discuss some of the physical and structural issues involved. Besides supplying an important historical perspective, this brief detour illustrates one of the reasons why certain types of star infinite directed sets and multidirected sets are worth studying in the discrete causal context, even if they are not ultimately included among the class of objects actually used to model classical spacetime.[18] As described in Chapter 3, most causal sets, and most discrete directed sets in general, bear little overall structural resemblance to relativistic spacetime manifolds, and no dynamical mechanism is yet known for actually *generating* "manifold-like" directed sets. Hence, an important method of obtaining "physically relevant" directed sets for study purposes is to "work backward" from pre-existing pseudo-Riemannian structure. A standard way to do this, in the context of causal set theory, is to begin with a pseudo-Riemannian manifold X satisfying the causal condition,[19] such as Minkowski spacetime $\mathbb{R}^{3+1}$, then randomly "sprinkle" elements onto X, as illustrated in the left-hand diagram in Figure 4.5.1. As usual, two spatial dimensions are suppressed in the diagram for the purposes of illustration. A few of the elements are labeled for future reference. In general, the number of elements sprinkled may be finite or infinite, and may involve

[18] Another such reason is that structurally similar objects arise at the quantum level, as discussed in Chapter 7. However, even in the classical context, idealized "limiting cases" or "simplified models" are important objects of study in general.

[19] The reason for invoking the causal condition in this context is simply to ensure that the sprinkled directed set will be acyclic, and hence a causal set. The extent to which this causal set is guaranteed to "faithfully approximate" the original geometry involves subtler conditions, as stated in the metric recovery theorems described in Section 2.8.

all or part of X. In the present setting, the sprinklings of principal interest are *global sprinklings into* $\mathbb{R}^{3+1}$, involving an infinite number of sprinkled elements. These sprinklings are constructed via a *Poisson process,* under which the probability of finding n elements in a spacetime region of volume V is $(\rho V)^n e^{-\rho V}/n!$, for some constant ρ.[20]

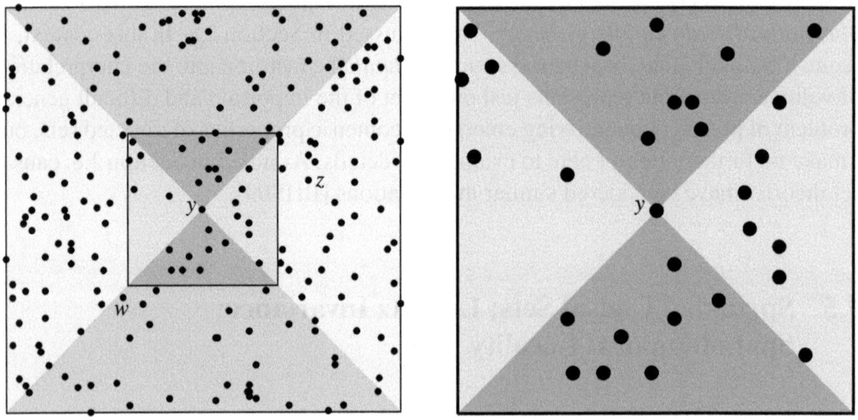

Figure 4.5.1. Sprinkling elements into Minkowski spacetime $\mathbb{R}^{3+1}$; "zooming in" near a particular element y in a particular frame of reference.

The set of sprinkled elements may be denoted by C, in anticipation of regarding this set as a causal set. The diagram is "centered" at the element y, and the total domain of influence of y in $\mathbb{R}^{3+1}$, bounded by its null cone, is indicated by the darker shading.[21] Given such a sprinkling, one may then define a binary relation $\prec_{\text{CS}}$ on C, induced by the relativistic causal relation $\prec_{\text{GR}}$ on $\mathbb{R}^{3+1}$. In particular, $\prec_{\text{CS}}$ includes the individual relation $w \prec_{\text{CS}} y$, because w belongs to the causal past of y, but does *not* include an individual relation between y and z, because these elements are spacelike separated. For future reference, it is useful to note that this construction automatically induces a morphism of directed sets from $(C, \prec_{\text{CS}})$ into $(\mathbb{R}^{3+1}, \prec_{\text{GR}})$. The small, darker-shaded box in the left-hand diagram represents a small neighborhood of y with respect to the manifold topology on $\mathbb{R}^{3+1}$, chosen in this case to correspond in an obvious way to the frame of reference in which constant-time sections are drawn horizontally. This is the frame labeled "A" in Figure 4.5.4 below. The right-hand diagram "zooms in" on this neighborhood. Naïvely, one might take it for granted that such zooming in would "resolve the local structure of the causal set $(C, \prec_{\text{CS}})$, near y," but the actual situation is subtler, as explained below.

An obvious limitation of the sprinkling method is that simply *copying* structure from pseudo-Riemannian geometry naturally suppresses a variety of interesting ways

[20] See [BHS09] for details.

[21] The reason for denoting a typical element by y, rather than x, in this section, is because the notation used here for spatial coordinates in $\mathbb{R}^{3+1}$ involves the symbol x, as in x_{A} or x_{B}.

in which actual spacetime structure may *differ* from the predictions of general relativity at scales beyond what is currently accessible to experiment. Discovering and understanding such differences is, of course, the central objective of studying quantum gravity in the first place. In some sense, then, sprinkled causal sets are "not very exciting," since their structure is, by construction, essentially featureless. By contrast, computer simulations in causal dynamical triangulations reveal interesting qualitative changes in structure across a range of scales, including *changes of dimension* [AJL05]. Further, if one wishes to model nontrivial *local* structure, such as emergent matter, under the strong interpretation of the causal metric hypothesis (CMH), then sprinkled causal sets have little to offer. Nevertheless, sprinkled causal sets, and other similar types of "reverse-engineered" directed sets, are indispensable at present for connecting discrete causal theory to established physics, and will likely remain so until the dynamics of the theory is better understood. Many important results in discrete causal theory involve the study of sprinkled causal sets; perhaps the most prominent of these is the prediction of the size of fluctuations of the "cosmological constant," described by Sorkin and his coauthors in the paper [ADGS04].

Returning to the issue of local causal structure of $(C, \prec_{\text{CS}})$, one might expect that any "reasonable" topology on $\mathbb{R}^{3+1}$; in this case, the manifold topology, would "resolve" this structure, in the sense that suitable small neighborhoods of y in $\mathbb{R}^{3+1}$ would capture all the maximal predecessors and minimal successors of y in C. The left-hand diagram in Figure 4.5.2 reproduces the "zoomed in" neighborhood of y first illustrated in the right-hand diagram in Figure 4.5.1 above, with maximal predecessors and minimal successors of y in C indicated by the black nodes. The right-hand diagram in Figure 4.5.2 illustrates the induced causal structure of these elements in C. In general, these two diagrams seem to present a "reasonable local picture," since the nearest neighbors of y in C captured by the "zoomed in" neighborhood in $\mathbb{R}^{3+1}$ mostly appear to be "close to y" in a spatiotemporal sense. Naïvely, one might assume that this neighborhood captures *all* the maximal predecessors and minimal successors of y in C. The one "eyebrow-raising" element is v, which lies near the boundary of the neighborhood. However, it is easy to see that v is indeed a minimal successor of y under the causal set relation $\prec_{\text{CS}}$ on C induced by the relativistic causal relation $\prec_{\text{GR}}$ on $\mathbb{R}^{3+1}$. Indeed, v is a successor of y, since it falls inside the null cone of y. On the other hand, v is *not* a successor of either of the other two successors of y shown in the diagram. In particular, the dashed lines in the diagram illustrate that v "just misses" being inside the null cone of u.

Zooming back out reveals that the element v is not the exception, as one might have expected, but the rule. The left-hand diagram in Figure 4.5.3 shows the maximal predecessors and minimal successors of y in the "zoomed out" neighborhood, indicated by the black nodes. The corresponding abstract causal structure near y is shown in the right-hand diagram. From this picture, it is intuitively clear that for a typical global sprinkling[22] into $\mathbb{R}^{3+1}$, one may identify infinite sequences of sprinkled elements in the causal future of y, approaching the null cone in a spatiotemporal

[22]The word "typical" here has a precise *measure-theoretic* meaning: the same meaning as the words "almost every" that often appear in proofs in analysis. In more detail, one may view a set Ω of

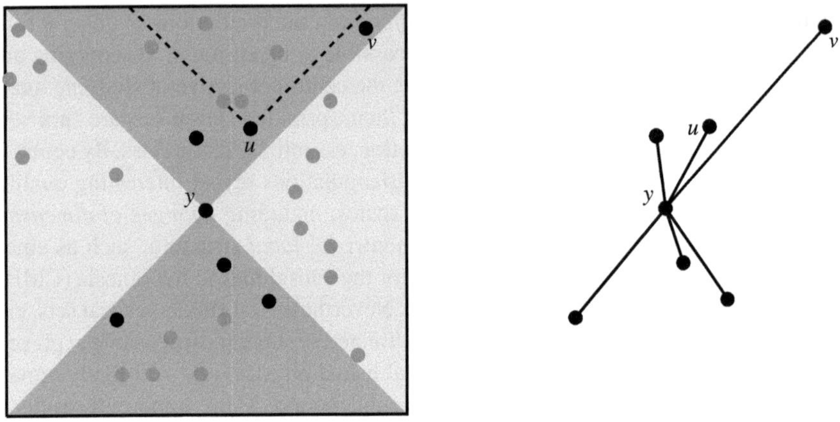

Figure 4.5.2. Zoomed-in sprinkling shows maximal predecessors and minimal successors of y mostly "close to" y in the diagram; abstract causal structure of these "nearest neighbors".

sense, as one goes further and further forward in time from y in a given frame of reference. Every element of such a sequence will lie outside the null cone of any preceding element, and will therefore be a minimal successor of y in $(C, \prec_{CS})$. A similar statement applies to maximal predecessors of y. The conclusion is that *a typical causal set* $(C, \prec_{CS})$ *defined via global sprinkling into* $\mathbb{R}^{3+1}$ *is star infinite at every element.* It is easy to make this statement rigorous and prove it; the result has been known in the causal sets community since the 1980s.

Sprinkling and systematic Lorentz invariance. One of the most important properties of causal sets defined via global sprinkling into Minkowski spacetime $\mathbb{R}^{3+1}$ is that a typical such set $C = (C, \prec_{CS})$ avoids, by construction,[23] *systematic violations of Lorentz invariance.* In a similar manner, causal sets defined via sprinkling into other pseudo-Riemannian manifolds, such as de Sitter spacetime, reproduce, by construction, some of the large-scale properties of these manifolds. The question of whether or not one *should* choose models that automatically preserve such properties in studying quantum gravity is a complicated one, since it remains a matter of debate which aspects of general relativity should be expected to break down in this context, and at what scales. Focusing on Lorentz invariance in particular, there are strong experimental constraints on the possible types and degrees of violation, which

(Footnote 22 continued)
sprinklings of a given type as an abstract space, as discussed below. One may then fix a measure μ on Ω, together with a *sigma algebra* Σ of measurable sets. The statement that a "typical" sprinkling satisfies a certain property then means that the subset of Ω consisting of sprinklings that do *not* satisfy this property has measure zero.

[23] Here, I mean that the source of properties such as systematic Lorentz invariance, exhibited by such causal sets, is the structure of $\mathbb{R}^{3+1}$ from which they are constructed, *not* that it is a priori obvious that they possess these properties. Otherwise, results such as the theorem in [BHS09] would be trivial.

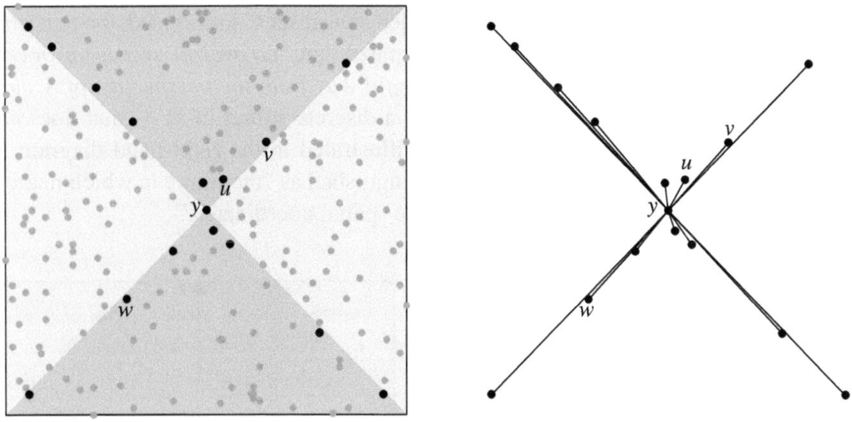

Figure 4.5.3. Zoomed-out sprinkling shows that maximal predecessors and minimal successors of y approach the *null cone* of y in a given frame, not necessarily y itself; abstract local structure of these "nearest neighbors".

provides an obvious motivation for considering models that avoid such violations as much as possible. On the other hand, some physicists *expect* systematic violations of Lorentz invariance in "extreme regimes;" for example, at sufficiently high energies. Even in more-ordinary settings, one may identify a number of candidates for a "preferred frame of reference;" for example, the "frame at rest with respect to the cosmic microwave background," called the *CMB frame.* These considerations suggest that it may be most reasonable to focus on models that enable "arbitrarily close approximation" to Lorentz invariance, in a suitable sense, without ruling out certain types of deviations.

Of course, Lorentz invariance is a continuous symmetry property of $\mathbb{R}^{3+1}$, so no discrete directed set is *actually* Lorentz invariant; this notion is not even meaningful, since it conflates symmetries of different objects. In fact, as discussed in Section 6.3, a typical discrete directed set is *rigid;* i.e., it has no symmetries at all. The meaning of the statement that C avoids *systematic* Lorentz invariance violation is, in the words of Bombelli, Dowker, Henson, and Sorkin [DHS04], [BHS09], that the structure of C does not *"serve to pick out a preferred frame (or frames) of reference in the approximating continuum,"* i.e., in $\mathbb{R}^{3+1}$. The left-hand diagram in Figure 4.5.4 illustrates two different frames of reference A and B in $\mathbb{R}^{3+1}$, with coordinates labeled (t_A, x_A) and (t_B, x_B), respectively.[24] Frame B is "boosted by a factor of $\beta = \frac{1}{2}$" with respect to frame A, which means that an object at rest in frame B moves to the right

[24] As usual, factors of the speed of light c are really involved here, if one wishes to obtain convenient ratios of scales between the coordinate axes in these diagrams. To be precise, I should either replace t_A and t_B with ct_A and ct_B, or else stipulate that I am "working with units in which $c = 1$."

at a velocity of $\frac{1}{2}c$ in frame A.[25] The sprinkled causal set C introduced above is also included in the diagram. Informally, one requires that *"no overall average property of C looks different in a physically meaningful way from the perspective of A than from the perspective of B."* An example of a discrete subset of $\mathbb{R}^{3+1}$ that does *not* possess this property is the regular lattice illustrated in the right-hand diagram in Figure 4.5.4. Here, the frame A may be distinguished as "the frame in which nearest timelike-separated neighbors share the same spatial coordinates."

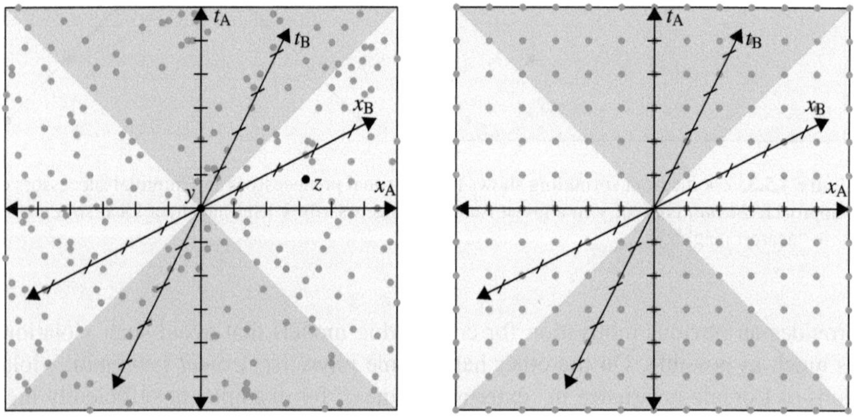

Figure 4.5.4. Different frames of reference A and B, with coordinates (t_A, x_A) and (t_B, x_B), on $\mathbb{R}^{3+1}$; a regular lattice selects a preferred frame.

The above description of "avoiding systematic violations" remains rather vague; a mathematical formulation is necessary to render it sufficiently precise to be useful. Such a formulation is supplied by the paper [BHS09]. The criterion of "picking out a preferred frame" is first generalized to "selecting a preferred temporal direction" at a chosen point of origin in $\mathbb{R}^{3+1}$, in this case, the point y. This is then formalized in terms of the existence of a suitable "direction map" $\Delta : \Omega \to H$ from an appropriate space Ω of sprinklings on $\mathbb{R}^{3+1}$ to the hyperboloid H of unit future-directed timelike vectors at y.[26] The Lorentz group L acts on both Ω and H; i.e., an active Lorentz transformation Λ converts each sprinkling $C \in \Omega$ to a new sprinkling $\Lambda(C)$, and each vector $\mathbf{u} \in H$ to a new vector $\Lambda(\mathbf{u})$. The "physically meaningful direction maps" are those which are *equivariant* with respect to the *proper orthochronous Lorentz group* L'', i.e., those for which $\Delta(\Lambda(C)) = \Lambda(\Delta(C))$ for all $\Lambda \in L''$ and all $C \in \Omega$. The authors prove a theorem showing that no such equivariant map exists. Among the conclusions they draw from this is that *the structure of a typical causal set defined via global sprinkling does not select a preferred frame of reference in* $\mathbb{R}^{3+1}$.

[25] Note that the scales on the axes in frame B differ by a factor of $\sqrt{\frac{1+\beta}{1-\beta}}$ from the corresponding scales for frame A. This, in particular, is why the "experiments" illustrated in Figure 4.5.6 below cover the same spacetime volumes. The same factor appears in the "cutoff equation" (4.5.7).

[26] In fact, the argument applies to $\mathbb{R}^{n+1}$ for any $n \geq 1$.

To understand the significance of the equivariance criterion, note that the abstract structures of C and $\Lambda(C)$ are isomorphic as causal sets; as the authors describe it, one wishes the direction chosen by Δ *"to have come from the sprinkling and nothing else."* The equivariance criterion formalizes this.

It is reasonable to question the scope of applicability of this result, given the fact that Minkowski spacetime $\mathbb{R}^{3+1}$ is the original source *all* of the "spacetime structure" involved in the theorem. Ultimately, one wishes to turn this picture around; discrete spacetime models should come first, and the "approximating continuum" should be secondary. This leads to a more general question; namely, *"which directed sets approximated at large scales by* $\mathbb{R}^{3+1}$ *serve to select a preferred frame of reference in* $\mathbb{R}^{3+1}$*?"* In general, one does *not* expect all the structure of such directed sets to be precisely *induced* by $\mathbb{R}^{3+1}$, as it is for sprinkled causal sets. In particular, such sets may exhibit interesting structural differences across a range of scales, such as the changes in dimension and nontrivial local features mentioned above. The authors of [BHS09] do not seem to make any specific claims about this more general question, although their result does go considerably further than merely ruling out a preferred frame of reference selected by a typical global sprinkling. In particular, they generalize "preferred frame of reference" to "preferred temporal direction at a point," and further to "family of preferred temporal directions at a point." Perhaps most notably, they prove that *"any method used to associate a finite-valency graph to such a sprinkling would have to violate Lorentz invariance."* This implies, in particular, that a typical causal set constructed via global sprinkling into $\mathbb{R}^{3+1}$ will be star infinite, as already explained informally above.

During the initial reception of causal set theory by the physics community, the fact that such global sprinklings are star infinite was viewed by some physicists as a disturbing and undesirable property. In particular, in a follow-up "comment" on the inaugural causal sets paper of Bombelli, Lee, Meyer, and Sorkin [BLMS87], Christopher Moore [MO88] pointed out this issue, first explaining why a typical global sprinkling into $\mathbb{R}^{3+1}$ is star infinite, then continuing:

> *...but of course one wants the real coordination number* [i.e., valence], *at the smallest scales, to be finite... there seem to be four possibilities:*
>
> *(1)* [Classical spacetime] *is not, in fact, discrete.*
>
> *(2)* [Classical spacetime] *is discrete, but it has the nasty property that every point is influenced by an infinity of "nearest neighbors" which, in a given frame, are arbitrarily far back in time.*
>
> *(3) Physics is macroscopically Lorentz invariant and isotropic but the underlying* [discrete structure] *is not...*
>
> *(4)* [Classical spacetime] *is discrete, but* [this discrete structure] *can only be strictly defined in the non-relativistic limit...* ([MO88], page 655)

The authors of [BLMS87] responded to Moore's remarks [BLMS88] by explaining that the star-infinite behavior of global sprinklings is a natural consequence of the structure of $\mathbb{R}^{3+1}$, writing,

> *Points in a Lorentzian manifold have metric neighborhoods which converge to the* [null cone] *rather than to the point itself. Causal sets simply reflect this characteristic of* [spacetime]. ([BLMS88], page 656)

They argued, however, that this behavior does not represent a serious obstacle to causal set theory, because such idealized causal sets are neither physically realistic nor necessary for performing causal set computations in quantum gravity:

> *The difficulty, of course, is that one appears to have lost some of the advantages gained by discretization (such as manifest freedom from ultraviolet divergences) if one is forced to allow* [star infinite causal sets]. *Our purpose, however, eventually using causal sets in a* [path summation] *approach to quantum gravity, does not force us into this situation. Curved* [spacetime] *is only locally Lorentz invariant; in a* [Friedman-Lemaître-Robertson-Walker] *universe, for example... each element would... have only a finite number of* [maximal predecessors]... *In the general situation, the* [path summation] *approach* [need involve] *histories extending only a finite distance into the future... Thus it suffices to consider* [star finite causal sets]. ([BLMS88], page 656)

Over the subsequent generation, the principal efforts to develop suitable dynamics for causal sets; for example, Sorkin and Rideout's theory of *sequential growth dynamics* [SR99], have indeed focused on star finite causal sets. The most notable phenomenological success of the theory; namely, offering a plausible explanation for the magnitude of the "cosmological constant" [ADGS04], is also based on computations involving star finite causal sets. Recent papers involving causal set cosmology, such as [KKSR12], also restrict attention to the star finite case. However, star *infinite* causal sets have retained a degree of *theoretical* prominence, due to their connection with systematic Lorentz invariance, and the importance of the sprinkling construction. While the study of such objects is desirable in its own right, it has likely helped to perpetuate the awkward axiomatic role of interval finiteness (IF) as a "local finiteness condition," while possibly obscuring some of the advantages to be gained by using star finiteness (SF) as an alternative.

Generalization in terms of refinements. Before exploring these issues further, it is perhaps illuminating to view the topic of systematic Lorentz invariance in a more abstract context, which serves to introduce the far-reaching considerations regarding emergent spacetime symmetry outlined in Section 8.4. The point of departure, beginning from the perspective of the paper [BHS09], is the authors' generalization from "preferred frame of reference" to "preferred temporal direction." What a choice of temporal direction in $\mathbb{R}^{3+1}$ corresponds to, in order-theoretic terms, is a *refinement* of the relativistic causal relation $\prec_{\text{GR}}$ on $\mathbb{R}^{3+1}$.[27] As discussed below, this is true for either a local or global choice of temporal direction, although the types of refinements involved may differ. In the present context, it suffices to examine some special cases of refinements illustrated by the left-hand diagram in Figure 4.5.4. Choosing a unit future-directed timelike vector $\mathbf{u}$ at y serves to "split space and time" at y; i.e., it identifies a local "constant-time section" σ passing through y.[28] For example, if

[27]The general topic of refinements of partial orders was discussed in Section 3.8; to review, a refinement of $\prec_{\text{GR}}$ is a new, "larger" binary relation which includes all the individual relations in $\prec_{\text{GR}}$.

[28]In the case of $\mathbb{R}^{3+1}$, σ extends to a global Cauchy surface, but one may wish to work with only local sections near y (in the manifold topology), to permit consideration of nonparallel temporal directions at different points in $\mathbb{R}^{3+1}$.

u is chosen to point along the t_A-axis at y, then σ is part of the three-dimensional Euclidean space represented in the figure by the x_A-axis. Similarly, if **u** is chosen to point along the t_B-axis at y, then σ is part of the space represented by the x_B-axis. The way in which such a choice refines the relation $\prec_{GR}$ on $\mathbb{R}^{3+1}$ is by choosing temporal orderings for pairs of events which are spacelike separated, i.e., unrelated under $\prec_{GR}$. For example, the events y and z are spacelike separated, but choosing **u** to point along the t_A-axis at y selects a temporal order under which y precedes z. On the other hand, choosing **u** to point along the t_B-axis at y selects a temporal order under which z precedes y. The fact that either ordering is consistent with the original relation $\prec_{GR}$ is a manifestation of the relativity of simultaneity. A *global* choice of temporal direction on $\mathbb{R}^{3+1}$ has a similar but stronger effect; it identifies a complete foliation of $\mathbb{R}^{3+1}$ by global constant-time sections, and a correspondingly stronger refinement.

The motivation for viewing a choice of temporal direction in terms of refinements is that *this viewpoint generalizes to the case of arbitrary directed sets.* This may be understood by comparing the following two questions:

> **Question of** [BHS09]: *"When does a global sprinkling into Minkowski spacetime $\mathbb{R}^{3+1}$ select a preferred temporal direction, or family of directions, in $\mathbb{R}^{3+1}$?"*

> **Generalization:** *"When does a directed set $(D, \prec)$ select a preferred refinement, or family of refinements, of the binary relation $\prec'$ on an "approximating set" $(D', \prec')$?"*

A *formally* attractive simplification of this generalized question focuses on the case where D and D' are related by a morphism of directed sets, analogous to the morphism $(C, \prec_{CS}) \rightarrow (\mathbb{R}^{3+1}, \prec_{GR})$ induced by a sprinkling, though in general the "approximating set" $(D', \prec')$ need not be a pseudo-Riemannian manifold. However, this version, like the sprinkling construction itself, ignores important cases in which D possesses interesting structural features that are not sufficiently reflected by D' to permit a morphism between the two. A further simplification involves the case in which $D = D'$; in this case, the question becomes *"when does a binary relation $\prec$ on a set D possess a preferred refinement, or family of refinements?"* At a naïve level, this seems like a relevant question to ask, given that the discrete models one is working with are regarded as more fundamental than their "continuum approximations." Also, the question obviously makes sense in certain special cases. For example, the causal set induced by the embedding of the regular lattice into $\mathbb{R}^{3+1}$ illustrated in the right-hand diagram in Figure 4.5.4 obviously possesses a preferred refinement in an intrinsic sense; this refinement could be defined as "the refinement in which minimal successors of a given element are taken to be simultaneous." However, one must re-define what the notion of "preferred" means in this context, since directly mimicking the approach of [BHS09] in terms of "direction maps" produces a trivial answer.[29]

In the study of emergent symmetries, however, questions like the one posed in [BHS09] are more interesting than questions about individual directed sets. The reason why this is true is because the properties under consideration are no longer exact properties of any of the specific objects involved. This is by no means an uncommon situation in physics; for example, in statistical thermodynamics, one is often interested not in the specifics of a particular "microstate," but in the entire class of "microstates" that satisfy some "average property." An outstanding issue regarding causal sets and systematic Lorentz invariance seems to be that it is unclear how to precisely describe the class of causal sets satisfying the "average property" of being *"well-approximated by Minkowski spacetime* $\mathbb{R}^{3+1}$*."* Global sprinklings into $\mathbb{R}^{3+1}$ belong to this class by construction, but there does not seem to be any definitive characterization of the class in general. The following quotation from [BHS09] is illuminating in this regard:

> *The theorem also shows that a finite-valency graph like a "spin-foam" cannot be equivariantly associated to a* [global] *sprinkling of Minkowski* [spacetime]. *If we wanted our discrete structure to involve such a graph, we could attempt to use some other continuum approximation scheme than sprinkling. The problem is that no other way is known that preserves* [systematic] *Lorentz symmetry, when Minkowski* [spacetime] *is to be the effective continuum description. Other* [approaches] *could be imagined (e.g., adding distance information to the relations in the causal set), but we would conjecture that the causal set is, in some sense, the minimal* [systematically] *Lorentz invariant discrete structure from which the continuum can be reconstructed at macroscopic scales...* (page 2586)

Of course, the ultimate goal of causal set theory, and of discrete causal theory in general, is not to "reconstruct" Minkowski spacetime, but to describe *new structure,* which is not a priori induced by pseudo-Riemannian geometry at all. In particular, it is desirable to work with a class of models that can accommodate different types of behavior across a range of scales, and that can exhibit interesting local features. As reflected by the above quotations of Moore [MO88], and Bombelli et al. [BLMS88], [BHS09], it has long been recognized that systematic Lorentz invariance, in a precise formal sense, is neither a realistic nor even a desirable property for such models to satisfy. In particular, the authors of [BHS09] explain in their conclusion that only *"approximate Lorentz invariance,"* or *"effective local Lorentz invariance,"* may be expected to hold in more realistic situations involving finite regions of flat spacetime, or regions involving curvature. Nevertheless, idealized global results such as the theorem in [BHS09] remain crucial for understanding the basic structural issues at stake.

---

[29]Indeed, the space Ω of sprinklings is replaced here by the singleton set $\{D\}$, because no approximation is involved. Further, arbitrary "direction maps" are typically equivariant in a vacuous manner, since a typical discrete directed set is rigid.

Systematic Lorentz invariance versus spatiotemporal locality. As the founders of causal set theory explain in [BLMS88], the star infinite behavior of causal sets induced by global sprinklings into Minkowski spacetime $\mathbb{R}^{3+1}$ reflects a certain obvious property of the Minkowski metric η; namely, that events close to the null cone of an event y are close to y itself with respect to η, however distant from y they may be, in a purely spatial or temporal sense, in a given frame of reference. In a basic qualitative capacity, however, such causal sets actually *emphasize one property of* $\mathbb{R}^{3+1}$ *over another.* The emphasized property is systematic Lorentz invariance. The deprecated property is a certain compatibility between causal structure and the manifold topology on $\mathbb{R}^{3+1}$, in which small neighborhoods[30] of y "capture" any influence reaching or leaving y. This property may be called **spatiotemporal locality.** In the relativistic setting, one does *not* see any analogue of the situation arising in the present context for sprinkled causal sets, in which maximal predecessors or minimal successors of an element y appear spatially and temporally distant from y in a given frame of reference. Of course, there *are* no maximal predecessors or minimal successors under the relativistic paradigm, but that is really the point; discrete causal theory presents new structural possibilities that can only seldom be adequately addressed by simply copying properties of emergent geometry, or by elevating one emergent property over another. The left-hand diagram in Figure 4.5.5 illustrates spatiotemporal locality in Minkowski spacetime $\mathbb{R}^{3+1}$: given any pair of elements y and v in $\mathbb{R}^{3+1}$, related under the relativistic causal relation $\prec_{\mathrm{GR}}$, and any open neighborhood U of y in the manifold topology on $\mathbb{R}^{3+1}$, there exists an "interpolating element" $u \in U$ such that $y \prec_{\mathrm{GR}} u \prec_{\mathrm{GR}} v$. The right-hand diagram in the figure illustrates how the interpolative property of $\mathbb{R}$ extends to spacelike sections of the causal future of y in $\mathbb{R}^{3+1}$. Any such section is impermeable; i.e., it captures all of the influence flowing from y to its future, and one may always find another such section nearer to y, as near as desired in a given frame of reference. Hence, spatiotemporal locality meshes harmoniously with causal locality in the relativistic context.

By contrast, *no* finite-volume neighborhood of y in a typical causal set induced by global sprinkling into $\mathbb{R}^{3+1}$ captures all of the maximal predecessors and minimal successors of y, as illustrated above. This is one of the reasons for statements in the literature expressing the idea that causal set theory is "nonlocal," or that discreteness "forces a choice" between locality and local Lorentz invariance. A good example of such a statement may be found in Sorkin's recent paper *Does locality fail at intermediate length scales?* [SO09A]. Sorkin writes:

[30] Of course, such neighborhoods have uncountable *cardinality,* but their discrete causal counterparts have finite cardinality in the cases of principal interest, due to the relationship between discrete combinatorial structure and emergent volume suggested by the metric recovery theorems.

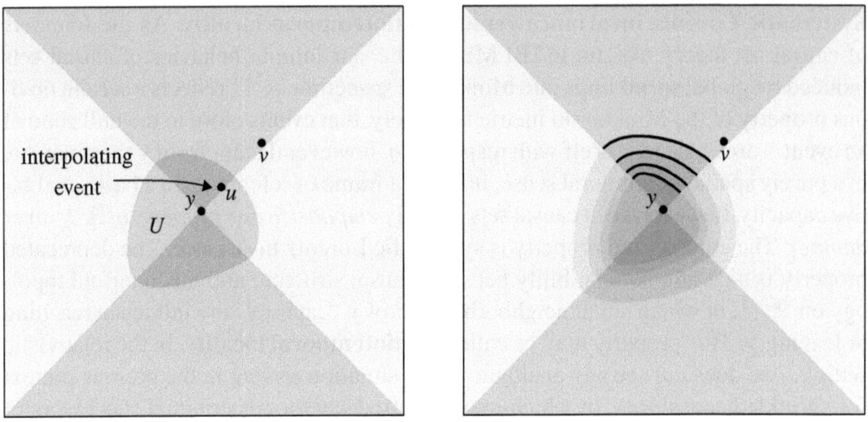

Figure 4.5.5. Finite-volume neighborhoods capture local causal structure in the relativistic setting; the interpolative property extends to spacelike sections of the causal past or future of y.

> *Moreover, the infinite volume of the Lorentz group implies that* [a causal set] *cannot be home to a local dynamics. Rather the "couplings" or "interactions" that describe physical processes occurring in the* [causal set] *are–of necessity–radically nonlocal.* (page 27)

An even sharper statement appears in Lisa Glaser's paper *A closed form expression for the causal set D'Alembertian* [GL14]. She writes:

> *Causal set theory keeps Lorentz invariance, thus sacrificing locality. Without locality the dynamics of the theory cannot follow the usual patterns...* (page 5008)

A third expression of the same idea may be found in Belenchia's recent preprint *Causal Set theory, non-locality and phenomenology* [BE15]. He writes:

> *The sprinkling process* [produces] *a random lattice that* [preserves] *Lorentz invariance in a well defined sense, in this way we have a kinematical randomness that preserves Lorentz invariance while working with a discrete structure. However, there is a price to pay: A fundamental nonlocality of causal sets.* (page 2)

The type of locality referred to in these statements is spatiotemporal locality, and the meaning of "Lorentz invariance" is, of course, "systematic Lorentz invariance." In general, I prefer not to describe the situation in the same terms as these quotations, because spatiotemporal locality is an emergent notion, which I would rather not emphasize at the expense of the more fundamental notion of causal locality. Indeed, as reflected by the quote of Sorkin appearing in Section 4.1, which is drawn from the same paper [SO09A] quoted in the present section, there seems to be no reasonable possibility of disagreement regarding what it means for two events to be local in a *causal* sense; Definition 4.1.1 is uncontroversial in this regard. By extension, there is nothing about the behavior of a star infinite causal set that should be considered *causally* nonlocal; it merely exhibits the "nasty property," as Moore [MO88] puts it, that the smallest open neighborhoods of typical elements in the star topology happen to be infinite, both in cardinality, and in emergent volume.[31]

None of the statements in the above quotations is intended to suggest that causal set theory implies a "fatal" degree of spatiotemporal nonlocality from the standpoint of available experimental evidence. Instead, these papers attempt to "describe," "quantify," and/or "control" such effects, in order to recover a reasonable degree of spatiotemporal locality in practice. In particular, Sorkin's paper [SO09A] posits the existence of an "intermediate length scale" below which nonlocal effects become important, and even hazards a rough guess of 10^{-14} m for the size of this scale.[32] These efforts are interesting, but remain limited in scope. First, they take for granted the existence of extreme spatiotemporal nonlocality in the *fundamental causal structure* of spacetime itself, and attempt to define background dependent data living on this structure, in the form of scalar fields and discrete analogues of differential operators such as the D'Alembertian, which somehow "do not see" the underlying nonlocality over a convenient range of scales. Second, the evidence that even this type of "effective locality recovery" is possible is based mostly on specialized constructions involving sprinkled causal sets and computer simulations.[33] Third, other potentially problematic notions play an important role in these constructions; for example, they involve "layering" of causal sets, which automatically raises issues involving transitivity (TR), the independence convention (IC), and the permeability problem.

What these papers *do* illustrate is that there is a nontrivial relationship between systematic Lorentz invariance and spatiotemporal locality in causal set theory, and there is every reason to believe that some version of this relationship extends to discrete causal theory in general. This relationship could be described as a "balance," or "trade-off," or "incompatibility." Although the context is purely classical at present, it is tempting to draw a comparison involving the uncertainty principle, in which classical histories such as sprinkled causal sets are something like *"Lorentz invariance eigenstates,"* and classical histories such as lattices are something like *"spatiotemporal locality eigenstates."* The purpose of introducing this analogy is to avoid falling prey to the general pitfall of assigning undue weight to any particular relic from a earlier theory when developing a new theory. For example, if one were to take the idea of a precisely-defined position too seriously in the context of ordinary quantum theory, then one might conclude that *"quantum theory retains the notion of a well-defined position, and sacrifices the notion of a well-defined momentum."* In fact, quantum theory does nothing of the sort; it merely reveals the existence of *relationships between pairs of physical properties that were not recognized in earlier theories.* Similarly, causal set theory itself does *not* sacrifice spatiotemporal locality in favor of systematic Lorentz invariance; it merely reveals the existence of a relation-

---

[31] From a basic structural standpoint, there is nothing particularly unusual about topologies in which basic open sets are "huge;" for example, in the Zariski topology familiar from algebraic geometry, open sets are generally so large that their complements are of smaller dimension.

[32] See [SO09A], page 42.

[33] Glaser [GL14] does go quite far with analytic methods.

ship between the two properties that was not recognized in the context of relativity. The theory actually accommodates a wide array of different types of objects, some of which preserve one or both properties to some degree of approximation, and most of which exhibit little resemblance to continuum-based geometry at all. The same is true of discrete causal theory in general. Of course, some individual researchers may choose to emphasize one emergent property over another. *Causal* locality, by contrast, is *not* an emergent property in this context; it is regarded as fundamental.

In any case, there are strong arguments against completely discarding spatiotemporal locality in order to preserve systematic Lorentz invariance in idealized situations. *True* Lorentz invariance in discrete causal theory makes no sense, due to discreteness itself, and obvious "discrete symmetry analogues" are doomed by the fact that a typical discrete directed set is rigid. Even *systematic* Lorentz invariance is doomed, in realistic scenarios, by emergent spacetime curvature. Hence, only an "approximate" or "effective" local notion of Lorentz invariance can survive, as explained by the authors of [BHS09], and this is all that anyone should expect or wish for. Further, approximate spatiotemporal locality is just as sacred from an experimental standpoint, at the classical level, as approximate local Lorentz invariance, and it is probably even harder to sacrifice in terms of actually doing physics. In particular, as described in Section 4.4, it would be difficult to formulate sensible conservation laws if one were always forced to worry about direct causes or effects occurring an arbitrary distance away in the laboratory frame. Fortunately, on the basis of rather naïve scale arguments, it does not seem difficult to retain a reasonable approximation of *both* local Lorentz invariance and spatiotemporal locality in discrete causal theory, although typical scenarios do not preserve either property precisely.

Approximate Lorentz invariance and spatiotemporal locality. Nature is the ultimate arbiter of "reasonable" structure in theoretical physics, and *both* local Lorentz invariance and spatiotemporal locality seem to hold, at the classical level, at currently accessible scales. Hence, whatever objects one chooses to use in modeling classical spacetime structure must accommodate a suitable degree of fidelity to both properties, even if they exhibit systematic deviations from one or both properties at extreme scales. In both causal set theory and in the version of discrete causal theory developed in this book, satisfactory models may be found among the class of star finite directed sets. What distinguishes these objects from certain problematic alternatives that have been examined previously in the discrete context is that their local structures are *random,* rather than uniform, like a regular lattice, and that their valences are allowed to take on *arbitrarily large values,* rather than being bounded above by some fixed positive integer. Such models may be chosen to approximate Lorentz invariance to an arbitrary degree of precision, or to approximate spatiotemporal locality to an arbitrary degree of precision, or to exhibit a balance between the two. Here I explain, mostly via informal examples, how, and in what sense, this is true, focusing on the case of Lorentz invariance.

In general terms, the geometry of any suitable pseudo-Riemannian manifold may be reconstructed from its causal and conformal structure, as established by the metric recovery theorems described in Section 2.8. Sprinklings represent one method of approximating this structure via random discrete directed sets. For any *finite-volume* region of such a manifold, however large, these directed sets are star finite, though their valences are not bounded by any specific positive integer. In practical terms, one generally works with such finite regions, even in the context of cosmology, as in the papers [BLMS88, ADGS04, KKSR12]. Methods for constructing such "reverse-engineered" approximations *other* than sprinkling have not been definitively ruled out, but sprinklings suffice for the present discussion. As always, the rationale for studying discrete approximations of *non-fundamental* continuum-based objects is to demonstrate that discrete structure can reproduce, up to experimental precision, whatever aspects of physical spacetime appear to be geometric.

I will illustrate "approximate Lorentz invariance" for star finite directed sets by examining some examples of how one could hypothetically detect *violations* of systematic Lorentz invariance in such sets, via arbitrarily precise and detailed experiments carried out in finite regions of spacetime. The directed sets involved in these particular examples are obtained by "modifying" star infinite causal sets constructed via global sprinklings into $\mathbb{R}^{3+1}$. The left-hand diagram in Figure 4.5.6 illustrates a pair of "idealized classical experiments," performed in the frame of reference A, first introduced in Figure 4.5.4. For the sake of the argument, I pretend here that both the copy of Minkowski spacetime $\mathbb{R}^{3+1}$ and the star infinite causal set $C = (C, \prec_{\mathrm{CS}})$ induced by the illustrated sprinkling "exist," in the same sense that the authors of [BHS09] do; i.e., I pretend that it makes sense to talk about the frames of reference A and B, even if the only events actually under consideration are those corresponding to elements of the sprinkled causal set C. The "experiments" occur inside the darker-shaded boxes[34] in the figure, one contained within the other. They are analogous to Feynman's [FE48] "ideal measurements," discussed in Section 6.6, in the sense that the only detail one knows, or needs to know, about them is that they reveal some information about "what happens in a specified region of spacetime." Under the strong interpretation of the causal metric hypothesis (CMH), "what happens" is completely specified by the causal structure, and this structure remains the notion of interest in the context of approximate Lorentz invariance, even under weaker versions of the hypothesis. Hence, an "idealized classical experiment" in this context is merely some unspecified procedure for determining part of the structure of the causal set $(C, \prec_{\mathrm{CS}})$, or modifications thereof. In this case, I assume that each experiment detects any maximal predecessors or minimal successors of the element y in C occurring inside the corresponding box.

As observed by Finkelstein [FI88], Moore [MO88], and others, it seems very dubious, a priori, that arbitrarily distant events in the frame A, or in any other given frame, should be directly related to y. Hence, it is natural to experiment with modifications of the sprinkling construction that do not exhibit this "nasty property." Perhaps the most obvious such modification is to impose a *cutoff,* under which events occurring before a certain time $t_{\mathrm{A}} = t_0$ preceding y, or after a certain time $t_{\mathrm{A}} = t_1$ succeeding

[34] Really, these are *cylinders,* i.e., certain regions of space observed for certain time intervals in the frame A. However, the two-dimensional nature of the diagrams reduces the spatial regions to horizontal line segments. Alternatively, one could use Alexandrov neighborhoods.

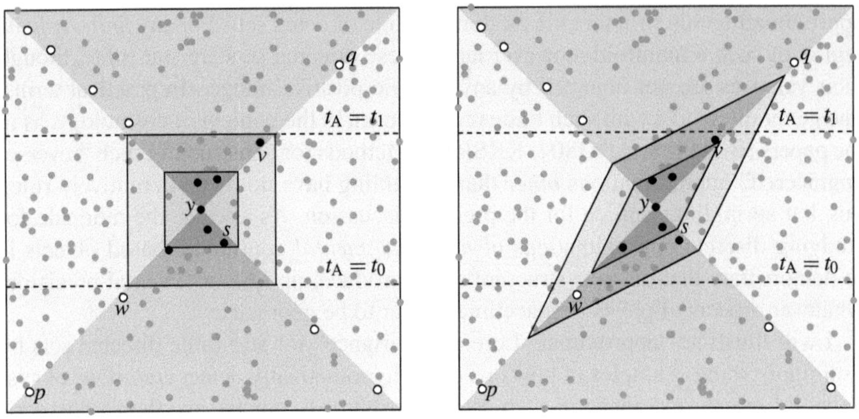

Figure 4.5.6. Systematic Lorentz invariance violation due to a "frame-dependent cutoff" is theoretically detectable, but only for sufficiently large and detailed experiments involving sufficiently high energies.

y, are prohibited from directly influencing y. These cutoff times are illustrated by the dashed horizontal lines in the figure. There exist any number of methods to design such cutoffs; for example, one may choose to "smooth the cutoff" by taking the probability of direct influence to decrease to zero in some "gentle manner," rather than dropping to zero abruptly. The simplest choice, however, is to impose a time-symmetric cutoff $t_0 = -T$ and $t_1 = T$ for some positive constant T, called simply "the cutoff," and this is the choice I follow in the rough estimates given below. Whatever the details, the qualitative structural consequence of such a cutoff is to "pare down" C to a causal set C' that is star finite in the skeleton, by eliminating relations that are "longer than the cutoff" in a temporal sense. In particular, the relations in C between y and the elements represented by the white nodes in the diagram, such as p, q, and w, are eliminated by the cutoff. This modification comes, of course, at the price of sacrificing systematic Lorentz invariance; in particular, A is now a "preferred frame." Further, as I have described the construction, the cutoff shrinks the overall past and future of y, not just its direct past and future. The reason for this is that the elements excluded by the cutoff include some of the maximal predecessors and minimal successors of y in C; in this case, those represented by the white nodes.[35] However, in this example, C' remains a "global approximation of $\mathbb{R}^{3+1}$." A simpler approach would be to consider a finite region of $\mathbb{R}^{3+1}$, but in this case the corresponding causal set would already be star finite.

The crucial observation to be drawn from this modified sprinkling construction is that *no conceivable small, low-energy, classical experiment can detect the cutoff.* The quantitative meaning of "small" and "low-energy" depends on the size of the cutoff. In particular, neither of the experiments illustrated in the left-hand diagram in

[35]One could attempt to preserve the original pasts and futures in a variety of ways, but I do not explore these details here.

Figure 4.5.6 can detect the cutoff for the element y, although a more complete experiment carried out in the larger box could presumably detect the cutoff for the element v, and for other elements near the times $t_A = t_0$ and $t_A = t_1$. However, provided that the cutoffs for different elements are of the same magnitudes, no experiment carried out in the smaller box can detect these cutoffs at all. The situation illustrated by the right-hand diagram is somewhat different. Here, the experiments under consideration are performed in the frame of reference B introduced in Figure 4.5.4, which is boosted by a factor of $\beta = \frac{1}{2}$ with respect to frame A. These are essentially "the same experiments performed in different frames," in the sense that the two parallelograms in the right-hand diagram look exactly like the squares in the left-hand diagram from the perspective of frame B. Note that the larger experiment in frame B *can* detect the cutoff for y, and this illustrates the sense in which the frame A, with respect to which the cutoff was chosen, is a preferred frame. The smaller experiment in frame B *cannot* detect the cutoff for y, although it "comes closer to doing so" than the corresponding smaller experiment in frame A. For any given boost, however, no matter how large, an experiment covering too small a spatial region for too short a time interval in the boosted frame cannot detect the cutoff.

It is easy enough to put rough numbers into such a scenario. Suppose that the frame B is boosted by a factor of β with respect to the preferred frame A, where $-1 < \beta < 1$, and suppose that a spherical region of space of radius r, centered at an event y in the boosted frame, is observed for a time interval $[-r/c, r/c]$. Then the largest cutoff for y that can be detected is[36]

$$T = \frac{r}{c}\sqrt{\frac{1+\beta}{1-\beta}}. \tag{4.5.7}$$

For example, for the boost $\beta = \frac{1}{2}$ illustrated in the right-hand diagram in Figure 4.5.6, the largest detectable cutoff is $T = \sqrt{3}r/c$, measured here in seconds. Hence, to detect a cutoff of one second, one would need to observe, in detail, for a little more than a second, a sphere of radius about 1.7×10^8 m in the boosted frame. This radius is a little less than half the distance from the Earth to the Moon, which does not make for a very practical experiment. Alternatively, for an ultrarelativistic boost $\beta = (1 - 10^{-17})$, involving, for example, a high-energy cosmic ray, the largest detectable cutoff is about $T \approx r$ seconds. If the cosmic ray is a proton, then the radius is about $r = 10^{-15}$ m, so one could theoretically detect a cutoff of around $T = 10^{-15}$ s by observing the cosmic ray cross the width of a bacterium in the preferred frame. In comparison, by colliding protons at the Large Hadron Collider, it would be difficult, even in theory, to detect a cutoff larger than 10^{-20} s, which corresponds to subatomic-scale nonlocality in a spatial sense. These naïve considerations suggest that spatiotemporal nonlocality in discrete causal theory need not be very pronounced,

---

[36]This is just the time value t_A in frame A corresponding to the coordinate values $t_B = r/c$ and $x_B = r$ in frame B, namely $t_A = \frac{1}{\sqrt{1-\beta^2}}(t_B + \frac{\beta x_B}{c}) = \frac{1}{\sqrt{1-\beta^2}}(\frac{r}{c} + \frac{\beta r}{c}) = \frac{r}{c}\frac{1+\beta}{\sqrt{1-\beta^2}} = \frac{r}{c}\sqrt{\frac{1+\beta}{1-\beta}}$.

from a conventional perspective, in order to preserve approximate Lorentz invariance to a reasonable degree.

In fact, it would almost certainly be impossible, at least by any classical means, to detect cutoffs as large as those cited above. The reason for this is that the "idealized classical experiments" involved are much more detailed than any experiment one could hope to perform using any existing or foreseeable technology. In particular, these experiments are naïvely assumed to reveal details of causal structure at the fundamental scale, thereby allowing detection of a cutoff as soon as individual relations start "dropping out" of the idealized spacetime regions involved. Realistically, any achievable experiment would detect only a much coarser version of this structure, which would have to differ rather drastically between the preferred frame and the boosted frame in order to produce any measurable effect. For example, using ballpark guesses regarding the fundamental scale, the cosmic ray experiment cited above might involve 10^{80} fundamental elements, and the entire portion of the causal future $J^+(y)$ of an element y delimited by a cutoff of 10^{-15} s might involve 10^{100} elements or more. The "outer shell" of this region, consisting of the "nearest neighbors" of y spared by the cutoff, would be smaller by perhaps 50 orders of magnitude,[37] and only this shell would be initially affected when the cutoff was reached. Nevertheless, the fact that these hypothetical experiments involve scales much larger than the Planck scale suggests that experimental searches for associated phenomena are worth considering. Moreover, the numbers of fundamental elements involved indicates that very large valences would not be surprising, as mentioned repeatedly beginning in Section 1.3. Such large valences may seem rather alien from the perspective of causal dynamical triangulations or spin foams, but less so from the perspective of neural networks.

Clarifications regarding diagrams; concluding remarks. Before concluding this section, it is prudent to clear up some possible sources of confusion associated with the diagrams I have been using to illustrate sprinklings, cutoffs, and related notions. First, all of the Minkowski diagrams appearing so far in this section have been drawn "from the perspective of the frame of reference A," even though information involving other frames also appears in some of these diagrams. For example, in the diagrams in Figure 4.5.4, coordinate axes are shown for both frame A and frame B, but both pairs of axes are drawn from the perspective of frame A, since the x_{A} and t_{A} axes appear "orthogonal," while the x_{B} and t_{B} axes appear "scissored together." If these diagrams were re-drawn "from the perspective of frame B," then the x_{B} and t_{B} axes would appear "orthogonal," while the x_{A} and t_{A} axes would appear "scissored apart." Such a diagram actually appears in Figure 7.3 of Chapter 7, where it is needed to discuss the relativity of simultaneity from both points of view.

The practice of drawing all the diagrams from the perspective of frame A may create the naïve impression that causal sets induced by global sprinkling do *not*

[37] In generalized Minkowski spacetime $\mathbb{R}^{n+1}$, this "outer shell" is represented by the region between the null cone and the hyperboloid one fundamental unit from y under the Minkowski metric η. The volume of this region scales like T^{n-2}, or like $\log T$ for $n = 2$. See [BLMS88] for a similar discussion. A cutoff of 10^{-15} s corresponds to about $T = 10^{28}$ in Planck units.

exhibit systematic Lorentz invariance. To understand why, consider again the right-hand diagram in Figure 4.5.6, which illustrates two "idealized classical experiments" performed in frame B, which is boosted by a factor of $\beta = \frac{1}{2}$ with respect to frame A. From the perspective of frame A, it appears as though the effect of the boost is to "capture more elements near one part of the null cone than another." In particular, in the two-dimensional version of the diagram actually shown, the two experiments in frame B seem to "capture more elements" near the line $ct = x$, which runs from the bottom left of the diagram to the top right of the diagram, than near the line $ct = -x$, which runs from the bottom right of the diagram to the top left of the diagram.[38] One might conclude that A *is* a preferred frame, even for the "full" causal set C, characterized by exhibiting a "balance of sprinkled elements around the null cone." The problem with this reasoning is that the property of "what part of the null cone an element is nearest to," in a spatiotemporal sense, depends on the frame of reference. For example, elements that appear "close to the right-hand side" of the future null cone in one frame may appear "close to the left-hand side" of the future null cone in another frame. This may be demonstrated by re-drawing the diagrams from the perspectives of other frames, as is done in Figure 4.5.8. Note that this involves "transforming elements instead of changing coordinates;" i.e., it involves *active* transformations of $\mathbb{R}^{3+1}$.

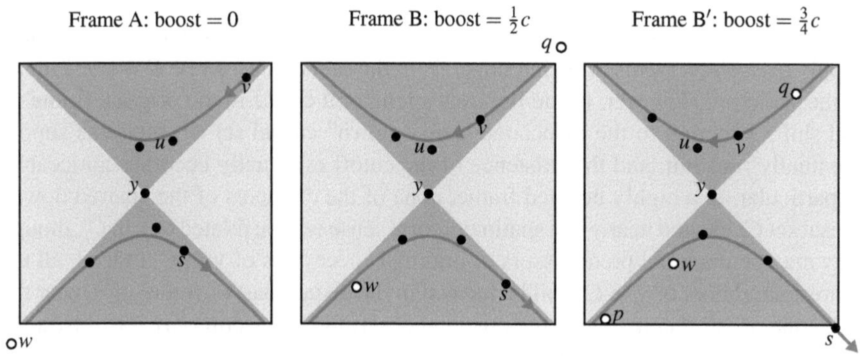

Figure 4.5.8. Spatiotemporal proximity of events to the "one side or the other" of the null cone of y is different in different frames of reference.

The left-hand diagram in the figure reproduces the results of the "larger experiment" in frame A, which detects the maximal predecessors and minimal successors of y located in the larger of the two darker-shaded boxes in the left-hand diagram in Figure 4.5.6. Also shown is the "off-screen" element w, which plays a role in the other two diagrams. The element p just above the caption is associated with the middle diagram, which reproduces the results of the "larger experiment" in frame B. This experiment detects the maximal predecessors and minimal successors of y located

[38]I eliminate subscripts here, because the null cone is invariant under Lorentz transformations.

in the larger of the two darker-shaded parallelograms in the right-hand diagram in Figure 4.5.6. In the present diagram, the same region of $\mathbb{R}^{3+1}$ is represented by a square, because the diagram is drawn from the perspective of frame B. Note how the element w, which was "off-screen" in the left-hand diagram, has now moved into the picture, just as it appears in the lower left corner of the parallelogram in Figure 4.5.6. The elements p and q are just visible "off-screen," at the extreme lower left and top right of the middle diagram, respectively. The right-hand diagram illustrates the "same experiment" performed in a third frame B′, which is boosted by a factor of $\beta' = \frac{3}{4}$ with respect to the frame A. If this experiment were illustrated in Figure 4.5.6, it would be represented by a "more stretched" parallelogram, extending all the way from the lower-left corner of the diagram to the top-right corner. In particular, this experiment captures the elements p and q, while the element s is "just moving off-screen." More generally, one may track the "motion" of elements as the boost changes continuously; each element traces out one branch of a hyperbola,[39] which approaches the null cone, in a spatiotemporal sense, as the boost approaches ± 1. In the two-dimensional diagrams used here, individual elements "move from left to right" in the causal past of y as the boost increases, and "move from right to left" in the corresponding causal future.

These diagrams provide a concrete illustration of one particular aspect of the systematic Lorentz invariance of causal sets induced by global sprinklings into $\mathbb{R}^{3+1}$. For example, although every minimal successor of y in such a causal set C eventually "moves toward one side of the null cone" as the boost increases, there is an endless supply of such elements "flowing down from the other side," since C is star infinite in the skeleton. However, if one imposes a temporal cutoff in the original frame A, and shifts attention to the associated "pared down" causal set C', then this supply eventually runs out, and the existence of the cutoff eventually becomes noticeable. In particular, in a highly boosted frame, none of the elements of the "paired down" causal set C' located near y in a spatiotemporal sense will be related to y in C', though they may be maximal predecessors or minimal successors of y in C. Indeed, all the nearest neighbors of y in C' will be located in the distant past or future of y, near the null cone of y. Of course, this hypothetical description is useful only for illustrative purposes, since it takes the ambient copy of $\mathbb{R}^{3+1}$, by which C and C' are induced, to be "real," in the sense that one may observe the arrangement of elements of C or C' from the perspective of different frames in $\mathbb{R}^{3+1}$. In the discrete causal context, the structure of $\mathbb{R}^{3+1}$ is regarded as merely an approximate way of describing certain aspects of discrete order-theoretic structure. Associated concepts, such as boosts, are emergent, and must be described, in suitable regimes, in order theoretic terms, rather than vice versa. The topic of emergent boosts is briefly revisited in Section 8.4.

---

[39]In the two-dimensional situation shown here, if the coordinates of an element are $(t_{\mathrm{A}}, x_{\mathrm{A}})$ in frame A, then this element traces out a branch of the hyperbola $X^2 - Y^2 = x_{\mathrm{A}}^2 - t_{\mathrm{A}}^2$ under the one-parameter family of active transformations parameterized by the boost value, where X and Y are fixed horizontal and vertical coordinates of $\mathbb{R}^{1+1}$. The right-hand side of this equation, of course, is just the invariant Minkowski spacetime interval between the origin y and the "moving" element, in units where $c = 1$.

To summarize, the star infinite structure of causal sets constructed, a posteriori, via global sprinklings into Minkowski spacetime $\mathbb{R}^{3+1}$, or other idealized pseudo-Riemannian geometries, provides one of several good reasons why star infinite directed sets are worth studying in the discrete causal context. In particular, global sprinklings into $\mathbb{R}^{3+1}$ provide the only known mechanism for constructing causal sets exhibiting systematic Lorentz invariance. This property itself is physically unrealistic, but experimental constraints on violation of *local* Lorentz invariance render "approximate Lorentz invariance" a necessary property for suitable classes of directed sets to accommodate. Fortunately, star finite directed sets can exhibit a suitable version of this property, while also retaining a reasonable degree of spatiotemporal locality. The latter property is completely sacrificed by star infinite directed sets at the level of fundamental causal structure. Strategies to recover "effective spatiotemporal locality" for background dependent *auxiliary* structure, examined in the causal set literature, are interesting but limited in scope. The generally undesirable properties of star infinite directed sets have been recognized since the inception of causal set theory; in particular, they were pointed out by Moore [MO88] in response to the inaugural paper on causal sets [BLMS87], and were subsequently acknowledged by the founders of causal set theory [BLMS88]. However, this acknowledgement was followed by an explanation that star finite structures suffice for modeling quantum gravity in the context of causal set theory. There are excellent reasons for extending this conclusion to apply to discrete causal theory in general. This subject is revisited once more in Sections 4.7 and 4.8.

4.6 Relative Multidirected Sets Over a Fixed Base

Infinite intervals. The incomparability of interval finiteness (IF) and star finiteness (SF) means that each condition permits behavior that the other condition rules out. As explained in Section 4.4 and elaborated in Section 4.5, interval finiteness permits models in which a given element possesses an infinite number of maximal predecessors and/or minimal successors. The resulting behavior ranges from awkward to cataclysmic, but is dubious enough, on the whole, to render such models undesirable, especially since they are unnecessary for maintaining the general connection to established physics afforded by the metric recovery theorems. However, the alternative condition of star finiteness permits "strange behavior" of a different sort; namely, infinite intervals between pairs of elements connected by finite chains of relations, or even single relations, as in the case of Jacob's ladder. What, if any, significance should be attached to these infinite intervals? If they are determined to be sufficiently pathological, then *neither* interval finiteness nor star finiteness may be considered adequate local finiteness conditions, and one should consider the options of either imposing both conditions together, or searching for some third alternative. At the opposite extreme, such intervals *could* be taken seriously as structures of possible physical interest. In the specific context of classical spacetime structure, I know of no positive physical evidence for the latter view, but there seem to be interesting *structural* grounds for considering such possibilities. In particular, the theory of *relative multidirected sets over a fixed base,* introduced in the present section, provides a perspective which suggests, among other conclusions, that censoring infinite intervals is unjustified. The terminology associated with this theory originates from Grothendieck's *relative viewpoint* (RV), introduced in Section 3.8.

Causal sets as relative directed sets over the integers. Every nonempty causal set $C = (C, \prec_{\mathrm{CS}})$ admits a bijective morphism of directed sets into a totally ordered subset of the integers: either $\mathbb{Z}$ itself, the natural numbers $\mathbb{N}$, the nonpositive integers $-\mathbb{N}$, or a *finite simplex* $[n] := \{0, 1, \ldots, n\}$ for some natural number n. Since $\mathbb{N}$, $-\mathbb{N}$, and $[n]$ all embed into $\mathbb{Z}$, every causal set may be viewed as a *relative directed set over* $\mathbb{Z}$, i.e., a directed set together with a distinguished morphism into $\mathbb{Z}$. Generally, there exist many such morphisms for a given causal set. Figure 4.6.1 illustrates two different such morphisms from a simple "v-shaped" causal set C into the simplex [2]. Such a morphism is equivalent to a refinement of the causal relation $\prec_{\mathrm{CS}}$ on C to yield an order $\prec_{\mathrm{TOT}}$ isomorphic to that of its totally ordered image. For example, the morphism illustrated in the left-hand diagram in the figure corresponds to the refinement of $\prec_{\mathrm{CS}}$ to yield the order $x \prec_{\mathrm{TOT}} y \prec_{\mathrm{TOT}} z$, while the morphism illustrated in the right-hand diagram corresponds to the refinement of $\prec_{\mathrm{CS}}$ to yield the order $x \prec'_{\mathrm{TOT}} z \prec'_{\mathrm{TOT}} y$.

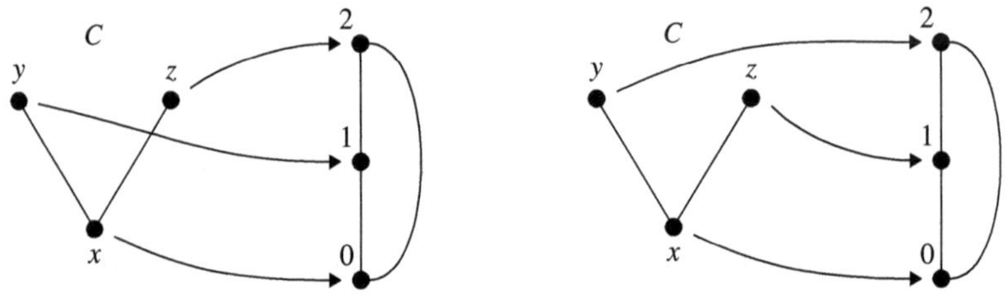

Figure 4.6.1. Two different ways of viewing a "v-shaped" causal set as a relative directed sent over the simplex $[2] = \{0, 1, 2\}$.

As mentioned in Section 3.4, such a refinement is analogous to a *frame of reference* in relativity. Figure 4.6.2 illustrates this analogy, using three spacetime events labeled x, y, and z, which correspond in an obvious way to the elements of the causal set $C = (C, \prec_{\mathrm{CS}})$ in Figure 4.6.1. The spacelike section σ represents a frame of reference under which y temporally precedes z; this frame corresponds to a refinement of the relativistic causal relation $\prec_{\mathrm{GR}}$ analogous to the refinement of $\prec_{\mathrm{CS}}$ associated with the morphism illustrated in the left-hand diagram in Figure 4.6.1. Similarly, the spacelike section τ represents a frame of reference under which z temporally precedes y; this frame corresponds to a refinement of $\prec_{\mathrm{GR}}$ analogous to the refinement of $\prec_{\mathrm{CS}}$ associated with the morphism illustrated in the right-hand diagram in Figure 4.6.1. Relationships between frames of reference and refinements of binary relations are discussed in much more detail in Chapter 7.

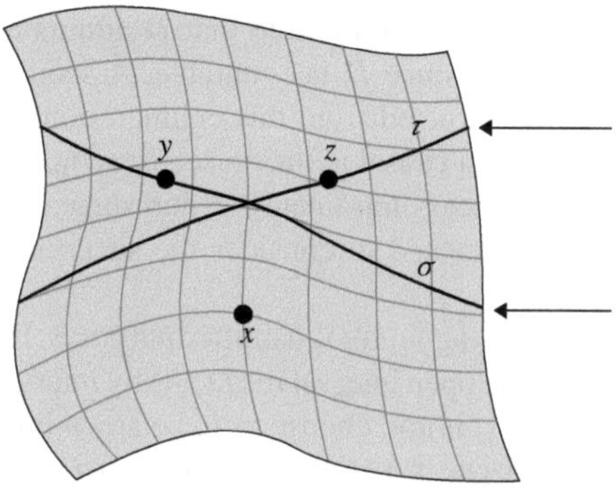

Figure 4.6.2. Relativistic frames of reference and refinements of the causal relation.

In a more general context, the existence of a bijective morphism from an arbitrary causal set into a totally ordered subset of the integers does *not* follow immediately from the *order refinement principle* (ORP), introduced in Section 3.8, since a total refinement of an interval finite strict partial order is generally not interval finite, as illustrated in Figure 4.6.6 below. However, such a refinement may be explicitly constructed. The proof I give of this result, which generalizes the picture somewhat by dropping the axiom of transitivity (TR), is essentially due to Joel David Hamkins,[40] though expressed in different language. It uses a special case of a technique I refer to as *atomic accretion,* in which an order-theoretic analogue of a convex set, which I refer to as a *causal atom* due to its physical interpretation in the context of classical spacetime structure, is built up via a sequential process. Although I present the proof of the theorem here, the reader may choose to postpone examining it until reading Section 5.6, where I introduce the theory of causal atoms more systematically. Although this particular "physically suggestive" method of proof is a recent construction, it seems that the *result* itself was first recognized in the context of pure order theory, in the work of Marcel Erné, circa 1979.[41]

Theorem 4.6.3. *Any nonempty, countable, interval finite, acyclic directed set admits a morphism into the integers.*

Proof. The proof has two steps. The first step is an induction argument establishing that any nonempty, countable, interval finite, acyclic directed set $D = (D, \prec)$ admits a **sequential atomic accretion** $\{x_i\}_{i\in I}$, where I is an interval in $\mathbb{Z}$ containing 0. The meaning of the words "atomic accretion" is that every finite, two-sided truncation $\{x_m, ..., x_n\}$ of $\{x_i\}_{i\in I}$ is a *causal atom* in D, i.e., a subset of D containing the intersection of its past and future. The second step is merely the observation that such an atomic accretion induces a morphism into $\mathbb{Z}$.

[40]Via informal communication, 2013.

[41]The earliest published references I could find are secondary remarks citing the unpublished lecture notes [ER79], which Erné was kind enough to send me copies of.

1. *D admits a sequential atomic accretion*. First note that D admits an *enumeration* $D = \{y_0, y_1, ...\}$ by hypothesis, since D is countable. A sequential atomic accretion $\{x_i\}_{i\in I}$ of D may be constructed from this enumeration by induction. By acyclicity, any singleton in D is a causal atom, so set $x_0 = y_0$. If D has only one element, then the accretion process terminates at this stage, and the result is obvious. Otherwise, one must proceed to consider the succeeding elements $y_1, y_2, ...$ in the enumeration.

   I explicitly describe the step involving y_1, for illustrative purposes. If x_0 precedes y_1 in $(D, \prec)$, then $\langle\langle x_0, y_1\rangle\rangle$ is a finite open interval in $(D, \prec)$ by interval finiteness. If this interval is empty, set $x_1 = y_1$. If not, choose x_1 to be any minimal element in $\langle\langle x_0, y_1\rangle\rangle$. In either case, the set $\{x_0, x_1\}$ is a causal atom in D. If the interval $\langle\langle x_0, y_1\rangle\rangle$ contains no other elements, set $x_2 = y_1$. Otherwise, choose x_2 to be a minimal element in $\langle\langle x_0, y_1\rangle\rangle - \{x_1\}$. In either case, the set $\{x_0, x_1, x_2\}$ is a causal atom in D. Continuing in this fashion, the interval $\langle\langle x_0, y_1\rangle\rangle$ is exhausted after a finite number of steps. A symmetric argument applies if x_0 succeeds y_1; in this case, the elements in the interval $\langle\langle y_1, x_0\rangle\rangle$ are chosen to be x_{-1}, x_{-2}, and so on. If y_1 is unrelated to x_0, then y_1 may be chosen to be either x_1 or x_{-1}, since in this case the intersection of the past and future of $\{x_0, y_1\}$ is empty. All of these cases lead to a process that terminates after a finite number of steps, yielding a causal atom containing y_0 and y_1.

   Now assume, by induction, that $\alpha := \{x_m, ..., x_n\}$ is a causal atom containing $y_0, ..., y_k$, and consider y_{k+1}. If $\alpha = D$, or if $y_{k+1} \in \alpha$, there is nothing to show. Suppose that y_{k+1} belongs to the future of α, but not to α itself. Since α is a causal atom, y_{k+1} does not belong to the past of α. Consider the generalized open interval $\langle\langle \alpha, y_{k+1}\rangle\rangle$ consisting of elements in the future of α and the past of y_{k+1}, but not belonging to α or equal to y_{k+1}. If $\langle\langle \alpha, y_{k+1}\rangle\rangle$ is empty, set $x_{n+1} = y_{k+1}$. Otherwise, choose x_{n+1} to be a minimal element in $\langle\langle \alpha, y_{k+1}\rangle\rangle$. In either case, $\alpha \cup \{x_{n+1}\}$ is a causal atom. Continuing in this fashion, the interval $\langle\langle \alpha, y_{k+1}\rangle\rangle$ is exhausted after a finite number of steps. A symmetric argument applies if y_{k+1} precedes α. If y_1 is unrelated to any element of α, then y_1 may be chosen to be either x_{n+1} or x_{m-1}. All of these cases lead to a process that terminates after a finite number of steps, yielding a causal atom containing $y_0, ..., y_{k+1}$. By induction, this process extends to all of D.

   The construction shows directly that the subset α of D obtained after adding the kth element y_k of the original enumeration $\{y_0, y_1, ...\}$ is a causal atom, and that each individual step involved in adding the elements "between α and y_{k+1}" preserves this property. However, it is actually the case that *every* finite, two-sided truncation $\beta = \{x_m, ..., x_n\}$ of $\{x_i\}_{i\in I}$ is a causal atom in D.[42] To see this, let w be an element of D in the complement of β. Then $w = x_i$ for some i, where either $i < m$ or $i > n$. If $i < m$, then by construction $w = x_i$ does not belong to the future of any of the elements $x_m, ..., x_n$. Similarly, if $i > n$, then by construction

[42]It would be interesting to create computer animations of the "motion" of a sequence of causal atoms $\{x_{m+j}, x_{m+j+1}, ..., x_{n+j}\}$ in a typical sequential atomic accretion, as j increases. At a naïve visual level, such an animation might resemble a "particle" moving from past to future in D.

$w = x_i$ does not belong to the past of any of the elements x_m, ..., x_n. Hence, the intersection $J^-(\beta) \cap J^+(\beta)$ does not contain w. Therefore, β is a causal atom, and the process defines a sequential atomic accretion.

2. *The set map $D \to \mathbb{Z}$ sending x_i to i is a morphism of directed sets.* Generalizing the above observation involving two-sided truncations, x_i does not belong to the future of any element with a larger label, or to the past of any element with a smaller label. Hence, the accretion process automatically respects the binary relation $\prec$ on D. □

Every causal set is a nonempty, countable, interval finite, acyclic directed set, and therefore admits a morphism into $\mathbb{Z}$ by the theorem. In fact, the proof of the theorem yields more specific results involving totally ordered subsets of $\mathbb{Z}$. In particular, if D is finite, with cardinality $n + 1$, then the construction described in the proof yields a bijective morphism into $[n]$, determined by the choice of enumeration, and the choices of how to order the maximal or minimal elements in the generalized open intervals involved in the construction. If D is infinite, bounded below,[43] and has a finite number of minimal elements, then the construction yields, after appropriate "shifts," a bijective morphism into $\mathbb{N}$. If D is infinite, bounded above, and has a finite number of maximal elements, then the construction yields a bijective morphism into $-\mathbb{N}$, in a similar manner. If D is infinite and unbounded above and below, then the construction yields a bijective morphism into $\mathbb{Z}$. The remaining cases may involve bijective morphisms into either $\mathbb{N}$, $-\mathbb{N}$, or $\mathbb{Z}$, depending on the details. For example, a countably infinite set with no relations admits bijective morphisms into all three of these target sets.

A directed set $(D, \prec)$, together with a *particular* morphism into $\mathbb{Z}$, bijective or otherwise, is called a **relative directed set over** $\mathbb{Z}$. Hence, every causal set may be viewed as a relative directed set over $\mathbb{Z}$. Conversely, every transitive relative directed set D of finite index over $\mathbb{Z}$ is a causal set,[44] since every interval in D is contained in the preimage of an interval in $\mathbb{Z}$, and since $\mathbb{Z}$ itself is irreflexive and interval finite under the familiar "less than" relation $<$. The axioms of causal set theory may therefore be reduced from six to four, with countability (C), interval finiteness (IF), and irreflexivity (IR) replaced by the existence of a morphism of finite index into $\mathbb{Z}$. This axiomatic streamlining is one advantage of the theory of relative multidirected sets over a fixed base. A more concrete application appears in Sorkin and Rideout's theory of sequential growth dynamics [SR99], in which a total labeling $\{x_0, x_1, x_2, \ldots\}$ of a causal set C exhibits C as a relative directed set over $\mathbb{Z}$, via the inclusion $\mathbb{N} \to \mathbb{Z}$.

[43] There are many different notions of boundedness for acyclic directed sets, both local and global. In this case, *bounded below* means that every chain in D has a minimal element. This is a relatively "weak" boundedness condition.

[44] Here, *index* refers to the index of the morphism $D \to \mathbb{Z}$, i.e., the supremum of the cardinalities of its fibers.

Arbitrary bases; foreshadowing "co-relative" constructions. The following definition generalizes the picture of relative directed sets over a fixed base to the case of arbitrary multidirected sets:

Definition 4.6.4. *Let $M = (M, R, i, t)$ be a fixed multidirected set. A* **relative multidirected set over** *M is a multidirected set $M' = (M', R', i', t')$, together with a morphism $M' \to M$. The* **category of relative multidirected sets over** *M is the category whose objects are multidirected sets over M, and whose morphisms are commutative triangles of the following form:*

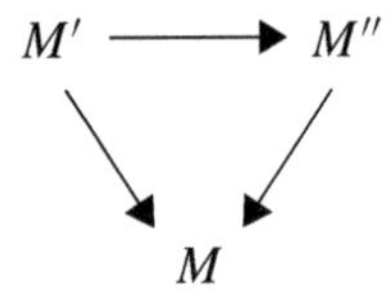

Here, the vertical arrows are the distinguished morphisms defining M' and M'' as multidirected sets over M. In this context, M is called the **base** *of M' and M''.*

Of particular interest in the study of classical spacetime structure is the case in which M, M' and M'' are directed sets $(D, \prec)$, $(D, \prec')$, and $(D, \prec'')$ sharing the same underlying set D, and where the morphisms are the identity set maps. In this case, the sequence of maps $(D, \prec') \to (D, \prec'') \to (D, \prec)$ corresponds to successive refinements of binary relations, with $\prec''$ refining $\prec'$, and $\prec$ further refining $\prec''$. The individual morphisms correspond to generalized frames of reference for their sources, adding "extraphysical information," analogous to temporal ordering of spacelike-separated events in relativity. Of even greater interest, however, are *dual diagrams,* given by "reversing the arrows," and considering morphisms from a single "cobase" into multiple "targets," rather than morphisms from multiple "sources" into a single "base." Such dual diagrams are central to the theory of *co-relative histories,* developed in Chapter 6, where the term "co-relative" is the category-theoretic abbreviation for "dual to relative." Another application of such dual diagrams involves consideration of just two different "targets" with the same underlying set; The relativistic analogue in this case is given by considering two different frames of reference on the same copy of spacetime. Such diagrams are analogous to "generalized elements" of the *Poincaré group P* of isometries of Minkowski spacetime $\mathbb{R}^{3+1}$, since elements of P *induce changes of reference frame on* $\mathbb{R}^{3+1}$, *under the passive viewpoint.* This idea is revisited in Section 8.4.

Democracy of bases. Consider now the following question, which bears directly on the plausibility of the axiom of interval finiteness (IF) as a local finiteness condition in discrete causal theory:

> **Question:** *Why should the causal structure of classical spacetime be modeled exclusively in terms of relative directed sets over $\mathbb{Z}$?*

Even causal sets, and in particular, connected[45] causal sets, generally admit bijective morphisms into totally ordered bases *not contained in the integers,* and it is difficult to argue that such bases should be ignored. The following example, illustrated in Figures 4.6.5 and 4.6.6, exhibits a connected causal set $C = (C, \prec_{CS})$ which may be viewed either as a relative directed set over $\mathbb{Z}$, *or* as a relative directed set over the transitive closure of the interval-infinite directed set called *Jacob's ladder,* introduced in Section 4.4. From a neutral perspective, these two alternative constructions represent radically different, but equally valid, "frames of reference" for C. The skeleton sk(C) of C, depicted in the left-hand diagram in Figure 4.6.5, consists of two infinite chains, one bounded below, with minimal element x; the other bounded above, with maximal element y. These chains are connected by a relation $x \prec y$. From the viewpoint of Sorkin and Rideout's theory of sequential growth dynamics [SR99], the goal is to provide a *"complete kinematic account of the evolution of C,"* which translates, at a mathematical level, to exhibiting C bijectively as a relative directed set over a totally ordered base. Augmenting C by the addition of two new relations $v \prec x$ and $y \prec u$, as shown in the right-hand diagram in the figure, uniquely determines a total order on C, since every pair of elements is now connected by a chain. This total order defines a bijective morphism $C \to \mathbb{Z}$.

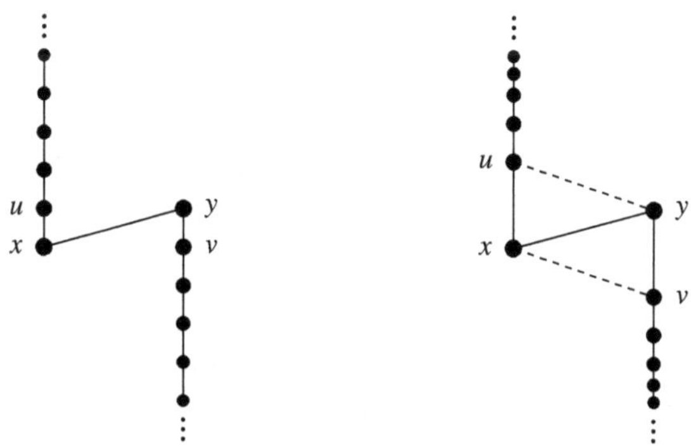

Figure 4.6.5. Skeleton sk(C) of a causal set $(C, \prec_{CS})$; adding relations $v \prec x$ and $y \prec u$, and taking the transitive closure, defines a bijective morphism $C \to \mathbb{Z}$.

However, an alternative total order on C, also refining $\prec_{CS}$, may be defined by taking the transitive closure of Jacob's ladder. Figure 4.6.6 illustrates how this refinement is constructed. The left-hand diagram in the figure, ignoring the relations indicated by the dashed line segments, depicts the skeleton sk(C) of C. This may not be obvious visually, since the left-hand diagrams of Figures 4.6.5 and 4.6.6 are arranged in very different ways. However, they do indeed represent the same directed set. Augmenting C by the addition of the "dashed" relations yields Jacob's ladder, and

[45]Connected, that is, in a graph-theoretic sense. The point of highlighting this property is that the existence of bijective morphisms into "strange" bases for sufficiently "sparse" causal sets, such as a countable set with no relations, is pedestrian.

taking the transitive closure yields the acyclic directed set illustrated in the right-hand diagram in Figure 4.6.6, which is neither interval finite, nor star finite. This last set is isomorphic to the union $\mathbb{N} \bigsqcup -\mathbb{N}$, endowed with the order in which every element of $\mathbb{N}$ is taken to precede every element of $-\mathbb{N}$.

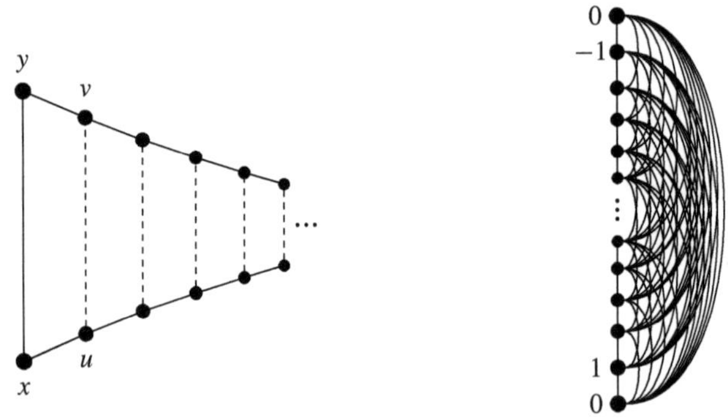

Figure 4.6.6. The transitive closure of Jacob's ladder gives an alternative refinement of $(C, \prec_{\text{CS}})$, which is not interval finite; this refinement defines a bijective morphism $C \to \mathbb{N} \bigsqcup -\mathbb{N}$.

Both of the totally ordered sets $\mathbb{Z}$ and $\mathbb{N} \bigsqcup -\mathbb{N}$ are examples of *discrete totally ordered sets.* Referring back to Section 3.5, a *discrete order* is a partial order in which every nonextremal element has a maximal predecessor and minimal successor. Hence, it is "almost correct" to say that a discrete *totally* ordered set $L = (L, <)$ is *"locally isomorphic to* $\mathbb{Z}$," in the sense that a typical small neighborhood of an element in L is isomorphic to a small neighborhood in $\mathbb{Z}$. Here, the letter L is chosen to suggest the word "linear," while the notation for the order $<$ is intended to evoke the "less than" relation on $\mathbb{Z}$. Of course, it is not true that a typical small neighborhood in L is isomorphic to $\mathbb{Z}$ *itself,* but one should not expect such a condition to hold in the discrete context, where one has the benefit of natural scale data. Indeed, from this perspective, the only reason why elements in n-dimensional real manifolds have "small neighborhoods" isomorphic to $\mathbb{R}^n$ itself, is because not enough natural structure is available in the continuum-based setting to "enforce" the intuition that *"small neighborhoods should be small, while* $\mathbb{R}^n$ *itself should be large."* The qualifier "locally" in the phrase "locally isomorphic to $\mathbb{Z}$," may be made precise by using the irreducible star topology on L, which represents one of the alternative notions of local structure discussed in Section 4.4. To avoid the possibility of confusion, I should point out that L is generally *not* locally isomorphic to $\mathbb{Z}$ with respect to the star topology, which does not "work well" in the transitive setting. Employing the *irreducible* star topology instead of the star topology in this context has essentially the same effect as first applying the skeleton operation, to "kill reducible relations" in L, then analyzing local structure in the usual way.

Discrete totally ordered sets may therefore be viewed, in rough terms, as *"one-dimensional discrete manifolds over the integers."* More precisely, they are analogous to manifolds *with boundary,* since they may include extremal elements at which the "usual" local conditions break down. The fact that such sets are defined in terms of *local* properties makes them particularly attractive as bases for relative directed sets, viewed as models of classical spacetime, since it is less objectionable to make modest assumptions about local structure than to prescribe global structure. From a concrete physical perspective, the relationship between $\mathbb{Z}$ and an arbitrary discrete totally ordered set is analogous to the relationship between Minkowski spacetime $\mathbb{R}^{3+1}$ and an arbitrary four-dimensional pseudo-Riemannian manifold of Lorentz signature X, which "looks like $\mathbb{R}^{3+1}$ locally," in the sense that its tangent spaces T_xX are isomorphic to $\mathbb{R}^{3+1}$ in the category of pseudo-Euclidean spaces.

Infinite analogue of the Kleitman-Rothschild pathology. In Section 3.9, I described how large finite causal sets are generically biased toward large "spatial size" and small "causal size," with the three-generation partially ordered sets described by Kleitman and Rothschild dominating asymptotically. Since such sets are manifestly irrelevant to the observed structure of classical spacetime, this asymptotic enumeration leads to basic structural concerns regarding causal set theory. This is particularly true in the quantum version of the theory, since the path summation approach must somehow incorporate these sets without being overwhelmed by them. In Section 3.9, I referred to this problem as the *Kleitman-Rothschild pathology.* Here, I briefly point out that an even more worrisome version of this problem arises for countably infinite interval finite acyclic directed sets, including infinite causal sets.

Order theory provides notions of "size" more nuanced than cardinality; namely, the isomorphism classes of totally ordered sets. In this refined sense, for example, Cantor's first transfinite ordinal ω is "larger" than $\mathbb{N}$, because it contains an "extra element," and the discrete totally ordered set $\mathbb{N}\bigsqcup -\mathbb{N}$ is "larger" still. More generally, one possible method of classifying the "size" of an acyclic directed set $D=(D,\prec)$ is to identify a *minimal totally ordered base for D*, i.e., a totally ordered set $L=(L,<)$ such that there exists a bijective morphism $D\rightarrow L$, but no bijective morphism from D into a proper totally ordered subobject of L not isomorphic to L itself. Such a minimal base may, of course, be non-unique. Special classes of acyclic directed sets may be defined in terms of restrictions imposed on such minimal totally ordered bases. Finiteness conditions, such as interval finiteness (IF) and star finiteness (SF), provide obvious examples of such restrictions. Physically, such conditions may be viewed as restrictions on "causal size," providing much more detail than the simple, coarse notion of cardinality.

In this sense, Theorem 4.6.3 demonstrates that interval finiteness limits acyclic directed sets to "causal size at most $\mathbb{Z}$," while star finiteness permits "much larger causal sizes." This viewpoint provides yet another way to understand how severely interval finiteness restricts the global structure of causal sets. The situation is already "bad" for large finite causal sets, as demonstrated by the Kleitman-Rothschild pathology, but is "even worse" for infinite acyclic directed sets, because the same restrictions apply to "causal size," while "spatial size" is unrestricted, except possibly by the coarse cardinality condition of countability (C). Figure 4.6.7 illustrates the skeleton of a causal set of "spatial size greater than $\mathbb{Z}$." A better understanding of this vague statement about spatial size may be given in terms of the theory of *relation space,* introduced in Chapter 5, but for now it is easy to see that the two connected subsets indicated by the braces in the figure are of "size $\mathbb{Z}$," in a heuristic sense.

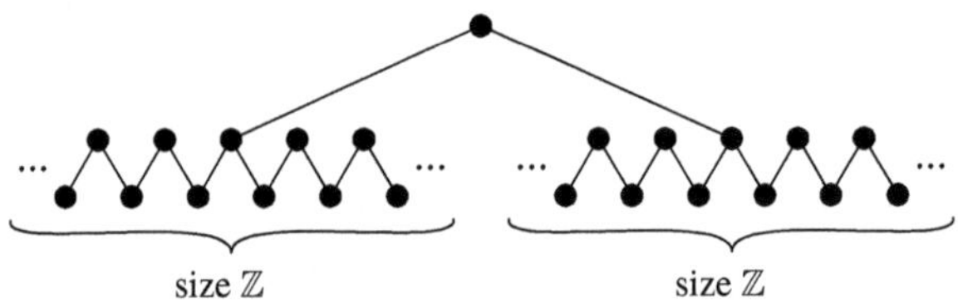

Figure 4.6.7. Skeleton of a causal set of "spatial size greater than $\mathbb{Z}$".

4.7 Seven Arguments Questioning Interval Finiteness

Preliminary remarks. I have now assembled the necessary background to fully explain the limitations of interval finiteness (IF) as a local finiteness condition in discrete causal theory. As stated in Section 4.4, a detailed explanation of these issues is necessary because of the prominence of interval finiteness in the literature. Some of the structural concerns discussed here, such as the fact that interval finiteness is not a local condition, have already been mentioned in previous sections. However, the broader limitations of the condition have yet to be demonstrated here. For example, the following arguments show that interval finiteness cannot even be *modified* to yield a genuinely local condition by working only with arbitrarily small open intervals, or with intersections of open intervals. In particular, topological local finiteness for the interval topology is also an unwieldy condition to use in this context. Further, the examples used here to illustrate these arguments involve only acyclic directed sets, so the arguments apply even under conservative assumptions. The conclusion is that *interval finiteness is not an optimal mathematical condition for formalizing the physical notion of local finiteness in discrete causal theory.* As explained in Sections 4.4 and 4.5, this does *not* mean that there is anything wrong with models of classical spacetime that *happen to be* interval finite, nor does it mean that star infinite objects arising under the condition of interval finiteness, such as causal sets induced by global sprinklings into Minkowski spacetime $\mathbb{R}^{3+1}$, are not worth studying. Rather, it means that the condition itself is not optimal for restricting attention to a class of directed sets suitable for modeling classical spacetime in the discrete causal context.

Seven arguments. The following arguments address a variety of different physical and mathematical issues associated with the use of interval finiteness as a local finiteness condition in discrete causal theory. Additional arguments could be formulated, but the list given here should suffice to illustrate the principal limitations of the condition.

1. *Interval finiteness fails to model a causally local condition for classical spacetime structure.* Referring to Definition 4.1.1, this means that interval finiteness generally cannot be checked by examining the independent causes and effects associated with each element in a directed set, interpreted as a model of classical spacetime under the classical causal metric hypothesis (CCMH). For example, consider the element x in the acyclic directed set illustrated in the left-hand diagram in Figure 4.7.1. The only open interval containing x is the interval $\langle\langle w, y\rangle\rangle$, which also contains all of the directed structure drawn in gray, which is causally unrelated to x. This is also the smallest open set containing x in the interval topology. Hence, even the smallest open neighborhoods of elements in the interval topology generally include causally irrelevant structure.[46] In this particular case, the directed set under consideration *happens to be* interval finite, and indeed, globally finite. However, Figure 4.7.3 below illustrates an acyclic directed set containing an element whose only open neighborhood in the interval topology is infinite, even though this element has only one predecessor and one successor!

Figure 4.7.1. The only open interval containing x also contains irrelevant directed structure; the singleton $\{x\}$ is open in the interval topology, but does not contain an open interval.

2. *Interval finiteness is not a local condition for the interval topology.* In particular, interval finiteness is not the same condition as topological local finiteness for the interval topology, as already mentioned in Section 4.2; it is a much stronger, "less local" condition. Of course, this is obvious, because interval finiteness involves open intervals of arbitrary size. More generally, however, *no* condition stated solely in terms of open intervals, even "small" open intervals, is local in the interval topology, because open intervals do not form a *basis* for the topology, in the sense of Section 4.2. This means that a neighborhood of an element need not contain an *open interval* containing the element. For example, consider the element x in the acyclic directed set illustrated in the right-hand diagram in Figure 4.7.1. The only nonempty open intervals in this set are $\langle\langle w, y\rangle\rangle$ and $\langle\langle w, z\rangle\rangle$, both of which contain x. Their intersection is the singleton $\{x\}$, which is therefore an open set in the interval topology, but does *not* contain an open interval.
3. *Topological local finiteness with respect to the interval topology does not imply that each element is contained in a finite open interval.* This perhaps-surprising

[46] As mentioned in Section 4.3, this is exactly what one would expect by comparing the intervals involved here to *Alexandrov neighborhoods* in the context of relativity.

fact provides a stark demonstration that interval finiteness is *not even close* to possessing a reasonable topological interpretation as a local finiteness condition. It is already clear from the right-hand diagram in Figure 4.7.1 that an element in a multidirected set may have neighborhoods in the interval topology that are smaller than any nonempty open interval. In the worst-case scenario, a minimal nonempty open interval containing a given element may be *infinite,* even though the element has a finite open neighborhood in the interval topology! For example, consider the acyclic directed set $D = (D, \prec)$ with elements

$$\{x\} \cup \{y_{p,q} \mid p, q \text{ prime}, q \neq p\} \cup \{z_l \mid l \text{ prime}\},$$

and relations

$$x \prec y_{p,q}, \quad y_{p,q} \prec z_p, \quad \text{and} \quad y_{p,q} \prec z_q.$$

A portion of D is illustrated in the left-hand diagram in Figure 4.7.2, with the elements $y_{p,q}$ grouped together for each p, and with the relations $y_{p,q} \prec z_q$ indicated by dark segments. The only nonempty open intervals in D are the infinite intervals $\langle\langle x, z_l\rangle\rangle = \{y_{p,q} \mid p = l \text{ or } q = l\}$ for each l. The intersection of two such intervals $\langle\langle x, z_l\rangle\rangle$ and $\langle\langle x, z_{l'}\rangle\rangle$ is the two-point set $\{y_{l,l'}, y_{l',l}\}$. Meanwhile, the open set $J^+(y_{p,q})$ in D is the two-point set $\{z_p\} \cup \{z_q\}$, and the open set $J^-(y_{p,q})$ is the singleton $\{x\}$ for all p, q. Hence, every element of D is contained in a finite open subset of D in the interval topology, but *no* element of D lies in a finite open interval! Informally, this example demonstrates that interval finiteness generally "has very little to do with local structure," even when "locality" is understood in terms of the topology most naturally-related to interval finiteness.

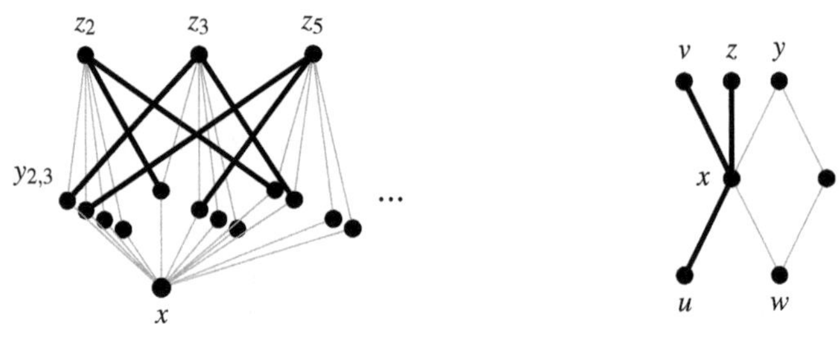

Figure 4.7.2. Every element lies in a finite open set in the interval topology, but not in a finite open interval; the open interval $\langle\langle w, y\rangle\rangle$ contains x, but does not capture local directed structure near x.

4. *Open intervals fail to even capture local multidirected structure, let alone isolate such structure.* Heuristically, this means that open intervals are not only "imprecise," but also "inaccurate," for measuring local structure. Despite being so "large" that they cannot *resolve* local multidirected structure, as shown in the previous example, open intervals generally "leave out" relevant information about local multidirected structure near each of their elements. For example, consider the

element x lying in the open interval $\langle\langle w, y\rangle\rangle$ in the acyclic directed set illustrated in the right-hand diagram in Figure 4.7.2. Three relations involving x; namely, $u \prec x$, $x \prec v$, and $x \prec z$, represented in the figure by dark segments, have terminal elements outside of $\langle\langle w, y\rangle\rangle$. Hence, $\langle\langle w, y\rangle\rangle$ contains relatively little local information relevant to x. Open sets in the interval topology are generally even worse in this regard, since they may be smaller than any open interval.

5. *Interval finiteness does not relate in a useful way to star finiteness.* The fact that interval finiteness does not *imply* star finiteness was demonstrated in Section 4.4, with the infinite bouquet, illustrated in the left-hand diagram in Figure 4.4.5, as a counterexample. Since open intervals are generally "too large" to resolve local multidirected structure, one might hope that requiring even such "large" subobjects of a directed set to be finite would *at least* constrain local causal structure, viewed as "small," to be finite as well. The fact that this is not true illustrates how badly open intervals fail to capture local causal structure. Topological local finiteness in the interval topology is even worse in this regard, since it is a weaker condition. Meanwhile, the fact that interval finiteness does not *follow from* star finiteness was demonstrated in Section 4.4, with Jacob's ladder, illustrated in Figure 4.4.6, as a counterexample. This fact itself is not very surprising, since conditions stated in terms of arbitrary open intervals are not local even in the interval topology. However, even topological local finiteness in the interval topology does not follow from star finiteness. To see this, one may merely "augment" Jacob's ladder by replacing the relation $x_0 \prec y_0$ with a 2-chain $x_0 \prec z_0 \prec y_0$, as illustrated in Figure 4.7.3. In the resulting star finite directed set, the only open set in the interval topology containing z_0 is the infinite interval $\langle\langle x_0, y_0\rangle\rangle$.

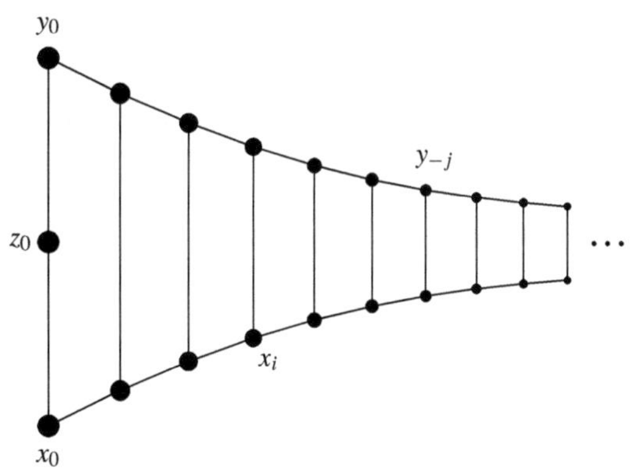

Figure 4.7.3. Augmented version of Jacob's ladder in which the only open interval containing z_0 is infinite.

6. *Interval finiteness permits physically absurd local behavior under any version of the classical causal metric hypothesis for which infinite sets are assigned infinite measure.* As repeated several times throughout the chapter, this includes

Sorkin's version of the classical causal metric hypothesis (CCMH). The physically absurd behavior permitted in this context consists of instantaneous expansion, or instantaneous collapse, of an infinite volume of spacetime, as explained in Section 4.4. Mixing finite and infinite local behavior seems dubious, even if infinite sets are permitted to have finite measure, but is potentially cataclysmic if these sets are assumed to have infinite measure. Topological local finiteness in the interval topology is equally problematic in this regard. As explained in Sections 4.4 and 4.5, some instances of locally infinite behavior may be assigned innocuous interpretations in this context, as occurs in the case of causal sets induced by global sprinklings into Minkowski spacetime $\mathbb{R}^{3+1}$. However, the fact that absurd local behavior is *permitted* under interval finiteness remains a cause for serious concern, especially in the quantum-theoretic context.

7. *Interval finiteness imposes unjustified restrictions on the global structure of classical spacetime.* This statement is made precise by Theorem 4.6.3, which demonstrates that every nonempty, countable, interval finite, acyclic directed set may be exhibited as a relative directed set over the integers. No solid cosmological evidence that I am aware of, either observational or theoretical, justifies this restriction. Topological local finiteness in the interval topology does not suffer from this problem, since it is at least a genuinely local condition. However, the absence of this particular limitation does not mean that topological local finiteness in the interval topology is a suitable local finiteness condition for discrete causal theory; it suffers from too many other issues.

4.8 Nine Arguments in Favor of Star Finiteness

Preliminary remarks. I now briefly review some of the principal arguments in favor of star finiteness (SF) as a local finiteness condition for directed sets, viewed as models of classical spacetime under the classical causal metric hypothesis (CCMH). To be clear, these arguments take for granted the general merits of discrete models, which, for basic physical reasons, require *some* type of local finiteness condition under any version of the hypothesis for which infinite sets are assigned infinite measure. Basic arguments in favor of discreteness itself, such as metric recovery, and the many problems inherent in continuum-based theories, have been covered in detail in previous chapters. The present discussion assumes this background as given, and focuses on establishing the more specific assertion that star finiteness is a "good" local finiteness condition. Most of the considerations involved here have played a role, at least implicitly, in the foregoing analysis.

Nine arguments. The following list of arguments serves both to contrast the limitations of interval finiteness (IF), described in Section 4.7, with some of the corresponding advantages of star finiteness, and to highlight some additional positive attributes of the latter condition. In particular, the first two arguments in favor of star finiteness in the following list correspond to the first two arguments against interval finiteness in the list in Section 4.7. Argument 3 in this section corresponds roughly to arguments 3 and 4 in Section 4.7, and arguments 4 and 5 in this section correspond to arguments 6 and 7 in Section 4.7. Arguments 6 through 9 in this section are not comparative; they describe additional advantages of star finiteness.

1. *Star finiteness models a causally local condition for classical spacetime structure.* Referring to Definition 4.1.1, this means that star finiteness may be checked by examining the independent causes and effects associated with each element in a directed set D, interpreted as a model of classical spacetime under the classical causal metric hypothesis. Under the independence convention (IC), these causes and effects are represented by the relation sets $R(x)$ at each element x in D, whose cardinalities determine whether or not D is star finite. If one chooses not to apply the independence convention, then the numbers of independent causes and effects represented by the relation sets $R(x)$ are generally smaller than their cardinalities; hence, star finiteness still ensures causal local finiteness in this case.

2. *Star finiteness is a local condition in the topological sense.* More precisely, star finiteness coincides with the condition of topological local finiteness for the star topology on the star model $M_\star$ of a multidirected set M, as proven in Lemma 4.4.2.

3. *Star finiteness both captures and isolates the basic cardinality behavior of local multidirected structure.* This means that consideration of the star $\mathrm{St}(x)$ at each element x in a multidirected set M is both necessary and sufficient to determine whether or not M is star finite. This follows immediately from the definition of star finiteness in Section 4.4.1.

4. *Star finiteness is physically compatible with any version of the classical causal metric hypothesis for which infinite sets are assigned infinite measure.* In particular, this includes Sorkin's version of the hypothesis. Star finiteness even allows the volumes assigned to singleton subsets of a directed set to be "bounded away from zero," without danger of fatal local behavior. Indeed, since the set $J_0^-(x) \cup J_0^+(x)$ of direct predecessors and successors of any element x in *any* star finite multidirected set $(M, \prec)$ is finite, this set has finite measure under *any* "generalized measure function" $\mu : \mathcal{P}(M) \rightarrow \mathbb{R} \cup \{\infty\}$, discrete or otherwise.

5. *Star finiteness does not impose unjustified restrictions on the global structure of classical spacetime.* For example, in the acyclic case, star finiteness admits relative directed sets over a wide variety of discrete totally ordered sets, not only totally ordered subsets of the integers $\mathbb{Z}$.

6. *Star finiteness is a natural condition for generally nontransitive multidirected structure under the independence convention.* Under the independence convention (IC), each element r of the relation set $R(x)$ at an element x of a multidirected set M possesses independent information-theoretic significance. In this context, star finiteness is determined precisely by "packets of information directly relevant to each x," as x ranges over M.

7. *Star finite directed sets suffice to realize metric recovery for relativistic spacetime to an arbitrary degree of precision.* In more detail, star finite models can

approximate the local geometry of any pseudo-Riemannian manifold to which the metric recovery theorems apply. This may be accomplished, for example, via local sprinkling constructions.

8. *Star finiteness permits a reasonable balance between local Lorentz invariance and spatiotemporal locality.* As explained in Section 4.5, star finite models may be chosen to exhibit the type of *"approximate Lorentz invariance,"* or *"effective local Lorentz invariance,"* described by the authors of [BHS09]. This may be accomplished without necessitating radical nonlocality of fundamental causal structure in a spatiotemporal sense.

9. *Star finite directed sets are convenient for formulating discrete causal dynamics.* This argument may be better understood after reading Chapters 5 and 6, but the conclusion is not surprising. In general, one expects discrete causal dynamics to involve discrete analogues of partial differential equations, which are much easier to construct and analyze when only a finite number of terms contribute to the local physics.

4.9 Hierarchy of Finiteness Conditions

Six additional finiteness conditions. In this section, I briefly introduce six more finiteness conditions, some local and some nonlocal, and prove two elementary theorems relating these conditions, for acyclic directed sets and multidirected sets, respectively. The corresponding conditions for directed sets may be easily inferred from the second theorem. The principal purpose of this section is to assist the curious or skeptical reader in evaluating how the basic structural implications for discrete causal theory change when one alters the choice of finiteness axiom.

Definition 4.9.1. *Let $M = (M, R, i, t)$ be a multidirected set. The following are alternative finiteness conditions for M:*

1. *M is called* **element finite** (EF), *if M has finite cardinality.*
2. *M is called* **locally element finite** (LEF), *if every element of M has a finite number of direct predecessors and successors.*
3. *M is called* **relation finite** (RF), *if R has finite cardinality, i.e., if M has a finite number of relations.*
4. *M is called* **pairwise relation finite** (PRF), *if every pair of elements in M has a finite number of relations between them.*
5. *M is called* **chain finite** (CF), *if every chain in M has finite length.*
6. *M is called* **antichain finite** (AF), *if every antichain in M has finite cardinality.*

The relationships among these finiteness conditions are spelled out in the statements of Theorems 4.9.2 and 4.9.3 below, but it is convenient to make a few qualitative remarks before stating and proving these theorems. First, some of these conditions acquire independent significance only in the general context of multidirected sets. In particular, local element finiteness coincides with star finiteness in the special case of directed sets, and pairwise relation finiteness follows directly from the definition in this case. Second, the remaining four conditions are all too restrictive to serve as axioms for directed sets, viewed as models of classical spacetime under the classical causal metric hypothesis (CCMH). Element finiteness is too restrictive because it prescribes finite classical histories, and relation finiteness is too restrictive because it prescribes classical histories with finite connected components. Chain finiteness is too restrictive because it prescribes classical histories that are causally bounded above and below, in the sense that every sequence of events has a "first cause" and "final effect." Finally, antichain finiteness is too restrictive because it prescribes classical histories that are "spatially finite" in any given "frame of reference." Each of these prescriptions rules out interesting classes of cosmological models.

Two theorems. The following two theorems describe relationships between pairs of finiteness conditions for acyclic directed sets, and for multidirected sets, respectively. The intent of Theorem 4.9.2 is to compare finiteness conditions for "conservative models" of discrete classical spacetime structure under the classical causal metric hypothesis. Theorem 4.9.3, meanwhile, is intended to apply primarily to the abstract structure of configuration spaces of directed sets, particularly kinematic preschemes and kinematic schemes, which arise in the path summation approach to discrete quantum causal theory. In both theorems, eight different conditions are compared, but not the *same* eight conditions. In particular, pairwise relation finiteness is not considered in Theorem 4.9.2, because it is part of the *definition* of a directed set; also, local element finiteness is not considered, because it coincides with star finiteness in this case. However, the variations on star finiteness introduced in Section 4.4; namely, star finiteness in the transitive closure, and star finiteness in the skeleton, *are* considered. Theorem 4.9.3 omits these last two conditions, due to the choice in Section 3.10 to restrict consideration of the transitive closure functor and the skeleton operation to the context of *directed* sets. However, the latter theorem *does* consider local element finiteness, which differs from star finiteness in the general case, as well as pairwise relation finiteness, which is not part of the definition of a multidirected set.

Note that only *pairwise* relationships, i.e., logical implications, or lack thereof, between *pairs* of conditions, are considered in this section. Theorems 4.9.2 and 4.9.3 do not address subtler relationships, in which a combination of two or more conditions may imply a third.[47] It is impossible to overlook the fact that the diagrams in the statements of these theorems possess the abstract structure of acyclic directed sets, with the arrows indicating logical implication, rather than causal influence, in this context. Of course, the general analogy between logical and causal structure is both interesting and important, but in the present setting, it would not be useful to emphasize this analogy. Hence, these diagrams are *not* drawn in the form of generalized Hasse diagrams.

Theorem 4.9.2. *For acyclic directed sets, the conditions of element finiteness* (EF), *relation finiteness* (RF), *star finiteness in the transitive closure* (SFT), *star finiteness*

[47] A familiar example of such a more-complicated relationship involves the three conditions of *irreflexivity* (IF), *transitivity* (TR), and *acyclicity* (AC), discussed in Sections 3.5 and 3.6. The first two conditions *together* imply the third, but neither does so independently.

(SF), *star finiteness in the skeleton* (SFS), *chain finiteness* (CF), *interval finiteness* (IF), *and antichain finiteness* (AF), *are related as follows, where arrows indicate logical implications:*

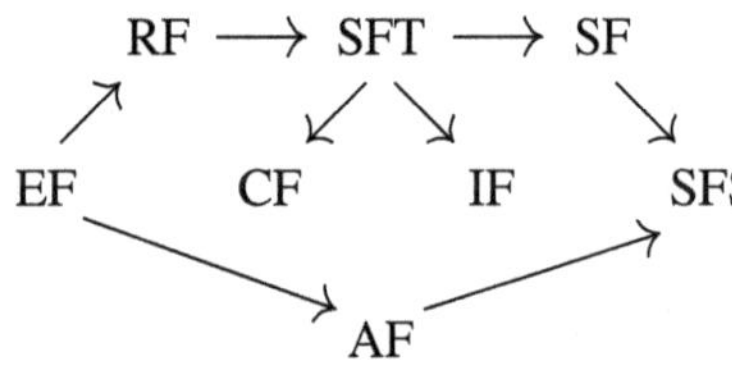

Moreover, these are the only pairwise logical implications among these conditions.

Proof. Let $D = (D, \prec)$ be an acyclic directed set. I break the proof into two parts, first showing that the claimed logical implications are valid, then showing that no other pairwise logical implications exist among these conditions. Some of the statements proven here are more general than what is actually necessary to establish the theorem. The logical implications in the diagram appearing in the statement of the theorem are drawn with ordinary arrows, to avoid cluttering the picture, but the corresponding statements of these implications in the proof use the conventional "implies" symbol $\Rightarrow$.

1. *The claimed logical implications are valid.*

   EF $\Rightarrow$ RF: For *any* directed set, relations may be identified with ordered pairs of elements. Hence, the number of relations is bounded by the square of the number of elements.

   EF $\Rightarrow$ AF: The number of elements in any antichain in *any* multidirected set is bounded by the total number of elements in the set.

   RF $\Rightarrow$ SFT: Suppose that D has n relations. The cardinality of the relation set $R(x)$ at any element x in $\mathrm{tr}(D)$ is bounded by the total number of relations in $\mathrm{tr}(D)$, which is bounded by the total number of positive-length chains in D. Since such chains may be identified with families of relations in the acyclic case, this number is bounded by the cardinality of the power set of the set $\prec$ of relations of D, which is 2^n.

   SFT $\Rightarrow$ SF: Every relation in D corresponds to a relation in $\mathrm{tr}(D)$, so any element with a finite relation set in $\mathrm{tr}(D)$ also has a finite relation set in D.

   SFT $\Rightarrow$ CF: I prove the contrapositive. Suppose that D contains an infinite chain $\gamma = \{... \prec x_0 \prec x_1 \prec ...\}$, and let $x = x_n$ be an element of D lying on γ. Then $\mathrm{tr}(D)$ includes a relation between x and every other element x_k lying on γ, in one direction or the other, and the elements x_k are all distinct by acyclicity. Hence, the relation set $R(x)$ at x in $\mathrm{tr}(D)$ is infinite.

   SFT $\Rightarrow$ IF: I prove the contrapositive. Suppose that D is not interval finite, and let $\langle\langle w, y\rangle\rangle$ be an infinite open interval in D. Every element in this interval is a direct successor of w in $\mathrm{tr}(D)$, so D is not star finite in the transitive closure.

SF ⇒ SFS: Every relation in $\mathrm{sk}(D)$ corresponds to a relation in D, so any element with a finite relation set in D also has a finite relation set in $\mathrm{sk}(D)$.

AF ⇒ SFS: I prove the contrapositive. Suppose that D is not star finite in the skeleton. Then there exists an element in D with either an infinite number of maximal predecessors or an infinite number of minimal successors, which comprise an infinite antichain in D by the definition of irreducibility.

2. *No other pairwise logical implications exist among these conditions.* Note that disproving a particular implication of the form A ⇒ B also disproves every implication of the form A ⇒ C, where C ⇒ B, and every implication of the form C ⇒ B, where A ⇒ C. For example, disproving the implication SFT ⇒ RF automatically disproves the implication SFT ⇒ EF, since EF ⇒ RF, and also automatically disproves the implication SF ⇒ RF, since SFT ⇒ SF.

   RF ⇏ AF: A countably infinite set with no relations is a counterexample.

   SFT ⇏ RF: A countably infinite number of disjoint 1-chains is a counterexample.

   SF ⇏ CF, IF: Jacob's ladder is a counterexample for both conditions.

   SFS ⇏ SF: The transitive closure of Jacob's ladder is a counterexample.

   CF ⇏ IF, SFS: The *mirror bouquet*, with elements x, $\{y_i\}_{i\in\mathbb{Z}}$, and z, and relations $x \prec y_i$ and $y_i \prec z$ for all i, is a counterexample for both conditions.

   IF ⇏ SFS, CF: The disjoint union of the infinite bouquet and $\mathbb{Z}$ is a counterexample for both conditions.

   AF ⇏ SF, CF, IF: Cantor's first transfinite ordinal is a counterexample for all three conditions. □

Theorem 4.9.3. *For multidirected sets, the conditions of element finiteness* (EF), *local element finiteness* (LEF), *relation finiteness* (RF), *pairwise relation finiteness* (PRF), *star finiteness* (SF), *chain finiteness* (CF), *interval finiteness* (IF), *and antichain finiteness* (AF), *are related as follows, where arrows indicate logical implications:*

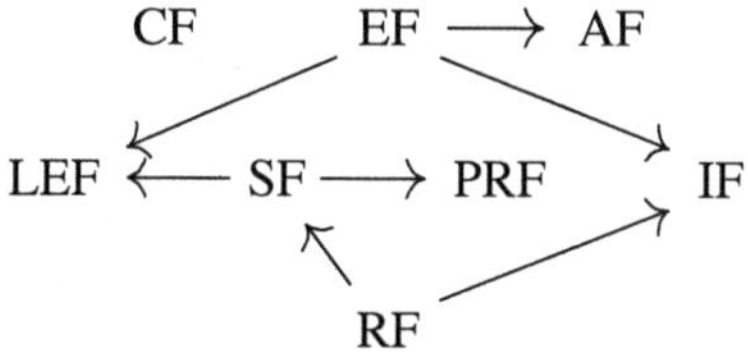

Moreover, these are the only pairwise logical implications among these conditions.

Proof. Let $M = (M, R, i, t)$ be a multidirected set. As in Theorem 4.9.2, I first prove the claimed logical implications, then prove that no other pairwise logical implications exist among these conditions. The remarks about notation also carry over from Theorem 4.9.2.

1. *The claimed logical implications are valid.*

   EF $\Rightarrow$ LEF: The number of direct predecessors and successors of each element in M is bounded by the total number of elements of M.

   EF $\Rightarrow$ IF: The number of elements in any open interval in M is bounded by the total number of elements of M.

   EF $\Rightarrow$ AF: The proof of this implication in Theorem 4.9.2 is for general multidirected sets.

   RF $\Rightarrow$ SF: The number of relations in the relation set $R(x)$ at any element x in M, which determines star finiteness at x, is bounded by the cardinality of the total relation set R of M.

   RF $\Rightarrow$ IF: I prove the contrapositive. Suppose that M possesses an infinite interval $\langle\langle w, y\rangle\rangle$, and let $\{x_i\}_{i\in\mathbb{Z}}$ be an infinite subfamily of $\langle\langle w, y\rangle\rangle$. By the definition of an open interval, there exist chains γ_i in M beginning at w and terminating at x_i for each i. The terminal relations in these chains are all distinct, since they possess distinct terminal elements. Hence, these relations form an infinite subset of the relation set R of M.

   SF $\Rightarrow$ LEF: The number of elements directly related to an element x of M is bounded by the cardinality of the relation set $R(x)$ at x. If M is star finite, then the latter number is finite by definition, so the former number is also finite.

   SF $\Rightarrow$ PRF: The number of relations between two elements x and y in M is bounded by the cardinalities of both of the relation sets $R(x)$ and $R(y)$. If either of the latter numbers is finite, then so is the former.

2. *No other pairwise logical implications exist among these conditions.* All of the counterexamples from Theorem 4.9.2 remain valid, since multidirected sets are more general than acyclic directed sets. The remaining necessary counterexamples are as follows:

   EF $\not\Rightarrow$ PRF, CF: A multidirected set with two elements and an infinite number of relations between them in both directions is a counterexample for both conditions.

   LEF $\not\Rightarrow$ CF, AF: The disjoint union of the previous counterexample and an infinite set of isolated elements is a counterexample for both conditions.

   RF $\not\Rightarrow$ CF: A reflexive relation $x \prec x$ is a counterexample.

   PRF $\not\Rightarrow$ LEF, CF, AF: The disjoint union of the infinite bouquet, $\mathbb{Z}$, and an infinite set of isolated elements is a counterexample for all three conditions.

CF $\nRightarrow$ LEF, PRF: Replacing each relation in the infinite bouquet with a countably infinite number of relations yields a counterexample for both conditions.

IF $\nRightarrow$ LEF: $\mathbb{Z}$ is a counterexample.

AF $\nRightarrow$ LEF: Again, $\mathbb{Z}$ is a counterexample. □

4.10 New Axioms for Discrete Classical Causal Theory

Summary of foregoing axiomatic analysis. After introducing the causal metric hypothesis (CMH) at the beginning of Chapter 2, I noted that the conceptual content of this hypothesis *"must be expressed mathematically before its physical consequences may be explored in a precise quantitative fashion."* The first step in this direction, as described in Section 2.2, was to identify *directed sets* as the general class of mathematical objects most appropriate for modeling classical spacetime structure. Following this identification, however, there remained a need to narrow the picture further, to focus attention on those *special classes of directed sets* most likely to prove physically relevant. To accomplish this in a convincing manner required the development of significant background material, laid out in Chapter 3 and this chapter. For both technical and historical reasons, these chapters include repeated references to causal set theory, which serves as a crucial point of reference for the somewhat different version of discrete causal theory developed in this book. Among the most notable differences between the two approaches are the choices to *abstain from the causal set axioms of transitivity* (TR) *and interval finiteness* (IF). These axioms lead, in my view, to serious limitations in discrete causal theory under any version of the classical causal metric hypothesis (CCMH) for which infinite subsets of a directed set are assigned infinite measure. Section 3.9 analyzed the limitations of transitivity, focusing on the case of acyclic directed sets. The conclusion of this analysis was that *"transitive binary relations are information-theoretically inadequate to model classical spacetime structure in discrete causal theory."* In this book, such relations are viewed as derivative constructs, generated by generally nontransitive binary relations, called *causal relations,* physically interpreted according to the *independence convention* (IC). Section 4.7 examined some of the limitations of interval finiteness as a local finiteness condition for directed sets and multidirected sets. The conclusion of this analysis was that *"interval finiteness is not an optimal mathematical condition for formalizing the physical notion of local finiteness in discrete causal theory."* An attractive alternative, pursued in this book, is to *replace interval finiteness with the axiom of star finiteness* (SF). It is important to reiterate that neither of these choices suggest that there are no useful applications for transitive, interval finite models, or that such models are unworthy of serious study.

Besides transitivity and interval finiteness, causal set theory uses the binary axiom, the measure axiom (M), countability, and irreflexivity (IR), as described in Section

3.5. None of these latter axioms raise concerns comparable to those raised by transitivity and interval finiteness, but modifications are nevertheless worth considering from a broad point of view. Although some of the theoretical background supporting the possibility of such modifications is not examined in detail until Part II of the book, I devote a few words to the subject in the present section. It is convenient here to discuss these axioms in the opposite order to the listing above, considering irreflexivity first, countability second, the measure axiom third, and the binary axiom fourth. In the nontransitive context, irreflexivity no longer rules out causal cycles, and this greatly diminishes its potential role in the version of discrete causal theory developed in this book. Hence, *irreflexivity is not included in the "suggested list" of axioms given below.* This leaves the choice to either permit cycles in directed sets, thereby "taking seriously" the existence of relativistic spacetime manifolds violating the causal condition, or to elevate acyclicity (AC) to the status of an axiom. Many of the results of this book do *not* depend on acyclicity, but some of the most important results of Chapters 6 and 7 are developed and presented only in the acyclic case. For this reason, *I include acyclicity in the "suggested list" of axioms given below, with the understanding that this is a "tentative" and "provisional" choice.* Good practical reasons exist to retain the axiom of countability, at least for most classes of directed sets involving a discrete measure and a reasonable notion of local finiteness. One obvious reason for this is that the prospect of events occurring "an infinite distance away" from the site of any conceivable local observation is of limited practical interest, even if such events might exist hypothetically. Imposing countability in this context reflects, at worst, a determination to examine *"the part of the universe relevant to the observer."* Of course, one could conceive of compelling *structural* arguments that might justify the consideration of such infinitely-distant events. Note that infinite distances do not necessarily arise in the consideration of relative directed sets over infinite bases other than $\mathbb{Z}$; for example, any pair of elements in Jacob's ladder are related by a finite chain. In any case, *countability is included in the "suggested list" of axioms.*

Turning to the measure axiom, I have already discussed the possibility of relaxing this axiom to admit consideration of *local causal structure* in the emergent definition of spacetime volume. In the special case of star finite directed sets, such a modification is not so radical as to annul the basic insight that *"directed structure plus natural scale equals geometry,"* which generalizes Sorkin's version of the classical causal metric hypothesis (CCMH). Hence, *I include a generalized version of the measure axiom in the "suggested list" of axioms.* Finally, the binary axiom admits some interesting generalizations, but proper understanding of these ideas requires some of the results of Chapter 5, which examines the general use of multidirected structure to encode relationships among entities more complicated than individual events. The principal aim of this development is to facilitate the subsequent study of relationships between pairs of classical histories, i.e., *co-relative histories,* appearing in Chapter 6, but an interesting tangential idea is to consider *holistic* relationships among subobjects of a given classical history. One possible framework for developing this concept is the theory of *power spaces,* discussed in Chapter 5. However, these details are not central to the goals or the philosophy of this book. For these reasons, *I retain the binary axiom among the "suggested list" of axioms,* although its wording is slightly altered to apply to directed sets in general.

Suggested list of axioms. The following five axioms for discrete classical causal theory represent, in my judgment, a reasonable balance among the criteria of physical relevance, structural integrity, and generality, in light of the analysis presented in Chapter 3 and this chapter.[48] While Part I of the book, in my view, already provides strong justification for the decision to focus on models satisfying these axioms, results from Part II play a significant role in demonstrating that the chosen axioms actually lead to interesting new methods and applications. These results also assist in evaluating the more "radical" axiomatic alternatives described below. As stated above, the inclusion of acyclicity among the five "suggested" axioms is a conservative, provisional choice, based on the lack of clear physical evidence for the existence of causal cycles, despite their theoretical prominence in general relativity.

Definition 4.10.1. *A new version of discrete classical causal theory may be defined by the following five axioms:*

1. **Binary axiom** (B): *Classical spacetime may be modeled via a directed set* $(D, \prec)$*, whose elements represent spacetime events, and whose relations represent causal relationships between pairs of spacetime events.*
2. **Generalized measure axiom** (M*): D *is equipped with a set function* $\mu : \mathcal{P}(D) \rightarrow \mathbb{R} \cup \{\infty\}$*, viewed as a "generalized measure," which assigns finite positive values to nonempty finite subsets of* D*, and infinite values to infinite subsets of* D*.*
3. **Countability** (C): D *is countable.*
4. **Star finiteness** (SF): *For every element* x *of* D*, the star* $\text{St}(x)$ *of* x *is finite.*
5. **Acyclicity** (AC): $(D, \prec)$ *possesses no cycles.*

The foremost of the existing theories to which these five axioms are intended to provide an alternative is, of course, causal set theory. For clarity, I include a few additional remarks about the particular way in which these axioms are stated. First, since the binary axiom is carried over essentially unchanged from causal set theory, the resulting formulation of discrete classical causal theory is not *irreducibly holistic,* in the sense of Chapter 5. However some of the objects studied in Part II of the book, including relation spaces, path spaces, and other natural directed or multidirected structures whose elements correspond to entities more complex than individual spacetime events, remain crucially important to the theory, even if they do not involve information "distributed nonlocally" in a physical sense. The *generalized measure axiom* (M*) appearing here is very different, and less specific, than the causal set measure axiom (M) appearing in Section 3.5, although it is based on the same general idea that discrete combinatorial data can serve as a proxy for the "missing conformal factor" in the metric recovery theorems, discussed in Section 2.8. The choice to use a generalized measure here is an exercise in caution, since it

---

[48]The hyperlinks for these axioms in the electronic version of the book link to the statements listed in this section.

is conceivable that the properties of a true measure may be inappropriate at the fundamental scale.[49] However, the generalized measure axiom (M*) does not attempt to specify exactly *how* scale data arises from discrete causal structure, as the causal set measure axiom (M) does, and a more specific version must be chosen in any specific incarnation of the theory. The most attractive versions I know of are very similar to the causal set version. In the present generalized setting, μ assigns positive "sizes" to singletons, but these "sizes" are not assumed to be constant, or even constant up to statistical fluctuations. This a priori allows for the "undesirable" possibility of sequences of singleton volumes decreasing to zero. If μ were *countably additive,* as in the case of a true measure, then the requirement that infinite subsets are assigned infinite values would rule out such sequences, because one could always identify subsequences whose corresponding series would converge. However, since much more specific methods of assigning "volumes" must ultimately be employed in any case, I choose not to add any more specifics to the present statement of this axiom.

Conservative alternative: "minimally modified causal set theory." The "suggested list" of axioms for discrete classical causal theory presented above is "already conservative," in the sense that it excludes both causal cycles and classical holism. For comparative purposes, however, it is worthwhile to outline an even more conservative approach, incorporating only those modifications *absolutely necessary* to alleviate the most serious structural concerns regarding classical causal set theory described in Chapter 3 and this chapter.[50] This ultra-conservative alternative permits only *irreducible* relations between pairs of elements, thus obviating the need for the independence convention (IC). The causal set version of the measure axiom (M) is retained, rather than being replaced by a generalized measure axiom, such as (M*). Interval finiteness is also retained, though it is no longer viewed as a local condition. Rather, it serves to narrow the focus to the class of relative acyclic directed sets over $\mathbb{Z}$, which is among the most conservative generalizations of the class of finite directed sets. Star finiteness is *added* to the picture, to rule out pathologies arising from locally infinite behavior. Irreflexivity is replaced by acyclicity, since the former axiom is insufficient to rule out cycles in the nontransitive setting. Using this "minimally modified" list of axioms, many existing methods and results from causal set theory carry over with little or no alteration. New concepts, such as *relation spaces, path spaces, co-relative histories,* and *kinematic schemes,* may then be brought to bear, without drastically altering the viewpoint or information content of the existing corpus of causal set theory.

---

[49]Technically, a *measure* μ on a set D is a set function $\mu : \Sigma \to \mathbb{R} \cup \{\infty\}$ which assigns non-negative values to elements of a *sigma algebra* Σ of subsets of D, and which satisfies the additional properties that $\mu(\emptyset) = 0$, and that $\mu\left(\bigcup_{k=0}^{\infty} S_k\right) = \sum_{k=1}^{\infty} \mu(S_k)$ for any countable family $\{S_k\}_{k=0}^{\infty}$ of pairwise disjoint subsets of D. It is this last property, called *countable additivity,* that seems possibly questionable at the fundamental scale. In this book, I generally take the sigma algebra Σ to be the entire power set $\mathcal{P}(D)$ of D, whether or not μ is actually a measure. However, this choice could be reconsidered, even in the countable case.

[50]The hyperlink for the axiom of irreducibility (IRR) in the electronic version of the book links here, since this axiom does not appear among the previous lists of axioms.

Definition 4.10.2. *A conservative version of discrete classical causal theory may be defined by the following seven axioms:*

1. **Binary axiom**: *Classical spacetime may be modeled via a directed set* $(D, \prec)$, *whose elements represent spacetime events, and whose relations represent causal relationships between pairs of spacetime events.*
2. **Measure axiom**: *D is also equipped with a discrete measure μ, which assigns to each subset of D a volume equal to its number of elements in fundamental units, up to Poisson-type fluctuations.*
3. **Countability**: *C is countable.*
4. **Irreducibility** (IRR): *Every element of the binary relation $\prec$ is irreducible, i.e., admits no reducing chain.*
5. **Interval finiteness**: *For every pair of elements x and z in D, the open interval*

$$\langle\!\langle x, z \rangle\!\rangle := \{y \in D \mid x \prec y \prec z\}$$

has finite cardinality.
6. **Star finiteness**: *For every element x of D, the star* $\mathrm{St}(x)$ *of x is finite.*
7. **Acyclicity**: $(D, \prec)$ *has no cycles.*

The three axioms of countability, interval finiteness, and acyclicity may be replaced by the single axiom that $(D, \prec)$ admits a morphism of finite index into the integers $\mathbb{Z}$.

Radical alternatives. Many, but not all, of the methods and results introduced in this book also apply to more general approaches to discrete classical causal theory, incorporating "conceptually radical" notions, such as causal cycles and classical holism. As mentioned more than once in Part I of the book, the consideration of cycles in directed sets has strong theoretical precedent arising from the rather generic appearance of closed causal curves in general relativity. Perhaps the most reasonable "radical alternative" to the "suggested" list of axioms above would be to simply drop the axiom of acyclicity, thereby reducing the number of axioms to four. Regarding classical holism, probably the simplest versions of this idea involve "families of events influencing other families of events" in an irreducible fashion. This notion admits straightforward formalization in terms of power spaces, as described in Section 5.6. Modified versions of the measure axiom, countability, star finiteness, and acyclicity still make sense in this context. Alternatively, one might prefer to regard "spacetime events," and the elements representing them, as mere proxies for aspects of structure defined in essentially different terms. The relation space viewpoint, developed in Chapter 5, opens the door to this way of thinking, without going far enough to necessitate any actual changes to the axioms. Penrose's *twistor theory,* briefly revisited in Chapter 5, provides a classical continuum-based precedent for this perspective. More generally, Grothendieck's relative viewpoint (RV), Isham's topos-theoretic approach to fundamental physics, and other structural notions from modern mathematics, provide many alternative generalizations of these ideas, besides those arising from discrete causal theory itself. Finally, one may choose to incorporate ideas from *nondiscrete* versions of classical causal theory, such as domain theory.

References

[SO05] Rafael Sorkin. *Causal Sets: Discrete Gravity*. Lectures on Quantum Gravity, edited by Andrés Gomberoff and Donald Marolf. Springer, 2005. arXiv preprint: http://arxiv.org/pdf/gr-qc/0309009v1.pdf.

[FI88] David Finkelstein. *"Superconducting" Causal Nets*. International Journal of Theoretical Physics, **27**, 4, pp. 473–519, 1988.

[RA00] Ioannis Raptis. *Algebraic Quantization of Causal Sets*. International Journal of Theoretical Physics, **39**, 5, pp. 1233–1240, 2000. arXiv preprint: http://arxiv.org/pdf/gr-qc/9906103v1.pdf.

[SO09A] Rafael Sorkin. *Does locality fail at intermediate length scales?* Approaches to Quantum Gravity, edited by Daniele Oriti. Cambridge University Press, 2009.

[BD10] Dionigi M. T. Benincasa and Fay Dowker. *Scalar Curvature of a Causal Set*. Physical Review Letters, **104**, 18, 181301, 2010. arXiv preprint: http://arxiv.org/pdf/1001.2725v4.pdf.

[SU11] Sumati Surya. *Directions in Causal Set Quantum Gravity*. In Recent Research in Quantum Gravity, edited by Arundhati Dasgupta. Nova Science Publishing Incorporated, 2012. arXiv preprint: http://arxiv.org/pdf/1103.6272v1.pdf.

[MP06] Keye Martin and Prakash Panangaden. *A Domain of Spacetime Intervals in General Relativity*. Communications in Mathematical Physics, **267**, 3, pp. 563–586, 2006.

[ER80] Marcel Erné. *Topologies on products of partially ordered sets I: Interval topologies*. Algebra Universalis, **11**, 1, pp. 295–311, 1980.

[BLMS87] Luca Bombelli, Joohan Lee, David Meyer, and Rafael Sorkin. *Space-Time as a Causal Set*. Physical Review Letters, **59**, 5, pp. 521–524, 1987.

[WA84] Robert M. Wald *General Relativity*. University of Chicago Press, 1984.

[HE73] S. W. Hawking and G. F. R. Ellis. *The large scale structure of space-time*. Cambridge Monographs on Mathematical Physics. Cambridge University Press, 1973.

[BPS08] Rick Blute, Prakash Panangaden, and Lutz Straßburger. *The Logic BV and Quantum Causality*. Preprint. http://www.lix.polytechnique.fr/Labo/Lutz.Strassburger/papers/BVlocative.pdf.

[KR75] D. J. Kleitman and B. L. Rothschild. *Asymptotic Enumeration of Partial Orders on a Finite Set*. Transactions of the American Mathematical Society, **205**, pp. 205–220, 1975.

[PE10] Roger Penrose. *Cycles of Time*. Vintage Books, New York, 2010.

[KKSR12] Dmitri Krioukov, Maksim Kitsak, Robert S. Sinkovits, David Rideout, David Meyer, and Marián Boguñá. *Network Cosmology*. Nature Scientific Reports **2**, 793, 2012.

[BHS09] Luca Bombelli, Joe Henson, and Rafael Sorkin. *Discreteness without symmetry breaking: a theorem*. Modern Physics Letters A, **24**, 32, pp. 2579–2587, 2009. arXiv preprint: http://arxiv.org/pdf/gr-qc/0605006v1.pdf.

[AJL05] Jan Ambjørn, J. Jurkiewicz, and R. Loll. *Spectral Dimension of the Universe*. Physical Review Letters, **95**, 171301, 2005. arXiv preprint: https://arxiv.org/pdf/hep-th/0505113v2.pdf.

[ADGS04] Maqbool Ahmed, Scott Dodelson, Patrick B. Greene, and Rafael Sorkin. *Everpresent* Λ. Physical Review D, **69**, 10, 103523, 2004. arXiv preprint: http://arxiv.org/pdf/astro-ph/0209274v1.pdf.

[DHS04] Fay Dowker, Joe Henson, and Rafael Sorkin. *Quantum gravity phenomenology, Lorentz invariance and discreteness*. Modern Physics Letters A, **19**, 24, pp. 1829–1840, 2004. arXiv preprint: http://arxiv.org/pdf/gr-qc/0311055v3.pdf.

[MO88] Christopher Moore. *Comment on "Space-Time as a Causal Set."* Physical Review Letters, **60**, 7, pp. 655, 1988.

[BLMS88] Luca Bombelli, Joohan Lee, David Meyer, and Rafael Sorkin. *Bombelli et al. Reply to Comment on "Space-Time as a Causal Set."* Physical Review Letters, **60**, 7, pp. 656, 1988.

[SR99] David Rideout and Rafael Sorkin. *Classical sequential growth dynamics for causal sets*. Physical Review D, **61**, 2, 024002, 2000. arXiv preprint: http://arxiv.org/pdf/gr-qc/9904062v3.pdf.

[GL14] Lisa Glaser. *A closed form expression for the causal set D'Alembertian.* Classical and Quantum Gravity, **31**, 9, pp. 5007, 2014. arXiv preprint: http://arxiv.org/pdf/1311.1701v3.pdf.

[BE15] Alessio Belenchia. *Causal Set theory, non-locality and phenomenology*. Preprint, 2015. arXiv preprint: http://arxiv.org/pdf/1512.08485v1.pdf.

[FE48] Richard Feynman. *Space-Time Approach to Non-Relativistic Quantum Mechanics.* Reviews of Modern Physics, **20**, 2, pp. 367–387, 1948.

[ER79] Marcel Erné, unpublished lecture notes, 1979.

References

[illegible]

Part II
Discrete Quantum Causal Theory

Chapter 5
Relation Space and Generalizations

Abstract. This chapter begins to develop the machinery necessary for a suitable version of quantum theory in the discrete causal setting. Section 5.1 introduces the *relation space* over a directed set or multidirected set, which is the space whose elements are relations between pairs of elements in the original set. Relation space represents the simplest discrete causal application of Grothendieck's *relative viewpoint,* in which the "objects" are elements, and the "natural relationships" are relations between pairs of elements. Section 5.2 introduces *abstract element space,* which is an "approximate inverse" to relation space. Its primary role is to facilitate the investigation of the properties of relation space. Section 5.3 explains how passage to relation space preserves the information in a directed set or multidirected set, except possibly at the causal boundary. Section 5.4 examines an important technical problem, called the *permeability problem,* which has obstructed "foliation-type" approaches to discrete causal dynamics appearing in the literature. This problem is elegantly solved via passage to relation space, due to the *impermeability theorem,* which states that "maximal antichains in relation space are impermeable." Section 5.5 describes analogies between relation spaces and *morphism categories* in abstract algebra. Section 5.6 introduces *power spaces,* which are generalizations of relation spaces, whose "elements" are nontrivial subobjects of directed or multidirected sets. Power spaces provide a natural context for studying the topic of *classical holism,* which plays a role in novel viewpoints regarding causal structure, such as *top–down causation.* Section 5.7 introduces a technical method called *causal atomic resolution,* which enables systematic "coarsening" of directed or multidirected structure, for the purposes of computation and approximation. Section 5.8 discusses top–down causation and classical holism specifically, along with related ideas involving theories such as *shape dynamics* and *twistor theory.* Section 5.9 introduces *causal path spaces,* which provide the basic structural scaffolding for the *path summation approach* to discrete quantum causal theory, adapted from Feynman's conventional path summation approach. Section 5.10 introduces *causal path algebras,* which augment the structure of the corresponding path spaces in a manner suitable for computation.

B.F. Dribus, *Discrete Causal Theory*, DOI 10.1007/978-3-319-50083-6_5

5.1 Relation Space Over a Multidirected Set

Generalized interpretation of multidirected structure. A simple way to express the distinction between classical and quantum physics, under the discrete causal paradigm, is to say that *"classical physics is about relationships between pairs of events; quantum physics is about relationships between pairs of histories."* This statement represents an important facet of the overall philosophical viewpoint underlying discrete causal theory, as outlined in Section 2.10. Due to the principle of *iteration of structure* (IS), the basic objects of both classical and quantum theory may be expressed in terms of directed or multidirected sets in this setting, and many additional applications of such sets arise in auxiliary roles throughout the theory. The specific physical interpretation of elements of a directed set $D = (D, \prec)$ as representing *spacetime events,* expressed by the binary axiom (B) of Section 4.10, applies only when D itself is interpreted as representing a classical history; other interpretations of such objects are often equally useful. In particular, as expressed in Definition 5.1.1 below, the *relation space* $\mathcal{R}(D)$ over D is *itself* a directed set, whose elements represent relationships between pairs of spacetime events, *not* individual events. At the quantum level, the elements of the *underlying multidirected set* of a *kinematic scheme* represent entire classical histories, and the elements of the corresponding relation space represent *co-relative histories.*

A variety of possible interpretations of directed sets and multidirected sets are examined in this chapter, and applications of these interpretations are explored. The intent of these developments is not to advocate additional axiomatic innovations with respect to existing versions of discrete causal theory, beyond the new set of axioms already suggested in Section 4.10, although possible alterations of the binary axiom to incorporate the unconventional possibility of *classical holism* are briefly outlined in Section 5.6. Instead, the goals of this chapter are to provide an expanded structural perspective, introduce new technical methods, and apply the developments of Chapters 3 and 4 to lay the groundwork for the background independent quantum theory introduced in Chapters 6 and 7. From a mathematical viewpoint, this chapter adds an increasingly *algebraic* flavor to the theory, which is further amplified in subsequent chapters.

Relation space preliminaries. In Section 2.3, I described in elementary terms the distinction between models of physical spacetime, and the "spaces" of various types that arise in purely mathematical settings, which may often be studied using methods partly attributable to physical intuition. Physicists are well-accustomed to working with such abstract "spaces," but the physical context itself introduces a new subtlety, namely, that all the "spaces" under consideration *do* possess actual physical significance of one type or another. In this setting, it is useful to make a finer distinction, separating the class of "spaces" that correspond to physical spacetime merely in the sense that the same methods are useful for studying both, from the class of "spaces" that represent *the same or very similar information* as physical spacetime. The first class includes Hilbert spaces in conventional quantum theory, multiparticle configuration spaces and phase spaces in mechanics and thermodynamics, measure spaces in physical applications of probability, and the various Lie groups, bundles, and sheaves arising in continuum-based field theories. The second class includes frequency spaces in electronics, momentum spaces in classical and quantum mechanics, twistor spaces in Penrose's twistor theory, and the conformal boundary in the AdS/CFT correspondence in string theory.

The *relation space* $\mathcal{R}(M)$ over a multidirected set $M = (M, R, i, t)$, formally introduced in Definition 5.1.1 below, falls into the second class, since classical histories in discrete causal theory "may be recovered from their relation spaces, up to information on the causal boundary." More generally and precisely, Theorem 5.3.1 demonstrates that *any pair of multidirected sets with isomorphic relation spaces have isomorphic interiors.* Hence, $\mathcal{R}(M)$ is "approximately equivalent" to M in an information-theoretic sense; however, it provides a superior viewpoint in several important ways. For example, $\mathcal{R}(M)$ is merely a directed set, rather than a multidirected set, so *passage to relation space reduces multidirected structure to directed structure.* Further, $\mathcal{R}(M)$ is immune to the *permeability problem* for maximal antichains in typical multidirected sets, introduced in Section 2.7. This property provides a technical advantage of immense importance in discrete causal dynamics. As an aside, I remark that spaces belonging to the second class mentioned above may often be associated in a precise manner with "ordinary space or spacetime" via an appropriate *duality* theory, such as *Pontryagin duality, Tannaka–Klein duality,* or *Maldacena duality.* The rudiments of such a duality theory for multidirected sets and their relation spaces may be discerned in Definition 5.2.1 and Theorem 5.3.1, but this viewpoint is not emphasized in this book. Indeed, this is only one of many important algebraic aspects of discrete causal theory that remain largely unexplored, at least in the context of fundamental spacetime structure.

The theory of relation space represents a comprehensive application of Grothendieck's relative viewpoint (RV) to the *internal* structure of a multidirected set $M = (M, R, i, t)$, where the "objects" under consideration are the elements of M, and the "natural relationships" are the elements of its relation set R. In the special case of a directed set, viewed as a model of classical spacetime under the classical causal metric hypothesis (CCMH), these "objects" and "relationships" represent spacetime events and independent causal relationships between pairs of events, respectively. For a multidirected set arising from the abstract structure of a kinematic scheme, "objects" are directed sets, viewed as classical histories, and "relationships" are co-relative histories. Ultimately, the quantum dynamics encoded in the *causal Schrödinger-type equation* (1.1.2) is described in terms of the *"relation space over a kinematic scheme."* This freedom to juggle interpretations among elements, relations, classical histories, co-relative histories, and other analogous objects arising in discrete causal theory is a crucial conceptual device, of which the theory of relation space provides the prototypical embodiment.

Definition of relation space. The relation space $\mathcal{R}(M)$ over a multidirected set $M = (M, R, i, t)$ is a *directed* set whose underlying set of elements is the set R of relations in M. This is true even if M itself is *genuinely multidirected;* i.e., the structure of $\mathcal{R}(M)$ may always be completely described in terms of a binary relation on the set R, rather than the a priori more general structure of initial and terminal element maps necessary to describe a typical multidirected set. I usually denote this binary relation, called the *induced relation,* by the same symbol $\prec$ used throughout the book to denote the causal relation on a directed set. The induced relation $\prec$ on $\mathcal{R}(M)$ is defined by taking one relation r_0 in R to precede another relation r_1 in R if and only if the terminal element $t(r_0)$ of r_0 in M coincides with the initial element $i(r_1)$ of r_1 in M. The fact that this description involves only *ordered pairs* of elements of R automatically implies the assertion that $\mathcal{R}(M)$ is merely a directed set. This is the precise meaning of the above statement that *"passage to relation space reduces multidirected structure to directed structure."* The following definition makes these ideas precise:

Definition 5.1.1. *Let $M = (M, R, i, t)$ be a multidirected set, and let r_0 and r_1 be elements of its relation set R.*

1. *The* **induced relation** $\prec$ *on R is the binary relation defined by setting $r_0 \prec r_1$ if and only if*[1] $t(r_0) = i(r_1)$.
2. *The directed set $\mathcal{R}(M) = (R, \prec)$ is called the* **relation space over** M.

In the parlance of graph theory, the directed set $\mathcal{R}(M) = (R, \prec)$ is called the *line digraph* of M, where M is viewed as a small directed multigraph. Here, the word "line" in the phrase "line digraph" means the same thing as "relation," and the word "digraph" means the same thing as "directed set." The first explicit appearance of line digraphs in the literature seems to be in Frank Harary and Robert Norman's 1960 paper *Some Properties of Line Digraphs* [HN60]. As far as I know, these objects have received relatively little attention in the context of fundamental spacetime structure, although they do appear somewhat routinely in recent articles on physically relevant topics, such as the *control theory of network dynamics* [NV12]. However, Markopoulou does discuss a special case of relation space, referred to as the *"edge set of a planar trivalent graph,"* in her inaugural paper on quantum causal histories [MA00]. More recently, in their paper *The Logic BV and Quantum Causality,* Blute, Panangaden, and Straßburger [BPS08] define the *"precedence relation on the edge set a pre-causal graph,"* which is essentially the same thing as the induced relation on relation space in Definition 5.1.1.

The construction of the relation space over a very small directed set $D = (D, \prec)$ was illustrated informally in Figure 1.5.2 of Chapter 1; an adaptation of this figure is reproduced here as Figure 5.1.2. Note, in particular, that the element y is both the terminal element of the relation r_0, and the initial element of the relation r_1.

[1] Anticipating the definition of *abstract element space* in Section 5.2, one might wonder whether or not the "if and only if" statement here is justified, i.e., if there might not exist additional "non-obvious" induced relations involving sequences of elements of R. The reason for the appearance of such "non-obvious" relations in the context of abstract element space is because an *equivalence relation* is involved in the definition. The "obvious" relations used to generate this equivalence relation are supplemented by additional relations implied by the transitive property, which is part of the definition of an equivalence relation. In the present context, one may freely stipulate that *only* the specified relations are involved, because no other properties are included in the definition.

Hence, there is an induced relation $r_0 \prec r_1$ in the relation space $\mathcal{R}(D)$. The only difference between this description and the description given in Chapter 1 is that the objects involved are now referred to as "directed sets," rather "classical histories," since the former term is neutral, while the latter term carries a particular physical interpretation.

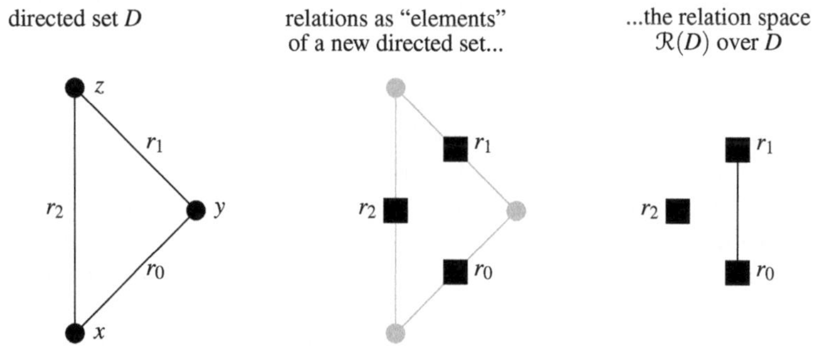

Figure 5.1.2. Construction of the relation space $\mathcal{R}(D)$ over a small directed set $D = (D, \prec)$.

Figure 5.1.3 illustrates the construction of the relation space over a more complicated directed set, which I also denote by D. The left-hand diagram in the figure highlights a particular pair of relations r_0 and r_1 in D, with an induced relation between them. The right-hand diagram shows the global structure of $\mathcal{R}(D)$ as a directed set. In this particular example, D is an *acyclic* directed set, so $\mathcal{R}(D)$ is an *irreducible* acyclic directed set, as proven in Theorem 5.1.4. Both Figures 5.1.2 and 5.1.3 follow the convention of representing elements of relation space by *square nodes,* while elements of the original "element space" are represented by circular nodes, in the usual manner. This is the default convention used throughout the remainder of the book.

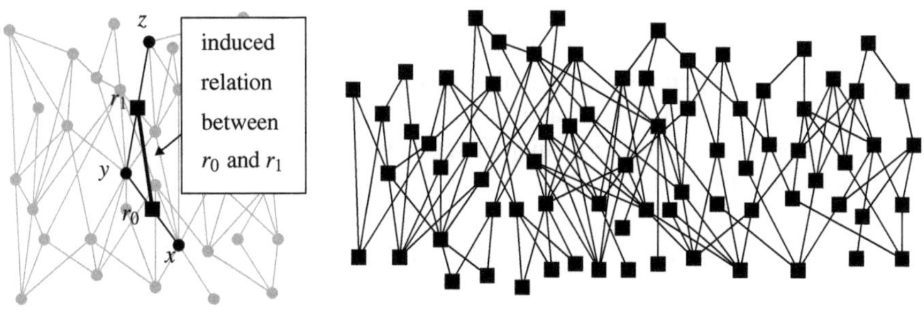

Figure 5.1.3. Induced relation between a pair of relations r_0 and r_1 in a directed set $D = (D \prec)$; global view of $\mathcal{R}(D)$.

Basic properties of relation space. Theorem 5.1.4 establishes the basic properties of the relation space $\mathcal{R}(M)$ over an arbitrary multidirected set $M = (M, R, i, t)$. It is convenient here to use the category-theoretic description that *"passage to relation space defines a functor from the category of multidirected sets to the category of directed sets."* This description "bundles together" in a single statement many of the properties established in the proof of the theorem. In Section 3.10, I mentioned that the transitive closure operation is a functor on the category of directed sets, but did not explain the meaning of this statement in any detail. Hence, I briefly outline the notion here. Informally, a functor is a "morphism-like entity at a higher level of algebraic hierarchy" than an ordinary morphism. It represents a *relationship between a pair of categories,* analogous to the relationship between a pair of objects represented by a morphism in a fixed category. Just as morphisms are generally defined so as to "preserve the structure" of the objects they relate, so functors are defined so as to "preserve the structure" of the categories they relate. At a technical level, this means that a functor *"converts each morphism in its source category to a morphism in its target category."* Establishing this property for relation space is the first item in the proof of Theorem 5.1.4. Further, this "conversion of morphisms" by a functor is required to preserve the properties of identity morphisms and compositions of morphisms. This is the second item in the proof of the theorem. The remaining items in the proof demonstrate that passage to relation space is a *very special type of functor,* which also preserves a number of other physically significant properties, such as star finiteness.

Theorem 5.1.4. *Passage to relation space defines a functor $\mathcal{R}$ from the category $\mathcal{M}$ of multidirected sets to the category $\mathcal{D}$ of directed sets. This functor sends acyclic multidirected sets to irreducible acyclic directed sets, and preserves star finiteness. Moreover, a cycle of length n in a multidirected set induces a cycle of length n in its relation space.*

Proof. Let $M = (M, R, i, t)$ be a multidirected set, and let $\mathcal{R}(M) = (R, \prec)$ be its relation space. The proof proceeds in five steps:

1. *A morphism $f : M \rightarrow M'$ in $\mathcal{M}$ induces a morphism $\mathcal{R}(f) : \mathcal{R}(M) \rightarrow \mathcal{R}(M')$ in $\mathcal{D}$.* Let $M' = (M', R', i', t')$ be a second multidirected set, and let $\mathcal{R}(M') = (R', \prec')$ be its relation space. Recall that a morphism $f : M \rightarrow M'$ consists of a *pair of maps* $f_{\text{ELT}} : M \rightarrow M'$ and $f_{\text{REL}} : R \rightarrow R'$, respecting i, t, i' and t'. Define $\mathcal{R}(f)$ to be the map f_{REL}, as a map of sets, and let $r_0 \prec r_1$ be a relation in $\mathcal{R}(M)$. Then by definition, $t(r_0) = i(r_1)$. Since f is a morphism,
$$t'(f_{\text{REL}}(r_0)) = f_{\text{ELT}}(t(r_0)) = f_{\text{ELT}}(i(r_1)) = i'(f_{\text{REL}}(r_1)).$$
Hence, $f_{\text{REL}}(r_0) \prec' f_{\text{REL}}(r_1)$ in $\mathcal{R}(M')$, so $\mathcal{R}(f)$ is a morphism in $\mathcal{D}$.
2. *$\mathcal{R}$ is a functor.* It remains to show that $\mathcal{R}$ preserves identity morphisms and compositions of morphisms. The identity morphism $\text{Id}_{(M,R,i,t)}$ consists of the identity maps Id_M and Id_R on the *sets* M and R, respectively, the latter of which coincides with the induced morphism $\mathcal{R}(\text{Id}_{(M,R,i,t)})$ on the relation space $\mathcal{R}(M)$,

by definition. Hence, $\mathcal{R}$ preserves identities. Similarly, given two morphisms $f : (M, R, i, t) \to (M', R', i', t')$ and $g : (M', R', i', t') \to (M'', R'', i'', t'')$, the map $(g \circ f)_{\text{REL}}$ is by definition equal to the composition $g_{\text{REL}} \circ f_{\text{REL}}$, so $\mathcal{R}$ preserves compositions.

3. *If M is acyclic, then so is $\mathcal{R}(M)$.* I prove the contrapositive. Suppose that $r_0 \prec r_1 \prec \cdots \prec r_n \prec r_0$ is a cycle in $\mathcal{R}(M)$. Then $t(r_j) = i(r_{j+1})$ for $0 \leq j < n$, and $t(r_n) = i(r_0)$, by the definition of $\mathcal{R}(M)$. Therefore $i(r_0) \prec i(r_1) \prec \cdots \prec i(r_n) \prec i(r_0)$ is a cycle in M.

4. *If M is acyclic, then $\mathcal{R}(M)$ is irreducible.* Again, I prove the contrapositive. Suppose that $r_0 \prec r_1 \prec \cdots \prec r_n$ is a reducing chain for a relation $r_0 \prec r_n$ in $\mathcal{R}(M)$. Then $t(r_0)$ coincides with both $i(r_1)$ and $i(r_n)$. By definition, the reducing chain has length at least 2. If $n = 2$, then $t(r_1)$ coincides with $i(r_n)$, so r_1 is a reflexive relation. If $n > 2$, then $t(r_1)$ precedes $i(r_n)$, so $i(r_1) \prec t(r_1) \prec \cdots \prec i(r_n) = i(r_1)$ is a cycle in M.

5. *If M is star finite, then so is $\mathcal{R}(M)$.* I prove the contrapositive. Suppose that $\mathcal{R}(M)$ is not star finite, and let r be an element of $\mathcal{R}(M)$ with infinite relation set. Since $\mathcal{R}(M)$ is a directed set, either the set of direct predecessors of r, or the set of direct successors of r, is infinite. Assume the former condition, and let $\{r_j\}_{j \in \mathbb{Z}}$ be an infinite family of direct predecessors of r. By the definition of $\mathcal{R}(M)$, the initial element $i(r)$ of r coincides with each terminal element $t(r_j)$ for all j. Therefore, the relation set $R(i(r))$ in M is infinite, so M is not star finite. A symmetric argument applies if the set of direct successors of r is infinite.

6. *A cycle of length n in M induces a cycle of length n in $\mathcal{R}(M)$.* Let $x_0 \prec x_1 \prec \cdots \prec x_n = x_0$ be a cycle of length n in M, where the relation $x_j \prec x_{j+1}$ is labeled as r_j.[2] Then $r_0 \prec r_1 \prec \cdots \prec r_{n-1} \prec r_0$ is a cycle of length n in $\mathcal{R}(M)$, by the definition of relation space. □

In the special case where M is an acyclic directed set, its relation space actually satisfies a stronger condition than irreducibility. For example, no acyclic directed set with multiple 2-chains between a given pair of elements is the relation space of an acyclic directed set $D = (D, \prec)$, since the existence of such chains implies multidirected structure on D. Indeed, if $r_0 \prec r_1 \prec r_2$ and $r_0 \prec \overline{r_1} \prec r_2$ are two 2-chains in the relation space $\mathcal{R}(M)$ over a multidirected set M, then r_1 and $\overline{r_1}$ must share the same initial element; namely, the terminal element of r_0, and must also share the same terminal element; namely, the initial element of r_2. Hence, M is genuinely multidirected, not merely directed. Relation space also satisfies "non-obvious" properties in the case where M contains cycles. For example, any pair of

[2] Of course, there are generally other relations between these two elements, since M is multidirected, but r_j is the relation belonging to the cycle under consideration.

cycles in $\mathcal{R}(M)$ connected by a relation are "subcycles" of a larger cycle in $\mathcal{R}(M)$ joined together by a chain "in the opposite direction." To be more precise, suppose that r and s are elements of $\mathcal{R}(M)$ such that $r \prec s$. Let $\gamma = \{r \prec r_0 \prec \cdots \prec r_m \prec r\}$ be a chain in $\mathcal{R}(M)$ beginning and terminating at r, and let $\delta = \{s \prec s_0 \prec \cdots \prec s_n \prec s\}$ be a chain in $\mathcal{R}(M)$ beginning and terminating at s. Then by the definition of relation space, $t(s_n) = i(s) = t(r) = i(r_0)$, so $s_n \prec r_0$ in $\mathcal{R}(M)$, and there is a chain $s \prec s_0 \prec \cdots \prec s_n \prec r_0 \prec \cdots \prec r_m \prec r$ in $\mathcal{R}(M)$. Harary and Norman [HN60] give a precise criterion identifying when a digraph is the line digraph of a multidigraph, but this result is not needed for the purposes of this book.

The relation space functor $\mathcal{R}$ is particularly well-suited to the discrete context. The fact that $\mathcal{R}$ preserves star finiteness, as shown in the proof of Theorem 5.1.4, is one indication of this. It is interesting to observe how strongly this contrasts with the behavior of $\mathcal{R}$ in the interpolative setting. For example, the relation space $\mathcal{R}(\mathbb{R})$ over the real line $\mathbb{R}$, equipped with the familiar "less than" relation $<$, is not only non-interpolative, but irreducible! To see this, observe that $\mathcal{R}(\mathbb{R})$ may be identified, as a set, with the half-plane above the diagonal $\{(x,x) \mid x \in \mathbb{R}\}$ in $\mathbb{R}^2$. Under the induced relation on $\mathcal{R}(\mathbb{R})$, each element (x, y) precedes precisely those elements of the form (y, z), and these relations are irreducible. Similar considerations apply to Minkowski spacetime $\mathbb{R}^{3+1}$; every relation directly precedes a "solid future timelike/null cone" of relations in a section of $\mathbb{R}^{3+1} \times \mathbb{R}^{3+1}$, and these relations are irreducible. It is worth noting that one may recognize *timelike* and *null* subspaces of $\mathcal{R}(\mathbb{R}^{3+1})$, whose elements are relations between timelike and null-separated elements of $\mathbb{R}^{3+1}$, respectively. Similar considerations apply to $\mathcal{R}(X)$ for any pseudo-Riemannian manifold X of Lorentz signature. In the special case of $\mathbb{R}^{3+1}$, one may define an interesting quotient space of the null subspace of $\mathcal{R}(\mathbb{R}^{3+1})$, by identifying relations between elements lying on common null lines in $\mathbb{R}^{3+1}$. This quotient space corresponds bijectively to Penrose's *twistor space* T over $\mathbb{R}^{3+1}$.[3]

5.2 Abstract Element Space

"Inverting" the relation space functor. I turn now to the task of explaining the precise sense in which the relation space $\mathcal{R}(M)$ over a multidirected set $M = (M, R, i, t)$ "represents the same information as M, except on the boundary ∂M of M." A conceptually useful way to approach this topic is by constructing an "approximate inverse" to the relation space functor $\mathcal{R}$, called the *abstract element space functor*, denoted by $\mathcal{E}$.[4] Heuristically, one imagines that the elements of some directed set $R = (R, \prec)$ are "actually relations," then constructs, in a canonical manner, a multidirected set $\mathcal{E}(R) = (M, R, i, t)$, called the *abstract element space over* R, whose relation space

[3]In this context, *twistor space* may be identified with a certain five-dimensional real manifold. See the footnote in Section 5.8 regarding the various conflicting uses of the term "twistor space."

[4]In an unfortunate coincidence of notation, Markopoulou [MA00] uses $\mathcal{E}C$ to denote the "edge set over a trivalent planar causal set C," which is essentially the *relation space* $\mathcal{R}(C)$ over C!

is "as close as possible" to being isomorphic to $(R, \prec)$. In the case where R actually is the relation space over a "preexisting" multidirected set, R and $\mathcal{RE}(R)$ really are isomorphic, as demonstrated below. More generally, however, the functor $\mathcal{E}$ may be applied to *any* directed set; in this context, the letter "R" is used only for suggestive purposes, since not every directed set is the relation space over a multidirected set, as proven by Harary and Norman [HN60]. Hence, *$\mathcal{E}$ is not quite a true inverse to* $\mathcal{R}$, and it is important to analyze in what manner, and to what extent, the inverse property of $\mathcal{E}$ fails. Informally, the conclusion of this analysis is that "everything that can go wrong is confined to the causal boundary." More precisely, as demonstrated in Theorem 5.3.1 below, $\mathcal{E}(R)$ possesses the "approximate inverse property" that *any multidirected set whose relation space is isomorphic to R has interior isomorphic to* $\mathrm{Int}(\mathcal{E}(R))$. Abstract element space is constructed via the following procedure:

Definition 5.2.1. *Let $R = (R, \prec)$ be a directed set. Construct a multidirected set $\mathcal{E}(R) = (M, R, i, t)$ as follows:*

1. *Define abstract sets $M^- := \{x_r \mid r \in R\}$ and $M^+ := \{y_r \mid r \in R\}$.*
2. *Form the quotient set $M := M^- \bigsqcup M^+/(y_{r_0} \sim x_{r_1}$ if $r_0 \prec r_1$ in $R)$. For each r in R, denote the equivalence classes of x_r and y_r by $\widetilde{x}_r$ and $\widetilde{y}_r$, respectively.*
3. *Define maps i and t from R to M by setting $i(r) = \widetilde{x}_r$ and $t(r) = \widetilde{y}_r$ for each r in R.*

The multidirected set $\mathcal{E}(R) = (M, R, i, t)$ is called the **abstract element space** *over R.*

It is important to point out that the statement "$y_{r_0} \sim x_{r_1}$ if $r_0 \prec r_1$ in R," appearing in part 2 of Definition 5.2.1, is *not* an "if and only if" statement for an arbitrary directed set $R = (R, \prec)$, because $\sim$ is required to be an *equivalence relation.* The transitive property of equivalence relations therefore generates "non-obvious" equivalences between certain pairs of elements. However, if R is *itself* the relation space over a "preexisting multidirected set," then it *will* be the case that $r_0 \prec r_1$ whenever $y_{r_0} \sim x_{r_1}$ in $\mathcal{E}(R)$. To see why this is true, note that the equivalence $y_{r_0} \sim x_{r_1}$ *does* imply the existence of a sequence of "back and forth" relations $r_0 \prec s_0 \succ s_1 \prec s_2 \succ \dots \succ s_n \prec r_1$ in R, which may be visualized as a "sawtooth" subobject of R. Now if R is the relation space over a "preexisting" multidirected set $M = (M, R, i, t)$, then by Definition 5.1.1, $t(r_0) = i(s_0) = t(s_1) = i(s_2) = \dots = t(s_n) = i(r_1)$ in M, which telescopes to $t(r_0) = i(r_1)$, and this *is* equivalent to the statement that $r_0 \prec r_1$ in R, again by the definition of relation space. To see that the "only if" direction of the statement generally fails if R is *not* a relation space, consider the directed set R appearing in the left-hand diagram in Figure 5.3.2. Here, $y_u \sim x_u$ in $\mathcal{E}(R)$, even though $u \not\prec u$ in R. It is now easy to spell out why the relation space $\mathcal{RE}(R)$ really is isomorphic to $R = (R, \prec)$ in the case where R is itself a relation space. First, both spaces share the same underlying set R by definition. Hence, the obvious candidate for an isomorphism between them is the identity set map Id_R. As noted above, $r_0 \prec r_1$ in $\mathcal{RE}(R)$ if and only if $t(r_0) = i(r_1)$ in $\mathcal{E}(R)$, by the definition of relation space. By the definition of abstract element space, $t(r_0) = \widetilde{y_{r_0}}$ and $i(r_1) = \widetilde{x_{r_1}}$, so this may be rewritten as $\widetilde{y_{r_0}} = \widetilde{x_{r_1}}$. If R is a relation space, then this is true if and only if $r_0 \prec r_1$ in R, which yields the desired result.

Figure 5.2.2 illustrates the construction of the abstract element space $\mathcal{E}(R)$ over a directed set R. The elements of R are represented by circular nodes in the left-hand diagram in the figure, to emphasize that R need not be a relation space, although applying the relation space functor $\mathcal{R}$ to $\mathcal{E}(R)$ actually *does* recover R in this particular example. The same elements of R are represented by directed edges in the middle diagram, where they are "pulled apart and considered individually." In the right-hand diagram, these directed edges are "stitched together" as relations in $\mathcal{E}(R)$, using the equivalence relation from part 2 of Definition 5.2.1. The elements of $M^{\pm}$ and $\mathcal{E}(R)$ are represented by open circular nodes.

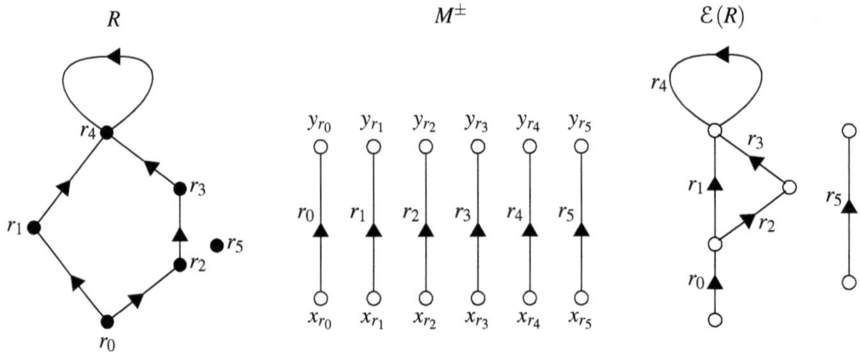

Figure 5.2.2. Directed set $R = (R, \prec)$; abstract sets $M^{\pm}$ (*open nodes*), where *arrows* are added in anticipation of relations in $\mathcal{E}(R)$; abstract element space $\mathcal{E}(R)$ over R. Note that appropriate pairs of elements of $M^{\pm}$ have been identified according to Definition 5.2.1.

As mentioned above, the directed set $R = (R, \prec)$ appearing in Definition 5.2.1 often arises, in practice, by application of the relation space functor $\mathcal{R}$ to some "pre-existing" multidirected set $M = (M, R, i, t)$. In such cases, there are *two different notions* of initial and terminal elements for an element r of R. The images of r under the initial and terminal element maps i and t of the preexisting multidirected set M are called the **concrete initial and terminal elements** of r, while the equivalence classes $\widetilde{x_r}$ and $\widetilde{y_r}$, arising from the construction in Definition 5.2.1, are called the **abstract initial and terminal elements** of r. Theorem 5.3.1 establishes that *concrete interior elements correspond bijectively to abstract interior elements,* while *concrete boundary elements correspond to families of abstract boundary elements.* Hence, any multidirected set with relation space isomorphic to R may be obtained from $\mathcal{E}(R)$, up to isomorphism, by identifying appropriate families of abstract boundary elements. To clarify the use of terminology, I remark that the phrase "element space viewpoint" *always* refers to the concrete context, rather than the abstract context, unless stated otherwise. In other words, it refers to a multidirected set M "before applying $\mathcal{R}$," *not* to a multidirected set $\mathcal{E}(R)$ "obtained by applying $\mathcal{E}$."

Examples involving familiar number systems. It is interesting to note that the abstract element space over the real line $\mathbb{R}$, equipped with the familiar "less than" relation $<$, is a multidirected set consisting of $2^{\aleph_0}$ reflexive relations at a singleton. In particular, the total order on $\mathbb{R}$ is completely lost after applying $\mathcal{E}$. To see why this is true, note that for any pair of real numbers r_0 and r_1, one may choose a third real number u such that $r_0 < u$ and $r_1 < u$. The letter u is chosen here to suggest "upper bound." It then follows by the definition of $\mathcal{E}(\mathbb{R})$ that the abstract initial element of u coincides with the abstract terminal elements of both r_0 and r_1. Hence, *all* terminal elements in $\mathcal{E}(\mathbb{R})$ coincide. Similarly, all initial elements in $\mathcal{E}(\mathbb{R})$ coincide, by applying a symmetric argument to a "lower bound" ℓ for each pair r_0 and r_1. An analogous argument applies to Minkowski spacetime $\mathbb{R}^{3+1}$. More generally, an analogous argument applies to any directed set $(D, \prec)$ satisfying the property that each pair of elements r_0 and r_1 possesses corresponding "upper and lower bounds" u and ℓ such that $\ell \prec r_0 \prec u$ and $\ell \prec r_1 \prec u$. The abstract element spaces over such sets depend only on their cardinalities.[5] It is also instructive to examine the interaction of $\mathcal{E}$ with the transitive closure functor and the skeleton operation. The set of integers $\mathbb{Z}$ and the set of rational numbers $\mathbb{Q}$, each equipped with the familiar "less than" relation $<$, provide useful examples. Recall that $\text{tr}(\mathbb{Z}) = \mathbb{Z}$ and $\text{tr}(\mathbb{Q}) = \mathbb{Q}$, since $<$ is transitive, but $\text{sk}(\mathbb{Z})$ is an irreducible unbounded chain, while $\text{sk}(\mathbb{Q})$ is just a set of cardinality $\aleph_0$, with no relations. The abstract element spaces $\mathcal{E}(\mathbb{Z})$ and $\mathcal{E}(\mathbb{Q})$ both consist of $\aleph_0$ reflexive relations at a singleton, but $\mathcal{E}\big(\text{sk}(\mathbb{Z})\big) \cong \text{sk}(\mathbb{Z})$, while $\mathcal{E}\big(\text{sk}(\mathbb{Q})\big)$ consists of $\aleph_0$ disjoint 1-chains. These examples provide another illustration of the structural clutter arising from the use of the axiom of transitivity (TR) in modeling discrete spacetime structure.

Basic properties of abstract element space. Theorem 5.2.3 establishes some of the basic properties of abstract element space. Included in the statement of the theorem are certain *shortcomings* of $\mathcal{E}$; i.e., the theorem also specifies some important properties *not* preserved by $\mathcal{E}$. This helps to distinguish $\mathcal{E}$ from the "better-behaved" relation space functor $\mathcal{R}$.

Theorem 5.2.3. *Passage to abstract element space defines a functor $\mathcal{E}$ from the category $\mathcal{D}$ of directed sets to the category $\mathcal{M}$ of multidirected sets. This functor fails to preserve either star finiteness or acyclicity, even in the irreducible case.*

Proof. Let $R = (R, \prec)$ and $R' = (R', \prec')$ be directed sets, with abstract element spaces $\mathcal{E}(R) = (M, R, i, t)$ and $\mathcal{E}(R') = (M', R', i', t')$, respectively. Let $f : R \to R'$ be a morphism. Recall that for elements r and r' in R and R', respectively,

$$i(r) = \widetilde{x_r}, \qquad t(r) = \widetilde{y_r}, \qquad i'(r') = \widetilde{x_{r'}}, \qquad \text{and} \qquad t'(r') = \widetilde{y_{r'}},$$

by part 3 of Definition 5.2.1. The induced morphism $\mathcal{E}(f)$ consists of a map of elements $f_{\text{ELT}} : M \to M'$, and a map of relations $f_{\text{REL}} : R \to R'$. Define f_{ELT} by setting $f_{\text{ELT}}(\widetilde{x_r}) = \widetilde{x_{f(r)}}$ and $f_{\text{ELT}}(\widetilde{y_r}) = \widetilde{y_{f(r)}}$. Define f_{REL} to coincide with the given morphism $f : R \to R'$. The proof proceeds in four steps:

---

[5]I thank Johnny Feng for pointing out this obvious fact to me. Such sets are generalized versions of sets that are both "upward directed" and "downward directed," under the conventional terminology.

1. *The map of elements* $f_{\mathrm{ELT}} : M \to M'$ *is well-defined.* A given element $z \in M$ may be the abstract initial element of many different relations in R, and the abstract terminal element of many others. It is necessary to show that the images of these elements under f_{ELT} all coincide in M'. For example, suppose that z is the abstract initial element of two different relations r_0 and r_1 in R; i.e., $z = \widetilde{x_{r_0}} = \widetilde{x_{r_1}}$, and consider the corresponding images $\widetilde{x_{f(r_0)}}$ and $\widetilde{x_{f(r_1)}}$ in M' under f_{ELT}. In this case, there is a sequence of "back-and-forth" relations

$$r_0 \succ s_0 \prec s_1 \succ s_2 \prec \cdots \succ s_n \prec r_1$$

in R, which generates the equivalence $x_{r_0} \sim x_{r_1}$ under transitivity of the equivalence relation $\sim$ in part 2 of Definition 5.2.1. Since $f : R \to R'$ is a morphism, this sequence induces a sequence

$$f(r_0) \succ f(s_0) \prec f(s_1) \succ f(s_2) \prec \cdots \succ f(s_n) \prec f(r_1)$$

in R'. Invoking part 2 of Definition 5.2.1 once more, there is a sequence of equivalences

$$x_{f(r_0)} \sim y_{f(s_0)} \sim x_{f(s_1)} \sim y_{f(s_2)} \sim \cdots \sim y_{f(s_n)} \sim x_{f(r_1)},$$

so that $\widetilde{x_{f(r_0)}} = \widetilde{x_{f(r_1)}}$ in M'. A symmetric argument applies if $z = \widetilde{y_{r_0}} = \widetilde{y_{r_1}}$, and an analogous argument applies if $z = \widetilde{y_{r_0}} = \widetilde{x_{r_1}}$.

2. *The pair* $\mathcal{E}(f) := (f_{\mathrm{ELT}}, f_{\mathrm{REL}})$ *is a morphism of multidirected sets.* It is necessary to show that f_{ELT} and f_{REL} respect initial and terminal element maps, in the sense that

$$f_{\mathrm{ELT}}\big(i(r)\big) = i'\big(f_{\mathrm{REL}}(r)\big) \quad \text{and} \quad f_{\mathrm{ELT}}\big(t(r)\big) = t'\big(f_{\mathrm{REL}}(r)\big),$$

for every r in R. Writing this in terms of equivalence classes, and recalling that f_{REL} is just the given morphism f, this becomes

$$f_{\mathrm{ELT}}(\widetilde{x_r}) = \widetilde{x_{f(r)}} \quad \text{and} \quad f_{\mathrm{ELT}}(\widetilde{y_r}) = \widetilde{y_{f(r)}},$$

which is the *definition* of f_{ELT}.

3. $\mathcal{E}$ *is a functor.* Part 2 of the proof shows that $\mathcal{E}$ sends morphisms to morphisms. It remains to show that $\mathcal{E}$ preserves identity morphisms and compositions of morphisms. If $f : R \to R$ is the identity morphism on R, then the map of relations f_{REL} contributing to the morphism $\mathcal{E}(f) = (f_{\mathrm{ELT}}, f_{\mathrm{REL}})$ is the identity morphism on R by definition, and the map of elements f_{ELT} sends each $\widetilde{x_r}$ to $\widetilde{x_{f(r)}} = \widetilde{x_r}$. A symmetric statement applies to abstract terminal elements. Hence, $\mathcal{E}$ preserves identity morphisms. Similarly, given two morphisms $f : R \to R'$ and $g : R' \to R''$, where $R'' = (R'', \prec'')$ is a third directed set, the map of relations $(g \circ f)_{\mathrm{REL}}$ in the induced morphism $\mathcal{E}(g \circ f)$ is the composition $g \circ f = g_{\mathrm{REL}} \circ f_{\mathrm{REL}}$ by definition, while the map of elements $(g \circ f)_{\mathrm{ELT}}$ sends $\widetilde{x_r}$

to $\widetilde{x_{g\circ f(r)}} = g_{\mathrm{ELT}}(\widetilde{x_{f(r)}}) = g_{\mathrm{ELT}} \circ f_{\mathrm{ELT}}(\widetilde{x_r})$, and similarly for terminal elements. Hence, $\mathcal{E}$ preserves compositions of morphisms.

4. *$\mathcal{E}$ fails to preserve either star finiteness or acyclicity, even in the irreducible case.* A counterexample for star finiteness is the *infinite sawtooth,* which has elements r_n^- and r_n^+, together with relations $r_n^- \prec r_n^+$ and $r_n^- \prec r_{n+1}^+$, for all integers n. Its abstract element space is isomorphic to the *infinite hourglass,* which has minimal elements u_n and maximal elements w_n, for all integers n, as well as a single "intermediate element" v, together with relations $u_n \prec v$ and $v \prec w_n$ for all n. A counterexample for acyclicity is the irreducible acyclic directed set illustrated in Figure 5.3.2. □

5.3 Preservation of Information

"Replacing" a multidirected set by its relation space. The relation space functor $\mathcal{R} : \mathcal{M} \to \mathcal{D}$ and the abstract element space functor $\mathcal{E} : \mathcal{D} \to \mathcal{M}$ interact in interesting and useful ways, especially in the discrete context. The most important general property of these functors is that they may be used, with a high degree of fidelity, to "shift viewpoints" between a given element space and its relation space. In particular, Theorem 5.3.1 renders precise the assertion, already repeated multiple times over the previous two sections, that the relation space $\mathcal{R}(M)$ over a multidirected set $M = (M, R, i, t)$ "represents the same information as M, except on the boundary ∂M of M." This theorem has the same mathematical substance as theorem 3 of Harary and Norman [HN60], proven more than 50 years ago in a graph-theoretic setting, but possesses important and heretofore unrecognized physical significance. In particular, the *replacement of certain badly-behaved multidirected sets by their well-behaved relation spaces* facilitates the background independent discrete causal adaptation of Feynman's *path summation approach to quantum theory* developed in Chapters 6 and 7. The most notable type of "bad behavior" eliminated by this procedure is *permeability of maximal antichains in element space,* mentioned periodically since Chapter 2, and revisited definitively in Section 5.4.

Theorem 5.3.1. *Any pair of multidirected sets with isomorphic relation spaces have isomorphic interiors.*

Proof. The proof actually establishes the more specific statement that if $R = (R, \prec)$ is a directed set, and if M is any multidirected set whose relation space $\mathcal{R}(M)$ is isomorphic to R, then $\mathrm{Int}(M) \cong \mathrm{Int}(\mathcal{E}(R))$. Since the relation space $\mathcal{R}(M)$ over M is isomorphic to $(R, \prec)$, M may be expressed, up to isomorphism, as a quadruple

(M, R, i, t), where the "R" appearing in the quadruple coincides, as a set, with the given directed set $(R, \prec)$, and where the maps i and t identify *concrete* initial and terminal elements in M. Also, the relation space $\mathcal{RE}(R)$ is isomorphic to the original set $R = (R, \prec)$ in this case, as explained in the discussion following Definition 5.2.1. Throughout the proof, I use the "tilde notation," appearing in part 2 of the same definition, to denote elements of $\text{Int}(\mathcal{E}(R))$, i.e., *abstract* initial and terminal elements.

The interior $\text{Int}(M)$ of M, viewed as a full subobject of M, is the multidirected set whose elements are nonextremal elements of M, whose relations are elements of R having nonextremal initial and terminal elements, and whose initial and terminal element maps are the restrictions of the initial and terminal element maps i and t of M. In particular, any relation between two elements in $\text{Int}(M)$ itself belongs to $\text{Int}(\mathcal{R}(M))$, by the definition of the induced relation on $\text{Int}(\mathcal{R}(M))$. Therefore, an isomorphism f between $\text{Int}(M)$ and $\text{Int}(\mathcal{E}(R))$, if it exists, may be expressed as a map of elements $f_{\text{ELT}} : \text{Int}(M) \to \text{Int}(\mathcal{E}(R))$, and a map of relations $f_{\text{REL}} :$ $\text{Int}(\mathcal{R}(M)) \to \text{Int}(R)$, where the expression for the latter map incorporates the fact that $\mathcal{RE}(R) \cong R$. I define this latter map f_{REL} to be the restriction of the given isomorphism $f : \mathcal{R}(M) \to R$ to the interior of $\mathcal{R}(M)$. Using the fact that any element of $\text{Int}(M)$ may be written as either an initial element or a terminal element, I define the former map f_{ELT} by setting $f_{\text{ELT}}(i(r)) = \widetilde{x_{f_{\text{REL}}(r)}}$ and $f_{\text{ELT}}(t(r)) = \widetilde{y_{f_{\text{REL}}(r)}}$. The proof then proceeds in four steps:

1. *The map f_{ELT} is well-defined.* This is similar to part 1 of the proof of Theorem 5.2.3 above, but is easier because the source elements are concrete rather than abstract. Hence, one does not have to worry about sequences of "back-and-forth" relations inducing "non-obvious" equivalences. In particular, suppose that $z = i(r_0) = i(r_1)$ for two elements r_0 and r_1 of R. Then it is necessary to show that the images $\widetilde{x_{f_{\text{REL}}(r_0)}}$ and $\widetilde{x_{f_{\text{REL}}(r_1)}}$ belong to $\text{Int}(\mathcal{E}(R))$ and are equal. Since z belongs to $\text{Int}(M)$, there exists an element r_2 of R such that $t(r_2) = z$ in M. Therefore, r_2 directly precedes both r_0 and r_1 in $\mathcal{R}(M)$, so $f_{\text{REL}}(r_2)$ directly precedes both $f_{\text{REL}}(r_0)$ and $f_{\text{REL}}(r_1)$ in R. Hence, the elements $\widetilde{x_{f_{\text{REL}}(r_0)}}$ and $\widetilde{x_{f_{\text{REL}}(r_1)}}$ coincide in $\mathcal{E}(R)$, by the Definition 5.2.1 of abstract element space. A symmetric argument applies if $z = t(r_0) = t(r_1)$. Finally, if $z = t(r_0) = i(r_1)$, then $f_{\text{REL}}(r_0)$ directly precedes $f_{\text{REL}}(r_1)$ in R, so $\widetilde{y_{f_{\text{REL}}(r_0)}} = \widetilde{x_{f_{\text{REL}}(r_1)}}$ in $\mathcal{E}(R)$, again by the definition of abstract element space.

2. *The pair of maps $(f_{\text{ELT}}, f_{\text{REL}})$ defines a morphism of multidirected sets.* The map f_{REL} simply identifies each relation in $\text{Int}(\mathcal{R}(M))$ with the corresponding relation in $\text{Int}(R)$, via the hypothesized isomorphism $f : \mathcal{R}(M) \to R$ of directed sets. If r belongs to $\text{Int}(\mathcal{R}(M))$, then its concrete initial and terminal elements $i(r)$ and $t(r)$ in M map to the abstract initial and terminal elements $\widetilde{x_{f_{\text{REL}}(r)}}$ and $\widetilde{y_{f_{\text{REL}}(r)}}$ of its image $f_{\text{REL}}(r)$ in $\text{Int}(\mathcal{E}(R))$, by the definition of f_{ELT}. Hence, $(f_{\text{ELT}}, f_{\text{REL}})$ respects initial and terminal element maps.

3. *The map f_{ELT} is bijective.* Suppose that $z = t(r_0) = i(r_1)$ and $w = t(\overline{r_0}) = i(\overline{r_1})$ are two elements of $\text{Int}(M)$ whose images under f_{ELT} coincide, where $r_0, r_1, \overline{r_0}$, and $\overline{r_1}$ are elements of R. Then by the definition of f_{ELT}, $\widetilde{x_{f_{\text{REL}}(r_1)}} = \widetilde{x_{f_{\text{REL}}(\overline{r_1})}} =$

$\widetilde{y_{f_{\mathrm{REL}}(r_0)}} = \widetilde{y_{f_{\mathrm{REL}}(\overline{r_0})}}$ in Int($\mathcal{E}(R)$). Therefore, $f_{\mathrm{REL}}(r_0)$ and $f_{\mathrm{REL}}(\overline{r_0})$ both directly precede $f_{\mathrm{REL}}(r_1)$ and $f_{\mathrm{REL}}(\overline{r_1})$ in Int(R), since R is a relation space.[6] Since f_{REL} is an isomorphism of directed sets, r_0 and $\overline{r_0}$ both directly precede r_1 and $\overline{r_1}$ in Int($\mathcal{R}(M)$). Hence, $t(r_0) = t(\overline{r_0}) = i(r_1) = i(\overline{r_1})$, so $z = w$, and f_{ELT} is injective. Next, observe that any element in Int($\mathcal{E}(R)$) may be written as $\widetilde{x_{f_{\mathrm{REL}}(r_1)}}$ for some relation r_1 in Int($\mathcal{R}(M)$). By definition of f_{ELT}, this element lifts to the element $i(r_1)$ in Int(M). Thus, f_{ELT} is surjective.

4. *The pair of inverse maps* $(f_{\mathrm{ELT}}^{-1}, f_{\mathrm{REL}}^{-1})$ *defines a morphism of multidirected sets.* Since $f_{\mathrm{REL}} = f$ is an isomorphism, any element of Int(R) may be expressed as $f_{\mathrm{REL}}(r)$ for a unique element r of $\mathcal{R}(M)$. The image of this element under the inverse relation map f_{REL}^{-1} is just r. By definition, the inverse element map f_{ELT}^{-1} carries the abstract initial and terminal elements of $f_{\mathrm{REL}}(r)$ in Int($\mathcal{E}(R)$) to the concrete initial and terminal elements $i(r)$ and $t(r)$ in Int(M). Hence, the pair $(f_{\mathrm{ELT}}^{-1}, f_{\mathrm{REL}}^{-1})$ respects initial and terminal element maps.

□

Weak information preservation of $\mathcal{E}$. As suggested by the statement of Theorem 5.2.3, the abstract element space functor $\mathcal{E}$ generally fails to preserve information as faithfully as the relation space functor $\mathcal{R}$. In particular, it is generally impossible to recover the interior of a directed set from its abstract element space. This can be true even for small finite irreducible acyclic directed sets, as illustrated in Figure 5.3.2. Here, application of $\mathcal{E}$, followed by application of $\mathcal{R}$, results in a directed set that includes the structure of the original directed set R, but also includes a large number of additional relations, represented by the "dashed lines" in the right-hand diagram in the figure. It is generally impossible to determine, given only $\mathcal{E}(R)$, which properties of its relation space "actually originate from R," and which are "spurious artifacts of the recovery process."

Fortunately, the principal application of the abstract element space functor $\mathcal{E}$ in this book is to facilitate formal analysis of the properties of the relation space functor $\mathcal{R}$. Preservation of information by $\mathcal{E}$ itself is generally of secondary importance in this context. Of course, one may imagine hypothetical situations in which $\mathcal{E}$ could play a larger role. For example, one might consider the concept of an *infinite* number of different levels of "horizontal structural hierarchy," parameterized by $\mathbb{Z}$, with application of $\mathcal{R}$ *"shifting levels in the positive direction,"* and application of $\mathcal{E}$ *"shifting levels in the negative direction."* Instead of individual multidirected sets, then, one could study sequences of multidirected sets related by $\mathcal{R}$ and $\mathcal{E}$, abstractly similar to the *spectra* of spaces studied in algebraic topology. At present, however, I know of no convincing *physical* reason to insist on shifting the focus to such structures. Of greater interest is the "vertical structural hierarchy" involved in the principle of *iteration of structure* (IS), formally introduced in Section 6.3.

[6] Note that this step fails for an arbitrary directed set R!

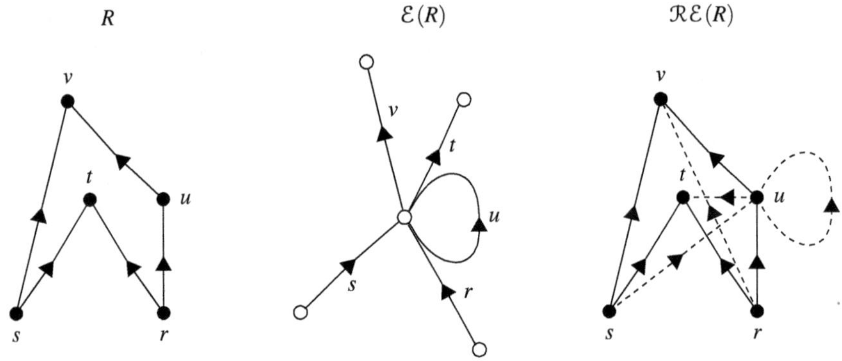

Figure 5.3.2. Irreducible acyclic directed set R; abstract element space $\mathcal{E}(R)$ over R; relation space $\mathcal{RE}(R)$ over $\mathcal{E}(R)$. Note "extra relations".

5.4 Permeability

Cauchy surfaces as filters of information flow. I now revisit in detail an information-theoretic device which has proven to be of great importance in conventional continuum-based physics, both in the classical and quantum settings, but whose adaptation to discrete causal theory has hitherto been hampered by a particular technical obstacle. This device involves the use of "spacelike sections of spacetime" to quantify the flow of information from "past to future." In continuum-based theories such as general relativity, this notion is made precise in terms of *Cauchy surfaces.* In particular, as discussed in Chapter 2, a Cauchy surface in a relativistic spacetime manifold X is defined to be a subset σ of X such that every inextensible causal curve in X intersects σ exactly once. A Cauchy surface may therefore be viewed as representing *"space at a particular instant in time, in a particular frame of reference."* From an information-theoretic viewpoint, a Cauchy surface *filters* the flow of information from past to future in X. Foliation of spacetime via Cauchy surfaces is a method of central importance in relativistic cosmology, and also plays a role in various approaches to quantum gravity, including loop quantum gravity, noncommutative geometry, and shape dynamics. The left-hand diagram in Figure 5.4.1 illustrates a Cauchy surface σ in a relativistic spacetime manifold X, with two representative causal curves passing through σ from past to future. By the definition of a Cauchy surface, these curves cannot "avoid" intersecting σ; in this case, I have labeled the two intersection points x and y. A descriptive way to express this situation is to say that *"Cauchy surfaces are impermeable to causal curves."*

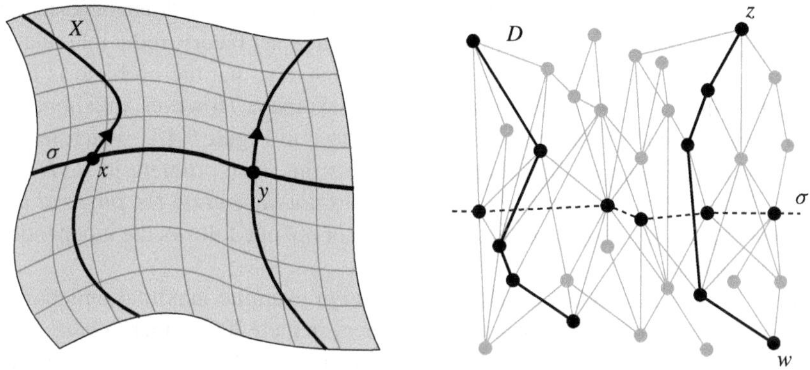

Figure 5.4.1. Cauchy surface σ in a relativistic spacetime manifold X, intersected by two causal curves; maximal antichain σ in a directed set $D = (D, \prec)$, permeated by two chains.

Maximal antichains; permeability problem. As discussed in Section 2.7, the naïve analogue of a Cauchy surface in a directed set $D = (D, \prec)$ is a *maximal antichain* in D, i.e., a subset σ of D such that no chain of nonzero length in D connects any pair of elements of σ, distinct or otherwise, but any subset of D properly containing σ exhibits at least one such connecting chain. It is *not* necessarily true that any element in the complement of σ is connected to an element of σ by a chain, since an element x lying on a cycle in D may belong to neither the past nor the future of σ. However, any element x of D in the compliment of σ and *not* belonging to a cycle does belong to either the past or the future of σ by Lemma 3.7.2, since otherwise x could be added to σ to obtain a larger antichain. A maximal antichain in a directed set D is illustrated in the right-hand diagram in Figure 5.4.1, which reproduces the right-hand diagram in Figure 2.7.6 of Section 2.7. The maximal antichain σ is represented by the "horizontal string" of dark nodes in the diagram. The dashed lines connecting these nodes are included merely as a visual aid; they are not part of the actual structure of D. This situation generalizes immediately to the case of an arbitrary multidirected set $M = (M, R, i, t)$.

A serious technical problem associated with maximal antichains in multidirected sets, also mentioned in the special case of directed sets in Section 2.7, is that they are generally **permeable.** As one would expect from the previous discussion, this means that given a maximal antichain σ in a multidirected set $M = (M, R, i, t)$, there generally exists at least one chain γ in M beginning at an element in the past of σ, and terminating at an element in the future of σ, *without intersecting* σ. It is sometimes useful to refer to such a chain as a **permeating chain.** Two permeating chains are illustrated in the right-hand diagram in Figure 5.4.1, which shows the special case of an acyclic directed set. Heuristically, the existence of permeating chains implies that *information can "leak through" the maximal antichain* σ, which leads to difficulties in describing the dynamics of any theory depending on this information. This problem is particularly severe in the transitive case, with numerous chains, and even individual relations, carrying information directly from the "distant past" to the "distant future" of σ. In the specific context of causal set theory, reducible relations are taken to be dependent, so one might hope, naïvely, that the problem is only a formal one that disappears upon passage to the skeleton. However, this hope fails, as illustrated by the permeating chain from w to z in Figure 5.4.1, which involves only irreducible relations, and therefore presents the same problem in $\mathrm{sk}(D)$. The technical obstacle of permeability therefore *does not depend on the physical interpretation of reducible relations,* i.e., on whether or not one follows the independence convention (IC).

Seth Major, David Rideout, and Sumati Surya examine maximal antichains as analogues of Cauchy surfaces in causal set theory in their 2006 paper *Spatial Hypersurfaces in Causal Set Cosmology* [MRS06]. Recognizing the permeability problem, they attempt to address it by "thickening" maximal antichains to "capture" permeating chains. Surya mentions the same problem in her 2011 review article [SU11]. The basic idea of thickening maximal antichains is reasonable and interesting for a variety of reasons. Perhaps the most obvious is that conventional continuum-based theories are described partly in terms of differential equations involving time derivatives, whose natural discrete causal analogues are discrete equations involving families of events of "small causal separation," which may be described in terms of such thickenings. However, the actual attempts at formulating "thickening procedures" appearing in the existing literature do not seem to supply a decisive solution to the permeability problem. A better approach, in my view, is to pass to relation space, and work with *maximal antichains of relations.* Remarkably, no additional device is needed: the permeability problem disappears in relation space, as demonstrated by Theorem 5.4.3. One may choose, a posteriori, to return to the element space viewpoint, and view maximal antichains of relations as *"optimal thickenings of maximal antichains of elements,"* but the relation space viewpoint is superior in many respects.[7]

Figure 5.4.2 illustrates a maximal antichain in the relation space over a multidirected set. In this particular example, the multidirected set under consideration is the familiar acyclic directed set $(D, \prec)$ appearing in Figure 5.4.1, and in previous figures, whose relation space was first illustrated in Figure 1.5.4, and later in Figure 5.1.3. To avoid proliferating symbols, *I use the same symbol σ for a third time* to denote this maximal antichain; however, it is entirely different from the Cauchy surface in relativistic spacetime, and from the maximal antichain in D itself, both also labeled σ in Figure 5.4.1. In the present case, σ is represented by the "horizontal string" of dark square nodes. At an informal level, it seems clear from a cursory examination of Figures 5.4.1 and 5.4.2 that a typical antichain in relation space is much "denser" than a typical antichain in element space, and is therefore "likely to be much less permeable." In particular, the maximal antichain in $\mathcal{R}(D)$ illustrated in Figure 5.4.2 contains five times as many elements as the maximal antichain in D illustrated in Figure 5.4.1!

[7] Blute, Panangaden, and Straßburger [BPS08] discuss information preservation involving structures that are "almost the same as" antichains in relation space, working in a somewhat different context

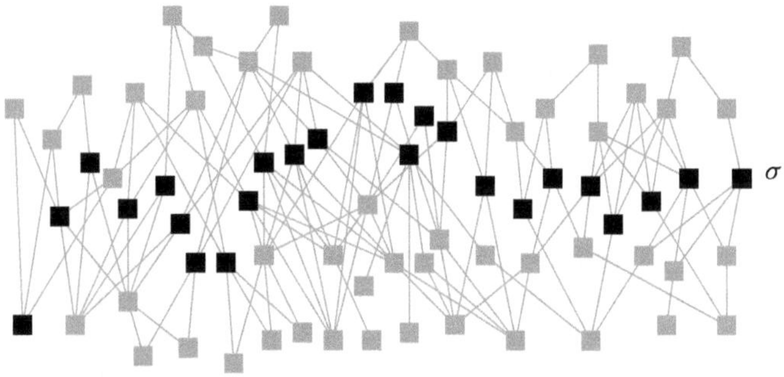

Figure 5.4.2. Maximal antichain in relation space.

Impermeability theorem. Theorem 5.4.3 establishes that a maximal antichain σ in the relation space $\mathcal{R}(M)$ over a multidirected set $M = (M, R, i, t)$ is impermeable, i.e., that any chain in $\mathcal{R}(M)$ connecting an element of the past $J^-(\sigma)$ of σ to an element of the future $J^+(\sigma)$ of σ also contains an element of σ. The physical interpretation of this result is that *maximal antichains in relation space can serve as effective "filters" for the flow of information from "past to future" in* M.

Theorem 5.4.3. *Maximal antichains in relation space are impermeable.*

Proof. The proof is by contradiction. Let $M = (M, R, i, t)$ be a multidirected set with relation space $\mathcal{R}(M) = (R, \prec)$, where $\prec$ is the induced relation of Definition 5.1.1, and let σ be a maximal antichain in $\mathcal{R}(M)$. Suppose that σ is permeable, and let γ be a chain in $\mathcal{R}(M)$ permeating σ. By definition, γ may be expressed as a finite sequence of elements and relations in $\mathcal{R}(M)$, with elements indexed by consecutive natural numbers, and with initial and terminal elements belonging to $J^-(\sigma)$ and $J^+(\sigma)$, respectively. Note that since $\mathcal{R}(M)$ is a directed set, γ is uniquely defined by its sequence of elements. Also note that a given element of $\mathcal{R}(M)$ may appear multiple times in the sequence defining γ, and may therefore be assigned multiple indices.[8] Let r be the element of γ of maximal index belonging to $J^-(\sigma)$, and let s be the element of γ of minimal index belonging to $J^+(\sigma)$. Then r and s, together with the elements and relations strictly between them along γ, make up a restriction γ' of γ, which is minimal among all restrictions of γ that are themselves chains permeating σ. I consider separately the cases $r \prec s$ and $r \nprec s$.

1. If $r \prec s$, then since r belongs to $J^-(\sigma)$, there exists a chain α of the form $r_0 \prec \cdots \prec r_m$ in $\mathcal{R}(M)$, with $r \prec r_0$ and $r_m \in \sigma$. This includes the case $m = 0$, i.e., the case in which r directly precedes an element of σ. Similarly, since s belongs to $J^+(\sigma)$, there exists a chain β of the form $s_0 \prec \cdots \prec s_n$ in $\mathcal{R}(M)$, with $s_n \prec s$

(Footnote 7 continued)
in which the information under consideration involves auxiliary algebraic structure in a background dependent quantum theory.

[8] See the discussion following the proof of the theorem for more details.

and $s_0 \in \sigma$. This includes the case $n = 0$, i.e., the case in which an element of σ directly precedes s. The properties of relation space enter the picture here; by Definition 5.1.1,

$$t(s_n) = i(s) = t(r) = i(r_0) \qquad \text{in } M,$$

since $s_n \prec s, r \prec s$, and $r \prec r_0$ in $\mathcal{R}(M)$. This sequence of equalities is illustrated in the left-hand diagram in Figure 5.4.4, which shows the element space viewpoint. Hence, $s_n \prec r_0$ in $\mathcal{R}(M)$, so the chains α and β "patch together" to yield a product chain $\alpha \sqcup \beta$ of the form $s_0 \prec \cdots \prec s_n \prec r_0 \prec \cdots \prec r_m$. This is an example of the *concatenation product* of chains, studied in detail in Section 5.9. Since the initial and terminal elements s_0 and r_m of $\alpha \sqcup \beta$ both belong to σ, this contradicts the hypothesis that σ is an antichain.

2. If $r \not\prec s$, then the elements of γ' strictly between r and s must belong to a common cycle in $\mathcal{R}(M)$. Indeed, any such element not belonging to a cycle would perforce belong to $J^-(\sigma)$ or $J^+(\sigma)$ by the maximality of σ, thus contradicting the minimality of γ', and any pair of cycles in relation space joined by a relation may be identified as restrictions of a larger cycle, as explained in the discussion following the proof of Theorem 5.1.4. Let r^+ be the minimal successor of r along γ', and let s^- be the maximal predecessor of r along γ'. This includes the case $r^+ = s^-$. Since r^+ and s^- belong to a common cycle in $\mathcal{R}(M)$, there exists a chain δ from s^- to r^+ in $\mathcal{R}(M)$ of the form $s^- \prec c_0 \prec \cdots \prec c_k \prec r^+$. For example, in the case where $r^+ = s^-$ and r^+ is related to itself via a reflexive relation, one may set $s^- = c_0 = \cdots = c_k = r^+$ and take δ to be the chain wrapping $k+2$ times around this relation. As in the proof of the case $r \prec s$ in part 1 of the theorem, there exist chains α and β of the forms $r_0 \prec \cdots \prec r_m$ and $s_0 \prec \cdots \prec s_n$ in $\mathcal{R}(M)$, respectively, such that $r \prec r_0$, $s_n \prec s$, and r_m and s_0 belong to σ. This setup is illustrated in the right-hand diagram of Figure 5.4.4, which shows the case where the elements $s^-, c_0, \ldots, c_k, r^+$ are distinct. By the definition of relation space, $t(s_n) = i(s) = t(s^-) = i(c_0)$ and $t(c_k) = i(r^+) = t(r) = i(r_0)$. Hence, the chains β, δ, and α "patch together" to yield a product chain $\beta \sqcup \delta \sqcup \alpha$ from σ to itself, contradicting the hypothesis that σ is an antichain.

□

Discussion of the theorem and proof. At the beginning of the proof of Theorem 5.4.3, note that γ' is minimal among restrictions of γ whose initial and terminal elements belong to $J^-(\sigma)$ and $J^+(\sigma)$, respectively, and whose elements are indexed by consecutive natural numbers *under the same indexing describing γ itself.* As explained in Section 5.9, chains may be viewed as special cases of *causal paths,* which are represented by morphisms whose sources are *linear directed sets.* In the present case, γ is represented by a morphism whose source is a set of consecutive natural numbers, connected by their nearest-neighbor relations; this is a more-sophisticated way of viewing the "indexing" described in the proof. Such a morphism need not

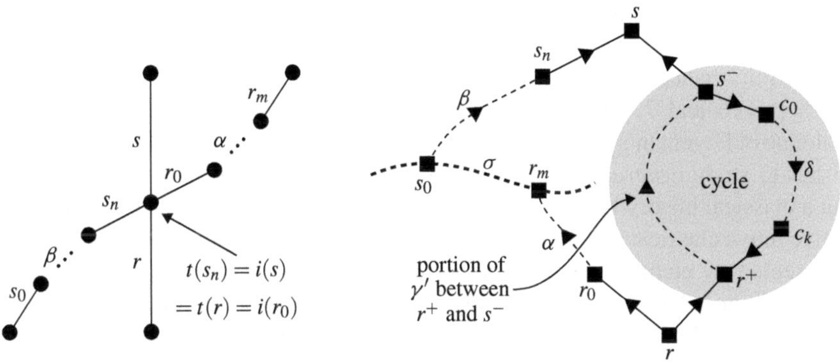

Figure 5.4.4. Maximal antichains in relation space are impermeable.

be injective, so γ may intersect itself, which is why chains are defined as *sequences* rather than *subsets* in Definition 3.7.1 of Chapter 3. In the present setting, multiple indices are assigned to the same element of $\mathcal{R}(M)$ if γ passes through this element multiple times; in particular, the set of indices of a given element is just its preimage under the morphism representing γ. It is sometimes possible to define an even smaller permeating chain than γ' involving a *non-consecutive* subsequence of the sequence of elements defining γ, by "cutting out cycles along γ'," but this requires re-indexing to avoid working with a disconnected source. A morphism with a disconnected source does not technically represent a causal path, even if one may easily engineer a connected source with the same sequence of image elements, for example, by reindexing. This is why γ' is characterized as the minimal restriction of γ *that is itself a chain permeating* σ.

Note also that the crucial step in the first part of the proof; namely, the conclusion that $s_n \prec r_0$, which enables the chains α and β to "patch together" to yield the product chain $\alpha \sqcup \beta$ beginning and terminating on σ, *fails* in general if the entire relation space $\mathcal{R}(M)$ is replaced by an arbitrary subobject of itself. This is because *the condition* $t(s_n) = i(r_0)$ *guarantees the existence of the relation* $s_n \prec r_0$ *only in* $\mathcal{R}(M)$ *itself, not in an arbitrary subobject of* $\mathcal{R}(M)$, *which may possess fewer relations.* For example, the directed set R appearing in the left-hand diagram in Figure 5.3.2 is a subobject of the relation space $\mathcal{RE}(R)$ appearing in the corresponding right-hand diagram; in this case, R is obtained from $\mathcal{RE}(R)$ by deleting the relations represented by the dashed lines. This ruins impermeability; in particular, the subset $\{t, u\}$ of R is a maximal antichain, but is permeated by the chain $s \prec v$. However, the conclusion of the theorem *does* apply if $\mathcal{R}(M)$ is replaced by a *full* subobject of itself, since the relation $s_n \prec r_0$ holds in a full subobject of $\mathcal{R}(M)$ whenever it holds in $\mathcal{R}(M)$ itself. Similarly, the special properties of relation space are crucial to the arguments in the second part of the proof. In particular, it is not at all obvious by merely glancing at the right-hand diagram of Figure 5.4.4 that the chains β, δ, and α "patch together" to yield a product chain. The properties of relation space are necessary to ensure the existence of the "connecting relations" $s_n \prec c_0$ and $c_k \prec r_0$, which are not shown in the figure.

The importance of Theorem 5.4.3 in discrete causal dynamics becomes clear in Chapters 6 and 7, where I develop a background independent discrete causal analogue of Feynman's path summation approach to quantum theory. Informally, one wishes to study *generalized quantum amplitudes,* which describe the "likelihood" that a classical history belonging to a given *cobase family* of classical histories will evolve into a classical history belonging to a given *target family.* These histories are represented by elements in the *underlying multidirected set* $\mathcal{M}(\mathbb{S})$ of a kinematic scheme $\mathbb{S}$, and "evolutionary pathways" between the two families, which consist of sequences of co-relative histories in $\mathbb{S}$, are represented by chains in $\mathcal{M}(\mathbb{S})$. The practical nature of experiment and observation leads one to formulate "initial value problems" in this context, where the cobase family is chosen to represent a variety of closely-related choices of initial conditions, while the target family is chosen to represent corresponding choices of terminal conditions. It would be natural, then, from a naïve perspective, to choose cobase and target families corresponding to maximal antichains in a subset of $\mathcal{M}(\mathbb{S})$. However, such antichains typically represent "deficient families," due to the permeability problem. This issue is resolved by passage to the relation space $\mathcal{R}\big(\mathcal{M}(\mathbb{S})\big)$.

At a general conceptual level, Theorem 5.4.3 provides evidence that the choice to view fundamental spacetime structure in terms of relations, rather than in terms of elements, is much more than just a philosophical preference, since the resulting perspective provides straightforward access to methods that are "hidden" under the element space viewpoint.[9] Additional motivations are evident for formulating emergent "spatial notions" in the context of relation space. One such motivation, at least in the acyclic case, is *spatial adjacency:* any pair of relations with the same initial or terminal element in an acyclic multidirected set M forms a minimal nontrivial antichain in $\mathcal{R}(M)$. Such pairs are unambiguously adjacent in the sense that they share a common element; i.e., they form a connected subobject of M in a graph-theoretic sense. They may therefore be viewed as "building blocks of spatial structure" in M. While it is not true that every maximal antichain in $\mathcal{R}(M)$ is connected in this sense, a "high degree of connectivity" is typical. For example, the maximal antichain in the relation space $\mathcal{R}(D)$ illustrated in Figure 5.4.2 has just two connected components. This is illustrated, from the element space viewpoint, in the left-hand diagram in Figure 5.4.5. It is worth remembering, as discussed in Sections 1.7 and 4.5, that there are good reasons to consider the possibility of very large valences in physically realistic settings, and large valences are generally associated with a high degree of connectivity.

Similarly, the maximal antichain consisting of the "first generation of relations" in the directed set illustrated in the right-hand diagram of Figure 5.4.5 has just two connected components from the element space viewpoint, while maximal antichains

[9] It would be interesting to explore continuum-based applications of results similar to Theorem 5.4.3; for example, in the context of partial differential equations and boundary value problems involving regions of non-globally hyperbolic pseudo-Riemannian manifolds of Lorentz signature.

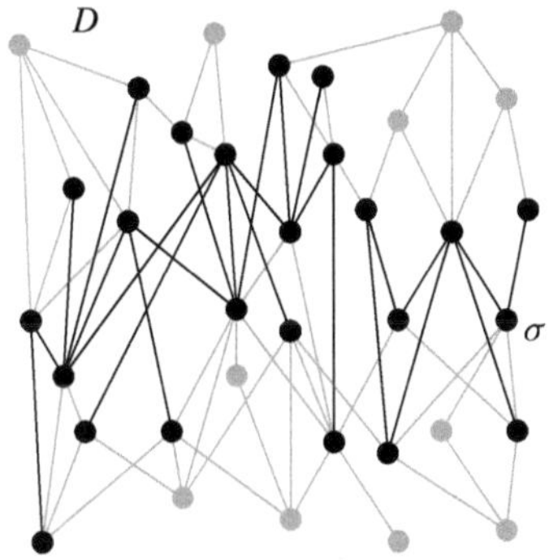

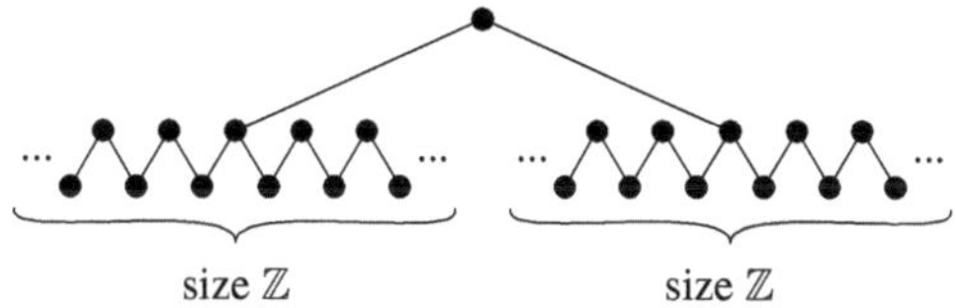

Figure 5.4.5. Maximal antichain σ from Figure 5.4.2 possesses a high degree of connectivity as a subobject of element space; skeleton of causal set from Figure 4.6.7 contains a highly-connected antichain of relations from the element space perspective.

in this set itself have an infinite number of connected components. This particular directed set appeared previously in Figure 4.6.7 of Section 4.6, where it was introduced as the skeleton of a causal set of "spatial size greater than $\mathbb{Z}$." Each connected component of its "first generation of relations" corresponds to a copy of $\mathbb{Z}$ in an obvious way, giving precise meaning to the statement regarding "spatial size." Ultimately, the *limited, but nontrivial, internal structure* of maximal antichains of relations, as seen from the element space viewpoint, has important dynamical consequences in the context of *generational dynamics,* discussed in Sections 7.8, 8.2, and 8.3. In particular, such antichains possess a nontrivial notion of *entropy,* expressed in terms of their element-space automorphism groups.

5.5 Analogy Involving Morphism Categories

Morphism categories. A *morphism category*, also called an *arrow category,* is a category whose class of objects is the class $\text{Mor}(\mathcal{C})$ of *morphisms* of some other category $\mathcal{C}$, and whose class of morphisms is the class $\text{Mor}\big(\text{Mor}(\mathcal{C})\big)$ of "morphisms between morphisms" of $\mathcal{C}$, also called 2*-morphisms,* described more precisely below. It is sometimes useful to "combine" $\mathcal{C}$, $\text{Mor}(\mathcal{C})$, and $\text{Mor}\big(\text{Mor}(\mathcal{C})\big)$ into a single "augmented category," called a 2*-category,* which possesses objects, morphisms, and 2-morphisms. This construction may be generalized to yield 3-categories, n-categories, and even ∞-categories. The general study of such structures is sometimes called *higher category theory.* In this brief section, however, I focus only on morphism categories and 2-categories. Let W, X, Y, and Z be objects of a category $\mathcal{C}$. Then a 2-morphism between morphisms $f : X \to Y$ and $g : W \to Z$ in $\mathcal{C}$ is usually[10] defined to be a *commutative square* of the form illustrated in the left-hand diagram

[10]Many other definitions are possible. A useful online resource for exploring such notions is nLab.

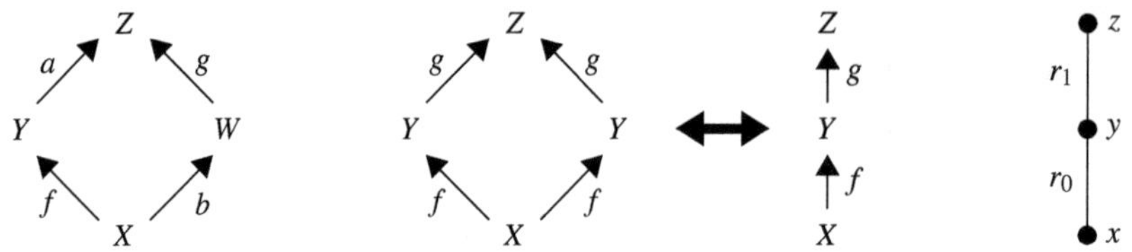

Figure 5.5.1. A 2-morphism; "commutativity" without composition; 2-chain analogue.

in Figure 5.5.1. In this context, commutativity means that *"one obtains the same answer by going either way around the edges of the square from* X to Z*,"* i.e., that the composition of morphisms $g \circ b$ is equal to the composition $a \circ f$.

Analogy between $\mathrm{Mor}(\mathcal{C})$ **and** $\mathcal{R}(M)$**.** Of course, a basic requirement that must be satisfied for the notion of commutativity of an arbitrary diagram of morphisms to make sense is the *composition axiom* of category theory, which guarantees that the ordered pairs of morphisms (f, a) and (b, g) in Figure 5.5.1 do indeed compose to define morphisms $a \circ f : X \to Z$ and $g \circ b : X \to Z$, which may then be compared for equality. In conventional settings where category theory plays a natural role, the justification for the composition axiom is usually obvious. However, in more general settings, in which certain structural features common to familiar categories may be absent, it is not at all obvious that a reasonable notion of composition exists, and if so, how it should be defined. For example, as noted in Section 3.9, the composition axiom is similar, at an abstract level, to the axiom of transitivity (TR), which is a problematic condition to impose on directed sets in the physical context of discrete classical causal theory. Hence, it is prudent to reconsider, from first principles, the questions of *if, when,* and *how* "commutative diagrams" and "morphisms between morphisms" generalize to settings in which the composition axiom is not taken for granted.

There is at least one obvious case in which the left-hand diagram in Figure 5.5.1 may reasonably be regarded as "commutative" even without the composition axiom; namely, when $W = Y$, $b = f$, and $a = g$. In this case, the square diagram is *redundant;* all the necessary information is encoded in the sequence of morphisms $X \xrightarrow{f} Y \xrightarrow{g} Z$, as illustrated in the middle diagram in Figure 5.5.1. This sequence is analogous to a 2-chain in a multidirected set $M = (M, R, i, t)$, consisting of relations r_0 and r_1, with initial and terminal elements $i(r_0) = x$, $t(r_0) = i(r_1) = y$, and $t(r_1) = z$, as illustrated in the right-hand diagram in the figure. By the definition of the relation space $\mathcal{R}(M)$ over M, such a 2-chain corresponds to a relation $r_0 \prec r_1$ in $\mathcal{R}(M)$. Under this correspondence, $\mathcal{R}(M)$ is analogous to the morphism category with object class $\mathrm{Mor}(\mathcal{C})$ discussed above. $\mathcal{R}(M)$ may therefore be regarded as a *"generalized morphism category,"* in which the composition axiom does not apply.

The analogy between $\mathrm{Mor}(\mathcal{C})$ and $\mathcal{R}(M)$ provides an example of a general structural phenomenon briefly discussed in Section 3.8, in the context of category-theoretic and topos-theoretic approaches to fundamental physics. In discrete causal theory, mathematical objects naturally arise that are *not* categories, but that play similar structural roles. Unmodified "appropriation" of category-theoretic notions can do more harm than good in this context. For example, the naïve discrete causal analogue of a category-theoretic 2-morphism is a "causal diamond," given by replacing the objects W, X, Y, and Z in the left-hand diagram in Figure 5.5.1 with elements of a multidirected set, and the morphisms a, b, f, and g with relations. However, this analogy leads in the wrong direction, in the sense that it obstructs the "discovery" of the induced relation on relation space. It is worth pointing out that the "correct" choice of analogy for composition of pairs of consecutive relations is much less obvious if one works under the transitive paradigm and without the *independence convention* (IC), since reducible relations can masquerade as "compositions" in this context. In the more general nontransitive setting, the only obvious analogue of composition for relations $x \prec y$ and $y \prec z$ is the 2-chain $x \prec y \prec z$, since one cannot take for granted the existence of a relation $x \prec z$.

The star model and 2-categories. In the star model $M_\star$ of a multidirected set $M = (M, R, i, t)$, introduced in Chapter 4 to facilitate the study of local conditions on M, elements of the relation set R of M play a dual role, as both *elements* of R and as *relations* between pairs of elements of M. It is sometimes useful to define an **enhanced star model**, by elevating $M_\star$ to a multidirected set $(M_\star, R_\star, i_\star, t_\star)$. This is accomplished by means of the induced relation $\prec$ on the relation space $\mathcal{R}(M)$ over M. As a set, the enhanced star model is just the union $M \cup R$. It is endowed with multidirected structure by defining its relation set $R_\star$ to be the union $R \cup \prec$, where elements of R relate pairs of elements of M, and elements of $\prec$ relate pairs of elements of R. Initial and terminal element maps $i_\star$ and $t_\star$ are then defined in the obvious way, by combining the maps i and t with the obvious maps arising from $\prec$, which send each induced relation $r_0 \prec r_1$ to its initial and terminal elements r_0 and r_1 in R, respectively. The enhanced star model of a directed set is itself a directed set, rather than a multidirected set, since both R and $\prec$ define binary relations in this case. Referring back to the discussion at the end of Section 5.3, the enhanced star model incorporates three different levels of "horizontal structural hierarchy," represented by elements, relations, and *"relations between pairs of relations,"* i.e., 2-chains in M. Invoking the category-theoretic analogy described above, elements correspond to objects, relations to morphisms, and 2-chains to 2-morphisms. Hence, the enhanced star model $M_\star$ is analogous to a 2-category.[11] Anticipating results developed later in the present chapter, the enhanced star model combines the "lowest-order graded pieces" of the *causal concatenation spaces* over M and $\mathcal{R}(M)$, which are themselves analogous to ∞-categories. Many other such "category-like" structures appear in succeeding sections.

[11] Another suggestive analogy, mentioned in Section 3.8, compares $M_\star$ to an *algebraic scheme* in the context of algebraic geometry. Under this analogy, the "non-obvious" or "hidden" elements of $M_\star$ corresponding to relations are viewed as analogues of *generic points* corresponding to positive-dimension subschemes of an algebraic scheme.

5.6 Power Spaces

Relationships between pairs of subsets of a multidirected set. From the element space viewpoint, the induced relation $\prec$ on the relation space $\mathcal{R}(D)$ over a directed set $D = (D, \prec)$ may be viewed as encoding relationships between *pairs of ordered pairs* of elements of D, since each individual relation $x \prec y$ in D is uniquely specified by its ordered pair (x, y) of initial and terminal elements. Of course, the only such relationships that can exist in this case involve ordered pairs (x, y) and (w, z) that "link up in the middle," in the sense that $y = w$. More generally, the induced relation $\prec$ on the relation space $\mathcal{R}(M)$ over a *multidirected* set $M = (M, R, i, t)$ usually *cannot* be described in terms of relationships between ordered pairs of elements of M, because there may exist many different relations in R sharing the same initial and terminal elements in M. However, it is sometimes useful to "condense" the information in $\prec$ into a "coarse relation," encoding only which "pairs of pairs" are related under the induced relation, while forgetting *how* they are related, or in how many different ways. Such a coarse relation is described in detail below.

More generally, one may ask under what circumstances it is useful to consider relationships between *pairs of nonsingleton subsets* of the underlying set of elements of a multidirected set $M = (M, R, i, t)$. In the broadest setting, one may consider arbitrary subsets, which are, by definition, elements of the *power set* $\mathcal{P}(M)$ over M. Endowing such a power set with a directed or multidirected structure yields a *power space,* defined formally in Definition 5.6.2 below. Power spaces are very general, and it is a priori unlikely that a randomly-selected power space over a given classical history or configuration space of classical histories will itself possess any compelling physical interpretation. However, the utility of certain *special types* of power spaces is not too difficult to recognize. In particular, besides the natural relationships between "pairs of pairs" of elements of M, mentioned above, repeated application of the relation space functor $\mathcal{R}$ yields natural relationships between pairs of special n-element subsets of M, for every positive integer n. Due to the existence of such examples, the general machinery of power spaces is worth outlining, even if physically-useful applications arise only in very special cases. The construction of this machinery pays other dividends as well. Unsurprisingly, it facilitates the study of radical alternative models of classical spacetime, as described in Section 5.8. At a more practical level, it leads to useful technical methods involving approximations in computational discrete causal theory, outlined in Section 5.7.

Coarse relation on two-element subsets. As mentioned above, the induced relation $\prec$ on the relation space $\mathcal{R}(M)$ over a multidirected set $M = (M, R, i, t)$ may be used to construct a binary relation on the set of two-element subsets of M, which I denote here by M_2. This relation is defined by taking subsets $\{x, y\}$ and $\{u, v\}$ to be related if and only if there exists a "first relation" $x \prec y$ or $y \prec x$, a "second relation" $u \prec v$ or $v \prec u$, and the terminal element of the first relation coincides with the initial element of the second relation. The resulting binary relation on M_2 is called the **coarse relation,** because its construction generally destroys a large quantity of information. For example, unlike passage to relation space, formation of the coarse relation fails to preserve information in the interior of M, since merely specifying a pair of elements of M neither determines which element is initial and which is terminal, nor distinguishes among different relations sharing the same initial and terminal elements. Further, information about reflexive relations is lost entirely. Less information loss occurs in the special case of an acyclic directed set D, since in this case each two-element subset $\{x, y\}$ of D corresponds to *at most one* relation in D involving x and y, and no reflexive relations exist. For these reasons, the coarse relation on two-element subsets of D may be viewed as a "trivial extension" of the induced relation on the relation space $\mathcal{R}(D)$ over D. To understand the meaning of this statement, note that in general, many elements of the set D_2 of two-element subsets of D will *not* correspond to relations in D, and the coarse relation on D_2 does not encode any relationships involving such elements. In this context, D_2, equipped with the coarse relation, consists of the relation space $\mathcal{R}(D)$, together with a large number of additional isolated elements.

More generally, most of the multidirected sets used as examples in this book induce very "sparse" coarse relations on their sets of two-element subsets, since most of these subsets do not consist of the initial and terminal elements of a relation. Figure 5.6.1 illustrates this sparseness, using the familiar 35-element acyclic directed set $D = (D, \prec)$ appearing repeatedly in previous examples. The left-hand diagram in the figure illustrates the obvious fact that a "typical" two-element subset of D does *not* correspond to a relation in this case; in particular, D has $\binom{35}{2} = 595$ two-element subsets, but only 75 relations. Hence, most of the elements of D_2 are isolated under the induced relation on D_2. The right-hand diagram illustrates D_2, equipped with the coarse relation; isolated elements of D_2 are represented by small square nodes, while non-isolated elements are represented by large square nodes. Since D is an acyclic directed set in this example, the coarse relation on D_2 may be viewed as a trivial extension of the induced relation on $\mathcal{R}(D)$; this may be understood by comparing Figure 5.6.1 to Figure 5.1.3.

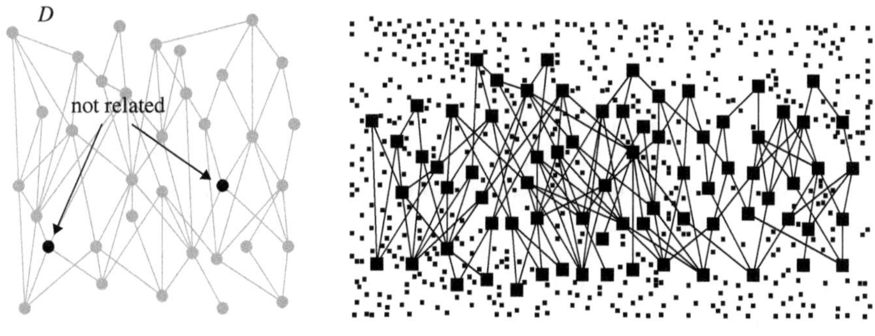

Figure 5.6.1. In the example shown here, most two-element subsets of the directed set D do not consist of the initial and terminal elements of a relation in D; coarse relation on D_2.

Power spaces; induced and holistic power spaces. The notion of relations between pairs of two-element subsets of a multidirected set $M = (M, R, i, t)$ naturally motivates consideration of relations between pairs of subsets of M of arbitrary cardinalities. Since subsets of M correspond to elements of the power set $\mathcal{P}(M)$ of M, families of relations between pairs of subsets of M correspond to multidirected structures on $\mathcal{P}(M)$, which I refer to as *power spaces* over M. The following general set-theoretic definition makes this idea precise:

Definition 5.6.2. *Let M be an arbitrary set. A* **power space** *over M is a multidirected set whose underlying set is the power set* $\mathcal{P}(M)$ *of M.*

The reason for taking M to be an arbitrary set in Definition 5.6.2 is that the general "power space viewpoint" takes multidirected structure on M to be "residual structure," left over after ignoring relations on $\mathcal{P}(M)$ involving nonsingleton subsets. This enables a unified perspective, under which "ordinary" multidirected sets are special cases of power spaces. As discussed below, it is generally not possible to recover a power space from this residual structure, but power spaces that *do* arise naturally from the structure of a "preexisting" multidirected set $M = (M, R, i, t)$, called **induced power spaces,** provide some of the most physically-interesting examples of power spaces. In fact, the majority of power spaces considered in this book *are* induced power spaces, but the reader should be aware that these are exceptional cases from an overall structural viewpoint.

Figure 5.6.3 illustrates the construction of an induced power space over a directed set D with three elements, labeled x, y, and z. The reader will recognize this directed set as one of the "toy histories" from Chapter 1, used to illustrate independent modes of influence in Chapter 3, and again to illustrate the construction of relation space in the present chapter. Since D has three elements, its power set $\mathcal{P}(D)$ has $2^3 = 8$ elements, one for each subset of D. These subsets are represented graphically inside the large nodes in the right-hand diagram in the figure, with the directed set D reproduced in light gray inside each node, to illustrate how each subset "fits into" D. Subsets are viewed as full subobjects of D, meaning that their relations are included in the picture. The light gray lines between pairs of large nodes illustrate "possible types of relations" between pairs of elements of $\mathcal{P}(D)$. An actual choice of power space structure on $\mathcal{P}(D)$ may include any or all of these types of relations, in either direction, and may include multiple instances of each type. The thick black lines indicate one particular choice of power space structure. This structure is induced by the structure of D, since it encodes *possible evolution of subsets of D under sequential growth.* For example, the empty set can only evolve sequentially into the singleton subset S_1 with element x, since the elements of the other singleton subsets of D have predecessors in D. S_1, meanwhile, can evolve into either of the two-element subsets S_2 or S_3, but the latter subset cannot evolve into D itself, since this would involve the appearance of a new element y "in the past of the existing element z."[12]

[12]This is one of the relatively few instances in the book where two isomorphic directed sets; namely, S_2 and S_3, are considered to be "different." The reason why is obvious: these sets *play different roles* as subsets of a "preexisting" directed set D, which is the object of principal interest.

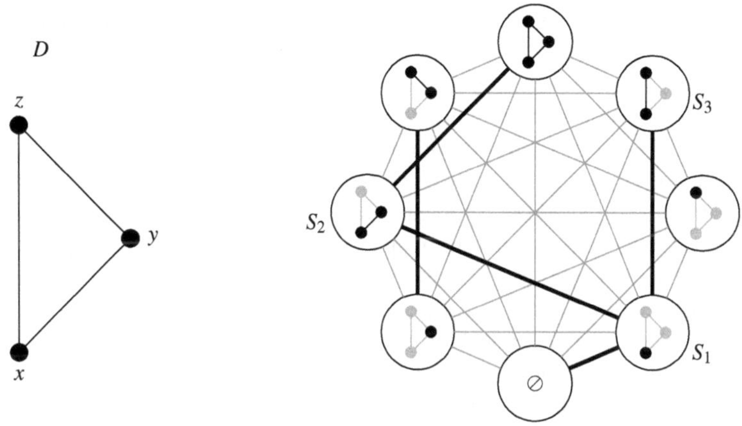

Figure 5.6.3. A power space over a three-element directed set D.

It is easy to exhibit many other examples of induced power spaces over a multidirected set $M = (M, R, i, t)$. A "trivial" example is given by taking the relation set over $\mathcal{P}(M)$ to be R, viewed as a set of relations between pairs of singleton subsets of M, with no relations involving subsets of higher cardinality. This example merely "realizes M as a power space." Other "trivial" examples may be constructed using only the set structure, and not the multidirected structure, of M. The most obvious of these is the *lattice of subsets* of M. The power space illustrated in Figure 5.6.3 is a subobject of the lattice of subsets of the underlying set of the directed set $(D, \prec)$, since a subset relationship is a necessary, though insufficient, condition for the possible evolution of one subset into another. A slightly more interesting example of an induced power space may be constructed by combining the relation set R of a multidirected set M with the coarse relation on the set M_2 of two-element subsets of M, introduced above. This defines a power space over M encoding relations between distinguished pairs of one and/or two-element subsets of M. If M is an acyclic directed set $D = (D, \prec)$, then the *enhanced star model* $D_\star$ of D, viewed as a "generalized 2-category," as described in Section 5.5, is the "nontrivial part" of this power space. However, the enhanced star model of a general multidirected set M does not embed naturally into a power space over M, since relations in M generally cannot be identified with two-element subsets of M, as explained above. A third induced power space over M is the space with one relation between each pair of subsets M' and M'' of M for each relation between an element of M' and an element of M''. At a conceptual level, this power space "lumps together families of elements in every possible way," while "keeping relations separate."

In contrast to these examples, most power spaces over a given set M are **holistic power spaces**, meaning that their information content consists at least partly of "irreducibly complex" relationships between pairs of nonsingleton subsets of M. In other words, such power spaces are *not* induced in any natural way by any "preexisting" multidirected structure on M.[13] The concept of structure "more general than multidirected structure" is already rather mainstream at the quantum level; for example, the recent paper of Oreshkov, Costa, and Brukner [OCB16] includes an interesting discussion of *"non-causal quantum correlations"* in ordinary quantum theory. The existence of such correlations provides strong evidence for the necessity of background independence in discrete quantum causal theory, because it implies that an individual directed set, or at least an individual *acyclic* directed set, is inadequate to model the causal structure of spacetime in the quantum setting. This topic is briefly revisited in Section 7.4. However, it has recently become popular in certain scientific fields, not limited to physics, to take holistic relationships among families of complex substructures seriously *even at the classical level.* I briefly discuss a few instances of this notion in Section 5.8, including the general concept of *top–down causation,* and some specific holistic approaches to fundamental physics.

Higher induced relations. Since relations in a multidirected set $M = (M, R, i, t)$ may be identified with 1-chains in M, an obvious way to define induced relations between pairs of distinguished nonsingleton subsets of M is to modify the induced relation $\prec$ on the relation space $\mathcal{R}(M)$ over M to apply to longer chains. Complications immediately arise in the general multidirected context. For example, an n-chain in a multidirected set M does not always correspond to an n-element subset of M, since chains may intersect themselves. Further, multiple chains may share the same family of elements. Hence, this approach generally involves a choice between accepting significant loss of information, as in the case of the coarse relation on M_2, and leaving the realm of power spaces, as in the theory of *causal path spaces,* introduced in Section 5.9. However, in the special case of an acyclic directed set $D = (D, \prec)$, each n-chain in D uniquely identifies an n-element subset of D, and this permits relations between pairs of chains in D to be viewed as relations between elements of the power set $\mathcal{P}(D)$ of D.

A straightforward way to define such relations between pairs of chains is to *iterate the process of forming relation spaces.* Let $D = (D, \prec)$ be an acyclic directed set, with relation space $\mathcal{R}(D)$, and set of 2-chains $\mathrm{Ch}_2(D)$. Here, it is convenient to denote the binary relation on D by $\prec_0$, and the induced binary relation on $\mathcal{R}(D)$ by $\prec_1$. Now let $w \prec_0 x \prec_0 y \prec_0 z$ be a 3-chain in D, and denote the relations $w \prec_0 x$, $x \prec_0 y$, and $y \prec_0 z$, by the ordered pairs (w, x), (x, y), and (y, z), respectively. Denote the 2-chains $w \prec_0 x \prec_0 y$ and $x \prec_0 y \prec_0 z$ by the ordered triples (w, x, y) and (x, y, z), respectively. Then there exists an *induced 2-chain* $(w, x) \prec_1 (x, y) \prec_1 (y, z)$ in $\mathcal{R}(D)$. The **second relation space** $\mathcal{R}^2(D) := \mathcal{R}(\mathcal{R}(D))$ of D corresponds to $\mathrm{Ch}_2(D)$ in an obvious way, and the **second induced relation** $\prec_2$ on $\mathcal{R}^2(D)$, which is induced by $\prec_1$, may be naturally identified with the binary relation on $\mathrm{Ch}_2(D)$ defined by taking one 2-chain to precede another if and only if these chains "overlap" in a single

[13] The caveat "naturally" is included here because one *can* define bijections between the set of *finite* multidirected structures on a *finite* set and the set of *finite* multidirected structures on its power set, since both sets of multidirected structures are countably infinite. Under such a bijection, a typical power space *could* be encoded, in an unnatural and arbitrary manner, by invoking large numbers of relations between typical pairs of elements of M.

relation in D. For example, in the case described above, $(w, x, y) \prec_2 (x, y, z)$, since these two chains overlap in the relation (x, y). One may proceed iteratively to define **higher induced relations** $\prec_n$ on the chain sets $\mathrm{Ch}_n(D)$ of D. Since D is assumed to be an acyclic directed set in this particular context, the higher induced relations $\prec_n$ descend unambiguously to define binary relations on the corresponding sets of n-element subsets of D. Any subfamily of these relations may be used to define a power space over D. To generalize this construction to the case of arbitrary multi-directed sets, one must specify *which* relations between pairs of elements are being used to construct chains, since chains no longer correspond uniquely to sequences of elements in this setting.

Figure 5.6.4 illustrates the construction of the first and second induced relations $\prec_1$ and $\prec_2$ for an acyclic directed set $D = (D, \prec)$. The left-hand diagram in the figure shows a chain in D. The middle diagram focuses on this chain, and shows how it induces a chain with one fewer element in the relation space $\mathcal{R}(D)$. As usual, square nodes represent elements of $\mathcal{R}(D)$, and relations between pairs of square nodes represent elements of the first induced relation $\prec_1$. The right-hand diagram illustrates the construction of the second induced relation $\prec_2$ on $\mathcal{R}^2(D)$. Triangular nodes represent elements of $\mathcal{R}^2(D)$, which correspond to relations between pairs of elements of $\mathcal{R}(D)$, and thence to 2-chains in D itself.

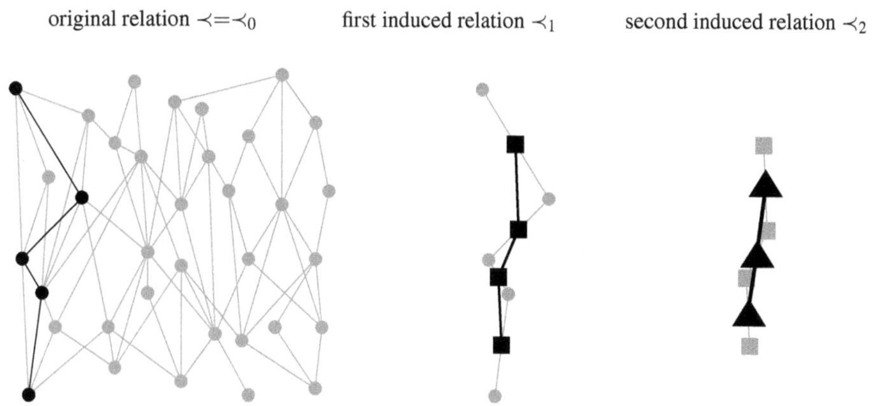

Figure 5.6.4. Higher induced relations in the acyclic directed case.

Simplicial power spaces. Whenever two overlapping n-chains $\gamma = \{x_0 \prec x_1 \prec \cdots \prec x_n\}$ and $\delta = \{x_1 \prec \cdots \prec x_n \prec x_{n+1}\}$ in an acyclic directed set $D = (D, \prec)$ each extend to n-simplices $\overline{\gamma}$ and $\overline{\delta}$ in D, i.e., totally ordered subsets of D, the relation $\gamma \prec_n \delta$ may be viewed as *"joining the maximal face of the initial simplex $\overline{\gamma}$ to the minimal face of the terminal simplex $\overline{\delta}$."* This viewpoint motivates the definition of alternative power spaces over D, "sparser" than those defined in terms of the higher induced relations described above. These **simplicial power spaces** are defined by recognizing only relations between pairs of n-element subsets of D that define simplices in D. The "nontrivial parts" of simplicial power spaces may be viewed as generalizations of the star model $D_\star$ of D, introduced in Section 4.3. Despite obvious complications, these ideas may be extended in interesting ways to apply to general multidirected sets. At a physical level, this is one possible way of viewing "triangulated spacetimes," such as those studied in the theory of causal dynamical triangulations.[14] Power spaces defined in terms of higher induced relations, and simplicial power spaces, are both examples of **binary power spaces,** i.e., power spaces involving only directed, rather than multidirected, structures on power sets.

Figure 5.6.5 illustrates chains of 2-simplices in an acyclic directed set D. The left-hand diagram in the figure shows two chains of 2-simplices, one consisting of three "consecutive" simplices, and the other consisting of two. Note that most 2-chains in D do *not* extend to 2-simplices in this particular case; i.e., D is "very nontransitive at the level of 2-chains." The right-hand diagram in the figure illustrates a different visualization of the same 2-simplices as "triangles," which is how such simplices are often represented in geometric and topological contexts. This representation emphasizes that the simplices are to be viewed as "individual entities" in the corresponding power space. From this viewpoint, a multidirected set may be regarded as a *directed simplicial complex of mixed dimension.* This provides a clear picture of how "triangulated spacetimes" may be modeled in the discrete causal context. For example, a causal dynamical triangulation involves a very special type of directed set, whose dimension as a simplicial complex is artificially fixed to agree with the apparent large-scale dimension of spacetime.[15] A recent paper of Bianconi and Rahmede [BR15] gives an interesting discussion of the properties of simplicial complexes of constant dimension in the context of quantum gravity.

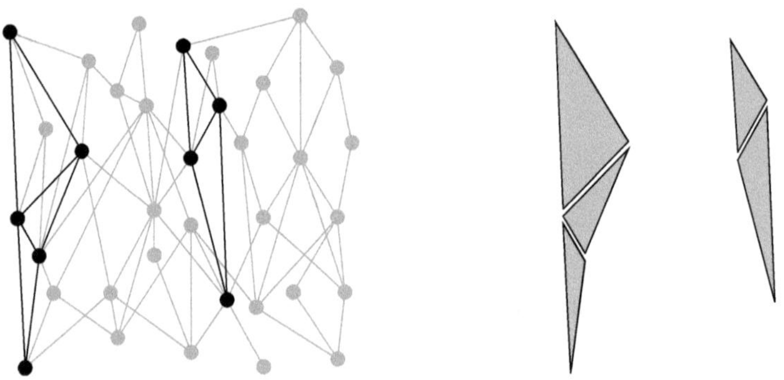

Figure 5.6.5. Chains of 2-simplices in a multidirected set.

[14]However, in causal dynamical triangulations, edges may be either "causal," i.e., timelike or null, or "spacelike." Here, all the edges are "causal."

[15]As described in the previous footnote, a causal dynamical triangulation also involves spatial information. Interestingly, computer simulations suggest that the dimension of spacetime in causal dynamical triangulations is scale-dependent, despite the assumption of four-dimensionality at the fundamental scale! See [AJL05] for details.

Splice relations. Higher induced relations, and the binary relations defining simplicial power spaces, are special cases of **splice relations**, in which pairs of chains, or more complicated "generalized elements" associated with a multidirected set M, are taken to be related if they may be "spliced together" in some specified way, to form a "generalized element of greater causal size." Since there is often more than one way to "splice together" a given pair of "generalized elements," splice relations often define multidirected, rather than merely directed, structures. "Generalized elements" may often be formalized in terms of *spaces of morphisms into* M, such as the *causal path spaces* introduced in Section 5.9. Such spaces of morphisms generally *cannot* be identified with power spaces over M in any obvious way, since morphisms into M generally involve information that cannot be recovered from the images of their element maps. Besides this basic information-theoretic distinction, typical "interesting" splice relations differ in an important *qualitative* respect from the relations featured in the present section, including higher induced relations and relations defining simplicial power spaces. The latter relations involve pairs of chains or simplices "overlapping in increasingly long subchains or subsimplices,"[16] and are therefore "highly redundant," in an obvious sense. It is often more useful to work with splice relations involving pairs of generalized elements of "small or empty intersection." These include the *concatenation relation* and the *directed product relation* for pairs of causal paths in a multidirected set, introduced in Section 5.9. These relations define special examples of causal path spaces, called the *causal concatenation space,* and the *causal directed product space,* respectively.

5.7 Causal Atomic Resolution

Causal atoms. A useful application of induced power spaces is to "organize information in a multidirected set" into families of subobjects, for the purposes of approximation and computation. Here I present a method of using "nested families of subobjects" of a multidirected set to approximate structure at different scales. Such approximation methods are important, for practical purposes, in the study of spacetime structure, since the scale of individual elements may be many orders of magnitude too small to be accessible to direct observation in the foreseeable future. The most interesting such approximation methods are those that organize information in a manner "compatible with the multidirected structure." The specific method outlined in this section involves grouping elements of a multidirected set into subsets, called *causal atoms,* or more generally, *local causal atoms,* which may be viewed as elements of a "coarser" multidirected set. The reader should compare this method to the "decimation approach to coarse-graining of causal sets," appearing, for example, in [SR01]. The utility of causal atoms has already been illustrated in the technique of *atomic accretion* used in the proof of Theorem 4.6.3.

[16]This is imprecise language; since the objects being spliced together are properly defined in terms of morphisms, one should really speak in terms of *restrictions,* not "subobjects." However, the meaning is reasonably clear in the present context.

A causal atom α in a multidirected set $M = (M, R, i, t)$ is a "convex subset" of M in a generalized order-theoretic sense, meaning that every element lying on a chain in M between a pair of elements in α also belongs to α. Under this description, the analogy between causal atoms and convex subsets of Euclidean space $\mathbb{R}^n$ is obvious, since the definition of convexity in the latter setting is that every point lying on a line segment connecting two points in a convex subset also belongs to the subset. Figure 5.7.1 illustrates this analogy. The left-hand diagram in the figure shows a convex subset U and a non-convex subset V of $\mathbb{R}^n$, while the right-hand diagram illustrates a causal atom and a non-atom in a multidirected set. In the non-atom, the chain $x \prec y \prec z$ violates order-theoretic convexity. The figure is drawn in such a way as to emphasize only the *elements* involved, but causal atoms may also be viewed as full subobjects of a directed or multidirected set, by "recognizing their internal structure."

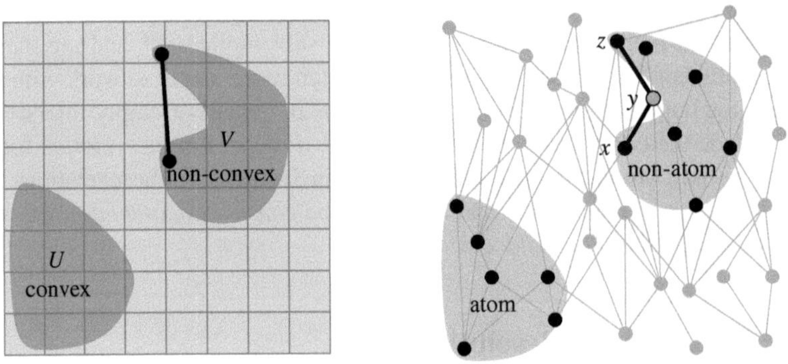

Figure 5.7.1. Analogy between convex subsets of $\mathbb{R}^n$ and causal atoms in a multidirected set.

It is convenient to state the definition of causal atoms in terms of pasts and futures, rather than in terms of chains. This definition obviously has the same meaning as the description given above, since pasts and futures are themselves defined in terms of chains.

Definition 5.7.2. *A subset α of a multidirected set $M = (M, R, i, t)$ is called a* **causal atom** *if every element in the intersection $J^-(\alpha) \cap J^+(\alpha)$ belongs to α.*

The definition implies that every element in the complement $M - \alpha$ of α is either exclusively in the past of α, exclusively in the future of α, or unrelated to any element of α. In particular, α *"absorbs any cycle it intersects."* Hence, the possible relationships between α and an element in $M - \alpha$ are analogous to the possible relationships between a pair of elements in an acyclic directed set. Causal atoms may therefore be viewed as "generalized elements," which explains the choice of terminology. To temper the "absorption property" of causal atoms in multidirected sets involving large cycles, one may define **local causal atoms,** which include all elements lying on "sufficiently short" chains between pairs of their elements. The motivation for defining local causal atoms is that they may be used to represent *"locally causally self-contained subsets of classical spacetime."* Since cycles in multidirected sets are analogous to closed causal curves in general relativity, relativistic scenarios may be used to help clarify how the idea of local causal atoms is relevant to local experimentation and observation. For example, in a relativistic spacetime manifold containing "large closed causal curves," it might be theoretically possible, but realistically impractical, for elements of a local region in an Earth-bound laboratory to communicate along a causal curve leaving the region. Actually achieving such communication might require sending messages along extended paths through wormholes connecting remote galaxies to Earth's distant past, rendering the effect of such communications negligible for practical purposes. It would be pointless to consider such gigantic structures as "generalized elements" in this context; it is more suitable to formulate such notions locally.

Atomic topologies. Relationships between pairs of causal atoms in a multidirected set $M = (M, R, i, t)$ are generally more complicated than relationships between pairs of elements of M, or between a causal atom in M and an element of M. If M is acyclic, then its individual elements are causal atoms, but in the general case, elements lying on cycles in M are *"smaller than any causal atom,"* due to the absorption property. The open interval $\langle\langle x, z\rangle\rangle$ between two elements x and z in M is a causal atom, since every chain between two elements in $\langle\langle x, z\rangle\rangle$ extends to a chain from x to z. Antichains are causal atoms, since they admit no chains between pairs of their elements. The images of maximal cycles[17] are causal atoms, since any element belonging to both the past and future of a cycle lies on an intersecting cycle. The family of causal atoms in M is closed under intersection, so appropriate subfamilies can serve as *bases,* rather than merely *subbases,* for topologies on the underlying set of M, called **atomic topologies**. In particular, intersections of open intervals are causal atoms, so the interval topology is an atomic topology. The discrete topology is atomic only in the acyclic case. Many other examples of atomic topologies exist. The star topology on the star model $D_\star$ of an acyclic directed set D is an atomic topology for the "2-category-type directed structure" that results from elevating $D_\star$ to the "enhanced star model," as described in Section 5.5. More generally, however, the star $\mathrm{St}(x)$ at an element x in a multidirected set M may be "smaller than any causal atom containing x." *Local atomic topologies* may also be defined, in terms of local causal atoms.

Causal atomic decomposition. The principal reason for studying causal atoms, as mentioned above, is that they may be viewed as "generalized elements" of a multidirected set $M = (M, R, i, t)$, whose relationships represent a "coarsening" of the structure of M. This provides a precise means of distinguishing "structure at different scales," in a context where metric notions are emergent.

---

[17] A cycle ω is regarded as "maximal" in this context if its image contains the image of any other cycle it intersects. One may generally define cycles with "larger sources" that "wrap more times around the same image," but only the images are relevant in this context.

Definition 5.7.3. *Let $M = (M, R, i, t)$ be a multidirected set. A* **causal atomic decomposition** *of M is a partitioning of M into causal atoms.*

It is sometimes useful to relabel M as M^0, while the causal atoms in the decomposition are viewed as elements of a "coarser" set M^1. This set may be endowed with multidirected structure in a variety of different ways. In particular, it is possible, though often undesirable, to define multidirected structure on M^1 in such a way that it contains "higher-level cycles," even though every cycle in the original multidirected set M is absorbed into a unique causal atom. For example, if a pair of causal atoms α and α' in M each contain elements in the past of the other, it might be reasonable to define relations in both directions between α and α' in M^1. *Local causal atomic decompositions* may also be constructed. In this case, it is often natural for "large cycles" in M to induce "smaller cycles" in M^1, since local causal atoms in M may not be large enough to absorb large cycles. Multidirected structures arising from either type of decomposition define power spaces over M, since causal atoms and local causal atoms in M correspond uniquely to subsets of M. Note that any family $\{\alpha_i\}_{i\in I}$ of causal atoms in a multidirected set M, where I is some index set, extends to a causal atomic decomposition of M. Indeed, suppose that x is an element of M not belonging to any atom α_i in the specified family. If x does not lie on a cycle in M, then $\{x\}$ is itself a causal atom. If x does lie on a cycle in M, then the image of the maximal cycle[18] in M containing x is a causal atom disjoint from $\bigcup_{i\in I}\alpha_i$, since any cycle intersecting a causal atom must lie entirely inside the atom. Hence, every element of M not belonging to a member of the family $\{\alpha_i\}_{i\in I}$ belongs to a causal atom in M disjoint from every atom in the family.

A causal atomic decomposition of a multidirected set $M = M^0$ is illustrated in the left-hand diagram in Figure 5.7.4, with the elements represented by the nodes in each light gray region grouped together as causal atoms. The middle diagram in the figure illustrates one possible method of assigning a multidirected structure to the set M^1 of causal atoms in this decomposition. In this case, the chosen structure is

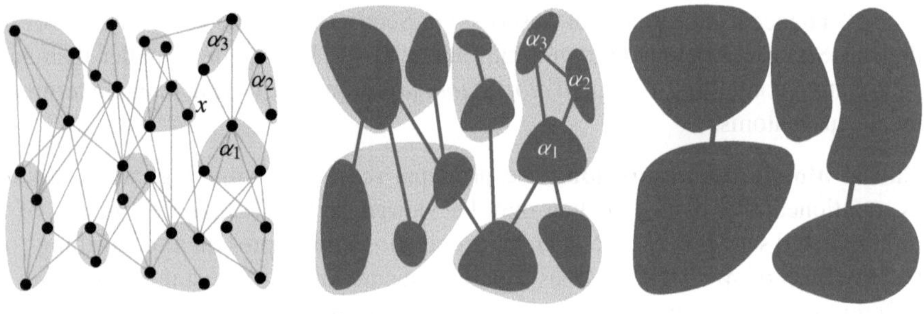

Figure 5.7.4. Grouping elements of a multidirected set $M = M^0$ into causal atoms; causal atomic decomposition M^1, with elements grouped for further decomposition; second decomposition M^2.

[18]More precisely, this means the "maximal subset of M containing x that is the image of a cycle in M."

represented by a binary relation $\prec$, where $\alpha \prec \alpha'$ in M^1 if and only if the following three conditions are satisfied: (1) every element of α precedes some element of α' in M; (2) no element of α succeeds any element of α' in M; (3) some element of α *directly* precedes some element of α' in M. There is nothing particularly special about this choice of binary relation, which yields a very "sparse" structure for M^1 in this case.

The further grouping of elements of M^1 indicated by the light gray regions in the middle diagram represents a second causal atomic decomposition, whose abstract structure is shown in the right-hand diagram. The use of "irregularly-shaped planar regions" to represent elements of M^1 and M^2 is for illustrative purposes only, and signifies nothing about the actual structure of these sets. Indeed, once one has passed from a multidirected set M to a causal atomic decomposition of M, the "internal structure" of the causal atoms involved in the decomposition is forgotten, and these atoms are viewed as abstract, structureless elements, just like the elements of any other multidirected set.

Foliations; generational dynamics. Although the foregoing discussion of causal atoms is deliberately tailored to the discrete context, the same basic idea remains both interesting and important in non-discrete settings, including general relativity. Here, the analogy involving convexity remains equally valid, but the role of chains is played by the appropriate continuum-based analogue; namely, causal curves.[19] An important instance of causal atomic decomposition in this context is the familiar method of foliating a region of a relativistic spacetime manifold X via Cauchy surfaces, as discussed in Section 2.7. The left-hand diagram in Figure 5.7.5 illustrates twelve "nearby" Cauchy surfaces in X, which serve as leaves of such a foliation. In fact, the leaves of a foliation *cover* the specified region of X, meaning that there are no "gaps" between them; every element in the region belongs to a leaf of the foliation. Sufficiently accurate intuition, however, is supplied by thinking of the leaves as

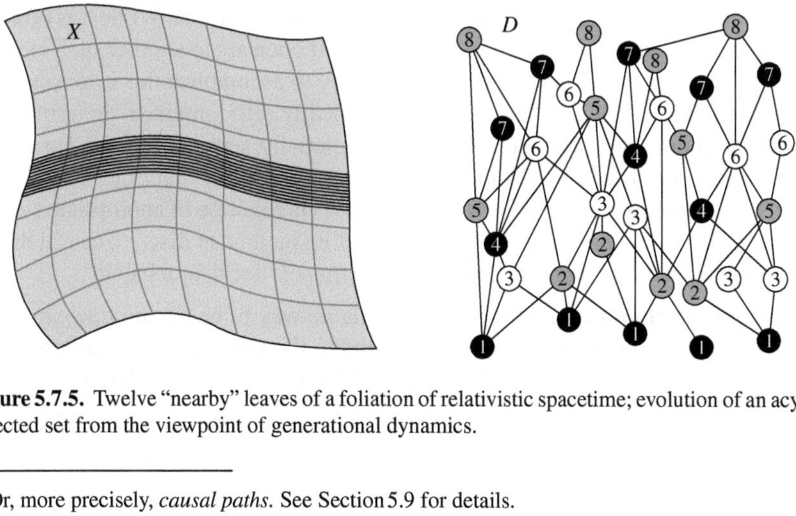

Figure 5.7.5. Twelve "nearby" leaves of a foliation of relativistic spacetime; evolution of an acyclic directed set from the viewpoint of generational dynamics.

[19]Or, more precisely, *causal paths*. See Section 5.9 for details.

"sheets lying on top of one another," as suggested by the diagram. Physically, each leaf represents an "instant in time" under suitable choices of local coordinates on X. Of course, a *global* foliation by Cauchy surfaces requires X to satisfy the globally hyperbolic condition.

A discrete causal analogue of such a foliation is given by a choice of *generational dynamics,* in which one *"builds up an acyclic directed set D by adding one layer of elements at a time."* This method, further discussed in Chapters 7 and 8, generalizes the theory of *sequential growth dynamics* for causal sets, introduced by Sorkin and Rideout [SR99]. The right-hand diagram in Figure 5.7.5 illustrates a choice of generational dynamics for an acyclic directed set D. Here, all the elements in the "first generation of D," labeled with numeral 1's, "appear simultaneously at the first stage of evolution." These elements make up an antichain in D. Next, all the elements in the "second generation," labeled with numeral 2's, appear, together with all relations between the first and second generations. The process then continues, adding the third, fourth, and subsequent generations. For a given acyclic directed set D, there generally exist many different generational dynamical processes leading to D, which is roughly analogous to the existence of many different frames of reference in relativistic spacetime.[20] Most of the antichains involved in this particular example are highly permeable, which is inevitable, due to the fact that the dynamics is described from the element space viewpoint. This would pose a technical problem if one wished to use this specific acyclic directed set D as a fixed background for a quantum theory. However, in the background independent context, path summation is performed at a "higher level of structure," in which specific directed sets such as D are viewed as merely "elements of a kinematic scheme." In this setting, the permeability of antichains in D, defined by a choice of generational dynamics, is innocuous, as explained in Section 7.7.

Local causal atoms and local causal atomic decompositions, like their corresponding global versions, may be generalized beyond the discrete context. A simple example of a local causal atomic decomposition in the cyclic interpolative case is the partitioning of a directed circle into a sequence of half-open arcs. While it is possible in this example to move from one element of a half open arc to another such element along a directed path exiting the arc, the only way to accomplish this is to travel "all the way around the circle." Hence, the arc is "locally self-contained" in terms of its directed structure, even though it fails to be "globally self-contained." Turning to a more physically relevant example, a relativistic spacetime manifold X including closed causal curves may often be studied locally, via foliation of appropriate subsets that "avoid global pathologies." The corresponding method in discrete causal theory is *local generational dynamics,* described in terms of "local antichains."

Causal atomic resolution. At present, there is no way to be certain how small the fundamental scale might actually be, compared to the scales presently accessible to direct observation. However, as described in Chapter 4, it is likely that these scales are separated by many orders of magnitude. Assuming, conservatively, a difference

[20] Closer analogues of relativistic frames of reference may be described in terms of *refinements of causal relations,* as described in Chapter 7.

of 15 orders of magnitude in both "spatial" and "temporal" scales, a causal atom of presently-observable size might contain 10^{60} fundamental elements, provided that "four-dimensionality" remains sufficiently predominant as scales decrease. This number is so large that such an atom might involve many different levels of qualitatively different structure, just as the known interactions manifest themselves in different degrees over similar ranges of scales. For this reason, a single causal atomic decomposition of a region of classical spacetime is unlikely to yield a coarsened structure that captures interesting new behavior in an effective manner. A straightforward response to this problem is to *iterate* the process of causal atomic decomposition, to progressively "resolve" the information in a directed set down to the fundamental scale. The same idea of resolution remains useful in the more general context of multidirected sets.

Definition 5.7.6. *Let $M = (M, R, i, t)$ be a multidirected set. A* **causal atomic resolution** *of M is a sequence*

$$\mathbf{M} := \{M^0, M^1, M^2, \ldots\}$$

of multidirected sets $M^n = (M^n, R^n, i^n, t^n)$, where $M^0 = M$, and where M^{n+1} is a causal atomic decomposition of M^n for each $n \geq 0$. M^n is called the nth **level** *of* $\mathbf{M}$.

Analogous constructions may be carried out locally. Note that this general definition does not discriminate among the many possible ways of defining multidirected structure on each M^n, but only requires that the "next set M^{n+1} at each stage of the resolution respects previously defined structure." An annoying technical point is that the construction of a causal atomic decomposition is generally *not* a transitive operation. This means that if $\alpha_1, \ldots, \alpha_k$ are elements of M^{n+1}, such that the subset $\{\alpha_1, \ldots, \alpha_k\}$ of M^{n+1} is a causal atom, then the union of the elements of $\alpha_1, \ldots, \alpha_k$, *viewed as subsets of the "previous" multidirected set M^n*, is generally not a causal atom in M^n. For example, the element x in the left-hand diagram in Figure 5.7.4 belongs to the intersection of the past and future of the seven-element set $\alpha_1 \cup \alpha_2 \cup \alpha_3$. Hence, this set is *not* a causal atom in $M = M^0$. However, the three-element set $\{\alpha_1, \alpha_2, \alpha_3\}$ *is* a causal atom in M^1, since the process of constructing M^1 "forgets the more-detailed structure of M^0." For a single causal atomic decomposition, this detail is immaterial, but in a causal atomic resolution, it can be inconvenient. This provides yet another example in which a notion analogous to category-theoretic composition fails to apply in discrete causal theory. More generally, "failure of composition" is not unusual in "information-filtering constructions." A familiar example is given by *composition series* of finite groups in the context of the *Jordan–Hölder theorem,* since formation of normal subgroups is nontransitive.

There exist a number of different ways to characterize a causal atomic resolution $\mathbf{M}$ in terms of the properties of its levels M^n, and the relationships among them. For example, in the case of finite atoms, the nth **growth factor** λ_n of $\mathbf{M}$ is the average number of elements of M^{n-1} making up an element of M^n. If the sequence $\Lambda(\mathbf{M}) := \{\lambda_1, \lambda_2, \ldots\}$ of growth factors of $\mathbf{M}$ is constant, or "near-constant," over a certain range of indices, then $\mathbf{M}$ is called **exponential** over this range, since the cardinalities in M^0 of typical atoms at nearby levels are related, on average, by powers of the common growth factor. Similarly, if the sequence $\Lambda(\mathbf{M})$ is decreasing or increasing, respectively, over a certain range of indices, then **M** is called **subexponential** or **superexponential**, respectively, over this range. The growth factors of the causal atomic resolution illustrated in Figure 5.7.4 are $\lambda_1 = 35/12 \approx 2.92$ and $\lambda_2 = 12/5 = 2.4$, so this particular resolution is "mildly subexponential." For comparison, an exponential causal atomic decomposition with a growth factor of 2.5 would require about 50 levels to "interpolate" between the Planck scale and the smallest scale presently accessible to direct observation.

There are also some obvious ways to characterize the *uniformity* of a given level M^n of a causal atomic resolution **M**. For example, one might use elementary statistics to measure the distribution of different "sizes" of causal atoms, described either in terms of raw cardinality, or in terms of local causal structure. One might also attempt to quantify the distribution of "shapes" of causal atoms in a causal atomic resolution. For example, generational dynamics involves "broad, flat" atoms. More generally, the order-theoretic convexity property implicit in the definition of a causal atom tends to limit the existence of "long, narrow" atoms in most multidirected sets. An interesting technical topic to consider is the notion of *"cyclic-to-non-cyclic phase transitions"* in a local causal atomic resolution of a multidirected set including cycles. The idea here is that cycles might exist across a limited range of scales, and such cycles would eventually be absorbed by increasingly large local causal atoms. Levels of the resolution "above the top level" of such an absorption process would be *global* causal atomic decompositions of their previous levels. More generally, causal atomic resolutions, along with their local counterparts, supply methods of describing other qualitative changes in structure across specific ranges of scales. These include changes in emergent dimension, arising in discrete causal analogues of the Kaluza–Klein mechanism, string compactifications, and similar continuum-based scenarios. Such changes in dimension have already been suggested by computer simulations in the context of causal dynamical triangulations.[21]

At a broad conceptual level, the theory of causal atomic resolution is analogous to *multiresolution analysis* in the study of wavelets in the field of harmonic analysis. The general aim of both techniques is to provide a unified method of dealing with aspects of structure existing at different scales. This viewpoint is of particular interest in physical or mathematical settings in which emergent phenomena are prominent, since emergence is often scale-dependent in some sense. On a very general basis, an important key to formulating successful versions of such methods is to make "natural" use of whatever type of structure is involved. Causal atomic resolution is a strikingly clear example of this. What is particularly pleasing about the method, however, is how conveniently it fits into the overall conceptual panorama of discrete causal theory, in which spaces of elements, relations, chains, causal atoms, and ultimately directed sets themselves, fall under a unified structural umbrella.

[21] More generally, see the recent paper of Carlip [CA15], already referenced in Chapter 2, which discusses possible "dimension reduction" in discrete gravity to yield an essentially 2-dimensional structure at small scales.

5.8 Top–Down Causation; Classical Holism

Radical possibilities regarding hidden structure. This section deviates from the main course of development of the chapter, by examining the possibility of certain variants of causal structure radically different from anything naturally suggested by the metric recovery results described in Chapter 2. In particular, this section examines scenarios under which elements of a directed or multidirected set, viewed as a model of classical spacetime, may represent *fundamental substructures larger and more complicated than individual events.* This viewpoint goes far beyond the pragmatic approach of *causal atomic resolution,* described in Section 5.7, in which nontrivial subobjects of multidirected sets are "collapsed" into individual elements, for the purposes of approximation and computation. In this section, directed relationships at the classical level between entities possessing complicated internal structure are deliberately interpreted as representing instances of *classical holism,* i.e., interactions which cannot be reduced to local descriptions in the usual sense. Of course, such interactions can, and should, be treated as "local" under the abstract structural viewpoint described in Chapter 4, where "influence independent of intervening influences" is local *by definition.* However, individual elements of the multidirected sets involved in such holistic scenarios are typically interpreted as representing "extended regions of classical spacetime," rather than "individual events." From an abstract perspective, the viewpoint here is very similar to Grothendieck's treatment of algebraic schemes in algebraic geometry, in which individual elements may represent complicated substructures, such as irreducible varieties. Hence, this section relies on the *principle of hidden structure* (HS) in an essential way.

Top–down causation. A number of prominent physicists, computer scientists, chemists, biologists, neuroscientists, and philosophers have recently drawn attention to the concept of classical **top–down causation**, which challenges the well-established philosophy of **classical reductionism**. The latter philosophy, applied to the discrete causal setting, assumes that classical causal structure may be described completely in terms of relations between pairs of individual events, as implied by the binary axiom (B) in the "suggested list" of axioms for discrete classical causal theory appearing in Section 4.10. Representative samplings of the top–down viewpoint appear in the recent collections *Top–Down Causation* [ENO09], and *Downward Causation and the Neurobiology of Free Will* [MEO09]. Certain ambitious versions of top–down causation either approach or cross the threshold of **classical holism**, which may be expressed, in the discrete causal context, by the assertion that classical causal structure *cannot* be described completely in terms of relations between pairs of individual events. As George Ellis [EL05] puts it, *"the higher levels in the hierarchy of complexity*[22] *have autonomous causal powers that are functionally independent of lower-level processes."* The implication is not that the physical laws invoked under the usual reductionist approach to classical physics are *wrong,* but rather that

[22] As Ellis' statement suggests, the "levels of hierarchy" involved here are distinguished in terms of *complexity,* not in terms of "classical versus quantum-theoretic interpretation," as in the principle of *iteration of structure* (IS) in discrete causal theory.

the reductionist viewpoint is *incomplete;* i.e., additional, independent influence is ascribed to holistic mechanisms.

To evaluate the legitimacy and implications of the top–down viewpoint, it is necessary to formulate the idea in precise mathematical terms. Since top–down causation involves influences exerted "irreducibly" by entire families of events, the theory of *power spaces,* developed in Section 5.6, is a natural tool to use in describing it. Different implementations are possible, with conservative versions remaining "essentially reductionist," while more radical versions are unambiguously holistic. Alternatives to the power space viewpoint for describing holistic behavior are also worth considering. In particular, even the most general power spaces preserve a certain *binary* character, by focusing on relationships *between pairs* of families of elements, rather than relationships *among families* of families of elements. There are excellent physical motivations for this focus, but one may still choose to generalize further, and allow for *ternary, N-ary,* and other types of relationships. Some of these ideas are briefly revisited in Section 8.7.

Conservative top–down scenarios. A conservative, and conceptually mundane, version of the top–down viewpoint may be adopted in the context of causal atomic resolution, described in Section 5.7, where individual influences are ascribed to entire families of events purely for convenience. Power spaces arise naturally in this setting, by "patching together" the various levels of a resolution, but these power spaces do not represent any essential physical information beyond that contained in the original multidirected set. Rather, they merely represent ways of organizing and filtering this information. They certainly do *not* involve any profound hypotheses about the nature of the physical universe. One way to emphasize this point is to note that the "approximate" multidirected structures given by causal atomic decompositions may "innocuously" exhibit properties that would arouse controversy in more fundamental contexts. For example, two causal atoms may be "related to each other," in the sense that each contains events in the other's past, yet this is irrelevant to the controversial issue of causal cycles. Similarly, one causal atom may be related to another, even though some of the corresponding events are *unrelated,* yet this is irrelevant to the controversial issue of "nonlocal" interaction.

More interesting, though still conservative, versions of the top–down philosophy may be invoked in the study of exact, rather than merely approximate, induced relationships between pairs of families of elements in a multidirected set $M = (M, R, i, t)$. Although the corresponding power spaces may still be recovered from the structure of M in this context, there may exist alternative representations that better organize the information contained in M. Choosing such a representation is analogous to choosing a "convenient basis" from among a linearly dependent set of vectors. For example, the relation space $\mathcal{R}(M)$ over a multidirected set M, together with its induced relation $\prec$, gives an alternative representation of the information contained in M, except at the boundary ∂M of M. The disappearance of the permeability problem upon passage to $\mathcal{R}(M)$, made precise by Theorem 5.4.3, illustrates the fact that advantages may sometimes be gained by working in terms of one representation rather than another. This example demonstrates that the reductionist viewpoint is not always the most *useful* one, even when reductionism is information-theoretically adequate. The argument that "emergence" and reductionism are not mutually exclusive has been recognized for some time; see, for example, the classic paper by Anderson [AN72], or the recent paper by Butterfield [BU11].

An important example of a classical continuum-based alternative representation of "spacetime information" is given by the *twistor space* T over Minkowski spacetime $\mathbb{R}^{3+1}$ in Roger Penrose's *twistor theory*, introduced in his 1967 paper *Twistor Algebra* [PE67]. T may be identified with a five-dimensional real hypersurface in the complex projective space $\mathbb{CP}^3$, which is six-dimensional as a real manifold.[23] Elements of T correspond to null lines in $\mathbb{R}^{3+1}$, physically interpreted in the usual manner as the spacetime paths of rays of light. As mentioned at the end of Section 5.1, twistor space corresponds bijectively to a quotient space of the *null subspace* of the relation space $\mathcal{R}(\mathbb{R}^{3+1})$. From the discrete causal perspective, however, it is perhaps more interesting to examine analogies between twistor space and the *full* relation space $\mathcal{R}(M)$ over a discrete multidirected set $M = (M, R, i, t)$. The relations incident at a given element x in M may be viewed as roughly analogous to "minimal null segments" at x, although there is no general distinction between "timelike" and "null" relations in this context. In causal set theory, *irreducible* relations i.e., "links," are sometimes treated as close analogues of null segments; see, for example, Sorkin's paper [SO09B]. The discussion involving sprinkled causal sets in Section 4.5 explains the sense in which irreducible relations in this context are "almost null." One might interpret this as an indication that a discrete causal analogue of twistor space should be something like a "quotient of $\mathcal{R}\big(\mathrm{sk}(D)\big)$ for a directed set $D = (D, \prec)$." However, even a typical star-infinite causal set induced by global sprinkling into $\mathbb{R}^{3+1}$ does not contain a single "truly null" relation in its relation space, in the sense that no pair of sprinkled elements lie on a common null line in $\mathbb{R}^{3+1}$.[24]

It is interesting to compare the basic role of "spatiotemporally nonlocal" null lines as *individual entities* in twistor theory, with the rather subtle status of spatiotemporal locality in discrete causal theory. In particular, the possibility that emergent spatiotemporal locality may remain sufficiently fuzzy to produce noticeable effects at scales far above the Planck scale, as suggested by Sorkin [SO09A] and others, indicates that the twistor space viewpoint may offer interesting complementary ideas to the Minkowski space viewpoint in this setting. It is tempting to hope that certain twistor space methods and insights, which usually represent "equivalent alternatives" in conventional continuum-based physics, may possess *essentially different* discrete causal analogues. One of the most attractive features of twistor theory in general is that it permits the conversion of certain important but intractable equations

[23] This is not standard notation, and in fact the term "twistor space" is used for a variety of different objects arising in twistor theory. For example, Ward and Wells [WW90] define "twistor space" to be a fixed four-dimensional complex vector space $\mathbb{T}$; the corresponding complex projective space is the copy of $\mathbb{CP}^3$ cited here. The subspace corresponding to the set of null lines in $\mathbb{R}^{3+1}$ is a real hypersurface in $\mathbb{CP}^3$, i.e., a five-dimensional real submanifold; it is this latter space that I denote by T, and call "twistor space."

[24] To see why this is true, note that the number of such pairs is countable, while the number of possible "slopes" in $\mathbb{R}^{3+1}$ is uncountable.

involving relativistic spacetime structure into more natural and manageable forms. In a roughly analogous manner, passage to relation space allows a natural and straightforward derivation of causal Schrödinger-type equations, such as Equation (1.1.2). These derivations are described in detail in the second half of Chapter 6.

Radical top–down scenarios. The most radical versions of top–down causation involve models of spacetime containing more information, or different information, than can be naturally encoded in any multidirected structure on their underlying sets. Holistic power spaces represent a broad class of such models. Such irreducible holism at the classical level is broadly agreed by the scientific community to be at least *strongly constrained* by empirical considerations, for example, by experimental restrictions on deviations from spatiotemporal locality. However, this view is not universal. A modern theory incorporating a radical degree of classical holism is Julian Barbour's *shape dynamics* [BA12]. Shape dynamics does *not* fall under the umbrella of causal theory, discrete or otherwise. Rather, it takes a nearly opposite, "space-first" approach to fundamental spacetime structure, beginning with *conformal geometries,* viewed as "purely spatial structures." As in any physical theory, however, consideration of evolutionary processes is inevitable in this context, and this suggests the possibility of interesting connections between shape dynamics and *generational dynamics* in discrete causal theory. This idea is briefly revisited in Section 8.8.

In shape dynamics, one spatial state of the universe is taken to succeed another along a unique evolution curve in *shape space,* which is identified as the *conformal superspace* i.e., configuration space, of all conformal 3-geometries.[25] In a broader context, conformal superspace is only one possible choice of shape space; for example, one may define *discrete shape spaces.* Underlying shape dynamics is the philosophy of Ernst Mach (1838–1916), which states that *global structure determines local effects.* Although Mach's ideas influenced Einstein in the development of general relativity, they have since become a "strange attractor" for dubious and incoherent theorizing. Carlo Rovelli [RO04] demonstrates the disunity of modern "Machian" ideas in an amusing manner, by listing *eight* different versions of "Mach's principle."[26] Shape dynamics represents the utmost extreme of classical holism: every aspect of each "state of the universe" is relevant to all subsequent behavior. Among other "eyebrow-raising" implications, this entails a *universal time parameter* and *absolute simultaneity.* Shape dynamics also favors a spatially closed universe, which is disfavored by certain conventional interpretations of current cosmological observations. Despite this, the theory succeeds in reproducing certain results of the *ADM formulation* of general relativity.

Figure 5.8.1 illustrates a simple example of a radically holistic discrete spacetime model S. The left-hand diagram in the figure shows the underlying set of S, which consists of four elements, labeled w, x, y, and z, interpreted as representing "events." The conceptual meaning of "event" in this context is the usual one: a "specific location at a specific time." What is radical about S is that much of its structure consists of relationships between pairs of subsets that are not singleton subsets. In particular,

[25] See Barbour, Koslowsi, and Mercati [BA14] for details.

[26] See [RO04], p. 76.

although the events represented by w and x each individually influence the events represented by y and z, holistic influences also play a role; these include the "influence of x on $\{y, z\}$," and the "influence of $\{w, x\}$ on z," illustrated in the middle diagram in the figure. Note that one may still recognize an "arrow of time" in this context. More generally, the structure of S may still be viewed as "causal structure," and familiar intuition remains useful in analyzing this structure. For example, it makes sense to think of w and x as representing the "first generation" of S, while y and z represent the "second generation." However, the binary axiom (B) of discrete causal theory, as stated in Section 4.10, fails to hold for S, because the "causal structure" of S is not described via a binary relation on S.

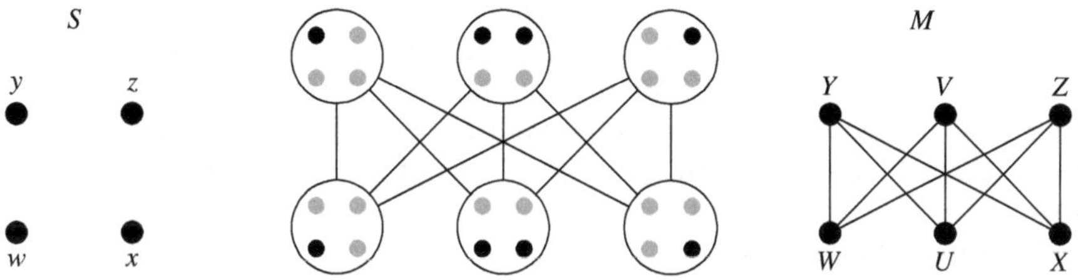

Figure 5.8.1. Radically holistic discrete spacetime model S; relationships among subsets of S; multidirected set M representing S.

It is still possible, however, to represent the relationships between subsets of S in terms of an **associated multidirected set** $M = (M, R, i, t)$, which arises as a power space over S. In this particular case, M is the acyclic directed set shown in the right-hand diagram in Figure 5.8.1. It represents, in an obvious way, an abstract version of the information encoded in the middle diagram. Some, but not all, of the elements of M represent "individual spacetime events." In particular, the elements of M labeled W, X, Y, and Z represent the same spacetime events as the elements of S labeled w, x, y, and z. However, the element U of M represents the two-element set $\{w, x\}$ of S, and the element V of M represents the two-element set $\{y, z\}$ of S. Any model of classical spacetime defined in terms of relations between pairs of families of events induces an associated multidirected set M in a similar fashion. However, for a typical "large, radically holistic classical universe" S, the vast majority of elements of the associated multidirected set M will *not* represent individual spacetime events, simply because most subsets of a large set are not singleton subsets. This means that the "naïve geometry" of M, understood in a "metric recovery-type context," will generally be related to the "physical geometry" of S only in a very indirect way. For this reason, the metric recovery results described in Chapter 2 do not seem to provide particularly strong motivation for a radically holistic universe. Further, it is generally impossible to recover the entire structure of S from M; for example, it is impossible to tell by looking at the right-hand diagram in Figure 5.8.1 which elements of M represent two-element subsets of S, and which elements of M represent singletons.

5.9 Causal Path Spaces

Path spaces in mathematics and theoretical physics. I now begin to develop the technical foundations for the theory of path summation over a multidirected set, which leads to a background independent discrete causal analogue of Feynman's path summation approach to ordinary quantum mechanics and quantum field theory, often called the *path integral approach* in conventional settings. Path summation over an arbitrary multidirected set is introduced in Section 6.2, and is applied in Sections 6.6–6.10 to construct the new quantum theory. Despite the vast differences in basic structural assumptions between Feynman's original setup and the novel scenarios studied in this book, both theories are described in terms of spaces of paths, either in models of classical spacetime, or in configuration spaces built from such models, such as kinematic schemes. The importance of such path spaces in mathematics and theoretical physics goes back too far, and originates from too many independent sources, for a historical description to be worth attempting here. However, it is useful for contextual purposes to briefly mention some of the conventional *uses* of path spaces.

In mathematics, path spaces form the basis for *homotopy theory*[27] in algebraic topology, via the fundamental groupoid and its higher-dimensional analogues. They also feature prominently in the theory of *topological groups,* and particularly Lie groups, partly due to the importance of *one-parameter subgroups.* In virtually every area of theoretical physics, conventional or otherwise, path spaces play a prominent role. At the simplest level, they represent trajectories of point-particles in Newtonian physics. More generally, they play a crucial part in the theory of phase space, with particularly iconic status in the field of nonlinear dynamics. Perhaps most importantly, they underlie the Lagrangian and Hamiltonian formulations of classical mechanics, and extend to analogous methods in quantum theory, culminating in Feynman's path summation approach. Moving beyond experimentally established physics, path spaces find additional applications in string theory, due to the one-dimensional nature of strings, and in loop quantum gravity, due to the use of *holonomy theory* in the construction of the *loop representation.* However, the *directed* nature of cause and effect renders the use of path spaces particularly natural, and especially important, in the context of discrete causal theory.

Representing paths via morphisms. In mathematical settings, the word "path" is often understood to mean essentially the same thing as the word "curve" means in the relativistic context of Chapter 2, i.e., a *map* from an interval in $\mathbb{R}$ into a topological space, often a real manifold. In theoretical physics, however, a "path" is usually defined to be an *equivalence class* of such maps. This is especially true in the context of general relativity and quantum gravity. For example, Rovelli [RO04] and Thiemann [TH07] both define a "path" as an equivalence class of curves, although with

---

[27] That is, "ordinary" homotopy theory. Other variants of the same idea are possible, such as Voevodsky's theory of $\mathbb{A}^1$-*homotopy* in algebraic geometry. "Causal homotopy" is a potentially interesting topic, for example, in the domain-theoretic setting, but I do not pursue it here.

somewhat different equivalence relations.[28] As mentioned in Section 2.6, a typical curve in a relativistic spacetime manifold X belongs to a very large equivalence class of curves representing the same physical data. The principal source of such large classes is *reparameterization;* for example, a given curve $\gamma : [0, 1] \rightarrow X$ may be pre-composed with an uncountable number of different smooth automorphisms of the unit interval $[0, 1]$, each of which produces a physically equivalent, but mathematically distinct, curve in X. All of these curves are viewed as representatives of the same "physical" path.

The conceptual meaning of "path" is similar in this book, but the objects involved are generally much more rigid; hence, *paths may often be represented by unique morphisms in discrete causal theory.* Discreteness tends to "freeze out" inconvenient ambiguities, such as those involved in reparameterization in the continuum-based setting. In particular, discrete morphisms encoding physical scenarios such as "trajectories" or "evolutionary processes" are often unique. For example, a finite chain L,[29] viewed as a discrete causal analogue of the unit interval $[0, 1]$, possesses no nontrivial automorphisms; hence, different morphisms from L into a multidirected set $M = (M, R, i, t)$ generally have different physical interpretations. If one replaces L with an *infinite chain,* the situation is somewhat different; for example, $\mathrm{sk}(\mathbb{Z})$ has a countably infinite number of automorphisms, each mapping 0 to n for some $n \in \mathbb{Z}$. A given morphism $\gamma : \mathrm{sk}(\mathbb{Z}) \rightarrow M$ may be pre-composed with any such automorphism of $\mathrm{sk}(\mathbb{Z})$, producing a countable number of physically equivalent, but mathematically distinct, morphisms into M, all of which are viewed as representatives of the same "physical" path. However, the uniform "shifting" involved in this construction carries much less structural significance than the "stretching and shrinking" involved in reparameterization. Perhaps the most convenient way to define paths in the discrete causal setting is as *"morphisms whose sources are defined only up to isomorphism,"* and this is the option followed in Definition 5.9.3 below. To be precise, this means that if $\beta : L' \rightarrow L$ is an isomorphism of "potential source objects for paths," then morphisms $\gamma : L \rightarrow M$ and $\gamma \circ \beta : L' \rightarrow M$ are understood to represent the same path in M. Here, the morphism β is analogous to a reparameterization $[0, 1] \rightarrow [0, 1]$ of the unit interval $[0, 1]$. With this understanding, it is usually innocuous to refer to paths as if they were individual morphisms, and I do so routinely throughout the remainder of the book. The source objects of these morphisms are required, like intervals in $\mathbb{R}$, to be "linear" in a suitable sense, but the context is much more general: arbitrary totally ordered sets are permitted, as are nontransitive directed sets whose transitive closures are totally ordered.

Linear directed sets: sources for paths. Totally ordered sets, or equivalently, linearly ordered sets, have already made a number of prominent appearances in Part I of this book. For example, Section 2.9 details the order-theoretic definition of $\mathbb{R}$ as the unique Archimedean complete totally ordered field, up to isomorphism. Section 3.8 includes a discussion of the order refinement principle (ORP), and Section 4.6 focuses

---

[28]See [RO04], pp. xxii–xxiii, and [TH07], p. 164.

[29]This choice of notation anticipates the more general use of the symbol "L" to denote a *linear directed set,* as described below.

on the theory of relative multidirected sets over totally ordered bases. In the present section, however, it is useful to expand the notion of "linearity" to include *generally nontransitive* directed sets whose transitive closures are totally ordered, such as acyclic chains. Such sets, which I refer to as *linear directed sets,* may be viewed as analogues of intervals on the real line, in the sense that they supply *sources* for paths in multidirected sets.[30]

Definition 5.9.1. *A directed set* $L = (L, \prec)$ *is called* **linear** *if its transitive closure* $\mathrm{tr}(L) = (L, \prec_{\mathrm{tr}})$ *is totally ordered.*

To be clear, this means that the binary relation $\prec_{\mathrm{tr}}$ on L is a *strict* total order, analogous to the familiar "less than" relation $<$ on $\mathbb{Z}$. Stated in a different way, L is linear if it is acyclic, and if every pair of elements of L is connected by a chain. A finite linear directed set has unique maximal and minimal elements, which play an important role in the definition of products of causal paths. Figure 5.9.2 illustrates some examples of linear directed sets. Perhaps the most notable feature of the figure is that only one of the directed sets shown here, namely, the chain appearing in the second diagram from the left, "looks linear," in a naïve intuitive sense. There is no doubt that Definition 5.9.1 is the "correct" definition of linearity from a structural perspective, since it properly encodes the idea that the information contained in the directed structure is sufficient to specify a linear order. The potential for confusion highlighted by Figure 5.9.2 involves a rather prosaic aspect of *visualizing* linearity, which is nonetheless worth clarifying. Discreteness enables one to simultaneously visualize both the elements and the relations of a multidirected set; the directed set illustrated in the far-right of the figure, for example, is just the set of natural numbers $\mathbb{N}$, which is "non-controversially" linear. However, one usually *draws* only the elements of $\mathbb{N}$, not the relations, and certainly not the reducible relations. For the sake of comparison, it is worth recalling from Section 5.1 that the relation set of the real numbers $\mathbb{R}$ is "essentially two-dimensional." More generally, familiar linearly ordered sets "look linear" only because it is conventional to leave their relations out of the picture.

The *length* of a finite linear directed set L is defined to be the chain length of its skeleton $\mathrm{sk}(L)$, in the sense of Definition 3.7.1, i.e., its maximal number of consecutive *relations*. Infinite linear directed sets may be "longer and more complicated than any acyclic chain." In particular, the image of any acyclic chain may be represented as a relative directed set over $\mathbb{Z}$, but a linear directed set may a "too large" to be represented in this manner. For example, Jacob's ladder is linear, but does not admit a bijective morphism into any subobject of $\mathbb{Z}$. The analogue of length for an infinite linear directed set L is the *isomorphism class of its transitive closure* $\mathrm{tr}(L)$ as a linearly ordered set.

[30]The reason why there is no need for this generalization in Chapter 4 is because any relative multidirected set over a linear base L may be expressed as a relative multidirected set over $\mathrm{tr}(L)$, via the canonical morphism $L \to \mathrm{tr}(L)$ induced by the identity set map.

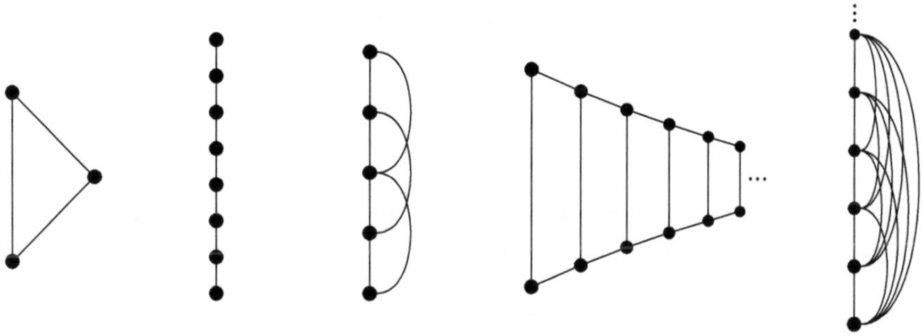

Figure 5.9.2. Linear directed sets: 2-simplex; acyclic 7-chain; chain of 2-simplices; Jacob's ladder; natural numbers $\mathbb{N}$.

Causal path sets. A *directed path* in a multidirected set $M = (M, R, i, t)$ is a morphism from a linear directed set into M, whose source is defined up to isomorphism. In the special case where M is a directed set, viewed as a model of classical spacetime structure under the classical causal metric hypothesis (CCMH), directed paths are physically interpreted as *causal paths,* since an event represented by an element x of M influences an event represented by an element y of M if and only if there exists a directed path "from x to y in M," i.e., a path whose minimal and maximal source elements map to x and y, respectively. The following definitions make it clear that this description meshes well with the definitions of domains of influence given in Section 3.7, which are expressed in terms of chains. The same physical intuition is useful for multidirected sets in general, so I use the adjective "causal" to refer to a number of general mathematical constructions involving directed paths. The conceptual intent of Definition 5.9.3 is to capture the *physical essence of an individual relationship between a linear directed set and a multidirected set,* which is the motivation for defining the source only up to isomorphism. A similar concept applies to the definition of *co-relative histories* in Section 6.4, although the latter context is subtler, and generally requires nontrivial equivalence classes of morphisms to adequately describe.

Definition 5.9.3. *Let $M = (M, R, i, t)$ be a multidirected set.*

1. *A* **directed path,** *or* **causal path,** *in M, is a morphism $\gamma : L \to M$ for some linear directed set L, defined up to isomorphism. L is called the* **source** *of γ, and M is called the* **target** *of γ.*
2. *If the source L of a causal path $\gamma : L \to M$ is finite, then the* **length** *of γ is defined to be the length of L. In this case, the images in M of the maximal and minimal elements of L are called the* **initial** *and* **terminal elements** *of γ, respectively.* If L is infinite, then the isomorphism class of $\mathrm{tr}(L)$ supersedes the usual notion of length.
3. *The* **causal path set** *$\Gamma(M)$ of M is the set whose elements are causal paths in M.*

4. *A causal path $\gamma : L \rightarrow M$ is called* **irreducible** *if its source L is irreducible. The* **irreducible causal path set** $\Gamma_{\mathrm{IRR}}(M)$ *of M is the subset of $\Gamma(M)$ whose elements are irreducible causal paths.*

The causal path set $\Gamma(M)$ of a multidirected set $M = (M, R, i, t)$ may be viewed as a generalization of the chain set $\mathrm{Ch}(M)$ of M, introduced in Definition 3.7.1, which itself generalizes the relation set R of M. In particular, a relation in M may be regarded as a causal path of length one, and an n-chain in M may be regarded as a causal path of length n, whose source is an irreducible linear directed set. Conversely, the image of an irreducible causal path $\gamma : L \rightarrow M$ is a chain in M, *not* necessarily irreducible, since pairs of elements "may acquire new relations between them in the image." Hence, the chain set $\mathrm{Ch}(M)$ of M may be identified with the irreducible causal path set $\Gamma_{\mathrm{IRR}}(M)$ of M. As in the special case of chains, the notions of initial and terminal elements generalize in the obvious way to infinite causal paths that are "bounded in only one direction;" for example, the natural inclusion $\mathbb{N} \rightarrow \mathbb{Z}$ represents a causal path in $\mathbb{Z}$ with initial element 0 in $\mathbb{Z}$, but no terminal element. In Definition 5.9.3, and elsewhere in this section, I have chosen to suppress explicit reference to certain structural details, such as the binary relations on sources of causal paths, and, of course, the fact that such sources are defined only up to isomorphism. These choices are made in order to avoid notational clutter, and should cause no serious confusion.

The remarks in Section 3.7 regarding the minimality or maximality of initial or terminal elements of chains generalize to the case of causal paths; in fact, these notions are easier to understand in the latter setting, due to the explicit description of causal paths in terms of morphisms. In particular, the terminal element z of a causal path $\gamma : L \rightarrow M$, if it possesses one, need not be maximal among the elements of $\gamma(L)$, in the sense that there may exist another causal path in M with initial element z and terminal element belonging to $\gamma(L)$. In fact, a stronger statement is true: z need not be maximal among the elements of $\gamma(L)$ even if one confines attention to the *minimal full subobject of M containing $\gamma(L)$*, and ignores the rest of the structure of M. A similar statement applies to initial elements. The simplest example of these possibilities is given by the unique causal path γ from a linear directed set L with one element x and no relations into the multidirected set M with one element y and one relation $y \prec y$. The element x is both maximal and minimal in L, so its image element y in M is both the initial and terminal element of γ. However, y is neither minimal nor maximal in the full subobject of M containing $\gamma(L)$, namely, M itself. Figure 5.9.4 illustrates two different causal paths in a multidirected set. The sources of these paths are the second and third linear directed sets from Figure 5.9.2. The causal path γ_1 illustrated in the left-hand diagram in Figure 5.9.4 has an irreducible source, so it may be viewed as a chain in M. The causal path γ_2 illustrated in the right-hand diagram has a more complicated source, consisting of a chain of three 2-simplices. In this book, I focus almost exclusively on the case of chains, i.e., causal paths with irreducible sources.

Causal paths in discrete causal theory differ from their relativistic analogues in at least two important respects, besides the obvious differences arising from the use of

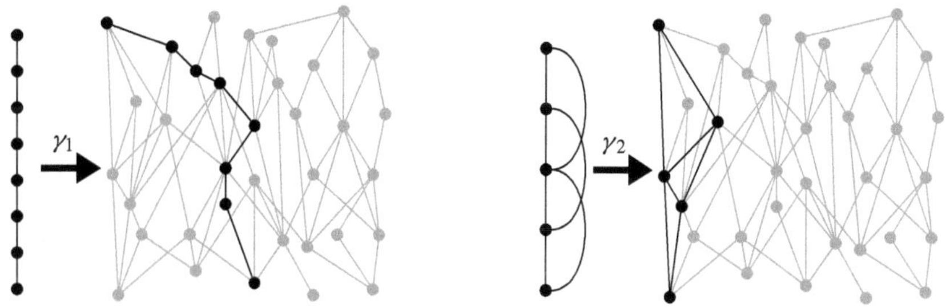

Figure 5.9.4. Two causal paths in a multidirected set.

discrete sources. First, in terms of their physical interpretation, they represent *actual* influence, not merely *potential* influence; i.e., following the philosophy outlined in Section 2.10, they *describe* causal structure, rather than *prescribing* it. Second, in terms of mathematical structure, there exist an immense variety of distinct types of linear directed sets that can serve as sources for causal paths in discrete causal theory, as illustrated in Figure 5.9.2. Causal paths in relativity, by contrast, always have real intervals as their sources.

Causal path spaces. A wide array of algebraic operations may be defined on causal path sets of directed sets and/or multidirected sets. Two of the most important such operations, for the purposes of discrete causal theory, are the *concatenation product* and the *directed product,* defined below. These are special cases of **splice products,** which are partially defined binary operations formalizing various notions of *"joining morphisms together end-to-end in a directed sense."* Russling [RU95] discusses essentially equivalent notions in the special case of chains, and applies these constructions to problems in algorithmic graph theory, but the general theory of splice products remains largely unexplored in physical contexts. These operations define natural directed structures on causal path sets, elevating them to directed sets called *causal path spaces,* which may be viewed as generalizations of the relation space $\mathcal{R}(M)$ over a multidirected set M. From an algebraic viewpoint, causal path spaces may be regarded as generalized "non-unital categories," or *semicategories,* called *causal path semicategories.* These algebraic objects, further examined in Section 5.10, supply the multiplicative structures of *causal path algebras,* which serve as a crucial tool for encoding the physical information represented by a multidirected set in a mathematically tractable manner.

Multidirected, rather than merely directed, structures, may also be defined on causal path sets. The most obvious such structures arise from the fact that two given causal paths may often be "spliced together" in many different ways. This idea is physically relevant, but is not explored in detail in this book. Another way to generalize the above notions is to consider sets whose elements are causal paths *together with* distinguished subpaths of length zero, physically representing "choices of present" in linear sequences of events. Such sets are structurally analogous to *flag varieties* in algebraic geometry. Extending the analogy between causal theory and twistor theory mentioned in Section 5.6, such "enriched causal path sets" may be used to define *correspondences,* analogous to those described by Ward and Wells [WW90]. However, besides a footnote in Chapter 7 mentioning an analogous "higher-level" construction in the context of kinematic schemes, I do not explore enriched causal path sets further in this book. It is not even feasible here to replicate, for the various spaces and algebraic objects introduced in this section, the modest examination of functorial, information-theoretic, and structural properties carried out for relation space in Section 5.1.

Causal concatenation space over a directed set. One of the most important causal path spaces associated with a directed set $D = (D, \prec)$ is the *causal concatenation space* over D. As stated above, this space may be described in terms of a partially defined product operation on the causal path set $\Gamma(D)$, called the *concatenation product,* which made a brief appearance in the proof of the impermeability theorem in Section 5.4. Informally, the concatenation product "splices together" pairs of causal paths γ and δ in D, whenever the terminal element of γ *directly precedes* the initial element of δ. The following definition makes this notion precise:

Definition 5.9.5. *Let $\gamma : L \rightarrow D$ and $\delta : L' \rightarrow D$ be causal paths in a directed set $D = (D, \prec)$. Suppose that L has a maximal element ℓ^+, and L' has a minimal element ℓ'^-, such that the terminal element $\gamma(\ell^+)$ of γ directly precedes the initial element $\delta(\ell'^-)$ of δ in D.*

1. *Denote by $L \sqcup L'$ the linear directed set whose underlying set is the disjoint union $L \bigsqcup L'$, and whose binary relation is given by augmenting the union of the binary relations on L and L' with the single relation $\ell^+ \prec \ell'^-$, called the* **connecting relation**. *Then the* **concatenation product** *$\gamma \sqcup \delta : L \sqcup L' \rightarrow D$ of γ and δ is the causal path in D defined by setting*

$$\gamma \sqcup \delta(\ell) = \begin{cases} \gamma(\ell) & \text{if } \ell \in L \\ \delta(\ell) & \text{if } \ell \in L' \end{cases}.$$

2. *The* **concatenation relation** *$\prec_\sqcup$ on the causal path set $\Gamma(D)$ of D is the binary relation given by setting $\gamma \prec_\sqcup \delta$ if and only if $\gamma \sqcup \delta$ exists. The* **causal concatenation space** *over $(D, \prec)$ is the directed set $\big(\Gamma(D), \prec_\sqcup\big)$.*

Figure 5.9.6 illustrates the construction of the concatenation product of two causal paths in a directed set D. The left-hand and middle diagrams in the figure show a pair of causal paths γ and δ in D, satisfying the hypotheses of Definition 5.9.5. The right-hand diagram shows the concatenation product $\gamma \sqcup \delta$. In all three diagrams, the linear directed sets serving as the sources for the causal paths involved are not shown, since this would only clutter the figure. The "correct" sources are obvious from the figure, however: the sources of γ, δ, and $\gamma \sqcup \delta$ are irreducible linear directed sets with three, four, and seven elements, respectively.

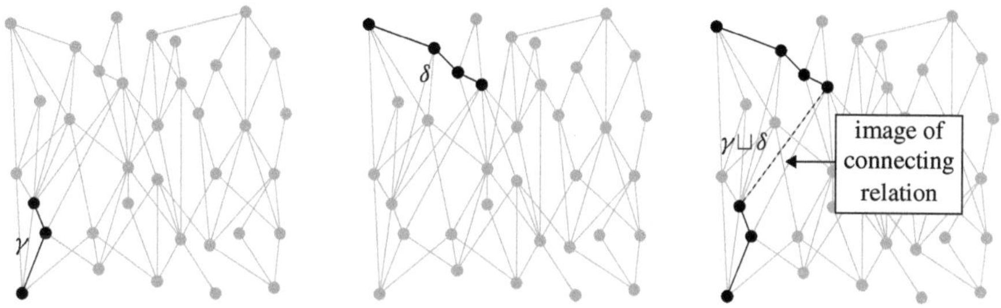

Figure 5.9.6. Causal path γ in a directed set; second causal path δ; concatenation product $\gamma \sqcup \delta$ (sources of paths suppressed to avoid clutter).

As implied by Definition 5.9.5, the concatenation product is uniquely defined only for directed sets. Indeed, in a multidirected set $M = (M, R, i, t)$, there may exist multiple connecting relations between $\gamma(\ell^+)$ and $\delta(\ell'^-)$, and there is no natural way to choose which of these relations is the "correct" choice for defining the product. Of course, one could choose to incorporate *all* connecting relations, but this would require either generalizing the definition of the product to include families of paths, or allowing sources of product morphisms to possess "linear" *multidirected* structure. The concatenation *relation* $\prec_\sqcup$ is easier to generalize to the multidirected context, because the "correct" definition is obvious: one should simply define a relation in $\Gamma(M)$ between γ and δ for every connecting relation between $\gamma(\ell^+)$ and $\delta(\ell'^-)$ in M. These generalizations may be interesting, but they are not pursued further in this book. The notation $\sqcup$ for the concatenation product is deliberately chosen to evoke the *disjoint union* $\bigsqcup$ *of sets.* The concatenation relation $\prec_\sqcup$ on $\Gamma(D)$ extends the binary relation $\prec$ on D. Hence, $(D, \prec)$ *embeds as a full subobject in its causal concatenation space.*

Causal directed product space over a multidirected set. A second causal path space of great importance in discrete causal theory is the *causal directed product space* over a multidirected set $M = (M, R, i, t)$. This space may be defined in terms of a partially defined product operation on the causal path set $\Gamma(M)$ of M, called the *directed product,* which "splices together" pairs of causal paths γ and δ in M, whenever the terminal element of γ *coincides with* the initial element of δ. The following definition makes this notion precise:

Definition 5.9.7. *Let $\gamma : L \to M$ and $\delta : L' \to M$ be causal paths in a multidirected set $M = (M, R, i, t)$. Suppose that L has a maximal element ℓ^+, and L' has a minimal element ℓ'^-, such that the terminal element $\gamma(\ell^+)$ of γ coincides with the initial element $\delta(\ell'^-)$ of δ in M.*

1. *Denote by $L \vee L'$ the linear directed set whose underlying set is the quotient space of the disjoint union $L \bigsqcup L'$, given by identifying ℓ^+ and ℓ'^-, and whose binary relation is the union of the binary relations on L and L'. Then the* **directed product** *$\gamma \vee \delta : L \vee L' \to M$ of γ and δ is the causal path in M defined by setting*

$$\gamma \vee \delta(\ell) = \begin{cases} \gamma(\ell) & \textit{if } \ell \in L \\ \delta(\ell) & \textit{if } \ell \in L' \end{cases}.$$

2. *The* **directed product relation** $\prec_\vee$ *on the causal path set* $\Gamma(M)$ *of* M *is the binary relation given by setting* $\gamma \prec_\vee \delta$ *if and only if* $\gamma \vee \delta$ *exists. The* **causal directed product space** *over* M *is the directed set* $\big(\Gamma(M), \prec_\vee\big)$.

Figure 5.9.8 illustrates the construction of the directed product of two causal paths in a multidirected set M. The left-hand and middle diagrams in the figure show a pair of causal paths γ and δ, satisfying the hypotheses of Definition 5.9.7. The right-hand diagram shows the directed product $\gamma \vee \delta$. As in Figure 5.9.6, the linear directed sets serving as the sources of these causal paths are not shown, but may be inferred from their images. The notation $\vee$ for the directed product is deliberately chosen to evoke the *wedge sum* $\bigvee$, familiar from topology, which joins topological spaces at distinguished basepoints. The directed product relation $\prec_\vee$ on $\Gamma(M)$ extends the induced relation on the relation space $\mathcal{R}(M)$ over M. Hence, $\mathcal{R}(M)$ *embeds as a full subobject in the causal directed product space over* M. Note that, unlike the definition of the concatenation product, the definition of the directed product does not rely upon the existence of directed or multidirected structure in the target object M. Hence, the notion of directed product extends to "non-causal" settings. In particular, it appears in the context of ordinary homotopy theory.

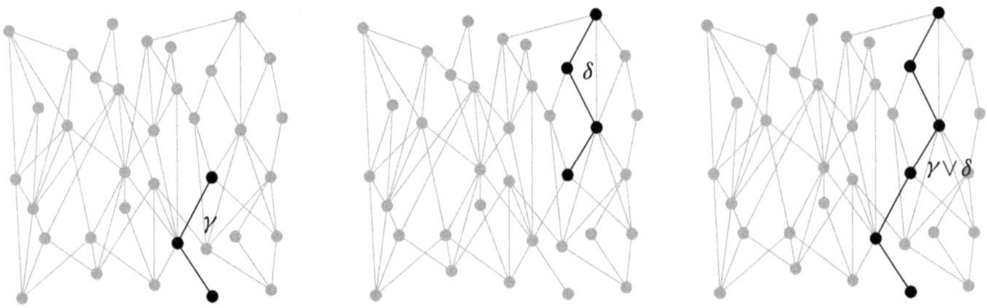

Figure 5.9.8. Causal path γ in a multidirected set; second causal path δ; directed product $\gamma \vee \delta$ (sources of paths suppressed to avoid clutter).

5.10 Causal Path Algebras

"Algebra" versus "algebras." To actually calculate quantities of physical interest in discrete causal theory, it is necessary to develop suitable algebraic tools; in particular, the notion of "algebras of causal paths" in a multidirected set. I refer to these algebras as *causal path algebras.* They are specific, though rather unconventional, examples of the general notion of an *algebra over a ring.* At the risk of outraging the mathematically sophisticated reader, I give here a very informal preliminary description of this idea, which will then be sharpened sufficiently for the purposes of this book. The reason for this approach is that a proper description of causal path algebras requires examination of rather abstract structural notions, such as "generalized semicategories," and it is prudent to outline the basic motivation of these considerations before undertaking them. The first point to clarify is that the word "algebra" itself has multiple meanings. The phrase "algebra over a ring" identifies a *special type of mathematical object* in which "one can multiply, add, and distinguish scalars." It is sometimes useful to regard such an algebra as the "end result of a progressive increase in structure," in the same spirit as the constructive view of $\mathbb{R}$, outlined in Section 2.9. For example, one might begin with a group, which is a mathematical object in which "one can multiply *or* add;" then increase the level of structure to obtain a ring, in which "one can multiply *and* add;" then add further structure to distinguish "scalars," or "constants," from "generators," or "vectors," or "variables," or whatever one chooses to call "non-scalars." In the context of causal path algebras, the "scalars" are numbers, i.e., elements of a suitable ring, and the "non-scalars" are paths in a multidirected set.

In a more general context, the term "algebra" refers to an *entire field of mathematics,* in which objects such as groups, rings, algebras over rings, and many other types of mathematical structures, are studied. More generally still, it refers to a *viewpoint,* or general approach, to doing mathematics. In this context, the adjective "algebraic" is used to qualify other mathematical topics examined via techniques or perspectives from algebra. For example, in "algebraic geometry," one studies the properties of certain geometric objects, such as the vanishing sets of polynomials, in terms of their algebraic counterparts, such as prime ideals in commutative rings. The language of category theory is ubiquitous in modern algebra. Of course, like most other complex and ancient fields of knowledge, algebra defies any definition that is both tidy and comprehensive. It is often viewed, to paraphrase a number of influential opinions, as *"the study of structural properties and relationships that apply to broad classes of mathematical objects, without undue concern for the idiosyncrasies of individual objects."* This rough description is far from definitive, but points in the right direction. However, as mathematics has matured as a whole, an increased emphasis on generality and structure has led to a situation in which other fields of mathematics; for example, order theory, may be described in similar terms. Hence, it is sometimes remarked that "mathematics as a whole is becoming more algebraic," and the same is certainly true of theoretical physics. In particular, the "study of structural properties and relationships" is absolutely central to discrete causal theory, and it is therefore not surprising that the algebraic viewpoint features so prominently in this book.

Multiplication via generalized semicategories. It happens that the operations defining "multiplication" in causal path algebras arise from a very general class of algebraic objects, called *generalized semicategories.* I often refer to these objects as just "semicategories;" the distinction is described below. Multiplication in "ordinary algebras," such as polynomial algebras, arises from less general, and more familiar, objects, such as groups or semigroups. The principal feature distinguishing semicategories from these familiar objects is that "multiplication" in a semicategory is only partially defined. Partially defined operations are "less popular in conventional algebra" than totally defined operations, which is why I describe causal path algebras as "unconventional" at the beginning of this section. However, this may be viewed as a mere shortcoming in the emphasis of "conventional" algebra, at least from a physical standpoint, since the information-theoretic reasons for studying partially defined operations are obvious. This is especially true in the context of discrete causal theory, where the algebraic operations of principal interest are splice products, such as the concatenation product and the directed product, defined in Section 5.9. These operations are only partially defined because *only special pairs of subobjects of a given multidirected set splice together in a natural manner.* This is exactly the type of mathematical structure needed to describe the flow of information from "past" to "future," and it would be ridiculous to deprecate such structure because of an aesthetic preference for the "good behavior" associated with totally defined operations.

From categories to semicategories and beyond. The terminological similarity between semicategories and categories is not an accident. As described below, a semicategory may be viewed as a "generalized category," in which a particular axiom of category theory; namely, the existence of identity morphisms, is no longer taken for granted. This represents yet another point of comparison between the mathematical apparatus of discrete causal theory and category theory. As discussed in previous chapters, these analogies arise because both theories incorporate certain common structural themes. Most importantly, they both involve *multidirected structure:* discrete causal "relationships," such as relations between pairs of elements, or co-relative histories between pairs of classical histories, "point from one element or history to another," while category-theoretic "relationships," such as morphisms or functors, "point from one object or category to another." Further, both theories emphasize *preservation of structure:* discrete causal representations of evolutionary processes, such as those appearing in sequential growth dynamics, require "descendant histories" to preserve the structure of "ancestor histories," while morphisms or functors are chosen to "mirror the structure of the source in the target." In addition, both theories feature similar approaches to *structural hierarchy,* as illustrated by the analogy between relation spaces and morphism categories outlined in Section 5.5. However, as I have repeatedly noted, the structural emphases of the two theories also diverge in important ways. The necessity of using generalized semicategories, rather than categories, to describe causal path algebras in discrete causal theory, is another manifestation of this fact.

As described in Section 1.5, category theory was originally developed to study certain special relationships in algebraic topology. In particular, categories were introduced to relate pairs of *cohomology theories*, which are special families of functors, often from a category of topological spaces to a category of groups. These relationships, called *natural transformations,* make a brief reappearance in Section 8.6. In this original context, the "causally suggestive" features of category theory, cited above, are naturally accompanied by other notions, such as composition of morphisms, associativity of morphisms, and the existence of identity morphisms. When "objects" are topological spaces, and "morphisms" are continuous maps, the relevance of these additional concepts is obvious. However, there is no justification for naïvely transporting such machinery to every other context in which multidirected structure, preservation of structure, and hierarchy play a central role.[31] This is particularly true in discrete causal theory. For example, as explained in Section 3.9, a pair of relations $x \prec y$ and $y \prec z$ in an arbitrary directed set $D = (D, \prec)$ generally do not "compose" to yield a third relation; such "composition" requires transitivity (TR). Associativity makes no sense without composition. Identity morphisms correspond to reflexive relations, which may or may not exist for general directed sets, and are actually forbidden in the acyclic case. Similar, though more complicated, issues arise when attempting to transport such notions to the theory of co-relative histories.

Semicategories, or "non-unital categories," are similar to categories, but generally lack identity morphisms. Causal path spaces over a directed or multidirected set may be viewed as *generalized* semicategories, which admit *"morphisms without sources and/or targets."* A useful heuristic may be drawn from classical electromagnetism, in which field lines "either terminate on charges or extend to infinity."[32] More precisely, a **semicategory** consists of a class of objects $\mathcal{S}$, together with a class of morphisms $\Gamma(\mathcal{S})$, such that every morphism γ has an initial object $\mathbf{i}(\gamma)$ and a terminal object $\mathbf{t}(\gamma)$. In the context of discrete causal theory, $\mathcal{S}$ may be viewed as an abstraction of the underlying set of elements of a multidirected set $M = (M, R, i, t)$, while $\Gamma(\mathcal{S})$ may be viewed as an abstraction of the class of *finite chains* in M. However, generalizations of these analogies, involving the entire causal path set $\Gamma(M)$ of M, are required below. The maps $\mathbf{i} : \Gamma(\mathcal{S}) \to \mathcal{S}$ and $\mathbf{t} : \Gamma(\mathcal{S}) \to \mathcal{S}$ may be viewed as abstractions of the initial and terminal element maps i and t of M, extended to apply to finite chains of arbitrary length, rather than merely relations. A semicategory $\mathcal{S}$ also admits *associative compositions* of morphisms, defined whenever the terminal object of the first morphism coincides with the initial object of the second morphism. This composition may be viewed as an abstraction of the directed product of chains. Note that although the "composition" of relations is not a *relation* in this context, it *is* a chain, more specifically, a 2-chain. The analogy between complex chains and morphisms should not be carried too far, however. In Section 5.5, I compared 2-chains to 2-morphisms, rather than morphisms, and this is often a "better" structural analogy. Similarly, the generalized semicategory structure of causal path spaces is perhaps better understood by comparing it to the category-theoretic approach to *groups*, already mentioned in Section 3.8, in which group elements are viewed as endomorphisms in a category with a single structureless object. In this sense, a generalized semicategory is a "grouplike structure," whose morphisms are its *elements,* while composition is viewed as a partially defined algebraic operation

---

[31] Note again the contrast with Christopher Isham's viewpoint, appearing, for example, in the papers [IS05, IS11]. Isham adds *additional* properties to categories to obtain topoi, whereas I abstain from certain axioms that apply to all categories. Of course, Isham works with topoi of many different kinds, not only those whose objects are intended to model causal structure.

[32] It is interesting to compare such structures with Blute Panangaden, and Straßburger's *pre-causal graphs* [BPS08], which are allowed to include "half-edges."

on these elements. From this perspective, a **generalized semicategory** is simply *a class of elements, endowed with an associative partially defined operation, generally noncommutative, and without identity or inverses.*

Although it is much easier to digest the motivation for studying such objects after examining physically relevant examples, such as the causal path spaces introduced in Section 5.9, it is worth noting that the appearance of such structures in mathematics is more common, and more conventional, than one might realize. Examples include the space of paths from an interval on the real line into a topological space, under the operation of directed product[33]; the class of *all* matrices over a given ring, under the operation of matrix multiplication; and the class of algebraic cycles on an algebraic variety, under the operation of *proper* intersection.[34] An interesting example of a similar structure arising naturally in a different area of theoretical physics is the *"category of points and paths"* appearing in Thiemann's book [TH07], in the context of the *holonomy flux algebra* for loop quantum gravity. In this setting, "objects" are elements in a manifold, and "morphisms" are equivalence classes of directed paths between pairs of points.[35] In this case, the structure really *is* a category; in fact, all morphisms are isomorphisms, essentially because the manifold itself is not directed; i.e., any path between two elements representing a particular morphism may be reversed to yield a representative of the inverse morphism.

Causal path semicategories. The causal path set $\Gamma(D)$ of a directed set $D = (D \prec)$, equipped with the partially defined operation of concatenation product $\sqcup$, is a generalized semicategory, called the **concatenation semicategory** over D. Similarly, the causal path set $\Gamma(M)$ of a multidirected set $M = (M, R, i, t)$, equipped with the partially defined operation of directed product $\vee$, is a generalized semicategory, called the **directed product semicategory** over M. Both of these structures, viewed as "group-like objects," are examples of what I refer to as **causal path semicategories**. They represent the "algebraic viewpoint" regarding the natural structure of causal path sets, in contrast to the "generalized order-theoretic viewpoint" represented by the corresponding causal path *spaces,* equipped with appropriate binary relations, such as $\prec_{\sqcup}$ and $\prec_{\vee}$. The "if and only if" statements relating the products $\sqcup$ and $\vee$ to the relations $\prec_{\sqcup}$ and $\prec_{\vee}$ in Definitions 5.9.5 and 5.9.7, ensure that the two viewpoints are equivalent in the present context. Of course, as explained in Section 5.9, the concatenation relation $\prec_{\sqcup}$ is easier to generalize to the multidirected context than the concatenation product $\sqcup$, but such details are unnecessary here. Since causal path semicategories are only *generalized* semicategories, their morphisms, i.e., causal

---

[33] As mentioned at the end of Section 5.9, the notion of directed product extends to situations in which the target of paths does not itself possess directed structure; one may still "join paths together end-to-end," provided that "the terminal element of the first path coincides with the initial element of the second path."

[34] In all three of these examples, information may be discarded in order to obtain a structure that is better-behaved mathematically, but is less complete and less natural; namely, *fundamental groupoids* or *groups, matrix groups,* and *Chow rings,* respectively.

[35] What Thiemann calls a "path"' in this context is a class of curves sharing common initial and terminal elements, as well as common holonomies for a specified class of connections. See [TH07], Definition 6.2.4, p. 165.

paths, do not necessarily possess initial or terminal objects, i.e., initial or terminal elements in the target set. This "structural novelty" is actually not very strange, since infinite paths with familiar sources, such as $\mathbb{Z}$, exhibit it. Further, the composition law for morphisms γ and δ in a causal path semicategory is not necessarily defined via the conventional semicategory condition $\mathbf{t}(\gamma) = \mathbf{i}(\delta)$, used in the directed product, since other splice products, such as the concatenation product, are sometimes of interest. I reemphasize, however, that I sometimes abbreviate phases such as "generalized semicategory," or "causal path semicategory," to just "semicategory," since the narrower conventional definition of a semicategory as a non-unital category plays no significant role in this book.

Interesting relationships exist among generalized semicategory operations on causal path sets, such as the concatenation product $\sqcup$ and the directed product $\vee$, and functors between categories of "underlying" directed or multidirected sets, such as the relation space functor $\mathcal{R}$ and the abstract element space functor $\mathcal{E}$. For example, the concatenation semicategory $\big(\Gamma\big(\mathcal{R}(M)\big), \sqcup\big)$ over the relation space $\mathcal{R}(M)$ over a multidirected set $M = (M, R, i, t)$ corresponds to a subsemicategory of the directed product semicategory $\big(\Gamma(M), \vee\big)$ over M itself, as illustrated by the following example: if r_0 and r_1 are elements of $\mathcal{R}(M)$, with initial and terminal elements $i(r_0) = x$, $t(r_0) = i(r_1) = y$, and $t(r_1) = z$, in M, then the directed product in $\Gamma(M)$ of r_0 and r_1, viewed as causal paths of length one, corresponds to the concatenation product in $\Gamma\big(\mathcal{R}(M)\big)$ of r_0 and r_1, viewed as causal paths of length zero. This relationship illustrates once again how the relation space functor $\mathcal{R}$ *reduces multidirected structure to directed structure,* since concatenation products are uniquely defined only in the directed case. The physical importance of *both* element space and relation space is the principal reason for studying *both* the concatenation semicategory and the directed product semicategory.

Since the operations $\sqcup$ and $\vee$ preserve irreducibility of causal paths, useful *subsemicategories* of $\big(\Gamma(D), \sqcup\big)$ and $\big(\Gamma(M), \vee\big)$ may be defined, by restricting attention to the irreducible causal path sets of M and D, which may be identified with the corresponding chain sets $\mathrm{Ch}(D)$ and $\mathrm{Ch}(M)$, respectively, as explained in Section 5.9. I refer to these subsemicategories as the **chain concatenation semicategory** over D, and the **chain directed product semicategory** over M, respectively. These **chain semicategories** $\big(\mathrm{Ch}(D), \sqcup\big)$ and $\big(\mathrm{Ch}(M), \vee\big)$ belong to a special class of causal path semicategories that may be studied without explicit reference to morphisms.[36] They sometimes appear implicitly in the literature as structural substrata for multiplicative operations in special types of path algebras, as elaborated below. Though I focus mostly on the special case of chains in this book, there are good reasons behind the choice to develop the theory in terms of general causal paths. For example, this

[36]Another case in which explicit reference to morphisms may be avoided is the case in which all causal paths under consideration are *isomorphisms onto their images,* viewed as subobjects M. However, this case does not behave well algebraically. In particular, splice products of such causal paths are generally *not* isomorphisms onto their images; for example, there may be relations in the target between the image of the first path and the image of the second path.

approach renders the relative viewpoint (RV) explicit, enables more refined treatment of target objects, and incorporates the case of causal paths of transfinite length. Special *quotient semicategories* also play a significant role in the theory of causal path semicategories. They may be viewed concretely as families of *congruence classes of causal paths.* A *congruence relation* $\sim$ on an arbitrary generalized semicategory $(\Gamma, \sqcup)$ is a reflexive, symmetric, transitive binary relation on Γ, preserving the semicategory operation $\sqcup$, in the sense that for any four elements $\alpha, \beta, \gamma, \delta$ of Γ, if $\alpha \sim \beta$ and $\gamma \sim \delta$, then $\alpha \sqcup \gamma \sim \beta \sqcup \delta$. Important congruence relations on $\big(\Gamma(D), \sqcup\big)$ and $\big(\Gamma(M), \vee\big)$ are given by taking two causal paths to be congruent if and only if they share the same initial and terminal elements. Such congruence relations arise in the theory of co-relative histories and kinematic schemes, developed in Chapters 6 and 7, where they identify *congruent families of kinematic accounts of evolutionary processes.*

Causal path algebras. Causal path semicategories are "primitive" algebraic objects, in the sense that they involve partially defined operations that fail to satisfy a number of familiar algebraic properties. This is part of the reason why such objects are relatively "unpopular in conventional algebra," as noted above. Richer objects, called *causal path algebras,* may be constructed from causal path semicategories, by incorporating the auxiliary structure of a ring, without sacrificing any information-theoretic content. It is important to note, however, that *adding* this "extrinsic ring structure" to the picture a priori threatens the strong interpretation of the causal metric hypothesis (CMH), since the information contained in an arbitrary ring cannot be attributed to "causal structure" in any natural way. There are at least two possible ways to address this difficulty. First, one might restrict attention to rings that "do not really introduce any essentially new information," but merely provide a device for encoding existing information. For example, the ring of integers $\mathbb{Z}$ may be used to encode quantities such as "how many causal paths in one subobject of a multidirected set splice together with causal paths in another subobject," which can hardly be regarded as extrinsic. Second, at a more sophisticated level, one may choose to view the formation of causal path algebras as a "functor-like operation," which takes rings as input, but does not depend on the choice of any specific ring. In particular, the specific choice of ring must not determine the physical predictions of the theory. In certain areas of mathematics, such as homological algebra, it is common to follow this type of reasoning, and choose different "coefficient rings" to compute invariants of objects that have no special relationship to one ring or another. It may be the case that different rings provide convenient descriptions of different quantities. However, it is worth keeping in mind that insisting on using a *particular special ring* to define causal path algebras risks elevating extrinsic information to an unjustified role.

An analogy from elementary algebra is useful to motivate the construction of a causal path algebra, which takes two distinct objects, namely, a ring and a causal path semicategory, as input. The analogy involves multiplication of monomials with real coefficients, which itself incorporates two distinct operations, namely, multiplication of coefficients in $\mathbb{R}$, and addition of exponents in $\mathbb{N}$. In this context, the additive structure of $\mathbb{N}$ may be viewed as a "substratum for multiplication" in the polynomial algebra $\mathbb{R}[x]$. More generally, group or semigroup algebras combine the structure of a ring, analogous to $\mathbb{R}$, with the structure of a group or semigroup, analogous to $\mathbb{N}$. In this setting, the group or semigroup participates only in the *multiplicative* operation of the resulting algebra, not the additive operation. Still more general are semicategory algebras, which derive their multiplicative structures from a semicategory operation. Finally, generalized semicategory algebras may be defined in terms of generalized semicategories, such as causal path semicategories.

As one might expect, two important types of generalized semicategory algebras arising naturally in discrete causal theory are the *concatenation algebra* over a directed set, and the *directed product algebra* over a multidirected set. These algebras, introduced formally in Definitions 5.10.1 and 5.10.3 below, are constructed by combining the structure of a ring T with the appropriate generalized semicategory operations on the corresponding causal path sets. I refer to both types of algebras as **causal path algebras**, since they are generated by causal path semicategories. The choice to use the symbol T in this context is intended to evoke the word "target," since T serves as a "target object for causal path functionals," as explain in Chapter 6. The target ring T of a causal path algebra may be replaced by an object possessing "less structure," such as an abelian group, or by an object at a different level of algebraic hierarchy, such as a category. Under such a replacement, the algebra itself may be reduced to a *causal path module*, or elevated to a higher-level structure such as a *"generalized enriched functor category,"* respectively. Generalizing beyond the context of causal paths, any splice product for a class of morphisms into a multidirected set M may be used to define a *causal splice module* or *causal splice algebra* over M, or a higher-level analogue. Alternatively, one may consider many distinct splice products on the same class of morphisms simultaneously.[37]

Concatenation algebras are defined in the context of *directed* sets, since their multiplicative structures arise from concatenation products:

Definition 5.10.1. *Let $D = (D, \prec)$ be a directed set, with causal path set $\Gamma(D)$. Let T be a ring, usually assumed to be commutative with unit.*

1. *The* **concatenation algebra** *$T^{\sqcup}(D)$ over D is the algebra whose elements are formal sums $\sum_\gamma t_\gamma \gamma$, where $\gamma \in \Gamma(D)$ and $t_\gamma \in T$, and whose multiplication is defined by multiplying coefficients in T, and forming concatenation products of pairs of causal paths in $\Gamma(D)$, whenever these products exist:*

$$\Big(\sum_{\gamma} t_\gamma \gamma\Big) \sqcup \Big(\sum_{\gamma'} t_{\gamma'} \gamma'\Big) := \sum_{\gamma \prec_{\sqcup} \gamma'} t_\gamma t_{\gamma'} \gamma \sqcup \gamma'. \qquad (5.10.2)$$

2. *The* **chain concatenation algebra** *$T^{\sqcup}[D]$ over D is the subalgebra of $T^{\sqcup}(D)$ given by restricting γ to $\Gamma_{\mathrm{IRR}}(D) = \mathrm{Ch}(D)$.*

The empty sum is identified with the additive identity 0_T of T in both algebras.

---

[37] *Vertex operator algebras*, used extensively in string theory, provide familiar examples of algebraic objects involving many distinct operations.

The index notation $\gamma \prec_{\sqcup} \gamma'$ appearing in the sum on the right-hand side of Equation (5.10.2) encodes, in generalized order-theoretic terms, the condition that a pair γ and γ' of causal paths contributes to this sum only when the concatenation product $\gamma \sqcup \gamma'$ exists, since the existence of this product in the concatenation semicategory $(\Gamma(D), \sqcup)$ is equivalent to the existence of the relation $\gamma \prec_{\sqcup} \gamma'$ in the corresponding causal path space $(\Gamma(D), \prec_{\sqcup})$. The target ring T may also be called the "coefficient ring" for the concatenation algebra $T^{\sqcup}(D)$. Similarly, the algebra itself may be referred to as the *"concatenation algebra over D, with coefficients in T."*

Directed product algebras are defined in the context of *multidirected* sets, since their multiplicative structures arise from directed products:

Definition 5.10.3. *Let $M = (M, R, i, t)$ be a multidirected set, with causal path set $\Gamma(M)$. Let T be a ring, usually assumed to be commutative with unit.*

1. *The* **directed product algebra** $T^{\vee}(M)$ *over M is the algebra whose elements are formal sums $\sum_{\gamma} t_{\gamma}\gamma$, where $\gamma \in \Gamma(M)$ and $t_{\gamma} \in T$, and whose multiplication is defined by multiplying coefficients in T, and forming directed products of causal paths in $\Gamma(M)$, whenever these products exist:*

$$\Big(\sum_{\gamma} t_{\gamma}\gamma\Big) \vee \Big(\sum_{\gamma'} t_{\gamma'}\gamma'\Big) := \sum_{\gamma \prec_{\vee} \gamma'} t_{\gamma}t_{\gamma'}\gamma \vee \gamma'. \tag{5.10.4}$$

2. *The* **chain directed product algebra** $T^{\vee}[M]$ *over M is the subalgebra of $T^{\vee}(M)$ given by restricting γ to $\Gamma_{\mathrm{IRR}}(M) = \mathrm{Ch}(M)$.*

The empty sum is identified with the additive identity 0_T of T in both algebras.

The remarks about notation and terminology following Definition 5.10.1 may be modified in the obvious way to apply to Definition 5.10.3. Both the concatenation algebra $T^{\sqcup}(D)$ and the directed product algebra $T^{\vee}(M)$ may be classified as "noncommutative generalized graded algebras," since the "length functionals" encoding the lengths of causal paths in D and M are "roughly additive." This vague description may be sharpened by examining the structural details involved. For example, the length of the concatenation product $\gamma \sqcup \delta$ of a pair of finite causal paths γ and δ, if it exists, is not equal to the sum of the lengths of γ and δ, but rather to the sum of these lengths *plus one.* Further, "additivity" acquires a generalized meaning, expressed in terms of isomorphism classes, for causal paths of infinite length. Subject to these qualifications, the additive subgroups of the causal path algebras $T^{\sqcup}(D)$ and $T^{\vee}(M)$ generated by causal paths of length n may be regarded as their "nth graded pieces," and similarly for the corresponding chain algebras. From an algebraic viewpoint, this generalized graded structure may be credited for many of the most useful properties of causal path algebras, but I do not emphasize this perspective in this book. It is sometimes useful to consider special subalgebras of causal path algebras, defined in terms of *finite* formal sums. In this book, I mostly sidestep this distinction, by working with *finite* subsets of causal path sets in the acyclic case. Among other simplifications, this choice of focus avoids the nuisance of proliferating notation, for example, by using expressions such as $T^{\sqcup}[[D]]$ to denote algebras of "power-series-like formal sums," *without* finiteness assumptions. However, such details must ultimately be addressed in the most general versions of the theory.

Conventional applications of path algebras and similar objects. Path algebras similar to causal path algebras, or even structurally equivalent to special cases of such algebras, arise naturally in graph theory, information theory, computer science, network theory, category theory, noncommutative geometry, algebraic K-theory, and, of course, causal set theory. For example, in the special case where the coefficient ring T is a field, the subalgebra of finite formal sums in the chain directed product algebra $T^{\vee}[M]$ is the "standard path algebra" appearing in Abrams and Pino [AP05]. The chain concatenation algebra $T^{\sqcup}[D]$ also appears in the literature, as a special type of *quiver algebra.* Similar quiver algebras play important roles in recent and ongoing research in noncommutative algebra and noncommutative geometry, particularly in the study of certain so-called "quantum groups," defined as deformations of Hopf algebras. Another closely-related type of algebra is the *Temperley-Lieb algebra,* which plays a role in knot theory, braid theory, and various physical applications.

Path algebras possess a number of attractive mathematical properties, besides the generalized graded structure noted above. The chain directed product algebra $T^{\vee}[D]$ over a finite acyclic directed set $D = (D, \prec)$ suffices for illustrative purposes. $T^{\vee}[D]$ possesses a multiplicative identity, given by the formal sum of the elements in D. Each element x in D defines an idempotent operator on $T^{\vee}[D]$, whose image is an *indecomposable projective* $T^{\vee}[D]$*-module* $P(x)$, called the *punctual module* of x.[38] Every other chain in D has square zero. $T^{\vee}[D]$ is a *hereditary algebra,* which means that any submodule of a projective module over $T^{\vee}[D]$ is itself projective; this follows from the fact that every projective module over $T^{\vee}[D]$ is "built up from the punctual modules $P(x)$." These properties play an important role in the study of related *module categories, derived categories,* and *cluster categories.* For further details regarding such properties, I recommend Bernhard Keller's recent paper *Cluster Algebras, Quiver Representations and Triangulated Categories* [KE10]. Some of these properties, or suitable analogues thereof, still apply to general multidirected sets, but others break down. For example, the formal sum of elements in a multidirected set M is not a multiplicative identity for $T^{\vee}[M]$ if M has unbounded chains. In the cyclic case, a reflexive relation in M does not have square zero, but rather generates an infinite cyclic multiplicative subgroup of $T^{\vee}[M]$.

Raptis' incidence algebra. In his 2000 paper *Algebraic Quantization of Causal Sets* [RA00], Ioannis Raptis employs a noncommutative algebra called an *incidence algebra,* which is a close transitive analogue of the directed product algebra over an acyclic directed set, with coefficients in $\mathbb{C}$. As described by Raptis, a suggestive "ket-bra" notation may be used to represent elements of this algebra. In particular, each relation $x \prec_{\text{CS}} y$ in a causal set C may be viewed as an *operator* $|x\rangle\langle y|$, and relations may be "multiplied" according to the formula

---

[38] Here, the word "punctual" means "related to a particular point;" it has nothing to do with time.

$$|x\rangle\langle y|u\rangle\langle v| = \langle y|u\rangle|x\rangle\langle v|, \quad \text{where} \quad \langle y|u\rangle := \delta_{yu} = \begin{cases} 1 & \text{if } y = u \\ 0 & \text{if } y \neq u, \end{cases} \tag{5.10.5}$$

where the expression δ_{yu} denotes the *Kronecker delta function*, defined by the cases listed on the right-hand side of Equation (5.10.5). In order to make sense, this definition requires the causal set axiom of transitivity (TR), since the relation $x \prec_{\text{CS}} v$ may not exist in the nontransitive case. Equation (5.10.5) defines a semicategory operation on the relation space over a causal set, with Raptis' incidence algebra arising as the corresponding semicategory algebra with complex coefficients. This operation suffers from *loss of information,* compared with the product operation in the corresponding directed product algebra, since there may be many distinct 2-chains between a given pair of elements x and v in a causal set. Hence, Raptis' incidence algebra may be viewed as a special type of **short-circuit algebra**, i.e., a "degenerate splice algebra" which discards some of the information associated with the objects that are being spliced together. There are instances, of course, in which not all of the information contained in a structure is of equal relevance; the general ideas of quotient objects in mathematics, and gauge theory in physics, are examples. For this reason, short-circuit algebras are not necessarily "bad" from a theoretical viewpoint, although they do not play a significant role in this book. Raptis describes addition in the incidence algebra over a causal set C as representing *quantum-theoretic superposition.* While this is a reasonable physical interpretation, it can only apply in the background dependent context, since it involves combining algebraic data associated with substructures of a *particular* causal set. More relevant to the approach developed in this book are the *monoids of arrow fields* used by Christopher Isham in his *quantization on a category* [IS05], which may be viewed as "higher-level multiplicative analogues" of Raptis' incidence algebras. Isham's construction is discussed in more detail in Chapter 6.

References

[HN60] Frank Harary and Robert Z. Norman. *Some Properties of Line Digraphs.* Rediconti del Circolo Matematico di Palermo, **9**, 2, pp. 161–168, 1960.

[NV12] Tamás Nepusz and Tamás Vicsek. *Controlling edge dynamics in complex networks.* Nature Physics, **8**, 7, pp. 568–573, 2012. arXiv preprint: http://arxiv.org/pdf/1112.5945v1.pdf.

[MA00] Fotini Markopoulou. *Quantum Causal Histories.* Classical and Quantum Gravity, **17**, 10, pp. 2059, 2000. arXiv preprint: https://arxiv.org/pdf/hep-th/9904009v5.pdf.

[BPS08] Rick Blute, Prakash Panangaden, and Lutz Straßburger. *The Logic BV and Quantum Causality.* Preprint. URL: http://www.lix.polytechnique.fr/Labo/Lutz.Strassburger/papers/BVlocative.pdf.

[MRS06] Seth A. Major, David Rideout, and Sumati Surya. *Spatial Hypersurfaces in Causal Set Cosmology.* Classical and Quantum Gravity, **23**, 14, pp. 4743–4751, 2006. arXiv preprint: http://arxiv.org/pdf/gr-qc/0506133v2.pdf.

[SU11] Sumati Surya. *Directions in Causal Set Quantum Gravity.* In Recent Research in Quantum Gravity, edited by Arundhati Dasgupta. Nova Science Publishing Incorporated, 2012. arXiv preprint: http://arxiv.org/pdf/1103.6272v1.pdf.

[OCB16] Ognyan Oreshkov, Fabio Costa, and Časlav Brukner *Quantum correlations with no causal order*. Nature Communications **3**, 1092, 2012.

[AJL05] Jan Ambjørn, J. Jurkiewicz, and R. Loll. *Spectral Dimension of the Universe*. Physical Review Letters, **95**, 171301, 2005. arXiv preprint: https://arxiv.org/pdf/hep-th/0505113v2.pdf.

[BR15] Ginestra Bianconi and Christopher Rahmede. *Complex Quantum Network Manifolds in Dimension d>2 are Scale-Free*. Preprint, 2015. arXiv preprint: https://arxiv.org/pdf/1506.02648v1.pdf.

[SR01] David Rideout and Rafael Sorkin. *Evidence for a continuum limit in causal set dynamics*. Physical Review D, **63**, 10, 104011, 2001. arXiv preprint (updated 2008): http://arxiv.org/pdf/gr-qc/0003117v4.pdf.

[SR99] David Rideout and Rafael Sorkin. *Classical sequential growth dynamics for causal sets*. Physical Review D, **61**, 2, 024002, 2000. arXiv preprint: http://arxiv.org/pdf/gr-qc/9904062v3.pdf.

[CA15] S. Carlip. *Dimensional reduction in causal set gravity*. Classical and Quantum Gravity, **32**, 23,232001, 2015. arXiv preprint: http://arxiv.org/pdf/1506.08775v3.pdf.

[ENO09] George Ellis, Denis Noble, Timothy O'Connor (Editors). *Top-Down Causation*. Interface Focus, Feburary 6, 2009.

[MEO09] Nancey Murphy, George Ellis, Timothy O'Connor (Editors). *Downward Causation and the Neurobiology of Free Will (Understanding Complex Systems)*. Springer, 2009.

[EL05] G. F. R. Ellis. *Physics, complexity and causality*. Nature, **435**, 9, pp. 743, 2005.

[AN72] P. W. Anderson. *More is Different*. Science, New Series, **177**, 404, pp. 393–396, 1972.

[BU11] J. Butterfield. *Emergence, Reduction, and Supervenience: a Varied Landscape*. Foundations of Physics, **41**, 6, pp. 920–959, 2011. arXiv preprint: http://arxiv.org/pdf/1106.0704v1.pdf.

[PE67] Roger Penrose. *Twistor Algebra*. Journal of Mathematical Physics, **8**, 2, 345–366, 1967.

[WW90] R. S. Ward and Raymond O. Wells. Twistor Geometry and Field Theory. Cambridge Monographs on Mathematical Physics, *Cambridge University Press*, 1990.

[SO09B] Rafael Sorkin. *Light, Links and Causal Sets*. Journal of Physics: Conference Series, **174**, 012018, 2009. arXiv preprint: http://arxiv.org/pdf/0910.0673v1.pdf.

[SO09A] Rafael Sorkin. *Does locality fail at intermediate length scales?* Approaches to Quantum Gravity, edited by Daniele Oriti. Cambridge University Press, 2009.

[BA12] Julian Barbour. *Shape Dynamics: an Introduction*. Quantum Field Theory and Gravity: Conceptual and Mathematical Advances in the Search for a Unified Framework, Edited by Felix Finster, Olaf Müller, Marc Hardmann, Jürgen Tolksdorf, and Eberhard Zeidler. arXiv preprint: http://arxiv.org/pdf/1105.0183v1.pdf.

[BA14] Julian Barbour, Tim Koslowski, amd Flavio Mercati. *The Solution to the Problem of Time in Shape Dynamics*. Classical and Quantum Gravity, **31**, 15, 155001, 2014. arXiv preprint: http://arxiv.org/pdf/1302.6264v1.pdf.

[RO04] Carlo Rovelli. *Quantum Gravity*. Cambridge Monographs on Mathematical Physics. Cambridge University Press, 2004.

[TH07] Thomas Thiemann. *Modern Canonical Quantum General Relativity*. Cambridge Monographs on Mathematical Physics. Cambridge University Press, 2007.

[RU95] Martin Russling. *A general scheme for breadth-first graph traversal*. Mathematics of Program Construction. Lecture Notes in Computer Science, **947**, pp. 380–398, 1995. Springer-Verlag.

[IS05] Christopher Isham. *Quantising on a Category*. Foundations of Physics, **35**, 2, pp. 271–297, 2005. arXiv preprint: http://arxiv.org/pdf/quant-ph/0401175v1.pdf.

[IS11] Christopher Isham. *Topos Methods in the Foundations of Physics*. In Deep Beauty: Understanding the Quantum World through Mathematical Innovation, edited by Hans Halvorson. Cambridge University Press, 2011. arXiv preprint: http://arxiv.org/pdf/1004.3564v1.pdf.

[AP05] Gene Abrams and Gonzalo Aranda Pino. *The Leavitt path algebra of a graph*. Journal of Algebra, **293**, 2, pp 319–334, 2005. arXiv preprint: http://arxiv.org/pdf/math/0509494v1.pdf.

[KE10] Bernhard Keller. *Cluster Algebras, Quiver Representations and Triangulated Categories*. Preprint, 2010. arXiv preprint: http://arxiv.org/pdf/0807.1960v11.pdf.

[RA00] Ioannis Raptis. *Algebraic Quantization of Causal Sets*. International Journal of Theoretical Physics, **39**, 5, pp. 1233–1240, 2000. arXiv preprint: http://arxiv.org/pdf/gr-qc/9906103v1.pdf.

Chapter 6
Quantum Spacetime

Abstract. This chapter applies the foundational material from previous chapters to construct a flexible new version of discrete quantum causal theory, which may be applied either in the background dependent context of particles and fields existing on a fixed directed set, or in the background independent context of quantum spacetime structure. Section 6.1 reviews basic quantum theoretic notions, such as the *superposition principle* and the *path summation approach to quantum theory,* and discusses some modern contributions to background independent quantum theory by Isham and Sorkin. Section 6.2 introduces the theory of *path summation over a multidirected set,* which is analogous to *path integral methods* in conventional quantum field theory. Section 6.3 describes a striking structural feature of discrete causal theory, called *iteration of structure,* in which the same abstract type of structure, expressed in terms of directed relationships, exists at both the classical and quantum levels of the theory. At the classical level, these directed relationships are represented by relations between pairs of events, while at the quantum level, they are represented by relationships between pairs of classical histories, called *co-relative histories.* Section 6.4 examines the theory of co-relative histories in more detail, and explains why multidirected structure arises at the quantum level. This represents a "higher-level application" of Grothendieck's *relative viewpoint,* in which the "objects" are directed sets, and the "natural relationships" are co-relative histories. Section 6.5 explains how the abstract theory of path summation over a multidirected set may be used to define quantum theories in a variety of different ways. Section 6.6 revisits the construction of Feynman's path integral, in order to motivate the discrete causal constructions examined in the remainder of the chapter. Section 6.7 adapts relevant notions from conventional path integration to construct a path summation version of discrete quantum causal theory in relation space. Section 6.8 reviews how Feynman re-derived Schrödinger's equation, in order to motivate the derivation of analogous equations in the discrete causal context. Section 6.9 presents the derivation of *causal Schrödinger-type equations,* which serve as dynamical laws for discrete quantum causal theory. Section 6.10 describes how *generalized quantum amplitudes* may be computed via causal path algebras.

B.F. Dribus, *Discrete Causal Theory*, DOI 10.1007/978-3-319-50083-6_6

6.1 Quantum Preliminaries

From "classical" to "quantum." The transition from classical physics to the quantum realm in discrete causal theory is accomplished via an appropriate adaptation of Feynman's *path summation approach to quantum theory,* expressed in terms of path summation over multidirected sets. This transition involves a change of emphasis from *relationships between pairs of individual events,* to *relationships between pairs of classical histories,* i.e., co-relative histories, as foreshadowed at the beginning of Chapter 5. Due to the complexity of the subject, the presentation in this chapter is far from comprehensive, and involves a number of simplifying assumptions; in particular, I focus mostly on the acyclic case. Despite these limitations, the approach introduced here is very general at a conceptual level, and admits ready adaptation to a variety of settings. In particular, as mentioned above, it may be used to construct either background dependent theories of fields and particles existing on a particular directed set, or background independent theories involving structured configuration spaces of directed sets; in particular, kinematic schemes. As explained in Section 1.7, and again in Section 3.3, the former theories are "conservative;" they are analogous to quantum field theories in curved spacetime. The latter theories are more ambitious; they attempt to describe the quantum nature of spacetime itself, and ultimately, all physical phenomena. As the title of this chapter suggests, I focus principally on the latter type of theory, which is both more interesting and more difficult.

The fact that the same abstract approach applies to both background dependent and background independent versions of discrete quantum causal theory is a consequence of the striking "self-similarity" across different levels of structural hierarchy in the theory, referred to in this book as the principle of *iteration of structure* (IS). Throughout the next two chapters, I make repeated use of this principle, focusing on the prototypical case, which involves the "neighboring hierarchical levels" occupied by directed sets and kinematic schemes of directed sets, respectively.[1] In this book, these two levels of structure are identified as the "classical" and "quantum" levels, respectively, since the former level contains objects suitable for modeling classical spacetime structure in the context of the metric recovery theorems, while the latter level exhibits the features necessary to support an appropriate version of the path summation approach to quantum theory. In this manner, discrete causal theory ties together the two ingredients of *"suitable structural notions"* and a *"suitable approach to quantum theory,"* cited in the preface of this book as necessary for solving the problems of fundamental spacetime structure, quantum gravity, and unification of physical law.

Superposition; path summation approach. A generic feature of quantum theory, regardless of the specific approach adopted, or the specific context in which it is applied, is the notion of *superposition*, in which multiple independent physical scenarios coexist. In ordinary quantum mechanics, each of these scenarios is a solution

[1] In particular, the hyperlink for iteration of structure (IS) in the electronic version of the book links to a formalization of this special case, even though the underlying idea is more general.

of Schrödinger's equation, and superposition is a consequence of the *linearity* of this equation. As mentioned in Chapter 2, it is possible to imagine combining different physical scenarios in a *nonlinear* fashion, and some serious physicists, such as Roger Penrose, have suggested that this might be necessary in the context of quantum gravity. However, a sufficiently broad notion of superposition can incorporate technical modifications of this nature without any essential change in viewpoint.[2] In Feynman's original version of the path summation approach to quantum theory, each "independent physical scenario" is represented by a path in a background spacetime manifold or configuration space. Such a path may be taken to encode a possible *history* of a physical system, since the background is static in this context; i.e., all the necessary "historical details," and nothing else, are contained in the path itself. Predictions are made by summing over these "paths" to obtain *quantum amplitudes,* which may be translated into probabilities. All "possible paths," even those that seem classically absurd, contribute to these amplitudes. Roughly speaking, they are determined by an underlying "classical theory," via an action principle, and the summation process "favors classical solutions," by damping out the contributions of other "paths" through destructive interference. The resulting behavior is not exactly the same as that predicted by the classical theory, but generally *approaches* classical behavior "in the classical limit," which usually means "for large values of quantum numbers." This "scale-dependent convergence to the classical theory" is another generic feature of quantum theory, called the *correspondence principle.*

In its original continuum-based context, the path summation approach is implemented via technically-obtuse limits of iterated integrals called *path integrals,* and the approach itself is therefore often referred to as *path integration.* Bringing fields and special relativity into the picture, path integration becomes one of the principal tools, and one of the principal technical stumbling blocks, of conventional quantum field theory. However, from a basic conceptual perspective, the path summation approach need not be chained to this specific and forbidding technical machinery; rather, it may be viewed as a much more general strategy of *"weighing contributions from members of a configuration space of evolutionary processes,"* using whatever natural structural features are available. In the background dependent context, these processes may be represented by particle trajectories, or field configurations, or some other auxiliary structure "inhabiting" spacetime. In the background independent setting, evolutionary processes involve entire "classical universes," which may be represented by geometries, or spin foams, or dynamical triangulations, or causal sets, or other models. The flexibility and generality of the path summation approach, viewed in this broad conceptual fashion, allows it to be applied in contexts where most other versions of quantum theory do not even make sense.

Path summation in the discrete causal context. In discrete causal theory, it is straightforward, at a formal level, to construct a background *dependent* quantum theory, by simply replacing the background spacetime manifold in Feynman's theory of path integration with a discrete directed set, and applying the general theory

[2] A recent article of Bassi, Singh, and Ulbricht [BSU15] describes some nonlinear generalizations of quantum theory.

of path summation developed in Section 6.2 below. In this setting, a history may be represented by a "particle trajectory," i.e., a causal path, or by some more complex auxiliary structure, "inhabiting" the chosen background set. This approach leads to flexible discrete causal alternatives to quantum field theories in curved spacetime, and to lattice field theories. It is conceptually similar to the existing treatments of particles and fields on causal set backgrounds described in Chapter 3, but is considerably strengthened by the systematic use of relation space, and by the other innovations and changes of viewpoint described in the intervening chapters. Much deeper, more difficult, and more interesting are background *independent* versions of discrete quantum causal theory, in which each history is represented by a *different* directed set. This requires consideration of entire configuration spaces of directed sets, which are elevated to the status of *kinematic schemes* via recognition of their natural "higher-level" multidirected structure, induced by *co-relative histories* between pairs of directed sets. Important analogues of these developments appear in Isham's *quantization on a category* [IS05] and Sorkin's *quantum measure theory* [SO12]. For contextual purposes, I briefly outline these two approaches below, before commencing on the technical apparatus of path summation.

Isham's quantization on a category. As mentioned in Section 3.8, Christopher Isham has explored a number of ideas relevant to quantum causal theory in his groundbreaking work on physical applications of category theory. For example, in his 2005 paper *Quantising on a Category* [IS05], Isham introduces a category-theoretic approach to "quantizing" a broad variety of physical models, including causal sets. This represents only a small facet of Isham's ongoing research program to develop a general *topos-theoretic* framework for physics. A *topos* is an "enriched" category with properties rendering it suitable to serve as a "mathematical universe." For example, there is a *topos of sets,* which serves as Isham's "topos for classical physics." Topos theory is yet another crucial structural idea arising from Alexander Grothendieck's approach to commutative algebraic geometry, where it was conceived by Grothendieck and Verdier as a way of organizing information about *presheaves* and *sheaves* on topological spaces, and on generalized topological spaces called *sites*. For the purposes of this book, it suffices to highlight only a few specific aspects of Isham's work.

Isham [IS05] writes that his "main topic of interest" is

> *...to construct a quantum framework for a system whose "history configuration space"... ...is a collection of possible space-times, such as... ...causal sets... ...or... topological spaces.* (page 277)

and later,

> *...when quantising a system whose configuration space... ...is a collection... ...of sets with structure, the starting point will be an association to each pair* [of such sets] *a set... ...of transformations* [between them] *that, in some suitable sense, 'respect,' or 'reflect,'* [their] *internal structures...* (page 278)

Isham then proceeds to explain that such a family of "structured sets" may be organized into a category $\mathcal{Q}$, whose morphisms are the aforementioned "transformations." Abstracting this, he proposes to quantize *general* categories by

> ...*representing* [morphisms] *with operators on a Hilbert space in some way.* (page 280)

Without going into the details, Isham's method involves a *presheaf of Hilbert spaces* and a *monoid of arrow fields* over a category $\mathcal{Q}$. A given arrow field assigns to each object Q of $\mathcal{Q}$ a morphism $Q \to Q'$ to some other object Q'. A natural monoid structure on the class of arrow fields over $\mathcal{Q}$ is given by composition of morphisms. As mentioned in Section 5.9, Isham's monoid of arrow fields may be viewed as a "higher-level, purely multiplicative" analogue of Raptis' *incidence algebras* [RA00]. Both types of objects contain part, but not all, of the structure in the corresponding directed product algebras, described in Section 5.10,[3] Isham gives a simple example of his construction, using a category consisting of two causal sets, which is both interesting and relevant to the broader discrete causal context.[4] However, for the purposes of this book, it is Isham's *general reasoning,* represented by the above quotations, that is most important. The technical details of the approach presented here are somewhat different, involving higher-level phase maps on kinematic schemes rather than functors on categories. In particular, I do not a priori assign Hilbert spaces, or any other extrinsic, "ready-made" quantum-theoretic devices, to directed sets. I also work with path algebras, rather than with incidence algebras, arrow fields, or similar constructions, since these latter objects generally "lose too much information" under composition for the desired applications.

Sorkin's quantum measure theory. Rafael Sorkin has also done very general work on the path summation approach to background independent quantum theory, and has applied these ideas in specific ways to causal set theory. In particular, in his 2012 paper *Toward a Fundamental Theorem of Quantal Measure Theory* [SO12], Sorkin describes quantum-theoretic measures in the context of the Sorkin–Rideout kinematic scheme $\mathbb{S}_{\rm SR}$ describing the sequential growth of causal sets, illustrated in Figure 3.4.1 of Chapter 3, and described in more detail in Chapter 7. Sorkin works mostly in terms of *labeled* causal sets in this paper, which leads to the replacement of $\mathbb{S}_{\rm SR}$ by an appropriate *tree,* in the graph-theoretic sense. Sorkin writes,

> ...*the collection of naturally labeled finite* [causal sets]... ...*has itself the structure of a* [tree], *because its elements are labeled...* ...*Clearly, a particular realization of the growth process...* ...*can be conceived of as an upward path through this tree.* (page 7)

Here, a "realization of the growth process" means a sequence of finite causal sets, each differing from the last by the "addition of a single element," together with a family of relations. The nth element added is assigned the natural number n as a label, and the resulting causal set is the nth member of an upward-directed path through the tree. The labeling process supplies a *kinematic account* of the evolution of each of these causal sets, i.e., a "partly fictional description of its evolution,"

[3] Isham mentions Raptis' incidence algebras in the preprint version of [IS05].

[4] See pp. 293–296 of [IS05].

roughly analogous to a frame of reference in relativity.[5] In Chapter 7, I revisit such evolutionary processes and kinematic accounts in a more general context.

6.2 Path Summation

Path functionals; motivation from continuum theory. I begin this section by briefly reviewing some standard notions from continuum-based classical mechanics, in preparation for analogous constructions in the context of discrete multidirected sets. The reader should be aware that some of the notation appearing in this section, such as the standard notation for the *classical action* and the *Lagrangian,* is similar to notation I have used for completely different purposes elsewhere in the book.[6] *Path functionals* are maps from a space of paths into a target object, such as an abelian group, a ring, or a field. In the continuum-based setting, a path is an equivalence class of maps from an interval in the real line $\mathbb{R}$ into a real manifold, as described in Section 5.9. In the study of particle motion in background dependent continuum-based classical mechanics, the *classical action* $\mathcal{S}$ is a real-valued path functional, given by integrating the *Lagrangian* $\mathcal{L}$, which encodes the difference between a particle's kinetic and potential energies, with respect to time, over a spacetime path γ, viewed as a possible particle trajectory:

$$\mathcal{S}(\gamma) = \int_\gamma \mathcal{L}dt. \tag{6.2.1}$$

Hamilton's principle states that *"the classical trajectory γ_{CL} of the particle renders the classical action stationary."* Heuristically, this means that the Lagrangian $\mathcal{L}$ "chooses" the specific trajectory γ_{CL}, determined by how $\mathcal{S}$ varies with γ. This is a "more holistic" interpretation of classical mechanics than the interpretation provided by Newton's equations of motion, since the value of the action $\mathcal{S}(\gamma)$ depends on the entire trajectory γ. Thus, Hamilton's principle provides a classical example of the general fact, noted in Chapter 5, that a holistic viewpoint may be *useful*, even when the same physics may be described via a reductionist approach.

The left-hand diagram in Figure 6.2.2 illustrates a classical trajectory γ_{CL} in a fixed background spacetime manifold, along with several "nearby" trajectories. The figure is deliberately drawn to "suggest" Euclidean or Minkowski spacetime, in order to make it clear that the background is *not* dynamical in this context. In the Euclidean case, Newton's equations of motion may be recovered via Hamilton's principle, as

---

[5]Actually, such a labeling is somewhat more specific than a kinematic account, which generally does not distinguish unambiguously between "old" and "new" elements. See Sections 6.4 and 7.3 for details.

[6]Actually, the mathscript symbol $\mathcal{L}$, which I use for the Lagrangian, is often used to denote a Lagrangian *density* in the context of field theory. In this book, however, mathscript symbols generally appear at the "category-functor level of hierarchy," which is appropriate for the Lagrangian. In any case, densities are generally irrelevant in the discrete context.

the *Euler-Lagrange equations* for the classical action functional $\mathcal{S}$. This derivation, which represents an iconic application of the *calculus of variations,* generalizes to a host of other physical and mathematical settings. However, the physical situation changes radically when one moves to the background independent context; for example, in general relativity. Here, particle trajectories *interact* with the spacetime they "inhabit," so it is no longer adequate to merely integrate over individual trajectories. Instead, each trajectory corresponds to a completely different spacetime. This is illustrated in the right-hand diagram in Figure 6.2.2, which shows a particle trajectory with spacetime "warping" around it.

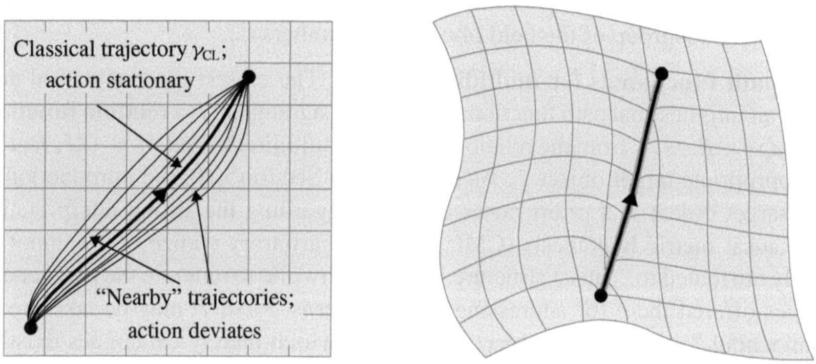

Figure 6.2.2. The classical action $\mathcal{S}$ "chooses" the classical trajectory γ_{CL}; in a background independent theory, *different trajectories involve different classical universes.*

Returning for the moment to the simple scenario of single-particle motion in the background dependent continuum-based setting, the Lagrangian $\mathcal{L}$ is an example of an *infinitesimal path functional*, which means that it depends only on the position and motion of the particle at a given time, not on the details of its trajectory. In fact, the Lagrangian is usually taken to depend only on "first-order information," i.e., on position, time, and velocity. As such, it may be viewed as a function on the *real tangent bundle* over a spacetime manifold. As described below, discrete causal analogues of such Lagrangians may be viewed as *relation functions,* i.e., functions assigning values to elements of the relation set R of a multidirected set $M = (M, R, i, t)$. The reader may recall here the comparison between R and the *tangent sheaf* in algebraic geometry, mentioned in Section 3.6. In any case, this "first-order" restriction makes computations easier; in particular, it simplifies the calculation of path integrals in ordinary quantum mechanics and quantum field theory. Generalizations of $\mathcal{L}$ and $\mathcal{S}$ are many and varied, but the primitive versions cited here are adequate for the purposes of illustration and analogy.

In his original formulation of the path summation approach to quantum theory, Feynman defined a path functional Θ, called a *phase map,* assigning to each particle trajectory γ in a region of classical spacetime a "phase" $\Theta(\gamma)$, equal to the *complex exponential of its classical action, in units of Planck's reduced constant:*

$$\Theta(\gamma) := e^{\frac{i}{\hbar}\mathcal{S}(\gamma)}. \tag{6.2.3}$$

The term "phase" reflects the fact that $\Theta(\gamma)$ always has unit *magnitude;* the differences between phases of different "paths" consist entirely of "directional information," which determines the extent to which these phases interfere constructively or destructively. The classical action $\mathcal{S}$ is additive for directed products of paths,[7] by the definition of integration. Hence, Feynman's phase map is multiplicative, since exponentiation converts addition to multiplication. In more sophisticated terms, $\mathcal{S}$ is a *generalized semicategory morphism* from an appropriate causal path semicategory over a spacetime manifold to the underlying additive group of $\mathbb{R}$, and Θ is a morphism from the same causal path semicategory into the unit circle S^1, viewed as a multiplicative subgroup of the field of complex numbers $\mathbb{C}$.

Causal path functionals for multidirected sets. The closest discrete causal analogue of an infinitesimal path functional, such as a Lagrangian, is a **relation function**, i.e., a map $\theta : R \to T$ from the relation set R of a multidirected set $M = (M, R, i, t)$ to an appropriate target object T. As discussed in Section 5.10, the introduction of such a target object T a priori raises concerns regarding the strong interpretation of the causal metric hypothesis (CMH), since an arbitrary choice of T cannot be naturally attributed to "causal structure." It is worthwhile to reiterate the two general strategies offered there for addressing such concerns. First, T may be taken to be an "innocuous," or "natural," object, such as $\mathbb{Z}$, which merely catalogues existing information, such as multiplicities, without introducing any essentially new structure. Second, the whole construction may be understood in a functorial sense, with different choices of T employed under different circumstances, on the basis of convenience, much like different choices of coefficient rings in cohomology theories. Under the second strategy, one must take care to ensure that the *physical* implications of the associated computations do not depend on the choice of T.

The corresponding discrete causal analogue of a "non-infinitesimal" path functional, such as the classical action $\mathcal{S}$, or Feynman's phase map Θ, is a map $\Theta : \Gamma(M) \to T$ from the causal path set $\Gamma(M)$ of M into T. Such a map may be referred to descriptively as a **causal path functional**, but I often use the physically suggestive term **phase map** instead, along with the notation Θ, since the examples of principal interest in this book play essentially the same role as Feynman's phase map. Under this choice of terminology, the individual value $\Theta(\gamma)$ is called the *phase* of the causal path γ. In abstract settings, the condition that "all phases have unit magnitude" may no longer apply in a conventional sense, but it is prudent to keep in mind the conceptual lesson that "different paths contribute equally" in Feynman's approach, whether or not one chooses to apply an analogous condition in more novel

---

[7] As mentioned at the end of Section 5.9, the notion of directed product extends to settings in which the target object is not a discrete multidirected set. Here, it means essentially the same thing as in Definition 5.9.7: a pair of paths in a real manifold has a directed product "joining the two paths together" if and only if "the terminal element of the first path coincides with the initial element of the second path." This product is familiar in *homotopy theory,* where it descends to multiplication in the *fundamental groupoid.*

theories. As hinted in Section 2.9, and later elaborated in Section 6.7, such a condition *could* be regarded as an expression of background independence. Of course, arbitrary causal path functionals need not inherit any assumptions regarding uniform "magnitudes" or "contributions."

When $\Gamma(M)$ is viewed as a generalized semicategory, via recognition of a natural operation such as the concatenation product or the directed product, the most interesting associated phase maps are usually generalized semicategory morphisms, i.e., maps $\Theta : \Gamma(M) \to T$ that "preserve the structure of $\Gamma(M)$ in the target object T." This preservation of structure generalizes the multiplicativity of Feynman's phase map. Depending on the context, the target object T may be a group, a ring, a field, or perhaps a "physically suggestive" algebraic object, such as a special type of operator algebra. For example, Isham [IS05] studies functionals whose values are Hilbert space operators. At higher levels of structural hierarchy, such as in the theory of kinematic schemes, phase maps may be viewed as "assignments," analogous to functors. For example, referring to notions developed later in the present chapter, and in Chapter 7, one may define "higher-level phase maps," assigning to each co-relative history in a kinematic scheme its *causal Galois groups.* It is often preferable, in fact, to view phase maps as generalized functors even when working over an individual directed set, rather than a kinematic scheme. It is not feasible to undertake a thorough exploration of such hierarchical considerations in this book, but a few general remarks on the topic appear in Section 8.6.

A phase map Θ on the causal path set $\Gamma(M)$ over a multidirected set M may be regarded as a *formal sum* $\sum_{\gamma\in\Gamma(M)}\Theta(\gamma)\gamma$, since this expression uniquely pairs each causal path with its corresponding phase. This viewpoint immediately evokes the causal path algebras described in Section 5.10. Indeed, in the case where T is a ring, a phase map Θ may be regarded as *"supplying the coefficients of a causal path algebra element."* Conversely, each element of a causal path algebra specifies a unique phase map. Hence, the causal path algebras $T^{\sqcup}(D)$ and $T^{\vee}(M)$, introduced in Definitions 5.10.1 and 5.10.3, respectively, supply multiplicative structures on the **mapping spaces** $T^{\Gamma(D)}$ and $T^{\Gamma(M)}$ of maps $\Gamma(D) \to T$ and $\Gamma(M) \to T$, respectively. For example, if $\Theta \leftrightarrow \sum\Theta(\gamma)\gamma$ and $\Psi \leftrightarrow \sum\Psi(\gamma)\gamma$ are elements of $T^{\Gamma(D)}$ or $T^{\Gamma(M)}$, i.e., phase maps, then the formulae

$$\Phi \sqcup \Psi = \sum_{\gamma \prec_{\sqcup} \gamma'} \Phi(\gamma)\Psi(\gamma')\gamma \sqcup \gamma' \tag{6.2.4}$$

and

$$\Phi \vee \Psi = \sum_{\gamma \prec_{\vee} \gamma'} \Phi(\gamma)\Psi(\gamma')\gamma \vee \gamma', \tag{6.2.5}$$

given by multiplication in $T^{\sqcup}(D)$ and $T^{\vee}(M)$, respectively, define analogues of *convolution* on $T^{\Gamma(D)}$ and $T^{\Gamma(M)}$, respectively. For similar reasons, more conventional path algebras are sometimes called *convolution algebras* in the literature. Using the concatenation algebra $T^{\sqcup}(D)$ as an example, the similarity of the product $\Phi \sqcup \Psi$ to ordinary convolution of real valued functions, given by the formula

$$f * g(t) = \int_\tau f(\tau)g(t-\tau)d\tau,$$

may be seen by expressing the value $\Phi \sqcup \Psi(\delta)$, for a particular causal path δ, as

$$\Phi \sqcup \Psi(\delta) = \sum_\gamma \Phi(\gamma)\Psi(\delta - \gamma),$$

where γ ranges over all causal paths that "may be extended in an appropriate way[8] to produce δ," and where the expression $\delta - \gamma$ means "the causal path such that $\gamma \sqcup (\delta - \gamma) = \delta$." Similarly, the chain algebras $T^{\sqcup}[D]$ and $T^{\vee}[M]$, also introduced in Definitions 5.10.1 and 5.10.3, supply multiplicative structures on the **chain mapping spaces** $T^{\mathrm{Ch}(D)}$ and $T^{\mathrm{Ch}(M)}$, while the corresponding causal path algebras of *finite* formal sums supply multiplicative structures on the spaces of *finitely supported* maps $\Gamma(D) \to T$ and $\Gamma(M) \to T$. Here, of course, the *support* of a phase map Θ is the set of all causal paths in $\Gamma(D)$ or $\Gamma(M)$ *not* mapping to the additive identity 0_T of T under Θ. More general mapping spaces T^{Γ} involving *arbitrary subsets* Γ of causal path sets enter the picture below, in the discussion of causal paths sums and evaluation maps.

Path summation over a multidirected set. Fortunately, there is no need to discuss the general theory of path integration in the continuum-based context before introducing the analogous theory of path summation over a multidirected set. Indeed, many of the difficulties of path integration are irrelevant to the latter theory, particularly in the star finite case. In Section 6.6, I examine in greater detail some *specific* facets of continuum-based path integration used by Feynman to formulate the path summation approach to ordinary quantum theory, since these particular fragments of the construction are directly relevant to the corresponding discrete quantum causal machinery. Here, however, I merely describe the rudiments of path summation over an arbitrary multidirected set, in the special case where the target object for the phase map is a commutative ring with unit. I treat phase maps and causal path algebra elements interchangeably, as needed. Note, however, that the same phase map could conceivably supply coefficients for elements in multiple *different* causal path algebras, defined in terms of different products. Hence, one must always keep in mind which product is involved when "multiplication of phase maps" is needed, as in the *generalized inner product formulas* of Section 6.10.

Definition 6.2.6. *Let $M = (M, R, i, t)$ be a multidirected set, and let Γ be a subset of the causal path set $\Gamma(M)$ of M. Let T be a commutative ring with unit, and let Θ be a T-valued phase map on $\Gamma(M)$. Then the* **causal path sum** *$\Theta(\Gamma)$ over Γ with respect to Θ, if it exists, is the T-valued sum*

$$\Theta(\Gamma) := \sum_{\gamma \in \Gamma} \Theta(\gamma). \tag{6.2.7}$$

[8]This may be formalized via the theory of *transitions,* developed in Section 6.3, since the image of γ embeds as a *full originary subobject* of the image of δ.

A convenient way to express causal path sums is in terms of *evaluation maps.* In the context of Definition 6.2.6, the **evaluation map** $e_\Gamma : T^\Gamma \dashrightarrow T$ is the partially-defined map given by "collapsing" the formal sum over paths in Γ with phases supplied by Θ, to yield the corresponding "concrete" causal path sum in T, whenever it exists:

$$e_\Gamma(\Theta) = e_\Gamma\Big(\sum_{\gamma\in\Gamma}\Theta(\gamma)\gamma\Big) = \sum_{\gamma\in\Gamma}\Theta(\gamma). \tag{6.2.8}$$

Note that the evaluation map e_Γ depends on T and Θ as well as Γ, but in the context of path summation, the target object and phase map are generally fixed, while the set of paths being summed over is allowed to vary. Hence, the notation chosen here emphasizes only the set of paths Γ. If M' is a subset of M, viewed as a full subobject, it is convenient to denote the evaluation map $e_{\Gamma(M')}$ by the more streamlined notation $e_{M'}$. In particular, e_M is the **global evaluation map**. In the important special case in which the multidirected set under consideration is a subobject R of an "ambient" relation space, the global evaluation map is denoted by e_R. This notation appears, for example, in Definitions 6.7.4, 6.9.5, and 6.9.9. In this context, the term "global" refers to the "entire object under present consideration," even if this object itself may be defined as a subobject of a larger object. The dashed arrow in the expression $e_\Gamma : T^\Gamma \dashrightarrow T$ is used to emphasize the fact that e_Γ is generally only partially defined, since causal path sums may not always converge to elements of T.

Evaluation maps may be expressed in terms of the **indicator functions** $e_\gamma : \Gamma(M) \to T$, defined by the formulae

$$e_\gamma(\gamma') = \delta_{\gamma\gamma'} := \begin{cases} 1_T & \text{if } \gamma' = \gamma \\ 0_T & \text{if } \gamma' \neq \gamma. \end{cases}$$

The indicator functions e_γ extend by T-linearity to yield totally-defined functions $T^{\Gamma(M)} \to T$, given by the formulae

$$e_\gamma(\Theta) = e_\gamma\Big(\sum_{\gamma'\in\Gamma(M)}\Theta(\gamma')\gamma'\Big) = \sum_{\gamma'\in\Gamma(M)}\Theta(\gamma')e_\gamma(\gamma') = \Theta(\gamma).$$

This particular application of e_γ is roughly analogous to "integrating a real-valued function against a Dirac delta function," to obtain a function value at a particular point. The evaluation map e_Γ may be expressed as a formal sum of indicator functions:

$$e_\Gamma = \sum_{\gamma\in\Gamma} e_\gamma.$$

In the case of finitely-supported phase maps, the evaluation map e_Γ is always totally defined, regardless of the choice of Γ. In the general case, however, the right-hand side of Equation (6.2.8) may diverge.

In many cases, it is appropriate to use physically suggestive notation and terminology for the causal path sum $\Theta(\Gamma)$ appearing in Equation (6.2.7). For example, in Section 6.7, the path set Γ is taken to be the set $\mathrm{Ch}_{\mathrm{MAX}}(R)$ of maximal chains in a finite subobject R of the relation space over a multidirected set, and the phase map Θ is assumed to be *generated* by a relation function θ, as described by Equation (6.7.3). In this context, the causal path sum $\Theta(\Gamma)$ is referred to as a *generalized quantum amplitude,* and is denoted by $\psi(R;\theta)$. This particular causal path sum is directly analogous to Feynman's path integral, as explained in Sections 6.6 and 6.7. More generally, sets of maximal chains in finite subsets of directed sets are particularly important index sets for path summation in discrete quantum causal theory, because such chains may be used to represent evolutionary processes between specified *cobase families* and *target families* of classical histories in a kinematic scheme, as mentioned in Section 5.4. Such sets of maximal chains play an important role in Sections 6.7, 6.9, and 6.10.

An interesting *abstract* use of causal path sums is to specify partially-defined *power set functions* $\mathcal{P}(M) \dashrightarrow T$, which may be viewed as T-valued *generalized measures* on M. To accomplish this, one may begin by assigning to every subset M' of M a particular set of causal paths in M', for example, the set of all maximal chains in M', as in the case of the generalized quantum amplitude $\psi(R;\theta)$ mentioned above. The "measure" assigned to the set M' is then defined to be the causal path sum given by summing the phases of these paths. It is worth noting that the machinery of phase maps, evaluation maps, and causal path sums generally provides a variety of different ways to express any given quantity. In particular, one often wishes to restrict the set of paths to be summed over. This may be accomplished either by applying a restricted evaluation map, or by restricting the support of the original phase map, or by defining a new "auxiliary phase map" with restricted support. Equation (6.7.5) of Section 6.7 provides an example of the third option, in which the generalized quantum amplitude $\psi(R;\theta)$ is given by applying the *global* evaluation map e_R to a specially-defined causal path functional $\Psi(R;\theta)$ with support restricted to the set $\Gamma = \mathrm{Ch}_{\mathrm{MAX}}(R)$. Following the first option instead, one *could* apply the evaluation map e_Γ to the original phase map Θ, to obtain the same sum.

6.3 Iteration of Structure

Iteration of structure as a general unifying principle. As stated in Section 6.1, the path summation approach to discrete quantum causal theory applies *both* to background dependent quantum theories involving fields and particles on a fixed directed set, *and* to background independent quantum theories, described in terms of *kinematic schemes* of directed sets. This fact is by no means obvious, since the machinery of path summation underlying the entire approach, outlined in Section 6.2, involves only the causal path sets of *individual* directed or multidirected sets, with no mention of kinematic schemes, or of configuration spaces of any kind. The link between the two types of theories is the principle of *iteration of structure* (IS), already described informally at several points in preceding chapters. This principle formalizes and generalizes the basic observation that configuration spaces of directed sets possess abstract "higher-level" multidirected structures, collectively induced by natural relationships between pairs of individual directed sets belonging to these spaces. In the discrete causal context, these natural relationships are interpreted as *co-relative histories,* and the corresponding structured configuration spaces are interpreted as *kinematic preschemes,* or, in "well-behaved" cases, as *kinematic schemes.* Since configuration spaces arise inevitably in the context of superposition, this iterative viewpoint regarding directed and multidirected structure provides a striking conceptual unification of classical theory and quantum theory. A prototypical example of iteration of structure in discrete causal theory is provided by the *positive sequential kinematic scheme* $\mathbb{S}_{PS}$ of finite acyclic directed sets, first illustrated in Figure 1.1.1 of Chapter 1. The first four generations of $\mathbb{S}_{PS}$ are reproduced in Figure 6.3.1.

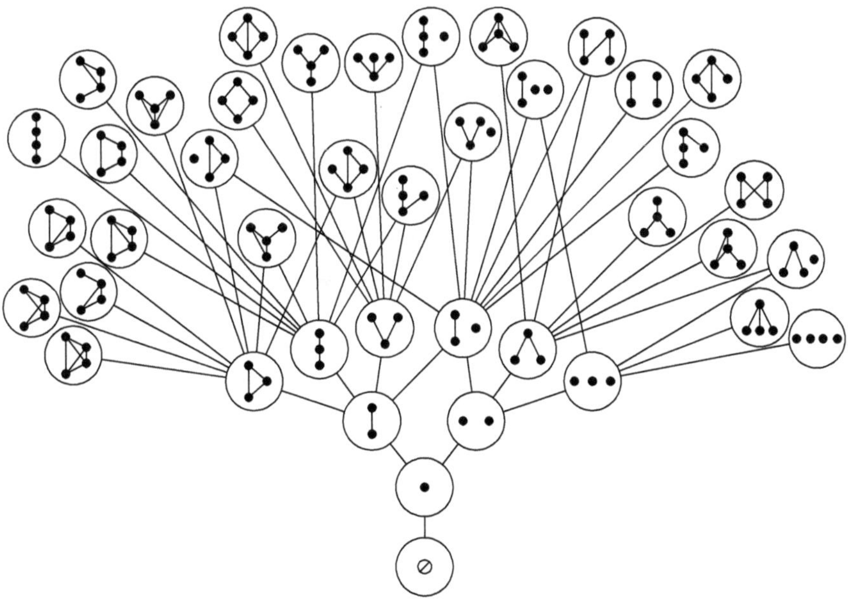

Figure 6.3.1. Iteration of structure via the positive sequential kinematic scheme $\mathbb{S}_{PS}$.

The reason why I have chosen the very general expression "iteration of structure" to describe this unifying principle is because the same basic idea applies to many other types of structures besides *"multidirected sets of multidirected sets."* Category theory, in particular, abounds with examples, including *"morphisms of morphisms"* and *"categories of categories."* However, due to the underlying multidirected structure of categories themselves, it is not a priori obvious that these examples actually represent a significant generalization. Hence, I briefly mention here a "nondirected setting" from pure mathematics in which essentially the same concept plays a critical role; namely, the theory of *moduli spaces* in geometry, particularly algebraic geometry. This analogy was already mentioned obliquely in Section 1.1. Without going into the details, moduli spaces are *"geometric objects of geometric objects,"* i.e., geometric objects whose individual elements represent geometric objects in their own right. Probably the simplest examples of such spaces are *projective spaces* and *Grassmannians,* which are *"spaces of linear subspaces of a vector space."* In particular, the projective space $\mathbb{P}^1(\mathbb{R})$ is defined to be the set of all lines through the origin in $\mathbb{R}^2$. Figure 6.3.2 illustrates the sense in which $\mathbb{P}^1(\mathbb{R})$ is a *"line of lines,"* which is why it is usually called the *real projective line*. Every line through the origin in $\mathbb{R}^2$, with the exception of the x-axis, intersects the "thick dark line" ℓ at a single point; conversely, every point on ℓ lies on a unique line through the origin. Hence, the elements of $\mathbb{P}^1(\mathbb{R})$ may be placed in one-to-one correspondence with the elements of the line ℓ, with exactly one element of $\mathbb{P}^1(\mathbb{R})$ left over. This element is usually called the "point at infinity," since "parallel lines intersect at infinity" in projective geometry.

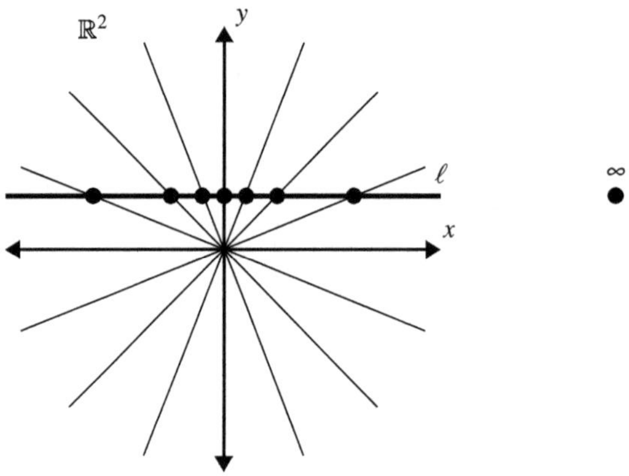

Figure 6.3.2. Projective space $\mathbb{P}^1(\mathbb{R})$ as a "line of lines".

More complicated moduli spaces include spaces of algebraic curves,[9] or "marked algebraic curves," such as *elliptic curves.* It is in the context of examples like these that the theory of moduli spaces begins to emerge as a major topic in its own right, since projective spaces and Grassmannians are generally viewed as part of "elementary algebraic geometry." Still more intricate examples from the same general field include *Hilbert schemes* and *Chow groups.* Iteration of structure is not just a coincidental property of moduli spaces, but one of the principal reasons why these spaces are mathematically interesting. Indeed, the close similarity between moduli spaces and the spaces represented by their individual points often allows the same technical machinery to be applied to both, leading to useful and far-reaching results. It is worth noting that, just as configuration spaces of directed sets are generally *multidirected,*

[9] As stated in the first footnote in Section 2.3, the word "curve" means simply "locus of points" in this setting; it does *not* mean a map into a manifold.

so moduli spaces tend to be "somewhat more general" than the spaces represented by their points. In particular, the theory of *algebraic stacks* is necessary for a suitable modern treatment of moduli spaces.

Iteration of structure in discrete causal theory. The following "narrow" and "physically-oriented" statement of the principle of iteration of structure is chosen in order to focus attention on the "neighboring hierarchical levels" occupied by directed sets and kinematic schemes of directed sets in discrete causal theory, as foreshadowed in Section 6.1. The foregoing discussion should make it clear, however, that much more general statements of this principle are possible; in fact, many additional examples of the same idea may be recognized even in the discrete causal context. For example, the relationship between kinematic schemes and *kinematic spaces,* briefly outlined in Chapter 7, is an example of iteration of structure "above the quantum level."

Definition 6.3.3. Iteration of structure (IS). *Background independent discrete quantum causal theory may be described in terms of multidirected sets whose elements represent directed sets.*

Understood in this specific context, iteration of structure represents perhaps the most important generalization of the binary axiom (B), i.e., the most important way of "assigning nontrivial internal structure to elements in a directed set or multidirected set." For future reference, it is worth noting that many different such structures exist, of which the positive sequential kinematic scheme $\mathbb{S}_{\mathrm{PS}}$ illustrated in Figure 6.3.1 is just one example. Another example, already mentioned periodically throughout the book, is *Sorkin and Rideout's kinematic scheme* $\mathbb{S}_{\mathrm{SR}}$ underlying the theory of *sequential growth dynamics* for causal sets [SR99], illustrated in Figure 3.4.1, and more thoroughly in Figure 7.5.1 of Chapter 7. Both of these kinematic schemes possess the abstract structures of countably infinite, star finite, acyclic multidirected sets. By contrast, the *generational kinematic scheme* $\mathbb{S}_{\mathrm{GEN}}$ and the *foliation scheme* $\mathbb{S}_{\mathrm{FOL}}$, which describe *generational growth* of finite acyclic directed sets, both possess the abstract structures of countably infinite, star *infinite,* acyclic multidirected sets. The properties of all four of these kinematic schemes are discussed in more detail in Chapter 7.

Despite the broad mathematical applicability of iteration of structure as a general organizing principle, its special *physical* role in discrete causal theory distinguishes the theory from other approaches to fundamental physics, since it allows classical theory to merge with quantum theory in a natural and inevitable way in this setting. This is of great interest in the context of unification. The paramount enterprise of modern theoretical physics, after all, involves the reconciliation of a particular classical theory; namely, general relativity, with a particular quantum theory; namely, the standard model of particle theory, and the complaint that these two theories involve "different languages" is a common refrain in the literature.[10] In discrete causal theory,

[10] As mentioned in the last footnote in Section 2.9, loop quantum gravity goes some way toward achieving a "common language" in this context.

however, classical and quantum physics do *not* involve different languages; they fit together harmoniously under the path summation approach to quantum theory, due to iteration of structure. For comparison, it is worth noting that the abstract structures of configuration spaces of relativistic spacetime manifolds, up to suitable equivalence, generally do not resemble individual spacetime geometries at all. For example, the moduli space of *Einstein metrics,* i.e., metrics solving the vacuum version of Einstein's equation, on the single diffeomorphism class of *K3 surfaces,* admits a local diffeomorphism into a 57-dimensional orbifold![11]

Transitions. Iteration of structure arises naturally in the search for suitable notions of *initial* and *terminal conditions* in discrete causal theory. As mentioned several times in previous chapters, it is useful to express this idea via Grothendieck's relative viewpoint (RV), in terms of *natural relationships between pairs of classical histories,* which I refer to as *co-relative histories.* These are studied in detail in Section 6.4. The category-theoretic origin of the term "co-relative" was briefly mentioned in Section 4.6, and is explained in more detail below. In this book, I focus on the special case of co-relative histories that formalize the *"evolution of one directed set into another."* As usual, only *isomorphism classes* of directed sets are significant in this context, but the necessary information may be described in terms of representatives of these classes. Co-relative histories are represented by *transitions,* which are special morphisms of directed sets.

Definition 6.3.4. *A* **transition** *in the category $\mathcal{D}$ of directed sets is a monomorphism $\tau : D_i \rightarrow D_t$, embedding its* **initial object**, *or* **source**, *D_i, into its* **terminal object**, *or* **target**, *D_t, as a proper, full, originary subobject. This means that $\tau(D_i)$ has nontrivial complement in D_t (proper), that $\tau(x) \prec \tau(y)$ in D_t if and only if $x \prec y$ in D_i (full), and that the isomorphic image $\tau(D_i)$* of *D_i in D_t contains its own past (originary).*

Informally, a transition τ "augments its source" D_i, by adding directed structure in a manner compatible with its existing structure, thereby "producing its target" D_t. There is no difficulty in extending the definition of a transition to multidirected sets, but such a generalization is not needed in this book. The "originary" condition limits the applicability of transitions, as defined here, when attempting to generalize to the context in which cycles are permitted. However, I do not examine these details here. Figure 6.3.5 illustrates a transition whose target is the familiar 35-element acyclic directed set used in previous figures, labeled as D_t in this context. The conditions specified in Definition 6.3.4 are evident from the figure. In particular, the image $\tau(D_i)$ of the source D_i has nontrivial complement in the target D_t, represented by the elements and relations drawn in gray, so τ is proper. Every relation in D_t between elements of the image $\tau(D_i)$ is "already present" in D_i itself, so τ is full. Finally, the complement $D_t - D_i$ of the image $\tau(D_i)$ contains no elements in the past of $\tau(D_i)$ in D_t, so τ is originary.

To avoid the possibility of confusion, I include a practical remark regarding the manner in which transitions are represented graphically in this chapter. In the interest of accurately representing "causal" analogies involving "directed relationships at

[11] See Anderson [AN09] for details.

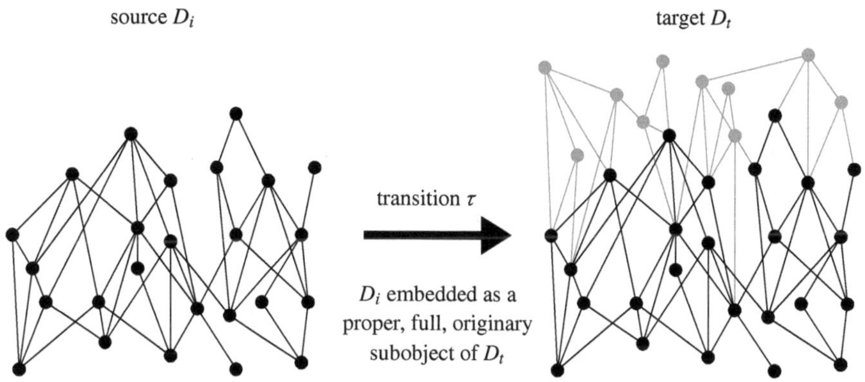

Figure 6.3.5. A transition $\tau : D_i \to D_t$.

different levels of hierarchy," understood in the context of iteration of structure (IS), it is preferable to draw transitions running "up the page," with the target drawn "above the source." However, to avoid wasting space, it is sometimes more convenient to "arrange transitions horizontally," as in Figure 6.3.5. This is particularly true when a chain of transitions is involved, as in Figure 6.3.6. For this reason, I follow the practice of drawing transitions in *whichever way seems best for representing the necessary information efficiently.* In particular, the graphical difference between "horizontally drawn" transitions and "vertically drawn" transitions does not signify any structural distinction whatsoever; it is merely a question of format.

Transitions and symmetry. Transitions are a priori too specific to be regarded as physically fundamental, since they generally contain "extraphysical information," due to symmetries, i.e., automorphisms, of their sources and targets. For example, the composition of a transition τ with an automorphism of its source defines a "physically equivalent" transition. The automorphism group of the target of τ also plays a role, whenever this group "mixes the image of τ with its complement." Given the "irregularity" of an arbitrary directed set, one might expect most such sets to lack "large-scale symmetries," i.e., automorphisms mapping large subsets to their complements. On the other hand, one might expect "small-scale symmetries," i.e., automorphisms permuting small numbers of elements locally, to be fairly common. An interesting result, contravening the latter intuition, is that *a typical finite acyclic directed set is rigid;* i.e., it has no nontrivial automorphisms at all. This result, already mentioned informally in Sections 2.3, 3.10, 4.5, and 5.9, was proven by Edward Bender and Robert Robinson, in their 1988 paper *The Asymptotic Number of Acyclic Digraphs II* [BR88].

The physical implications of this rigidity result in the context of discrete causal theory depend on a variety of factors. For example, a *typical* finite directed set has a "large number of relations," due to elementary counting considerations. This tends to favor rigidity, since any given relation can break a symmetry. However, dynamical considerations may prove to favor "sparser" directed sets, allowing nontrivial automorphism groups to play a significant role. For example, the discrete causal alternative to the *inflationary hypothesis* in cosmology, discussed in Section 1.9, involves a "phase transition to sparser structure." Then again, directed sets that are *too* sparse may encounter issues involving *systematic Lorentz invariance violation,* as discussed in Section 4.5. This suggests that the directed sets of greatest interest as models of classical spacetime may exhibit very large average valences, and hence, may be very rigid. However, it may be possible to accommodate approximate local Lorentz invariance at the *quantum* level, using directed sets with small average valences, in a manner analogous to the treatment of the "length spectrum" in loop quantum gravity [RS03]. This possibility is briefly revisited in Section 8.8. Questions of this nature are of considerable physical importance for several different reasons. For example, automorphisms of directed sets are analogous to "external" or "spacetime" symmetries in conventional physics, such as those encoded by the *Poincaré group* P of isometries of Minkowski spacetime $\mathbb{R}^{3+1}$. Such external symmetries impose basic constraints on the properties of elementary particles in conventional quantum field theory, via the representation theory of Lie groups, and similar considerations are inevitable in the dynamical theory of particles and fields on discrete directed sets, such as those arising in quantum causal set theory. Section 8.4 briefly discusses the *emergence of the Poincaré group,* focusing on the case in which the typical directed sets under consideration *are* assumed to be rigid. However, one is often interested in "less-rigid" subobjects of directed sets, such as those arising in the study of *generational growth,* outlined in Sections 7.8, 8.2, and 8.3.

Symmetry preservation, extension, breaking, generation. Given the possibility that exact "external" symmetries may prove relevant, at some level, in discrete causal theory, despite Bender and Robinson's rigidity result [BR88], it is worthwhile to briefly consider how such symmetries interact with transitions. In this context, a *symmetry* means simply an automorphism of a particular directed set. A transition may *preserve, extend,* or *break* symmetries of its source, or it may *generate* symmetries of its target, or some combination of these four possibilities. Care is required in the physical interpretation of these notions, particularly because the too-specific nature of individual transitions adds extraphysical information to the picture. For example, *different but physically equivalent* transitions mapping a common source to different subobjects of a common target generally preserve, extend, break, and/or generate *different but physically equivalent* families of symmetries.

Let $\tau : D_i \rightarrow D_t$ be a transition of directed sets, and let $\text{Aut}(D_i)$ and $\text{Aut}(D_t)$ be the automorphism groups of the source D_i and the target D_t of τ, respectively. Let $F_\tau := \text{Aut}(D_t; \tau(D_i))$ be the subgroup of $\text{Aut}(D_t)$ whose elements *permute* $\tau(D_i)$ *and its complement separately.* In other words, each element of F_τ maps elements of $\tau(D_i)$ exclusively to elements of $\tau(D_i)$, and elements of $D_t - \tau(D_i)$ exclusively to elements of $D_t - \tau(D_i)$. Each element of F_τ defines an automorphism of D_i, via restriction and conjugation by τ. Hence, I refer to F_τ as the **group of extensions** of symmetries of D_i. More precisely, there is a group homomorphism

$$\rho_\tau : F_\tau \longrightarrow \text{Aut}(D_i),$$

$$\beta \mapsto \tau^{-1} \circ \beta \circ \tau,$$

which I refer to as the **restriction homomorphism**, where the factors τ^{-1} and β in the composition are understood to be restricted to the image $\tau(D_i)$ of τ. If ρ_τ is surjective, then τ **preserves the symmetries** of the source D_i of τ. The kernel K_τ of ρ_τ is the normal subgroup of F_τ whose elements consist of automorphisms of D_t fixing $\tau(D_i)$. Hence, if τ preserves the symmetries of D_i, there is a short exact sequence

$$1 \to K_\tau \overset{\iota}{\to} F_\tau \overset{\rho_\tau}{\to} \text{Aut}(D_i) \to 1,$$

where ι is the inclusion of the kernel K_τ of ρ_τ into $\text{Aut}(D_i)$. Hence, F_τ is a *group extension* of $\text{Aut}(D_i)$ by K_τ if and only if τ preserves the symmetries of D_i. The quotient group $G_\tau := F_\tau/K_\tau$ may be identified with the normal subgroup of F_τ fixing the *complement* of $\tau(D_i)$ in D_t. The group F_τ is isomorphic to the *group direct product* of K_τ and G_τ:

$$F_\tau \cong K_\tau \times G_\tau.$$

Whether or not ρ_τ is surjective, it induces a quotient monomorphism $\overline{\rho_\tau} : G_\tau \to \text{Aut}(D_i)$, identifying G_τ with the *subgroup of automorphisms of D_i preserved by τ*. In the finite case, the *subgroup index* $[F_\tau : G_\tau]$ of G_τ in F_τ measures the number of different ways in which automorphisms of D_i preserved by τ are extended by τ. If ρ_τ is surjective, then $\overline{\rho_\tau}$ is an isomorphism. If ρ_τ is not surjective, then every element of the complement $\text{Aut}(D_i) - \overline{\rho_\tau}(G_\tau)$ represents a **symmetry broken by** τ. In the finite case, the subgroup index $[\text{Aut}(D_i) : \overline{\rho_\tau}(G_\tau)]$ measures the extent of this symmetry breaking. The groups K_τ and G_τ may be called **causal Galois groups**, since they are groups of automorphisms fixing distinguished subobjects of directed sets in the context of discrete causal theory. Many other interesting analogies exist between causal theory and Galois theory, which cannot be properly explored in this book. For example, there is no room here to discuss the obviously relevant theory of *Galois connections.* It is useful to generalize the notion of causal Galois groups to include analogous groups involving specialized subobjects of D_i and D_t, such as the "structural increments" arising in *generational dynamics,* discussed in Sections 7.8, 8.2 and 8.3. For example, progressively larger subobjects of such structural increments tend to be fixed as more generations are taken into account. This means that families of elements permuted by the *generational automorphism groups* introduced in Section 8.2 "become distinguishable" as the level of detail is increased. This leads to interesting *entropic* considerations, under which a specified level of detail is analogous to a *coarse-graining of phase space* in statistical thermodynamics. The remark regarding "filtrations of pasts and futures" preceding Definition 3.7.8 is relevant in this context.

Figure 6.3.6 illustrates simple examples of symmetry preservation, extension, and breaking. Symmetry generation, also depicted in the figure, is discussed below. Let $\tau_0 : D_0 \to D_1$ be the transition sending the elements t_0, u_0, and v_0 of D_0 to the elements t_1, u_1, and v_1 of D_1, respectively. The automorphism group $\text{Aut}(D_0)$ of the source D_0 of τ_0 is isomorphic to $\mathbb{Z}_2$, generated by the unique automorphism α of D_0 interchanging u_0 and v_0. The automorphism group $\mathrm{Aut}(D_1)$ of the target D_1 of τ_0 is trivial. Hence, the subgroups F_{τ_0}, K_{τ_0}, and G_{τ_0} of $\mathrm{Aut}(D_1)$ are trivial. The index $[F_{\tau_0} : G_{\tau_0}]$ is therefore equal to 1, which means that τ_0 extends the trivial automorphism of D_0 to the trivial automorphism of D_1, with no other extensions. The index $[\mathrm{Aut}(D_0) : \overline{\rho_\tau}(G_{\tau_0})]$ is equal to 2, which means that τ_0 breaks the unique nontrivial symmetry α of D_0.

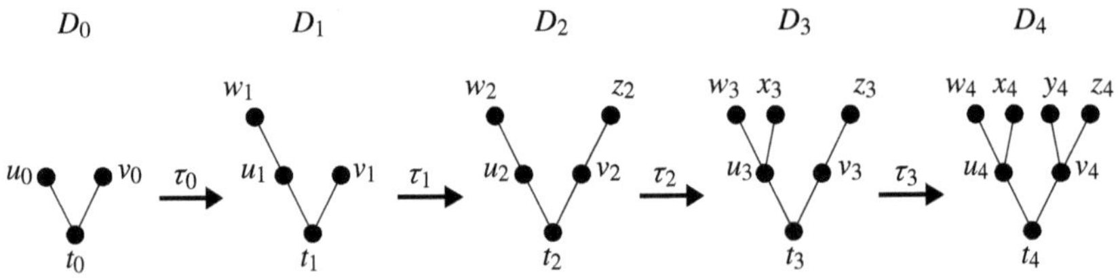

Figure 6.3.6. Symmetry preservation, extension, breaking, and generation, via transitions.

It is instructive to examine the significance of this broken $\mathbb{Z}_2$ symmetry in an informal sense, first "from the perspective of the source D_0," then "from the perspective of the target D_1." There are obvious physical motivations for considering such viewpoints. In particular, one might begin with D_0, and examine the possible ways in which it "could evolve into D_1;" conversely, one might begin with D_1, and investigate how it "could have evolved from D_0." From the perspective of the source D_0 of τ_0, the broken $\mathbb{Z}_2$ symmetry reflects the existence of two "equivalent" ways to obtain a directed set isomorphic to D_1 by adding a single element to D_0. This "new element" may be added either in the future of u_0, or in the future of v_0. These two constructions yield transitions from D_0 into two *different, but isomorphic,* targets. It is important to note, however, that the existence of multiple isomorphic targets is not always associated with a broken symmetry of the common source. In particular, there may be multiple "inequivalent" ways to construct the isomorphism class of a specified target from a specified source *even if both source and target have trivial automorphism groups.* This curious phenomenon, explained in more detail below, can occur even in cases where only a single new element is "added" by the transition. From the perspective of the target D_1 of τ_0, there is a unique subobject of D_1 that may be identified with the source D_0. From this perspective, the broken $\mathbb{Z}_2$ symmetry reflects the fact that the unique nontrivial symmetry of this subobject does not extend to a symmetry of D_1.

Returning for the moment to the case of an arbitrary transition $\tau : D_i \to D_t$, the complement of the subgroup F_τ in $\mathrm{Aut}(D_t)$ consists of all automorphisms of D_t that *mix* the image $\tau(D_i)$ of τ with its complement in D_t. These automorphisms obviously do *not* form a subgroup of $\mathrm{Aut}(D_t)$, since they do not include the identity.

Further, F_τ is generally not a *normal* subgroup of $\text{Aut}(D_t)$,[12] so there is generally no quotient group to study. These automorphisms are new "from the perspective of the transition τ," since they do not restrict to automorphisms of the image $\tau(D_i)$. Hence, I refer to them as symmetries **generated by** τ. However, these automorphisms are not necessarily new "from the perspective of the source D_i of τ." The meaning of this distinction is that there may exist *other* transitions $\tau' : D_i \to D_t$, corresponding to different, but isomorphic, subgroups $F_{\tau'}$ of $\text{Aut}(D_t)$, whose elements extend symmetries of D_i broken by τ, and which are *equivalent to* τ, in the sense that their images are mapped isomorphically onto the image of τ by automorphisms of D_t. This notion of equivalence is made precise in Definition 6.4.3 below. This distinction reflects the fact that transitions are *too specific* to be considered physically fundamental.

Referring again to Figure 6.3.6, let $\tau_1 : D_1 \to D_2$ be the transition sending the elements t_1, u_1, v_1, and w_1 of D_1 to the elements t_2, u_2, v_2, and w_2 of D_2, respectively. The automorphism group $\text{Aut}(D_1)$ of the source D_1 of τ_1 is trivial, while the automorphism group $\text{Aut}(D_2)$ of the target D_2 of τ_1 is isomorphic to $\mathbb{Z}_2$, generated by the unique automorphism β of D_2 interchanging u_2 and v_2, and w_2 and z_2, respectively. The extension group F_{τ_1} is trivial, since β mixes $\tau_1(D_1)$ with its complement. Hence, the causal Galois groups K_{τ_1} and G_{τ_1} are also trivial. The complement of F_{τ_1} in $\text{Aut}(D_2)$ is the singleton $\{\beta\}$; therefore, β is a symmetry generated by τ_1. From the perspective of the source D_1 of τ_1, there is only one way to add an extra element to D_1 to obtain a directed set isomorphic to the target D_2. However, from the perspective of the target D_2 of τ_1, there are *two different subobjects* that may be identified with the source D_1. This reflects the fact that there is an equivalent transition $\tau_1{}' : D_1 \to D_2$, sending the elements t_1, u_1, v_1, and w_1 of D_1 to the elements t_2, v_2, u_2, and z_2 of D_2, respectively. The transition $\tau_1{}'$ *also* generates the unique nontrivial symmetry β of D_2.

The remaining transitions $\tau_2 : D_2 \to D_3$ and $\tau_3 : D_3 \to D_4$ appearing in Figure 6.3.6 involve more complex phenomena. The transition τ_2 breaks the $\mathbb{Z}_2$ symmetry of D_2, but generates a *new* $\mathbb{Z}_2$ symmetry in D_3. Hence, the symmetry groups of D_2 and D_3 are "accidentally" isomorphic. The transition τ_3, meanwhile, generates multiple new symmetries. One may also consider compositions of the transitions in the figure; for example, the composition $\tau_1 \circ \tau_0 : D_0 \to D_2$ preserves the symmetries of D_0, without generating any new symmetries. In this case, the fact that the symmetry groups of the source and target are isomorphic is *not* "accidental."

6.4 Co-Relative Histories

Equivalence of transitions. In defining co-relative histories, the objective is to precisely capture the *physical essence of individual relationships between pairs of classical histories,* represented by directed sets, following Grothendieck's relative

[12]For example, consider the automorphism β of D_4 in Figure 6.3.6 interchanging u_4, w_4, and x_4 with v_4, y_4, and z_4, respectively. It is easy to see that $\beta^{-1} F_{\tau_3} \beta \neq F_{\tau_3}$.

viewpoint (RV). The naïve approach, in which one attempts to represent such relationships via *specific morphisms* in the category $\mathcal{D}$ of directed sets, for example, specific transitions, leads in the wrong direction, at least in the general case in which the directed sets under consideration may possess nontrivial automorphism groups. A better choice is to represent such relationships via *equivalence classes of morphisms, whose images are related by automorphisms of their target sets.* Here, I focus on equivalence classes of transitions, since the principal application is to formalize *evolutionary processes* in discrete causal theory. These equivalence classes define *proper, full, originary co-relative histories.* More general types of co-relative histories may be defined, but maximal generality is not the goal in the present context.

Examples of physically equivalent and inequivalent transitions between pairs of directed sets are illustrated in Figure 6.4.1, with the labels on some of the elements suppressed to avoid clutter. Three copies of the same target set D_t appear in the figure, along with three different transitions τ, τ', and τ'', from the source set D_i into D_t. The transitions τ and τ' are equivalent, since their images are related by the unique automorphism α of D_t interchanging the elements x and y. The transition τ'' is *not* equivalent to τ or τ', since α is the only nontrivial automorphism of D_t, and α fixes the element z.

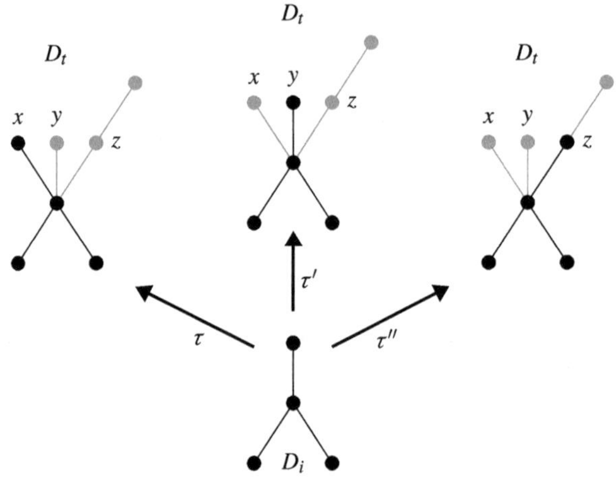

Figure 6.4.1. Transitions τ and τ' are equivalent to each other, but not to τ''.

McKay's example. An interesting fact, already mentioned in Section 6.3 in the context of symmetry breaking, is that there may exist multiple "inequivalent" ways to construct the isomorphism class of a specified target set D_t from a specified source set D_i, even if both $\text{Aut}(D_i)$ and $\text{Aut}(D_t)$ are trivial, and even if D_i and D_t differ by only a single element. Equivalently, viewing D_i and D_t as "given," rather than approaching the problem constructively, there may exist multiple inequivalent transitions between D_i and D_t. Figure 6.4.2 illustrates an example of this phenomenon, supplied by Brendan McKay.[13] The element labels are included to make manifest the *existence* of the transitions τ and τ', since the images of D_i in D_t under τ and τ' "look different" than D_i in the figure. The elements x and y in the complements of τ and τ', respectively, are called *pseudosimilar vertices* in the language of graph theory.[14] Viewing D_i and D_t as classical histories, it is "reasonably obvious" that the relationships between them represented by τ and τ' differ in a physically significant way. The existence of this type of behavior has significant structural consequences in discrete causal theory. For example, it implies that the abstract structures of kinematic schemes are generally *multidirected,* rather than merely directed. For example, *Sorkin and Rideout's kinematic scheme* $\mathbb{S}_{\mathrm{SR}}$ describing the sequential growth of causal sets possesses the abstract structure of a multidirected set, *not* merely a partially ordered set, or the skeleton of a partially ordered set, as suggested in some of the literature.

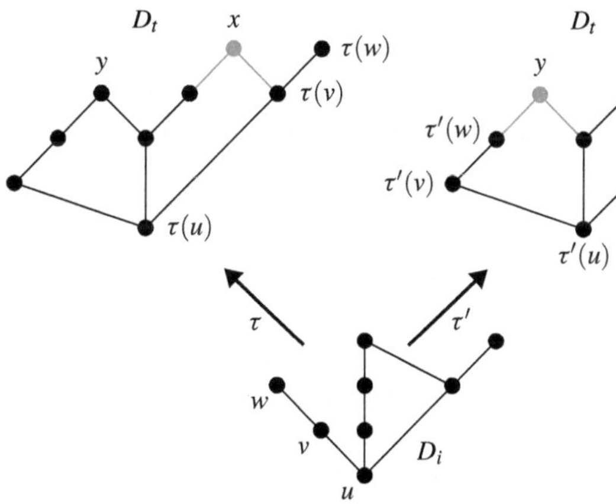

Figure 6.4.2. McKay's example of inequivalent single-element transitions between the same pair of directed sets.

Co-relative histories between pairs of directed sets. The foregoing discussion motivates the following formal definition of co-relative histories, suitably general for the purposes of this book. The terminology is chosen to suggest a *fixed source,* reflecting the intended evolutionary viewpoint. However, other perspectives are possible.

Definition 6.4.3. *A* **proper, full, originary co-relative history** $h : D_i \Rightarrow D_t$ *between directed sets* D_i *and* D_t *is an equivalence class of transitions from* D_i *to* D_t*, where two transitions* τ and τ' *are equivalent if and only if there exists an automorphism* β *of* D_t *mapping* $\tau(D_i)$ *onto* $\tau'(D_i)$*. The common initial set* D_i *of*

---

[13] Via private communication.

[14] The theory of pseudosimilarity is of interest in the *graph reconstruction problem.* McKay tells me that little work has been done on pseudosimilarity in the case of *directed* graphs.

the transitions making up h is called the **cobase** *of h, and the common terminal set D_t of these transitions is called the* **target** *of h.*

The physical intuition underlying Definition 6.4.3 is reasonably clear when the cobase D_i and target D_t are both finite. In the infinite case, "cycles of co-relative histories" are possible under this definition, even when the directed sets involved are acyclic, and this raises interesting interpretive challenges. The hyphen in the word "co-relative" is added to avoid confusion with the similar word "correlative." The "double-arrow notation" $\Rightarrow$ in the definition is intended to convey the idea that a co-relative history consists of a *family* of morphisms, "bundled together to encode a single physical relationship." More generally, one may specify an entire **cobase family $\mathbf{D}_i$** of directed sets, and study families $\mathbf{D}_t$ of *histories co-relative to* $\mathbf{D}_i$, called **target families**. The reason for using the terms "co-relative" and "cobase," rather than "relative" and "base," is that D_i is taken to represent *initial* conditions, while D_t is allowed to vary over possible *terminal conditions.* This viewpoint *reverses* the viewpoint accompanying the theory of *relative multidirected sets over a fixed base,* introduced in Section 4.6, in which the base is a fixed *target* set, while the *source* is allowed to vary. The terminology in both cases originates from the category-theoretic notions of *covariant* and *contravariant* constructions, related to one another by "reversing the arrows."[15] The interpretation of D_i and D_t as representing initial and terminal conditions immediately suggests a naïve *S-matrix-like approach* to discrete causal theory, in which one merely assigns an "amplitude" to every "permissible" co-relative history. These "amplitudes" may then be interpreted as encoding the *likelihoods that any possible initial state will evolve into any possible terminal state.* However, the path summation approach to discrete quantum causal theory developed in the second half the present chapter, and elaborated in Chapter 7, accomplishes the same purpose in a much more elegant and explanatory manner.[16]

Note that Definition 6.4.3 does *not* require that $\tau' = \beta \circ \tau$ as *morphisms;* it is only the *images* of τ and τ' that must be related by an automorphism of the target. For example, if τ breaks a symmetry α of its cobase D_i, then the transition $\tau' = \tau \circ \alpha$ cannot be recovered by applying an automorphism to its target D_t *after* application of τ. The simplest nontrivial example of this situation is illustrated in Figure 6.4.4, in which D_i consists of a pair of unrelated elements x and y, and D_t consists of three elements x', y', and z', with one relation $y' \prec z'$. In this case, $\text{Aut}(D_i)$ is isomorphic to $\mathbb{Z}_2$, generated by the automorphism interchanging x and y, while $\text{Aut}(D_t)$ is trivial. There is a unique co-relative history $h : D_i \Rightarrow D_t$, represented by two equivalent transitions τ and τ', where τ sends the elements x and y of D_i to the elements x' and y' of D_t, respectively, and where τ' reverses these images. Both transitions break the

---

[15] The category-theoretic meaning of the term *covariant* has only loose relationships with the various physics-related meanings of this term. Unfortunately, both of the terms *covariant* and *contravariant* carry many different and mutually contradictory meanings in both mathematics and physics.

[16] In particular, one is generally interested in *how* a given initial history D_i may evolve into a given terminal history D_t, and this leads to the consideration of *evolutionary pathways.* Also, S-matrix-like approaches require compatibility conditions for consistency purposes, and such conditions are easier to formulate using the path summation approach.

unique nontrivial symmetry of D_i. These transitions are equivalent, via the identity automorphism of D_t, since they share the same image, but there is no automorphism of D_t relating them as morphisms.

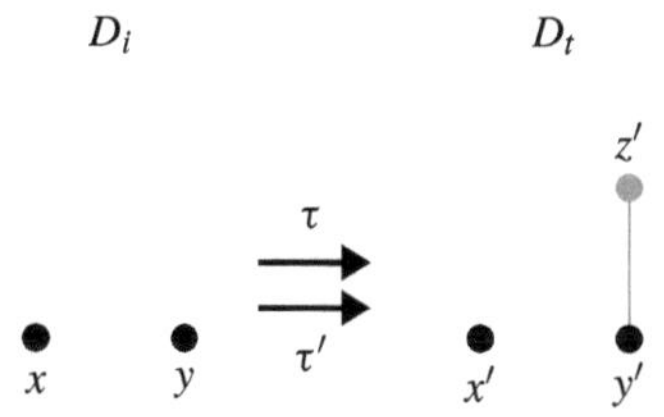

Figure 6.4.4. Equivalent transitions may be unrelated as *morphisms,* due to symmetry breaking.

Co-relative histories and category theory. As the previous example suggests, co-relative histories are subtle from a category-theoretic perspective. For example, equivalence of transitions is *not* a congruence relation under composition in the category $\mathfrak{D}$ of directed sets. This means that, given equivalent transitions from a common source D_0 to a common target D_1, and from D_1 to a third directed set D_2, it is generally *not* true that the compositions of these transitions are equivalent. This is illustrated in Figure 6.4.5. The transitions τ_0 and τ_0' from D_0 to D_1 are related by any automorphism of D_1 interchanging its two terminal elements, but the compositions $\tau_1 \circ \tau_0$ and $\tau_1 \circ \tau_0'$ represent *different* co-relative histories between D_0 and D_2. This example demonstrates that equivalence of transitions is not even *cancellative,* since the "left-hand factor" τ_1 is the same in both cases. More generally, pairs of "consecutive" co-relative histories generally do not compose to give unique "products." Instead, the "composition" of a pair of "consecutive" co-relative histories is generally a *family* of co-relative histories.

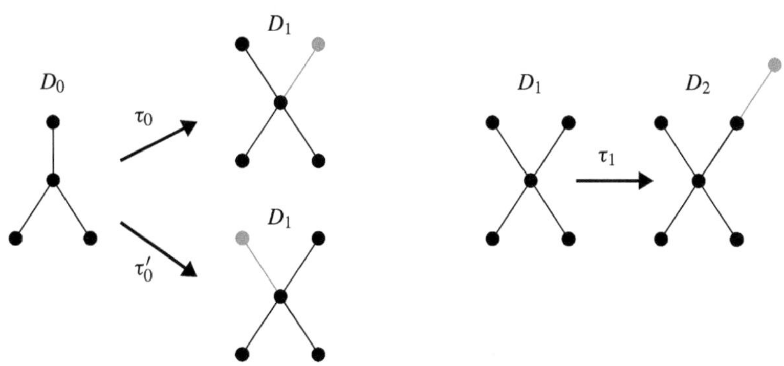

Figure 6.4.5. Equivalence of transitions is not a congruence relation under composition in $\mathfrak{D}$.

Congruence relations are so convenient that it is tempting to try to work with congruence classes of transitions, rather than the equivalence classes specified in Definition 6.4.3. One way to do this is to define a stricter equivalence relation, in which two transitions τ and τ' from D_i to D_t are taken to be equivalent if and only if they *share the same image*. This equivalence is obviously a congruence; the corresponding congruence classes may be called **image-fixed co-relative histories**. A co-relative history $h : D_i \Rightarrow D_t$ may then be viewed as an equivalence class of image-fixed co-relative histories, with members indexed by an appropriate subgroup of the automorphism group $\mathrm{Aut}(D_t)$; namely, the subgroup that "moves the image of a representative transition around its orbit in D_t." Heuristically, transitions *freeze the actions* of both the automorphism groups $\mathrm{Aut}(D_i)$ and $\mathrm{Aut}(D_t)$, while image-fixed co-relative histories allow $\mathrm{Aut}(D_i)$ to act, and arbitrary co-relative histories allow both groups to act.

Co-relative histories help to embody the specific statement of the principle of *iteration of structure* (IS) given in Definition 6.3.3, by supplying "higher-level relations" between pairs of directed sets in a configuration space of directed sets. Such a configuration space is elevated to a kinematic scheme by specifying an appropriate class of co-relative histories, "suitable for the evolutionary viewpoint." Figure 6.4.6 illustrates how a particular family of four co-relative histories sharing a common cobase fits into the *positive sequential kinematic scheme* $\mathbb{S}_{\mathrm{PS}}$, studied in more detail in Chapter 7. The directed sets contributing to these co-relative histories are none other than the "parent history" and four "child histories" featured at the very beginning of the book; for example, in Figure 1.2.1.

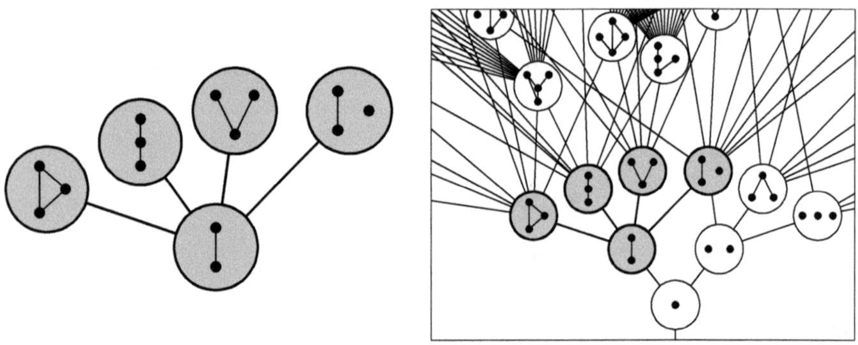

Figure 6.4.6. Four irreducible co-relative histories sharing a common cobase; co-relative histories embedded in the positive sequential kinematic scheme $\mathbb{S}_{\mathrm{PS}}$.

As noted above in the context of McKay's example of inequivalent transitions, the nonuniqueness of co-relative histories between a given cobase and target implies that the abstract structure of a kinematic scheme is generally multidirected, rather than merely directed. "Higher-level analogues" of *irreducibility* and *independence* are of interest in this context. For example, the four co-relative histories illustrated in Figure 6.4.6 are irreducible, since each target differs from the common cobase by a "minimal amount," i.e., by a single element. McKay's example implies that *irreducible co-relative histories are generally nonunique,* which is merely another way of restating the fact that multidirected structure is unavoidable in kinematic schemes. Turning to the subject of independence, a basic requirement built into the definition of a kinematic scheme is that it must include "enough independent co-relative histories" to admit "evolutionary pathways for every possible history." This requirement is formalized by the condition of *accessibility* (A), defined in Section 7.4.[17]

6.5 Quantum Causal Generalities

Formal path summation approach. The path summation approach to quantum theory, first implemented in the continuum-based setting via Feynman's path integral, may be adapted without *formal* difficulty to the discrete causal context, by means of the theory of *path summation over a multidirected set,* as outlined in Sections 6.1 and 6.2. To accomplish this, one assigns phases $\Theta(\gamma)$ to causal paths γ in a suitable causal path set Γ, constructs formal sums of the form $\sum_{\gamma\in\Gamma}\Theta(\gamma)\gamma$, then "evaluates" these formal sums, if possible, to yield causal path sums $\sum_{\gamma\in\Gamma}\Theta(\gamma)$ in the target object T of the phase map Θ. This is a very general construction, which may be applied to any multidirected set and any phase map whatsoever. In particular, the formal sum $\sum_{\gamma\in\Gamma}\Theta(\gamma)\gamma$ retains a perfectly valid meaning even if its evaluation fails to converge in the target object T. Of course, even more general constructions are possible. For example, one might consider *essentially nonlinear* methods of combining the phases associated with various causal paths, for example, methods involving *products* of phases in the target object, as well as sums. The version of the theory I choose to focus on in this book takes it for granted that such further generalization is unnecessary; i.e., the theory is "linear" in a suitable generalized sense. Further, following the principle of *iteration of structure* (IS), introduced in Section 6.3, the *same* assortment of abstract quantum theoretic methods are available for application to *every* directed or multidirected set arising in the theory, regardless of its physical interpretation as a model of classical spacetime, a relation space, a power space, the underlying structure of a kinematic scheme, or something else. These choices may be expressed via the following *path summation principle:*

Definition 6.5.1. Path summation principle (PSP). *Discrete quantum causal theory, whether background dependent or background independent, may be described in terms of path summation over multidirected sets.*

In the remaining sections of this chapter, I develop this abstract general theory, focusing on the important special case of *finite regions of acyclic multidirected sets.* The first task, undertaken in Section 6.6, is to revisit Feynman's original version of the path summation approach to ordinary quantum theory, from a perspective

---

[17] In fact, several *different* accessibility conditions are worth considering. These details are also discussed in Section 7.4.

convenient for abstraction and generalization to the discrete causal context. The second step, carried out in Section 6.7, is to actually develop the discrete causal analogue of this theory in a precise way. Here, I make some simplifying assumptions; for example, I do not investigate the implications of causal cycles. These assumptions are not intended to imply that more general behavior is physically unrealistic; they are made purely for the purposes of convenience and manageability. Further, while these developments do not entail any particular choice of physical interpretation of the mathematic objects involved, I do choose to adopt certain perspectives, and implement certain methods, that are specifically targeted to the study of fundamental spacetime structure. In particular, I make frequent use of relation space methods. The final step, taken in Sections 6.8–6.10, is to deduce from the abstract theory dynamical laws and "concrete quantities" such as generalized quantum amplitudes, which are ultimately necessary for experimental physics.

Preliminary remarks regarding applications. Following these developments, the context ceases to be purely formal, and applications become the primary concern. Conceptually, one must choose which interpretations are most appropriate for the mathematical structures appearing in the theory. Technically, one must ensure that the chosen methods correctly process the physical information involved, and produce "intelligible results." The most basic choice regarding interpretation is whether the objects over which path summation is performed are to be regarded as models of classical spacetime, or as configuration spaces of such models. This distinction determines whether the resulting theory is background dependent or background independent. The question of information processing involves both conceptual issues, such as *independent modes of influence,* and technical issues, such as the *permeability problem.* The general theory developed in preceding chapters has been carefully designed to avoid pitfalls involving these issues. In particular, the choice to model classical spacetime in terms of *generally nontransitive* binary relations enables suitable treatment of independent modes of influence, as discussed in Section 3.9, while the choice to work in relation space eliminates the permeability problem, as discussed in Section 5.4.

The question of "intelligible results" is more open-ended, since it ultimately involves experimental physics. A basic necessity in this context is to assign *concrete meanings to path sums.* If one begins with no assumptions beyond the abstract theory itself, then this remains a very general problem. For example, one might ask if these sums should be familiar complex amplitudes, interpreted in the usual way, or if quantum gravity might require the use of *"generalized truth values,"* as suggested by Isham [IS11]. If the latter choice proves necessary, then obvious secondary questions include what mathematical objects should be used to represent these values, and what they should be taken to *mean* in terms of measurements. Basic technical questions also arise. For example, the absence of a global finiteness axiom, or even a local finiteness axiom in the case of causal set theory, raises the generic question of when the "concrete" causal path sums associated with formal sums actually converge. In practice, it might be most reasonable to begin by studying simplified scenarios, for example, scenarios involving finite subobjects of acyclic multidirected sets, interpreted as representing configuration spaces of finite regions of classical spacetime. Similarly, complex-valued path sums, viewed as ordinary quantum amplitudes, could be studied before invoking "generalized truth values." One could then investigate the extent to which the resulting simplified theory might already suggest interesting phenomenology, before making any serious effort to generalize.

It may not be necessary to "get all the details right" in order to make useful progress in this manner; for example, it is instructive to recall the successes of "small-scale physics" in the early 1900s, before anyone understood the rudiments of ordinary quantum theory. "Correct" structural assertions, such as the fact that the blackbody spectrum *is* quantized, and the fact that electrons *do* occupy discrete energy levels in the hydrogen atom, produced significant results long before these structural properties were adequately explained. Similarly, even at its present early stage of development, discrete causal theory has insights to offer in terms of *where to search* for phenomena beyond the scope of relativity and the standard model, and even *what to search for.* Perhaps the most obvious such phenomena involve deviations from Lorentz invariance, but many other approaches to "new physics" predict such deviations as well, and their observation might not favor discrete causal theory specifically, depending on the details. Modest deviation from spatiotemporal locality also appears to be a generic feature of the theory, and rough estimates suggest that this property might lead to effects far above the Planck scale, as discussed in Section 4.5. A third general type of effect to look for is *small statistical deviations from "sensible" continuum-based values* for basic physical quantities such as the dimension of spacetime, which may be expected to result from replacing uncountable manifolds with finite or locally finite sets. The causal set explanation of the size of the "cosmological constant" falls into this category. A fourth class of effects to consider are those arising from *graph-dynamical behavior,* such as the *phase transitions* suggested in Section 1.9 as an alternative to the *inflationary hypothesis* in cosmology. All four of these general types of effects might be accessible at a rather crude, "semiclassical" level, since they depend to a large degree on the properties of broad classes of classical histories.

Background independence revisited. In previous chapters, I have described some of the qualitative differences between background dependent and background independent versions of discrete quantum causal theory. Section 1.6 explains the basic conceptual distinction between such theories in a general setting: background dependent theories involve a fixed choice of spacetime, "inhabited by material entities," such as particles and fields, while background independent theories endow spacetime itself with dynamical properties. Narrowing the focus to the discrete causal context, there remain different types and degrees of background independence. For example, one may choose to retain certain "auxiliary information," in the form of "non-gravitational matter," as in the dynamics of quantum causal set theory, described in Section 3.4. In this setting, one follows the example of general relativity, by allowing matter to "interact with underlying causal structure." Such "partially background independent" approaches are preferable to the naïve approach of *prescribing* a fixed background, because they "explain" the origin of the structure involved. However, they still present philosophical difficulties, because one may always inquire as to *why* a particular configuration of causal structure and matter arises in the "actual universe." Such theories also fail to completely eliminate consistency issues involving clashes between different types of structure, such as those occurring in "time-travel paradoxes." Further, they offer no automatic assistance in building suitable quantum theories. In fact, if general relativity is any guide, "partial background independence" renders the transition to the quantum realm especially difficult.

The alternative pursued in this book is to try to build a *perfectly* background independent theory, by incorporating *all* "permissible" models of classical spacetime. Under this approach, no questions remain regarding why a particular structure is observed in the "actual universe," since every classical history plays a role. Such a theory can no longer be "classical" itself, because it does not isolate specific classical histories as physical "solutions." In fact, the very device of incorporating multiple classical histories is exactly what is needed to construct a quantum theory. Hence, one achieves perfect background independence *and* a quantum-theoretic description at a single stroke. Additional benefits of this approach are readily apparent. For example, in background dependent theories, there is often no obvious fundamental reason why the "past to future" direction should be more relevant, from a physical standpoint, than the "future to past" direction. In particular, the continuum-based constructions involved in Feynman's path summation approach to ordinary quantum theory, sketched in Sections 6.6 and 6.8, are remarkably symmetric with respect to time. The same is generally true of background dependent discrete quantum causal theories, such as theories of particles and fields on causal set backgrounds. However, as explained in Section 7.5, the "past to future" direction is *automatically selected in the background independent context,* by the asymmetry of kinematic schemes. For example, the "past to future" direction in Sorkin and Rideout's kinematic scheme $\mathbb{S}_{SR}$, illustrated in Figure 7.5.1, or in the positive sequential kinematic scheme $\mathbb{S}_{PS}$, first illustrated in Figure 1.1.1, are readily apparent.

Improved background dependent theories. As described in Sections 1.6 and 1.7, the least-complicated versions of discrete quantum causal theory are background dependent theories involving a specific choice of directed set D. Such a set plays essentially the same role as Minkowski spacetime $\mathbb{R}^{3+1}$ in special relativity and ordinary quantum field theory. Temporarily ignoring the inherent limitations of background dependence itself, as well as the results of intervening chapters, one may attempt to construct and analyze such theories in terms of path summation over D. Many examples of this strategy may be found in the literature, particularly in the special case where D is a causal set. Such efforts, a few of which were briefly discussed in Sections 3.3, 3.4, and 4.5, can potentially be improved by addressing a few specific problems, including the issues associated with *transitivity* (TR) and *interval finiteness* (IF), described in Sections 3.9 and 4.7, the difficulties of working in element space, highlighted in Sections 2.7 and 5.4, and the limitations of certain algebraic devices, such as the *incidence algebras* discussed at the end of Section 5.10. Many of these problems are either resolved or significantly ameliorated by the results and methods offered in this book, without requiring any radical changes in perspective.

This points the way toward flexible new alternatives to quantum field theories in curved spacetime and lattice field theories, expressed in terms of *path summation over relation spaces over fixed star finite directed sets.*

New background independent theory. The ultimate need for a background independent quantum theory of fundamental spacetime structure is the final catalyst for adopting the novel perspective represented by the principle of *iteration of structure* (IS) and the theory of *co-relative histories,* introduced in Sections 6.3 and 6.4. These developments enable the abstract theory of path summation over a multidirected set to be applied in a context involving *no arbitrary choices,* except for the modest choices inherent in the axioms. While there is much existing precedent for these ideas in the work of Sorkin, Isham, and others, it seems that a truly clear structural picture emerges only via a combination of conceptual material drawn from a wide variety of technical fields, including order theory, graph theory, category theory, algebraic geometry, and information theory, in addition to the obvious guiding influences from general relativity and quantum theory. Full realization of this approach requires, at a minimum, the simultaneous application of the *causal metric hypothesis* (CMH), *discreteness,* the *independence convention* (IC), *star finiteness* (SF), *hidden structure* (HS), the *relative viewpoint* (RV), *background independence,* and the *path summation principle* (PSP). The "correct type of object over which to perform path summation," under the new theory, is the *relation space* $\mathcal{R}(\mathcal{M}(\mathbb{S}))$ *over the underlying multidirected set* $\mathcal{M}(\mathbb{S})$ *of a kinematic scheme* $\mathbb{S}$ *of directed sets.* This relation space represents the abstract directed structure of the class of co-relative histories representing individual relationships between pairs of directed sets belonging to $\mathbb{S}$. In more concrete, physically-descriptive language, the fundamental quantum-theoretic properties of spacetime are described not in terms of a space of individual events, or a space of relationships between pairs of events, or even a space of classical histories, but rather *a space of relationships between pairs of classical histories.*

6.6 Feynman Revisited

Path integration. For contextual purposes, I briefly review Feynman's original formulation of the path summation approach to quantum theory in the continuum-based setting, before developing its discrete causal analogue. Here, spacetime is a fixed real manifold, often Euclidean spacetime $\mathbb{R}^4$ or Minkowski spacetime $\mathbb{R}^{3+1}$, and paths are equivalence classes of future-directed continuous curves in a finite spacetime region X, viewed as particle trajectories. Path sums are limits of multiple integrals involving these paths, taken as the number of integrals goes to infinity. Following Feynman's 1948 paper [FE48] on the subject, such *path integrals* have found broad use in the computation of complex-valued *quantum amplitudes,* interpreted as encoding relative probabilities for the values of physical measurements in ordinary quantum mechanics and quantum field theory. At a technical level, path integrals present significant challenges regarding rigorous definition, convergence, and computation. *Wick rotation* is a "very non-causally-motivated" technique for dealing with certain existence and convergence issues. The iconic *Feynman diagrams,* which form a centerpiece of the standard computational toolkit of quantum field theory, may be viewed as shorthand devices for calculating contributions to path integrals.

Although the credit for the path summation approach, as a general formulation of quantum theory, rightly belongs to Feynman, it has important antecedents. In particular, Feynman explicitly credits the 1935 edition of Dirac's classic book *The Principles of Quantum Mechanics* [DI35] as contributing to his ideas on the subject, while Norbert Wiener's earlier work on Brownian motion, published in the early 1920s, just before the introduction of "modern quantum theory" by Schrödinger and Heisenberg, seems to represent the first appearance of path integrals as a technical device. In the present section, I give a very brief sketch of path integration, using the case originally considered by Feynman: the quantum-theoretic behavior of a single point-particle in a finite region of Euclidean spacetime $\mathbb{R}^4$, equipped with a choice of temporal direction.[18] The sole purpose of this sketch is to motivate the corresponding discrete quantum causal constructions to follow, so technical details incidental to the main concepts are ignored. Surprisingly, there is little conceptual advantage to be gained, for this specific application, by considering more "realistic" or elaborate continuum-based scenarios involving, for instance, special relativity or field theory. In particular, most of the familiar machinery for evaluating or estimating path integrals in quantum field theory is wholly irrelevant in this context.

Feynman's setup and postulates. To describe Feyman's path integral, I consider a finite subset X of Euclidean spacetime $\mathbb{R}^4$, bounded between two spatial sections defined by setting $t = t^{\pm}$, where t is the universal time parameter associated with a choice of temporal direction on $\mathbb{R}^4$, and where $t^{\pm}$ are specific time values. Euclidean spacetimes $\mathbb{R}^n$ of different dimensions may be considered without altering the procedure, and of course the accompanying figures suggest $\mathbb{R}^2$, but *the intended context is ordinary, nonrelativistic quantum mechanics in three space dimensions and one time dimension.* The left-hand diagram in Figure 6.6.1 illustrates the setup for Feynman's path integral. The symbols appearing in the diagram are identified below. The right-hand diagram in the figure shows the naïve discrete causal analogue of Feynman's setup, which involves a causal path γ in a directed set D.

The "lower boundary" ∂^- of the region X may be taken, in this case, to be the set of all lower endpoints of temporal intervals in X at fixed spatial points, while the corresponding "upper boundary" ∂^+ may be taken to be the set of all upper endpoints of such intervals. Mathematically, one may imagine all manner of perverse spacetime regions, but for the present illustrative purposes, it does no harm to assume that X is bounded by a smooth hypersurface, or even that X is convex. Let $\Gamma = \Gamma_{\text{MAX}}(X)$ be the set of all "maximal directed paths," or "maximal causal paths," in X, i.e., equivalence classes of future-directed continuous curves in X beginning somewhere

[18] In the paper *On the Axioms of Causal Set Theory* [DR13], which provides much of the background for this book, I denote Euclidean spacetime by $\mathbb{R}^{3+1}$, to highlight the distinguished time direction. Here, of course, I use $\mathbb{R}^{3+1}$ to denote Minkowski spacetime. The latter convention is preferable, because there is nothing intrinsic to the geometry of Euclidean spacetime that distinguishes a particular time direction.

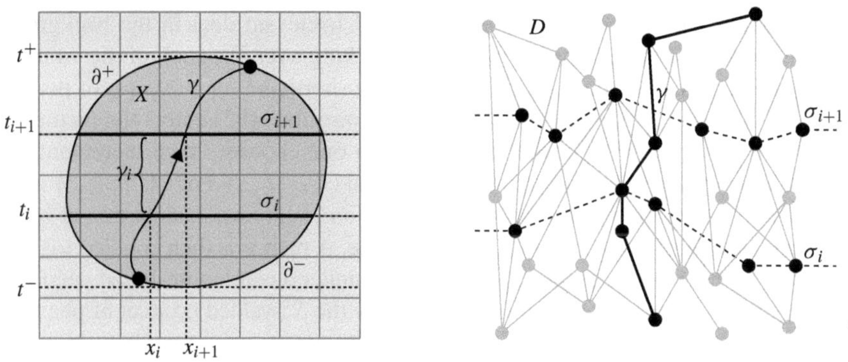

Figure 6.6.1. Setup for Feynman's path integral; naïve discrete causal analogue.

on the lower boundary ∂^- and terminating somewhere on the upper boundary ∂^+. Note that in Euclidean spacetime, *any* future-directed curve is taken to be "causal;" i.e., velocity is unlimited in magnitude. Consider now the motion of a single particle, with Lagrangian $\mathcal{L}$, along a path γ belonging to Γ, with classical action $\mathcal{S} = \int_\gamma \mathcal{L}dt$. Feynman's approach to studying the quantum behavior of such a particle is based on the following two "postulates:"

Definition 6.6.2. Feynman's Postulates.

1. (F1) *If an "ideal measurement" is performed, to determine whether the trajectory of the particle belongs to $\Gamma = \Gamma_{\text{MAX}}(X)$, then the probability of an affirmative result is the square modulus of a complex-valued* **quantum amplitude** $\psi(X; \mathcal{L})$, *given by summing complex contributions from each path $\gamma \in \Gamma$.*[19]
2. (F2) *Each path $\gamma \in \Gamma$ contributes equally in magnitude to the amplitude $\psi(X; \mathcal{L})$, with complex* **phase** $\Theta(\gamma) = e^{\frac{i}{\hbar}\mathcal{S}(\gamma)}$, *where $\mathcal{S}$ is the classical action corresponding to $\mathcal{L}$, and where $\hbar = h/2\pi$ is Planck's reduced constant.*[20]

Feynman's path integral. What these "postulates" really describe is the *heuristic intent of a limiting procedure,* in which the interval $[t^-, t^+]$ is subjected to a sequence of increasingly fine *partitions* $\Delta = \{t^- = t_0, t_1, \ldots, t_N = t^+\}$, and in which a "path" γ is specified by a sequence $\mathbf{x} := \{x_0, x_1, \ldots, x_N\}$ of position values, one for each time in Δ. Hence, the construction is actually rather crude; one does not begin with arbitrary maximal causal paths, as Feynman's first postulate (F1) suggests, but with finite sequences of spacetime points, through which such paths might hypothetically pass on their way from the lower boundary ∂^- of X to the upper boundary ∂^+ of X. In fact, there is a large degree of flexibility in the construction, in the sense that one may derive the same values for quantum amplitudes under a variety of different strategies for representing paths. From the discrete causal viewpoint, this may be

[19] Adapted from [FE48], p. 8.
[20] Adapted from [FE48], p. 9.

interpreted as an indication of the presence of irrelevant data in the background continuum manifold. Assuming that one is working over $\mathbb{R}^4$, each of the values x_i represents a position vector in $\mathbb{R}^3$, so $\mathbf{x}$ is really shorthand for a "vector of position vectors." The phase map in Feynman's second postulate (F2) enters the picture via a suitable "interpolating convention," in which one chooses "path increments" γ_i between each pair of spacetime points (x_i, t_i) and (x_{i+1}, t_{i+1}). For example, γ_i might be taken to be the classical trajectory between its endpoints, as determined by the Lagrangian $\mathfrak{L}$, according to Hamilton's principle. A path γ is then cobbled together by "joining these path increments end-to-end," under a continuum-based version of the directed product. The phase of γ is given by the S^1-valued product of phases of its increments:

$$\Theta(\gamma) = \prod_{i=1}^{N} \Theta(\gamma_i) = \prod_{i=1}^{N} e^{\frac{i}{\hbar}\mathcal{S}(\gamma_i)} = \exp\left(\sum_{i=1}^{N} \frac{i}{\hbar}\mathcal{S}(\gamma_i)\right). \tag{6.6.3}$$

Now denote by $\sigma_i = \sigma_i(\Delta)$ the spatial section of X at time $t = t_i$ in the partition Δ. Each spatial section represents a temporal choice of "present," which is successively incremented as i increases from 0 to N. Let $|\Delta|$ be the *norm* of the partition Δ, i.e., the supremum of the values $t_{i+1} - t_i$. Then Feynman's quantum amplitude $\psi(X; \mathfrak{L})$ is *defined* to be the limit:

$$\psi(X; \mathfrak{L}) := \lim_{|\Delta|\to 0} \int_{\mathbf{X}} C \cdot \Theta(\gamma) d\mathbf{x}$$

$$= \lim_{|\Delta|\to 0} \int_{\sigma_N} \cdots \int_{\sigma_1} \int_{\sigma_0} C \cdot \exp\left(\sum_{i=1}^{N} \frac{i}{\hbar}\mathcal{S}(\gamma_i)\right) dx_0 dx_1 \ldots dx_N, \tag{6.6.4}$$

if this limit exists. The top expression in Equation (6.6.4) is mostly symbolic, meaning *"integrate* $N+1$ times, *over all values of* $x_0, \ldots, x_N$, *a complex quantity* $C \cdot \Theta(\gamma)$, *depending on the partition* Δ *and the Lagrangian* $\mathfrak{L}$, *then take the limit as* $|\Delta| \to 0$." The second expression makes this more explicit at a particular level of approximation. As the norm $|\Delta|$ of the partition approaches zero, the number N of intervals in the partition approaches infinity, so the number of iterated integrals approaches infinity. The factor C appearing in the integrand is a proportionality constant which may be ignored here, given the informal overall level of description. No such constant appears in the discrete causal analogue of Feynman's construction.

Referring to previous discussions regarding the "flow of information from past to future through spacetime," particularly those in Sections 2.7 and 5.4, one may observe that the spatial sections σ_i serve as *Cauchy surfaces* for X, i.e., subsets of X that uniquely intersect every inextensible causal curve in X. Indeed, the "inextensible causal curves" in this context are just future-directed continuous curves from the lower boundary ∂^- to the upper boundary ∂^+ of X, and such curves intersect each spatial section σ_i in a unique point, under the presently-innocuous convexity assumption. This includes the piecewise-defined curves representing the paths that actually contribute to the construction of Feynman's path integral. The physical significance of this observation is that *Feynman's path integral processes information in a manner suitable to describe dynamics in* X; i.e., it captures all information flowing from ∂^- to ∂^+ in X, and more generally, all information flowing from the past of X, through X, to the future of X.

In particular, Feynman's path integral does not suffer from any information-theoretic deficiency resembling the *permeability problem* studied by Major, Rideout, and Surya [MRS06] in the context of causal set theory. The right-hand diagram in Figure 6.6.1, which shows the naïve discrete causal analogue of Feynman's setup, illustrates this permeability problem. The dashed lines in the diagram are a visual aid, indicating membership in the maximal antichains σ_i and σ_{i+1}, respectively; they have nothing to do with the structure of D. In this particular example, the causal path γ intersects the first maximal antichain σ_i, but permeates the second maximum antichain σ_{i+1}. This precludes defining a suitable analogue of the "path increment" γ_i for such a causal path. Further, causal paths originating in the past of σ_i may permeate *both* maximal antichains, while still passing through intervening elements. Or, they may cross directly from the "distant past" to the "distant future," without interacting with the region at all. Hence, naïve discrete causal analogues of Feynman's path integral are very poorly behaved in terms of information processing. Fortunately, the *impermeability theorem* of Section 5.4 enables a straightforward resolution of this problem, via passage to relation space.

6.7 Quantum Theory in Relation Space

Discrete causal analogues of Feynman's path integral. Using the theory of path summation over a multidirected set, described in Section 6.2, it is straightforward to define *abstract* analogues of Feynman's path integral, with maximal antichains playing the role of spatial sections of continuum spacetime. However, as demonstrated in previous chapters, new issues, as well as new choices, arise in the discrete causal context. First, one must select a suitable class of directed sets to serve as models of classical spacetime. Much of the existing literature involves causal sets, but the results of Chapters 3 and 4 strongly suggest using certain classes of *generally non-transitive star finite directed sets* as an alternative. Second, one must decide whether to work in the background dependent or background independent context. This book focuses mostly on the background independent case, which means that the directed or multidirected sets over which path summation is actually performed are generally *not* models of classical spacetime, but "higher-level" objects representing the structures of kinematic schemes, described more fully in Chapter 7. The abstract properties of these objects are *derived from* the properties of the corresponding classes of classical histories, but are not exactly the same properties in general. For example, the underlying multidirected set $\mathcal{M}(\mathbb{S})$ of a kinematic scheme $\mathbb{S}$ is generally *not* star finite, even when the individual classical histories in $\mathbb{S}$ *are* star finite.[21] In the present context, I simplify matters by *working with finite subobjects* of the multidirected sets over which path summation is performed. Third, one must decide what kinds of causal paths to consider. Since the information carried by a causal path with a reducible source may "generally be expressed in terms of chains along the path" in the cases of principal interest, I work *exclusively with chains* in this section. However, I do not attempt to argue rigorously that causal paths with reducible sources are always irrelevant. Fourth, regardless of these choices, the generic problem of permeability of maximal antichains in multidirected sets renders naïve path summation problematic from an information-processing standpoint. It is this fourth issue that requires a *technical* solution. As already explained in Chapter 5, and repeated at the end of Section 6.6, the strategy pursued here is to simply *pass to relation space,* where the permeability problem disappears by Theorem 5.4.3.

Choices involving causal cycles. One basic structural issue that arises in attempting to adapt the path summation approach to the discrete causal context is the question of whether or not to permit directed or multidirected sets containing cycles, and if so, how to treat path summation when cycles are involved. This complication is absent in both Feynman's original construction, and in the modern context of background dependent quantum causal set theory. However, even a small finite region of a multidirected set containing a cycle admits an infinite number of distinct chains wrapping around this cycle different numbers of times, and this introduces potential convergence issues involving the corresponding path sums. This problem cannot necessarily be avoided by restricting attention to acyclic directed sets as models of classical spacetime, since "higher-level cycles" may still arise in the corresponding theory of co-relative histories and kinematic schemes, as already mentioned in Section 6.1. Hence, the issue of cycles remains relevant in the background independent context, even for relatively conservative models of classical spacetime.

A number of possible approaches to this problem may be devised. First, one may simply apply the same formal methods as in the acyclic case, and try to show that the resulting path sums *do* converge in some suitable sense. Second, one may attempt to treat cycles in a "holistic" manner. A simple method is to define an equivalence relation $\sim$ on causal paths in a multidirected set M, in which $\gamma \sim \gamma'$ if and only if γ "coincides with γ' outside of cycles," then assign a single phase to each equivalence class. A more elaborate version of the same idea is to perform a causal atomic decomposition of M, choosing the constituent causal atoms to be the images of maximal cycles in M. The resulting directed set M^1 is a *"maximal acyclic approximation"* of M, in an obvious sense. Causal paths in M^1 then correspond to the above equivalence classes, and the phases assigned to such paths may be attributed partly to the internal cyclic structures of the "elements" in their images. A complementary method, based on the relative viewpoint (RV), is to generalize cycles to equivalence classes of *morphisms* into M, whose sources, defined up to isomorphism, are *cyclic directed sets,* i.e., directed sets in which there exist chains in both directions between any pair

[21] For example, the kinematic schemes describing generational growth of finite acyclic directed sets in Section 7.8 are not star finite, since each generation may be of arbitrary finite size.

of elements. This generalization of cycles is analogous to the familiar generalization of chains to causal paths, i.e., equivalence classes of morphisms whose sources are linear directed sets. Finally, if all cycles in M happen to be "large," then one may investigate local problems without considering cycles at all. A similar situation arises in certain solutions to general relativity, in which causal cycles exist only at cosmic scales. A modest elaboration on the subject of cyclic structure appears in Section 8.7. In the remainder of the present chapter, however, I *sidestep the issue of cycles entirely, restricting attention to the abstract acyclic case.*

Discrete causal setup for path summation. Applying the simplifying assumptions described above, let $R = (R, \prec)$ be a finite subobject of the relation space $\mathcal{R}(M')$ over an acyclic multidirected set $M' = (M', R', i', t')$. Here, primes are used to denote aspects of structure associated with the "ambient" set M' and its relation space $\mathcal{R}(M')$, so as to leave the "unprimed" symbol R free to denote the object of principal interest. Similarly, the binary relation on R is denoted by $\prec$, even though it is the subset relation on R induced by the relation $\prec'$ on $\mathcal{R}(M') = (R', \prec')$, and would therefore ordinarily be denoted by the same symbol $\prec'$. The relation $\prec'$, in turn, is induced in a familiar manner by the initial and terminal element maps i' and t' on M', as described in Definition 5.1.1. These conditions, as I have stated them, are *not sufficient to guarantee that R is immune to the permeability problem,* because arbitrary subobjects of relation spaces may exhibit permeable maximal antichains, as noted in the discussion following Theorem 5.4.3. The simplest way to avoid this issue is to take R to be a *full* subobject of $\mathcal{R}(M')$, but one may sometimes wish instead to first specify a finite subobject M of M', "working at the element space level," then define R to be the relation space over M. In this case, R is not necessarily a full subobject of $\mathcal{R}(M')$, but is a relation space in its own right, and is therefore immune to permeability. The latter procedure might be summarized by the phrase *"localize first, then move to relation space."* In what follows, I simply assume that R is immune to the permeability problem, since this can be easily arranged.

Let $\big(\mathrm{Ch}(R), \sqcup\big)$ be the chain concatenation semicategory over R, and let $\Gamma = \mathrm{Ch}_{\mathrm{MAX}}(R)$ be the subset[22] of $\big(\mathrm{Ch}(R), \sqcup\big)$ consisting of all maximal chains in R. Let σ be a maximal antichain in R, separating R into the disjoint union $R = R^- \bigsqcup \sigma \bigsqcup R^+$. In this context, σ represents a choice of "present" in R, while R^- and R^+ are **past** and **future regions** with respect to σ. Let $\gamma \in \Gamma$ be a maximal chain in R, whose unique intersection with σ occurs at the element r. Comparing these designations to Feynman's original continuum-based setup, $\mathcal{R}(M')$ corresponds to Euclidean spacetime $\mathbb{R}^4$, at least, in terms of how it relates to the construction of the path sum,[23] R corresponds to the finite spacetime region X, Γ corresponds to the set of maximal

[22]From the algebraic viewpoint, the concatenation product of any pair of elements of Γ is zero, since these elements are maximal chains. Similarly, from the order-theoretic viewpoint, no element of Γ directly precedes any other element of Γ under the causal concatenation relation. Hence, it makes no difference whether Γ is called a "subset," a "subsemicategory," a "subobject," or a "subspace" in this context. I choose the first option.

[23]Of course, M' itself may be regarded as a closer analogue of $\mathbb{R}^4$ in Feynman's setup, in the sense that "it is an element space, instead of a relation space." In the cases of principal interest, however, M' represents the abstract structure of a kinematic scheme, so *neither* M' nor $\mathcal{R}(M')$ are particularly

causal paths in X, γ corresponds to an individual such path, and σ corresponds to a spatial section of X. This setup is illustrated in Figure 6.7.1. As in Figure 6.6.1, the dashed lines in the figure are merely a visual aid to indicate membership in the maximal antichain σ. The shaded region is also a visual aid, to indicate membership in the subobject R of $\mathcal{R}(M')$.

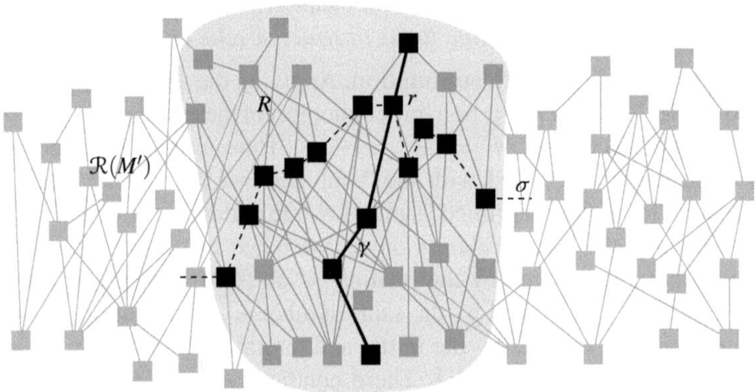

Figure 6.7.1. Setup for path summation over a finite subobject R of the relation space $\mathcal{R}(M')$ over an acyclic multidirected set $M' = (M', R', i', t')$.

Before proceeding, it is instructive to consider the same setup from the perspective of element space, i.e., to examine it in terms of the original multidirected set $M' = (M', R', i', t')$. This viewpoint is illustrated in Figure 6.7.2. Here, the significance of the diagram is not very easy to decipher, because the antichain of relations σ does not "visually resemble a spatial section of spacetime." Instead, σ visually resembles a cluttered mess, with complicated, irregular, and arbitrary-looking structure. For example, the elements w and z shown in the figure are endpoints of relations belonging to σ, but these elements are connected by multiple complex chains in M'. Due to the presence of such features, there is little obvious motivation to recommend σ as a discrete causal analogue of a Cauchy surface from the naïve element space viewpoint. This provides a good illustration of why relation space is so important in discrete quantum causal dynamics. Without it, one must choose between conceptually suggestive objects that are structurally deficient, such as maximal antichains of elements, and conceptually opaque objects whose actual structural advantages are virtually unrecognizable, such as the element-space incarnation of σ appearing in Figure 6.7.2. It is not very surprising that dynamical approaches based on the element space viewpoint have thus far enjoyed only limited success.

At the end of Section 6.6, I explained how the permeability problem interferes with defining suitable element-space analogues of Feynman's "path increments,"

(Footnote 23 continued)
close analogues of $\mathbb{R}^4$. Ultimately, this is because path integration over subsets of $\mathbb{R}^4$ yields only a very primitive theory.

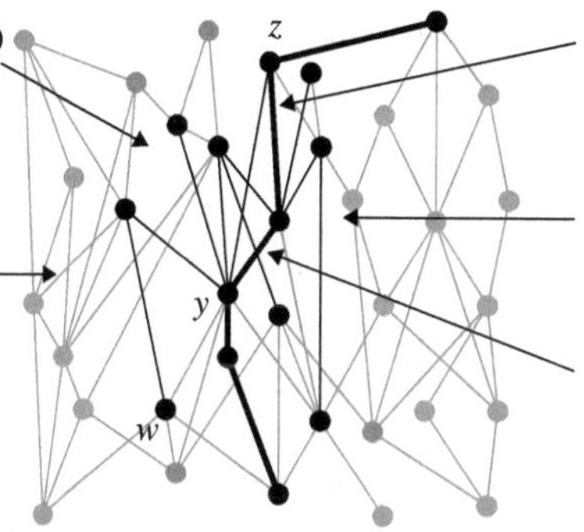

Figure 6.7.2. Setup for path summation from the element space viewpoint. From this perspective, the structural advantages of σ are completely obscure.

such as the portion γ_i of the path γ between the "consecutive" spatial sections σ_i and σ_{i+1} illustrated in Figure 6.6.1. Of course, the reason why path increments between *pairs* of spatial sections are necessary in Feynman's construction is because a single spatial section of continuum spacetime possesses "no thickness," and therefore cannot encode any finite "information flow" in its own right. In the discrete causal setup illustrated in Figure 6.7.1, however, the *single* maximal antichain of relations σ serves essentially the same purpose as the pair σ_i and σ_{i+1}, together with the family of all path increments between them, since each relation r belonging to σ *itself* constitutes an "increment" of any chain $\ldots \prec r^- \prec r \prec r^+ \prec \ldots$ passing through r. This highlights one of the advantages of working with "spatial sections" possessing *limited, but nontrivial, internal structure,* as described in Section 5.4. For more general causal paths, of course, the notion of increments may be more complicated, since a chain containing r may be "short-circuited" by a reducible relation in the image of the path, also belonging to σ by the impermeability theorem. A candidate for such a relation is the relation $y \prec z$ in Figure 6.7.2. This subtlety is one reason for choosing to work only with chains in the present setting.

Target objects and phase maps. Having specified the necessary data for path summation *intrinsic to* R, I now consider the target object T in which the values of causal path sums involving R are taken to lie, along with the phase map $\Theta : \mathrm{Ch}(R) \rightarrow T$ that supplies the individual terms of the particular path sums analogous to Feynman's path integral. Both T and Θ are a priori independent of R, but physical and philosophical considerations narrow the field of reasonable possibilities. For example, the target object T is strongly constrained if one wishes to interpret path summation in terms of probabilities. More abstract alternatives may be proposed; for example, as noted in Sections 3.8 and 6.5, Isham suggests replacing probabilities with *"generalized truth values."*[24] As suggested in Section 5.10, and again in Section 6.2, it may be fruitful to admit a "democracy of targets," just as one considers different rings of coefficients in homological algebra. In both cases, the "coefficients" chosen are

[24] See, for example, [IS11], p. 6.

generally not intrinsic to the mathematical objects of principal interest, but may be used as convenient tools to encode, organize, or isolate structural features of these objects.

It is worthwhile to revisit the precedent of "equal magnitudes" set by Feymman's phase map $\Theta(\gamma) = e^{\frac{i}{\hbar}S(\gamma)}$, which formalizes the intuitive notion that *"each path contributes equally to the path integral."* This topic was mentioned obliquely in Section 2.9, while analyzing the general quantum-theoretic role of the complex numbers $\mathbb{C}$, and again in Section 6.2, in the context of causal path functionals. However, the material developed in intervening sections enables a somewhat more focused treatment. A suitable generalization of the idea of "equal magnitudes" might be viewed as an expression of background independence, since assigning different magnitudes to different paths might be considered a form of "discrimination" among evolutionary processes for classical histories, "usurping the proper role of interference effects." This reasoning, together with the path summation principle (PSP) itself, might be used to narrow the choices of target objects of phase maps considerably. A "permissible" target object T would be required to possess a sensible notion of magnitude, under which the images of individual causal paths would share the same measure. This common value could be "set to unity, via a suitable choice of units." The subobject T_0 of T consisting of *all* elements of "unit magnitude" could then be viewed as the "unit sphere in T," just as all elements of the form $e^{i\theta}$ make up the unit circle S^1 in $\mathbb{C}$. The set Γ of causal paths under consideration would then map into the subobject T_0 of T, mirroring the behavior of Feymman's phase map. Further, T would be required to possess an "additive" operation for combining contributions from different paths, and there are good reasons to expect that an actual abelian group operation, such as addition in $\mathbb{C}$, would be most suitable. Finally, one could make strong arguments for the existence of a "multiplicative" operation on T, or at least on T_0, to reflect the algebraic structure of causal path spaces. Such an operation would be analogous to complex multiplication, restricted to S^1. Conventional knowledge regarding "well-behaved" algebraic structures on spheres suggests that very few reasonable possibilities for T might remain after imposing these requirements; in fact, the first examples that come to mind are countable subrings of $\mathbb{C}$, or similar subobjects of the quaternions $\mathbb{H}$ or the octonions $\mathbb{O}$.

It is certainly intriguing that the very general notions of background independence, path summation, and causal structure alone might lead in such a specific direction; in particular, suggesting why the complex numbers play such a pivotal role in quantum theory. However, the reasoning involved here is of a type particularly vulnerable to wishful thinking. The principal motivation behind these loose arguments is the pre-existing knowledge that Feynman's phase map *does* "work," and that the complex numbers *are* the "correct" target for this map. Such a posteriori reconstruction and generalization of a desired result is risky, especially because there are subtle differences in the meanings of the analogous structures involved. In Feynman's original setup, different paths represent physically distinct scenarios, in which a particle follows different trajectories in the same background spacetime manifold. However, as explained in Chapter 7, "higher-level causal paths" in background independent discrete quantum causal theory sharing the same initial and terminal elements do *not* represent physically distinct scenarios, but merely different *kinematic accounts* describing the same scenario, roughly analogous to different frames of reference in relativity. For this reason, it is not obvious that the precedent of "equal magnitudes" should still apply to individual paths in this setting; perhaps the situation calls for a more complicated condition involving *congruence classes* of paths representing alternative kinematic accounts of evolution between fixed pairs of classical histories. It is also worth noting that certain interesting kinematic schemes, such as the foliation scheme $\mathbb{S}_{\mathrm{FOL}}$ introduced in Section 7.8, *do* "prefer certain evolutionary processes over others," in a sense.

Returning to the discrete causal setup for path summation illustrated in Figure 6.7.1, the phase map $\Theta : \mathrm{Ch}(R) \rightarrow T$ specifies the dynamics of the theory, just as Feynman's phase map specifies dynamics in the conventional continuum-based setting. By Equation (6.2.3), the information-theoretic content of Feynman's phase map is encoded in the classical action $\mathcal{S}$, which serves as a real-valued "classical precursor" to the phase map. In particular, it is certainly *not* true that the classical actions of different paths share the same magnitude; the transition from "classical" to "quantum" produces this balanced relationship by mapping the values $\mathcal{S}(\gamma)$ to the circle S^1. The action, in turn, is determined by the Lagrangian $\mathcal{L}$. Identifying unique "correct" analogues of $\mathcal{L}$ and $\mathcal{S}$ in the discrete causal context, or at least unique physically-equivalent classes of such, is a very deep and subtle problem, and one that this book does not make any serious attempt to solve, despite the suggestive constructions in Sections 8.2 and 8.3.[25] The most obvious choices for target objects of such functionals are countable subsets of the real numbers, although "higher-level" functor-like interpretations are also possible. Under the strong interpretation of the causal metric hypothesis (CMH), these functionals ought to be "somehow determined by underlying directed structure," perhaps via natural "mediating structures," such as the *valence fields* introduced in Section 4.3, or the *causal Galois groups* introduced in Section 6.3. As discussed in Section 6.2, the most obvious discrete causal analogues of $\mathcal{L}$ are relation functions, whose arguments represent co-relative histories in the background independent setting. The entire internal structures of the source and target sets of these co-relative histories must a priori be taken into consideration in determining the correct discrete causal analogue of the Lagrangian, which illustrates why the problem is so formidable. It is reasonable to surmise, however, that a suitable notion of *entropy* plays a significant role. This subject is revisited, and some tentative guesses are proposed, in Section 8.2.

Additional simplifying assumptions. In the present abstract context, I make only modest simplifying assumptions regarding the target object T and the phase map Θ. The principal purpose of these assumptions is to enable the derivation of interesting and relevant dynamical laws; namely, the *causal Schrödinger-type equations* introduced in Section 6.9. In particular, I assume that the target object T is a commutative ring with unit, and that the phase map Θ is *generated under the concatenation product on* $\mathrm{Ch}(R)$ by a relation function $\theta : R \rightarrow T$. The latter assumption means that

[25] As noted in Chapter 3, this problem has been studied in the case of causal sets by Fay Dowker and a few others [BD10].

the phase $\Theta(\gamma)$ of any chain γ of the form $\gamma = r_1 \sqcup \ldots \sqcup r_N$ in R may be expressed as a product of the form

$$\Theta(\gamma) = \prod_{i=1}^{N} \theta(r_i) \tag{6.7.3}$$

in T. Since R is assumed to be a finite directed set in the present context, every chain γ in R has a unique factorization of this form under the concatenation product,[26] so every phase $\Theta(\gamma)$ may be expressed uniquely in the form of Equation (6.7.3) for such a phase map. At a more technical level, such a phase map Θ may be regarded as a *generalized semicategory morphism* from the chain concatenation semicategory $\big(\mathrm{Ch}(R), \sqcup\big)$ over R to the target object T. The product formula for $\Theta(\gamma)$ in Equation (6.7.3) is analogous to Feynman's formula (6.6.3), reproduced here for convenience,

$$\Theta(\gamma) = \prod_{i=1}^{N} \Theta(\gamma_i) = \prod_{i=1}^{N} e^{\frac{i}{\hbar}\mathcal{S}(\gamma_i)} = \exp\bigg(\sum_{i=1}^{N} \frac{i}{\hbar}\mathcal{S}(\gamma_i)\bigg),$$

which expresses the phase of a path γ, defined via a sequence of spacetime points, in terms of the classical actions of "interpolating path increments." Of course, much more general treatments are possible in the present context, even under the assumption that the phase map Θ preserves the structure of the source in the target. For example, the target object T might be replaced by some "higher-level" algebraic object, such as a *monoidal category*.

Chain functionals and amplitudes. Under these assumptions, the discrete causal analogue of Feynman's quantum amplitude $\psi(X; \mathfrak{L})$ is a causal path sum $\psi(R; \theta)$ in the target object T, given by summing phases over the set $\Gamma = \mathrm{Ch}_{\mathrm{MAX}}(R)$ of maximal chains in R. In the present simplified setting, the existence of $\psi(R; \theta)$ is automatic. The choice of notation for this *generalized quantum amplitude* reflects the fact that the relation function θ is information-theoretically analogous to a Lagrangian. It is useful to define an element $\Psi(R; \theta)$ of the chain concatenation algebra $T^{\sqcup}[R]$ to serve as an "algebraic precursor" to $\psi(R; \theta)$.[27] As a causal path functional, $\Psi(R; \theta)$ is merely the *restriction to* Γ of the phase map Θ generated by the relation function θ. The corresponding amplitude $\psi(R; \theta)$ is then given by applying the global evaluation map e_R to $\Psi(R; \theta)$. The following definition makes these notions precise:

Definition 6.7.4. *Let $M' = (M', R', i', t')$ be an acyclic multidirected set with relation space $\mathcal{R}(M')$, and let R be a finite subobject of $\mathcal{R}(M')$. Let $\Gamma = \mathrm{Ch}_{\mathrm{MAX}}(R)$ be the set of maximal chains in R, viewed as a subset of the chain concatenation semicategory $\big(\mathrm{Ch}(R), \sqcup\big)$. Let T be a commutative ring with unit, and let $\Theta : \mathrm{Ch}(R) \to T$ be a phase map generated under the concatenation product $\sqcup$ by a relation function $\theta : R \to T$, as expressed by Equation* (6.7.3).

[26] Again, this simplification does not apply to general causal paths in R.

[27] There is nothing to prevent consideration of such precursor functionals in continuum-based theories, but they are a priori cumbersome as formal sums, due to the large cardinalities of their sets of summands. However, they represent a radical method of treating certain divergence issues.

1. *The* **maximal chain functional** $\Psi(R;\theta)$ *of* R *with respect to* θ *is the causal path functional* $\Gamma \rightarrow T$ *given by restricting the phase map* Θ *to* Γ*:*

$$\Psi(R;\theta) = \Theta|_{\Gamma} = \sum_{\gamma\in\Gamma} \Theta(\gamma)\gamma. \tag{6.7.4}$$

2. *The* **(generalized quantum) amplitude** $\psi(R;\theta)$ *of* R *with respect to* θ *is the causal path sum given by applying the global evaluation map* e_R to $\Psi(R;\theta)$*:*

$$\psi(R;\theta) = e_R\big(\Psi(R;\theta)\big) = \sum_{\gamma\in\Gamma} \Theta(\gamma) = \sum_{\gamma\in\Gamma}\left(\prod_{i=1}^{N_\gamma}\theta(r_{\gamma,i})\right), \tag{6.7.5}$$

where $\gamma = r_{\gamma,1} \sqcup \ldots \sqcup r_{\gamma,N_\gamma}$ *is the unique factorization of* γ *in* $R \subset \big(\text{Ch}(R), \sqcup\big)$.

To understand Equation (6.7.4), recall from Section 6.2 that functionals $\text{Ch}(R) \rightarrow T$ and elements of $T^{\sqcup}[R]$ are interchangeable, since the coefficients t_γ of causal paths γ in a formal sum $\sum_\gamma t_\gamma \gamma$ denoting an element of $T^{\sqcup}[R]$ may be regarded as images of these paths under a functional $\text{Ch}(R) \rightarrow T$, and vice versa. This is the sense in which the formal sum $\sum_{\gamma\in\Gamma} \Theta(\gamma)\gamma$ represents the restriction $\Theta|_\Gamma$. As described at the end of Section 6.2, there exist a variety of different ways to express the generalized quantum amplitude $\psi(R;\theta)$ in terms of phase maps, evaluation maps, and causal path sums. In particular, under the "generic causal path sum" notation of Equation (6.2.7), one would write $\psi(R;\theta) = \Theta(\Gamma)$. Alternatively, in terms of the "restricted evaluation map" e_Γ, one would write $\psi(R;\theta) = e_\Gamma(\Theta)$. The sum $\sum_{\gamma\in\Gamma}()$ appearing on the far right of Equation (6.7.5) is analogous to the integral $\int_{\mathbf{x}}()d\mathbf{x}$ appearing in Equation (6.6.4) in Feynman's continuum-based construction. However, in the discrete causal context, this sum is *exact;* no limiting process is necessary.

6.8 Schrödinger's Equation via Path Summation

General Schrödinger-type equations. Schrödinger's equation, first appearing in this book as Equation (1.1.3), and repeated as Equation (6.8.1) below, describes the dynamics of nonrelativistic quantum systems, modeled via Hilbert spaces of complex-valued functions on Euclidean spacetime. It may be derived from general principles via path integral methods, as demonstrated by Feynman [FE48]. Since these methods lead directly to the vastly more general path summation approach to quantum theory, Schrödinger's equation represents much more than merely a "low-energy approximation of quantum field theory." So far as the general reasoning behind the path summation approach itself is valid, the descriptive limitations of Schrödinger's equation arise entirely from shortcomings in the choices of history configuration spaces used in nonrelativistic quantum theory; namely, spaces of paths traced out on a fixed copy of Euclidean spacetime. Superior models of classical histories may be expected to produce superior dynamical laws of the same general type, without altering the conceptual setting in any significant way. This expectation is already borne out, to some extent, in modern implementations of path integration in quantum field theory, and also in various approaches to quantum gravity, such as loop quantum gravity and causal dynamical triangulations. In Section 6.9, I carry out an analogous procedure in the discrete causal setting, under the simplifying assumptions described in Section 6.7. The "punchline" of this undertaking is the causal Schrödinger-type equation (1.1.2) appearing at the very beginning of the book, which is reproduced as Equation (6.9.10) below. The present section provides conceptual background for this development, by reviewing some details of Feynman's derivation.

Ordinary Schrödinger equation. One of the principal achievements of Feynman's original paper [FE48] on the path summation approach to quantum theory is the recovery, via path integral methods, of the **nonrelativistic Schrödinger equation**

$$i\hbar\frac{\partial\psi^-}{\partial t} = \mathbf{H}\psi^-, \tag{6.8.1}$$

which describes the mathematical behavior of the *past state function* ψ^-, associated with the motion of a particle through a distinguished region X of Euclidean spacetime $\mathbb{R}^4$. This function ψ^- is the usual "state function," or "wave function," encountered in elementary quantum theory; the adjective "past," and the minus sign in the superscript, indicate that it depends only on events *preceding* a specified choice of "present," represented by an appropriate spatial section of X. As discussed below, Feynman also defined a *future state function* ψ^+, which may be regarded as a "time-reversed version of ψ^-;" it satisfies an equation analogous to Equation (6.8.1), but depending only on events *succeeding* the specified choice of "present." More generally, as mentioned in Section 6.5, there is nothing intrinsic to Feynman's setup that explains the manifest fact that *the forward causal direction is more relevant than the backward causal direction* from a physical perspective. Ultimately, a background *independent* quantum theory is necessary to explain this asymmetry. I choose the term "state function," rather than "wave function," for ψ^-, because the latter term is an archaism, evoking unhelpful analogies involving the classical wave equation. Of course, the physical state of a system in ordinary quantum theory is not exactly represented by ψ^- either; rather, it is represented by a *Hilbert space ray containing* ψ^-. The symbol $\mathbf{H}$ denotes the **Hamiltonian operator**, defined in the single-particle case by the formula

$$\mathbf{H}\psi^- = \frac{-\hbar^2}{2m}\nabla^2\psi^- + V\psi^-, \tag{6.8.2}$$

where m is the mass of the particle, and V is its potential energy. Schrödinger's equation serves as a dynamical law for nonrelativistic quantum theory, enabling highly successful quantitative predictions regarding phenomenology. The concrete "dynamical content" of the equation is supplied by the details of the Hamiltonian $\mathbf{H}$, with different choices of m and V corresponding to different types of interactions. In particular, different physical fields on spacetime, such as electromagnetic fields and "gravitational fields," contribute in different ways to the potential energy function V. Complete or partial knowledge of the Hamiltonian is necessary in order to distill from Schrödinger's equation concrete information about the behavior of the particle. Hence, in an information-theoretic sense, **H** "repackages" data contained in Feynman's phase map Θ.[28]

Setup to derive Schrödinger's equation. I return now to the context of Feynman's path integral, as described in Section 6.6. To derive Schrödinger's equation in this setting, Feynman first expressed the quantum amplitude $\psi(X;\mathfrak{L})$ as the complex inner product of the past and future state functions $\psi^{\pm}$ mentioned above:

$$\psi(X;\mathfrak{L}) = \langle \psi^+(x',t'), \psi^-(x',t') \rangle := \int_{\sigma'} \psi^+(x',t')^* \psi^-(x',t') dx'. \qquad (6.8.3)$$

Informal definitions of these state functions are given below. Note that the domain of integration of the variable x' in Equation (6.8.3) is a *single* spatial section σ' of X at time $t = t'$. This spatial section partitions X into a disjoint union $X = X^- \bigsqcup \sigma' \bigsqcup X^+$, where $X^{\pm}$ are the **past** and **future regions** with respect to the choice of "present" represented by σ'. The past and future state functions $\psi^{\pm}$ depend on *both position and time*; hence, different choices of σ' produce different values of $\psi^{\pm}$. However, the quantum amplitude $\psi(X;\mathfrak{L})$ is a complex number depending only on the spacetime region X and the Lagrangian $\mathfrak{L}$, so different choices of σ' produce the same inner product in Equation (6.8.3). Figure 6.8.4 illustrates the partition $X = X^- \bigsqcup \sigma' \bigsqcup X^+$.

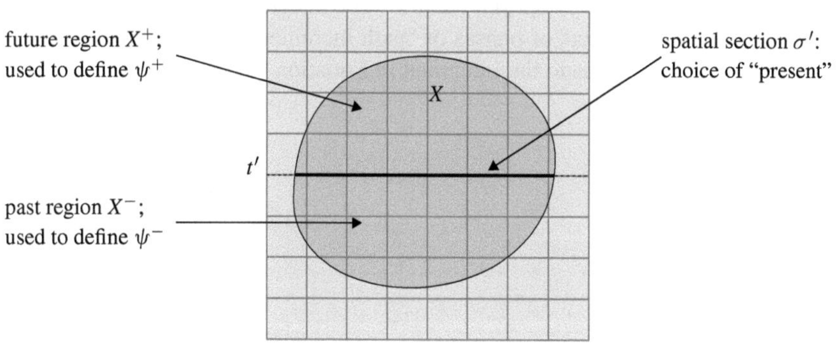

Figure 6.8.4. Partition $X^- \bigsqcup \sigma' \bigsqcup X^+$ of a region X of Euclidean space time $\mathbb{R}^4$, used to define past and future state functions $\psi^{\pm}$ in Feynman's derivation of Schrödinger's equation.

Defining the past and future state functions $\psi^{\pm}$. Without belaboring the details, the value $\psi^-(x',t')$ of the past state function ψ^- at a point (x',t') in X is determined

[28] It is not feasible to undertake here an explanation of the relationships between Lagrangian and Hamiltonian methods, or to describe the circumstances under which the two are equivalent. In the discrete causal context, the motivations for working with relation functions, and the manner in which they resemble Lagrangians, are both quite clear.

by *"integrating over the set of all maximal causal paths in the past region X^- of X terminating at (x', t')."* Here, the word "maximal" means that the paths begin on the lower boundary of X, and terminate on the spatial section σ'; in this case, at the specific point (x', t'). A family of such paths is illustrated in the left-hand diagram in Figure 6.8.6. Similarly, the value of the future state function at (x', t') is determined by *"integrating over all maximal causal paths in the future region X^+ of X beginning at (x', t')."* A family of such paths is illustrated in the right-hand diagram in the figure. As in the definition of the quantum amplitude $\psi(X; \mathfrak{L})$, given in Feynman's first "postulate" (F1), the instructions to "integrate over sets of paths" to obtain values of the past and future state functions $\psi^{\pm}$, are really heuristic descriptions of limiting processes. Note that every path contributing to the quantum amplitude $\psi(X; \mathfrak{L})$ is given by *"joining paths in X^- and X^+ at a common point (x', t') in the spatial section σ',"* since σ' is a Cauchy surface. Hence, by combining the state functions $\psi^{\pm}$, via the inner product formula in Equation (6.8.3), one may recover $\psi(X; \mathfrak{L})$, with the x'-independence "integrated out."

To be a bit more precise, the **past state function** ψ^- may be defined as a limit:

$$\psi^-(x', t') := \lim_{|\Delta^-| \to 0} \int_{\mathbf{x}^-} C \cdot \Theta(\gamma) d\mathbf{x}^-, \tag{6.8.5}$$

where $\{\Delta^-\}$ is a sequence of partitions of the time interval $[t^-, t']$ bounding the past region X^- of X, with the upper boundary $t = t'$ defining the spatial section σ'. The integration is performed over all sequences $\mathbf{x}^-$ of values of spatial coordinates corresponding to such partitions, subject to the additional condition that *these sequences must terminate at the specific spatial value x'*. The integrand $C \cdot \Theta(\gamma)$ is given, up to the constant C, by products of phases of "path increments" determined by Δ^- and $\mathbf{x}^-$, in a manner analogous to the integrand in Equation (6.6.4).

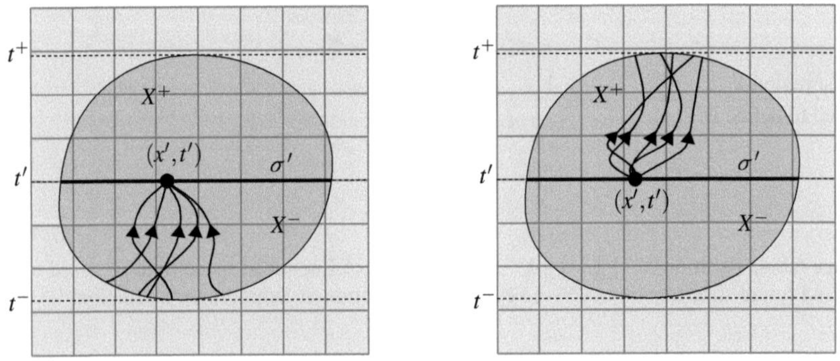

Figure 6.8.6. Family of paths contributing to the value $\psi^-(x', t')$ of the past state function ψ^-; family of paths contributing to the value $\psi^+(x', t')$ of the corresponding future state function ψ^+.

Derivation of Schrödinger's equation. Schrödinger's Equation (6.8.1) is derived by comparing the values of the past state function ψ^- for "nearby" time values

t' and t'', where $t' < t''$, then taking the limit as the difference $t'' - t'$ approaches zero. Here, I give an "adequate" sketch of the derivation, without describing all the details. Feynman [FE48] provides a good explanation, and the same result is recapitulated elsewhere in the literature. For simplicity, I replace the limit notation with the "approximation symbol" $\approx$ below, since no limiting process is needed when carrying out the discrete causal analogue of this derivation in Section 6.9. Let t' and t'' be a pair of "nearby" time values, and let σ' and σ'' be the corresponding spatial sections of the spacetime region X. Let Δ^- be a partition of the time interval $[t^-, t']$ bounding X, as described in Section 6.6. This setup is illustrated in the left-hand diagram in Figure 6.8.7. The right-hand diagram illustrates the naïve discrete causal analogue. As one would expect, it suffers from structural deficiencies. In particular, as described at the end of Section 6.6, there is generally no sensible way to define element-space analogues of continuum "path increments," such as the increment $\delta\gamma$ appearing in the left-hand diagram, due to the permeability problem. For example, the chain γ illustrated in the right-hand diagram permeates the maximal antichain σ''. The fact that element space is generally non-irreducible makes this problem even worse; one cannot even define a "preceding spatial section" σ' in the naïve way, as *"the set of all elements directly preceding σ'',"* since this set is generally not an antichain. For example, the consecutive elements x and y in γ both directly precede the same element z in σ''. Problems of this nature are cured upon passage to relation space, as described in Section 6.9.

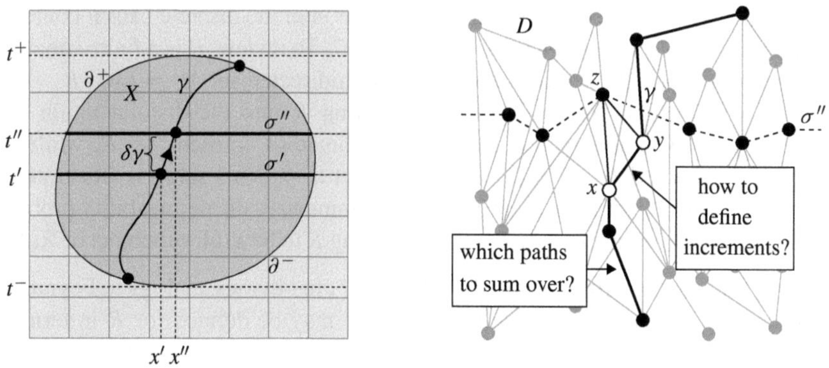

Figure 6.8.7. Deriving Schrödinger's equation via Feynman's path integral; naïve discrete causal analogue, illustrating structural problems of working in element space.

The approximation

$$\psi^-(x', t') \approx \int_{\mathbf{X}^-} C \cdot \Theta(\gamma) d\mathbf{x}^-,$$

computed at time $t = t'$, is given by dropping the limit in Equation (6.8.5) above. The "subsequent" approximation, computed at time $t = t''$, may be written as

$$\psi^-(x'', t'') \approx \int_{\sigma'} \int_{\mathbf{x}^-} C \cdot \Theta(\gamma) d\mathbf{x}^- \exp\Big(\frac{i}{\hbar} \mathcal{S}(\delta\gamma)\Big) dx',$$

where $\delta\gamma$ ranges over all "path increments" beginning somewhere on the spatial section σ', and terminating at the specific point (x'', t'') of the spatial section σ''. These expressions lead to the "approximate recursion"

$$\psi^-(x'', t'') \approx \int_{\sigma'} \psi^-(x', t') \exp\Big(\frac{i}{\hbar} \mathcal{S}(\delta\gamma)\Big) dx', \qquad (6.8.8)$$

which yields Schrödinger's equation (6.8.1) in the limit. The justification for giving such a terse description here is that the details of the derivation are irrelevant to what follows. The philosophy here is essentially the same as in the description of metric recovery in Chapter 2; the objective is not to reproduce rigorous derivations of standard results in continuum-based theories, but merely to explain the basic reasoning involved, and to adapt whichever features are relevant to the discrete causal context.

6.9 Causal Schrödinger-Type Equations

Setup to derive discrete causal analogue of Schrödinger's equation. To derive a dynamical law analogous to Schrödinger's equation in the discrete causal context, I now return to the scenario described in Section 6.7. To review, R is a finite subobject of the relation space $\mathcal{R}(M')$ over an acyclic multidirected set $M' = (M', R', i', t')$, while σ is a maximal antichain in R, separating R into the disjoint union $R = R^- \bigsqcup \sigma \bigsqcup R^+$. The antichain σ represents a choice of "present" in R, while R^- and R^+ are *past* and *future regions* with respect to σ. This setup is illustrated in Figure 6.9.1. Recall that I am assuming R to be immune to the permeability problem. This can easily be arranged; for example, by taking R to be a full subobject of $\mathcal{R}(M')$.

Defining the past and future state functions $\psi^{\pm}$. In obvious analogy to Feynman's construction, past and future state functions $\psi^{\pm}$ may be defined for R in terms of the decomposition $R = R^- \bigsqcup \sigma \bigsqcup R^+$. The algebraic expressions for these state functions, given in Definition 6.9.5, involve phases of maximal chains either terminating or beginning at the maximal antichain σ. Hence, a choice of phase map Θ is also required to define these functions. For the convenience of the reader, I repeat the simplifying assumptions I am imposing on Θ in the present context, originally stated in Section 6.7: Θ is assumed to be a generalized semicategory morphism from the chain concatenation semicategory $\big(\mathrm{Ch}(R), \sqcup\big)$ of R into a commutative ring T with unit, generated under the concatenation product $\sqcup$ by a relation function $\theta : R \to T$.

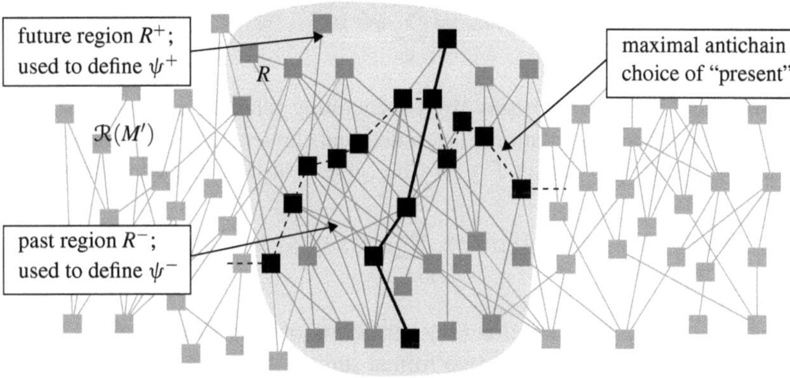

Figure 6.9.1. Setup for deriving discrete causal analogue of Schrödinger's equation.

This means that the phase $\Theta(\gamma)$ of each chain $\gamma = r_1 \sqcup \ldots \sqcup r_N$ in R is equal to the product $\prod_{i=1}^{N} \theta(r_i)$, as first expressed in Equation (6.7.3) above.

Since the past and future state functions $\psi^{\pm}$ are of central importance to discrete quantum causal theory, I describe their derivation in detail. I focus on the case of ψ^-; the case of ψ^+ is similar. Let γ be a maximal chain in the past region R^- of R, admitting extension by r, i.e., a maximal chain in R^- such that $\gamma' = \gamma \sqcup r$ is a maximal chain in $R^- \bigsqcup \sigma$. Figure 6.9.2 illustrates this situation. In what follows, r, γ, and γ' will be allowed to vary as needed. For example, the relation r will be allowed to vary over σ, and later, over all of R. In the present construction, r may be viewed as the relation-space analogue of the continuum path increment $\delta\gamma$ appearing in the left-hand diagram in Figure 6.8.7.

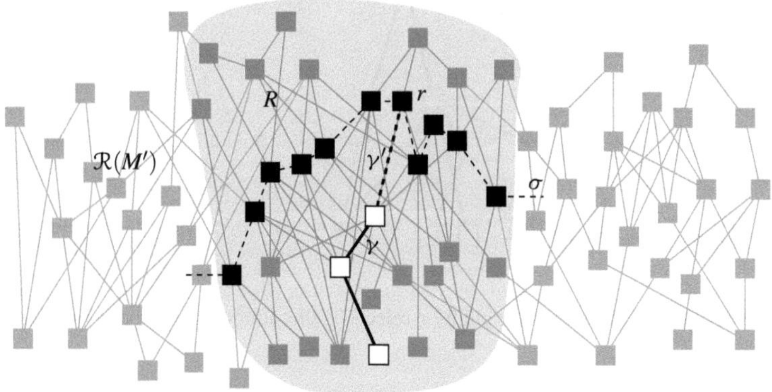

Figure 6.9.2. Incrementing a chain in relation space: γ is incremented by r to yield $\gamma' = \gamma \sqcup r$.

Now consider the restriction $\Theta|_{\sigma}$ of the phase map $\Theta : \mathrm{Ch}(R) \to T$ to the maximal antichain σ, viewed as an element of the chain concatenation algebra $T^{\sqcup}[R]$ over R:

$$\Theta|_\sigma = \sum_{r\in\sigma} \theta(r)r.$$

Here, I have replaced Θ in the sum by its generating relation function θ, since the two coincide on individual relations, i.e., individual elements of R. Consider also the restriction $\Theta|_{\Gamma^-}$ of Θ to the set $\Gamma^- := \mathrm{Ch}_{\mathrm{MAX}}(R^-)$ of maximal chains in the past region R^-:

$$\Theta|_{\Gamma^-} = \sum_{\gamma\in\Gamma^-} \Theta(\gamma)\gamma,$$

where I have chosen *not* to factor $\Theta(\gamma)$ in terms of θ in the summand. I claim that the product in $T^{\sqcup}[R]$ of these two formal sums is the formal sum representing the restriction $\Theta|_{\Gamma_\sigma^-}$ of Θ to the set $\Gamma_\sigma^- := \mathrm{Ch}_{\mathrm{MAX}}(R^- \bigsqcup \sigma)$ of maximal chains in $R^- \bigsqcup \sigma$:

$$\Theta|_{\Gamma^-} \sqcup \Theta|_\sigma = \Theta|_{\Gamma_\sigma^-} = \sum_{\gamma'\in\Gamma_\sigma^-} \Theta(\gamma')\gamma'. \tag{6.9.3}$$

To see that the formula in Equation (6.9.3) actually holds, observe that for any chain γ in Γ^-, and for any relation r belonging to σ, the concatenation product $\gamma' = \gamma \sqcup r$ is nonzero in $T^{\sqcup}[R]$ if and only if γ admits extension by r. An equivalent condition is that the terminal relation r^- of γ directly precedes r in R. Hence, the summand $\theta(r)r$ in $\Theta|_\sigma$, multiplying the element $\Theta|_{\Gamma^-}$ on the right, annihilates all chains γ in Γ^- *except those admitting extension by* r, as illustrated in Figure 6.9.4.

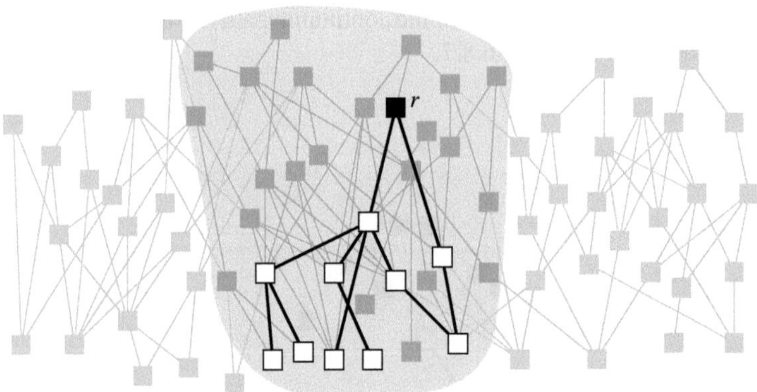

Figure 6.9.4. Multiplication on the right by r selects chains admitting extension by r, annihilating all others.

The surviving terms in the product $\Theta|_{\Gamma^-} \sqcup \theta(r)r$ are therefore chains γ' in Γ_σ^- terminating at r, weighted by their appropriate phases, since Θ is generated by θ. The qualifier "weighted by their appropriate phases" makes this statement true even when $\theta(r) = 0_T$. Letting r vary over σ, it remains to show that *every* such chain γ' is included in the product $\Theta|_{\Gamma^-} \sqcup \Theta|_\sigma$ on the left-hand side of Equation (6.9.3). This follows from the impermeability of the maximal antichain σ, which is guaranteed by Theorem 5.4.3 and the choice of R as a "suitable" subobject of relation space. Indeed, impermeability implies that every maximal chain γ in R^- terminates in a relation r^- directly preceding some relation r belonging to σ.

Now let r vary over all of R. Denote by $\Gamma_r^- := \mathrm{Ch}_{\mathrm{MAX}}^-(R; r)$ the set of all maximal chains in R terminating at r, and by $\Gamma_r^+ := \mathrm{Ch}_{\mathrm{MAX}}^+(R; r)$ the set of all maximal chains in R beginning at r. Proceeding as in the construction of the maximal chain functional $\Psi(R; \theta)$ and the generalized quantum amplitude $\psi(R; \theta)$ in Section 6.7, I define *past and future chain functionals* $\Psi_{R;\theta}^{\pm}(r)$ on the chain sets $\Gamma_r^{\pm}$, viewed as elements of the chain concatenation algebra $T^{\sqcup}[R]$, which serve as algebraic precursors to *past and future state functions* $\psi_{R;\theta}^{\pm}(r)$, obtained from the functionals $\Psi_{R;\theta}^{\pm}(r)$ by application of the global evaluation map e_R. The following definition makes these notions precise:

Definition 6.9.5. *Let $M' = (M', R', i', t')$ be an acyclic multidirected set, with relation space $\mathcal{R}(M')$, and let R be a finite subobject of $\mathcal{R}(M')$. Let r be an element of R, and let $\Gamma_r^- = \mathrm{Ch}_{\mathrm{MAX}}^-(R; r)$ and $\Gamma_r^+ = \mathrm{Ch}_{\mathrm{MAX}}^+(R; r)$ be the sets of maximal chains in R terminating at r and beginning at r, respectively, viewed as subsets of the chain concatenation semicategory $\big(\mathrm{Ch}(R), \sqcup\big)$. Let T be a commutative ring with unit, and let $\Theta : \mathrm{Ch}(R) \to T$ be a phase map generated under the concatenation product $\sqcup$ by a relation function $\theta : R \to T$, as expressed by Equation* (6.7.3).

1. *The elements*

$$\Psi_{R;\theta}^-(r) := \sum_{\gamma \in \Gamma_r^-} \Theta(\gamma)\gamma \quad \text{and} \quad \Psi_{R;\theta}^+(r) := \sum_{\gamma \in \Gamma_r^+} \Theta(\gamma)\gamma, \tag{6.9.6}$$

   of the chain concatenation algebra $T^{\sqcup}[R]$, are called the **past** *and* **future chain functionals** *of r in R with respect to θ, respectively.*

2. *The relation functions*

$$\psi_{R;\theta}^-(r) := \sum_{\gamma \in \Gamma_r^-} \Theta(\gamma) \quad \text{and} \quad \psi_{R;\theta}^+(r) := \sum_{\gamma \in \Gamma_r^+} \Theta(\gamma), \tag{6.9.7}$$

   given by applying the global evaluation map e_R to the past and future chain functionals $\Psi_{R;\theta}^-(r)$ and $\Psi_{R;\theta}^+(r)$, are called the **past** *and* **future state functions** *over R with respect to θ, respectively.*

Given two different elements r_1 and r_2 of R, the functionals $\Psi_{R;\theta}^-(r_1)$ and $\Psi_{R;\theta}^-(r_2)$ are *two different functionals,* mapping two different chain sets $\Gamma_{r_1}^-$ and $\Gamma_{r_2}^-$ to the common target object T, respectively, while $\psi_{R;\theta}^-(r_1)$ and $\psi_{R;\theta}^-(r_2)$ are merely *two different values of the same past state function* $\psi_{R;\theta}^-$, mapping R to T. This is because application of the global evaluation map e_R to a causal path functional "collapses" the functional to a T-valued causal path sum. Similar statements apply to the future chain functionals $\Psi_{R;\theta}^+(r_1)$ and $\Psi_{R;\theta}^+(r_2)$, and the corresponding values $\psi_{R;\theta}^+(r_1)$ and $\psi_{R;\theta}^+(r_2)$ of the future state function $\psi_{R;\theta}^+$. Heuristically, passage from past and future chain functionals to values of past and future state functions may be viewed as a "hierarchy-reducing process," analogous to evaluating an ordinary function to obtain a specific value, or taking a category-theoretic limit of a functor to obtain a specific object. Such processes are often "diagonal," in the sense that a "functor-like entity" at one level of hierarchy is reduced to an "object-like entity" at the next lower level of hierarchy. The overall structural viewpoint behind these comparisons is briefly described in Section 8.6.

As in the case of Feynman's past and future state functions $\psi^{\pm}$, defined in Section 6.8, there is nothing intrinsic to the construction of the state functions $\psi^{\pm}_{R;\theta}$ that singles out the past state function $\psi^{-}_{R;\theta}$ as more relevant than the future state function $\psi^{+}_{R;\theta}$. Instead, the physical priority of the "past to future" direction over the "future to past" direction is specific to the background independent context, arising from the asymmetry of the higher-level structure of kinematic schemes, as described in Section 7.5. In the abstract theory developed in the present chapter, quantities defined in terms of "past" and "future" information stand on essentially equal footing. In this context, the reason for deriving a causal Schrödinger-type equation only for the past state function $\psi^{-}_{R;\theta}$ is simply that the steps for deriving a corresponding equation for the future state function $\psi^{+}_{R;\theta}$ are essentially the same, and therefore need not be repeated.

Derivation of the causal Schrödinger-type equation for $\psi^{-}_{R;\theta}$. Now consider in more detail the past chain functional $\Psi^{-}_{R;\theta}(r)$ of r in R with respect to θ. If r is a minimal element of R, then $\Psi^{-}_{R;\theta}(r)$ consists of the single summand $\theta(r)r$. Otherwise, $\Psi^{-}_{R;\theta}(r)$ may be *decomposed* in terms of the set $J_0^{-}(R;r)$ of direct predecessors of r in R, i.e., those elements r^{-} in R such that $r^{-} \prec r$.[29] Indeed, every maximal chain γ' in R terminating at r may be expressed as a concatenation product $\gamma \sqcup r$ for an appropriate maximal chain γ in R^{-} terminating at some $r^{-} \in J_0^{-}(R;r)$. Fixing r^{-}, and summing in $T^{\sqcup}[R]$ over all phase-weighted chains γ of this form, yields the past chain functional $\Psi^{-}_{R;\theta}(r^{-})$ of r^{-} in R with respect to θ. Now letting r^{-} vary over *all* direct predecessors of r in R, the past chain functional $\Psi^{-}_{R;\theta}(r)$ of r in R with respect to θ may be written as a product in $T^{\sqcup}[R]$:

$$\Psi^{-}_{R;\theta}(r) = \Big(\sum_{r^{-}\prec r} \Psi^{-}_{R;\theta}(r^{-})\Big) \sqcup \theta(r)r. \tag{6.9.8}$$

Equation (6.9.8) serves as an algebraic precursor to the causal Schrödinger-type equation introduced in Definition 6.9.9, just as the functionals $\Psi(R;\theta)$ and $\Psi^{\pm}_{R;\theta}$ are algebraic precursors to the generalized quantum amplitude $\psi(R;\theta)$, and the past and future state functions $\psi^{\pm}_{R;\theta}$, respectively.

[29] The reason for using the notation $J_0^{-}(R;r)$, rather than merely $J_0^{-}(r)$, is to make it absolutely clear that only direct predecessors of r belonging to R are considered here. Other direct predecessors of r in the "ambient relation space" $\mathcal{R}(M')$ are excluded.

Definition 6.9.9. *Under the hypotheses of Definition 6.9.5, the equation*

$$\psi^{-}_{R;\theta}(r) = \theta(r) \sum_{r^{-} \prec r} \psi^{-}_{R;\theta}(r^{-}), \tag{6.9.10}$$

given by applying the global evaluation map e_R to Equation (6.9.8), is called the **causal Schrödinger-type equation** *over R with respect to θ.*

Note that care is required in applying an evaluation map to a product in a causal path algebra, because some of the products of individual paths in the underlying causal path semicategory may not exist. In this case, the corresponding contributions to the product in the causal path algebra will be zero, thereby annihilating the associated products of phases, whether or not these products vanish in the target object. In other words, *evaluation maps generally do not commute with products in causal path algebras*. However, in the case of Equation (6.9.8), every chain contributing to each past chain functional $\Psi^{-}_{R;\theta}(r^{-})$ terminates in a relation r^{-} directly preceding r, as indicated by the index of summation. Hence, each of the products of paths in the underlying semicategory contributing to the overall product exist, and the global evaluation map *does* commute with products in this special case. In the corresponding causal Schrödinger-type equation (6.9.10), I have moved the factor $\theta(r)$ to the left of the sum for legibility, since T is commutative. However, it is worth noting that the factor $\theta(r)$ "naturally" appears to the right of the sum, and this makes a difference if T is replaced by a noncommutative target object. The situation is reversed in the corresponding Schrödinger-type equation for the *future* state function $\psi^{+}_{R;\theta}$, in which the factor $\theta(r)$ naturally appears to the *left* of the sum.

Fractal decomposition; higher-order equations. The method used to derive Equation (6.9.8), in which the past chain functional of a relation r is decomposed in terms of the direct predecessors of r, may be iterated, in a process that may be referred to as *past fractal decomposition,* or *past dendritic decomposition.* The adjective "fractal" refers to the structural similarity of the nested chain sets appearing at various levels of the decomposition. The alternative adjective "dendritic" refers to the tree-like structure of the decomposition itself. Past fractal decomposition may be used to define *higher-order discrete causal analogues of Schrödinger's equation.* Figure 6.9.11 illustrates past fractal decomposition at the element r of R appearing in previous figures. In particular, the reader will recognize the directed set appearing in the top node of the decomposition as the subobject of R defined by all chains admitting extension by r, first illustrated in Figure 6.9.4.

6.10 Generalized Quantum Amplitudes via Causal Path Functionals

"Inner product formula" for the concatenation product. To conclude this chapter, I briefly discuss two discrete causal analogues of Feynman's complex inner product formula

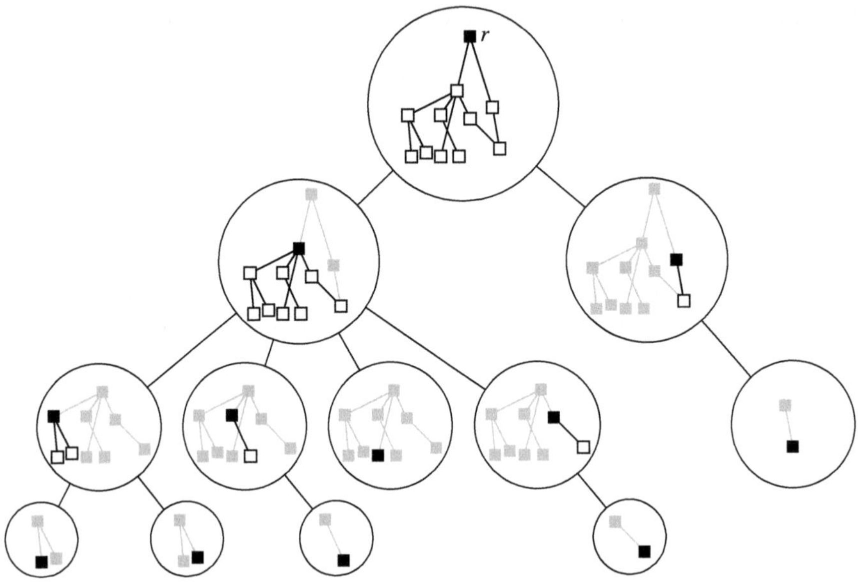

Figure 6.9.11. Example of a past fractal decomposition leading to a higher-order causal Schrödinger-type equation.

$$\psi(X;\mathfrak{L}) = \langle \psi^+(x',t'), \psi^-(x',t') \rangle := \int_{\sigma'} \psi^+(x',t')^* \psi^-(x',t') dx',$$

first appearing in Equation (6.8.3), which expresses the quantum amplitude $\psi(X;\mathfrak{L})$ in terms of the past and future state functions $\psi^{\pm}$. To review, the interpretation of the amplitude $\psi(X;\mathfrak{L})$, under Feynman's first postulate (F1), is that its square modulus measures the probability that the trajectory of a particle belongs to the set of maximal directed paths in a region X of continuum spacetime, as illustrated in Figure 6.6.1. The value $\psi^-(x',t')$ of the past state function ψ^- encodes information about maximal causal paths in X terminating at the point (x',t'), while the value $\psi^+(x',t')$ of the future state function ψ^+ encodes information about maximal causal paths in X beginning at (x',t'). The integral in the inner product formula is taken over the spatial section σ' of X given by setting $t = t'$. Different choices of t' produce the same amplitude, because the same maximal causal paths in X are always involved; they are merely "constructed from different subpaths" for different choices of t'. In other words, *changes in ψ^- are "balanced" by changes in ψ^+* in the inner product formula when a different value of t' is chosen. Spatial dependence, of course, is "integrated out" in the formula. Discrete causal analogues of this formula enable computation of generalized quantum amplitudes, such as the amplitude $\psi(R;\theta)$ defined in Equation (6.7.5), and related causal path functionals, which are interpreted as measuring "generalized likelihoods" for certain families of evolutionary processes. For example, in the background independent context, these amplitudes and functionals encode the "likelihood" that a classical history belonging to a given *cobase family* will evolve into a classical history belonging to a given *target family*. Whether or not these "likelihoods" always can, or always should, be viewed as "conventional values," such as complex amplitudes or probabilities, is debatable.

To construct the first such "inner product formula," I begin by expressing the maximal chain functional $\Psi(R;\theta)$ defined in Equation (6.7.4) as a causal path algebra element in the usual way, i.e., as the formal sum $\sum_{\gamma\in\Gamma}\Theta(\gamma)\gamma$ of all chains γ in $\Gamma=\mathrm{Ch}_{\mathrm{MAX}}(R)$, weighted by their appropriate phases $\Theta(\gamma)$. Next, I consider a maximal antichain σ in R, and assume for the sake of simplicity that σ contains no extremal elements. Then, by the impermeability of σ, every term in the sum for $\Psi(R;\theta)$ may be uniquely expressed as a concatenation product of the form $\Theta(\gamma^-)\gamma^-\sqcup\theta(r)r\sqcup\Theta(\gamma^+)\gamma^+$, for appropriate chains $\gamma^\pm$ in $\Gamma^\pm=\mathrm{Ch}_{\mathrm{MAX}}(R^\pm)$, and an appropriate relation r in σ. The sum of all such terms is therefore given by the generalized "inner product" formula

$$\Psi(R;\theta)=\Psi(R^-;\theta)\sqcup\Theta|_\sigma\sqcup\Psi(R^+;\theta). \tag{6.10.1}$$

In the case where σ *does* contain extremal elements, it is necessary to add to the formula some terms of the forms $\Theta(\gamma^-)\gamma^-\sqcup\theta(r)r$ and/or $\theta(r)r\sqcup\Theta(\gamma^+)\gamma^+$ and/or $\theta(r)r$, since the product in Equation (6.10.1) annihilates terms lacking nontrivial contributions from all three factors $\Psi(R^-;\theta)$, $\Theta|_\sigma$, and $\Psi(R^+;\theta)$. Hence, although the value of the functional $\Psi(R;\theta)$ itself does not depend upon the choice of maximal antichain σ, the actual form of the corresponding "inner product formula" in terms of the concatenation product *does* depend on σ.

"Inner product formula" for the directed product. To construct a second analogue of Feynman's inner product formula, I begin with the past and future chain functionals $\Psi^-_{R;\theta}(r)$ and $\Psi^+_{R;\theta}(r)$, defined in Equation (6.9.6). To review, these chain functionals serve as "algebraic precursors" to the past and future state functions $\psi^-_{R;\theta}$ and $\psi^+_{R;\theta}$, defined in Equation (6.9.7), whose values $\psi^-_{R;\theta}(r)$ and $\psi^+_{R;\theta}(r)$ encode information about chains in R beginning and terminating at r, respectively. Observe that for any relation r in R, the directed product $\Psi^-_{R;\theta}(r)\vee\Psi^+_{R;\theta}(r)$ *almost* coincides with the sum of phase-weighted maximal chains in the region $J^-(R;r)\cup\{r\}\cup J^+(R;r)$, which is the "total domain of influence of r in R, augmented by r itself." The only difference between the product and the sum is that the phase of r is counted twice in the product. To see why this is true, recall the simplifying assumption that the phase map Θ is generated under the *concatenation* product by the relation function θ; i.e., if $\gamma=r_1\sqcup\ldots\sqcup r_N$ is a chain in R, then $\Theta(\gamma)=\prod_{i=1}^N\theta(r_i)$, as expressed in Equation (6.7.3). The formula $\gamma=r_1\sqcup\ldots\sqcup r_N$ expresses the chain γ from the "semicategory viewpoint," i.e., in terms of a product. From the alternative "causal path space viewpoint," γ may be rewritten as $r_1\prec_\sqcup\ldots\prec_\sqcup r_N$; or, since all the factors are individual relations, as just $r_1\prec\ldots\prec r_N$. Using the more sophisticated language of Section 5.9, this is a reflection of the fact that $(R,\prec)$ embeds as a full subobject in its causal concatenation space. Now, suppose that the chain γ terminates at the relation r; i.e., $r_N=r$, and let γ' be another chain $\gamma'=r_1'\prec\ldots\prec r_M'$ beginning at r; i.e., $r_1'=r$. Then the *directed* product $\gamma\vee\gamma'$ of γ and γ' exists, and $\big(\Theta(\gamma)\gamma\big)\vee\big(\Theta(\gamma')\gamma'\big)=\Theta(\gamma)\Theta(\gamma')\gamma\vee\gamma'$ by the definition of the chain directed product algebra. The product $\Theta(\gamma)\Theta(\gamma')$ in the target object T of Θ may be written as

$$\Theta(\gamma)\Theta(\gamma') = \prod_{i=1}^{N}\theta(r_i)\prod_{j=1}^{M}\theta(r'_j) = \theta(r)\prod_{i=1}^{N}\theta(r_i)\prod_{j=2}^{M}\theta(r'_j) = \theta(r)\Theta(\gamma \vee \gamma'),$$

where in the second equality I have pulled a factor of $\theta(r)$ out of the second product, since $r'_1 = r$, and where the third equality follows from the fact that $\gamma \vee \gamma' = r_1 \prec \ldots \prec r_N \prec r'_2 \prec \ldots \prec r'_M$, since $r_N = r'_1 = r$. This conclusion also uses the simplifying assumption that T is a *commutative* ring.

Now given a maximal antichain σ in R, every maximal causal path in R passes through a unique relation r in σ. Therefore, $\Psi(R;\theta)$ is given by the formal sum

$$\Psi(R;\theta) = \sum_{r\in\sigma}\frac{1}{\theta(r)}\Psi^-_{R;\theta}(r) \vee \Psi^+_{R;\theta}(r), \tag{6.10.2}$$

where the factors of $\theta(r)$ in the denominators "correct for the corresponding extra factors in the products." In particular, one does not encounter "division by zero" in case $\theta(r) = 0_T$, because the factor $\theta(r)$ in the denominator of each term cancels with one of the two corresponding factors in the numerator. Equation (6.10.2) "collapses" to an expression for the corresponding amplitude $\psi(R;\theta)$. To obtain this expression, note that since every causal path contributing to $\Psi^-_{R;\theta}(r)$ precedes every causal path contributing to $\Psi^+_{R;\theta}(r)$ under the directed product relation, the global evaluation map e_R commutes with the directed product in this case. Hence,

$$\psi(R;\theta) = \sum_{r\in\sigma}\frac{1}{\theta(r)}\psi^-_{R;\theta}(r)\psi^+_{R;\theta}(r), \tag{6.10.3}$$

which is directly analogous to the inner product formula in Equation (6.8.3).

References

[BSU15] Angelo Bassi, Tejinder Singh, and Hendrik Ulbricht. *Is Quantum Linear Superposition an Exact Principle of Nature?* Questioning the Foundations of Physics, The Frontiers Collection, Springer, pp. 131–138, 2015.

[IS05] Christopher Isham. *Quantising on a Category*. Foundations of Physics, **35**, 2, pp. 271–297, 2005. arXiv preprint: http://arxiv.org/pdf/quant-ph/0401175v1.pdf.

[SO12] Rafael Sorkin. *Toward a Fundamental Theorem of Quantal Measure Theory*. Mathematical Structures in Computer Science, **22**, 05 (special issue), pp. 816–852, 2012. arXiv preprint: http://arxiv.org/pdf/1104.0997v2.pdf.

[RA00] Ioannis Raptis. *Algebraic Quantization of Causal Sets*. International Journal of Theoretical Physics, **39**, 5, pp. 1233–1240, 2000. arXiv preprint: http://arxiv.org/pdf/gr-qc/9906103v1.pdf.

[SR99] David Rideout and Rafael Sorkin. *Classical sequential growth dynamics for causal sets*. Physical Review D, **61**, 2, 024002, 2000. arXiv preprint: http://arxiv.org/pdf/gr-qc/9904062v3.pdf.

[AN09] Michael T. Anderson. *A Survey of Einstein Metrics on* 4 *-Manifolds*. Preprint, 2009. arXiv preprint: http://arxiv.org/pdf/0810.4830v2.pdf.

[BR88] Edward A. Bender and Robert W. Robinson. *The Asymptotic Number of Acyclic Digraphs II*. Journal of Combinatorial Theory, Series B, **44**, pp. 363–369, 1988.

[RS03] Carlo Rovelli and Simone Speziale. *Reconcile Planck-scale discreteness and the Lorentz-Fitzgerald contraction*. Physical Review D, **67**, 6, 064019, 2003. arXiv preprint: http://arxiv.org/pdf/gr-qc/0205108v1.pdf.

[IS11] Christopher Isham. *Topos Methods in the Foundations of Physics*. In Deep Beauty: Understanding the Quantum World through Mathematical Innovation, edited by Hans Halvorson. Cambridge University Press, 2011. arXiv preprint: http://arxiv.org/pdf/1004.3564v1.pdf.

[FE48] Richard Feynman. *Space-Time Approach to Non-Relativistic Quantum Mechanics*. Reviews of Modern Physics, **20**, 2, pp. 367–387, 1948.

[DI35] P. A. M. Dirac. *The Principles of Quantum Mechanics, Second Edition*. Oxford University Press. 1935.

[DR13] Benjamin F. Dribus. *On the Axioms of Causal Set Theory*. Preprint. arXiv link: http://arxiv-web3.library.cornell.edu/pdf/1311.2148v3.pdf.

[MRS06] Seth A. Major, David Rideout, and Sumati Surya. *Spatial Hypersurfaces in Causal Set Cosmology*. Classical and Quantum Gravity, **23**, 14, pp. 4743–4751, 2006. arXiv preprint: http://arxiv.org/pdf/gr-qc/0506133v2.pdf.

[BD10] Dionigi M. T. Benincasa and Fay Dowker. *Scalar Curvature of a Causal Set*. Physical Review Letters, **104**, 18, 181301, 2010. arXiv preprint: http://arxiv.org/pdf/1001.2725v4.pdf.

Chapter 7
Kinematics and Dynamics

Abstract. This chapter continues the development of discrete quantum causal theory from Chapter 6, focusing on the background independent context. Section 7.1 reviews the notion of *frames of reference* in relativity, and describes how each frame defines a *refinement of the causal relation,* by specifying *temporal orders* between pairs of unrelated events. Section 7.2 examines the general theory of refinements in more detail. Section 7.3 introduces the concept of *kinematic accounts,* which are "partially fictional" descriptions of the evolution of directed sets. The "extraphysical information" contained in a kinematic account is analogous to the "fictional" ordering of spacelike separated events in relativistic spacetime under a particular choice of temporal order. Section 7.4 introduces the theory of *kinematic schemes,* which are "higher-level" structures organizing families of kinematic accounts. A kinematic scheme provides evolutionary pathways for every "permissible" classical history in a specified class of classical histories. It possesses the abstract structure of a multidirected set, whose relations represent co-relative histories. Section 7.5 focuses on the *positive sequential kinematic scheme* $\mathbb{S}_{PS}$, which organizes all evolutionary processes "adding one element at a time" to a directed set. The positive sequential kinematic scheme contains, as a subobject, the kinematic scheme $\mathbb{S}_{SR}$ appearing in Sorkin and Rideout's theory of *sequential growth dynamics* for causal sets. Section 7.6 introduces the *quantum causal metric hypothesis,* which states that *"quantum spacetime may be modeled in terms of kinematic schemes of directed sets."* Section 7.7 describes *path summation over kinematic schemes,* which leads to a background independent version of discrete quantum causal theory. Section 7.8 discusses an alternative to sequential growth in describing discrete causal dynamics; namely, *generational growth.* The *generational kinematic scheme* and the *foliation scheme* are two different kinematic schemes in which an entire generation of elements, rather than an individual element, is added to each directed set at each "stage of evolution." Section 7.9 discusses *completions of kinematic schemes,* which are constructed by considering "limits" of evolutionary processes involving infinite numbers of stages. Section 7.10 introduces the idea of *hyperquantum theory,* which involves additional levels of hierarchy above the level of kinematic schemes.

B.F. Dribus, *Discrete Causal Theory*, DOI 10.1007/978-3-319-50083-6_7

7.1 Frames of Reference

Physical versus extraphysical information. Before the general methods of discrete quantum causal theory developed in Chapter 6 can be applied to the practical problems of modern physics, it is necessary to flesh out a more specific and detailed version of the theory. Any serious effort in this direction requires a background independent approach, so it is necessary to work in terms of the "higher-level" structure of *kinematic schemes* of directed sets, rather than in terms of individual directed sets. Much of the necessary preparation for this approach appears in the previous two chapters, but at least two important facets remain largely untouched. First, few details have been offered thus far regarding *specific* multidirected structures of interest in this setting. Indeed, the only two kinematic schemes described even superficially up to this point have been the *positive sequential kinematic scheme* $\mathbb{S}_{\rm PS}$ of finite acyclic directed sets, and *Sorkin and Rideout's kinematic scheme* $\mathbb{S}_{\rm SR}$ of finite causal sets. The actual properties of these kinematic schemes have not been examined in any detail. Second, issues involving uniqueness have not been addressed, even though it is already clear that different kinematic schemes, and different "evolutionary pathways" in a given kinematic scheme, may be used to describe relationships involving the same classical histories. Satisfactory treatment of both topics requires consideration of the broad question of how to organize and interpret different viewpoints regarding the evolution of directed sets. This leads to the *general theory of kinematics and dynamics* in discrete causal theory.

From a basic information theoretic perspective, the problem of dealing with multiple "evolutionary viewpoints" revolves around the subtle general issue of distinguishing "absolute," or "physical," information, from "arbitrary," or "gauge-like," or "extraphysical" information. An important example of this distinction, arising in relativity, is the difference between the physically absolute causal ordering of events, and the partly-extraphysical temporal ordering associated with a given frame of reference. The fact that different frames yield different temporal orderings is popularized via the phrase *"relativity of simultaneity,"* an idea whose real meaning and significance remains underappreciated in conventional physics. Before discussing this point, however, it is instructive to revisit an example from Chapter 1, which provides a simple illustration of the distinction between physical and extraphysical information in the discrete causal context. This example involves the classical stochastic "toy dynamics" discussed in Section 1.8. Figure 7.1.1, which reproduces and highlights part of Figure 1.8.2, illustrates "evolutionary pathways" in the positive sequential kinematic scheme $\mathbb{S}_{\rm PS}$. The probabilistic information appearing in the original figure is irrelevant here; the feature of interest is the existence of *two separate* pathways in $\mathbb{S}_{\rm PS}$ from the empty set $\emptyset$ to the directed set D_t, indicated in the figure by the darker shading. One of these pathways passes through the directed set D_i, while the other passes through D_i'. This means that there exist two separate, equally valid, *kinematic accounts* of "how D_t could have evolved." The alternative between these accounts is extraphysical; i.e., it is devoid of any absolute physical significance. All of the absolute physical information pertaining to D_t is *contained* *in the internal directed structure of D_t itself.* However, one must still decide how to handle the existence of such evolutionary alternatives in order to build a precise theory; at least, under the general requirement that this theory should supply notions of "initial and terminal conditions" suitable for predictive purposes.

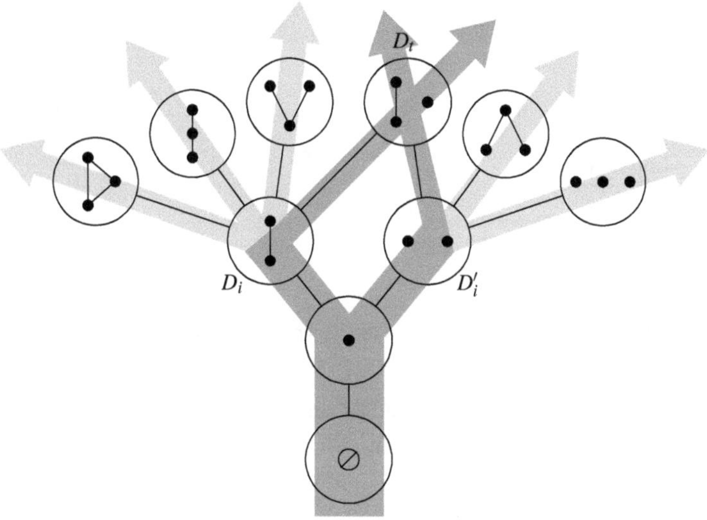

Figure 7.1.1. Physically equivalent kinematic accounts of the evolution of D_t via two different directed sets D_i and D'_i.

Relativity of simultaneity. As mentioned above, it is useful to compare the existence of "extraphysical alternatives" between evolutionary pathways in a kinematic scheme, such as those illustrated in Figure 7.1.1, to the existence of different frames of reference in relativity. Here, *special* relativity is adequate for the purposes of illustration; the first chapter of Carroll's book [CA04] gives an easy introduction. The feature of principal interest at present is the *relativity of simultaneity,* under which spacelike separated pairs of events in Minkowski spacetime $\mathbb{R}^{3+1}$ are assigned different temporal orders in different frames of reference. Relativity of simultaneity played a limited role in the discussion of sprinkled causal sets and systematic Lorentz invariance in Section 4.5. It is useful here to briefly review this familiar topic, since many modern treatments of special relativity ignore some of the most important general lessons to be drawn from it. Adopting the physical presuppositions underlying special relativity, I consider a physical scenario involving two "nearby objects" A and B, in "uniform rectilinear motion" with respect to each other. In particular, one may consider object A to be "at rest," while object B moves at a constant velocity v, or one may consider object B to be "at rest," while object A moves at a constant velocity $-v$. Alternatively, one may consider *neither* object to be "at rest," but this perspective need not be considered at present. Various philosophical issues may be raised regarding the scenarios examined here, but these are well-documented elsewhere, and involve considerations that are largely irrelevant to discrete causal theory. The left-hand diagram in Figure 7.1.2 illustrates the situation from the "viewpoint of object A."

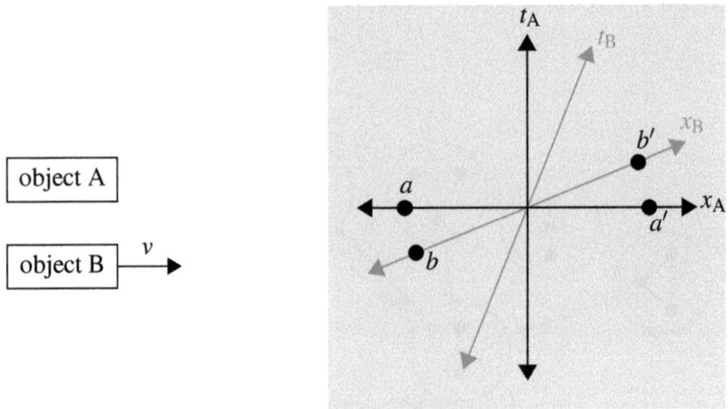

Figure 7.1.2. Relativity of simultaneity from the viewpoint of object A.

The term "viewpoint" is made precise, in this context, in terms of an *inertial frame of reference* centered at object A, which is described via a special choice of coordinates on Minkowski spacetime $\mathbb{R}^{3+1}$. Here, I follow the usual simplifying procedure of aligning a particular spatial axis with the velocity vector v. This reduces the problem to a two-dimensional one, which may be studied in $\mathbb{R}^{1+1}$. I label the temporal coordinate t_A, and the spatial coordinate x_A. The right-hand diagram in the figure illustrates this choice of coordinates, with the (t_A, x_A) coordinate system drawn in black.[1] For simplicity, no "coordinate grid" or domain of influence is shown in the diagram, in contrast to the more-detailed diagrams appearing in Section 4.5. An alternative coordinate system (t_B, x_B), drawn in gray, encodes the corresponding temporal and spatial coordinates associated with an inertial frame of reference centered at object B. For convenience, I have drawn the objects A and B in the left-hand diagram as merely "passing close to each other," rather than actually "colliding," which is the literal implication of the common origin of the coordinate systems in the right-hand diagram. This detail is unimportant for the purposes of the present discussion.

The "scissoring together" of the temporal and spatial coordinates t_B and x_B in the "moving frame of reference" associated with object B is an iconic feature of spacetime diagrams in special relativity, often called *Minkowski diagrams.* Superficially, this scissoring effect seems to belie the relativity of uniform rectilinear motion, since one could argue that the "true rest frame" is the one that "looks orthogonal." However, as explained in Section 4.5, it is not the frames of reference that are biased here, but merely the diagram; which shows *both* coordinate systems (t_A, x_A) and (t_B, x_B)

---

[1]It is traditional in this context to list the temporal coordinate t_A first in the expression (t_A, x_A), and to associate this coordinate with the *vertical* axis. This reverses the relationship between abscissa and ordinate conventional in elementary calculus.

"from the viewpoint of object A." Taking the object B to be at rest leads to a different diagram, illustrated in Figure 7.1.3, in which the coordinate system (t_B, x_B) "looks orthogonal," while the coordinate system (t_A, x_A) is "scissored apart."

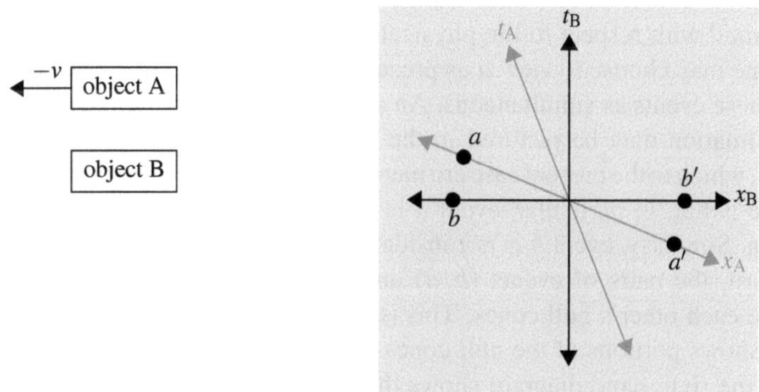

Figure 7.1.3. Relativity of simultaneity from the viewpoint of object B.

Either figure may be used to illustrate the relativity of simultaneity. Referring, for instance, to Figure 7.1.2, one may consider the two spacetime events a and a' lying on the x_A-axis. These events are spatially distant from the origin of the coordinate system (t_A, x_A); one has a positive x_A-value, while the other has a negative x_A-value. The two events share the same null value of the temporal coordinate t_A; i.e., $t_A(a) = t_A(a') = 0$. The physical interpretation of these values is that a and a' are *simultaneous* events from the viewpoint of object A. However, from the viewpoint of object B, the event a' temporally precedes the event a, since the value of the temporal coordinate $t_B(a')$ at a' is negative, while $t_B(a)$ is positive. Similar remarks apply to the pair of events b and b'. From the viewpoint of object B, b and b' are simultaneous, but from the viewpoint of object A, b temporally precedes b'. The fact that different choices of inertial frames of reference can lead to different conclusions about the temporal order of pairs of events demonstrates that *the notion of temporal order itself is based at least partly on extraphysical information.* In particular, the statement that "a' temporally precedes a" is not an absolute, definitive statement about physical reality, but merely an expression of one way in which data about these events may be arranged from a particular viewpoint.

7.2 Refinements

Relativistic causal relation revisited. What *is* physically absolute about Minkowski spacetime $\mathbb{R}^{3+1}$ is its relativistic causal relation $\prec_{GR}$. This relation was introduced in Section 2.6 in the broader context of *general* relativity, which is the reason for the "GR" subscript. In the special case of Minkowski spacetime, the relativistic causal relation is a *strict partial order* in the traditional sense of the term, i.e., an irreflexive, transitive binary relation, defined as a subset of the Cartesian product $\mathbb{R}^{3+1} \times \mathbb{R}^{3+1}$. The reason why the temporal ordering of certain pairs of events, such as the pair a and a' illustrated in Figure 7.1.2, and the pair b and b' illustrated in Figure 7.1.3, depend on the choice of reference frame, is because these particular pairs of events are *unrelated* with respect to the physically-absolute relation $\prec_{\mathrm{GR}}$. This leads to a choice: one may choose to view a as preceding a', or vice versa, or one may choose to view these events as simultaneous. An analogous statement applies to b and b'.

This situation may be pictured in the familiar manner in terms of domains of influence, which in the present case are merely "filled in" null cones, since Minkowski spacetime is flat. In particular, event a is *not* inside the null cone of event a', and vice versa. Similarly, event b is *not* inside the null cone of event b', and vice versa. By contrast, the pairs of events (b, a) and (a', b') *are* causally related; i.e., they fall inside each other's null cones. This is illustrated in Figure 7.2.1. The left-hand diagram shows portions of the null cones of a and a' from the viewpoint of object A, while the right-hand diagram shows the same null cones from the viewpoint of object B.

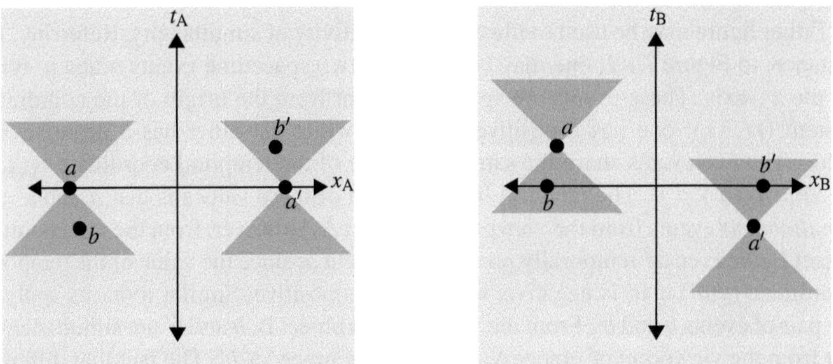

Figure 7.2.1. Causal structure from the viewpoint of object A; causal structure from the viewpoint of object B.

Temporal relation as a refinement of the relativistic causal relation. The specific choice to view event b as temporally preceding event b' involves adding a "new" individual relation, with initial element b and terminal element b', to the relativistic causal relation $\prec_{\mathrm{GR}}$. To avoid confusion, I remind the reader that the word "relation" is being used here in two different ways. The first use refers to a binary relation, such as $\prec_{\mathrm{GR}}$, while the second use refers to an individual relation between a specific pair of events or elements, such as $b \prec_{\mathrm{GR}} a$. The elements of the binary relation $\prec_{\mathrm{GR}}$, viewed as a subset of $\mathbb{R}^{3+1} \times \mathbb{R}^{3+1}$, are just its individual relations. Returning to the discussion at hand, the "new" individual relation from b to b', added "from the viewpoint of object A," is extraphysical; one could just as easily add a relation pointing in the opposite direction, or leave the pair of events unrelated. The frame of reference centered at object A, illustrated in Figure 7.1.2, incorporates the choice that b precedes b', and also incorporates an uncountable number of other such choices, encoded via the coordinate system (t_A, x_A). Note, however, that the pair of events a and a' are left unrelated under this choice of coordinates. In an analogous manner, the frame of reference centered at object B, illustrated in Figure 7.1.3, incorporates the choice that a' precedes a, along with an uncountable number of other such choices, encoded via the coordinate system (t_B, x_B). In this case, the pair of events b and b' are left unrelated.

From an order-theoretic perspective, the choice of coordinates (t_A, x_A) represents a new binary relation $\prec_A$, called the *temporal relation* with respect to A, which refines the relativistic causal relation $\prec_{GR}$ in the sense of Section 3.8, meaning that $\prec_A$ contains $\prec_{GR}$ as a subset of the Cartesian product $\mathbb{R}^{3+1} \times \mathbb{R}^{3+1}$. In more concrete terms, this means that any pair of events related under $\prec_{GR}$ is also related under $\prec_A$. The converse, of course, is false; in particular, $b \prec_A b'$, even though b and b' are unrelated with respect to $\prec_{GR}$. In an analogous manner, the choice of coordinates (t_B, x_B) represents an *alternative* temporal relation $\prec_B$, which refines $\prec_{GR}$ in a different way; for example, $a' \prec_B a$. Note that these temporal relations are *not* the same as the relativistic chronological relation $\prec\prec_{GR}$ discussed in Chapter 2, nor close analogues of it, even though both involve "time" in a certain sense. In particular, the chronological relation is a *subset* of $\prec_{GR}$, while temporal relations are *supersets* of $\prec_{GR}$; i.e., they contain $\prec_{GR}$ as subsets.

It is sometimes useful to adopt a "constructive viewpoint" regarding refinements of $\prec_{GR}$, such as $\prec_A$ and $\prec_B$. From this viewpoint, one "begins with $\prec_{GR}$," then "adds new individual relations" between appropriate pairs of events to obtain $\prec_A$, or alternatively, $\prec_B$. In the present case, this involves adding an uncountable number of new individual relations. However, it suffices for the purposes of illustration to focus on new relations involving the four events a, a', b, and b' illustrated in the previous figures. In particular, as stated above, the refinement $\prec_A$ adds one new individual relation involving these elements; namely, the relation $b \prec_A b'$, while the refinement $\prec_B$ adds a different new individual relation $a' \prec_B a$. These new individual relations are illustrated in Figure 7.2.2.

Refinements of discrete causal relations. In discrete causal theory, the absolute, physically-significant information associated with a directed set $(D, \prec)$, viewed as a model of classical spacetime under the classical causal metric hypothesis (CCMH), is contained in its causal relation $\prec$. Abstracting and generalizing the relativistic discussion above, a "frame of reference" in this context is represented by a *refinement* R of $\prec$, i.e., a subset R of the Cartesian product $D \times D$, containing $\prec$. The familiar "relation set" symbol R from Chapters 5 and 6 is used here to denote such a refinement for two reasons: first, R *is* the relation set of the "refined directed set" (D, R); second, the letter "R" suggests the word "refinement." Often, additional conditions, such as acyclicity, are imposed on such a refinement R. Generally, R represents extraphysical information added to $\prec$ for pragmatic reasons, particularly to facilitate the description of evolutionary processes. Figure 7.2.3 illustrates a refinement R of the causal relation

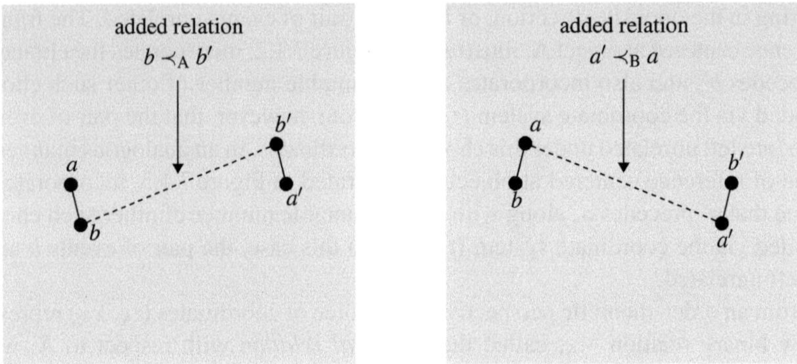

Figure 7.2.2. The temporal relations $\prec_A$ and $\prec_B$ refine the relativistic causal relation $\prec_{GR}$ in two different ways.

on a four-element acyclic directed set $D = (D, \prec)$. The left-hand diagram in the figure shows D itself; for future reference, it is worth noting that D may be obtained by adding an additional step to the "evolutionary process" producing the directed set D_t appearing in Figure 7.1.1. The right-hand diagram shows the refined set (D, R), given by adding three new individual relations to $\prec$, represented by the dashed lines in the diagram. In this case, R is a *maximal refinement* of $\prec$, under the assumption of acyclicity, since it extends $\prec$ to a strict total order.

Figure 7.2.3. A refinement R of the causal relation $\prec$ on a four-element acyclic directed set D.

Generational refinement types. I now introduce two convenient, though imprecise, additional methods of representing such refinements, in the special case where $D = (D, \prec)$ is an acyclic directed set, and where the chosen refinement R of $\prec$ partitions the elements of D into well-defined "generations." Such "generational refinements" are exceedingly rare among the class of all refinements of a typical causal relation. However, they provide useful sources of examples, due to their close resemblance to coordinate systems in relativity. They are also convenient for computational purposes. In particular, since the generations specified by R are necessarily antichains in $(D, \prec)$, R defines an associated causal atomic decomposition of D

to yield a linear directed set[2] whose elements are the generations of (D, R). This decomposition is analogous to a *foliation* of a relativistic spacetime manifold, as described in Section 5.7. The first method of representing such a refinement is to assign positive integer-valued labels to the elements of D, where each element of the "first generation" is labeled with a numeral 1, each element of the "second generation" is labeled with a numeral 2, and so on. This method is illustrated in the left-hand diagram in Figure 7.2.4. A much more complicated example of the same type of construction appeared in Figure 5.7.5 of Chapter 5, and another example appears in Figure 7.8.2. In the present example, the fact that each generation has precisely one element indicates that the refined directed set (D, R) is itself linear. Similarly, the labelings arising in the theory of *sequential growth dynamics* of causal sets, mentioned in Section 3.4, represent special cases of such linear refinements. The second method is to use shaded planar regions as a visual aid to convey the same information. This method, which works particularly well for "small, simple" acyclic directed sets, is illustrated in the right-hand diagram in Figure 7.2.4. Figures 5.7.5 and 7.8.2 "combine both methods" in an obvious sense, by using shaded nodes to "reinforce" the information represented by the integer labels. In the present context, the conceptual meaning of the shading is that *elements in darker-shaded regions precede elements in lighter-shaded regions under the refined relation R.*

Figure 7.2.4. Alternative representations of the refinement type appearing in Figure 7.2.3.

Both of these methods have limitations. Besides the fact that they apply only to generational refinements, they are generally ambiguous; in particular, they do not specify which, if any, new individual relations are added between non-consecutive generations. For example, one could begin with a three-element directed set, with no relations, then add relations to obtain a refinement with three generations. The resulting "shaded representation" for such a refinement is illustrated in the left-hand diagram in Figure 7.2.5. This representation corresponds to *either* of the two refinements shown in the corresponding right-hand diagram. For the specific "numbered" or "shaded" representations appearing in this chapter, such ambiguities are generally insignificant, but it is nevertheless prudent to mention them. One may use phrases

[2] As discussed below, such a refinement is not necessarily totally ordered, because it may be "missing relations between nonconsecutive generations." Of course, its transitive closure is totally ordered, by the definition of a linear directed set.

such as "refinements up to generation," or "generational refinement types," to indicate that one is deliberately ignoring such ambiguities.

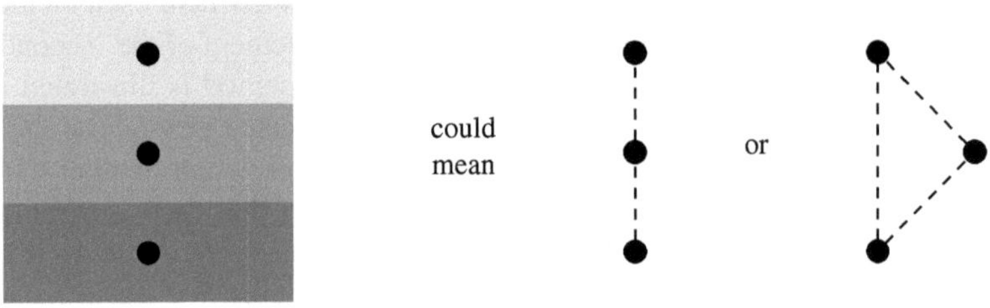

Figure 7.2.5. Ambiguity of "shaded representations".

The directed set $D = (D, \prec)$ illustrated in Figures 7.2.3 and 7.2.4 admits a total of eleven different nontrivial types of generational refinements of its causal relation. Shaded representations of these refinement types are illustrated in Figure 7.2.6. Some of these representations are ambiguous in the manner discussed above; for example, the third representation in the top row corresponds to two different refinements, one including a new individual relation from the "bottom right element" to the "top left element," and the other omitting this relation. The representations in the figure are arranged according to the number of generations appearing in the corresponding refinements. Since the causal relation $\prec$ on D is nonempty, any generational refinement R of $\prec$ must include at least two generations, because the initial element of each individual relation in $(D, \prec)$ must precede the corresponding terminal element under R. There exists only one nontrivial generational refinement of $\prec$ with two generations; this is the refinement illustrated on the bottom left in the figure. In terms of "adding new individual relations," this refinement is given by adding one individual relation from the "bottom right element" to the "top left element." The middle row in the figure illustrates the five types of generational refinements of $\prec$ with three generations, while the top row illustrates the five types with four generations, i.e. the linear refinement types of $\prec$. Also shown, in the bottom row, are two specific examples of refinements which are *non-generational,* and which therefore cannot be represented in a convenient way by shaded diagrams. In particular, the top left element in the first of these two refinements could be taken either to precede or succeed either of the elements on the right, while the bottom right element in the second refinement could be taken either to precede or succeed either of the elements on the left.

In general, it is rather complicated to determine how many refinements of a particular type a given causal relation will admit. The pivotal factor in determining this number is the question of "how refined the causal relation already is." In the acyclic context, the extreme cases are totally ordered sets, which admit no nontrivial acyclic refinements, and totally unordered sets, i.e., directed sets with empty relation sets, which admit "as many refinements as possible." In the finite case, the latter number may be viewed as the number of *labeled acyclic directed graphs* of a given size, which grows very rapidly [BR88]. After all, there already exist $n!$ different total

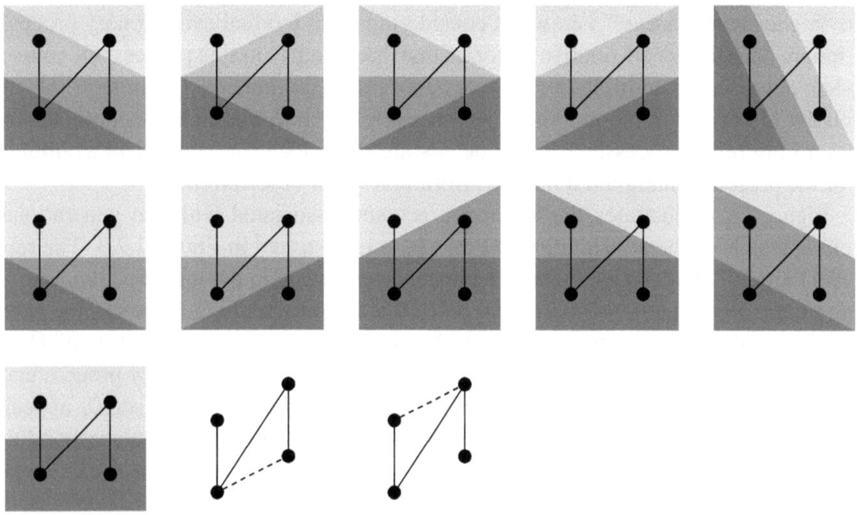

Figure 7.2.6. Eleven types of generational refinements for a 4-element directed set, together with two specific non-generational refinements.

orders on a set of cardinality n, before even considering refinements with multiple elements belonging to one or more generations!

7.3 Kinematic Accounts

Kinematic accounts via refinements. Part of the interpretation of a refinement R of a causal relation $\prec$ on a directed set $D = (D, \prec)$, such as the refinement illustrated in Figure 7.2.3, is that it "tells a story," generally both incomplete and partly fictional, about "the order in which events occur" in the classical history represented by D. Referring back to the terminology associated with the classical stochastic "toy dynamics" introduced in Section 1.8, and revisited in Section 7.1, it is convenient to refer to such a "story" as a *kinematic account.* One of the purposes of this toy dynamics was to illustrate, at an informal level, the complementary roles of kinematics and dynamics. To reiterate, the kinematics of a theory describes what behavior is *permissible,* without spelling out what behavior is *determined or favored under specified conditions.* In this spirit, a kinematic account furnishes a somewhat arbitrary description of *how* a particular physical system evolves, without attempting to explain *why* such evolution occurs, i.e., what specific laws govern its behavior. The latter question is a dynamical issue. At present, I am being deliberately vague about what exactly constitutes a "kinematic account." Ultimately, the term means a "higher-level causal path" in a subobject of a kinematic prescheme or kinematic scheme. This meaning is made precise in Definition 7.7.2. In practice, I generally restrict attention to "higher-level *chains;*" i.e., when considering an individual evolutionary process, I ignore "auxiliary" co-relative histories that "skip steps" in this process. Of course, when considering multiple evolutionary processes simultaneously, some of the co-relative histories corresponding to one process may "short-circuit higher-level chains describing other processes." These details are made clear in subsequent examples. For the present, I maintain a mostly informal level of description.

Figure 7.3.1 illustrates the kinematic account associated with the generational refinement R of the acyclic directed set $(D, \prec)$ illustrated in Figure 7.2.3. The representation of R via shaded planar regions is reproduced in the left-hand diagram in the figure. Note that in this particular case, R is the only refinement belonging to its generational refinement type. The right-hand diagram shows the evolution of D, as described from the perspective of R. As noted above, this evolutionary process may be regarded as a continuation of the process leading to the directed set D_t appearing in Figure 7.1.1, since D_t is the penultimate directed set in the corresponding kinematic account. The fact that this process may be described in terms of a linear sequence of steps, i.e., a "higher-level chain," whose "elements" are directed sets, is a consequence of the fact that R is a generational refinement. The evolutionary description derived from R, expressed in ordinary English, is as follows: *"Beginning with the empty set, the bottom-left element appears first, followed by the bottom-right element. Next, the top-left element appears, related to the bottom-left element. Finally, the top-right element appears, related to the two bottom elements."*

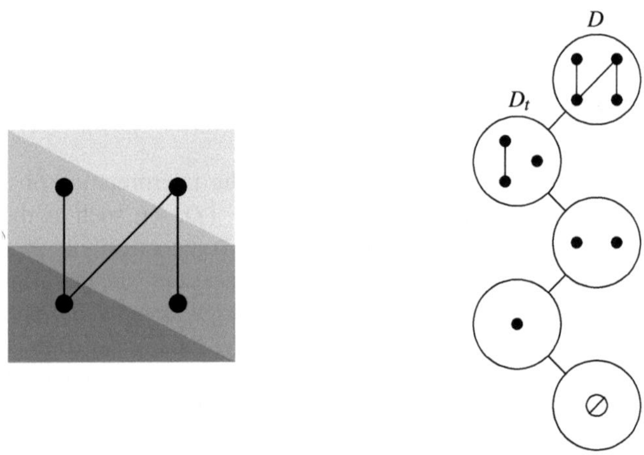

Figure 7.3.1. "Shaded representation" of the generational refinement R from Figure 7.2.3; kinematic account associated with R.

It should be immediately obvious, however, that this description is open to criticism. The difficulty arises not from the fact that extraphysical information is introduced, which is a deliberate aspect of the construction, but from a subtler issue: the conclusions drawn from the shaded representation illustrated in the left-hand diagram in the figure depend on viewing the evolution of D a posteriori, rather than allowing D to evolve via an evolutionary process that could conceivably pursue a different course. Indeed, the only reason why the "bottom-left element" of D is even *physically distinguishable* from the "bottom-right element" of D is because the subsequent addition of the two "top elements" breaks the $\mathbb{Z}_2$ symmetry of the pair of unrelated "bottom elements." Hence, the kinematic account illustrated in the right-hand diagram in the figure is actually *less specific* than the naïve description associated with the refinement in the left-hand diagram. What this kinematic account *really* says, in its own right, is as follows: *"Beginning with the empty set, an element appears, followed by a second element, unrelated to the first. Next, a third element appears, related to one of the first two elements, but not the other. Finally, a fourth element appears, related to both of the first two elements."*

Referring to the table of eleven generational refinement types of the causal relation $\prec$ on $(D, \prec)$, illustrated in Figure 7.2.6, one may observe that there is *another* such refinement type that leads to the same kinematic account; namely, the third type appearing in the top row of the table. Indeed, this refinement type *also* describes the consecutive appearance of two unrelated elements, followed by a third element related to one of the first two elements, followed by a fourth element related to both of the first two elements. This ambiguity is highlighted in Figure 7.3.2.

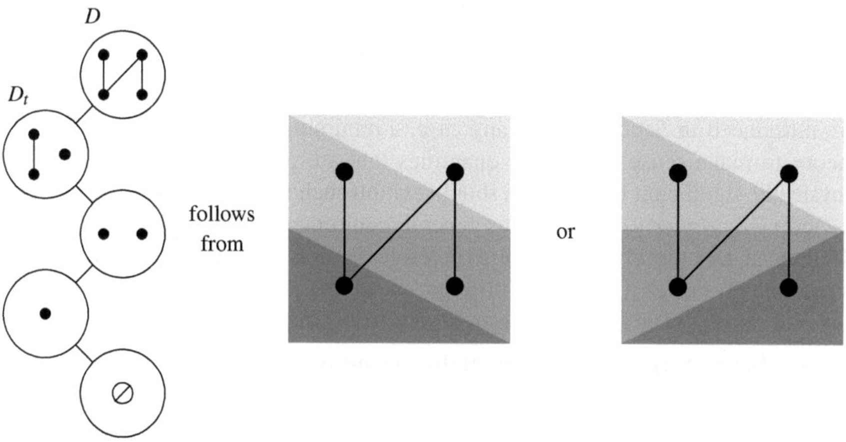

Figure 7.3.2. Distinct refinement types may lead to the same kinematic account.

Extraphysical information; covariance. It should be clear from the foregoing discussion that discrete causal analogues of relativistic frames of reference are rather subtle from an information-theoretic perspective. It is perhaps prudent, therefore, to review *why one should be interested* in such constructions, since the information they introduce is extraphysical. The most important reason, already clear from the context, is that such constructions are useful for describing evolutionary processes, and are therefore automatically relevant to applications of discrete causal theory. More generally, however, the systematic practice of invoking "arbitrary" or "conditional" information as a tool for studying "natural" or "absolute" information is important throughout mathematics and theoretical physics. A basic example of this practice is to invoke a *choice of local coordinates* on a manifold, as one routinely does in relativity. Another example of particular importance in theoretical physics is to fix a *choice of gauge* in a field theory. An elementary mathematical example is to specify a *choice of basis* in a vector space. In all such cases, one *deliberately imposes arbitrary choices as a means to a computational or descriptive end,* then discards these choices once the end is achieved.

One of the costs of this practice is the necessity of verifying that the results one derives from a theory *do not depend on arbitrary choices.* Turning this around, one may ask which quantities in a theory remain *invariant* when one switches to a different set of arbitrary choices. In some cases, the physical building blocks of a theory may be largely determined by such invariance properties. For example, in conventional quantum field theory, "possible particle types" may be classified via representations of the *Poincaré group* P, which may be viewed, from the "passive viewpoint," as effecting changes of reference frame on Minkowski spacetime $\mathbb{R}^{3+1}$.[3] It is possible to formulate interesting discrete causal analogues of this approach using the ideas introduced above. In particular, since each frame of reference corresponds to a refinement of the relativistic causal relation $\prec_{\text{GR}}$ on $\mathbb{R}^{3+1}$, elements of the Poincaré group may be associated with ordered pairs of refinements, or more precisely, with *equivalence classes of ordered pairs of refinements.* This suggests an avenue toward generalizing the Poincaré group, along with the associated phenomenology, in the discrete causal context.[4] The details are rather messy, but the rudiments of this idea are introduced in Section 8.4. In any case, a minimum requirement for a physical theory to make sense is that the quantities that are assumed to encode absolute, physically-significant information should exhibit such invariance. In this book, I use the term *covariance* to describe this general requirement. However, as mentioned in Section 1.8, this terminology requires careful attention to context, because the word "covariance" carries a variety of different meanings in different mathematical and physical settings. The subject of discrete causal covariance is revisited in Section 8.3.

Organizing refinements and kinematic accounts. The first step toward systematically describing evolutionary processes in discrete causal theory, and ultimately predicting the observable consequences of such processes, is to organize the various kinematic accounts describing the evolution of each "permissible" directed set. This leads to the theory of *kinematic preschemes,* which are *"arbitrary families of kinematic accounts,"* and *kinematic schemes,* which are *"well-behaved families of kinematic accounts."* The notion of "good behavior" may be made precise by imposing technical conditions on kinematic preschemes. In this book, I choose to impose the conditions of *accessibility* (A) and the *hereditary property* (H), introduced in Definition 7.4.7. However, these conditions are not very restrictive, and much stronger

[3] To be precise, one is a priori interested in *projective* representations, since physical states are represented by *Hilbert space rays.*

[4] Of course, following this recipe for a *particular* directed set leads to a background *dependent* theory.

conditions are satisfied in most of the cases of principal interest. In particular, the *generational property* (G) is an attractive condition to impose on kinematic preschemes of *acyclic* directed sets. Also, several alternative accessibility conditions are worth considering in the general case. These details are explained in Section 7.4. In the present section, it is advantageous to build up the intuition underlying these notions by focusing on a simple concrete example; in this case, the four-element acyclic directed set $D = (D, \prec)$ whose generational refinement types were illustrated in Figure 7.2.6. One general observation is that the set of refinements of $\prec$ possesses the abstract structure of a partially ordered set, where one refinement R of $\prec$ is taken to precede another refinement R' of $\prec$ if and only if "the second refinement further refines the first," i.e., if and only if R' may be obtained from R by adding new individual relations. Even a causal relation as simple as the one under consideration here possesses a surprising number of refinements, so I will illustrate a coarser "induced partial order" on the corresponding set of generational refinement types. The left-hand diagram in Figure 7.3.3 identifies a pair of such refinement types that are related under this partial order. In this particular example, each refinement type is represented by only one refinement; these specific refinements are illustrated in the right-hand diagram in the figure.

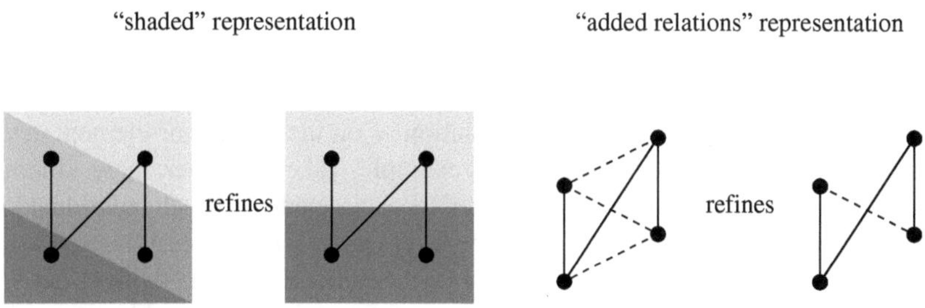

Figure 7.3.3. Refinements of refinements.

Figure 7.3.4 shows all eleven generational refinement types of $\prec$, together with the trivial refinement, organized into a partially ordered set. To avoid clutter, only the skeleton of this set is illustrated. Since the "elements" of this set possess internal structure, it is a "higher-level partially ordered set," analogous to a category, where the analogue of a morphism is an individual relation between two refinement types. Focusing on the discrete causal context, it is also analogous to a kinematic scheme, in the sense that it "exists at the hierarchical level above directed sets." However, it is not as useful as a kinematic scheme for studying evolutionary processes for directed sets. The main reason why this is true, as already mentioned above, is that such refinement types may only be examined a posteriori, i.e., only after the directed set in question has "already evolved."

A different, and more useful, type of "higher-level" structure, which leads directly to the notion of kinematic preschemes and kinematic schemes, may be assembled by organizing families of kinematic accounts into a single diagram. For example,

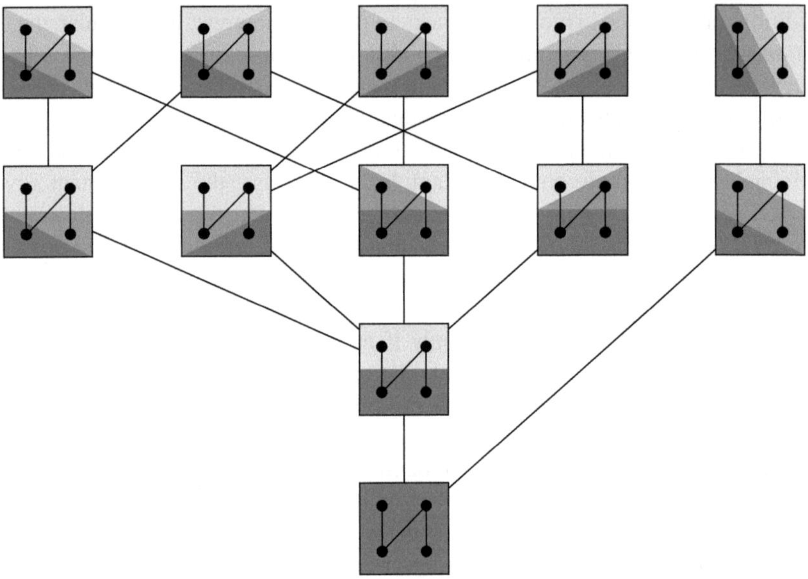

Figure 7.3.4. Partial order on the set of generational refinement types of a causal relation.

Figure 7.3.5 illustrates the family of kinematic accounts associated with the generational refinement types of the causal relation $\prec$ on the same four-element acyclic directed set D appearing in the previous example. Taken together, these kinematic accounts generate the structure of a "higher-level directed set." Each individual kinematic account is represented by a "higher-level causal path" in this directed set. As in the case of refinement types, a category-theoretic analogy is evident here: the "objects" under consideration are subobjects of D, while the "morphisms" are *relationships describing which subobjects may evolve directly into others.* These "morphisms," of course, are special examples of *co-relative histories,* already studied in some detail in Chapter 6. What is new in the present context is the systematic study of *entire families* of co-relative histories. In general, "higher-level *multidirected* sets" are necessary to provide an adequate description of such families, because there generally exist multiple different ways in which one directed set may "evolve directly into another." A simple case of this phenomena is provided by McKay's example of two distinct co-relative histories between an eight-element directed set and a nine-element directed set, illustrated at the level of transitions in Figure 6.4.2 of Chapter 6.

The higher-level directed set appearing in Figure 7.3.5 may be regarded as a "toy example" of a kinematic scheme. Unlike the set of generational refinement types of D, it is *not* a partially ordered set. In particular, it is nontransitive, since a sequence of two co-relative histories, each representing a possible instance of "direct evolution," generally will *not* itself represent "direct evolution." For example, the singleton acyclic directed set cannot "evolve directly" into D, because this would involve the "appearance of a new individual relation between two new elements." On the other

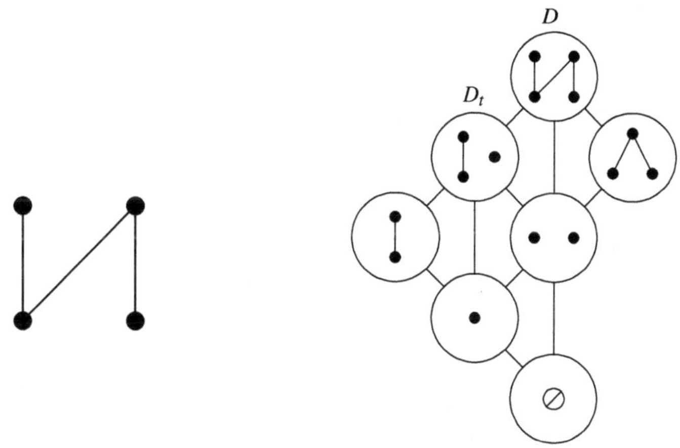

Figure 7.3.5. Directed set $D = (D, \prec)$; kinematic accounts for D, organized into a "higher-level directed set".

hand, *some* nontrivial sequences of co-relative histories can represent "direct evolution," because more than one new element may be taken to appear "simultaneously." For example, the two-element directed set with no relations can "evolve directly" into D, because all of the new individual relations involved in this evolutionary step connect "old elements" to "new elements." The same example illustrates the fact that the higher-level directed set encoding the kinematic accounts of the evolution of D is *not irreducible,* since one could "interpolate" this particular co-relative history with an intermediate stage; for instance, involving D_t.

Up to this point, the discussion has been deliberately informal, but it may easily be made more precise. For example, in the acyclic case, the technical meaning of the statement that a co-relative history $h : D_i \Rightarrow D_t$ represents "direct evolution" is that the set of elements in the complement $D_t - \tau(D_i)$ of a transition τ representing h form an *antichain* in D_t. This statement appears in the definition of the *generational property* (G) in Section 7.4, which is an attractive condition to impose on kinematic preschemes of acyclic directed sets. The motivation for focusing on such "direct" co-relative histories ultimately goes back to the causal metric hypothesis (CMH). Indeed, from the evolutionary viewpoint, one wishes to describe the "relative likelihoods" that different configurations of "new elements" will appear in the "next evolutionary step" in a given process. Since these "new elements" are "supposed to be caused by the existing structure," it is desirable to avoid the "unexplained appearance of new relations between pairs of new elements." More succinctly, it is better to think of "old structure causing new structure," rather than "new structure partially causing itself." Of course, the picture becomes more complicated when causal cycles are permitted at the classical level, but the present chapter continues the practice of Chapter 6, focusing mostly on the acyclic case.

Figure 7.3.6 illustrates the eight different kinematic accounts of D, each represented by a "higher-level chain" in the set of kinematic accounts illustrated in

Figure 7.3.5. Each account is associated with either one or two types of generational refinements of the causal relation $\prec$ on D. The reason why some of the kinematic accounts correspond to two different refinement types is that distinctions between these refinement types arise "before certain symmetries are broken in the evolutionary process." In particular, the two "lower elements" in D are distinguishable, in the sense that there exists no automorphism of D interchanging them; indeed, D is completely rigid. However, there exist three different kinematic accounts of the evolution of D involving the co-relative history from the singleton directed set to the directed set with two elements and no relations, and this target set *does* possess a nontrivial symmetry, which is broken by any evolutionary process leading to D.

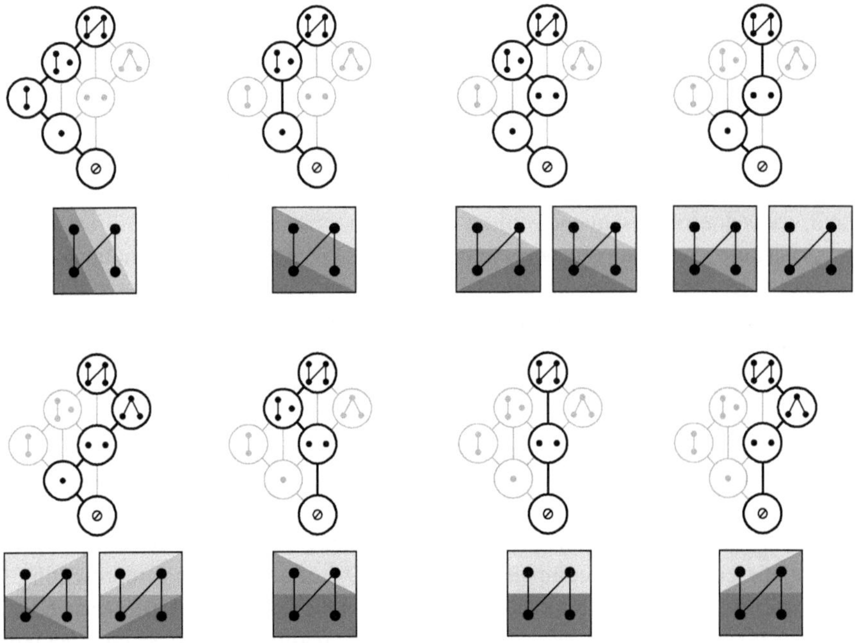

Figure 7.3.6. Eight different kinematic accounts of the evolution of $(D, \prec)$.

It is prudent at this point to dispose of a tempting fallacy suggested by naïve interpretation of the family of "higher-level chains" from $\emptyset$ to D illustrated in Figure 7.3.6, already mentioned obliquely in Section 6.7 in the context of target objects for phase maps. This fallacy arises from a spurious analogy between conventional path integration over a static background model of spacetime, and path summation in the present "higher-level" background independent setting. In particular, one might compare the family of "higher-level chains" in the "higher-level directed set" illustrated in Figure 7.3.5 to the family of particle trajectories depicted in the left-hand diagram in Figure 6.6.2, and might thereby conclude that each individual "higher-level chain" is analogous to such a trajectory. This reasoning is not quite correct, however, because different particle trajectories represent *physically distinct scenarios,* whereas the different kinematic accounts illustrated in Figure 7.3.6 represent *precisely the same physics,* viewed from different, but equally valid, perspectives. The root of this conceptual pitfall lies in the background dependence of conventional quantum theory, which ignores the general-relativistic notion that the "ambient spacetime" interacts in different ways with different particle trajectories. After incorporating these interactions, each trajectory is associated with a *different* spacetime manifold, and the kinematic accounts associated with these trajectories diverge. What *is* correct about the above reasoning, at least from the perspective of this book, is that one *should* assign generalized phases to kinematic accounts, and one *should* apply an appropriate version of path summation to compute generalized quantum amplitudes and related quantities. However, the purpose of this procedure is not to discriminate among "higher-level chains" in a kinematic scheme sharing common initial and terminal histories, since distinctions between such chains are always extraphysical. Rather, the objective is to compare evolutionary processes with *physically distinct* outcomes.

7.4 Kinematic Schemes

Path summation in the background independent context. At this point, the discussion becomes more technical, with earlier concepts revisited in a more precise fashion. According to the *path summation principle* (PSP), introduced in Section 6.5, the theory of *path summation over multidirected sets* is the proper tool for constructing a discrete causal adaptation of Feynman's path summation approach to quantum theory. Path summation over multidirected sets was studied in Chapter 6, but without describing in any detail *which specific types* of multidirected sets are most interesting in applications. In the background independent context, the proper sets over which to perform path summation are "higher-level relation spaces" over "higher-level multidirected sets" called *kinematic schemes,* whose "elements" are directed sets, viewed as models of classical spacetime under the classical causal metric hypothesis (CCMH). This approach combines the theory of *relation space,* developed in Chapter 5, with the principle of *iteration of structure* (IS), introduced in Section 6.3. The elements of these "higher-level relation spaces" are *co-relative histories,* i.e., natural relationships between pairs of classical histories. The theory of kinematic schemes, together with the associated dynamics, formalizes the study of *evolutionary processes* in discrete causal theory, generalizing Sorkin and Rideout's theory of *sequential growth dynamics* for causal sets [SR99]. Kinematic schemes play a structural role analogous to that of categories, while their "relation spaces" play a role analogous to that of morphism categories. The abstract structure of a kinematic scheme is generally more complicated than the structures of its individual constituent histories. For example, kinematic schemes of finite acyclic directed sets generally possess *multidirected,* rather than merely directed, structures, as demonstrated by McKay's example of multiple distinct co-relative histories between a pair of acyclic directed sets differing by a single element, illustrated in Figure 6.4.2 of

Section 6.4. Kinematic schemes including countably infinite acyclic directed sets are even more complicated, generally possessing "higher-level cycles."

Kinematic preschemes; classes of co-relative histories. It is useful here to revisit the distinction between kinematics and dynamics, previously mentioned in Sections 1.8 and 7.3. *Kinematics* is the study of the general types of behavior *permissible* within a theory, while *dynamics* is the study of which behaviors are *determined or favored under specified conditions.* Under the strong interpretation of the causal metric hypothesis (CMH), all physical phenomena are ultimately attributable to causal structure, which is encoded mathematically via directed sets. "Permissible behavior" is therefore defined by specifying a distinguished class $\mathcal{K}$ of directed sets.[5] This class, viewed as a whole, may be endowed with a "higher-level multidirected structure," by specifying a class $\mathscr{H}$ of distinguished co-relative histories between pairs of members of $\mathcal{K}$, which represent possible instances of evolution from one directed set into another. As discussed in Section 7.3, it is advantageous to focus on *direct* evolution whenever possible. An arbitrary choice of $\mathscr{H}$ and $\mathcal{K}$ yields an object $\mathbb{S}$, called a *kinematic prescheme,* which possesses the structure of a "higher-level multidirected set." Each "directed path" in this set represents a partial kinematic account of the evolution of its terminal directed set. Kinematic accounts sharing common initial and terminal histories describe the same physics, and this must be carefully accounted for in the theory. Imposing additional conditions on $\mathscr{H}$ and $\mathcal{K}$ yields a *kinematic scheme,* which is a "well-behaved kinematic prescheme." A variety of possible methods exist for quantifying "good behavior" in this context, and the conditions given in this section are intended only to rule out certain obviously pathological kinematic preschemes that would be a waste of time to study. Much stronger conditions apply in most of the cases of obvious interest. Once a kinematic scheme $\mathbb{S}$ is chosen, dynamical information is supplied by assigning an appropriate phase to each kinematic account in $\mathbb{S}$. Under the strong interpretation of the causal metric hypothesis, the ultimate *expectation* is that there will prove to be a unique "correct" choice of phase map, or family of physically equivalent phase maps, for each "well-behaved" choice of $\mathscr{H}$ and $\mathcal{K}$, thereby "unifying kinematics and dynamics."

Definition 7.4.1. *A* **kinematic prescheme** *is a pair* $\mathbb{S} = (\mathcal{K}, \mathscr{H})$*, where* $\mathcal{K}$ *is a distinguished class of directed sets, and* $\mathscr{H}$ *is a distinguished class of co-relative histories between pairs of members of* $\mathcal{K}$.

The class $\mathcal{K}$ is called the **object class** of $\mathbb{S}$, and the class $\mathscr{H}$ is called the **class of co-relative histories** of $\mathbb{S}$. This definition may be generalized to define kinematic preschemes of *multidirected* sets, but the physical role of such kinematic preschemes is less obvious, since there seems to be no clear physical motivation for multidirected models of classical spacetime. Following the precedent of Section 6.4, I focus attention on *proper, full, originary* co-relative histories, which are represented by equivalence classes of *transitions,* as expressed in Definition 6.4.3. This choice of focus is

[5] As usual, only the isomorphism classes of directed sets are significant. For simplicity, I work with representatives, with the implicit understanding that at most one representative of each isomorphism class appears in $\mathcal{K}$.

"obviously appropriate" for acyclic models of classical spacetime, but may require reconsideration in more general contexts. A kinematic prescheme whose class of co-relative histories $\mathscr{H}$ includes only proper, full, originary co-relative histories is called a *proper, full, originary kinematic prescheme.* All kinematic preschemes in the present section are assumed to be proper, full, and originary, unless stated otherwise. In fact, for reasons already mentioned in Section 7.3, it is often convenient to narrow the focus even further, to consider only kinematic preschemes assembled from co-relative histories involving a "single generation of new elements." Such kinematic preschemes satisfy the *generational property* (G) defined below, which I choose to exclude from the actual definition of a kinematic scheme only to accommodate a broader context in which cycles are permitted at the classical level.

Figure 7.4.2 illustrates a kinematic prescheme $\mathbb{S} = (\mathcal{K}, \mathscr{H})$, whose object class $\mathcal{K}$ consists of all the acyclic directed sets of cardinality at most three, with the exception of the directed set with two elements and no relations. The class $\mathscr{H}$ of co-relative histories of $\mathbb{S}$ exhibits some specific features that are worth pointing out for future reference. Perhaps the most noticeable such feature is the existence of two distinct co-relative histories between the directed sets labeled D_0 and D_3. One of these co-relative histories is represented by a transition mapping the unique element in D_0 to the isolated element in D_3, while the other is represented by a transition mapping this element to the non-isolated minimal element in D_3. These two relationships should clearly be considered "different" from an evolutionary perspective. While this particular instance of multidirected structure may be avoided by choosing a "better" class of co-relative histories, a typical kinematic prescheme will exhibit many instances of multidirected structure that *cannot be directly eliminated in any practical way, without loss of information.*[6]

An obvious "irregularity" in $\mathbb{S}$ is the presence of an isolated directed set, labeled D_2. This is troubling from an evolutionary perspective, because $\mathbb{S}$ provides no nontrivial kinematic account of the evolution of this directed set. This problem may be eliminated by imposing a technical condition called *accessibility* (A). A second questionable property of $\mathbb{S}$ is that some of its member sets have proper, full, originary subsets that do *not* appear in $\mathbb{S}$. In particular, the directed set with two elements and no relations is a proper, full, originary subset of D_3, D_4, and D_5. This is troubling because it suggests dubious scenarios under which certain "evolutionary outcomes" are permitted, while plausible "prior stages of evolution" of these outcomes are forbidden. This problem may be eliminated by imposing a technical condition called the *hereditary property* (H). A third "undesirable" feature of $\mathbb{S}$ is the presence of co-relative histories that "skip generations;" in this case, co-relative histories from the empty set $\varnothing$ to the directed sets D_1 and D_5, which possess nonempty relation sets. As mentioned in Section 7.3, allowing multiple generations of elements to "appear simultaneously" in a kinematic account is uncomfortable, from an evolutionary viewpoint, under the causal metric hypothesis (CMH). The co-relative history $\varnothing \Rightarrow D_1$ is reducible in $\mathbb{S}$; in particular, it may be expressed in terms of a sequence of simpler

[6] Of course, passage to "higher-level relation space" converts this multidirected structure to directed structure.

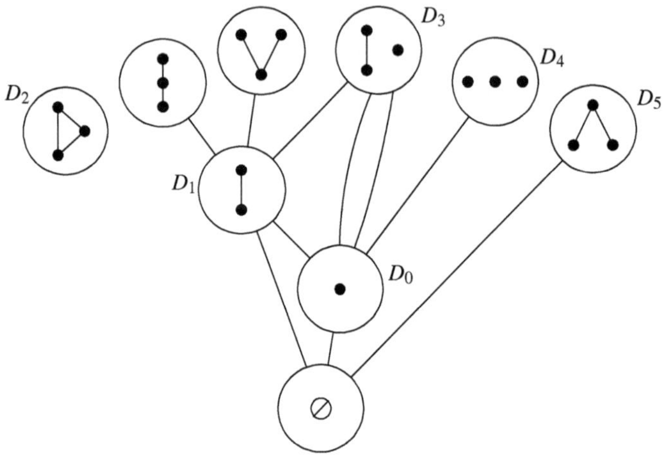

Figure 7.4.2. A kinematic prescheme $\mathbb{S} = (\mathcal{K}, \mathscr{H})$ exhibiting "bad," or "irregular," behavior.

co-relative histories $\oslash \Rrightarrow D_0$ and $D_0 \Rrightarrow D_1$, defining a "reducing chain" for the co-relative history $\oslash \Rrightarrow D_1$ in the higher-level structure of $\mathbb{S}$. From an evolutionary perspective, it is often preferable to discard reducible co-relative histories whenever they "skip generations." Note, however, that the two co-relative histories $D_0 \Rrightarrow D_3$ in the figure are also reducible, but only one of them "skips generations;" namely, the co-relative history mapping the single element in D_0 to the isolated element in D_1. Hence, reducibility itself is not always a good criterion for eliminating a particular co-relative history from consideration. A "worse" problem than "reducible co-relative histories" is posed by *irreducible* co-relative histories that "skip generations," such as the co-relative history from $\oslash$ to D_5.[7] The presence of such co-relative histories allows new, nontrivial causal structure to appear "without explanation." This problem may be eliminated, at least in the acyclic case, by imposing a technical condition called the *generational property* (G).

Each co-relative history in the class $\mathscr{H}$ of co-relative histories of a kinematic prescheme $\mathbb{S} = (\mathcal{K}, \mathscr{H})$ *bundles equivalent transitions together into a single morphism-like entity*, for the purpose of distinguishing physical information from extraphysical "coordinate-like" or "gauge-like" information. Despite this, it is often useful to "remember the internal structures" of co-relative histories, i.e., to preserve explicit knowledge of their constituent transitions, rather than simply treating them as "abstract arrows." One motivation for maintaining this viewpoint is that co-relative histories exhibit "unfamiliar behavior" from a category-theoretic perspective, as discussed in Section 6.4. In particular, the "composition" of two co-relative histories $h : D \Rrightarrow D'$ and $h' : D' \Rrightarrow D''$ is generally *not* a co-relative history $D \Rrightarrow D''$, but rather a *family* of co-relative histories $D \Rrightarrow D''$. The reason for this is that different

[7] Irreducible, that is, within the structure of $\mathbb{S}$ itself. One could add additional histories and co-relative histories to $\mathbb{S}$ that would render this co-relative history reducible, but this would change the kinematics of the theory.

representatives of h and h' may compose to yield inequivalent transitions $D \rightarrow D''$; i.e., equivalence of transitions is not a *congruence relation* under composition in the category $\mathcal{D}$ of directed sets. An example of this phenomenon appeared in Figure 6.4.5 of Section 6.4. Transitions, on the other hand, *are* morphisms in $\mathcal{D}$, so they compose to yield unique transitions. For such reasons, it is best to regard a kinematic prescheme as a "category-like entity" whose "objects" and "morphisms" both encode nontrivial internal structure. Whenever any aspect of this structure is not needed, it may be systematically discarded via a "quotient operation," or a "decategorification-like procedure," as explained below.

Figure 7.4.3 illustrates how a co-relative history in a kinematic prescheme $\mathbb{S} = (\mathcal{K}, \mathscr{H})$ may be regarded as more than merely an "abstract arrow." The left-hand diagram in the figure shows how integer weights may be assigned to the co-relative histories in $\mathbb{S}$, indicating how many different transitions represent each co-relative history. The right-hand diagram shows the two transitions representing the unique co-relative history from the 1-chain in $\mathcal{K}$ to the "v-shaped" directed set drawn directly above it. There exists another co-relative history in $\mathbb{S}$ represented by multiple transitions; namely, the co-relative history from the singleton directed set to the directed set with three elements and no relations. Since the elements of the target set are indistinguishable in this case, there exist three different transitions representing this co-relative history, each mapping the unique element of the cobase to one of the three elements of the target.

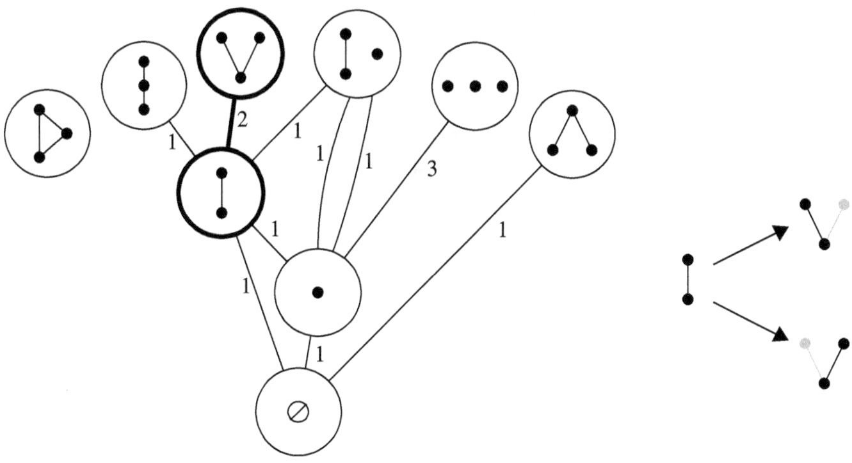

Figure 7.4.3. Remembering the structure of transitions representing co-relative histories in a kinematic prescheme.

Underlying directed sets and multidirected sets of kinematic preschemes. Throughout the book, I have repeatedly referred to kinematic schemes, and more generally, kinematic preschemes, as "higher-level multidirected sets." In fact, a variety of different "higher-level directed or multidirected structures" may be ascribed to a kinematic prescheme $\mathbb{S} = (\mathcal{K}, \mathscr{H})$, with each such structure defined by taking "relations" between members of its object class $\mathcal{K}$ to correspond to one of several types of "morphism-like entities" associated with its class $\mathscr{H}$ of co-relative histories. The "preferred" notion of a "morphism-like entity" in this context is simply a member of $\mathscr{H}$ itself, but variations of this notion are sometimes useful. Here, I describe three such "higher-level structures," in increasing order of complexity.

Definition 7.4.4. *Let $\mathbb{S} = (\mathcal{K}, \mathscr{H})$ be a kinematic prescheme.*

1. *The* **underlying directed set** $\mathcal{D}(\mathbb{S})$ *of $\mathbb{S}$ is the directed set possessing one element $x(D)$ for each directed set D in $\mathcal{K}$, and one relation $x(D_i) \prec x(D_t)$ for each pair of directed sets D_i and D_t in $\mathcal{K}$ with a co-relative history between them in $\mathscr{H}$.*
2. *The* **underlying multidirected set** $\mathcal{M}(\mathbb{S})$ *of $\mathbb{S}$ is the multidirected set possessing one element $x(D)$ for each directed set D in $\mathcal{K}$, one relation $r(h)$ for each co-relative history $h : D_i \Rightarrow D_t$ in $\mathscr{H}$, and initial and terminal element maps i and t sending $r(h)$ to $x(D_i)$ and $x(D_t)$, respectively.*
3. *The* **underlying transition structure** $\mathcal{T}(\mathbb{S})$ *of $\mathbb{S}$ is the multidirected set possessing one element $x(D)$ for each directed set D in $\mathcal{K}$, one relation $r(\tau)$ for each transition τ representing a co-relative history $h : D_i \Rightarrow D_t$ in $\mathscr{H}$, and initial and terminal element maps i and t sending $r(\tau)$* to *$x(D_i)$ and $x(D_t)$, respectively.*

Figure 7.4.5 illustrates the underlying directed set $\mathcal{D}(\mathbb{S})$, the underlying multidirected set $\mathcal{M}(\mathbb{S})$, and the underlying transition structure $\mathcal{T}(\mathbb{S})$ of the kinematic prescheme $\mathbb{S} = (\mathcal{K}, \mathscr{H})$ from Figure 7.4.2.

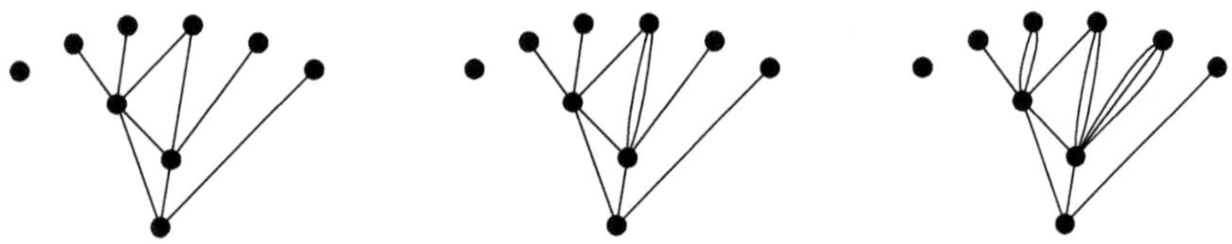

Figure 7.4.5. Underlying directed set, multidirected set, and transition structure of a kinematic prescheme.

Additional "higher-level" directed or multidirected structures may be associated with a kinematic prescheme $\mathbb{S} = (\mathcal{K}, \mathscr{H})$, each defined by preserving or ignoring different amounts and/or types of structure. For example, relations may be chosen to represent *image-fixed co-relative histories,* defined in Section 6.4. However, the three alternatives appearing in Definition 7.4.4 suffice for the purposes of this book. The underlying multidirected set $\mathcal{M}(\mathbb{S})$ of $\mathbb{S}$ is a quotient of its underlying transition structure $\mathcal{T}(\mathbb{S})$, given by identifying relations corresponding to transitions representing the same co-relative history in $\mathscr{H}$. The underlying directed set $\mathcal{D}(\mathbb{S})$ of $\mathbb{S}$, in turn, is a quotient of $\mathcal{M}(\mathbb{S})$, given by identifying all relations corresponding to co-relative histories in $\mathbb{S}$ sharing the same cobase and target. All three sets possess the same abstract element set $\{x(D)|D \in \mathcal{K}\}$. The assignments $\mathbb{S} \mapsto \mathcal{D}(\mathbb{S})$, $\mathbb{S} \mapsto \mathcal{M}(\mathbb{S})$, and

$\mathbb{S} \mapsto \mathcal{T}(\mathbb{S})$ may be viewed more formally in a functorial sense, but I do not elaborate on this viewpoint in this book.

Beginning with a kinematic prescheme $\mathbb{S} = (\mathcal{K}, \mathscr{H})$, passage to the directed set $\mathcal{D}(\mathbb{S})$, or to one of the multidirected sets $\mathcal{M}(\mathbb{S})$ or $\mathcal{T}(\mathbb{S})$, involves "forgetting internal structure" associated with members of $\mathcal{K}$ and $\mathscr{H}$. This may be viewed as a *decategorification-like procedure,* systematically reducing the "category-like entity" $\mathbb{S}$ to an "object-like entity." It is important to emphasize that $\mathcal{D}(\mathbb{S})$, $\mathcal{M}(\mathbb{S})$, and $\mathcal{T}(\mathbb{S})$ all derive their relations from the class of co-relative histories $\mathscr{H}$ of $\mathbb{S}$, *not* from the ambient category $\mathcal{D}$ of directed sets. In particular, there may be pairs of directed sets D and D' in the object class $\mathcal{K}$ of $\mathbb{S}$ related by transitions in $\mathcal{D}$, whose corresponding co-relative histories are *not chosen* to be members of the distinguished class of co-relative histories $\mathscr{H}$, and whose underlying elements $x(D)$ and $x(D')$ are therefore not related in $\mathcal{D}(\mathbb{S})$, $\mathcal{M}(\mathbb{S})$, or $\mathcal{T}(\mathbb{S})$. The abstract structural similarities between underlying directed sets or multidirected sets of kinematic preschemes, and the individual members of their object classes, is the principal motivating example leading to the identification of the principle of *iteration of structure* (IS) as a crucial aspect of discrete quantum causal theory. Application of the relation space functor $\mathcal{R}$ to such underlying directed or multidirected sets renders this similarity "perfect," in the sense that *all structures under consideration are then directed sets "at different levels of structural hierarchy,"* since $\mathcal{R}$ reduces multidirected structure to directed structure. However, a more practical reason for applying $\mathcal{R}$ in this context is that it circumvents the *permeability problem* for maximal antichains in multidirected sets, as proven in Theorem 5.4.3. This enables effective implementation of the abstract quantum causal theory developed in Chapter 6 in terms of path summation over multidirected sets.

It is worth emphasizing that there is generally no sense in discussing "causal relationships," or lack thereof, between pairs of elements y and y' belonging to *different* classical histories $D = (D, \prec)$ and $D' = (D', \prec')$ in a kinematic prescheme $\mathbb{S} = (\mathcal{K}, \mathscr{H})$. Of course, if there happens to exist a proper, full, originary co-relative history $h : D \Rightarrow D'$ in $\mathscr{H}$, represented by a transition $\tau : D \rightarrow D'$, then it makes sense to describe $\tau(y)$ and $\tau(y')$ as being either related or unrelated under $\prec'$, by identifying D with the subobject $\tau(D) \subset D'$. However, for an arbitrary choice of D and D', there generally exists no natural binary relation by which to compare y and y'. One may analyze the situation further, and consider classical histories $D'' = (D'', \prec'')$ whose underlying elements $x(D'')$ are common predecessors or successors of $x(D)$ and $x(D')$ in the underlying multidirected set $\mathcal{M}(\mathbb{S})$ of $\mathbb{S}$, and may thereby compare elements in these histories corresponding to y and y',[8] but I will not elaborate the details. One reason why this subject is interesting is because of certain results pertaining to global causal structure in *ordinary* quantum theory. For example, Oreshkov, Costa, and Brukner [OCB16] have recently proved a result that they describe as demonstrating the existence of *"quantum correlations that cannot be understood in terms of any definite* [global] *causal order."* On the experimental side, Ringbauer and five co-authors [RGCC16] have ruled out a wide variety of "non-local

[8]This idea is abstractly similar to certain methods involving *filtered systems* in category theory.

causal models" involving direct influences between components of a system that are spacelike separated in a conventional sense. As mentioned in Section 5.6, results of this nature might raise serious questions regarding any background *dependent* quantum causal theory; for example, a theory of particles and fields on a fixed causal set. In particular, such results do not suggest merely that the causal structure of relativistic spacetime fails to explain quantum correlations, but that a broad class of acyclic directed models is similarly inadequate. However, in the background independent context, it is not surprising that there is no natural way to compare contributions from different classical histories under a single causal structure. This is ultimately just an issue of hierarchy; quantum theory is about relationships between pairs of *histories,* not pairs of *events.*

Kinematic schemes. As stated in Section 7.3, kinematic schemes are *"well-behaved families of kinematic accounts."* This description reflects the fact that arbitrary kinematic preschemes are somewhat too general to conveniently describe evolutionary processes for directed sets. In particular, they can exhibit certain types of inconvenient behavior, some of which may be easily eliminated without compromising their scope of physical description. Some examples of such behavior in a kinematic prescheme were illustrated in Figure 7.4.2. Here, I elaborate a bit further on what types of behavior are considered undesirable, and offer some technical conditions that may be imposed in order to limit such behavior. Two of these conditions, called *accessibility* (A) and the *hereditary property* (H), are actually built into the definition of a kinematic scheme in Definition 7.4.7, since they are sufficiently general that they may be reasonably imposed in most cases of conceivable physical interest. Additional conditions may be safely imposed in more specific cases, but it is prudent to leave such conditions out of the definition of a kinematic scheme, even though this permits some rather unwieldy kinematic preschemes to qualify as kinematic schemes. In particular, for kinematic preschemes of acyclic directed sets, a technical property called the *generational property* (G) offers some attractive features, as described below.

Figure 7.4.6 elaborates on three specific types of deficiencies from which an arbitrary kinematic prescheme $\mathbb{S} = (\mathcal{K}, \mathscr{H})$ may suffer, all exhibited by the kinematic prescheme illustrated in Figure 7.4.2. First, there may exist no nontrivial kinematic accounts for the evolution of some of the directed sets in the object class $\mathcal{K}$ of $\mathbb{S}$, i.e., no nontrivial "evolutionary pathways" in $\mathbb{S}$ along which such sets may evolve. For example, there is no nontrivial kinematic account for the directed set D_2 in the figure. Part of the physical interpretation of a kinematic prescheme is that the directed sets belonging to its object class are "permissible" according to the theory; this is automatically implied by the use of the adjective "kinematic." It is dubious, however, to regard a directed set as "permissible" if it is not also "realizable." In more detail, it makes no physical sense to include a particular directed set in $\mathcal{K}$ without also including at least one nontrivial kinematic account in $\mathbb{S}$ that "explains the evolution" of this set. Second, $\mathbb{S}$ may contain a given directed set D without containing "earlier stages of evolution" of D, which may be identified in the acyclic case as proper, full originary subobjects of D. For example, the directed set D_5 appearing in the figure may be viewed as the "next stage of evolution" of the gray-shaded directed set with two elements and no relations, but this set does not appear in the object class of $\mathbb{S}$. Third, even if $\mathbb{S}$ contains a kinematic account for a given directed set, it may be a "deficient" account, in which part of the causal structure of the set "appears without any reasonable explanation." For example, the co-relative history between $\oslash$ and D_5 in the figure "skips a generation," in the sense that elements *and* relations appear "simultaneously," and in an irreducible fashion, from a kinematic perspective.

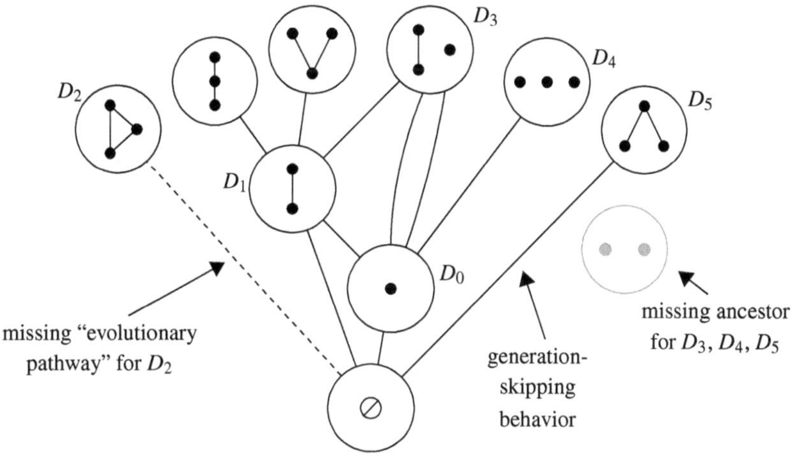

Figure 7.4.6. Pathologies cured by *accessibility* (A), the *hereditary property* (H), and the *generational property* (G).

These three particular types of pathologies may be eliminated by imposing the technical conditions of *accessibility* (A), the *hereditary property* (H), and the *generational property* (G), respectively. The first two of these properties complete the definition of a kinematic scheme in the general case, while the third is offered as a useful auxiliary condition to impose on kinematic preschemes of *acyclic* directed sets. The subsequent discussion connects the rather technical statement of these conditions with the preceding conceptual explanation of what types of behavior they are intended to eliminate. Regarding the notion of accessibility, it is not completely obvious, at least to me, what is the "best" choice of condition to impose, so a number of alternative accessibility conditions are briefly examined below.

Definition 7.4.7. *A kinematic prescheme* $\mathbb{S} = (\mathcal{K}, \mathscr{H})$ *is called a* **kinematic scheme** *if it satisfies the following two properties:*

1. **Accessibility** (A): *Suppose that* D *is a directed set belonging to the object class* $\mathcal{K}$ *of* $\mathbb{S}$. *Then there exists a chain from* $x(\oslash)$ *to* $x(D)$ *in* $\mathcal{T}(\mathbb{S})$.
2. **Hereditary property** (H): *The object class* $\mathcal{K}$ *of* $\mathbb{S}$ *is closed under the formation of proper, full, originary subobjects.*

As discussed above, the intuition underlying the accessibility condition (A) is that *"any permissible directed set D should be realizable in* $\mathbb{S}$*,"* in the sense that $\mathbb{S}$ should exhibit at least one nontrivial kinematic account explaining the evolution of D. Note that accessibility does not depend on the choice to use the underlying transition structure $\mathcal{T}(\mathbb{S})$ of $\mathbb{S}$ in the definition, rather than one of its quotients $\mathcal{D}(\mathbb{S})$ or $\mathcal{M}(\mathbb{S})$. The intuition underlying the hereditary property (H) is that *"if a given directed set D is a permissible model of causal structure, then so are its ancestors,"* i.e., its earlier stages of evolution. As explained in Chapter 6, these "earlier stages" are formalized in terms of proper, full, originary subobjects of D. The condition "proper" could be dropped here, because it merely means that one is actually thinking about "ancestors" of D, rather than D itself. The condition "full" means that *"new structure is compatible with old structure;"* in particular, it is illegal to add "new relations between pairs of existing elements." The condition "originary" means that *"new structure really is new from a causal perspective,"* i.e., "new structure does not affect the past of old structure." It may prove necessary to reexamine this latter condition in the general case in which cycles are permitted, but any resulting modification of the hereditary property would at least *include* proper, full, originary subobjects; hence, this condition seems safe to impose in the general case.

Elimination of "generation-skipping" behavior is feasible only in cases in which well-defined generations may be identified in the first place, which is why the generational property defined below is not necessarily appropriate for kinematic preschemes of directed sets containing cycles. In the acyclic case, however, it offers obvious advantages for modeling evolutionary processes. In particular, it rules out certain "pathological" kinematic schemes permitted by Definition 7.4.7. One example is the kinematic scheme whose objects are finite acyclic directed sets, and whose only co-relative histories are the trivial co-relative histories $\emptyset \Rightarrow D$ for each finite acyclic directed set D.

Definition 7.4.8. *A kinematic prescheme* $\mathbb{S} = (\mathcal{K}, \mathscr{H})$ *satisfies the* **generational property** (G) *if every irreducible relation* $r(\tau)$ *in the underlying transition structure* $\mathcal{T}(\mathbb{S})$ *of* $\mathbb{S}$ *represents a transition adding an antichain of elements to its target.*

In more detail, the generational property means that if $\tau : D_i \rightarrow D_t$ is a transition representing a co-relative history in $\mathbb{S}$, such that the relation $r(\tau)$ in the underlying transition structure $\mathcal{T}(\mathbb{S})$ of $\mathbb{S}$ is irreducible, then the elements belonging to the complement $D_t - \tau(D_i)$ of $\tau(D_i)$ in D_t form an antichain in D_t. The intuition underlying this property is that *"at each stage of evolution, new elements should be caused by existing causal structure, not by new causal structure."* Although I have chosen not to include the generational property in the definition of a kinematic scheme, the kinematic schemes of principal interest in this book all satisfy this property. These include the *Sorkin-Rideout kinematic scheme* $\mathbb{S}_{\text{SR}}$, the *positive sequential kinematic scheme* $\mathbb{S}_{\text{PS}}$, the *generational kinematic scheme* $\mathbb{S}_{\text{GEN}}$, the *foliation scheme* $\mathbb{S}_{\text{FOL}}$, and the *universal kinematic scheme* $\mathbb{S}(\mathcal{A}_{\text{FIN}})$ over the class $\mathcal{A}_{\text{FIN}}$ of finite acyclic directed sets.

It should be understood that the two classes $\mathcal{K}$ and $\mathscr{H}$ making up a kinematic scheme $\mathbb{S} = (\mathcal{K}, \mathscr{H})$ stand on very different footings. The choice of an object class $\mathcal{K}$ embodies a *a hypothesis about the structure of the physical universe,* while the corresponding choice of a class $\mathscr{H}$ of co-relative histories merely represents a convenient perspective for viewing this structure, selected from among a wide variety of equally valid alternatives. This difference is analogous to the difference in general relativity between choosing a class of pseudo-Riemannian manifolds as models of classical spacetime, and choosing particular families of coordinate systems on these manifolds. The arbitrary choice of the class of co-relative histories in a kinematic scheme requires an appropriate notion of *covariance,* already mentioned in Section 7.3. Such a notion is necessary in order to ensure that the physical predictions of the theory do not depend on this choice. This topic is rather subtle, partly because the "frames of reference" encoded by kinematic schemes tend to be much more general and varied than relativistic coordinate systems.[9]

Alternative accessibility conditions. Most of the directed sets studied in this book as possible models of classical spacetime structure belong to the special class $\mathcal{A}_{\text{FIN}}$ of finite acyclic directed sets. There are good reasons to consider imposing a stronger accessibility condition called *relative accessibility* (RA) on kinematic schemes of such sets, despite the fact that at least one interesting example of such a kinematic scheme, namely, the *foliation scheme* $\mathbb{S}_{\text{FOL}}$ introduced in Section 7.8, fails to satisfy this condition. Of course, one may also choose to study kinematic preschemes involving classes of infinite directed sets and/or directed sets containing causal cycles. In particular, the suggested list of axioms for directed sets appearing in Section 4.10 does *not* include any global finiteness condition, and includes the axiom of acyclicity (AC) only in a "provisional" capacity. In situations involving infinite directed sets, alternative accessibility conditions involving only *finite* chains are worth considering. The following definition introduces two such "finite accessibility conditions," along with relative accessibility:

Definition 7.4.9. *Let $\mathbb{S} = (\mathcal{K}, \mathscr{H})$ be a kinematic prescheme. The following are alternative accessibility conditions for $\mathbb{S}$.*

1. **Relative accessibility** (RA): *Suppose that D and D' are members of the object class $\mathcal{K}$ of $\mathbb{S}$. If there exists a transition $\tau : D \rightarrow D'$ in the category $\mathcal{D}$ of directed sets, then there exists a chain from $x(D)$ to $x(D')$ in $\mathcal{T}(\mathbb{S})$.*
2. **Finite accessibility** (FA): *Suppose that D is a finite directed set belonging to the object class $\mathcal{K}$ of $\mathbb{S}$. Then there exists a chain from $x(\emptyset)$ to $x(D)$ in $\mathcal{T}(\mathbb{S})$.*
3. **Finite relative accessibility** (FRA): *Suppose that D and D' are members of the object class $\mathcal{K}$ of $\mathbb{S}$. If there exists a transition $\tau : D \rightarrow D'$ in the category $\mathcal{D}$ of directed sets, such that $\tau(D)$ has finite complement in D', then there exists a chain from $x(D)$ to $x(D')$ in $\mathcal{T}(\mathbb{S})$.*

The relative accessibility condition (RA) is a stronger condition than accessibility (A), because it requires the existence of more evolutionary pathways in $\mathbb{S}$. Indeed, for

[9] Sorkin and Rideout [SR99] discuss *"discrete general covariance"* in the context of sequential growth dynamics. This notion is conceptually related to, but different than, the notion of covariance necessary for the version of discrete causal theory developed in this book.

any "target set" D_t in $\mathbb{S}$, the unique "empty function" $\emptyset \rightarrow D_t$ vacuously satisfies the conditions to be a transition, so relative accessibility guarantees the existence of a chain from $x(\emptyset)$ to $x(D_t)$ in $\mathcal{T}(\mathbb{S})$. The intuition underlying relative accessibility is that *"if D is an ancestor of D_t, then $\mathbb{S}$ should provide an evolutionary pathway between D and D_t."* It is instructive to note how this statement differs from the hereditary property (H), which "accommodates ancestors" in a different manner. In one way, the hereditary property is a stronger condition than relative accessibility, because it guarantees that all directed sets that could be considered "ancestors" of D_t belong to $\mathbb{S}$ whenever D_t belongs to $\mathbb{S}$. In another way, however, it is a weaker condition, because it does not guarantee that D_t may actually be reached in $\mathbb{S}$ from each of these "ancestors." The intuition underlying relative accessibility is rather compelling, since the notion of "inaccessible descendants" could be considered oxymoronic. Of course, the ordinary accessibility condition (A) only guarantees that each directed set D belonging to $\mathbb{S}$ is accessible from $\emptyset$, *not* that D is accessible from each of its ancestors in $\mathbb{S}$.

Another advantage of relative accessibility is that, in concert with the hereditary property (H), it implies the generational property (G) in the acyclic case. Indeed, suppose that $\tau : D_i \rightarrow D_t$ is a transition representing a co-relative history in a kinematic prescheme $\mathbb{S}$ of acyclic directed sets, such that the elements belonging to $D_t - \tau(D_i)$ do not form an antichain in D_t. Let x be a non-maximal element in $D_t - \tau(D_i)$, and consider the directed set $D' = D_t - J^+(x)$, viewed as a full subobject of D_t. Then D_i is a proper, full, originary subobject of D', which is itself a proper, full, originary subobject of D_t. By the hereditary property, D_t belongs to $\mathbb{S}$, and by relative accessibility, there exist sequences of co-relative histories in $\mathbb{S}$ between D_i and D' and between D' and D_t. Therefore, the relation $r(\tau)$ is reducible in $\mathcal{T}(\mathbb{S})$. Despite these advantages, the existence of certain interesting kinematic preschemes that do *not* satisfy relative accessibility is enough, in my view, to exclude this condition as part of the actual definition of a kinematic scheme. For example, as mentioned above, the *foliation scheme* $\mathbb{S}_{\text{FOL}}$, introduced in Section 7.8, satisfies accessibility, but not relative accessibility. This indicates that the viewpoint offered by the foliation scheme incorporates a "preference for some evolutionary pathways over others." For example, $\mathbb{S}_{\text{FOL}}$ "insists on adding isolated elements at the beginning of a kinematic account," rather than at later stages.

The conditions of finite accessibility (FA) and finite relative accessibility (FRA) differ from the corresponding accessibility (A) and relative accessibility (RA) conditions only for kinematic preschemes involving infinite directed sets. In this context, they are strictly weaker conditions, since they require the existence of evolutionary pathways only between pairs of directed sets differing by finite numbers of elements. The reason for considering such conditions is that any chain from $x(D)$ to $x(D')$ in $\mathcal{T}(\mathbb{S})$ is itself finite; hence, ordinary accessibility (A) generally requires certain pairs of directed sets differing by an infinite number of elements to be "connected by a finite number of evolutionary steps" in $\mathbb{S}$. At a naïve level, this might seem like an undesirable property. However, these "evolutionary steps" are not assumed to be irreducible, so ordinary accessibility (A) does not imply actual "generation-skipping" behavior in this context; at worst, it implies a certain redundancy in kinematic schemes involving infinite directed sets, which may not be removable without sacrificing important aspects of structure. A similar situation arises in the simpler context of individual directed sets satisfying the interpolative property, such as $\mathbb{Q}$, whose reducible relations cannot be removed from the picture without sacrificing its essential attributes. It is illuminating to compare accessibility conditions for kinematic preschemes to more familiar notions of accessibility in conventional order theory. For example, one may access any positive integer in a finite number of steps by taking minimal successors, which are connected by irreducible relations, but may *not* access Cantor's first transfinite ordinal in this manner. If one were to insist on modeling accessibility exclusively in terms of irreducible relationships, then the resulting modification of the ordinary accessibility condition (A) could well be considered "too strong" to apply to kinematic preschemes involving infinite directed sets. However, it seems advantageous to allow a degree of reducibility in this context. This approach seems to sacrifice nothing of importance, since interesting kinematic preschemes satisfying finite accessibility (FA) or finite relative accessibility (FRA) may generally be viewed as natural subobjects of kinematic schemes satisfying ordinary accessibility (A).

7.5 Positive Sequential Kinematic Scheme

Sorkin and Rideout's kinematic scheme. The foremost example of a kinematic scheme appearing in the mainstream literature, though described in a manner very different than that of Section 7.4, appears in Sorkin and Rideout's theory of *sequential growth dynamics* of causal sets [SR99]. I denote this kinematic scheme by $\mathbb{S}_{\mathrm{SR}}$, for "Sorkin and Rideout." Under the terminology used in this book, the object class of $\mathbb{S}_{\mathrm{SR}}$ is the class $\mathcal{C}_{\mathrm{FIN}}$ of finite causal sets, and the corresponding class of co-relative histories is the class $\mathscr{H}_1$ of co-relative histories $h : C_i \Rightarrow C_t$ such that C_t *differs from* C_i *by the "addition of a single element."* I am taking some liberties with this description, however. In particular, co-relative histories do not appear in the causal set literature; rather, evolutionary steps are described in terms of transitions in this context. Further, the higher-level structure of $\mathbb{S}_{\mathrm{SR}}$ is described in the literature as a "partially ordered set." However, these conventions seem to overlook the physical equivalence of certain families of transitions representing single co-relative histories in $\mathscr{H}_1$, as well as the essentially multidirected structure of $\mathscr{H}_1$ itself. In what follows, I amend these conventions, while still referring to $\mathbb{S}_{\mathrm{SR}}$ as "Sorkin and Rideout's kinematic scheme," or the "Sorkin-Rideout kinematic scheme."

It is prudent to check that $\mathbb{S}_{\mathrm{SR}}$ really *is* a kinematic scheme, since it is the first serious formal example of such appearing in the book. This is a very easy exercise. The hereditary property (H) holds because any proper, full, originary subobject of a finite causal set is itself a finite causal set. Accessibility (A) follows from the stronger condition of relative accessibility (RA), which may be established by a straightforward induction argument: if $\tau : D \to D'$ is a transition between finite causal sets, then the complement $D' - \tau(D)$ of $\tau(D)$ in D' contains a finite number of elements, none of which belong to the past of $\tau(D)$. Hence, $D' - \tau(D)$ contains at least one maximal element x, so there exists a transition representing a co-relative history h in $\mathscr{H}_1$ that "adds x to its source" to obtain D'. This procedure may then be repeated, replacing D' with the source of this transition. Continuing in this manner, one obtains the individual relations of a chain from $x(D)$ to $x(D')$ in the underlying transition structure $\mathcal{T}(\mathbb{S}_{\mathrm{SR}})$ of $\mathbb{S}_{\mathrm{SR}}$, in reverse order. Since causal sets are acyclic, relative accessibility and the hereditary property together imply that $\mathbb{S}_{\mathrm{SR}}$ also satisfies the generational property (G), but this is easy to see directly. Indeed, each co-relative history in $\mathbb{S}_{\mathrm{SR}}$ adds a single element to its source, which must be an antichain in its target, since causal sets are acyclic. Figure 7.5.1 illustrates a portion of Sorkin and Rideout's kinematic scheme $\mathbb{S}_{\mathrm{SR}}$, showing only causal sets with at most four elements. The odd, sparse appearance of the diagram is due to the fact that it is deliberately constructed by "culling" nontransitive acyclic directed sets from the "more complete" positive sequential kinematic scheme $\mathbb{S}_{\mathrm{PS}}$, described below. Figure 7.5.1 is very similar to Figure 1 in Sorkin and Rideout's paper *Classical sequential growth dynamics for causal sets* [SR99], which is one of the most important references on the subject.

It is important to emphasize that the edges between the large open nodes in Figure 7.5.1 represent co-relative histories, *not* transitions. The underlying transition structure $\mathcal{T}(\mathbb{S}_{\mathrm{SR}})$ of $\mathbb{S}_{\mathrm{SR}}$ is more complicated, since many of the co-relative histories illustrated in the figure are represented by multiple transitions. For example, the co-relative history between D_i' and D_t' is represented by three different transitions, corresponding to the $\binom{3}{2} = 3$ different ways of mapping the two elements of D_i' to two out of the three elements of D_t'. It is necessary to work with $\mathcal{T}(\mathbb{S}_{\mathrm{SR}})$ if one wishes to construct unambiguous sequential labelings of the elements of a given causal set C in $\mathcal{C}_{\mathrm{FIN}}$, since co-relative histories do not specify which elements are "new" and which elements are "old." This is a case in which it is prudent to "remember the internal structures of co-relative histories," by preserving explicit knowledge of their constituent transitions, as discussed in Section 7.4.

An important property of Sorkin and Rideout's kinematic scheme $\mathbb{S}_{\mathrm{SR}}$, which turns out to be a common feature of kinematic schemes in general, is that the overall structure of $\mathbb{S}_{\mathrm{SR}}$ "exhibits," or "acquires" an *asymmetry* that cannot be ascribed to any of its individual constituent histories. This asymmetry, mentioned briefly in Section 6.5, neatly answers the question of why the "past to future" direction is more relevant, from a physical standpoint, than the "future to past" direction, i.e., why one direction for the flow of causal influence is preferred over the other. Any causal set C belonging to $\mathbb{S}_{\mathrm{SR}}$ may be "dualized" to yield a causal set in which "causal influence flows in the opposite direction," and there is nothing about the structure of C itself that allows one to determine "which direction is correct." One may identify pairs of such *causal duals* in Figure 7.5.1; for example, the causal sets labeled D'' and D''' are dual to each other. Some of the sets appearing in the figure are *self-dual;* for example, both D_i' and D_t' are self-dual. By contrast, the higher-level structure of $\mathbb{S}_{\mathrm{SR}}$ clearly distinguishes the "past to future" direction from the "future to past" direction. This property, suitably abstracted, provides very general evidence that *certain fundamental attributes of time are essentially quantum theoretic and background independent* in nature. In particular, this idea does not depend in any

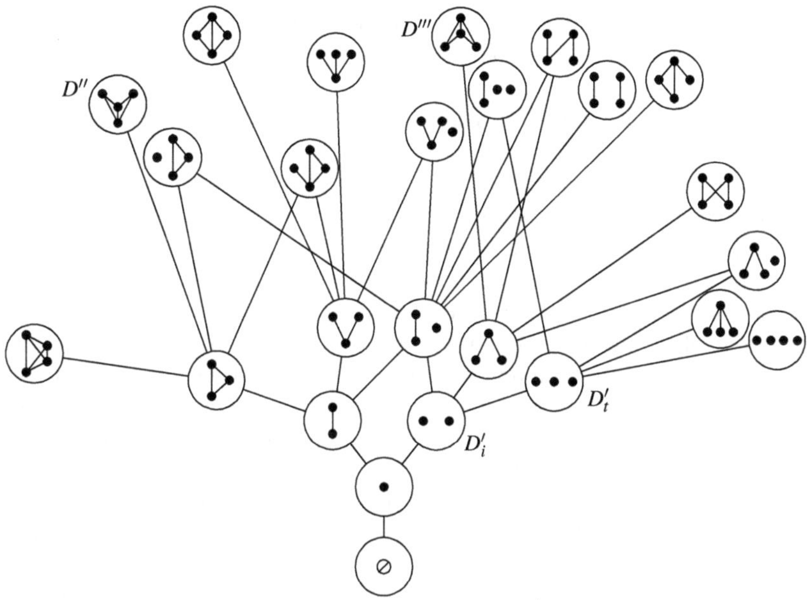

Figure 7.5.1. Portion of Sorkin and Rideout's kinematic scheme $\mathbb{S}_{SR} = (\mathcal{C}_{FIN}, \mathcal{H}_1)$, describing sequential growth of causal sets.

essential way on the fact that individual classical histories are modeled via directed sets; the same general conclusion may be drawn from a variety of other types of configuration spaces.

Positive sequential kinematic scheme. Due to the basic structural differences between causal set theory and the version of discrete causal theory developed in this book, it is necessary here to substitute an "enlarged" kinematic scheme to play a theoretical role analogous to that of $\mathbb{S}_{SR}$; namely, the role of describing a "conservative" version of background independent discrete quantum causal theory. This kinematic scheme, discussed repeatedly at an informal level in previous chapters, is constructed by replacing the class $\mathcal{C}_{FIN}$ of finite causal sets with the larger class $\mathcal{A}_{FIN}$ of finite acyclic directed sets, and enlarging the class of co-relative histories $\mathcal{H}_1$ to accommodate this larger class of objects, without changing its abstract definition or its notation. To be clear, this means that each target set D_t of a co-relative history in $\mathcal{H}_1$ differs from its cobase C_i by a single element, along with a family of relations from "existing elements" to the "new element." I refer to $(\mathcal{A}_{FIN}, \mathcal{H}_1)$ as the **positive sequential kinematic scheme** for finite acyclic directed sets, and denote it by $\mathbb{S}_{PS}$. The terminology comes from the fact that each chain in the underlying transition structure $\mathcal{T}(\mathbb{S}_{PS})$ of $\mathbb{S}_{PS}$ defines a partial sequential labeling of the elements of its terminal acyclic directed set.[10] In fact, this sequential description may be carried

[10]The analogous statement for the underlying multidirected set $\mathcal{M}(\mathbb{S}_{PS})$ of $\mathbb{S}_{PS}$ is not quite true, because co-relative histories generally do not distinguish between "old" and "new" elements.

further: since the object class $\mathcal{A}_{\mathrm{FIN}}$ of $\mathbb{S}_{\mathrm{PS}}$ includes finite acyclic directed sets of arbitrary size, one may also use the same scheme to construct sequential labelings of the elements of countably infinite "limit sets" defined via "infinite chains in $\mathbb{S}_{\mathrm{PS}}$," in a sense made precise in Section 7.9. This includes certain types of star infinite directed sets; in particular, it includes all causal sets. $\mathbb{S}_{\mathrm{PS}}$ is a kinematic scheme by essentially the same arguments used above for Sorkin and Rideout's kinematic scheme $\mathbb{S}_{\mathrm{SR}}$. In a similar manner, it also satisfies the generational property (G).

The reason for the statement that the positive sequential kinematic scheme $\mathbb{S}_{\mathrm{PS}}$ describes a "conservative" version of quantum theory is that its class of objects $\mathcal{A}_{\mathrm{FIN}}$ is the "smallest" class that may be used without significantly changing the underlying assumptions of the theory, or imposing unwieldy additional axioms. For example, the smaller object class of finite causal sets has already been shown to raise concerns in this context, due to the limitations associated with the axiom of transitivity (TR). Alternative choices of object classes may defined in terms of the "minimally modified causal set theory" of Definition 4.10.2, but this definition employs a cumbersome catalogue of axioms, some lacking any compelling justification. Even smaller object classes may be defined by explicitly restricting the valences of elements; for example, one may recover theories similar to causal dynamical triangulations by fixing a uniform valence. However, such choices raise immediate concerns regarding approximate local Lorentz invariance in the classical theory. In the other direction, one may generalize $\mathbb{S}_{\mathrm{PS}}$ by enlarging its object class to include infinite directed sets, or directed sets containing cycles, or even star-infinite directed sets, such as causal sets induced by global sprinklings into Minkowski spacetime $\mathbb{R}^{3+1}$. Of course, some such sets are already included in the picture as limits of $\mathbb{S}_{\mathrm{PS}}$, as described above. In any case, the choice to focus on $\mathbb{S}_{\mathrm{PS}}$ as a source of examples throughout the book should not be viewed as definitive, but merely "within the realm of obvious interest." A portion of $\mathbb{S}_{\mathrm{PS}}$ is illustrated in Figure 7.5.2. As in the case of Sorkin and Rideout's kinematic scheme $\mathbb{S}_{\mathrm{SR}}$, only sets with at most four elements are shown. The nontransitive acyclic directed sets "missing" from Sorkin and Rideout's kinematic scheme are highlighted in gray.

The large numbers 0 through 4, and the dashed curves appearing in the figure, illustrate how each member set D of $\mathbb{S}_{\mathrm{PS}}$ may be assigned a natural-number label matching its cardinality, or equivalently, matching the common length of all originary chains terminating at $x(D)$ in the underlying multidirected set $\mathcal{M}(\mathbb{S}_{\mathrm{PS}})$ of $\mathbb{S}_{\mathrm{PS}}$. The usefulness of such labelings partly explains the emphasis on generational refinement types in Sections 7.2 and 7.3. In terms of abstract "higher-level structure," these same numbers signify generations in $\mathcal{M}(\mathbb{S}_{\mathrm{PS}})$. Generations 0 through 3 differ from the corresponding portion of Sorkin and Rideout's kinematic scheme $\mathbb{S}_{\mathrm{SR}}$ by only a single element, corresponding to the nontransitive chain of length three. However,

(Footnote 10 continued)
Figure 7.4.3 provides an example of this; *either* of the maximal elements in the "v-shaped" target history could be viewed as the "new element" in this case.

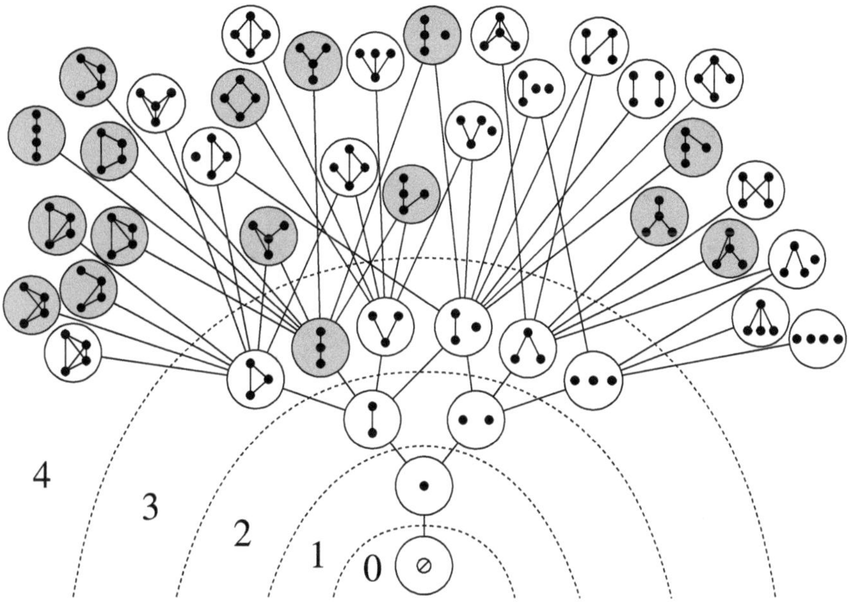

Figure 7.5.2. Portion of positive sequential kinematic scheme $\mathbb{S}_{PS} = (\mathcal{A}_{FIN}, \mathcal{H}_1)$.

about half of the fourth generation of $\mathcal{M}(\mathbb{S}_{PS})$ consists of elements representing nontransitive acyclic directed sets, and subsequent generations are dominated by them. The addition of these sets "cures" the *Kleitman–Rothschild pathology* [KR75] of $\mathbb{S}_{SR}$, mentioned repeatedly since Chapter 3.

Families of kinematic accounts in $\mathbb{S}_{PS}$. Due to the technical conditions of *accessibility* (A) and the *hereditary property* (H), a kinematic scheme is "large enough" and "possesses enough structure" to describe the evolution of all "permissible" directed sets, i.e., all directed sets contained in its object class. For example, Figure 7.5.3 illustrates how the positive sequential kinematic scheme $\mathbb{S}_{PS}$ encodes the eight kinematic accounts of the evolution of the four-element directed set D studied in Sections 7.2 and 7.3. The left-hand diagram in the figure reproduces the family of kinematic accounts for D, first illustrated in Figure 7.3.5. In the present context, however, the "reducible co-relative histories" contributing to the family are indicated by dashed lines. The right-hand diagram illustrates the sense in which these kinematic accounts are encoded by $\mathbb{S}_{PS}$. The reducible co-relative histories involved here are *not* considered to be part of the basic structure of $\mathbb{S}_{PS}$, since they are not sequential in nature; i.e., each of these co-relative histories "adds more than one new element" to its cobase. However, they are *generated* by $\mathbb{S}_{PS}$, in a manner analogous to the generation of a transitive relation $\prec_{tr}$ by the causal relation $\prec$ on an acyclic directed set $(D, \prec)$. More succinctly, $\mathbb{S}_{PS}$ *"contains the skeleton of the family in the left-hand diagram."*

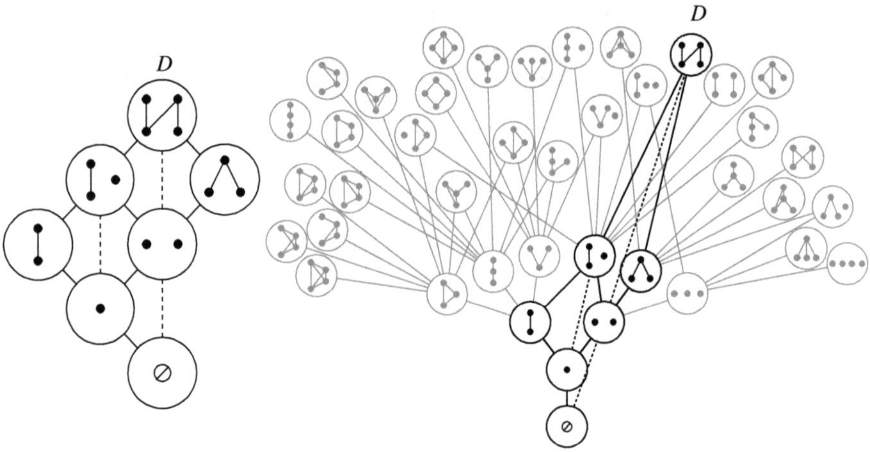

Figure 7.5.3. Family of kinematic accounts for D; the skeleton of this family appears as a subobject in the positive sequential kinematic scheme $\mathbb{S}_{\mathrm{PS}}$.

Underlying multidirected structure of $\mathbb{S}_{\mathrm{PS}}$. Statements making implicit reference to the "higher-level structure" of $\mathbb{S}_{\mathrm{PS}}$, such as the remarks about "reducible co-relative histories" in the previous paragraph, may be made precise in terms of the underlying multidirected set $\mathcal{M}(\mathbb{S}_{\mathrm{PS}})$ of $\mathbb{S}_{\mathrm{PS}}$, or its corresponding underlying directed set $\mathcal{D}(\mathbb{S}_{\mathrm{PS}})$ or transition structure $\mathcal{T}(\mathbb{S}_{\mathrm{PS}})$. Here, I focus on $\mathcal{M}(\mathbb{S}_{\mathrm{PS}})$, which is illustrated in the left-hand diagram in Figure 7.5.4. The diagram shows the portion of $\mathcal{M}(\mathbb{S}_{\mathrm{PS}})$ corresponding to the portion of $\mathbb{S}_{\mathrm{PS}}$ appearing in Figure 7.5.2.

The multidirected set $\mathcal{M}(\mathbb{S}_{\mathrm{PS}})$ is countably infinite, star finite, and acyclic. The portion of $\mathcal{M}(\mathbb{S}_{\mathrm{PS}})$ shown in the figure does not exhibit any instances of multiple relations between individual pairs of elements, since these only begin to appear around the seventh and eighth generations. The fact that the object class $\mathcal{A}_{\mathrm{FIN}}$ of $\mathbb{S}_{\mathrm{PS}}$, and hence the set of elements of $\mathcal{M}(\mathbb{S}_{\mathrm{PS}})$, is countable, was proven in Section 3.8. Star finiteness and acyclicity of $\mathcal{M}(\mathbb{S}_{\mathrm{PS}})$ follow from the finite cardinality of the member sets of $\mathcal{A}_{\mathrm{FIN}}$, together with the definition of a co-relative history. Underlying multidirected sets of kinematic schemes typically suffer from the same structural inconveniences that plague "element spaces" in general; in particular, the problem of permeability of maximal antichains. $\mathcal{M}(\mathbb{S}_{\mathrm{PS}})$ is no exception to this rule; for example, the three black nodes connected by the dashed lines in the left-hand diagram in Figure 7.5.4 represent a maximal antichain in $\mathcal{M}(\mathbb{S}_{\mathrm{PS}})$, which is permeated by the chain from $x(\emptyset)$ to $x(D)$. As usual, the dashed lines are merely a visual aid, not part of the structure of $\mathcal{M}(\mathbb{S}_{\mathrm{PS}})$. This indicates that passage to relation space remains a crucial tool in the context of kinematic schemes. The right-hand diagram in the figure illustrates a portion of the relation space $\mathcal{R}\big(\mathcal{M}(\mathbb{S}_{\mathrm{PS}})\big)$ over $\mathcal{M}(\mathbb{S}_{\mathrm{PS}})$. To reiterate

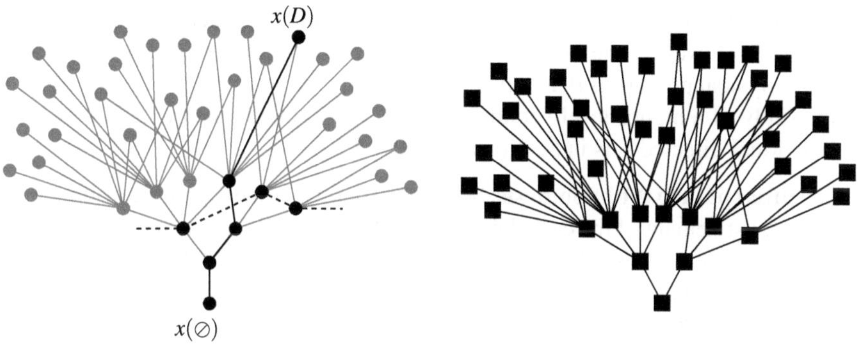

Figure 7.5.4. Portion of the underlying multidirected set $\mathcal{M}(\mathbb{S}_{\mathrm{PS}})$ of the positive sequential kinematic scheme $\mathbb{S}_{\mathrm{PS}} = (\mathcal{A}_{\mathrm{FIN}}, \mathscr{H}_1)$, illustrating the permeability problem; corresponding portion of the relation space $\mathcal{R}(\mathcal{M}(\mathbb{S}_{\mathrm{PS}}))$ over $\mathcal{M}(\mathbb{S}_{\mathrm{PS}})$.

the physical interpretation involved here, each element of $\mathcal{R}(\mathcal{M}(\mathbb{S}_{\mathrm{PS}}))$ represents a co-relative history $h : D_i \Rrightarrow D_t$ between two acyclic directed sets D_i and D_t, which "adds a single element" to its cobase D_i.

7.6 Quantum Causal Metric Hypothesis

Statement of the hypothesis. Just as the classical causal metric hypothesis (CCMH) takes the properties of classical spacetime to arise from directed structure, so the quantum causal metric hypothesis (QCMH) takes the properties of quantum spacetime to arise from *higher-level multidirected structure,* following the principle of *iteration of structure* (IS) described in Section 6.3. The theory of kinematic schemes enables a precise statement of this idea.

Definition 7.6.1. Quantum causal metric hypothesis (QCMH). *Quantum spacetime may be modeled via kinematic schemes of directed sets.*

As noted in Section 7.4, the class $\mathscr{H}$ of co-relative histories of a kinematic scheme $\mathbb{S} = (\mathcal{K}, \mathscr{H})$ involves a *choice of viewpoint,* analogous to a choice of reference frame in relativity. Hence, if one can identify the "correct" class $\mathcal{K}$ of classical histories, the quantum causal metric hypothesis may be realized via *any* suitable kinematic scheme with object class $\mathcal{K}$.[11] To achieve uniqueness, and to avoid inconsistency, one must then specify how to transform information back and forth between different kinematic schemes. Alternatively, uniqueness may be achieved by fixing a particular

[11]The qualifier "suitable" reflects the fact that the conditions of *accessibility* (A) and the *hereditary property* (H) appearing in the definition of a kinematic scheme are not very restrictive; in practice, one might wish to impose stronger conditions.

kinematic scheme, such as the positive sequential kinematic scheme $\mathbb{S}_{PS}$ described in Section 7.5. However, one must be very careful not to ascribe physical significance to extraphysical properties that depend on the choice of scheme. This problem leads to the study of a higher-level version of *covariance,* already mentioned at the level of individual directed sets in Sections 7.3 and 7.4.

Conceptual path to the hypothesis. Section 2.2 describes the reasoning leading from the general statement of the causal metric hypothesis (CMH) to the more precise and specific statement of the classical causal metric hypothesis (CCMH), expressed in terms of directed sets. To summarize, if one accepts the philosophy of the general hypothesis, i.e., that *"physics is about cause and effect,"* then the classical version of the hypothesis follows more or less inevitably from the self-evident observation that individual instances of cause and effect involve ordered pairs of events.[12] The class of directed sets is the "obvious" choice of mathematical objects for modeling families of such ordered pairs, without making additional assumptions regarding global structure. In the present section, I chart out a similar conceptual path to the quantum causal metric hypothesis (QCMH), with kinematic schemes subsuming the role of individual directed sets. The basic features of the resulting theory are already clear: kinematic schemes possess natural "higher-level multidirected structure," to which one may apply the abstract quantum theory developed in Chapter 6, which is formulated in terms of path summation over multidirected sets. This leads to a background independent discrete causal analogue of Feynman's path summation approach to ordinary quantum theory.

Since many other possible approaches to quantum theory exist, and since the essence of quantum physics remains much more obscure than that of classical physics, it is worthwhile to revisit some of the factors favoring this particular approach over others. In particular, the question of how to mathematically express the conceptual content of the causal metric hypothesis in a manner suitable for quantum physics is much less straightforward than the corresponding classical question, and the version given above is not the only conceivable alternative. What seems certain from the outset is that any reasonable attempt to answer this question must involve some version of *superposition.* As explained in Section 1.1, classical physics involves *specific permissible physical scenarios,* which are identified with *specific directed sets* in the discrete causal context. Quantum physics, on the other hand, involves *all permissible physical scenarios.* Hence, on very broad grounds, one begins the quest for a suitable quantum-theoretic version of the causal metric hypothesis by considering *families of directed sets.* The general idea of superposition calls for *combining the information in these families* in a harmonious way, and this leads to the consideration of natural relationships between pairs of their members, i.e., to the *co-relative histories* defined in Section 6.4. As explained in Section 6.1, this broad idea remains recognizable even under radical technical departures, such as nonlinearity. Co-relative histories combine naturally to yield kinematic accounts, which fit together naturally to yield kinematic schemes, as described in Sections 7.3 and 7.4. Hence, even without further

[12] This is true, at least, under the reductionist paradigm.

embellishment, superposition in discrete causal theory brings one very close to the theory of kinematic schemes.

However, what seems to single out kinematic schemes as the *only* reasonable choice for describing basic structure in discrete quantum causal theory is the *need to model evolutionary processes* in a framework involving superposition. The general precedent for the theory of co-relative histories as a natural application of Grothendieck's *relative viewpoint* (RV) renders kinematic schemes attractive from an abstract perspective, but it is the *physical* motivation for co-relative histories that is truly compelling in the discrete causal context. No other obvious mechanism for encoding the steps of evolutionary processes, or for properly modeling the "if-then" questions of principal interest in applications, seems evident in this setting. Indeed, as described in Section 6.4, the cobase of a proper, full, originary co-relative history supplies a natural discrete causal representation of "initial conditions," while the target supplies the corresponding representation of "terminal conditions." Numerical or algebraic quantities measuring the "likelihood" or "favorability" of co-relative histories are therefore roughly analogous to the entries of the *S-matrix* in conventional quantum field theory; in particular, such quantities are always relevant, regardless of what specific dynamical details are assumed to guide the intervening evolutionary processes. If one accepts this general reasoning, then the basic structural features of discrete quantum causal theory are uniquely determined, and the only remaining questions are technical in nature. These include the questions of exactly *how one should define co-relative histories in terms of morphisms,* and exactly *what types of behavior should distinguish kinematic schemes from among the larger class of kinematic preschemes.* Sections 6.4 and 7.4 offer, in my opinion, a plausible answer to the first of these questions, and a treatment of the second sufficiently broad to include all cases of obvious interest.

Conceptual pitfalls involving "paths" in kinematic schemes. However, the very structural properties that render kinematic schemes so appealing also offer tempting interpretive fallacies. Perhaps the most obvious such fallacy, already mentioned in Section 7.3, is to regard different "paths between a pair of directed sets D and D' in a kinematic scheme $\mathbb{S}$" as describing *different physics,* when they actually describe *different points of view regarding the same physics.* A closely-related fallacy is to view the abstract "past" and "future" of a given directed set D in $\mathbb{S}$ as representing its "past and future" in some physical sense, when in fact *D is itself a classical history,* with its own internal notions of past and future. For example, the directed sets represented by the elements in $J^-(x(D))$ in the underlying multidirected set $\mathcal{M}(\mathbb{S})$ of $\mathbb{S}$, do not "influence D," except in a metaphorical sense; rather, they are special subobjects of D. Paths in $\mathcal{M}(\mathbb{S})$ terminating at $x(D)$ do not represent "different histories of D," or different "influences reaching D;" indeed, these notions do not even make sense. What such paths *do* represent are *different kinematic accounts of the evolution of D*, as described in Sections 7.3 and 7.4.

Another related conceptual pitfall involves the *motivation* for performing path summation over kinematic schemes, as outlined in Section 7.7. Feynman's path summation approach to quantum theory is often referred to as the "sum over histories approach," which is a reasonable description in the background dependent setting. This is because different paths in a fixed spacetime background *do* represent physically distinct scenarios, i.e., different classical histories, unlike different kinematic accounts of a particular classical history in a kinematic scheme. On this basis, one might ask why the "correct" discrete causal analogue of Feynman's path integral is not merely a *sum over a configuration space of directed sets,* without consideration of "higher-level multidirected structure." After all, sums over more conventional models of classical spacetime, such as geometries, are common in other approaches to fundamental spacetime structure.

A number of considerations contribute to answering this question. First, Feynman's path integral could just as easily be called a *"sum over kinematic accounts that happen to represent distinct histories due to background dependence."* In particular, there is nothing about Feynman's theory that suggests that it is *wrong* to employ a more general summation procedure involving paths that may or may not represent distinct histories. Second, under any reasonable set of assumptions, summation over a "non-directed" configuration space $\mathcal{K}$ of directed sets may always be expressed *in terms of* path summation over a kinematic scheme $\mathbb{S}$ with object class $\mathcal{K}$, so there is no danger of actually losing descriptive power by working with kinematic schemes. Third, path summation over a kinematic scheme is *more conducive to answering physically interesting questions* than summation over a "non-directed" configuration space. As described above, it is especially well-suited for describing evolutionary processes, and for answering "if-then" questions involving initial and terminal conditions. Indeed, a common conceptual problem arising in various other approaches to quantum gravity is the awkwardness of dealing with entire histories as "complete objects," without convenient notions of "change," or "incrementation." This is one of the principal motivations for Sorkin and Rideout's theory of *sequential growth dynamics* for causal sets [SR99], which the theory of path summation over kinematic schemes generalizes. In this sense, the *S-matrix-like approach* to discrete quantum causal theory mentioned in Section 6.4 is "better than summation over structureless configuration spaces," since it enables treatment of initial and terminal conditions, but is "not as good as path summation over a kinematic scheme," because it fails to facilitate detailed description of complex evolutionary processes. Fourth, the *basic structural appeal* of iteration of structure (IS) and co-relative histories must be accorded some weight, even though aesthetic considerations are of secondary importance.

7.7 Path Summation Over a Kinematic Scheme

Path functionals for kinematic schemes. A preliminary step in building a background independent version of discrete quantum causal theory via "path summation over a kinematic scheme $\mathbb{S}$" is to define *appropriate path-dependent quantities* to "sum," or more generally, to combine, to obtain generalized quantum amplitudes, values of state functions, and related quantities. For clarity, I pause to spell out an "obvious" technical detail: the "paths" involved in this context are *not* morphisms of directed sets with target $\mathbb{S}$. This would make no sense, because $\mathbb{S}$ exists at a higher level of structural hierarchy than individual directed sets. Rather, these paths are morphisms into one of the natural abstract directed or multidirected sets associated with $\mathbb{S}$. For example, an important special case of obvious physical interest involves actual summation, in a target ring T, of phases associated with chains in the relation space $\mathcal{R}(\mathcal{M}(\mathbb{S}))$ over the underlying multidirected set $\mathcal{M}(\mathbb{S})$ of $\mathbb{S}$. It may sometimes be convenient to use a slightly different object, such as the relation space $\mathcal{R}(\mathcal{T}(\mathbb{S}))$ over the underlying transition structure $\mathcal{T}(\mathbb{S})$ of $\mathbb{S}$. It is natural to begin by studying cases in which the specified phase map respects the algebraic structures involved, in the sense that *the phase of a product of paths is equal to the product of the phases.* One may also restrict attention to finite subobjects of the directed or multidirected sets under consideration. For example, a simple and interesting case is given by focusing on multiplicative phase maps from the chain concatenation semicategory over a finite subobject of $\mathcal{R}(\mathcal{M}(\mathbb{S}))$ into T. In this context, any chain γ may be expressed uniquely as a product $r_1 \sqcup r_2 \sqcup \ldots \sqcup r_n$ of individual relations $r_1, r_2, \ldots, r_n$, each representing a co-relative history, and any phase map Θ is generated by a relation function $\theta : \mathcal{R}(\mathcal{M}(\mathbb{S})) \to T$. I focus mostly on such simple cases in what follows.

Toy relation functions. A variety of natural relation functions may be associated with directed or multidirected sets encoding the higher-level structure of kinematic schemes. Perhaps the most obvious choices for such relation functions are "toy functions" that measure some simple form of *entropy* associated with co-relative histories, in the sense that they count straightforward *multiplicities* of some kind. One such toy function is the relation function θ_1 that counts the *number of different ways in which the cobase associated with a given relation may evolve into the corresponding target.* The left-hand diagram in Figure 7.7.1 illustrates the values of θ_1 for the first three generations of the positive sequential kinematic scheme $\mathbb{S}_{PS}$. The labels h and k are included for future reference.

Figure 7.7.1. Simple examples of relation functions for the positive sequential kinematic scheme $\mathbb{S}_{PS}$.

To be precise, I reiterate the detail that θ_1 is really a function on the relation set $\mathcal{R}(\mathcal{M}(\mathbb{S}_{\mathrm{PS}}))$, even though the diagram shows the relevant part of $\mathbb{S}_{\mathrm{PS}}$ itself. At least for the small portion of $\mathbb{S}_{\mathrm{PS}}$ shown here, θ_1 does not look very interesting. In particular, only one of the relations illustrated in the diagram maps to a value other than 1; namely, the relation corresponding to the co-relative history h between the directed sets D_i' and D_t. This relation is assigned a multiplicity of 2, because the "new element added by h" may be taken to appear in the future of either of the two elements of D_i'. From a more sophisticated viewpoint, this multiplicity represents the $\mathbb{Z}_2$ symmetry-breaking of a transition representing h, as described in Section 6.3. Ultimately, this "global-symmetry-related origin" of θ_1 explains its relatively "uninteresting" nature in a much more general context, due to the generic *rigidity* of directed sets. This topic is revisited in more detail below.

Another toy relation function defined in terms of multiplicities is the relation function θ_2 that counts the *number of different ways in which the cobase of a co-relative history may be recognized as a proper, full, originary subobject of its target.* The right-hand diagram in Figure 7.7.1 illustrates the values of θ_2 for the first three generations of $\mathbb{S}_{\mathrm{PS}}$. Here, the picture looks slightly more interesting, essentially because some of the target sets involved, such as U, V, and D_t', possess nontrivial automorphism groups, which permute the images of the corresponding cobases under the various transitions representing the co-relative histories inducing the relations in $\mathcal{R}(\mathcal{M}(\mathbb{S}_{\mathrm{PS}}))$. For example, there are three different transitions representing the co-relative history $D_i' \Rightarrow D_t'$, each mapping the two elements of D_i' to two of the three elements of D_t'. These three images form a single orbit in the power set of D_t' under the action of the automorphism group $\mathrm{Aut}(D_t')$ of D_t', which is isomorphic to the symmetric group S_3. As in the case of θ_1, however, the relation function θ_2 involves global symmetries of the directed sets contributing to each co-relative history in $\mathbb{S}_{\mathrm{PS}}$, and therefore encounters the same basic rigidity issues.

A third such relation function θ_3, also involving global symmetries, simply assigns to each relation in $\mathcal{R}(\mathcal{M}(\mathbb{S}_{\mathrm{PS}}))$ the *number of transitions representing the corresponding co-relative history.* For example, the co-relative history $h : D_i' \Rightarrow D_t$ is represented by two different transitions, corresponding to the $\mathbb{Z}_2$ symmetry of the cobase D_i', so the value $\theta_3(r(h))$ assigned to the relation $r(h) \in \mathcal{R}(\mathcal{M}(\mathbb{S}_{\mathrm{PS}}))$ is 2. However, $\theta_2(r(h)) = 1$, because the two transitions representing h share the same image in D_t. Alternative relation functions may be defined to encode subtler properties of the underlying transition structure of a kinematic scheme. For example, the co-relative histories $h : D_i' \Rightarrow D_t$ and $k : D_i' \Rightarrow V$ are each represented by two transitions, but the transitions representing h assign physically distinct roles to the images of the two elements of D_i', while both transitions representing k assign the same roles to the images of each element. From a mathematical perspective, this is reflected in the fact that h breaks the $\mathbb{Z}_2$ symmetry of D_i', while k preserves this symmetry. Note that all three of the relation functions θ_1, θ_2, and θ_3 take positive integer values, and therefore do not satisfy the "equal magnitudes" condition suggested by Feynman's phase map. However, one may easily construct "associated relation functions," such as $r \mapsto e^{i\theta_1(r)}$ or $r \mapsto e^{i/\theta_1(r)}$, that *do* satisfy this condition, just as Feynman's phase map "re-targets" real-valued functionals to the circle S^1.

Relation functions and rigidity. A notable feature of the relation functions θ_1, θ_2, and θ_3 is that they depend on the structure of the *causal Galois groups* discussed in Section 6.3. In the simple cases examined in the present section, the groups involved are the causal Galois groups of the entire directed sets in each co-relative history, so the relation functions under consideration involve the global symmetries of these sets. From a physical standpoint, this means that information "reaching back to the beginning of time" is associated with each "evolutionary step," which is not necessarily realistic.[13] Nevertheless, relation functions of this kind provide reasonable toy examples, for at least three reasons. First, they depend only on the class $\mathscr{H}$ of co-relative histories of the kinematic scheme $\mathbb{S} = (\mathcal{K}, \mathscr{H})$ under consideration, and are therefore consistent with the quantum causal metric hypothesis (QCMH). Ultimately, the information contained in the phase maps generated by these relation functions derives from the causal relations of the individual directed sets belonging to $\mathcal{K}$. Second, these functions are easy to compute for small finite directed sets. Third, although their values are simply elements of a ring, these functions depend in a direct and obvious way on more-detailed algebraic structures; namely, the causal Galois groups. This points the way toward illustrative examples of "higher-level relation functions," defined in this case by assigning the *causal Galois groups themselves* to relations in $\mathcal{R}(\mathcal{M}(\mathbb{S}))$, rather than merely the cardinalities of these groups. Such "higher-level relation functions" are of interest in more general settings, because there is no obvious justification for believing that the intricate information content of co-relative histories can be suitably reduced to the individual ring elements produced by ordinary path summation. Instead, relation functions, and their corresponding phase maps, may ultimately be replaced by *generalized functors,* while literal path sums may be replaced by "higher-level evaluations," such as *generalized category-theoretic limits.* This brings into focus Isham's suggestion of considering *"generalized truth values"* [IS11], rather than mere amplitudes or probabilities, in the context of quantum gravity.

From a purely mathematical perspective, an obvious reason why the particular relation functions θ_1, θ_2, and θ_3 are unlikely to play more than an illustrative role in serious versions of discrete quantum causal theory is because of the generic rigidity of directed sets, demonstrated in the finite acyclic case by Bender and Robinson [BR88]. This property implies that in many cases, the values of such functions will be equal to 1 "almost everywhere," since the corresponding automorphism groups are "almost always trivial." The same is true for any relation function defined in terms of global symmetries. In particular, such functions have values equal to 1 almost everywhere in the special case of the positive sequential kinematic scheme $\mathbb{S}_{\mathrm{PS}}$. However, rigidity does not apply to every interesting class of directed sets. For example, directed sets of "large spatial but small causal size" are more likely to possess nontrivial automorphism groups, particularly under certain reasonable

[13] Of course, as discussed in Section 4.5, causal sets induced by global sprinklings into pseudo-Riemannian manifolds typically exhibit influences "reaching back to the beginning of time." However, such idealized objects are not viewed as physically realistic. In Section 8.2, I discuss the possibility that multiple generations of elements in a directed set may "remain relevant" in a more-limited role; namely, by breaking symmetries of the "most-recent" generation.

assumptions about spatiotemporal locality. In Section 8.2, I briefly describe how special subobjects of directed sets chosen to fit this description may be used to define more realistic relation functions, and hence, more realistic phase maps. Fortuitously, these phase maps involve a notion of entropy closely related to the physical entropy involved in conventional formulations of the second law of thermodynamics.

Co-relative kinematics; path summation. Given a phase map Θ generated by a relation function θ on the relation space $\mathcal{R}(\mathcal{M}(\mathbb{S}))$ over the underlying multidirected set $\mathcal{M}(\mathbb{S})$ of a kinematic scheme of directed sets $\mathbb{S}$, one may proceed to construct path sums, and related quantities of physical interest, as described in Chapter 6. This provides a precise realization of the quantum causal metric hypothesis (QCMH), by describing *how* the properties of quantum spacetime arise from the structure of $\mathbb{S}$. The formal aspects of this approach are more or less automatic, but faithful organization and interpretation of the corresponding physical information is a subtle matter. Here, I sacrifice some generality in the interest of manageability. The picture may be easily broadened to incorporate arbitrary phase maps, or to utilize other directed or multidirected sets associated with $\mathbb{S}$, such as its underlying transition structure. I also continue to abuse terminology by referring to "path summation over $\mathbb{S}$," when I actually mean path summation over an associated directed or multidirected set. I actually "codify" this abuse of terminology in Definition 7.7.3.

It is necessary to adapt the abstract description of path summation presented in Chapter 6 to the more specific present context of kinematics and dynamics for families of directed sets. This involves assigning a precise meaning to the conceptual idea of "kinematic accounts" discussed in Section 7.3. Let $\mathbb{S} = (\mathcal{K}, \mathscr{H})$ be a kinematic scheme, and let D and D' be directed sets belonging to the object class $\mathcal{K}$ of $\mathbb{S}$, such that there exists a chain γ from $x(D)$ to $x(D')$ in the underlying multidirected set $\mathcal{M}(\mathbb{S})$ of $\mathbb{S}$.[14] Such a chain defines a "partial kinematic account," which supplies a partly extraphysical description of the evolution of D into D'. The special case where $D = \emptyset$ gives a "complete kinematic account" of the evolution of D', from the perspective of $\mathbb{S}$.

Definition 7.7.2. *Let $\mathbb{S} = (\mathcal{K}, \mathscr{H})$ be a kinematic scheme, with underlying multidirected set $\mathcal{M}(\mathbb{S})$, and let D and D' be directed sets belonging to $\mathcal{K}$.*

1. *A chain γ in $\mathcal{M}(\mathbb{S})$ with initial element $x(D)$ and terminal element $x(D')$ is called a* **co-relative kinematics** *in $\mathbb{S}$ from D to D'.*
2. *A co-relative kinematics from $\emptyset$ to D' is called a* **(complete) kinematic account** *of the evolution of D.*

The reason why a co-relative kinematics γ from $\emptyset$ to D' is called a "complete" kinematic account is because it describes the evolution of D' "from beginning to end," in an obvious sense. This does *not* automatically imply that γ provides a "good" or "adequate" evolutionary description; for example, if $\mathbb{S}$ contains the trivial co-relative

[14] If $\mathbb{S}$ satisfies the condition of *relative accessibility* (RA) then this is true whenever there exists a transition $D \to D'$ in the ambient category $\mathcal{D}$, since the intervening directed sets belong to $\mathbb{S}$ by the hereditary property (H).

history $\emptyset \Rightarrow D'$, then the single relation $x(\emptyset) \prec (D')$ qualifies as a kinematic account under Definition 7.7.2. A equivalent way to describe a kinematic account is as an *originary chain in $\mathcal{M}(\mathbb{S})$ terminating at D'*. In this context, "originary" means "beginning at the element $x(\emptyset)$ corresponding to the empty set $\emptyset$." At a rough intuitive level, a co-relative kinematics from D to D' provides a generalized "frame of reference" for a particular co-relative history $h : D \Rightarrow D'$.[15] Of course, the class of refinements of the causal relation on D' provides more detailed "frames of reference," as described in Section 7.2. An alternative is to work with the underlying transition structure $\mathcal{T}(\mathbb{S})$ of $\mathbb{S}$, instead of the underlying multidirected set $\mathcal{M}(\mathbb{S})$ of $\mathbb{S}$. Each chain γ in $\mathcal{M}(\mathbb{S})$ corresponds to a family of chains in $\mathcal{T}(\mathbb{S})$, since each co-relative history contributing to γ is represented by a family of transitions. While it is "safest" work with $\mathcal{M}(\mathbb{S})$, from the perspective of avoiding physically meaningless distinctions, it may sometimes be convenient to work with $\mathcal{T}(\mathbb{S})$ from a computational perspective, since morphisms compose in a more convenient manner than do co-relative histories. A chain from $x(D)$ to $x(D')$ in $\mathcal{T}(\mathbb{S})$ corresponds to a generational refinement type of the causal relation on D', in which "relativity of simultaneity" is broken by arbitrary choices of succession among causally unrelated subsets of directed sets "lying between D and D' in $\mathbb{S}$." It is important to note that multiple such chains may correspond to a given refinement type. For example, the chain of co-relative histories illustrated in Figure 7.3.2 corresponds to four different chains of transitions, since the second and third co-relative histories in the chain are each represented by two different transitions. However, this chain corresponds to only two different generational refinement types. The following definition renders precise the notion of "path summation over a kinematic scheme." As described above, this is really an abuse of terminology, but it is convenient, and is unlikely to cause serious confusion.

Definition 7.7.3. *Let $\mathbb{S} = (\mathcal{K}, \mathscr{H})$ be a kinematic scheme, and let $\Theta : \mathcal{R}(\mathcal{M}(\mathbb{S})) \rightarrow T$ be a phase map from the relation space over the underlying multidirected set $\mathcal{M}(\mathbb{S})$ of $\mathbb{S}$ into a target object T possessing an additive operation. Then a T-valued* **path sum** *over $\mathbb{S}$ is a sum*

$$\sum_{\gamma \in \Gamma} \Theta(\gamma)$$

over a class Γ of co-relative kinematics in $\mathbb{S}$.

As explained in Section 7.6, it is important to emphasize the distinction between path sums and "sums over histories." Feynman's path integrals may be described in *either* manner, since distinct paths on a fixed background spacetime represent distinct classical histories. However, in background independent discrete quantum causal theory, classical histories are merely individual directed sets, not sequences of co-relative histories. Hence, a "path sum over $\mathbb{S}$" is better characterized as a "sum over

[15]There is an interesting analogy here arising from algebraic geometry: a co-relative kinematics may be compared to a point in a *flag variety,* since it involves a "distinguished sequence of subobjects in a space."

kinematics" than as a "sum over histories." This is because any *particular* "path in $\mathbb{S}$" involves extraphysical information, analogous to the arbitrary information treated by *gauge theory* in more conventional settings. Sorkin's quote regarding "upward directed paths in a tree of causal sets" in the context of quantum measure theory [SO12], cited in Section 6.1, foreshadows the general idea of co-relative kinematics. Generalizing this picture, any directed path γ in any underlying directed or multidirected set associated with $\mathbb{S}$ may be called a "kinematics" in $\mathbb{S}$. In this broader context, γ may not possess initial or terminal elements, so a general "kinematics" need not be "co-relative;" i.e., it need not correspond to a pair of directed sets. Path summation may be extended in an obvious manner to this more general setting. Of course, one must always consider practical issues such as convergence when dealing with path sums.

7.8 Generational Kinematics and Dynamics

Generational kinematics for $\mathcal{A}_{\text{FIN}}$. The positive sequential kinematic scheme $\mathbb{S}_{\text{PS}} = (\mathcal{A}_{\text{FIN}}, \mathscr{H}_1)$ is not the only interesting kinematic scheme whose object class is the set $\mathcal{A}_{\text{FIN}}$ of isomorphism classes of finite acyclic directed sets. A "larger" kinematic scheme with the same object class, which provides closer analogues of relativistic frames of reference than $\mathbb{S}_{\text{PS}}$, is the **generational kinematic scheme** $\mathbb{S}_{\text{GEN}} = (\mathcal{A}_{\text{FIN}}, \mathscr{H}_{\text{GEN}})$, whose class $\mathscr{H}_{\text{GEN}}$ of co-relative histories includes every co-relative history that adds a single *generation,* i.e., an antichain of "new elements," to its cobase, rather than merely a single element. As elaborated later in Sections 8.2 and 8.3, it is often useful to consider the entire "structural increments" of elements *and* relations added by such co-relative histories, up to suitable equivalence.[16] In particular, the automorphism groups of such structural increments, called *generational automorphism groups,* play an important role in the *phase theory* introduced in Section 8.2, under which phase maps are defined in terms of *entropic* considerations.

The generational kinematic scheme $\mathbb{S}_{\text{GEN}}$ provides natural analogues of "spacelike foliations" of members of $\mathcal{A}_{\text{FIN}}$, and also of countably infinite "limit sets" of $\mathcal{A}_{\text{FIN}}$, discussed in more detail in Section 7.9. The underlying multidirected set $\mathcal{M}(\mathbb{S}_{\text{GEN}})$ of $\mathbb{S}_{\text{GEN}}$ is acyclic, but is *not* star finite, since each new generation may possess any positive-integer cardinality. The reason why I refer to $\mathbb{S}_{\text{GEN}}$ as "larger" than the positive sequential kinematic scheme $\mathbb{S}_{\text{PS}}$ is because the class of co-relative histories $\mathscr{H}_1$ of $\mathbb{S}_{\text{PS}}$ is a subclass of $\mathscr{H}_{\text{GEN}}$. It is easy to see that $\mathbb{S}_{\text{GEN}}$ is a kinematic scheme satisfying the generational property (G); indeed, accessibility (A) and the hereditary property (H) are always preserved under the admission of new co-relative histories between members of a fixed object class, and the generational property follows directly from the definition of $\mathscr{H}_{\text{GEN}}$. At the level of underlying multidirected sets,

---

[16] As discussed in Section 8.2, different transitions representing the same co-relative history generally "add" different elements to their common target, so such structural increments are defined only up to isomorphism.

$\mathcal{M}(\mathbb{S}_{\text{PS}})$ is a subobject of $\mathcal{M}(\mathbb{S}_{\text{GEN}})$. The structure of $\mathcal{M}(\mathbb{S}_{\text{GEN}})$ is highly reducible, because the elements of a "new" antichain may be "added" to a given directed set in many different ways; either in a single step, or in a complex sequence of steps. This "excessive reducibility" can be inconvenient, so it is sometimes useful to consider *subschemes* of $\mathbb{S}_{\text{GEN}}$ that are "less reducible." The most obvious such subscheme is the positive sequential kinematic scheme $\mathbb{S}_{\text{PS}}$ itself; another is the *foliation scheme* $\mathbb{S}_{\text{FOL}}$, introduced below.[17]

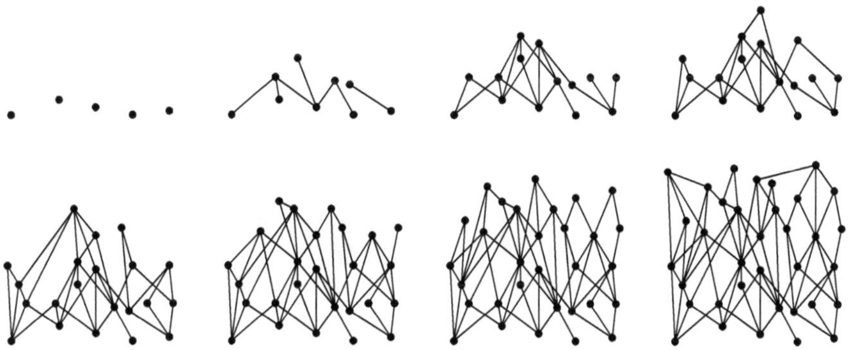

Figure 7.8.1. Generational growth of an acyclic directed set $D = (D, \prec)$.

Figure 7.8.1 illustrates an example of *generational growth* of an acyclic directed set $D = (D, \prec)$; i.e., growth along an evolutionary pathway in the generational kinematic scheme $\mathbb{S}_{\text{GEN}}$. The growth process is arranged graphically "from left to right and from top to bottom," with D appearing at the bottom right. Informally, the figure indicates that five mutually unrelated elements appear in the first generation of growth. These are followed by four elements in the second generation, each of which happens to be related to at least one element in the first generation. However, the structure of $\mathbb{S}_{\text{GEN}}$ does not mandate this type of relationship between consecutive generations in general; rather, "new elements" may be unrelated to any "old element," or they may be related to elements from "much older" generations, without being related to "more recent" generations. For example, the third generation illustrated in Figure 7.8.1 adds five new elements, some of which are related only to elements of the first generation, thereby "skipping" the second generation. As discussed in Section 7.4, it is often desirable to restrict this type of "generation-skipping" behavior; this topic is revisited below. The left-hand diagram in Figure 7.8.2 illustrates another way of representing generational growth, by assigning positive integer labels to the elements in an acyclic directed set D, to indicate the generation to which these elements belong. The reader may notice that this diagram is a reproduction of the right-hand diagram in Figure 5.7.5 of Section 5.7, where generational dynamics was

[17] I do not develop the theory of subschemes formally in this book, but the meaning is simply a *"smaller kinematic scheme embedded in the original kinematic scheme."* This topic is revisited briefly in Section 7.10.

briefly discussed in the context of causal atomic resolution. Such labeling also defines a generational refinement type of D, as described in Section 7.2. The right-hand diagram in Figure 7.8.2 illustrates an alternative description of generational growth for the same directed set, for future reference.

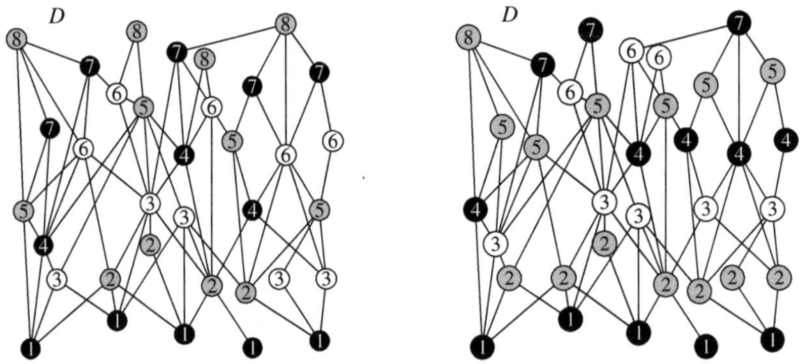

Figure 7.8.2. Two different partitions of an acyclic directed set D into generations.

Foliation scheme. While the generational kinematic scheme $\mathbb{S}_{\mathrm{GEN}}$ has the advantage of a simple definition, it also exhibits inconvenient structural aspects, described in the foregoing discussion as "excessive reducibility" and "generation-skipping." $\mathbb{S}_{\mathrm{GEN}}$ suffers from no actual information-theoretic *deficiency;* indeed, any kinematic scheme with object class $\mathcal{A}_{\mathrm{FIN}}$ may be used to describe evolutionary processes for members of $\mathcal{A}_{\mathrm{FIN}}$. Instead, $\mathbb{S}_{\mathrm{GEN}}$ contains "too much information," in the sense that it offers many redundant descriptions of essentially the same processes. For example, the "first generation" of elements appearing in Figure 7.8.1 could just as well be described as five *different* generations, each containing one element, and this alternative description also fits into $\mathbb{S}_{\mathrm{GEN}}$. Informally, one might express this situation by saying that $\mathbb{S}_{\mathrm{GEN}}$ is "not sufficiently specific to isolate the natural meaning of the term *generation.*" In particular, one might expect that every element of a given generation in a directed set $D = (D, \prec)$ should *actually succeed* some element in the previous generation under the causal relation $\prec$ on D. This particular type of awkwardness may be ameliorated by working with a subscheme $\mathbb{S}_{\mathrm{FOL}} = (\mathcal{A}_{\mathrm{FIN}}, \mathscr{H}_{\mathrm{FOL}})$ of $\mathbb{S}_{\mathrm{GEN}}$, where $\mathscr{H}_{\mathrm{FOL}}$ is the **foliation class**, which includes only co-relative histories that *"add elements related to maximal elements of their cobases."* These "new elements" may *also* be related to non-maximal elements of their cobases, but may not "skip the previous generation." As the notation suggests, the kinematic scheme $\mathbb{S}_{\mathrm{FOL}}$ is called the **foliation scheme,** because its "generations" actually resemble the leaves of foliations of relativistic spacetime in a sense.[18] Like $\mathbb{S}_{\mathrm{GEN}}$, $\mathbb{S}_{\mathrm{FOL}}$ is acyclic, but not star finite. Figure 7.8.3 illustrates a small portion of $\mathbb{S}_{\mathrm{FOL}}$.

[18]Of course, as mentioned in Section 2.7, these generations are generally *not* impermeable; i.e., they are not "discrete Cauchy surfaces."

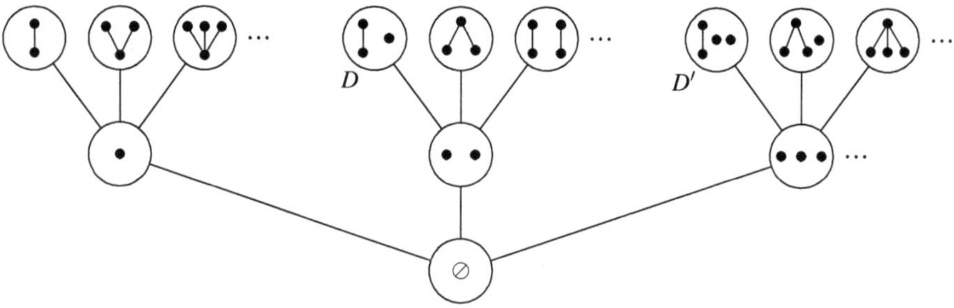

Figure 7.8.3. Portion of the foliation scheme $\mathbb{S}_{\mathrm{FOL}}$.

$\mathbb{S}_{\mathrm{FOL}}$ is a kinematic scheme satisfying the generational property (G), but this may not be a priori obvious from the definition. The hereditary property (H) holds because $\mathbb{S}_{\mathrm{FOL}}$ possesses the same object class as $\mathbb{S}_{\mathrm{GEN}}$, and the generational property follows directly from the definition of $\mathscr{H}_{\mathrm{FOL}}$, but accessibility (A) is worth checking. Let D be a finite acyclic directed set. As one might expect, a chain from $x(\oslash)$ to $x(D)$ in $\mathcal{T}(\mathbb{S}_{\mathrm{FOL}})$ may be constructed by induction. Assign to each element y of D a natural number $n(y)$ encoding the maximal length of all chains in D terminating at y. Assuming that D is nonempty, the set of such natural numbers $n(y)$ consists of a finite interval $[1, \dots, N]$ in $\mathbb{N}$. In the example illustrated in the right-hand diagram of Figure 7.8.2, for instance, $N = 8$. If $N = 1$, then D is an antichain, and there is nothing to prove. In any case, the set σ_n of all elements y in D such that $n(y) = n$ is an antichain in D; indeed, if $n(y) = n$ and there is a chain of length $m > 0$ from y to y' in D, then $n(y') \geq n + m$. It is convenient at this point to re-label D as D_N. Let D_{N-1} be the directed set obtained by removing σ_N from D_N, along with all relations terminating at elements of σ_N. Then D_{N-1} is a proper, full, originary subobject of D, and every element of $\sigma_N = D - D_{N-1}$ is related to a maximal element of D_{N-1} by construction. Hence, by the definition of $\mathbb{S}_{\mathrm{FOL}}$, there exists a transition representing a co-relative history $D_{N-1} \Rightarrow D$ in $\mathscr{H}_{\mathrm{FOL}}$, "adding σ_N to D." Now repeat the process, with $D = D_N$ replaced by D_{N-1}. Since D is finite, this process terminates after a finite number of steps, yielding a chain from $x(\oslash)$ to $x(D)$ in $\mathcal{T}(\mathbb{S}_{\mathrm{FOL}})$. It is worth noting that $\mathbb{S}_{\mathrm{FOL}}$ does *not* satisfy the stronger condition of *relative accessibility* (RA). For example, the directed sets D and D' appearing in Figure 7.8.3 are related by a transition, but there is no evolutionary pathway from D to D' in $\mathbb{S}_{\mathrm{FOL}}$. This suggests that $\mathbb{S}_{\mathrm{FOL}}$, in a sense, "prefers some evolutionary pathways over others." In particular, directed sets appearing at the Nth generation in $\mathbb{S}_{\mathrm{FOL}}$ are $N - 1$ chronons deep.

Figure 7.8.4 illustrates an example of the special type of generational growth described by $\mathbb{S}_{\mathrm{FOL}}$. The "result" of this growth process is the same acyclic directed set $D = (D, \prec)$ arising from the "less well behaved" generational growth process illustrated in Figure 7.8.1. Note how much more "uniform" the generations appear in the new process. This is the same process illustrated via labeling in the right-hand diagram in Figure 7.8.2.

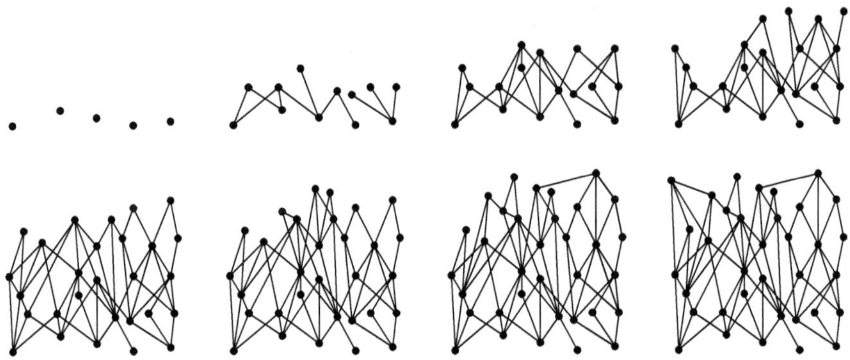

Figure 7.8.4. Generational growth in the foliation scheme $\mathbb{S}_{\mathrm{FOL}}$.

7.9 Completions of Kinematic Schemes

Chain completion of $\mathbb{S}_{\mathrm{PS}}$. In addition to organizing the set of isomorphism classes of finite acyclic directed sets into a higher-level multidirected structure, the positive sequential kinematic scheme $\mathbb{S}_{\mathrm{PS}}$ also encodes the structure of certain isomorphism classes of *countably infinite* acyclic directed sets, corresponding to *equivalence classes of maximal chains* in the underlying transition structure $\mathcal{T}(\mathbb{S}_{\mathrm{PS}})$ of $\mathbb{S}_{\mathrm{PS}}$. These countably infinite sets may be viewed as *limits,* in a generalized category-theoretic sense. At a more elementary level, they are analogous to limits of Cauchy sequences of rational numbers. More precisely, let

$$x(\emptyset) \prec x(D_1) \prec x(D_2) \prec \cdots \prec x(D_n) \prec \cdots$$

be a maximal chain in $\mathcal{T}(\mathbb{S}_{\mathrm{PS}})$, where in this context each precursor symbol $\prec$ is taken to represent a *specific* transition. Under these conditions, each finite acyclic directed set D_n in this chain may be regarded as a subset of the succeeding set D_{n+1}, and each relation in D_n may be regarded as a relation in D_{n+1}. Hence, the union $D_\infty := \cup_{n=1}^{\infty} D_n$ is a well-defined, countably infinite, acyclic directed set. The maximal chain in $\mathcal{T}(\mathbb{S}_{\mathrm{PS}})$ corresponding to D_∞ determines a total labeling of D_∞, which corresponds to a bijective morphism $D_\infty \to \mathbb{N}$ in the category of acyclic directed sets. Conversely, every bijective morphism from a directed set[19] into $\mathbb{N}$ determines a unique maximal chain in $\mathcal{T}(\mathbb{S}_{\mathrm{PS}})$. Hence, the class of maximal chains in $\mathcal{T}(\mathbb{S}_{\mathrm{PS}})$ corresponds bijectively to the class of countably infinite relative directed sets over $\mathbb{N}$.[20] As explained in Section 4.6, such a relative directed set consists of a directed set *together with a morphism* into $\mathbb{N}$; in particular, there are generally many such morphisms for a given source set. Fixing an appropriate equivalence relation

[19]The existence of such a morphism guarantees that the source is in fact a countably infinite acyclic directed set.

[20]As usual, only isomorphism classes are significant.

on maximal chains in $\mathcal{T}(\mathbb{S}_{\mathrm{PS}})$, corresponding to the equivalence relation identifying relative directed sets over $\mathbb{N}$ sharing isomorphic sources, one may define a class of countably infinite acyclic directed sets called *limit sets,* or simply *limits,* of the positive sequential kinematic scheme $\mathbb{S}_{\mathrm{PS}}$. Augmenting $\mathbb{S}_{\mathrm{PS}}$ by its class of limits yields a kinematic prescheme $\overline{\mathbb{S}_{\mathrm{PS}}}$, called the *chain completion* of $\mathbb{S}_{\mathrm{PS}}$. The object class of $\overline{\mathbb{S}_{\mathrm{PS}}}$ is the set of isomorphism classes of countable acyclic directed sets representable as relative directed sets over $\mathbb{N}$, which has cardinality $2^{\aleph_0}$. $\overline{\mathbb{S}_{\mathrm{PS}}}$ is related to $\mathbb{S}_{\mathrm{PS}}$ in roughly the same way as $\mathbb{R}$ is related to $\mathbb{Q}$, due to the analogy between maximal chains and Cauchy sequences, as well as cardinality considerations.

More generally, let $\mathbb{S} = (\mathcal{K}, \mathscr{H})$ be a kinematic scheme, and let $\widetilde{\mathrm{Ch}_{\mathrm{MAX}}}(\mathbb{S})$ be the class of **limits** of $\mathbb{S}$, i.e., the class of directed sets represented by equivalence classes of maximal chains in $\mathcal{T}(\mathbb{S})$.[21] Then the **chain completion** of $\mathbb{S}$ is defined to be the kinematic prescheme $\overline{\mathbb{S}} := \big(\mathcal{K} \cup \widetilde{\mathrm{Ch}_{\mathrm{MAX}}}(\mathbb{S}), \mathscr{H}\big)$, given by enlarging the object class $\mathcal{K}$ of $\mathbb{S}$ to include equivalence classes of maximal chains, without adding any new co-relative histories. The chain completion of a kinematic scheme $\mathbb{S}$ is generally *not* a kinematic scheme under Definition 7.4.7. The addition to $\mathcal{K}$ of directed sets corresponding to maximal chains preserves the *hereditary property* (H) by construction, but fails to preserve *accessibility* (A). Indeed, the limiting objects added to $\mathbb{S}$ in order to obtain $\overline{\mathbb{S}}$ are generally not accessible by any chain in $\mathcal{T}(\overline{\mathbb{S}})$.[22] A generally nontrivial problem is the question of whether or not a given directed set belongs to the chain completion of a given kinematic scheme. For example, the *foliation theorem* of Bleybel and Zaiour [BZ15] may be interpreted in this context, in the special case of partially ordered sets. More generally, this problem is roughly analogous to the determination of whether a given real number is rational, algebraic, or transcendental.

Kinematic preschemes of countably infinite acyclic directed sets. The chain completion $\overline{\mathbb{S}_{\mathrm{PS}}}$ of the positive sequential kinematic scheme $\mathbb{S}_{\mathrm{PS}}$ provides a simple example of a kinematic prescheme including infinite directed sets in its object class. However, these infinite sets are of a very restricted type, even in the star finite context, since they are representable as relative directed sets over $\mathbb{N}$. Further, they are "mere limits," in the sense that no co-relative histories involving these sets, either as cobases or targets, are admitted into the higher-level multidirected structure of $\overline{\mathbb{S}_{\mathrm{PS}}}$. There are good physical reasons to consider more complicated kinematic preschemes including countably infinite directed sets; for example, kinematic preschemes whose object classes are the classes $\mathcal{C}$ and $\mathcal{A}_{\aleph_0,\mathrm{SF}}$ of *all* causal sets, and *all* countable star finite acyclic directed sets, respectively. Such kinematic preschemes possess the same cardinality $2^{\aleph_0}$ as $\overline{\mathbb{S}_{\mathrm{PS}}}$, but are much more complicated in general. For example, they typically contain cycles, such as the "self-transition" of causal sets $-\mathbb{N} \rightarrow -\mathbb{N}$ sending $-n$ to $-n-1$, whose "new" element is the maximal element 0 in the second copy of $-\mathbb{N}$.

---

[21] Care is required when $\mathcal{T}(\mathbb{S})$ includes cycles.

[22] If the condition of accessibility (A) were replaced with the weaker condition of finite accessibility (FA) in Definition 7.4.7, then $\overline{\mathbb{S}}$ *would* be a kinematic scheme. This is one possible argument for imposing finite accessibility instead of accessibility.

7.10 Hyperquantum Theory

"Functors" of kinematic preschemes. In this final section of the chapter, I offer brief sketches of structural entities existing at "higher levels" of hierarchy than the quantum-theoretic level occupied by kinematic preschemes and kinematic schemes. The principle of *iteration of structure* (IS) extends to this more general context, with at least one higher level of hierarchy possessing an obvious claim to physical significance. This *first hyperquantum level* involves *natural relationships between pairs of kinematic preschemes,* understood in the context of Grothendieck's *relative viewpoint* (RV). Such relationships may be viewed as "higher-level morphisms," or "higher-level co-relative histories." Returning to the relativistic analogy, they represent an abstraction and generalization of *coordinate transformations.* From the algebraic perspective, under which kinematic preschemes may be viewed as "generalized categories," natural relationships between pairs of kinematic preschemes may be viewed as "generalized functors." In this book, I do not attempt to undertake any systematic analysis of which structural aspects of such "functor-like entities" are physically relevant in the discrete causal setting, as I did for "morphism-like entities" in Chapters 5 and 6, nor do I give formal definitions. Instead, I briefly discuss a particular class of examples that illustrate why these ideas are interesting. It is worth remarking that one may also posit *subclassical structure,* existing *below* the level of classical theory, by assigning "internal structure" to elements and relations of directed sets, viewed as models of classical spacetime. From this viewpoint, auxiliary structures such as particles and fields on directed sets in background dependent or "partially background independent" versions of discrete quantum causal theory may be viewed as subclassical in nature. However, the strong interpretation of the causal metric hypothesis (CMH) motivates the consideration of *directed* subclassical structure, such as "internal cycles." This idea is briefly revisited in Section 8.7.

Let $\mathbb{S} := (\mathcal{K}, \mathscr{H})$ and $\mathbb{S}' := (\mathcal{K}', \mathscr{H}')$ be kinematic preschemes, such that $\mathcal{K} \subset \mathcal{K}'$ and $\mathscr{H} \subset \mathscr{H}'$. In this setting, there exists a natural "inclusion functor" $\mathscr{I} : (\mathcal{K}, \mathscr{H}) \rightarrow (\mathcal{K}', \mathscr{H}')$. Since only an illustrative sketch is attempted here, I make some rather restrictive simplifying assumptions. First, I assume that the inclusion of object classes is proper, and the inclusion of classes of co-relative histories is full. From a physical perspective, this means that $\mathbb{S}'$ permits a broader variety of classical histories than $\mathbb{S}$, but that "new co-relative histories must involve new histories." Second, I narrow the picture much further by assuming that every object of $\mathcal{K}$ is finite, and every object of $\mathcal{K}'$ is countable. In this special case, the cardinalities of $\mathcal{K}$ and $\mathcal{K}'$ are "typically at most $\aleph_0$ and $2^{\aleph_0}$," respectively. A prototypical example of such an "inclusion functor" is given by the inclusion of the positive sequential kinematic scheme $\mathbb{S}_{\mathrm{PS}}$ into its chain completion $\overline{\mathbb{S}_{\mathrm{PS}}}$. Simpler "toy examples" may be exhibited, for which both $\mathbb{S}$ and $\mathbb{S}'$ are finite. Such an example is illustrated in Figure 7.10.1, where both $\mathbb{S}$ and $\mathbb{S}'$ are finite subschemes of $\mathbb{S}_{\mathrm{PS}}$.

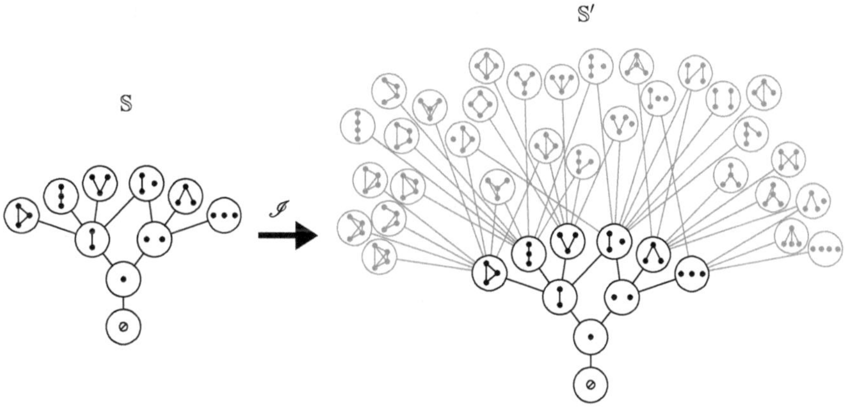

Figure 7.10.1. Inclusion functor $\mathscr{I} : \mathbb{S} \to \mathbb{S}'$ between two finite kinematic schemes.

Under the simplifying assumptions stated above, the "inclusion functor" $\mathscr{I}$ is a higher-level analogue of a transition, since it embeds $\mathbb{S}$ into $\mathbb{S}'$ as a "proper, full, originary subobject." The physical significance of such a relationship is that it encodes an *entire family of evolutionary stages, with varying initial and terminal conditions,* rather than merely encoding a single stage in the evolution of an individual classical history. Such families arise in computational scenarios of obvious importance in background independent discrete quantum causal theory; for example, in determining *generalized quantum amplitudes* of the type defined in Section 6.7. In particular, in the finite case, the "maximal histories" belonging to the source of $\mathscr{I}$ define a *cobase family,* as described in Section 6.4, while the "maximal histories" belonging to the target of $\mathscr{I}$ supply a corresponding *target family.*

Analogies involving number systems. Interesting analogies exist between "functors" $\mathscr{I} : \mathbb{S} \to \mathbb{S}'$ of kinematic preschemes and inclusions of familiar number systems, such as the inclusion of $\mathbb{Q}$ into $\mathbb{R}$. For simplicity, is convenient to restrict attention to the case in which the "source number system" is of cardinality $\aleph_0$; for example, $\mathbb{Z}$ or $\mathbb{Q}$, and the "target number system" is of cardinality $2^{\aleph_0}$; for example, $\mathbb{R}$, $\mathbb{C}$, $\mathbb{H}$, $\mathbb{O}$, or $\mathbb{Q}_p$. One such analogy involves the set $\mathbb{Q} \cap [0, 1]$ of rational numbers in the unit interval $[0, 1]$, represented by irreducible fractions, with binary relation $\prec$ induced by the usual strict total order $<$ on $\mathbb{Q}$, via the so-called *Farey sequences,*[23] or equivalently, via the closely-related *Stern–Brocot sequences,* in which each fraction is related to "the next two irreducible fractions bracketing it under $<$."[24] The resulting acyclic directed set $(T, \prec)$ is a *tree* from its "second generation" onward, and may be identified as a portion of the *Stern–Brocot tree,* with the exception of the single element 0, which is often excluded from the construction in conventional

[23]Farey, in fact, neither discovered the Farey sequences, nor proved their essential properties.

[24]Interestingly, the Riemann hypothesis may be re-stated in terms of Farey sequences; see, for example, the 1924 paper of Franel [FR24].

settings. The Stern–Brocot tree extends to include all positive rational numbers, but only the portion inside $[0, 1]$ is needed for the present illustration. A small portion of $(T, \prec)$ is illustrated in the left-hand diagram in Figure 7.10.2. A maximal chain in $(T, \prec)$ represents a Cauchy sequence of rational numbers in the interval $[0, 1]$, and hence a real number in this interval. Moreover, every real number in $[0, 1]$ is represented by an originary chain in $(T, \prec)$; finite originary chains may be identified with their terminal rational numbers in this context. The real interval $[0, 1]$ may thereby be viewed, roughly speaking, as the "chain completion of its rational subset." It is worthwhile to re-emphasize that the directed structures of both the "source" $\mathbb{Q} \cap [0, 1]$ and "target" $[0, 1]$ in this example are *not* the standard orders induced by $(\mathbb{R}, <)$, although they are induced by $<$ in a "nonstandard way."

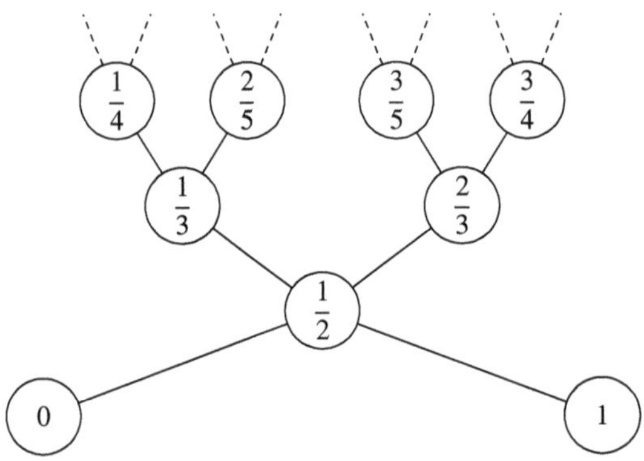

$\mathbb{S} = (\mathcal{K}, \mathscr{H})$	$\mathbb{Q}$
minimal nonempty objects	$\mathbb{Z}$
objects	$\mathbb{Q}$
equivalence classes of originary chains	$\mathbb{R}$

Figure 7.10.2. Portion of the Stern–Brocot tree $(T, \prec)$, with chain completion $[0, 1] \subset \mathbb{R}$; analogies between a countable kinematic prescheme $\mathbb{S}$ and the rational numbers.

In an analogous manner, a maximal chain in the underlying transition structure $\mathcal{T}(\mathbb{S})$ of a kinematic prescheme $\mathbb{S} := (\mathcal{K}, \mathscr{H})$ represents an object in its chain completion $\overline{\mathbb{S}}$. The right-hand diagram in Figure 7.10.2 lists some analogies between a countable kinematic prescheme $\mathbb{S}$, viewed as a "source prescheme for its chain completion," and the rational numbers $\mathbb{Q}$, viewed as a "source number system for the real numbers $\mathbb{R}$." The minimal nonempty objects in $\mathcal{K}$ are analogous to the integers, since integers are the minimal elements in $(T, \prec)$. In some cases, these objects "really are" integers in a sense; for example, in the foliation scheme $\mathbb{S}_{\mathrm{FOL}}$, illustrated in Figure 7.8.3, the minimal nonempty objects are just finite antichains, which may be identified with their positive integer cardinalities. Meanwhile, the objects of $\mathcal{K}$ itself are analogous to rational numbers, since the elements of $(T, \prec)$ are rational. Finally, suitable equivalence classes of originary chains in $\mathbb{S}$ are analogous to real numbers. Finite originary chains may be identified with their terminal objects, and maximal chains represent limit objects belonging to $\overline{\mathbb{S}}$. One obvious distinction between $\mathbb{Q}$ and $\mathbb{S}$ in this context is that a maximal chain in $(T, \prec)$ *may* represent a rational number, while limit objects of $\mathbb{S}$ generally do not belong to $\mathbb{S}$ itself. For example, the maximal chain $0 \prec \frac{1}{2} \prec \frac{2}{3} \prec \cdots \prec \frac{n}{n+1} \prec \cdots$ in $(T, \prec)$ represents the rational number 1, but no maximal chain in $\mathbb{S}_{PS}$ represents a finite acyclic directed set. However, the two constructions remain very similar in many respects.

Kinematic spaces. Having briefly outlined some examples of relationships between pairs of kinematic preschemes, the next logical step is to consider how entire *families* of such relationships may be assembled to yield *hyperquantum structures,* i.e., structures at a higher level of hierarchy than kinematic preschemes. There exist a number of interesting alternatives for defining such structures, but here I sketch just one of these. In Section 7.4, I noted that the two classes $\mathcal{K}$ and $\mathscr{H}$ defining a kinematic scheme $\mathbb{S} = (\mathcal{K}, \mathscr{H})$ do not share equal physical significance. The class $\mathcal{K}$ of objects of $\mathbb{S}$ represents a critical choice of how to model the physical universe, with different choices necessarily implying different physics. The class $\mathscr{H}$ of co-relative histories of $\mathbb{S}$, on the other hand, merely represents a choice of viewpoint, and may be selected in many different ways, subject only to the modest technical conditions of *accessibility* (A) and the *hereditary property* (H), imposed in order to facilitate convenient descriptions of evolutionary processes. The ultimate reason for this difference in status between $\mathcal{K}$ and $\mathscr{H}$ is that *all* co-relative histories between pairs of objects in $\mathcal{K}$ may be reconstructed by examining $\mathcal{K}$ itself; hence, a choice of $\mathscr{H}$ merely distinguishes certain information already present in $\mathcal{K}$. Similar reasoning applies to kinematic preschemes in general.

It is natural in this context to try to *organize* the class of all possible choices of $\mathscr{H}$ corresponding to a fixed choice of $\mathcal{K}$. This is roughly analogous to organizing all frames of reference for Minkowski spacetime $\mathbb{R}^{3+1}$, or all refinements of the causal relation $\prec$ on a directed set $D = (D, \prec)$. However, an important difference in the present case is that there exists one particular choice of $\mathscr{H}$ that "includes all the others;" namely, the class $\mathscr{H}(\mathcal{K})$ of *all* proper, full, originary co-relative histories between pairs of members of $\mathcal{K}$.[25] The underlying directed set of the resulting kinematic prescheme $\mathbb{S}(\mathcal{K}) := \big(\mathcal{K}, \mathscr{H}(\mathcal{K})\big)$ is *transitive,* since a composition of transitions is a transition. In particular, $\mathbb{S}(\mathcal{K})$ is automatically a kinematic scheme, assuming of course that $\mathcal{K}$ satisfies the hereditary property. Given any other kinematic prescheme $\mathbb{S}$ with object class $\mathcal{K}$, there exists a natural "inclusion functor" $\mathscr{I} : \mathbb{S} \rightarrow \mathbb{S}(\mathcal{K})$ of the type described at the beginning of this section; i.e., inclusion of object classes is proper, and inclusion of classes of co-relative histories is full. Hence, the kinematic scheme $\mathbb{S}(\mathcal{K})$ is "maximal" in the class of kinematic preschemes with object class $\mathcal{K}$. From a generalized category-theoretic perspective, $\mathbb{S}(\mathcal{K})$ possesses the "universal property" of being a "terminal object" for the class of such kinematic preschemes, and it is the unique kinematic prescheme with this property.[26] For this reason, I refer to $\mathbb{S}(\mathcal{K})$ as the **universal kinematic scheme** over $\mathcal{K}$. Despite the advantage of uniqueness, $\mathbb{S}(\mathcal{K})$ is generally very cumbersome to work with, since it encodes "many redundant accounts of each evolutionary process."

---

[25] As usual, the reason for confining attention to proper, full originary co-relative histories is that these are most convenient for describing evolutionary processes.

[26] By contrast, total refinements of a nonlinear acyclic causal relation are maximal, but not terminal, because different total refinements do not refine each other.

At the opposite extreme from universal kinematic schemes are "irreducible" kinematic preschemes, in which descriptions of evolutionary processes are "as unique as possible." In some cases, a kinematic scheme may be reduced to an irreducible kinematic scheme by "taking the skeleton," in the sense of discarding reducible transitions. For example, the "skeleton" of the universal kinematic scheme $\mathbb{S}(\mathcal{A}_{\mathrm{FIN}})$ over the set $\mathcal{A}_{\mathrm{FIN}}$ of isomorphism classes of finite acyclic directed sets, is just the familiar positive sequential kinematic scheme $\mathbb{S}_{\mathrm{PS}}$. Since "transitive kinematic schemes" over classes of *finite* directed sets are recoverable from their skeleta, $\mathbb{S}_{\mathrm{PS}}$ inherits a universal property of sorts from $\mathbb{S}(\mathcal{A}_{\mathrm{FIN}})$. This provides a purely structural reason for regarding $\mathbb{S}_{\mathrm{PS}}$ as important, aside from its conceptual convenience, and from the prominence in the literature of closely related structures, such as Sorkin and Rideout's kinematic scheme $\mathbb{S}_{\mathrm{SR}}$. The left-hand diagram in Figure 7.10.3 illustrates a portion of the underlying multidirected set $\mathcal{M}(\mathbb{S}_{\mathrm{PS}})$ of $\mathbb{S}_{\mathrm{PS}}$, while the corresponding right-hand diagram illustrates the corresponding portion of the underlying multidirected set $\mathcal{M}\big(\mathbb{S}(\mathcal{A}_{\mathrm{FIN}})\big)$ of the universal kinematic scheme $\mathbb{S}(\mathcal{A}_{\mathrm{FIN}})$. The cluttered appearance of the latter diagram demonstrates the computational inconvenience of working with universal kinematic schemes in general.

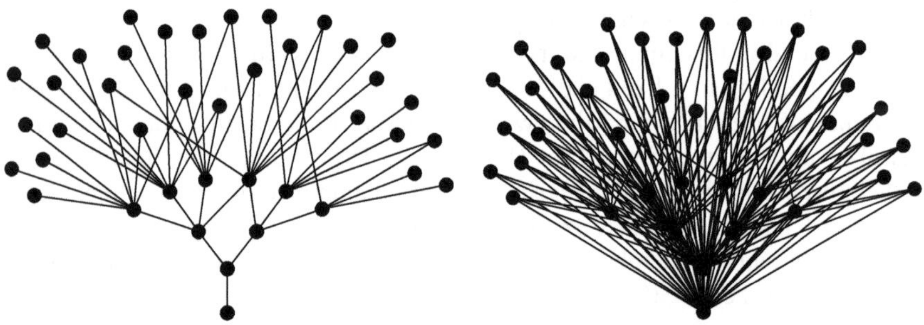

Figure 7.10.3. The positive sequential kinematic scheme $\mathbb{S}_{\mathrm{PS}}$ is the "skeleton" of the universal kinematic scheme $\mathbb{S}(\mathcal{A}_{\mathrm{FIN}})$.

The class of kinematic preschemes with object class $\mathcal{K}$ possesses a natural strict partial order $\prec_{\mathcal{K}}$, defined via proper inclusion of classes of co-relative histories, with the universal kinematic scheme $\mathbb{S}(\mathcal{K})$ as its unique maximal element. This defines a hyperquantum structure $\mathbf{Kin}(\mathcal{K})$, called the **kinematic space** over $\mathcal{K}$, whose "elements" are kinematic preschemes with object class $\mathcal{K}$, and whose "relations" are "functors" between pairs of such kinematic preschemes. At a conceptual level, the kinematic space $\mathbf{Kin}(\mathcal{K})$ *organizes all possible viewpoints regarding all possible evolutionary processes leading to all possible classical histories,* under the assumption that $\mathcal{K}$ represents the "correct choice" of objects for modeling classical spacetime.

References

[CA04] Sean Carroll. *Spacetime and Geometry: an Introduction to General Relativity*. Pearson, Addison Wesley, 2004.

[BR88] Edward A. Bender and Robert W. Robinson. *The Asymptotic Number of Acyclic Digraphs II*. Journal of Combinatorial Theory, Series B, **44**, pp. 363–369, 1988.

[SR99] David Rideout and Rafael Sorkin. *Classical sequential growth dynamics for causal sets*. Physical Review D, **61**, 2, 024002, 2000. arXiv preprint: http://arxiv.org/pdf/gr-qc/9904062v3.pdf.

[OCB16] Ognyan Oreshkov, Fabio Costa , and Časlav Brukner. *Quantum correlations with no causal order*. Nature Communications **3**, 1092, 2012.

[RGCC16] Martin Ringbauer, Christina Giarmatzi, Rafael Chaves, Fabio Costa, Andrew G. White, and Alessandro Fedrizzi. *Experimental test of nonlocal causality*. Science Advances, **2,** 8, 2016.

[KR75] D. J. Kleitman and B. L. Rothschild. *Asymptotic Enumeration of Partial Orders on a Finite Set*. Transactions of the American Mathematical Society, **205**, pp. 205–220, 1975.

[IS11] Christopher Isham. *Topos Methods in the Foundations of Physics*. In Deep Beauty: Understanding the Quantum World through Mathematical Innovation, edited by Hans Halvorson. Cambridge University Press, 2011. arXiv preprint: http://arxiv.org/pdf/1004.3564v1.pdf.

[SO12] Rafael Sorkin. *Toward a Fundamental Theorem of Quantal Measure Theory*. Mathematical Structures in Computer Science, **22**, 05 (special issue), pp. 816-852, 2012. arXiv preprint: http://arxiv.org/pdf/1104.0997v2.pdf.

[BZ15] Antonio N. Bernal and Miguel Sanchez. *A general theorem on temporal foliation of causal sets*. Preprint, 2015. arXiv preprint: http://arxiv.org/pdf/1508.01052v1.pdf.

[FR24] J. Franel. *Les suites de Farey et le problme des nombres premiers*.Nachrichten von der Gesellschaft der Wissenschaften zu Göttingen, Mathematisch-Physikalische, pp.198-201, 1924. URL: http://www.digizeitschriften.de/dms/img/?PID=GDZPPN00250653X.

Chapter 8
Toward Phenomenology

Abstract. This chapter summarizes the highlights of previous chapters, and describes further ideas necessary for studying discrete causal phenomenology. It also sketches relationships between discrete causal theory and other modern approaches to fundamental physics, and offers a few remarks about the genesis of the book. Section 8.1 summarizes the analysis of the foundations of discrete causal theory carried out in Part I of the book, with the benefit of the new perspectives afforded by Part II. This section also summarizes new technical methods developed throughout the book. Section 8.2 gives an overview of the important topic of *phase theory* for evolutionary processes in discrete causal theory, and suggests methods of constructing phase maps in terms of a generalized version of the second law of thermodynamics. Section 8.3 examines the subject of *covariance,* which involves the management of "extraphysical information," such as the information associated with an arbitrary choice of coordinate system or gauge in conventional physics. In the discrete causal context, arbitrary choices are involved in the selection of a kinematic scheme, and in choosing among different kinematic accounts describing the evolution of a particular classical history. Section 8.4 discusses the discrete causal emergence of the *Poincaré group* of isometries of Minkowski spacetime. The *active* and *passive viewpoints* regarding this group are shown to generalize in completely different ways. The notion of *rotation* is shown to emerge via an analogy involving *Thomas–Wigner rotation* in relativity. Section 8.5 discusses the rudiments of *random graph dynamics,* which provides natural mechanisms for the emergence of "sharp behavior" in discrete causal theory. Section 8.6 gives a brief overview of *algebraic structure and hierarchy* in discrete causal theory, supplemented by category-theoretic analogies. Section 8.7 gives snapshots of possible alternative formulations of discrete causal theory, including those involving *causal cycles* and/or *classical holism*. Section 8.8 describes relationships between discrete causal theory and other modern theories, such as *loop quantum gravity, causal dynamical triangulations,* and *domain theory*. Section 8.9 provides a similar discussion regarding related topics in physics-oriented mathematics, including *noncommutative geometry* and the theory of *loops* and *gyrogroups.* Section 8.10 offers some final remarks, and provides a few details about how this book came to be written.

B.F. Dribus, *Discrete Causal Theory*, DOI 10.1007/978-3-319-50083-6_8

8.1 Axioms and Technical Methods Revisited

Looking ahead. The purpose of this section is to summarize the overall approach to discrete causal theory presented in the preceding chapters, viewing the entire development as a coherent whole. It is much easier to understand the motivations behind the various choices made, and the various conventions adopted, after having seen some of their applications. For example, the choice to generalize the study of directed and multidirected structure far beyond a "minimal modification" of causal set theory is evident as early as Chapter 3, but the actual necessity of this expansive viewpoint cannot be fully appreciated without substantial knowledge of the material developed in Part II of the book. This summary begins with a restatement of the *general, classical,* and *quantum causal metric hypotheses,* since their original statements are separated by hundreds of pages in the book. This is followed by a brief recapitulation of the axiomatic analysis carried out in Chapters 3 and 4, with additional remarks based on the material developed in Part II. Finally, a concise global view is given of the new perspectives and technical methods presented in the book. In this section, I include only a modest number of references to specific sections in preceding chapters, in order to streamline the description.

Causal metric hypothesis revisited. The central motivating idea behind discrete causal theory is the causal metric hypothesis (CMH), which is mentioned, described, and/or analyzed hundreds of times throughout the book. Informally, the hypothesis says that "physics is about cause and effect." At a formal level, it is presented in three different ways: a general version, a classical version, and a quantum version. Here, I repeat these three statements back-to-back, for convenience and completeness:

> **Causal metric hypothesis** (CMH). *The properties of the physical universe are manifestations of causal structure.*

> **Classical causal metric hypothesis** (CCMH). *Classical histories may be modeled via directed sets.*

> **Quantum causal metric hypothesis** (QCMH). *Quantum spacetime may be modeled via kinematic schemes of directed sets.*

As observed in Chapter 2, the causal metric hypothesis is completely distinct from the assumption of fundamental discreteness. However, the hypothesis is more convincing in the discrete case, because discreteness supplies a potential combinatorial origin for natural scale, as suggested by the metric recovery theorems, described in Section 2.8. The proper scope of the causal metric hypothesis is also debatable. The general version of the hypothesis is deliberately stated so as to suggest the *strong interpretation,* which attributes *all of physics* to causal structure. More conservative formulations, such as the one used in causal set theory, restrict the scope of description to "spacetime," while attributing "matter and energy" to auxiliary structure. The strong interpretation is perhaps the most "pleasing" one from an aesthetic and philosophical perspective. In particular, it enables a version of discrete causal theory that enjoys *perfect background independence,* by eliminating the possibility of clashes between different types of structure, such as those occurring in relativistic "time travel paradoxes." However, the strong interpretation is very ambitious and optimistic, and goes far beyond what is directly supported by the metric recovery theorems. Most of the material in the book does not *depend* on the strong interpretation, but is developed and presented in a way that focuses on this interpretation as most desirable.

Axiomatic analysis revisited. Building the technical foundation for a precise version of discrete causal theory requires a suitable list of axioms for directed sets, viewed as models of classical spacetime under the classical causal metric hypothesis (CCMH). Chapters 3 and 4 of this book are largely devoted to enabling a judicious choice of such axioms, with Chapter 4 culminating in the presentation of a "suggested list" of axioms, followed by "conservative" and "radical" alternatives. From a neutral perspective, these two chapters may seem to devote a disproportionate amount of attention to causal set theory, but this emphasis is unavoidable for at least three reasons: first, causal set theory incorporates many of the ingredients necessary for an effective modern version of discrete causal theory; second, causal set theory is prominent in the current literature; third, significant departures from the causal set approach are necessary in order to enable many of the innovations developed in this book. Hence, it is impossible either to de-emphasize causal set theory, or to work within its existing framework. One of the major tasks of Chapters 3 and 4 is to explain why the causal set axioms of *transitivity* (TR) and *interval finiteness* (IF) are problematic from the viewpoint of this book. The conclusions of this analysis are that transitivity should be discarded entirely, and interval finiteness should be replaced by *star finiteness* (SF).

Other axioms for directed sets studied in Chapters 3 and 4 include the *binary axiom* (B), the *measure axiom* (M), *countability* (C), *irreflexivity* (IR), *acyclicity* (AC), *irreducibility* (IRR), and the *generalized measure axiom* (M*). Causal set theory incorporates the first four of these axioms, and consequently also satisfies acyclicity, which follows from the tandem of irreflexivity and transitivity. Unlike transitivity and interval finiteness, none of these first five axioms lead to sufficiently severe structural limitations to render them untenable, but the new viewpoint offered by the nontransitive, star finite paradigm developed in this book justifies their re-evaluation, and hints at possible modifications. In particular, one must decide how to treat acyclicity, since it depends on *both* transitivity and irreflexivity. One possibility is to regard it as a mere artifact of the transitive paradigm, to be discarded when considering more general causal relations. This viewpoint is supported circumstantially by the possibility of closed causal curves in general relativity. The other alternative is to view acyclicity as an *essential* property, and to enforce it as an axiom in its own right. Many of the results and techniques introduced in this book do *not* depend on acyclicity, but some of the most important results of Chapters 6 and 7 are developed and presented only in the acyclic case. Hence, acyclicity is tentatively included among the "suggested list" of axioms in Section 4.10.

The principal role of a measure axiom, such as (M) or (M*), is to accommodate a suitable version of metric recovery. Under the simplest measure axiom, invoked implicitly in Sorkin's statement *"order plus number equals geometry,"* the volume of a region of classical spacetime is determined by *counting* its fundamental elements. The causal set version (M) of the measure axiom modifies this recipe only slightly, incorporating statistical fluctuations, so as to better preserve "effective" or "approximate" Lorentz invariance. However, one may conceive of many other possible methods of deriving scale data from discrete causal structure. Although the topic of discrete causal measures is not studied in detail in this book, Chapters 2–4 include oblique remarks about the possibility of weakening the measure axiom to admit consideration of *local causal structure* as a contributing factor in the emergent notion of scale. In the star finite case, this change does not affect the basic insight provided by the metric recovery theorems; namely, that discrete combinatorial information can supply the "missing scale data" necessary to specify emergent geometry. The generalized measure axiom (M*) encodes "minimal requirements" that such scale data should satisfy. However, a more specific statement must ultimately be chosen in practice. Turning to the binary axiom (B), Chapter 5 provides ample illustrations of how this axiom *could* be altered; for example, by invoking holistic power spaces. However, the attitude toward the binary axiom adopted in this book is to *expand* the catalogue of structures that may be viewed in terms of "elements and relations," without insisting on a new interpretation of the correspondence between "events" and "elements" in the specific context of classical spacetime. There are excellent reasons to retain the axiom of countability (C), at least in the star finite case. Finally, irreducibility (IRR) seems to be too strong of a condition to impose in the general case, but it is worth considering if one restricts attention to "modified versions of causal set theory." Further remarks about these axioms appear below.

Suggested axioms revisited. It is convenient here to restate the "suggested list" of axioms for discrete classical causal theory presented in Section 4.10. Additional comments are added after each axiom, incorporating new perspectives from Part II of the book. A slightly-expanded discussion of alternative axiomatic systems follows.

1. **Binary axiom** (B): *Classical spacetime may be modeled via a directed set* $(D, \prec)$, *whose elements represent spacetime events, and whose relations represent causal relationships between pairs of spacetime events.*

   Inclusion of the binary axiom in this list indicates that the resulting formulation of discrete classical causal theory is *not* irreducibly holistic in the sense of Chapter 5. However, many other important structures in discrete causal theory are also modeled in terms of "elements and relations," with varying interpretations. In particular, *relation spaces, causal path spaces, kinematic schemes,* families of *refinements* or *refinement types,* and *kinematic spaces,* all possess "elements" that do not correspond to spacetime events.

2. **Generalized measure axiom** (M*): *D is equipped with a set function* $\mu : \mathcal{P}(D) \rightarrow \mathbb{R} \cup \{\infty\}$, *viewed as a "generalized measure," which assigns finite positive values to nonempty finite subsets of D, and infinite values to infinite subsets of D.*

For convenience, I remind the reader that $\mathcal{P}(D)$ denotes the power set of D, i.e., the set of all subsets of D, where subsets are viewed in this context as full subobjects. This version of the measure axiom is sufficiently general to admit consideration of local causal structure in the determination of emergent "scale," or "spacetime volume," while preserving the basic conceptual connection between discrete combinatorial data and scale suggested by the metric recovery theorems. As explained in Chapter 3, the real numbers appear in this version of the measure axiom purely for convenience; any "sufficiently large" extension of the positive integers suffices for assigning "sizes" to subsets, and no essential properties of the continuum are necessary.

3. **Countability** (C): *D is countable.*

   Most of the results of this book do not require any global cardinality assumptions, but countability is a reasonable condition to impose in most of the cases of principal interest. For example, countability is a rather mild restriction under any version of the measure axiom assigning infinite values to infinite subsets of directed sets, since any local observation in this context involves, at most, arbitrarily large finite sets of elements. For similar reasons, countability is natural in the star-finite context, or even the star-countable context. Of course, one could imagine the possibility of structural arguments in favor of "uncountable classical histories" that would be sufficiently compelling to motivate dropping countability from the list of axioms, leading to a more systematic study of cardinality conditions on directed sets and multidirected sets. Such a study might include the following: (1) application of Grothendieck's relative viewpoint (RV) to the topic of cardinality, to explain why cardinality conditions are less objectionable as global constraints than generalized order-theoretic conditions; (2) elaboration of the distinguished role of *finite* directed sets in the theory of experimentation and measurement; (3) further investigation of "entropy issues," such as the *Kleitman–Rothschild pathology;* (4) discussion of the *continuum hypothesis* and related set-theoretic and order-theoretic topics; (5) fuller treatment of qualitative structural differences between classes of directed sets and multidirected sets satisfying different cardinality conditions.

4. **Star finiteness** (SF): *For every element x of D, the star* $\mathrm{St}(x)$ *of x is finite.*

   Sorkin identifies "local finiteness" as a formal way of expressing the idea of discreteness. Of course, one may consider discrete topologies or discrete measures on *any* directed or multidirected set, but the most promising physical interpretations of these structures exist only under appropriate local finiteness conditions. As explained in Chapter 4, star finiteness seems to represent a suitable notion of local finiteness for directed sets, viewed as models of classical spacetime under any version of the classical causal metric hypothesis (CCMH) for which infinite sets are assigned infinite measure. As explained in Section 4.5, this does not mean that certain star infinite directed sets, such as causal sets induced by global sprinkling into Minkowski spacetime $\mathbb{R}^{3+1}$, are not worth studying as well. In the case of directed sets, star finiteness may be translated to the statement that each element possesses a finite number of direct predecessors and successors. In the context of kinematic schemes, star finiteness is a less-reasonable assumption;

for example, the underlying multidirected set of the *foliation scheme* $\mathbb{S}_{\mathrm{FOL}}$ is not star finite. Most of the quantum-theoretic techniques developed in Part II of this book do not ultimately depend on star finiteness, though it is often included as a simplifying assumption.

5. **Acyclicity** (AC): *$(D, \prec)$ possesses no cycles.*

   The inclusion of acyclicity in this list of axioms is a conservative, provisional choice, based on the lack of clear physical evidence for the existence of causal cycles. However, in many instances throughout the book, I have chosen to abstain from assumptions about causal cycles, even at the classical level. This reflects my belief that the status of causal cycles is far from settled. General relativity admits many solutions involving closed causal curves, and it seems premature to entirely dismiss these solutions as pathologies that may be expected to disappear in a suitable quantum theory of gravity. After all, general relativity is the only well-supported theory available which *could* shed any light on the status of causal cycles, since it is the only such theory that leaves their existence to be determined dynamically. The argument tipping the balance in favor of acyclicity in this book is simply that it is best to explore simple alternatives first, although this choice arguably contravenes the "anti-prescriptive philosophy" outlined in Section 2.10. Some supplementary remarks on the subject of causal cycles appear in Section 8.7.

Alternative formulations revisited. In Section 4.10, I discussed two alternative ways of choosing axioms for discrete causal theory. The first is very conservative, and leads to a "minimally modified" version of causal set theory. At a formal level, it is more cumbersome than the version of discrete causal theory based on the "suggested list." It is also much more prescriptive of classical causal structure, both locally and globally. It includes a total of seven axioms, of which four are copied from the suggested list; namely, the binary axiom, countability, star finiteness, and acyclicity. A version of the measure axiom is also included, which differs from the causal set version only in its choice of terminology. Two additional axioms are also included. The first of these is interval finiteness, copied from causal set theory. The other is irreducibility, included in order to avoid structural difficulties associated with transitivity.

The second alternative involves versions of causal theory, discrete or otherwise, which incorporate certain types of "radical" qualitative behavior. Its description is much more open-ended than the description of the "conservative alternative," and does not include a specific list of axioms. Rather, it proposes discarding or relaxing one or more axioms from the "suggested list." The most reasonable step in this direction is probably to simply drop the axiom of acyclicity, thereby taking seriously the existence of closed causal curves in general relativity. The resulting formulations of discrete causal theory are sufficiently interesting that I include a brief discussion of some of their properties in Section 8.7. More radical and speculative formulations involve modifying or discarding the binary axiom to admit classical holism. Section 8.7 also includes some basic ideas in this direction, which do not fall under the umbrella of power spaces. A final possibility is to drop the assumption of discreteness altogether, and study versions of causal theory, such as *domain theory,* that allow "infinitesimal structure" of some type. At an axiomatic level, this involves modifying the measure axiom, and discarding the entire idea of local finiteness.

Summary of new perspectives and technical methods. The "suggested list" of axioms repeated above serves both to move beyond limitations in existing versions of discrete causal theory, and to further enhance the theory by making room for a more general structural viewpoint and powerful new technical methods. While the specific idea that *"spacetime geometry is a way of describing causal structure"* has been previously articulated in causal set theory and related approaches, this book studies the causal metric hypothesis (CMH) as a much more general concept, working from first principles, and independent of any specific choice of mathematical model. This enables a clear, general, and flexible approach to discrete causal theory, incorporating many new features. In particular, it facilitates the explicit recognition of distinct classical and quantum versions of the causal metric hypothesis, restated above, which represent different hierarchical aspects of the theory. Other basic structural concepts appearing in Part I of the book include the *independence convention* (IC), the *order refinement principle* (ORP), the principle of *hidden structure* (HS), and Grothendieck's *relative viewpoint* (RV). These concepts are relatively straightforward on an individual basis, but may be combined in novel ways in the discrete causal context to yield unexpected insights. The guiding principles of natural philosophy expressed at the end of Chapter 2 provide a broad context for the interpretation and application of these ideas.

New material is systematically developed beginning in Chapter 3, where the fundamental physical role of transitive binary relations to model causal structure in relativity and causal set theory is reassigned to generally nontransitive binary relations called *causal relations,* which generate transitive relations under the operation of transitive closure. In the discrete setting, causal relations provide a fundamentally more detailed and nuanced view of causal structure than the transitive relations familiar from earlier version of discrete causal theory. In particular, they accommodate the possibilities of *irreducibility* and/or *independence* of relations between pairs of events, features which are completely absent in the interpolative continuum-based setting where the transitive paradigm originated. Previous work of Finkelstein, Raptis, and a few others examines structures formally equivalent to discrete directed sets, but with different physical interpretations. One of the many advantages afforded by the choice to elevate nontransitive causal relations to a central role is rectification of the *Kleitman–Rothschild pathology,* under which configuration spaces of causal sets are dominated by members of "negligible causal size."

Another major conceptual innovation introduced in Part I of the book is a new approach to *local* spacetime structure, developed in Chapter 4. The perspective involved here is much different than the interval-theoretic viewpoint of causal set theory, encoded in the axiom of interval finiteness (IF). The new approach is based on the physical concept of *causal locality,* under which a condition is taken to be local if and only if it may be checked by examining the independent causes and effects associated with each event. Here again, formal precursors for these ideas may be found in the work of Finkelstein, Raptis, and a few others. The foremost application of causal locality is the replacement of interval finiteness with the alternative axiom of *star finiteness* (SF). The *star topology* expresses this new local viewpoint in a standard mathematical manner. Star finite directed sets are adequate for realizing the general idea that *"directed structure plus natural scale equals geometry,"* implied by the metric recovery theorems. In particular, star finite directed sets suffice to adequately approximate both *local Lorentz invariance* and *spatiotemporal locality*. The theory of *relative multidirected sets over a fixed base,* appearing later in the chapter, is the first major application of Grothendieck's *relative viewpoint* (RV) in the book. The interplay of relative multidirected sets and causal locality provides an exceptionally clear picture of the structural limitations of interval finiteness, revealing the drastic global restrictions imposed by this supposedly "local" condition.

Part II of the book amplifies the new perspectives introduced in Part I, and applies them to the development of new technical methods for discrete causal theory. An overarching structural theme is to expand the viewpoint embodied in the binary axiom (B) to encompass abstract directed relationships between pairs of structures more complex than individual events, following Grothendieck's relative viewpoint (RV). The simplest manifestation of this theme is the theory of *relation space,* developed in Chapter 5, which formalizes the idea of relationships between pairs of relations in a multidirected set. The mathematical object defined by assembling all such relationships is a directed set in its own right, called the *relation space* over the original multidirected set. The resulting viewpoint leads to the solution of important technical problems in discrete causal theory, including the dynamical problem of *permeability of maximal antichains* in multidirected sets. Formal precursors to relation space exist in mathematical contexts, but have not been systematically applied to the study of fundamental spacetime structure.

Comparison of relation space with the closely-related concept of *abstract element space* leads to the striking realization that *typical multidirected sets are nearly interchangeable with their relation spaces*. This is an informal way of expressing the fact that passage from a multidirected set M to its relation space $\mathcal{R}(M)$ preserves all of the information contained in M, except possibly at the causal boundary ∂M of M. The second half of Chapter 5 introduces generalizations of relation space, providing further applications of natural relationships between pairs of nontrivial structures in discrete causal theory. Examples include *causal atoms,* which are useful for computation and approximation, *power spaces,* which serve as a natural framework for the study of *classical holism* in the discrete causal setting, and *causal path spaces,* which formalize the flow of information and influence in a broad context. Modern algebraic methods provide powerful techniques for studying the properties of these spaces, and for examining their physical consequences. In particular, *causal path algebras* play a central role in the theory of *path summation over a multidirected set,* which provides the formal machinery for discrete quantum causal theory.

Chapters 6 and 7 bring these methods to bear on the quantum theory of spacetime. The crucial link between the classical and quantum realms in discrete causal theory is the principle of *iteration of structure* (IS), which enables effective adaptation of Feynman's path summation approach to the background independent context. Iteration of structure formalizes the observation that configuration spaces of directed sets possess natural "higher-level multidirected structures," collectively induced by relationships between pairs of their member sets, called *co-relative histories.* Of particular interest are *proper, full, originary* co-relative histories, which may be viewed as equivalence classes of special morphisms called *transitions,* physically interpreted as representing the "evolution of one classical history into another." Most of Chapter 6 is devoted to developing the theory of path summation in the discrete causal context, with conceptual background and motivation from Feynman's continuum-based *path integral.* Near the end of the chapter, I apply this theory to derive *causal Schrödinger-type equations,* which serve as dynamical laws for discrete quantum causal theory. Chapter 7 explores the properties of *kinematic schemes,* which are special configuration spaces of directed sets encoding "evolutionary pathways for permissible classical histories." The theory of kinematic schemes generalizes Sorkin and Rideout's theory of *sequential growth dynamics* for causal sets.

8.2 Phase Theory

Basic phenomenological remarks. To achieve long-term success, discrete causal theory must eventually predict and explain phenomena beyond the scope of general relativity and the standard model of particle theory. Ideally, any substantial work detailing a new theoretical development should conclude with a robust and convincing discussion of phenomenology; however, at present, discrete causal theory can produce only a modest offering in this direction, as described in Sections 1.9, 3.4, 4.5, and 6.5. Of course, modern theoretical physics as a whole finds itself in the unenviable position of measuring each new contribution relative to its rather unspectacular competitors. In particular, the honeymoon for string theory is approaching its fiftieth year without exhibiting any tangible phenomenological success whatsoever.[1] Discrete causal theory at least has something to contribute regarding the "cosmological constant," and offers some prospect of eventually predicting accessible effects involving Lorentz invariance violation and spatiotemporal nonlocality. This is true despite the fact that the theory has only very recently attracted sufficient attention, or even existed in a sufficiently mature and flexible form, to enable serious evaluation.

The next four sections briefly outline some of the remaining prerequisites for embarking on a more rigorous and comprehensive phenomenological program in discrete causal theory. The present section focuses attention on the *phase maps* that supply the dynamical content of causal Schrödinger-type equations, such as Equation (1.1.2). Identification of suitable phase maps is an obvious requirement for deriving precise quantitative predictions from these equations. The highlight of this discussion is the topic of *entropic phase maps,* which enjoy the dual advantage of natural definition in terms of discrete causal structure, and incorporation of the *second law of thermodynamics,* which is perhaps the only well-established physical law sufficiently general to play a central role in any conceivable physical theory. The

[1] Thirty-five years might be more charitable; string theory was "non-mainstream" prior to 1984.

three subsequent sections briefly introduce three topics crucial for recovering and extending the results of experimentally established theories. These topics are discrete causal *covariance,* emergence of the *Poincaré group* of isometries of Minkowski spacetime $\mathbb{R}^{3+1}$, and *random graph dynamics.*

Undetermined dynamical content. The information contained in the causal Schrödinger-type equation (1.1.2)

$$\psi^-_{R;\theta}(r) = \theta(r) \sum_{r^- \prec r} \psi^-_{R;\theta}(r^-),$$

arises from two sources: the relation space R containing the relations r and r^-, and the relation function $\theta : R \to T$ determining the values of the individual phases $\theta(r)$. Regardless of whether or not one adopts the strong interpretation of the causal metric hypothesis (CMH), the structure of R arises from purely *kinematic* information. In particular, in the background independent context, R is a subset of the relation space $\mathcal{R}\big(\mathcal{M}(\mathbb{S})\big)$ over the underlying multidirected set $\mathcal{M}(\mathbb{S})$ of a kinematic scheme $\mathbb{S} = (\mathcal{K}, \mathscr{H})$. In this setting, the relations r and r^- represent co-relative histories h and h^- between pairs of directed sets belonging to the object class $\mathcal{K}$ of $\mathbb{S}$. This object class arises directly from a choice of axioms for discrete classical causal theory, and represents the most fundamental information regarding what the theorist "believes the physical universe to be like." The higher-level multidirected structure of $\mathbb{S}$ is specified by its class of co-relative histories $\mathscr{H}$, which is chosen for convenience from among a variety of different classes that provide "enough evolutionary pathways" for objects in $\mathcal{K}$. Of course, different choices of $\mathscr{H}$ lead to different equations, but these are related in a precise way under a suitable notion of covariance. Hence, provided that one is confident in the choice of axioms, one may readily formulate a version of Equation (1.1.2) whose kinematic information is completely understood.

However, Equation (1.1.2) also contains *a priori independent dynamical information,* determined by the relation function θ, which generates the phase map $\Theta : \Gamma(R) \to T$ under the simplifying assumptions described in Section 6.7. The values of θ, and hence of Θ, *may or may not be exclusively attributable to causal structure,* depending on how strongly one wishes to interpret the causal metric hypothesis. Under the strong interpretation, there *must* exist a unique "correct" choice of phase map Θ for each kinematic scheme $\mathbb{S}$, or at least a unique class of such maps that produce the same physics. In this sense, the strong interpretation "unifies kinematics and dynamics," as described in Sections 1.8 and 7.4. However, precise identification of this map, or class of maps, is generally a formidable problem. Even the individual values of relation functions such as θ a priori depend on the entire structures of co-relative histories in $\mathbb{S}$, which are generally very complicated. Without imposing simplifying assumptions, the values of Θ may depend on even more complicated information, involving "higher-level causal paths" of arbitrary length. In the general case, one must consider the "correct" way of defining Θ itself, since it is Θ that directly supplies the phases of the causal paths contributing to the values of the past state function ψ^-, as indicated by Definition 6.9.5. This leads to the following question:

> **Fundamental question of phase theory:** *Given a kinematic scheme* $\mathbb{S} = (\mathcal{K}, \mathscr{H})$*, and a subobject* R *of* $\mathcal{R}(\mathcal{M}(\mathbb{S}))$*, how is the "correct" phase map* $\Theta : \Gamma(R) \rightarrow T$ *determined, and what is its target object* T*?*

Without a precise answer to this question, the theory lacks a precise dynamics, and its predictive potential cannot be fully realized. Of course, it may still be possible, under these circumstances, to work out useful *qualitative* implications of the theory on basic structural grounds. To a limited extent this has already been done, as noted at the beginning of the section. One may also attempt to formulate general conclusions involving broad classes of phase maps. Or, one may try to gain insight by studying the much-less-complicated problem of defining phase maps for path summation over *individual* directed or multidirected sets, *not* viewed as representing "higher-level structure." This problem, too, has already been studied, although the literature does not describe it in exactly these terms. However, a much more desirable achievement would be to identify natural phase maps in the background independent context that are "obviously the correct ones." The reason for relegating this crucial topic to the final chapter of the book is that *at present I can offer only some interesting general ideas regarding how this problem might be approached.* Of these, the most promising involve attempts to define phase maps in terms of certain special "non-global" automorphism groups called *generational automorphism groups*, which are *not* automatically rendered trivial by the generic rigidity result of Bender and Robinson [BR88]. A striking physical motivation for this approach, already mentioned in Section 7.7, is that it raises the prospect of expressing discrete causal dynamics in terms of a *generalized version of the second law of thermodynamics.* This idea is described in more detail below.

Actions and Lagrangians. To begin the study of phase theory, it is natural to revisit Feynman's phase map Θ in the background dependent continuum-based context, reproduced here from Equation (6.2.3):

$$\Theta(\gamma) = e^{\frac{i}{\hbar}\mathcal{S}(\gamma)}.$$

As a reminder regarding the notation, i is the imaginary unit $\sqrt{-1}$, $\hbar$ is Planck's reduced constant $h/2\pi$, and $\mathcal{S}(\gamma)$ is the classical action for a particle trajectory γ, given by integrating a Lagrangian $\mathcal{L}$ over γ. Here, $\mathcal{L}$ encodes the dynamical content of the theory. In Feynman's conventional setting, $\mathcal{L}$ is physically interpreted as the difference between the particle's kinetic and potential energies, with physical fields such as the electromagnetic field making contributions to the potential energy portion. More abstractly, $\mathcal{L}$ may be regarded as a function on the real tangent bundle over spacetime, as explained in Section 6.2. This latter viewpoint is the one that generalizes naturally to the discrete causal context, since it enables relation functions to be regarded as "discrete causal Lagrangians." More precisely, the relation function θ appearing in Equation (1.1.2) is analogous to the "complex exponential of a Lagrangian," since it is a "quantum-theoretic entity," rather than a "classical precursor to a quantum-theoretic entity."

The task at hand, then, is to identify suitable "Lagrangians" and "actions," i.e., suitable relation functions and phase maps, in the discrete causal context. Interesting work has already been done in this direction in the special case of causal sets; for example, the recent paper *Scalar Curvature of a Causal Set* [BD10], by Benincasa and Dowker, already mentioned in Section 3.4, discusses discrete causal analogues of the *Einstein–Hilbert action* in general relativity, defined in terms of *generalized D'Alembertian operators* like those mentioned in Sections 3.3 and 4.5. I do not undertake the substantial task of attempting to evaluate, adapt, and generalize these particular efforts here. In general, however, it seems that it would require incredible luck to achieve substantial success in developing an action principle for discrete causal theory by attempting to mimic the Einstein–Hilbert action, except in the most abstract and general sense. The reason why this is true is because general relativity takes so much structure for granted; for example, the dimension of spacetime is fixed. In the discrete causal context, where every aspect of continuum-based geometry is emergent, the dynamical aspects of curvature described by Einstein's field equation almost certainly represent only a small part of the overall dynamical picture. Hence, it seems unlikely that Einstein's equation, or any alternative equation involving specific geometric attributes of specific pseudo-Riemannian manifolds of fixed dimension, will emerge in an easily-recognizable form as a "classical limit" of a causal Schrödinger-type equation. A more likely scenario, a priori, is that an explicit form of the latter type of equation will include many terms, with only a few "higher-order terms" involving quantities such as curvature.

Preliminary considerations regarding phase maps. Before proceeding, it is useful to make a few general observations regarding phase theory, based on elementary mathematical and information-theoretic considerations. Feynman's phase map takes values in the unit circle S^1, viewed as a subset of the complex numbers $\mathbb{C}$. These values are summed in $\mathbb{C}$ to yield quantum amplitudes, which are then translated into real-valued probabilities. As discussed in Section 2.9, and again in Section 6.7, there are good *general* reasons to consider "sphere-like" objects, such as S^1, and larger algebraic objects containing them, such as $\mathbb{C}$, as target objects for phase maps. This is especially true in the background independent setting, as described in Section 6.7; indeed, the three notions of *background independence, path summation,* and *directed structure* seem to recommend target objects possessing notions of *equal magnitudes, addition,* and *multiplication.* However, the *manifold* structures of S^1 and $\mathbb{C}$ themselves play no role in these considerations, and create a host of other problems if they are introduced artificially. Hence, in discrete causal theory, one expects objects such as S^1 and $\mathbb{C}$ to *emerge only as continuum limits of "more primitive" target objects.*

For example, working with a kinematic scheme whose object class is the class $\mathcal{A}_{\mathrm{FIN}}$ of finite acyclic directed sets, it would be reasonable to expect phases to belong to some finite algebraic object. Speculating further, different phases might belong to different natural targets, perhaps organized into something like a "filtered system," understood in a generalized category-theoretic sense, whose "limit object" might resemble a countable subset of $\mathbb{C}$. Alternatively, it might be appropriate to work at a higher level of structure throughout, and to view phase maps as "generalized functors." It is automatic, in the background independent setting, that the information determining phase maps *originates* at the level of kinematic schemes; the question is whether or not this information may be reasonably "condensed" into a form that may be described in terms of values in a "lower-level object," such as an individual ring. In either case, "entropic considerations," involving enumeration of certain types of directed sets, combinatorial properties of subobjects of these sets, and the structure of corresponding symmetry groups, may be expected to play a central role. A suitable notion of *covariance* is also necessary, since different kinematic schemes a priori require different families of targets.

Entropic phase maps. In Section 7.7, I briefly mentioned the possibility of defining phase maps for kinematic schemes in "entropic" terms, without explaining in any detail the technical reasoning behind this choice of terminology. In conventional statistical thermodynamics, entropy is understood as a way of describing "multiplicities of states," considered up to a specified level of detail. This idea may be generalized to any situation involving the enumeration of objects that may be regarded as "equivalent" up to such a specified level, where equivalence is defined in terms of whatever natural structure these objects happen to possess. In the discrete causal context, the most obvious sources of multiplicities are *automorphism groups of directed sets,* and related groups, such as the *causal Galois groups* associated with co-relative histories. A fundamental obstacle to the naïve application of such algebraic data is Bender and Robinson's result [BR88] that "most acyclic directed sets are rigid;" i.e., they possess trivial automorphism groups. However, *special classes* of directed sets of "large spatial and small causal size" include a larger proportion of objects with nontrivial automorphism groups, essentially because the valences of elements in these sets tend to be relatively small. In particular, such sets contain many maximal and minimal elements with at most one nearest neighbor. Informally, this means that *"there exist fewer relations to bind elements in place locally."* What is especially interesting about such directed sets from a physical perspective is that they arise naturally as subobjects of classical histories representing "spatial sections," or "thickened spatial sections," interpreted as encoding "recent behavior" in a particular frame of reference. The importance of this observation is that the "entropy" of such subobjects is closely related to the types of entropy involved in familiar formulations of the second law of thermodynamics. In particular, a rough abstraction of the informal physical idea that "entropy should increase with time" suggests the statement that *"later sections should possess larger automorphism groups than earlier sections."*

Of course, the most naïve notion of a "spatial section" in discrete causal theory is simply an antichain of cardinality n in a directed set, whose automorphism group is isomorphic to the symmetric group S_n. Hence, the most naïve "second law of thermodynamics" in this context simply states that *the universe tends to expand,* i.e., that "later antichains tend to be larger than earlier antichains." Even this unrealistically simple viewpoint may not be completely irrelevant as a "first-order approximation," since it seems to fit the observed cosmological data. However, it is much more interesting to consider *antichains in relation space,* which already play a distinguished dynamical role in discrete causal theory due to the impermeability result of Theorem 5.4.3. From the perspective of element space, such antichains possess *limited, but nontrivial, internal structure.* This suggests an intriguing balance, in which one may expect to encounter "interesting" automorphism groups, i.e., automorphism groups that are neither trivial, nor merely copies of S_n. An example of such an antichain σ, viewed in element space, is illustrated in Figure 8.2.1, which is adapted from Figure 5.4.5. The "ambient directed set" D is rigid in this case, but the subobject of D corresponding to σ possesses a nontrivial automorphism group; for example, this group contains an automorphism interchanging the elements x and y.

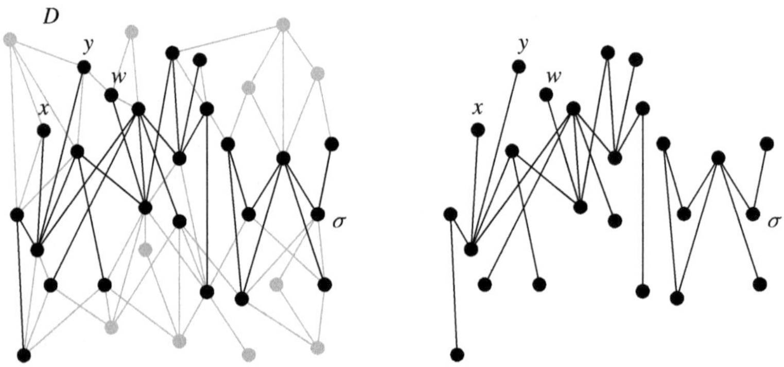

Figure 8.2.1. Subobjects of directed sets corresponding to antichains in relation space are more likely to be nonrigid than arbitrary directed sets.

Entropic phase maps in generational dynamics. A technical subtlety of antichains in relation space is that the corresponding element space subobjects are generally not *full* subobjects. In other words, given an antichain σ in the relation space over a directed set D, "extraneous relations" may exist in D between certain pairs of elements in the corresponding subobject. For example, for the maximal antichain σ illustrated in Figure 8.2.1, there exists a relation in D between the elements w and y, but this relation does not belong to σ. For this reason, it generally does not make sense to speak of a co-relative history "adding" a particular antichain of relations to a classical history, provided that one limits consideration to co-relative histories defined in terms of full subobjects. In the present example, the problem one encounters arises from the necessity of "adding a new relation between the two existing elements w and y," *after* the subobject corresponding to σ has already been "added." This conflicts with any reasonable evolutionary viewpoint, since it calls for the paradoxical appearance of "new causal structure in the past." Under deeper examination, this problem seems to rule out straightforward use of *maximal* antichains in relation space to construct realistic phase maps, since the elements contributing to such antichains generally involve intricate extraneous structure in the ambient directed set D. In particular, these elements may lie on chains of any length in D, as mentioned in Section 6.7.

Fortunately, the setting in which *maximal* antichains in relation space are actually needed, to solve the permeability problem, occurs at the level of kinematic schemes, since this is the level at which path summation must be performed in the background independent context. At the classical level, there is no need to insist on working with maximal antichains. A natural way to define generally non-maximal antichains of relations in this context is to "increment causal structure" by adding "new" antichains of *elements,* along with suitable families of relations from "existing elements" to "new elements," at each evolutionary step. These families of relations are automatically antichains in relation space. This, of course, is exactly the evolutionary approach involved in *generational dynamics,* as described in Section 7.8. In general, such antichains seem to exhibit attractive structural features for defining relation functions, and hence, multiplicative phase maps, for kinematic schemes. On the one hand, they possess nontrivial internal structure from the element space viewpoint, which allows for the possibility of meaningful phase values. On the other hand, this internal structure is "minimal" in a way that circumvents the generic problem of rigidity. At a physical level, generational measures of entropy are exactly what one needs in order to define relevant analogues of the second law of thermodynamics, since the entropy involved is associated with a "particular time," rather than with the "entire history of the universe." As an incidental note, this direction of thought seems to mesh well with the proposed alternative to the *inflationary hypothesis* in cosmology described in Section 1.9, i.e., the idea that causal structure grew sparser in the early universe. The reason why, at a basic qualitative level, is that sparser structure tends to "free up local symmetries" in subobjects of directed sets corresponding to maximal antichains in their relation spaces, thereby increasing entropy. Whether or not this rough idea is borne out *quantitatively* remains, as far as I know, undetermined.

Figure 8.2.2 reproduces the example of generational growth of an acyclic directed set D first appearing in Figure 7.8.1 of Chapter 7, along with the automorphism groups of the "structural increments" added in each generation of this process, viewed from the perspective of element space. More precisely, if $h : D_i \Rightarrow D_t$ is a co-relative history in the generational kinematic scheme $\mathbb{S}_{\mathrm{GEN}}$, then the "structural increment" associated with h is the isomorphism class of the element space subobject of D_t corresponding to the antichain of relations $\sigma(\tau)$ "added" by a transition τ representing h. Different transitions representing h generally add different subobjects to their sources, but these subobjects are isomorphic, by the definition of co-relative histories Definition 6.4.3. It is convenient to denote their common automorphism group by $\mathrm{Aut}(h)$, and refer to it as a **generational automorphism group**. The cardinalities $|\mathrm{Aut}(h)|$ of generational automorphism groups $\mathrm{Aut}(h)$ are the multiplicities that define the notion of "entropy" involved in the evolutionary process leading to D. No multiplicity is associated with the first generation appearing in the figure, because a single generation involves no relations.

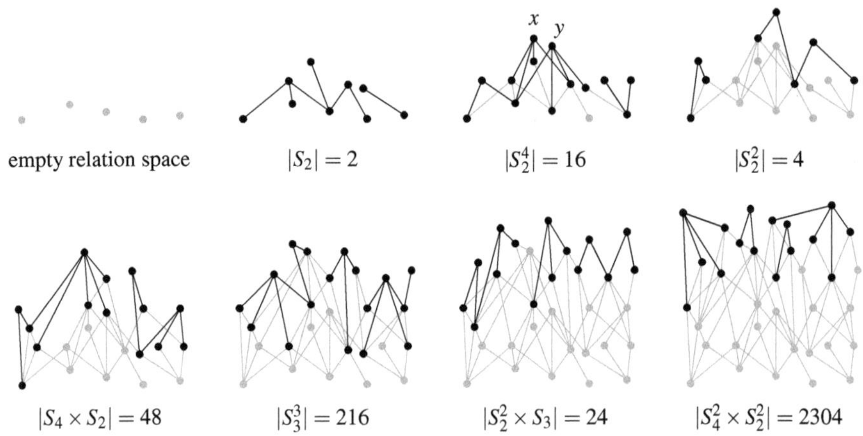

Figure 8.2.2. An example of multiplicities in generational dynamics.

It is not always easy to determine generational automorphism groups by inspection. For example, there exist "non-obvious" automorphisms interchanging the elements labeled x and y in the third generation illustrated in Figure 8.2.2. For large directed sets, actually computing these automorphism groups may prove cumbersome, although the examples illustrated here suggest that many of the symmetries involved tend to permute small families of elements locally.[2] Of course, "small" is a relative term in this context, since the valences one might expect to see in realistic settings may be much larger than those appearing in such "toy examples." As explained in Chapter 7, sequences of directed sets, such as the sequence illustrated in the figure, correspond to chains, or more generally, to *families* of chains, in the underlying multidirected set $\mathcal{M}(\mathbb{S}_{\text{GEN}})$ of the generational kinematic scheme $\mathbb{S}_{\text{GEN}}$. The reason why families of chains in $\mathcal{M}(\mathbb{S}_{\text{GEN}})$ may be involved is because of the possibility of multiple non-equivalent co-relative histories between a given pair of directed sets in $\mathbb{S}_{\text{GEN}}$, i.e., because of the multidirected structure of $\mathcal{M}(\mathbb{S}_{\text{GEN}})$.[3] The goal is to define an entropic phase map assigning values to such chains. The resulting data may then be fed into the machinery of path summation over multidirected sets, to determine, roughly speaking, which chains represent "favored evolutionary pathways."

Before proceeding, I should note that it is *not* obvious, at least to me, that the notion of entropy arising from the cardinalities of generational automorphism groups is exactly the right one from a physical point of view. Many alternatives are pos-

[2]It is worth pointing out that the contribution of "nonlocal symmetries" in this context; for example, symmetries involving the interchange of disjoint pairs of "wedge-shaped" or "vee-shaped" subobjects, introduces a type of holism at the quantum level. This is an "expected" form of holism, however.

[3]In the present example, the color-coding identifies specific *transitions* at each stage of evolution, by specifying which elements are considered "old," and which elements are considered "new."

sible. In particular, as noted earlier in the section, the familiar notion of entropy from statistical thermodynamics is defined not by counting states that are physically indistinguishable, i.e., related by a symmetry, but by counting states that are *indistinguishable only up to a specified level of detail.* Technically, this means that such states belong to the *same cell in a coarse-graining of phase space.*[4] It is obvious, based on simple counting, *why* the second law of thermodynamics holds in this context. Indeed, its qualitative meaning is merely that *an evolving system is more likely to move from a small cell to a large cell in a coarse-graining of phase space than vice versa.* In other words, the system is more likely to transition to a "macrostate" represented by a larger number of "microstates" than to a "macrostate" represented by a smaller number of "microstates." For analogous reasoning to hold in the context of generational automorphism groups, one must count the "structural increment" corresponding to a co-relative history $h : D_i \Rightarrow D_t$ in $\mathbb{S}_{\text{GEN}}$ as representing $|\text{Aut}(h)|$ *different* "microstates," rather than a single "microstate." This implies that its elements should be regarded as physically distinguishable at the finest level of detail, even if they are permuted by an element of $\text{Aut}(h)$. Of course, such elements typically *are* physically distinguishable when considers the contribution of previous generations, because the entire classical history D_t is typically rigid. In more detail, a given pair of elements might be indistinguishable when one takes into account, say, the most recent N generations, but might "become distinguishable" when one considers the most recent $N + 1$ generations.

Due to this same rigidity property, however, there are generally *not* $|\text{Aut}(h)|$ different ways for D_i to evolve into D_t via h in the sense of multiplicities; i.e., the elements of $\text{Aut}(h)$ generally do not extend to automorphisms of D_t. Hence, one could argue that "the proper multiplicities are already built into $\mathbb{S}_{\text{GEN}}$," and that the "favorability" of co-relative histories should *not* be taken to depend on their generational automorphism groups. On the other hand, experimental physics suggests that it is unnecessary to know the details of the distant past of a system in order to predict its future evolution; in particular, as discussed in Section 6.2, the Lagrangian is usually taken to depend only on "first-order information." The obvious analogue of such "first-order information" in generational dynamics is precisely the "structural increment" associated with a co-relative history h. On this basis, one might take the view that $|\text{Aut}(h)|$ *does* represent a physically relevant multiplicity for h. Or, one could consider symmetries involving multiple generations in an integrated fashion, since elements tend to "acquire distinguishability" as more and more generations are added to the picture. One would expect "older" generations to make smaller contributions in this context. This makes sense on naïve physical grounds, because the versions of entropy appearing in ordinary formulations of the second law of thermodynamics correspond to a given time, not to the entire history of the universe. Finally,

[4]Here, "phase space" generally means the abstract space whose coordinates correspond to the positions and momenta of particles in a system. In particular, it is not the same as the space of phases of paths involved in path summation. Since this "phase space" encodes "first-order data" about a system, it is roughly analogous to a space of "structural increments" associated with generational automorphisms. However, as discussed below, a given "structural increment" generally corresponds to more than one "microstate."

it is possible that generational automorphism groups may turn out to be very small, or even trivial, in the generic case, if the valence of a typical element in a "physically relevant" directed set turns out to be sufficiently large. It is conceivable that one could evaluate this possibility via experimental constraints on Lorentz invariance violation, together with ballpark figures regarding the fundamental scale. If these groups *do* turn out to be generically trivial, then one might attempt to define entropy via a coarse-graining procedure appropriate to the discrete causal setting, such as the method of *causal atomic resolution* studied in Section 5.7. On the other hand, if the dynamical picture arising from the use of generational automorphism groups turns out to be sufficiently compelling, one might consider focusing deliberately on the case of small valences in order to avoid rigidity, in spite of Lorentz invariance issues, in the hope that the latter problem might be resolved at the *quantum* level, as in loop quantum gravity [RS03]. This topic is briefly revisited in Section 8.8.

Examples of entropic phase maps. Returning to the subject of generational automorphism groups, there are several possible ways to define entropic phase maps in terms of such groups. Perhaps the most obvious of these is to begin with a positive-integer-valued "naïve relation function" $\overline{\theta}$, which simply measures the cardinalities of these groups. More precisely, the value $\overline{\theta}\big(r(h)\big)$, for any given co-relative history h in $\mathbb{S}_{\text{GEN}}$, is defined to be

$$\overline{\theta}\big(r(h)\big) = \big|\text{Aut}(h)\big|.$$

This "naïve relation function" may then be extended to a "naïve phase map" $\overline{\Theta}$, by multiplicativity. Products in this context merely represent a way of "combining favorabilities for consecutive stages of evolution;" they do *not* represent "integrated measures of multiplicity," since combining multiple generations tends to produce increasingly rigid subobjects.

An alternative relation function, equivalent to $\overline{\theta}$ from an information-theoretic perspective, is the "Boltzmann-type" **entropy function** $\log\overline{\theta}$, which is naturally additive. One might choose to view the corresponding "negative entropy function" $-\log\overline{\theta} = \log(1/\overline{\theta})$ as a "generalized Lagrangian," and define the "action" along a chain $\gamma = r(h_1) \prec \cdots \prec r(h_n)$ of relations $r(h_k)$ representing co-relative histories h_k in $\mathbb{S}_{\text{GEN}}$, or in a convenient subscheme, to be the sum $\mathbb{S}(\gamma) := -\sum_{k=1}^{n}\log\overline{\theta}\big(r(h_k)\big)$. The reason for the minus sign, which ultimately makes no information-theoretic difference, is a naïve version of *Hamilton's principle,* discussed briefly in Section 6.2. Hamilton's principle is often referred to, imprecisely, as the *principle of least action,* although it actually only invokes *stationary* action. In any case, the minus sign enables the intuitively familiar statement that *"favored chains γ minimize the action $\mathbb{S}(\gamma)$."* One may then "recover multiplicativity" and "impose equal magnitudes," by copying Feynman's recipe, and define the phase $\Theta(\gamma)$ of the chain γ to be the complex exponential

$$\Theta(\gamma) = e^{i\mathbb{S}(\gamma)} = \exp\Big(-i\sum_{k=1}^{n}\log\overline{\theta}\big(r(h_k)\big)\Big)$$

$$= \exp\Big(-i\sum_{k=1}^{n}\log\big|\text{Aut}(h_k)\big|\Big) = \prod_{k=1}^{n}\big|\text{Aut}(h_k)\big|^{-i}. \tag{8.2.3}$$

Under this recipe, the "actual relation function" θ generating Θ under multiplication has the form $\theta\big(r(h)\big) = \big|\text{Aut}(h)\big|^{-i}$.

In information theory, quantum information theory, and related subjects, one is accustomed to seeing more-sophisticated entropy functions of the form $H = -\sum_k p_k \log p_k$, where p_k is a "probabilistic quantity," such as the probability of the kth outcome among a list of possible outcomes of a measurement, or the *"probability of a system being in cell k of its phase space,"* as described by Shannon [SH48]. In particular, *Shannon entropy* and *von Neumann entropy* are both expressed in this form. Hence, one might wonder why the simpler-looking "Boltzmann-type" function $\log\overline{\theta} = \log\big|\text{Aut}(h)\big|$ is referred to above as the "entropy function." The reason why is because this function is associated with a *particular* co-relative history, rather than a family of different co-relative histories with different "favorabilities." Of course, one may also define "Shannon-type" entropy functions. For example, working in the classical stochastic context, one may begin with a fixed directed set D in a kinematic scheme $\mathbb{S}$, which I will assume to be a star finite subscheme of $\mathbb{S}_{\text{GEN}}$ for simplicity, and consider the set h_k of co-relative histories with source D. One may then define "probabilities"

$$p_k = \big|\text{Aut}(h_k)\big| / \sum_k \big|\text{Aut}(h_k). \tag{8.2.4}$$

The reason for assuming star finiteness of $\mathbb{S}$, of course, is because of the sum in the denominator. Extending this recipe to all of $\mathbb{S}$ defines a classical stochastic dynamics, similar to the "toy dynamics" discussed in Section 1.8. The reasoning behind Equation (8.2.4) is that if $\big|\text{Aut}(h_k)\big|$ encodes the "favorability of h_k" relative to other possible co-relative histories with source D, then p_k should be the "normalized" quantity measuring the associated probability. The **Shannon entropy** at D in $\mathbb{S}$ is then defined to be $H(D) = -\sum_k p_k \log p_k$, which is much different than the "Boltzmann-type" entropy function defined above. The construction reveals *why* the two functions are different; namely, because the Shannon entropy is a weighted sum associated with an entire family of co-relative histories, while the entropy function $\log\overline{\theta} = \log\big|\text{Aut}(h)\big|$ is associated with a single co-relative history.

Of course, "Shannon-type" entropy formulas might enter the picture even in the case of a single co-relative history, provided that one chooses to take into account how many generations must be considered before given pairs of elements in D "become distinguishable." For example, one might consider elements of $\text{Aut}(h_k)$ that are broken by the previous generation to represent a different "macrostate" than elements of $\text{Aut}(h_k)$ that extend to automorphisms of multiple generations. However, even this increase in detail remains rather conventional. A less-simplistic alternative, already mentioned above, is to work entirely at the higher level of structure represented by kinematic schemes. Following this approach, phase maps are viewed as "generalized functors," and the corresponding path sums as "generalized limits." In the present context of generational dynamics, one might define a "relation functor" ϑ that assigns to each co-relative history h the corresponding generational automorphism group $\mathrm{Aut}(h)$ itself. Phase maps arising from ϑ might involve category-theoretic products, or possibly ordered products, of the groups involved. Of course, a basic constraint on all these considerations is that one must eventually *interpret* the resulting generalized quantum amplitudes physically, in terms of probabilities, or some generalization thereof.

Entropic phase maps and local physics. As noted above, the generational automorphism groups involved in the simple example illustrated in Figure 8.2.2 tend to "permute small families of elements locally," rather than "mixing entire generations." The extent to which this behavior is actually representative of "physically interesting" evolutionary processes depends on a number of factors. In particular, it depends on the typical values of valences for elements in "favored" types of classical histories. What can be said, however, is that a general tendency for entropy to "depend mostly on local structure" would seem to provide an advantage in attempting to model realistic physical processes, which display a significant degree of locality. Of course, processes that are considered "local" in conventional particle theory may still be very "large" compared to the fundamental scale. However, "elementary particles" themselves are at least *approximately* local phenomena, and it is interesting to consider the question of whether or not entropic phase maps of the type introduced above might shed any light on the emergence of such particles. For example, it is natural to wonder if the rather artificial-looking view of a "particle trajectory" appearing in Figure 1.7.3 can be replaced with something similar, but more compelling. In particular, one could investigate whether or not there is any reason to expect the emergence of "self-replicating structures" in generational dynamics.

A purely speculative idea along these lines involves the possibility of *local entropic traps* involving large *simple groups.* In more detail, there might arise rare situations leading to symmetries permuting relatively large numbers of elements in a given generation of a classical history in an irreducible manner, meaning that these permutations cannot be broken up into products of permutations involving smaller sets of elements. In this context, a typical "small alteration" in local structure from one generation to the next might lead to a large decrease in local entropy, and this might encourage "perpetuation" or "replication" of certain aspects of local structure over a lengthy sequence of generations. In this sense, the local structure would be "trapped" in a particular configuration. The generational automorphism groups associated with such self-replicating structures would tend to possess large simple factors, so that a breakdown in any given aspect of structure would be highly disfavored from an entropic perspective.[5] Though I offer no quantitative justification whatsoever for this idea, it does suggest an interesting and completely novel way of attempting to associate groups with particles. It may also be interesting from the perspective of particle interactions, because the "collision" of two such self-replicating structures would provide an opportunity for the factors of the corresponding local groups

[5]The *classification of finite simple groups* is relevant in this context. Particularly interesting are the *sporadic simple groups,* such as the so-called *Monster group,* which has about 8×10^{53} elements.

to recombine in different ways, or for the two structures to annihilate each other. However, there are obvious objections to this idea; for example, these hypothetical self-replicating structures would have to exhibit similar properties in different "frames of reference." At present, I have no idea if such structures exist.

8.3 Covariance

Extraphysical data in entropic phase maps. Aside from concerns about the "correct" definition of entropy, the most obvious objection to the constructions sketched in Section 8.2 is that they involve a large quantity of extraphysical information. In particular, there generally exist many different chains terminating at a given directed set in a kinematic scheme, and many different kinematic schemes that could be used to represent evolutionary processes for the same class of directed sets. In this section, I give a very brief and informal sketch of the issues arising from such considerations, and describe in conceptual terms how they might ultimately be addressed. Figure 8.3.1 illustrates the evolution of the same directed set D used for illustrative purposes in Section 8.2, this time along an evolutionary pathway in the foliation scheme $\mathbb{S}_{\mathrm{FOL}}$. In this case, the generational automorphism groups are generally smaller, but there is no other obvious pattern relating the computations of entropy associated with the two descriptions. In particular, under both descriptions, entropy increases at some stages of evolution, and decreases at others, but these increases and decreases occur in different orders, and involve different aspects of the structure of D. The "actual favorability" of the evolution of D, relative to that of other similar directed sets, should not depend on these different points of view.

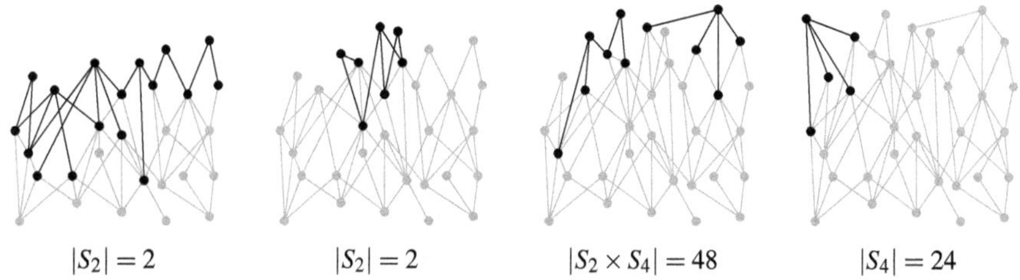

Figure 8.3.1. An evolutionary pathway for D in the foliation scheme $\mathbb{S}_{\mathrm{FOL}}$ involves completely different generational automorphism groups than the groups appearing in Figure 8.2.2.

It is instructive to consider yet another perspective regarding the evolution of D, illustrated in Figure 8.3.2. This latest viewpoint remedies an obvious shortcoming in the entropy values associated with the evolutionary process illustrated in Figure 8.3.1. This shortcoming arises from the "shrinking of generations" near the causal boundary ∂D of D. For example, the "final stage of evolution" in Figure 8.3.1 seems to be "entropically disfavored" for the prosaic reason that there is only one element in the final generation in the chosen "frame of reference." Clearly, this low value of entropy should be regarded as an artifact, arising from the choice of a "highly boosted frame" in a bounded classical history. One way to circumvent this problem, without considering more-complicated descendants of D, is to incorporate in the computation of entropy *all* relations terminating at maximal elements, at each stage in the evolutionary process, whether these relations are "new" or not. To be more precise, such relations need not be "new" in a *kinematic* sense, but they *are* "new" in a *physical* sense, because they involve the most recent behavior encoded in the causal structure. This modification, which results in more "physically relevant" multiplicities, may be carried out without altering the choice of generational growth. In particular, Figure 8.3.2 illustrates the resulting computations for the *same evolutionary pathway* in the foliation scheme illustrated in Figure 8.3.1. The same modification may be applied to the example of generational growth illustrated in Figure 8.2.2 of Section 8.2; in this case, only the entropy value associated with the fourth generation of growth is affected.

Besides correcting such artifacts involving causal boundaries, this modification of "including the contributions of all physically new relations" in the computation of entropy seems to produce intelligible results for a variety of different types of kinematic schemes. For example, if one wishes to work with the positive sequential kinematic scheme $\mathbb{S}_{\text{PS}}$, then the naïve generational automorphism groups corresponding to those illustrated in Figure 8.3.1 are just copies of S_n, where in this context, n is the number of relations terminating at the "new element." The version of the second law of thermodynamics given by using the cardinalities of these groups as multiplicities does not make much sense; it says that "new elements are related to as many existing elements as possible." This leads to essentially the opposite of the mechanism involved in the proposed alternative to the inflationary hypothesis outlined in Section 1.9. Indeed, rather than favoring a transition from highly interconnected structure to "approximately geometric structure," such a law encourages evolution toward a strict total order, in which each "new" element is related to all previous elements. This is manifestly nonphysical. However, if one works with modified generational automorphism groups of the type illustrated in Figure 8.3.2, then a more-balanced situation arises: each relation terminating at the "new element" *may* add new symmetries[6] involving this element, but it may also break existing symmetries involving other maximal elements. The "favorability" of an evolutionary step in this context is determined by the relationship between the "most recent physi-

[6]That is, symmetries of the element space subobject corresponding to all physically new relations, *not* of the entire terminal history. The latter is generally rigid.

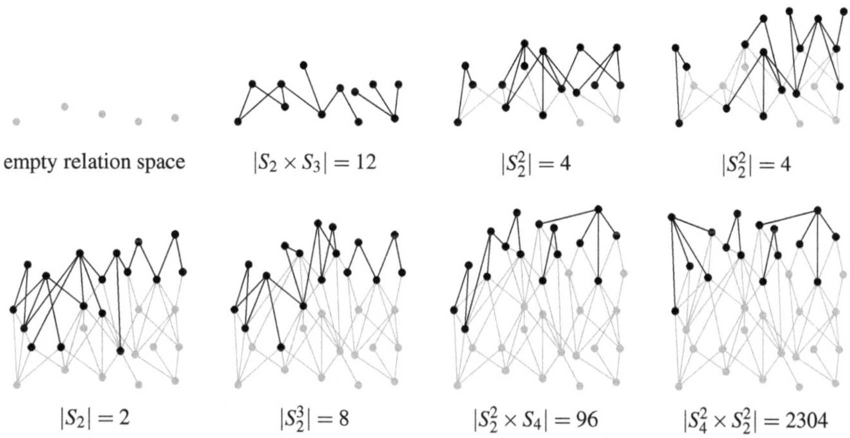

Figure 8.3.2. Modified computation of multiplicities for a generational growth process in $\mathbb{S}_{\mathrm{FOL}}$, involving "most recent" behavior in a causal sense.

cally significant data beforehand" and the "most recent physically significant data afterwards." This offers at least a *superficially* sensible way of modeling evolution.

Covariance. These examples illustrate a specific manifestation of a general source of subtlety in modern physics; namely, the notion of *covariance,* already discussed briefly in Sections 1.8 and 7.3. To review, *whenever a physical theory involves arbitrary choices that specify extraphysical data, it must provide mechanisms to ensure that its substance remains unchanged if these choices are altered.* This the essence of **covariance**. As explained in Section 7.3, this meaning of "covariance" is much different, and much more general, than the category-theoretic meaning involving functors. Also, the term *general covariance* is often used in physics to describe something *else* rather specific, namely, invariance of the mathematical forms of particular physical laws, such as Einstein's equation (1.3.1), under "suitable changes of coordinates."[7] General covariance is briefly revisited below. By contrast, the notion of "covariance" used in this book is very broad; it applies to the general study of organizing and managing physical and extraphysical information in a theory. A conventional example of this idea arises in *gauge theory,* in which a Lagrangian is invariant under a particular group of transformations. By extension, terms such as "gauge-like," or "pure gauge," are sometimes used to describe extraphysical information in the general sense needed in this book. However, the word "gauge" often carries group-theoretic connotations that I would rather avoid. As noted in Section 1.8, one benefit of the choice to use the term "covariance" in this context is that the word itself suggests a helpful heuristic; namely, that *varying one's point of view demands a corresponding variation in*

[7]The annoying synonymity of "covariance" and "invariance" in this context is a historical artifact, which is too deeply-ingrained to be erased.

associated quantities. In particular, the choice of phase map *varies along with* the choice of kinematic scheme in discrete causal dynamics.

Returning briefly to the continuum-based context of conventional physics, covariance is closely related to the theory of *automorphism groups* in various categories. For example, physical laws expressible as tensor equations, such as Einstein's equation (1.3.1), are often called *diffeomorphism invariant,* due to the relationship between the *active viewpoint* represented by diffeomorphisms, and the *passive viewpoint* represented by smooth coordinate transformations.[8] This diffeomorphism invariance is sometimes referred to as *general covariance.* While the passive and active viewpoints are essentially equivalent in many conventional settings, they generalize in completely different ways in the discrete causal context. This important distinction plays a major role in the emergence of the Poincaré group P of symmetries of Minkowski spacetime $\mathbb{R}^{3+1}$, discussed in Section 8.4. In particular, the version of "classical covariance" represented by coordinate independence generalizes in terms of refinements of causal relations and their associated kinematic accounts, which have little or nothing to do with automorphism groups of directed sets. An existing attempt to formalize the idea of covariance in the discrete causal context is embodied in the principle of *"discrete general covariance"* introduced in Sorkin and Rideout's theory of *sequential growth dynamics* for causal sets [SR99]. The purpose of this principle is to encode the idea that "all kinematic accounts of the evolution of a causal set are equally valid," which may be regarded as an analogue of Feynman's notion of "equal magnitudes," albeit in a classical stochastic setting. This effort is worthy of additional analysis, but it is not pursued further here. A somewhat different technical implementation of this idea is necessary for the version of discrete causal theory developed in this book.

Two different sources of extraphysical information require particular care in the context of discrete causal theory. The first source is the arbitrary choice of a kinematic account of the evolution of a particular directed set in a particular kinematic scheme $\mathbb{S} = (\mathcal{K}, \mathscr{H})$, such as the kinematic account of the evolution of the directed set D in the foliation scheme $\mathbb{S}_{\mathrm{FOL}}$, illustrated in Figure 8.3.2. This is a "low-level" source of extraphysical information, because it involves the structure of a specific classical history. The second source is the arbitrary choice of $\mathscr{H}$ itself, necessary to define $\mathbb{S}$. This is a "high-level" source of extraphysical information. At the lower level of structure, covariance means that *the "favorability" of evolution a given directed set incorporates all possible kinematic accounts in a "reasonable" manner.* This does not *necessarily* require that the same "favorability" must be assigned to each kinematic account, although the general merits of "equal magnitudes" are attractive. At the higher level of structure, covariance means that *the essential physical properties of the theory do not depend on the choice of the class of co-relative histories*

[8] To be clear, the form of Einstein's equation (1.3.1) *is* invariant under diffeomorphisms; the subtlety here is that the *essential* meaning of "general covariance" involves the passive viewpoint, and the equivalence of the passive and active viewpoints generally breaks down outside the continuum-based setting. See Rovelli [RO04], Chapter 2, for details.

$\mathscr{H}$. In practical terms, this means that if one changes kinematic schemes, one must change phase maps to compensate. The technical details involved in making these important principles precise are not developed further here.

8.4 Emergence of the Poincaré Group

Symmetry, covariance, and the Poincaré group. The symmetry properties of Minkowski spacetime $\mathbb{R}^{3+1}$ are crucial to every area of physics that incorporates special relativity, whether in an exact or an approximate sense. In particular, the "possible types of particles" arising in conventional quantum field theory, including those appearing in the standard model of particle theory, are classified in terms of these symmetries.[9] The **Poincaré group**, denoted by P, is the symmetry group of $\mathbb{R}^{3+1}$, i.e., the group of automorphisms of $\mathbb{R}^{3+1}$ in the pseudo-Euclidean category, also called *isometries*. An obvious facet in the recovery of the standard model via discrete causal theory is a description of the *emergence of* P, i.e., a description of how "near-symmetries" resembling Poincaré symmetries may arise from causal structure at the fundamental scale. In this section, I sketch some tentative progress in this direction, based on the study of refinements of causal relations, already discussed in Section 7.2, in the context of discrete causal kinematics. The resulting analysis may be generalized to apply to multidirected sets, but for simplicity I confine the discussion to directed sets in this section. A restricted type of covariance, called *Poincaré covariance,* may also be described in terms of P. Poincaré covariance requires that the mathematical forms of physical laws should be invariant under changes of coordinates "related by elements of P," as described below. A "slightly more restricted" type of covariance, of course, is "Lorentz covariance," usually called Lorentz *invariance,* already familiar from the discussion of sprinkled causal sets in Section 4.5. The Lorentz group L is a subgroup of the Poincaré group P, as described below. Lorentz invariance formalizes the special-relativistic notion that *"all inertial frames of reference are created equal."* As mentioned in Section 1.3, many physicists reject out of hand any theory that violates "local Lorentz invariance," while others *expect* such violations at high energies.

At a mathematical level, the Poincaré group P is a ten-dimensional *real Lie group.* Although the theory of Lie groups, Lie algebras, and their representations, is absolutely central to conventional physics, I have mentioned these objects only obliquely and informally in this book. This is because Lie theory depends in a crucial way on properties of real manifolds that are irrelevant in the discrete causal context, except in an emergent sense. A **real Lie group** is a real manifold which also possesses a group law, defining the composition of ordered pairs of elements in such a way that the manifold structure and group structure are compatible, in a suitable sense. The details of this compatibility are tangential to the present discussion, and good

[9] As mentioned in Section 7.3, one is actually a priori interested in *projective* representations of the Poincaré group in this context.

expositions on the subject are ubiquitous. Many other Lie groups besides the Poincaré group are crucial to the conventional modern understanding of theoretical physics; in particular, the *gauge groups* U(1), SU(2), and SU(3) of the standard model are real Lie groups. Generalizations of Lie theory exist in other settings; for example, one may define *complex Lie groups,* or *p-adic Lie groups.* An important facet of algebraic geometry is *algebraic Lie theory,* which studies the properties of algebraic varieties or schemes which are also groups. A simple example is the theory of *elliptic curves.* More generally, the idea of "multiple compatible structures" existing on a given mathematical object is useful in almost every conceivable mathematical or physical setting. However, the manifold structure of Lie groups places these particular objects beyond the pale in the discrete causal context.

The **Lorentz group** L is the six-dimensional subgroup of P representing isometries of $\mathbb{R}^{3+1}$ fixing the origin. The remaining structure of P arises from translations of the origin, and is therefore described in terms of a four-dimensional subgroup of P, isomorphic to $\mathbb{R}^{3+1}$ itself. For this reason, L is often considered to be the "interesting part" of P. As a manifold, L possesses four connected components; two comprise the **orthochronous Lorentz group** L', which preserves time orientation. One of these, containing the identity, is the **proper orthochronous Lorentz group** L'', which also preserves parity, i.e., left or right-handedness. L'' is generated as a group by *Lorentz boosts,* which are the familiar "Lorentz transformations" relating frames of reference in constant relative motion, such as those discussed in Sections 4.5 and 7.1, and illustrated in Figures 4.5.4 and 7.1.2. L'' also includes rotations, which may be expressed in terms of boosts, as *Thomas–Wigner rotations.* In particular, boosts do not form a subgroup of L'', as rotations do: composition of non-collinear boosts introduces rotation. Mathematically, this property of Lorentz boosts provides one of the principal motivating examples in the theory of *loops* and *gyrogroups,* which are generally noncommutative, nonassociative "group-like objects" possessing a "generalized associative property." The "information-theoretic preeminence" of boosts over rotations, reflected by the fact that the former generate the latter, is particularly interesting from the perspective of discrete causal theory. This is because rotations are "purely spatial," while boosts involve causal structure. This provides yet another hint that the causal structure of classical spacetime is more fundamental than any independent notion of spatial structure.

Individual elements of the Poincaré group P may be viewed either *actively* or *passively.* From the active viewpoint, each element p of P rearranges the elements of $\mathbb{R}^{3+1}$ via its corresponding isometry f_p. From this viewpoint, elements of P represent automorphisms of $\mathbb{R}^{3+1}$ in the pseudo-Euclidean category. From the passive viewpoint, p rearranges labels on the elements of $\mathbb{R}^{3+1}$, via a change of inertial frames of reference, represented by special coordinate systems. Poincaré covariance means that the laws of physics are invariant under such changes. A realization of significant importance, particularly in the context of discrete causal theory, is that *this dual "active-passive" role of the Poincaré group P is largely an artifact of continuum theory.* As discussed below, the active and passive viewpoints regarding P generalize in completely different ways when one departs the realm of real manifolds.

Active viewpoint: Poincaré group and causality group. Under the active viewpoint, P may be generalized by comparing it to the group of automorphisms of Minkowski spacetime $\mathbb{R}^{3+1}$ in a *different* category; namely, the category of partially ordered sets. The partial order in this context is the order defined by the relativistic causal relation $\prec_{\text{GR}}$ on $\mathbb{R}^{3+1}$, which encodes potential influences between pairs of events. In this specific context, of course, the subscript "GR" may be regarded as "overkill," because only special-relativistic considerations are involved. However, the conceptual lessons to be drawn from the discussion are much more general in scope. As described in Sections 2.6 and 3.5, $\prec_{\text{GR}}$ is an irreflexive,[10] transitive binary relation on $\mathbb{R}^{3+1}$; a partial order $\preceq$ in the conventional sense may be recovered by setting $x \preceq y$ if $x \prec_{\text{GR}} y$ or $x = y$.

Zeeman [ZE64] defines the **causality group** G of $\mathbb{R}^{3+1}$ to be the group of bijections $f : \mathbb{R}^{3+1} \to \mathbb{R}^{3+1}$ such that both f and f^{-1} preserve the causal relation $\prec_{\text{GR}}$.[11] He proves the following theorem relating P and G:

Theorem 8.4.1. (Zeeman). *The causality group G is generated by the orthochronous Lorentz group L', together with translations and dilations.*

Proof. See Zeeman [ZE64]. □

From the viewpoint of this book, Zeeman's theorem represents early motivation for introducing a fundamental scale in classical physics, since G coincides with the orthochronous subgroup of P up to scale. Indeed, the fact that G is somewhat larger than L' indicates that causal structure alone is slightly deficient for determining the "correct" symmetry properties in the continuum-based context. In particular, G includes dilations, under which the laws of physics are manifestly *not* invariant. However, as discussed in Chapter 2, the only "missing information" needed to correct this problem is a notion of natural scale, which may be ascribed to combinatorial sources in the discrete causal context. This enables the following analogy:

> **Symmetry analogy:** *The Poincaré group P is analogous to the group of automorphisms of a directed set.*

Of course, this analogy is not as useful for discrete causal models of spacetime as one might a priori expect, due to the now-familiar rigidity result of Bender and Robinson [BR88]. This suggests that not only the Poincaré group P itself, but the entire conventional apparatus of continuous symmetries underlying much of modern particle theory, should be viewed in an emergent sense.

Passive viewpoint: Poincaré group and refinements of the causal relation. Under the passive viewpoint, the Poincaré group P may be generalized in terms of the

---

[10] As explained in Section 2.6, $\prec_{\text{GR}}$ is irreflexive, as I have defined it, for any relativistic spacetime manifold satisfying the causal condition. Some references use reflexive relations in this context.

[11] Actually, as noted in Section 2.3, Zeeman works in terms of *timelike* structure. However, this makes no difference in the present context. Indeed, Malament's first lemma [MA77] states that such bijections preserve future-directed continuous null geodesics in a much more general context. In particular, they preserve the null segments joining null-separated elements of $\mathbb{R}^{3+1}$, and therefore preserve $\prec_{\text{GR}}$ in this special case.

relativity of simultaneity, discussed in some detail at the beginning of Chapter 7. Events x and y in Minkowski spacetime $\mathbb{R}^{3+1}$ are causally unrelated if $x \not\prec_{\text{GR}} y$ and $y \not\prec_{\text{GR}} x$. In this case, an inertial frame of reference F for $\mathbb{R}^{3+1}$ may be chosen under which x *temporally* precedes y; this may be done in many different ways. Each frame F therefore induces a special refinement R_F of $\prec_{\text{GR}}$, encoding the temporal order with respect to F. In this context, Poincaré covariance says that *"the laws of physics do not depend on a choice of refinement of the causal relation."* This viewpoint admits straightforward generalization to the case of arbitrary refinements in the discrete causal context, and has no necessary relationship to group theory. It is worth noting that the assignment $F \mapsto R_F$ is certainly *not* invertible; in particular, frames of reference related by a translation or rotation induce the same refinement. However, frames related by a boost induce different refinements. Hence, boosts are roughly analogous, at an abstract level, to ordered pairs of refinements of a partial order. Translations may be brought into the picture by working with *pointed* refinements, i.e., pairs (R, x), where R is a refinement of $\prec_{\text{GR}}$, and x is an element of $\mathbb{R}^{3+1}$. Rotations, as described above, may be expressed as sequences of boosts. This leads to the following analogy:

> **Refinement analogy:** *The Poincaré group P is analogous to the set of sequences of ordered pairs of pointed refinements of a binary relation.*

This analogy motivates the definition of the *Poincaré semigroup* below. Though unfamiliar from a conventional standpoint, it circumvents the issue of rigidity. It is also interesting to note that this particular analogy omits changes of *parity,* which again are "purely spatial."

Active generalization of the Poincaré group. Abstraction of Zeeman's causality group G provides one possible approach to generalizing the Poincaré group P:

Definition 8.4.2. *Let $D = (D, \prec)$ be a directed set, viewed as a model of classical spacetime under the classical causal metric hypothesis* (CCMH). *The* **(total) causality group** *$G(D)$ of D is the group* Aut(D) *of automorphisms of D in the category $\mathcal{D}$ of directed sets.*

This definition is very general; for example, it applies equally well in the "non-discrete" context of *domain theory,* and generalizes further to the multidirected setting. However, the rigidity result of Bender and Robinson [BR88] suggests that $G(D)$ does not provide a very useful generalization of P in the discrete causal context. There are at least two obvious ways of trying to circumvent this problem. First, one might consider automorphism groups of *subobjects* of directed sets that tend to be *nonrigid,* such as the *generational automorphism groups* described in Section 8.2. Of course, this would eliminate some of the traditional uses of P; for example, in classifying "particle types" in the background dependent context. Second, if one wishes to try to preserve a more conventional version of the active viewpoint, one might attempt to construct methods of defining and recognizing "near-automorphisms," by systematically ignoring local asymmetries in a scale-dependent fashion. This might allow discrete analogues of the Poincaré group to "emerge with increasing scale." This is yet another possible application for *causal atomic resolution,* described in Section 5.7. In particular, given a causal atomic resolution $\mathbf{D} = \{D = D^0, D^1, D^2, \ldots\}$ of a directed set $D = (D, \prec)$, the causality groups $G(D^0), G(D^1), G(D^2), \ldots$ may be interpreted as groups of "near-automorphisms" of D. For a judicious choice of resolution, one might hope that these causality groups would be nontrivial, and would resemble P over an appropriate range of scales. The general idea of "near-automorphisms" is important for trying to generalize results about systematic Lorentz invariance to apply to directed sets *not* induced by global sprinklings in $\mathbb{R}^{3+1}$, as discussed in Section 4.5. This might facilitate the study of interesting structural features, such as small-scale *dimension reduction,* which are invisible to sprinkling constructions. However, rigidity is so pervasive in this setting that it is not at all clear if one could construct meaningful groups using these methods.

Passive generalization of the Poincaré group. A completely different, and more natural, generalization of the Poincaré group P arises from the passive viewpoint.

Definition 8.4.3. *Let $D = (D, \prec)$ be a directed set, viewed as a model of classical spacetime under the classical causal metric hypothesis* (CCMH). *The* **Poincaré semigroup** $\mathbf{P}(D)$ *of D is the free semigroup generated by ordered pairs $\big[(R, x), (R', x')\big]$ of pointed refinements of the causal relation $\prec$ on D.*

This definition is also very general, and applies equally well to discrete and non-discrete models. It may also be generalized to the multidirected context, although this would require generalization of the theory of refinements discussed in Section 7.2. For the sake of brevity, I focus on the discrete directed case. From the foregoing discussion, it is clear how elements of $\mathbf{P}(D)$ may be compared to elements of P in certain special cases:

1. An ordered pair of the form $\big[(R, x), (R, x')\big]$ represents a "translation."
2. An ordered pair of the form $\big[(R, x), (R', x)\big]$ represents a "boost."
3. A sequence of ordered pairs of the form

$$\big[(R_1, x), (R_1', x)\big] \ldots \big[(R_n, x), (R_n', x)\big], \tag{8.4.4}$$

represents a "boost and rotation."

Obvious questions arise from these analogies. For example, how is a "pure rotation" represented? A reasonable answer, at the same rough analogical level of discourse, is that a "pure rotation" is represented by a sequence of the form (8.4.4), with the added condition that $R_1 = R_n'$. This follows the relativistic analogy of Thomas–Wigner rotation, in which a sequence of boosts "returns to the initial velocity." An interesting way of thinking about these analogies, at least from a mathematical perspective, is in terms of the theory of *relative directed sets over a fixed base,* introduced in Section 4.6 in the more general multidirected context. In particular, a refinement R of the causal relation $\prec$ on D induces a bijective morphism $(D, \prec) \rightarrow (D, R)$, whereby $(D, \prec)$ may be viewed as a relative directed set over (D, R). Similarly, an ordered pair of pointed refinements $\big[(R, x), (R', x')\big]$ may be viewed as an ordered pair of relative *pointed* directed sets sharing the same source. Figure 8.4.5 illustrates the resulting diagrams of morphisms for a "translation" and a "boost," as described above. Here, pairs of ordered *triples* are necessary, because one must specify a directed set, a refinement of its causal relation, and a distinguished element, for both the source and target of these morphisms. One must also remember that these pairs of morphisms are *ordered.*

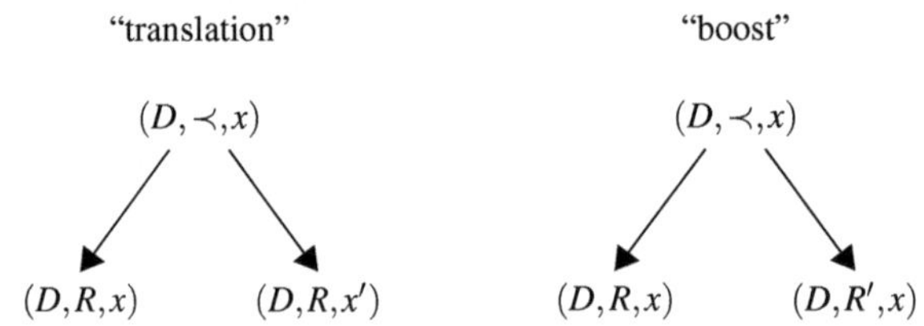

Figure 8.4.5. "Translations" and "boosts" in terms of morphisms.

The right-hand diagram, representing a "boost," is more attractive in terms of its abstract mathematical structure, since the morphisms involved are morphisms of *pointed* directed sets, i.e., morphisms that map the distinguished element of the first directed set to the distinguished element of the second directed set. If one chooses to focus on the Lorentz group, translations are removed from the picture, and one may work exclusively with morphisms of pointed directed sets. One of the advantages of working with such diagrams is that it provides useful intuition regarding the *composition of emergent group elements.* In particular, it is natural to try to "patch together" such diagrams to form larger diagrams, although I make no effort here to undertake any analysis of such diagrammatical calculus. It is also worth noting that these diagrams are abstractly similar to the *correspondences* arising in areas of physics that incorporate ideas from algebraic geometry and category theory, such as *twistor theory.* They also evoke certain constructions in *algebraic K-theory.*

Emergence of group elements under the passive viewpoint. The next logical question is how to characterize "physical equivalence" or "near-equivalence" of elements of the Poincaré semigroup $\mathbf{P}(D)$. This question is motivated by an ambiguity omitted from the discussion above. In particular, a given element p of the ordinary Poincaré group P represents *many different changes of inertial frames of reference,* one for each choice of initial frame. A boost, for example, really corresponds to an equivalence class of ordered pairs of pointed refinements, rather than a single pair. In the discrete causal context, one should not expect "exact equivalence" of pairs of pointed refinements. Rather, one should expect elements of P to emerge from large classes of elements of $\mathbf{P}(D)$ in an appropriate sense. To make this notion precise might require "something like a metric," or more precisely, a sequence of scale-dependent metrics, on $\mathbf{P}(D)$. These might be induced, in turn, by metrics on the space of refinements of the causal relation $\prec$ on D.

This is another area in which systematic methods of approximation such as causal atomic resolution might play a useful role. Given such a resolution $\mathbf{D} = \{D = D^0, D^1, D^2, \ldots\}$ of a directed set $D = (D, \prec)$, together with "metrics" $d = d^0, d^1, d^2, \ldots$ of some sort, on the spaces of refinements of the corresponding causal relations $\prec = \prec^0, \prec^1, \prec^2, \ldots$, one might characterize the "scale-dependent distance" between two refinements R and R' of $\prec$ in terms of the distances $d^n(R^n, R'^n)$ for corresponding refinements R^n and R'^n of $\prec^n$. This vague recipe raises many additional issues, such as the definition of the "metrics" d_n, the dependence of the procedure on the choice of resolution, the choice of the refinements R^n, and so on. One might hope, however, that the elements of $\mathbf{P}(D)$ would progressively separate into "near-equivalence classes" with increasing scale, at least for "dynamically favored" directed sets D. Despite the sketchiness of this description, and the a priori unwieldy nature of the Poincaré semigroup $\mathbf{P}(D)$ in general, the passive viewpoint at least offers the advantage of beginning with a rich algebraic structure in attempting to recover P. The active viewpoint begins, in the generic case, with no algebraic structure at all, due to rigidity.

Emergence of the group law. The free operation on the Poincaré semigroup $\mathbf{P}(D)$ may be viewed as a "precursor" to the group law on the Poincaré group P. Two special cases from relativity provide insight into the expected emergence of the group law under the appropriate dynamics. These two cases are *composition of translations,* and *composition of collinear boosts.* For translations, let x, x', and x'' be three elements of Minkowski spacetime $\mathbb{R}^{3+1}$, and let F be an inertial frame of reference at x. Let p be the translation changing F to the frame F' at x', and q the translation changing F' to the frame F'' at x''. Then the composition qp changes F to F''. By analogy, let D be a directed set with causal relation $\prec$, and let R be a refinement of $\prec$. Then in an appropriate quotient of $\mathbf{P}(D)$, it is physically reasonable to impose the relation

$$\big[(R, x), (R, x')\big]\big[(R, x'), (R, x'')\big] = \big[(R, x), (R, x'')\big]. \tag{8.4.6}$$

For *collinear* boosts p and q, the composition qp is again a boost, and this suggests the relation

$$\big[(R, x), (R', x)\big]\big[(R', x), (R'', x)\big] = \big[(R, x), (R'', x)\big]. \tag{8.4.7}$$

However, "collinearity" has no obvious fundamental meaning for arbitrary directed sets. Further reflection reveals that the relation (8.4.7) *cannot* be valid in general, because it ignores Thomas–Wigner rotation; in particular, it implies that the "composition of boosts"

$$\big[(R, x), (R', x)\big]\big[(R', x), (R'', x)\big]\big[(R'', x), (R, x)\big],$$

is equivalent to the pair $\big[(R, x), (R, x)\big]$, which should coincide with the identity in the emergent group law. One possible response to this complication might be to turn the logic around and use "deviation from Equation (8.4.7)" as a "measure of noncollinearity." This would require defining "something like a metric" on $\mathbf{P}(D)$, to measure the "sizes" of such deviations. Equations (8.4.6) and (8.4.7) involve sequences of ordered pairs of a special type, where each pair fixes either the refinement or the basepoint, and where the terminal pointed refinement of a given pair coincides with the initial pointed refinement of the next pair. For Minkowski spacetime, one may restrict attention to such sequences, because a given translation or boost may be specified by its action on any choice of frame. In particular, given an arbitrary sequence of pairs of pointed refinements, one may exchange pairs for equivalent pairs in such a way that these conditions are satisfied. In discrete causal theory, however, such equivalence is emergent, and arbitrary elements of $\mathbf{P}(D)$ cannot be expressed exactly in this special form. Instead, the general case seems to involve complex sequences of pairs, with no obvious relationships between consecutive pairs. Given the level of generality at which I have defined $\mathbf{P}(D)$, this rather obtuse behavior is not very surprising, since a typical directed set simply does not possess structure that supports anything resembling Poincaré symmetries. To go further, one might need to seek "assistance from the dynamics," to narrow down the class of directed sets to those favored under path summation, which should resemble relativistic spacetime over a large range of scales. In the meantime, one might hope to acquire insight by examining "artificially restricted" classes of directed sets, such as sprinkled causal sets, in this context.

Emergence of "rotation;" further remarks. An interesting specific feature of the foregoing discussion is the curious and indirect manner in which the notion of *rotation* arises in discrete causal theory, as an abstraction of Thomas–Wigner rotation in relativity. In the latter context, Thomas–Wigner rotation is often viewed as a "relativistic correction," rather than as a hint that rotation itself may not be fundamental. However, it seems to provide the only obvious generalization of what rotation should *mean* for an arbitrary directed set, and this obviates the need for any independent rotational notion, since a "rotation" may always be expressed as a "composition of boosts." The sketch of the emergence of the Poincaré group given in this section does not provide examples of directed sets for which a Lie group is actually proven to emerge from the Poincaré semigroup, nor does it even describe how to quantify such emergence. As mentioned above, these considerations are bound together with the problem of "how manifold-like directed sets arise dynamically in discrete causal theory." On the one hand, the natural emergence of Lie groups cannot be expected to arise without suitable dynamics; on the other hand, such emergence is a critical requirement for any viable dynamical approach. The hope is that the ideas described in this section will furnish a helpful point of reference for the further study and solution of these problems.

8.5 Elementary Graph Dynamical Considerations

Background: Erdös and Rényi. In the late 1950s, mathematicians Paul Erdös and Alfréd Rényi introduced the theory of *random graphs* in a long series of papers, beginning with [ER59].[12] The original context of this development was pure graph theory, with information-theoretic and physical applications appearing only afterwards. A simple example of an *undirected* random graph may be constructed via the following process: begin with a finite set V of cardinality N, called the *vertex set.* Next, consider each pair of elements (x, y) of V, in some order, adding an edge between x and y with some *fixed* probability p. When every pair of vertices has been considered, the process stops, and the result is a **random graph** of type (N, p). This constructive process is called a *graph dynamical process,* because the graph "develops" or "evolves" in *process time.* Figure 8.5.1 illustrates a random graph of type $(3, \frac{1}{2})$. The process begins with a set of three isolated vertices x_1, x_2, and x_3, shown in the far-left diagram. Next, a "coin is flipped" to determine whether an edge will be added between vertices x_1 and x_2. In this case, the answer is *yes;* the diagram second from the left shows the resulting graph. The process then continues, with each remaining pair of vertices considered independently. In this case, no edge is added between vertices x_1 and x_3, but an edge *is* added between vertices x_2 and x_3. At this point, the process stops, with the completed random graph illustrated in the far-right diagram. There is nothing about the graph itself that identifies it as "random;" rather, it is *considered to be random in this context because of the random process chosen to generate it.* From an abstract mathematical perspective, the vertex set V is merely a finite set, and the set of "potential edges" is merely the second symmetric power $V^{(2)}$ of V, i.e., the set of *unordered* pairs of elements of V. Hence, from this viewpoint, a random graph of type (N, p) is merely a random subset of a symmetric product.

Basic conceptual connections to discrete causal theory. One of the most important general aspects of discrete causal theory is the notion of the *evolution* of directed sets, which are equivalent to directed graphs from an abstract mathematical perspective. Hence, the use of dynamical "processes" in graph theory is naturally relevant

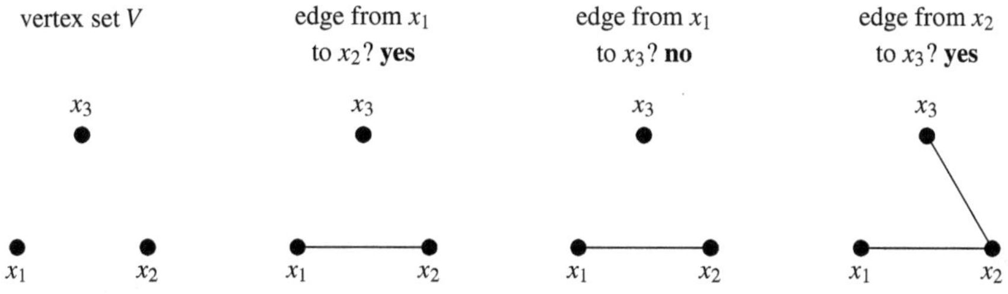

Figure 8.5.1. A simple graph dynamical process.

[12] A relatively modern reference on the subject is *Random Graphs* [BO01] by Béla Bollobás, one of the long-term leaders in the field.

to discrete causal theory, even in cases where the original motivation for adopting a dynamical perspective has nothing to do with physics. In particular, the sequential nature of the simple graph dynamical process illustrated in Figure 8.5.1 is closely akin to the process of evolution in the positive sequential kinematic scheme $\mathbb{S}_{PS}$. Hence, it is not surprising that this connection has already been examined in the context of Sorkin and Rideout's theory of *sequential growth dynamics* for causal sets [SR99]. In particular, Sorkin and Rideout explicitly mention the graph-dynamical viewpoint in their 2000 paper *Evidence of a Continuum Limit in Causal Set Dynamics* [SR01]. At a more sophisticated level, the theory of *transitions, co-relative histories,* and *kinematic schemes,* developed in Chapters 6 and 7 of this book, involve generalized versions of conventional graph dynamical processes. A number of complicating factors distinguish these physically-oriented constructions from the simple models originally studied by Erdös and Rényi, which still dominate the random graph literature. For example, the random graphs of interest in discrete causal theory are *directed* graphs, their vertex sets have no fixed cardinality, more complex structures than single edges are added at each step of the process, and the "probabilities" at each step depend on the details of the previous steps.

Suggestive graph-dynamical behavior. An important reason why the natural relationships between random graph dynamics and evolutionary aspects of discrete causal theory are so interesting is because of "peculiar" and physically suggestive phenomena occurring in random graph dynamics, phenomena apparently unimagined prior to Erdös and Rényi's papers. One class of phenomena of particular interest is graph-dynamical *phase transitions,*[13] already highlighted in Section 1.9. In very general terms, such a **phase transition** is an abrupt change in some qualitative property of a class of random graphs that occurs at some critical value of an associated parameter. The most obvious such parameter is the *process time* for the specified graph dynamical process producing a random graph in the class. However, other parameters may be considered as well, such as the probability p of the existence of an edge between a given pair of vertices. A simple and historically influential example of a family of graph-dynamical processes exhibiting such phase transitions may be described as follows: begin with the class of random graphs of type (N, p), as described above, and let N and p vary in such a way that $p = p(N)$ is a function of N. Now consider the limit $L := \lim_{N\to\infty} Np(N)$; conceptually, L measures the "combined influence" of two different factors; namely, the overall size of the graph, and the "average valence" of vertices in the graph. To obtain interesting cases, one must invoke a probability function p that "shrinks sufficiently rapidly" as N approaches infinity; otherwise, the limit L will be infinite. Hence, the classes of graphs under consideration have the general property that "an arbitrary pair of vertices in a large graph is less likely to be connected by an edge than an arbitrary pair in a smaller graph." However, this does not mean that larger graphs themselves are less likely to have strong connectivity properties, because each vertex in a large graph has many

[13]Of course, the meaning of the term "phase" here is entirely distinct from its meaning in the context of *phase theory* in Section 8.2.

more "potential partners" than each vertex in a small graph. It is interesting to compare this situation with the "exponentially decaying valences" in the recent paper [KKSR12], which involves causal set cosmology.

Given this setup, one may choose to gradually increase the function $p = p(N)$ in such a way that the limit L varies from less than 1 to greater than $\log N$. A sequence of phase transitions is then observed as p increases. These phase transitions encode abrupt changes in a specific measure of "connectivity" of a *typical random graph G constructed according to the specified process;* namely, the number of vertices belonging to the largest connected component of G. For sufficiently small values of p, each connected component of G typically possesses approximately $\log N$ vertices or fewer. As the limit of Np reaches the "critical value" of 1, a single *giant component* G_1 with roughly $N^{\frac{2}{3}}$ vertices suddenly emerges, and rapidly increases in size until it contains an appreciable fraction of all the vertices, while still leaving many small components and isolated vertices. This qualitative picture remains unchanged until the limit passes another "critical value" at $\log N$, at which point G_1 suddenly "swallows all the remaining components," and G becomes completely connected. Many other examples of such striking behavior exist throughout random graph dynamics.

It is almost unavoidable to ask if certain properties of fundamental physics might be explained in terms of such behavior in the discrete causal context. A particularly strong motivation for this idea is the desire to explain certain "sharp" and "precise" features of nature, such as the apparent integer dimension of spacetime, the apparently identical structure of certain types of "elementary particles," and the apparent manifestation of certain exact symmetries. At a naïve level, such uniformities might seem difficult to reconcile with the local irregularity at the heart of discrete causal theory, which contrasts so markedly with the local regularity of manifolds. However, statistical thermodynamics already provides a precedent for the possibility of probabilistic or entropic origins for behavior so sharp that its deviation from exact, deterministic behavior is virtually undetectable, and random graph dynamics provides natural mechanisms for describing how such processes may manifest themselves in discrete causal theory. Based on the most reasonable guesses about the size of the fundamental scale, there is "plenty of room" for such factors to achieve dominance; for instance, it would not be surprising to observe sharp behavior in random graphs modeling "particle interactions" involving perhaps 10^{100} fundamental elements, and an even larger number of fundamental relations. Of course, one of the most interesting potential applications of these ideas arises in a vastly larger context; namely, discrete causal alternatives to the inflationary hypothesis in the cosmology of the early universe, as outlined in Section 1.9.

8.6 Algebraic Structure and Hierarchy

Preliminary remarks. The remaining sections of this chapter gather together some miscellaneous material that could not be adequately covered in the main body of the book. The present section outlines some generalities regarding the subject of algebraic structure and hierarchy in discrete causal theory, which is encountered in previous chapters only via special cases. For example, some effort is made in Chapter 5 to explore the basic algebraic properties of *relation space,* but similar analyses could and should be carried out for *causal path spaces, kinematic schemes,* spaces of *co-relative histories, kinematic spaces,* and so on. Subsequent sections briefly detail some "more radical" possibilities for formulating discrete causal theory, sketch relationships between discrete causal theory and other fields of theoretical physics and modern mathematics, and give a few details regarding the genesis of the book.

Informal encounters with hierarchical topics. The analogy between kinematic schemes and categories naturally prompts a closer examination of their similarities and differences. This, in turn, leads to a broader consideration of algebraic structure in discrete causal theory, and in theoretical physics more generally, with a particular emphasis on questions of *hierarchy*. Isham's topos-theoretic framework for physics provides valuable and wide-ranging contributions in this direction. The principle of *iteration of structure* (IS), stated in Section 6.3, introduces the subject of hierarchy explicitly, but in a limited context, closely related to the conventional superposition principle of ordinary quantum theory. However, much more general notions of hierarchy are hinted at throughout Part II of the book, particularly in the study of relation space, causal path spaces, and kinematic spaces. In particular, *one may discern at least four distinct "vertical" levels of hierarchy, represented by elements, directed or multidirected sets, kinematic preschemes or schemes, and kinematic spaces.* Meanwhile, an infinite number of distinct "horizontal" divisions of hierarchy are readily apparent, represented, for example, by spaces of paths of various lengths in directed or multidirected sets.

Toward a systematic treatment of structure and hierarchy. It is not possible in this book to investigate the overall algebraic structure of discrete causal theory in an adequate manner. Even the structural details of many critically important objects, such as causal path spaces, are described here in a mostly informal way. Only the case of relation space is studied in a somewhat systematic fashion, and even this treatment leaves many elementary questions unaddressed. At higher levels of structure, the treatment given here amounts to only a brief glimpse of a few interesting examples. For instance, the theory of kinematic schemes is approached mostly via just four examples; namely, *Sorkin and Rideout's kinematic scheme* $\mathbb{S}_{\mathrm{SR}}$, the *positive sequential kinematic scheme* $\mathbb{S}_{\mathrm{PS}}$, the *generational kinematic scheme* $\mathbb{S}_{\mathrm{GEN}}$, and the *foliation scheme* $\mathbb{S}_{\mathrm{FOL}}$, all involving finite acyclic directed sets. Kinematic spaces are introduced without even a cursory examination of a single concrete example.

However, it is possible to briefly state some of the topics that *might* be addressed by a deeper examination of algebraic structure and hierarchy in discrete causal theory. These include the following: (1) more detailed comparison of discrete causal theory and category theory, particularly topos theory; (2) further explanation of the relevance and utility, or lack thereof, of various category-theoretic notions in physical contexts; (3) general discussion of the relationship between local and global aspects of hierarchy; (4) detailed comparison of categories endowed with operations, such as *monoidal categories,* with analogous structures in discrete causal theory; (5) systematic examination of discrete causal analogues of the theory of *categorification* and *decategorification,* generalizing the relationship between a kinematic scheme and its underlying multidirected set; (6) discussion of possible physical interpretations of additional levels of hierarchy, including further examination of *subclassical* and *hyperquantum* theories; (7) deeper analysis of the fate of Lie representation theory in theoretical physics under the causal metric hypothesis (CMH), including its bifurcation into *group-theoretic* and *non-group-theoretic* aspects, as in the case of the Poincaré group discussed in Section 8.4.

Local-global view of hierarchy. The remainder of this section is devoted to briefly examining just one of the points mentioned above; namely, the relationship between local and global aspects of hierarchy in discrete causal theory. This enables a more expansive view regarding the category-theoretic analogy than that adopted in previous chapters. As a byproduct, it leads to some interesting applications involving discrete causal analogues of familiar constructions, such as category-theoretic *limits.* Without attempting to be precise, one may describe the *"local hierarchy near a particular object in a class of similar objects"* as consisting of the natural relationships between the distinguished object and other individual objects in the class, while the corresponding *"global hierarchy"* involves the holistic properties of the class. Such local and global considerations may be recognized at each "level of hierarchy," where the nth level consists of objects constructed via n "globalizations." For example, in the specific context of discrete causal theory, the passage from elements to directed or multidirected sets, and thence to kinematic schemes, involves two globalizations. Hence, kinematic schemes exist at the second level of hierarchy. This, of course, is the "quantum level," under the physical interpretation used in this book.

One way to arrange these local and global relationships is in the form of a *binary tree,* where the "left branches" represent *"passage from objects to relationships between pairs of objects,"* which involves local structure, while the "right branches" represent *"passage from objects to classes of objects,"* which involves global structure. Such a binary tree encoding some of the hierarchical aspects of discrete causal theory is illustrated in Figure 8.6.1. Each rectangular node of the tree is labeled with a class of objects and a level of hierarchy, the latter indicated by the numbers in parentheses. As a visual aid, levels of hierarchy are color-coded as well as labeled, with darker shades indicating higher levels.[14] Vertical passage along each edge corresponds to "zooming out" on the structure of the theory, in either a local or global sense. The edge labels ℓ and g stand for "local" and "global," respectively. The most fundamental class of objects included in the tree is the class of elements, represented by the initial node of the tree. For simplicity, the "objects" at the first level of hierarchy are taken to be directed sets, rather than multidirected sets, but this choice may be immediately generalized. As indicated by the dashed upward edges in the figure, the tree may be extended. For example, n-chains occupy the left-most node at the

[14]Note, in particular, that the hierarchical level of a given class of objects appearing in the figure does *not* correspond to the "height" of the corresponding rectangular node on the page in this context.

nth row of the tree. Like elements and relations, however, they exist at the zeroth level of hierarchy, since they describe the internal structure of directed sets.

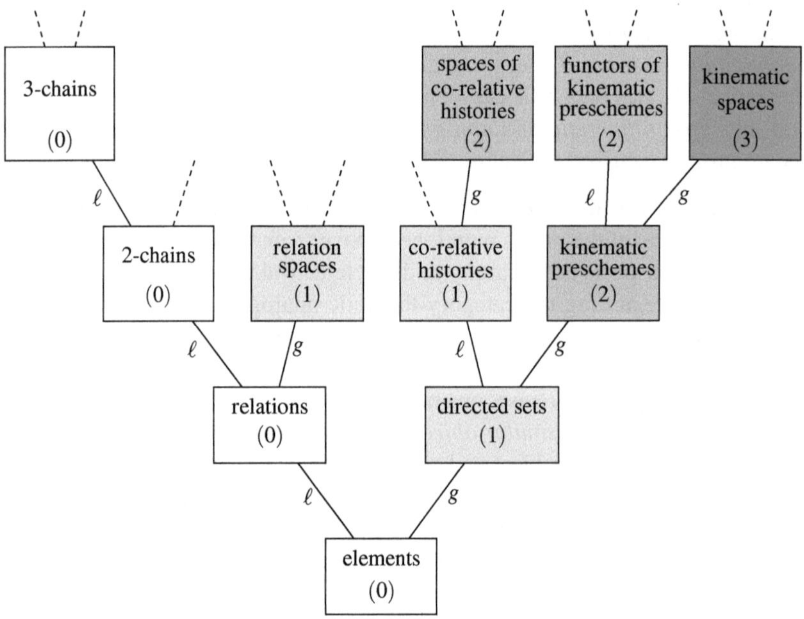

Figure 8.6.1. Partial view of hierarchy in discrete causal theory.

For the purpose of comparison, an analogous binary tree describing certain aspects of hierarchy in category theory is illustrated in Figure 8.6.2. Here, labels are suppressed to avoid clutter; however, color coding is again included as a visual aid, to indicate levels of hierarchy. Elements, when they exist, again occupy the initial node of the tree, but objects of a category need not possess elements. More generally, the lowest level of hierarchy, which represents the internal structure of objects, is not specified in any uniform way across classes of categories. In particular, some categories have objects with no internal structure at all; a familiar example is given by the category-theoretic description of a group, which possesses a single structureless object. This is why elements and relations appear in dashed boxes in the figure, while 2-chains and other zeroth-level constituents of internal directed structure are not shown at all. Of course, elements, relations, 2-chains, and so on, *do* exist for certain classes of categories; for example, the category of partially ordered sets. As in the discrete causal case, the binary tree for category theory illustrated here may be extended to include familiar higher-level structural entities, such as n-categories. All of the structures above the lowest level of hierarchy shown in the figure are both well-known and of central importance to category theory. For example, as noted in Section 1.5, *natural transformations* provided much of the early motivation for the development of category theory by Eilenberg and MacLane.

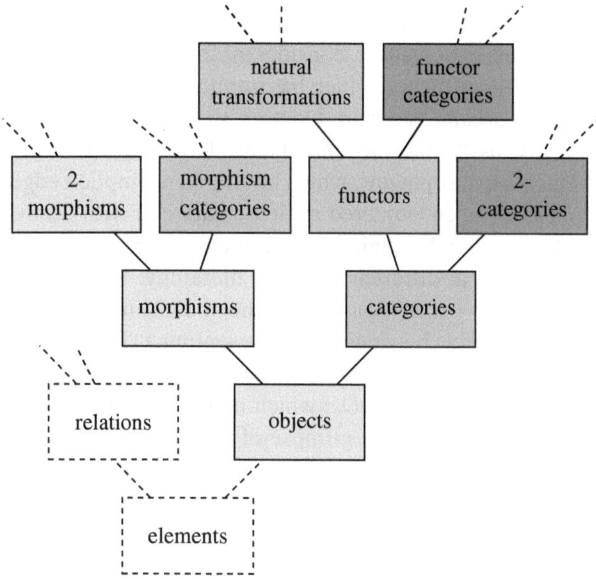

Figure 8.6.2. Partial view of hierarchy in category theory.

The binary trees illustrated in Figures 8.6.1 and 8.6.2 *themselves* possess the abstract structure of acyclic directed sets, with binary relations defined by the familiar "up the page" convention. The theory developed in this book may therefore be applied to these sets in interesting ways. For example, given any element x in an infinite binary tree $(T, \prec)$, there exists an isomorphism from the union $\{x\} \cup J^+(x)$ to $(T, \prec)$, which may be visualized as a *"translation of coordinates, with the new origin given by the distinguished element x."* In fact, there are an infinite number of such isomorphisms, given by rearranging the branches of the tree. However, in the present case, the edge labels ℓ and g, which encode passage to local or global hierarchy, select a unique isomorphism of this type. At a more formal level, one may replace ℓ and g with the elements 0 and 1 of $\mathbb{Z}_2$, and view the edge labels as values of a $\mathbb{Z}_2$-valued "relation function" on $(T, \prec)$. The distinguished "translation" is then elevated to an isomorphism in the enriched category of partially ordered sets equipped with such a function. This allows one to formalize the viewpoint that objects of a certain structural class, such as relations, are "elements" of an induced structure; in this case, a relation space. This procedure may be described as *"demoting objects to elements by forgetting their internal structure."* Similarly, given any element x of T, there exists an injective morphism of T sending the initial vertex of T to x. Again, edge labels select a unique morphism of this type. This allows one to formalize the notion of "promoting" classes of objects to a higher level of hierarchy. Together, these notions of "demotion and promotion" lead naturally to the broader subject of *generalized categorification* and *decategorification,* which has found important recent applications in a variety of algebraic settings.

The view of hierarchy sketched here suggests interesting unifications of familiar classes of constructions. For example, consider the 2-chain from objects to functors in the binary tree for category theory, which has implicit edge labels g and ℓ. Category-theoretic *limits* and *colimits* "convert functors to objects," and may therefore be viewed, heuristically, as "collapsing the chain." Now consider the corresponding 2-chain from elements to morphisms, which has the same implicit edge labels. Limits of functions, such as those encountered in elementary calculus, "convert morphisms to elements" by "collapsing the chain;" hence, the two constructions may be viewed as analogous procedures at different levels of hierarchy. In a similar way, a chain $\gamma := x(\emptyset) \prec x(D_1) \prec \cdots \prec x(D_n) \prec \ldots$ in the underlying multidirected set $\mathcal{M}(\mathbb{S})$ of a kinematic scheme $\mathbb{S}$ may be viewed as representing a "linear" kinematic scheme in its own right, and the "limit" of the "inclusion functor" mapping this kinematic scheme into $\mathbb{S}$ is the directed set $\cup_n D_n$, which may or may not belong to the object class of $\mathbb{S}$. This example provides a glimpse of how to better formalize the idea of *limits in kinematic schemes,* discussed at a mostly informal level in Section 7.9.

8.7 Miscellany Regarding Alternative Formulations

Further remarks regarding causal cycles. Substantial arguments exist both for and against the consideration of cycles in discrete causal theory. General *types* of objections to cycles include: (1) lack of observational evidence for their existence; (2) logical "problems," such as the *grandfather paradox;* (3) the fact that certain existing theories "conspire in unforeseen ways to avoid cycles," as illustrated by the *no-cloning theorem* in quantum information theory; and (4) technical complications arising in theories permitting cycles, such as the convergence issues for path sums mentioned in Section 6.7. The strong interpretation of the causal metric hypothesis (CMH) completely resolves the second type of objection, since it leads to theories with perfect background independence. The other types of objections, however, remain relevant in discrete causal theory. Arguments in *favor* of considering cycles include the following: (1*) general relativity permits closed causal curves; (2*) regardless of their classical status, cycles are generally unavoidable at higher levels of hierarchy, appearing, for example, in the underlying multidirected structures of kinematic schemes of countable acyclic directed sets; (3*) abstract element spaces over acyclic multidirected sets generally include cycles; (4*) cycles are generally nonlocal, and therefore do not necessarily lead to local pathologies; (5*) cycles are reasonable candidates for *subclassical structure,* i.e., "internal structure" of elements and relations in directed sets interpreted as models of classical spacetime.

Given this balance of positive and negative factors, it is worthwhile to briefly sketch some of the necessary groundwork for exploring the theory of causal cycles more systematically. In Section 3.7, I defined a cycle in a multidirected set $M = (M, R, i, t)$ to be a sequence of relations in M with the same initial and terminal element. Under this definition, cycles are special cases of chains of nonzero length in M, whose "ends join together." In Section 5.9, I generalized chains to causal paths, defined in terms of morphisms from linear directed sets into M. This greater level of generality is useful for a number of purposes; for example, it is necessary for studying *linear subobjects of transfinite length* in M. In a similar manner, it is useful to generalize cycles to *loops*, represented by morphisms from **cyclic directed sets** into M, where a directed set C is called *cyclic* if every pair of elements of C is connected by chains in both directions. As in the case of paths, the source of a loop is viewed as an isomorphism class, since pre-composition with an automorphism of the source produces a physically equivalent morphism into the target. Arbitrary cyclic directed sets generally contain many different cycles that are "distinct at the level of subobjects," in the sense that they involve different families of elements and relations. A cyclic directed set C is called **simple** if the continuum topology on the continuum model $C_{\mathbb{R}}$ of C coincides with the topology of S^1, i.e., if distinct cycles in C differ only in their "starting points and winding numbers." Any relation in a cyclic directed set C is reducible via chains "wrapping around cycles in C."

Definition 8.7.1. *Let $M = (M, R, i, t)$ be a multidirected set.*

1. *A* **loop** *in M is a morphism $\omega : C \rightarrow M$ for some cyclic directed set C, defined up to isomorphism. C is called the* **source** *of ω, and M is called the* **target** *of ω.*
2. *The* **loop set** *$\Omega(M)$ over M is the space whose elements are loops in M.*
3. *A loop $\omega : C \rightarrow M$ is called* **simple** *if its source is simple. The* **set of simple loops** *$\Omega_{\mathrm{SIM}}(M)$ over M is the subset of $\Omega(M)$ whose elements are simple.*

The image of a simple loop $\omega : C \rightarrow M$ is a cycle in M, whose underlying subobject in M is *not* necessarily simple; for example, it may have "self-intersections." Hence, the **set of cycles** in M, denoted by $\mathrm{Cyc}(M)$, may be identified with the set of simple loops $\Omega_{\mathrm{SIM}}(M)$ over M, just as the chain set $\mathrm{Ch}(M)$ of M may be identified with the irreducible causal path set $\Gamma_{\mathrm{IRR}}(M)$ over M, as described in Section 5.9. The theory of path summation over a multidirected set is generally much more complicated when cycles are permitted, and I make no attempt to describe its intricacies here. Instead, I briefly illustrate a simplifying procedure, which clears up much of the difficulty in certain interesting cases, while providing an intriguing window into how subclassical "internal structure" might be ascribed to elements and relations in a directed set, without discarding the strong interpretation of the causal metric hypothesis (CMH). Figure 8.7.2 illustrates this procedure, in the case of a directed set $(D, \prec)$ containing two disjoint maximal cyclic subobjects. The left-hand diagram shows D itself, with arrows indicating directions of influence. The two cyclic subobjects are highlighted in black. Informally, the procedure works by *"collapsing each cyclic subobject to a single element."* There is no actual dynamical aspect to this process; it is merely a method of partitioning information. The middle diagram shows these subobjects "nearly collapsed," while the right-hand diagram shows each subobject replaced by a single element. The result is an acyclic directed set, which is why no arrows appear in the right-hand diagram. At a more formal level, the two cyclic subobjects are *causal atoms* in D. One may perform a minimal *causal atomic decomposition* on D, whose elements are the "smallest possible" atoms. Under this decomposition, the two cyclic subobjects are the only two non-singleton atoms; every other element in D is an atom in its own right. In this manner, one may "isolate cyclic behavior" in D in such a way that methods such as path summation may be applied in a more straightforward manner.

Figure 8.7.2. Reduction of a causal relation containing cycles. Each connected family of cycles defines a causal atom, which may be viewed as a generalized element possessing subclassical "internal structure."

Alternatives to power spaces. Power spaces are only one of many ways to generalize binary relations. For example, the *causal path spaces* studied in Section 5.9 are not power spaces, since a morphism representing a causal path is generally not information-theoretically reducible to its image. However, even if one restricts consideration to *subobjects of individual structured sets,* rather than incorporating information about relationships among families of structured sets, a variety of alternatives to power spaces remain. In particular, the basic "binary" structure of relationships between *pairs* of objects is retained in the theory of power spaces; it is only the individual *members* of these pairs that are allowed greater complexity. An alternative viewpoint is to exchange pairs for N*-tuples.* For example, a **ternary relation** on a set S is a subset of the third Cartesian power $S \times S \times S$; its elements are *ordered triples*. Similarly, an N**-ary relation** on S is a subset of the Nth Cartesian power S^N. A "multidirected version" of a set equipped with an N-ary relation is an $(N+2)$-tuple of the form $\mathcal{S} = (S, R, e_1, \ldots, e_N)$, where S and R are sets, and where the maps $e_i : R \to S$ are "generalized initial and terminal element maps." In this context, an element of R may be viewed as an $(N-1)$-dimensional *simplex,* and the total structure of $\mathcal{S}$ may be viewed as an $(N-1)$-dimensional *simplicial complex.* One may also allow N to vary, thereby obtaining simplicial complexes of nonconstant dimension. In either case, random graph theory is subsumed by *random simplicial theory,* and mathematical problems such as *Whitehead's conjecture* come into play. The associated physical theory may be viewed as a generalization of *simplicial power spaces,* briefly mentioned in Section 5.6.

Another way to generalize binary relations is to consider *multiple independent* binary or N-ary relations on a single set. This may be accomplished in the context of multidirected sets, or N-ary analogues thereof, by adding *colors* to elements of relation sets or generalized relation sets. One obvious motivation for such a construction in the context of discrete spacetime structure would be the belief that the strong interpretation of the causal metric hypothesis (CMH) is *false*. For example, one might regard *spatial* as well as causal structure as fundamental. One might then invoke *two* relations: a *symmetric* relation encoding spatial structure, and an *antisymmetric* relation encoding causal structure. Of course, there already exist familiar theories involving combinations of spatial and causal structure on discrete manifolds at the fundamental scale; for example, causal dynamical triangulations and loop quantum gravity. At a broad conceptual level, there seems to be at least a partial *duality* involving spatial/causal structure and symmetric/antisymmetric binary relations in physics, observed to some extent even in the context of relativity. As far as I know, the scope and significance of this duality has not been definitively analyzed. In particular, one finds "time-first" theories, such as causal set theory, "space-first" theories, such as shape dynamics, and hybrid theories, such as causal dynamical triangulations, all competing for the same theoretical ground, with only modest interactions among them. It is conceivable that many of the basic mechanisms for encoding physical information represented by these theories are to some degree interchangeable.[15] Another possible application of multiple binary or N-ary relations might be to encode *different types of spatial or temporal structure.* For example, a pair of independent antisymmetric relations on a set might be interpreted as a *pair of timelike dimensions.* More generally, power spaces, N-ary relations, and colors may be combined in various ways as a "theme and variations." If nothing else, this concept creates fertile ground for speculation; for example, one might invoke whole families of independent relations, with "time" emerging as a "residual drift." If history is any judge, however, the simplest nontrivial versions of ideas are usually the most important.

8.8 Connections with Other Physical Theories

Preliminary remarks. Interesting analogies and potential connections exist between the version of discrete causal theory developed in this book and a variety of other approaches to physics at the fundamental scale. Although some of these connections have been partially explored in the context of previous versions of discrete causal theory, the new perspectives and methods presented here justify their re-examination in many cases. Below, I give an incomplete and informal rundown of some of these connections, not arranged in any deliberate order. Some of the theories involved have already been mentioned at various points in the book. Conspicuously absent is string theory, which has too little in common with the discrete causal perspective to warrant further discussion in this context, besides a footnote to the discussion of loop quantum gravity, and an oblique remark involving "tubular diagrams" at the very end of the section. This does not mean that string theory is not interesting in its own right, but if string theory is even "close to being correct" as a theory

[15] I have previously referred to this idea as *"shape-causal duality,"* for example, in forum discussions with other members of the Foundational Questions Institute.

of fundamental physics, then discrete causal theory is almost certainly "completely wrong," except as an operational tool in applied fields such as quantum information theory. Conversely, if discrete causal theory is even "close to being correct" in a fundamental sense, then string theory is probably useful only for its contributions to mathematics; for example, to algebraic geometry. By contrast, there are many aspects of causal set theory, loop quantum gravity, causal dynamical triangulations, and the other theories mentioned below, that will remain interesting and relevant even if the version of discrete causal theory developed in this book proves to be thoroughly successful. Causal set theory is omitted from the present discussion for a different reason; namely, because it has already been discussed in some detail at various points throughout the book.

Canonical quantum general relativity; loop quantum gravity. The oldest, and perhaps the most "obvious," avenue toward quantizing general relativity is to re-express the Einstein–Hilbert action in Hamiltonian form, then attempt to apply familiar, if forbidding, quantization machinery. It is traditional throughout physics to associate the word "canonical" with the Hamiltonian approach. Technically, the canonical approach to quantizing general relativity involves foliating a relativistic spacetime manifold X into "spatial sections," at least locally,[16] and thereby decomposing the Einstein–Hilbert action, as in the *ADM formalism.*[17] One may then apply the *Legendre transform* to obtain a family of *Hamiltonian constraints,* and attempt to quantize the resulting constrained system. This task is so difficult that it was largely abandoned after the 1960s, until the emergence of loop quantum gravity in the late 1980s. Loop quantum gravity has made considerable progress toward canonical quantization via a variety of new methods and insights. Rovelli [RO04] and Thiemann [TH07] are standard references on the subject. At present, loop quantum gravity is the leading mainstream alternative to string theory as a candidate for describing physics beyond the scope of general relativity and the standard model of particle theory.

The principal manner in which loop quantum gravity is encountered in this book is as an example of a theory that *arrives at a form of fundamental discreteness via a quantization procedure,* beginning from continuum-based assumptions, rather than building discreteness into the picture from the beginning. A posteriori, the continuum may be discarded, and spacetime may be viewed as a superposition of combinatorial objects called *spin foams,* as described below. This curious, indirect path to fundamental discreteness is actually viewed as a conceptual *strength* of the theory by some of its leading proponents. For example, in his seminal book on the subject, Carlo Rovelli remarks:

> *The program of* [loop quantum gravity] *is therefore conservative, and of small ambition. The physical inputs of the theory are just* [quantum mechanics] *and* [general relativity], *well-tested physical theories. No major additional physical hypothesis or assumption is made (such as: elementary objects are strings, space is made of individual discrete points, quan-*

---

[16]Of course, if X is globally hyperbolic, the foliation extends to cover X; see Thiemann [TH07], p. 40, for a discussion in the context of loop quantum gravity.

[17]Here, ADM stands for "Arnowitt, Deser, and Misner." See Thiemann [TH07], Chapter 1, for details of this decomposition.

> *tum mechanics is wrong,* [general relativity] *is wrong, supersymmetry, extra dimensions,...)* ([RO04], page 14)

In the resulting theory, however,

> *...loop quantum gravity leads to a picture of the short scale structure of spacetime extremely different from that of a smooth background geometry... Spacetime turns out to have a non-perturbative, quantized, discrete structure at the Planck scale...* ([RO04], page 8)

Similarly, in his definitive mathematical treatise on canonical quantization, Thomas Thiemann states:

> *The successes of* [loop quantum gravity] *are a mathematically rigorous framework, manifest background independence, a manifestly non-perturbative language, an inherent notion of quantum discreteness of spacetime which is derived rather than postulated, certain* [ultraviolet] *finiteness results, a promising path integral formulation (spin foams), and finally a consistent formulation of quantum black hole physics.* ([TH07], page 17)

From an objective viewpoint, this process of beginning with structural notions that are *not* believed to be fundamental; namely, continuum-based geometry, in order to arrive at structural conclusions that *are* believed to be fundamental; namely, discrete microstructure, must be regarded as odd, to say the least. Thiemann acknowledges this point, writing:

> [Non-orthodox approaches such as discrete causal theory] *start by questioning standard quantum field theory at an even more elementary level. Namely, if the ideas about spacetime foam (discrete structure of spacetime) are indeed true, then one should not even start formulating quantum field theory on a differentiable manifold, but rather something intrinsically discrete... These approaches are... the most difficult ones because the contact with standard quantum field theory is, a priori, very small... However, the ideas spelt out in these programmes could well reappear in the former approaches as well, once these have reached a sufficiently high degree of maturity in order to take the quantum leap to a more fundamental formulation.* ([TH07], page 22)

This quotation provides interesting context for Rovelli's remark, cited above, that loop quantum gravity is *"conservative, and of small ambition;"* the implication is that the discreteness emerging in the theory might later be reassigned as an *assumption* in more advanced versions of the theory, or in subsequent theories. In any case, "continuum first" approaches seems far less unusual to anyone steeped in the history of modern physics, and particularly in the practice of constructing quantum theories by "quantizing" classical theories. In particular, since general relativity, which is classical in nature, *does* involve a *"smooth background geometry,"* Rovelli's statement that loop quantum gravity *"does not assume that general relativity is wrong"* must be interpreted with the implicit addendum *"as a classical theory."* As discussed below, this viewpoint is based on a perspective regarding the relationship between classical and quantum theory that seems to differ significantly from the one I have chosen to adopt in this book.

In general, I agree with Rovelli and Thiemann that loop quantum gravity is a "good theory," in the sense that it is likely to make additional contributions to the understanding of fundamental physics beyond the scope of relativity and the standard model. This is partly because the assumptions that the theory *does* make are "at least no worse" than those of the most advanced experimentally established theories. For example, one would ultimately prefer not to make any assumption at all regarding the dimension of spacetime, but if one does make such an assumption, it is better not to assume what all evidence suggests to be the *wrong* dimension, as is done in string theory.[18] An excellent reason for appreciating loop quantum gravity is that its results suggest certain aspects of fundamental structure that seem very plausible from a variety of different viewpoints, including the viewpoint of this book. Having once discovered these structural principles, one may implicitly "kick away the ladder," and may take the view that *general relativity has merely provided crucial clues regarding what types of discrete combinatorial objects should be taken to contribute to quantum theory.*

However, there are at least two, and possibly three, areas in which I question Rovelli's and Thiemann's viewpoints, as I understand them. The same reservations apply to the perspective of the loop quantum gravity community in general. First, I do *not* view the discrete causal assumption that spacetime is *"made of individual discrete points* [and relations]*"* as an *"additional physical hypothesis"* supplementing the other basic *"physical inputs"* of the theory. Rather, I view discreteness as a *much simpler and more plausible replacement for the suspect hypothesis of continuum-based structure.* In other words, I disagree that the assumption of discreteness adds "baggage" to the theory; rather, I think that it *removes* "baggage" of the worst sort. Second, I believe that general relativity *is* wrong as a classical theory, in a sense similar to the manner in which Newtonian physics is wrong; i.e., I believe that it provides an ingenious but imperfect description of classical physics, invalid beyond a certain level of detail. I do not mean this statement as a mere platitude, to the effect that "future physicists may always tweak existing theories." Instead, I think that it is possible to make reasonable guesses about what this critical level of detail is, and I am confident that the consequences are relevant and important to both present and future physics. As Newtonian physics proved to harbor two entirely different sources of imprecision, provisionally addressed by general relativity and quantum theory, respectively, it is not unreasonable to be concerned that a theory of such *"small ambition"* as to assume that both of the latter theories are correct is simply not ambitious enough. In any case, what I believe *is* correct about general relativity is its qualitative insight that *spacetime is dynamical in nature,* and in this regard I agree with the loop quantum gravity community wholeheartedly.

A third *possible* area of disagreement with Rovelli, Thiemann, and other loop theorists concerns the basic nature of quantum theory, and how it should be expected to relate to classical theory. In particular, I feel that most of the quantization procedures used to construct quantum theories from classical theories are conceptually suspect,

[18]From the string-theoretic perspective, the necessity of working in 10, or 11, or 26 dimensions, depending on the context, is often represented in a favorable light, as a *prediction* of the fundamental dimensionality of spacetime, which general relativity fails to supply. Unfortunately, these values are greatly at odds with observation, and must be hidden via baroque and dubious "compactification" methods.

and work only because of very special assumptions about basic structure implicit in the types of classical theories to which they are usually applied. In particular, I think that most of the familiar machinery of quantization via operator algebras on function spaces over real manifolds is ultimately irrelevant, except in an emergent sense. Instead, I think that the proper way to view quantization is by combining families of evolutionary processes for classical histories, with Feynman's path summation approach to ordinary quantum theory representing the first reasonably mature realization of this idea. From this viewpoint, a model of "classical spacetime" ought to represent one of the "physical alternatives" involved in the corresponding quantum theory, not a completely different type of object. In particular, if hypothetical Planck-scale measurements are expected to return discrete values for "spacetime-related quantities" such as lengths and areas, as in loop quantum gravity, then the "classical objects" involved in building up a suitable quantum theory *should be discrete structures themselves, not manifolds.* Of course, such a potential disagreement may ultimately reduce to unimportant differences in terminology; for example, one might refer to a spin foam as a "quantum history," while placing it in essentially the same structural role as a "classical history" in this book. Similarly, "classical spacetime" might be taken to mean what "emergent geometry" means in this book.

However, this discussion brings to light a generic conceptual difficulty associated with most, if not all, attempts to "directly quantize" general relativity; namely, the "phantom role" that the original spacetime manifold plays in the theory. For example, in operator algebra-based approaches to conventional quantum theories or quantum field theories, one begins with a background real manifold, regarded as either "spacetime," or as some closely-related object, such as a fiber bundle over spacetime. One then considers function spaces over this manifold, along with algebras of operators on these function spaces, and proceeds to examine the spectra of these operators.[19] Manifestations of discreteness in conventional quantum theory are often understood in terms of the fact that some of these spectra happen to be discrete. In this context, the background manifold remains in the picture throughout the procedure, and is accorded physical existence, *regardless of the discreteness of the quantities involved in the resulting quantum theory.* When attempting to quantize general relativity in this manner, however, the *manifold itself* is supposed to "disappear from the picture," even though its underlying structure is the source of much of the information involved in the algebraic objects that lead to the "quantized spacetime" in the first place! Hence, one is expected to believe that certain objects; in this case, pseudo-Riemannian manifolds of Lorentz signature, which are *not believed to exist,* either as backgrounds for a quantum theory, or as "physical alternatives" contributing to the quantum theory itself, should nonetheless be relied upon to supply most of the basic information-theoretic content of the theory! There is nothing *contradictory* about this

[19] At a more sophisticated level, one may begin with abstract algebras of special types, and may consider *representations* of these algebras, which associate their elements with operators on various spaces. Conceptually, this viewpoint moves in the direction of *noncommutative geometry,* in which special algebraic structures, rather than special geometric structures, are isolated and elevated in significance.

viewpoint, but consideration of the historical and sociological environment in which the desire to quantize general relativity arose leads to the inevitable conclusion that such quantization methods would never have enjoyed such prominence if physicists had not *already experienced success* in using algebraic quantization techniques in the very different setting of background dependent theories, and if it were not *already known* that Lorentzian geometry is "the correct low-energy classical limit in quantum gravity."

In any case, quibbles involving the historical and conceptual road to loop quantum gravity do not disqualify the integrity of its technical apparatus, nor its results, which have led to an interesting combinatorial picture of spacetime microstructure, similar in important ways to the viewpoint of discrete causal theory. In particular, the *"nonperturbative, quantized, discrete structure"* of loop quantum gravity, mentioned in the above quote of Rovelli, is represented mathematically via objects called *spin networks* and *spin foams,* which may be viewed as "decorated graphs," and "decorated cell complexes," respectively. The "decorated structure" of these objects is defined in terms of algebraic quantities, such as group representations, associated with their vertices and higher-dimensional faces. Spin networks predate loop quantum gravity; in particular, Roger Penrose had adopted their use by the early 1970s, and the general practice of using decorated graphs to represent local quantum interactions goes back to Feymman diagrams, introduced to mainstream physics in the 1940s. A spin network may be used to represent an "instantaneous quantum state of the gravitational field," while a spin foam represents a "history of evolving spin networks." Quantum spacetime may therefore be viewed as a *superposition of spin foams.* "Non-gravitational matter" involves auxiliary structure in this context.

The basic similarities between spin foams and directed sets leads to many potential intersections between loop quantum gravity and discrete causal theory. The most obvious place to begin exploring such connections is by comparing the similarities and differences in how spin foams and directed sets encode information. In particular, a spin foam certainly *contains* causal structure, but it also contains a large quantity of "extra" spatial structure and algebraic structure. In light of the metric recovery theorems, it is reasonable to wonder if all this structure is really necessary for recovering "just gravity." If so, then the strong interpretation of the causal metric hypothesis (CMH) is *much* too ambitious. Marina Cortês and Lee Smolin have recently explored exactly this type of connection between loop quantum gravity and causal set theory, under the moniker *"energetic causal sets"* [CS14A, CS14B, CS14C]. Another interesting general topic to consider in the context of loop quantum gravity is the theory of co-relative histories and kinematic schemes. In particular, spin foams may be regarded as representing discrete histories, and one may therefore consider co-relative histories between pairs of spin foams. Adopting a global viewpoint, one ought to obtain kinematic schemes that look very similar to those arising in discrete causal theory, despite the significant structural differences between individual histories in the two theories. Hence, one ought to be able to relate the corresponding dynamical approaches. In particular, ideas such as relation space might prove very useful in loop quantum gravity at higher levels of structure. Spin foams lead to the "path integral viewpoint" regarding canonical quantization, which is analogous to the path summation approach to discrete quantum causal theory.

An interesting technical point of comparison between discrete causal theory and loop quantum gravity involves the interplay between *systematic Lorentz invariance* and *spatiotemporal locality* discussed in Section 4.5. In particular, spin foams involve graphs of small valence, and there is no known method of embedding such a graph into Minkowski spacetime $\mathbb{R}^{3+1}$ in a manner that preserves systematic Lorentz invariance. Many of the papers involving sprinkled causal sets appearing in the recent literature explicitly mention this issue as a point in favor of causal set theory and against loop quantum gravity. For example, the paper [BHS09], cited repeatedly in Section 4.5, notes that *"...the theorem also shows that a finite-valency graph like a "spin-foam" cannot be equivariantly associated to a* [global] *sprinkling of Minkowski* [spacetime]*."* This problem has been seriously considered by loop theorists, whose contention seems to be that loop quantum gravity recovers a suitable form of local Lorentz invariance at the *quantum* level. For example, the 2003 paper *Reconcile Planck-scale discreteness and the Lorentz–Fitzgerald contraction* [RS03], by Rovelli and Speziale, puts it in the following way:

> *A Planck-scale minimal observable length appears in many approaches to quantum gravity. It is sometimes argued that this minimal length might conflict with Lorentz invariance, because a boosted observer can see the minimal length further Lorentz contracted. We show that this is not the case within loop quantum gravity. In loop quantum gravity the minimal length (more precisely, minimal area) does not appear as a fixed property of geometry, but rather as the minimal (nonzero) eigenvalue of a quantum observable. The boosted observer can see the same observable spectrum, with the same minimal area. What changes continuously in the boost transformation is not the value of the minimal length: it is the probability distribution of seeing one or the other of the discrete eigenvalues of the area.* ([RS03], page 14)

Whatever the details underlying this particular explanation, the general lesson that *essentially quantum-theoretic properties* must be taken into account in evaluating the integrity of spacetime models is an important one. This provides a degree of flexibility in the consideration of models that seem untenable at the classical level. In particular, in the discrete causal context, one might investigate the viability of directed sets with very *small valences,* which might combine at the quantum level to preserve certain desirable properties that are sacrificed by any individual such set. However, it is an undeniable advantage of discrete causal theory that it *can* accommodate reasonable approximations of systematic Lorentz invariance and spatiotemporal locality even at the classical level.

Causal dynamical triangulations. I have already mentioned causal dynamical triangulations at several points throughout the book; in particular, in Sections 1.9, 1.10, 4.1, 4.5, 5.6, and 5.7. In many cases, these remarks have merely explained how the viewpoint and methods developed in this book may be applied to "triangulated spacetimes" in general, such as *simplicial power spaces,* but I have also highlighted some striking results arising in causal dynamical triangulations itself, including *phase transitions from non-geometric to geometric phases* [AJJR12], and *dimension reduction* [AJL05]. Overall, causal dynamical triangulations provides a very worthwhile and promising viewpoint, partly *because* it reduces the level of generality to focus on certain special cases that are especially likely to be physically interesting. In particular, even if one wishes to pursue the version of discrete causal theory developed in this book as a fundamental approach, one might very well choose to make simplifying assumptions for the purposes of testing the basic plausibility of the theory in simulations. Hence, more "pragmatic" approaches such as causal dynamical triangulations should not be viewed as "competing theories," but as facets of the same broad program. From a historical standpoint, the theory of causal dynamical triangulations is closely related to loop quantum gravity, and many of the same researchers work on both theories.

Causal dynamical triangulations are "discrete Lorentzian spacetimes," built from four-dimensional simplices, sometimes called *pentachora* in the literature. For the purpose of visualization, a pentachoron is often projected into 3-dimensional space as a tetrahedron with an additional vertex added at its center, and with edges connecting this vertex to its four other vertices. The five three-dimensional faces of the pentachoron are then identified as the four "internal cells" resulting from this construction, together with the "outer tetrahedron" itself. The "Lorentzian" nature of spacetime in causal dynamical triangulations arises by *assuming* a Lorentzian structure on each pentachoron, and by requiring that pentachora glue together in a "causally consistent" manner, i.e., that "timelike" or "null" edges of neighboring pentachora share the same orientation. Given a causal dynamical triangulation, one may construct a directed set by assigning an element to each pentachoron, and a relation between each pair of pentachora sharing a lower-dimensional sub-simplex through which a causal curve passes in the "glued geometry." The resulting directed set generally does not preserve all the information included in the original triangulation, because its binary relation ignores "spatial adjacency." Hence, only part of the spatial structure of the original "glued geometry" may be recovered from this relation. For this reason, there is a *degeneracy* involved in relating causal dynamical triangulations to directed sets. However, causal dynamical triangulations are "much closer to plain directed sets" than are spin foams.

Significant work has been done on the path summation approach to quantum theory for causal dynamical triangulations, and it would be interesting to carry out a detailed comparison of this work to the very general version of path summation developed in this book. For example, obvious questions regarding the relationship between spatial and causal structure in causal dynamical triangulations could be revisited with the benefit of relation space methods, the theory of co-relative histories and kinematic schemes, and the overall structural viewpoint offered in Part II of the book. In particular, one could undertake a deeper study of degeneracy issues, i.e., the problem of how path sums in causal dynamical triangulations are altered if one ignores certain aspects of spatial structure. Further, one could analyze the precise relationships between causal dynamical triangulations and the simplicial power spaces introduced in Section 5.6, which have "only causal edges."

Quantum information theory; quantum automaton theory. A curious interplay exists between the "synthetic" theory of quantum computing, in which one applies ordinary quantum theory to relatively large-scale information processing, and the fundamental theory of *how information is actually encoded and distributed at the fundamental scale.* Under the familiar Hilbert-space approach to quantum mechanics, quantum circuits may be modeled abstractly as special directed sets, equipped with matrix-valued relation functions. Applying the methods and perspectives developed in this book, one may then regard quantum computers as *analogues* of spacetime. This automatically introduces ideas of a *relativistic* flavor, such as *frames of reference, relativity of simultaneity,* and *domains of influence,* even though the underlying quantum theory is nonrelativistic. Naïvely, one might hope to turn this analogy around, and use quantum computers as "virtual Planck-scale laboratories," in which properties of discrete causal theory might be precisely simulated, thereby avoiding the seemingly insurmountable engineering obstacles involved in probing such scales directly. However, it would be very ironic if one could actually leverage elementary quantum theory to discover new information about a more fundamental theory in this manner. An alternative scenario is that the actual behavior of quantum computers may eventually reveal *defects* in their conventional modeling, and this may point the way to more fundamental explanations, in which discrete causal theory could play a role. It might be possible to search for such defects deliberately. For example, there has long been speculation about whether or not quantum evolution is actually precisely *unitary,* and it would be natural to ask if sufficiently complex quantum computations might be designed to "amplify" small deviations from unitarity in such a way as to shed light on this question. In the meantime, the structural similarities between directed sets and quantum circuits are worth exploring in more detail.

In the closely-related field of *quantum automaton theory,* G.M. D'Ariano and Paolo Perinotti [DP14] have recovered a version of the *Dirac equation* via fixed lattice models of spacetime, together with a list of simple evolutionary rules. It would be interesting to compare their approach with the path-summation method for deriving causal Schrödinger-type equations presented in Section 6.9. Ultimately, the use of a fixed background in quantum automaton theory carries with it all of the usual philosophical and structural baggage associated with conventional quantum theory and quantum field theory. However, it is possible that the advantages of discreteness may enable this approach to uncover glimpses of new physics in spite of its limitations. From this viewpoint, quantum automaton theory may be regarded as a "pragmatic theory" aimed at short-term progress, involving assumptions even more restrictive than those of loop quantum gravity or causal dynamical triangulations, with the possible payoff of more readily-accessible results. While it may be the case that this approach does not incorporate enough of the conceptual insights of general relativity to seriously contribute to the study of quantum gravity, it may still lead to new results involving interactions in particle theory.

Twistor theory. As described in Sections 5.1 and 5.8, Penrose's *twistor space* T over Minkowski spacetime $\mathbb{R}^{3+1}$ corresponds bijectively to a quotient space of the *null subspace* of the relation space $\mathcal{R}(\mathbb{R}^{3+1})$. Since relation spaces play such an important role in discrete causal theory, this relationship suggests that analogies between twistor theory and discrete causal theory may be interesting to explore in a broader context.

However, the absence of a clear distinction between "timelike relations" and "null relations" in arbitrary directed sets creates an immediate challenge in formulating such analogies in a precise manner. Further, passage to relation space in the discrete causal setting generally "keeps information local," in the sense that typical relations are of "small causal extent,"[20] while null rays "extend throughout spacetime." In this sense, twistor spaces might be considered more similar to *causal path spaces* than to relation spaces, although "most" causal paths should be considered "timelike," rather than "null." In any case, even rough analogies between the two theories might be worth examining further, since both theories encode basic aspects of spacetime structure in at least somewhat similar ways. There are also potential *algebraic* connections between the two theories, involving, for example, constructions such as *correspondences.*

Entropic gravity. Discrete causal theory is *at least* a theory of gravity, with broader applicability depending on which interpretation of the causal metric hypothesis (CMH) one wishes to adopt. As illustrated in Section 8.2, entropic considerations are of great interest in the determination of phase maps for path summation in discrete causal theory. Hence, there exists at least an abstract connection to Verlinde's notion of entropic gravity, which is based on the general idea that gravitation is essentially a statistical phenomena. However, if the strong interpretation of the causal metric hypothesis is valid, and if the "correct" choices for phase maps really are defined in entropic terms, then the second law of thermodynamics is ultimately the *only* fundamental physical notion, aside from the basic structural assumptions necessary to identify a suitable class of classical histories on which to build the quantum theory. Under this scenario, "entropic gravity" is subsumed by an "entropic unified theory."

Shape dynamics. As mentioned in Section 5.6, shape dynamics adopts a "space-first" viewpoint, nearly opposite to the "time-first" approach of discrete causal theory. Interesting connections between shape dynamics and generational dynamics in discrete causal theory arise when one considers evolutionary processes; conformal 3-geometries in shape dynamics are analogous to generations in discrete causal theory, while evolution curves in the shape space of all such geometries are analogous to kinematic accounts in kinematic schemes. Despite these obvious analogies, the evolutionary viewpoints involved in the two theories are very different. In discrete causal theory, generations are modeled by antichains of elements, and are therefore structureless, although nontrivial information such as generational automorphism groups may be associated with individual generations, by considering "recent" aspects of causal structure. In shape dynamics, by contrast, intricate geometric structure is ascribed to "generations" a priori, while the "causal" or "temporal" structure describing evolution is as simple as possible. As mentioned at the end of Section 8.7, it might be interesting to examine the question of whether or not *dual relationships* might exist between appropriate versions of shape dynamics and discrete causal theory, as sometimes occurs for other pairs of "opposite approaches," such as certain homology and cohomology theories. The importance of *conformal* structure in shape dynamics

[20] This is true, at least, in the star finite context.

increases the motivation for comparing the two theories, since conformal structure features prominently in the metric recovery theorems underlying the causal metric hypothesis (CMH). It is possible to formulate discrete versions of shape dynamics, but the fact that shape dynamics has been studied mostly in the continuum-based setting suggests that one might begin by looking for "duality-type" results between shape dynamics and *domain theory,* which may be characterized as a version of "continuous causal theory." This idea is briefly revisited below.

Other theories. More tenuous "connections" involving other physical theories may be exhibited. For example, it is mathematically tempting to "promote" directed sets to *tubular diagrams,* thus obtaining a "background independent causal theory of interacting closed spatial loops." Diagrams of this nature appear in string theory and topological quantum field theory. In this context, the direct future $J_0^+(x)$ of an element x in a directed set D might be replaced by a *cobordism of* S^1 *to a disjoint union of copies of* S^1 in the corresponding tubular diagram. Beyond this preliminary setup, however, the analogy seems difficult to extend in an obvious way. For example, it seems unclear how to define suitable analogues of individual relations between pairs of elements, and a natural analogue of *iteration of structure* (IS) does not seem evident.

8.9 Connections with Physics-Related Mathematics

Preliminary remarks on mathematical topics. This book draws material from many different mathematical fields, and a number of reflections on purely mathematical topics appear as "asides" throughout the book, particularly in the footnotes. "Interesting physics" has always shared a symbiotic relationship with pure mathematics, and the broad structural themes arising in discrete causal theory lead to particularly rich interactions of this nature, some of which remain largely unexplored. Among the mathematical topics discussed in the present section, some have already made sufficient contributions to theoretical physics to be listed among "approaches to fundamental physics" in some of the physics literature. For example, Connes' *noncommutative geometry* has contributed in profound ways to fundamental quantum-theoretic notions, and Connes' own physical ideas on the subject are sufficiently developed to be considered, in many sources, on an equal footing with theories of purely physical origin. In particular, Thiemann [TH07] lists noncommutative geometry alongside Sorkin's causal set theory, Penrose's twistor theory, Isham's topos theory, and 't Hooft's *superdeterminism* as an "unorthodox approach" to quantum gravity. Moreover, "orthodox" approaches such as string theory and loop quantum gravity have incorporated many of these same ideas, particularly in the last few years. However, noncommutative geometry itself is properly understood as a field of mathematics. Similarly, domain theory has made contributions to quantum information theory and general relativity, but is properly understood as a subfield of order theory. The division of topics between Section 8.8 and the present section is not intended to reflect any judgment regarding the *physical interest* of the ideas involved; it merely distinguishes between theories conceived with *specific physical interpretations,* and mathematical topics that have been *incidentally applied* to physics, however fruitfully.

Domain theory. Domains are special types of partially ordered sets satisfying a "continuity" property. They arose in the context of theoretical computer science, but have since been applied to a host of problems in other areas of mathematics and theoretical physics. Domains are perhaps the most natural choice for constructing a *nondiscrete* version of causal theory, and axiomatic systems based on domains are among the most interesting alternatives to the "suggested list" of axioms for directed sets appearing in Section 4.10. Of course, a vast "gray area" exists between discrete theories and continuum-based theories, including, for example, theories involving "interpolative but noncontinuous structures," such as the rational numbers $\mathbb{Q}$. An interesting general research project would be to examine in detail which of the concepts developed in this book admit useful analogues in the interpolative context, and which of these analogues extend further to apply to situations involving "limits" or "completeness properties," embodying various notions of continuity. Some of these concepts are obviously *not* useful in nondiscrete settings; for example, many of the advantages of working with nontransitive binary relations evaporate in the interpolative context, where every individual relation is reducible. Other ideas in the book extend in a straightforward but cumbersome way; for example, relation spaces over interpolative directed sets tend to be very "large" and "cluttered." Similarly, path sums in the interpolative context are generally very intricate, and are fraught with convergence issues and problems of definition, just as in the familiar context of conventional quantum field theory.

However, much of the material introduced in this book remains interesting even after discarding discreteness assumptions. In particular, the status and interpretation of the causal metric hypothesis (CMH) is worth examining more generally, especially in the context of metric recovery, where early-modern versions of the hypothesis, such as Sorkin's version, first arose. Recent domain-theoretic extensions of the original theorems of Hawking [HA76] and Malament [MA77], such as the 2006 paper of Martin and Panangaden [MP06], provide a different perspective regarding the general idea that *"directed structure plus natural scale equals geometry."* In the pseudo-Riemannian context, scale data is imposed extraneously, but it does not follow that *every* interpolative directed structure must be devoid of natural scale. It is interesting to consider whether there might exist some viable alternative structural paradigm that incorporates scale data in a natural way without resorting to discreteness. Another general idea that remains relevant in domain theory is Grothendieck's *relative viewpoint* (RV), and the consequent physical notion of *co-relative histories. Iteration of structure* (IS) also retains some degree of validity under any approach involving order-theoretic classical histories, although "higher-level structures" in this context tend to be "enormous." Evolutionary processes are a priori more difficult to analyze without the ability to decompose them precisely into discrete stages, but the general idea of *kinematic schemes* may still be studied under appropriate generalizations. The topic of *emergent Lie theory* remains interesting for domains and other nondiscrete order-theoretic structures. Finally, one may examine connections between domain theory and other continuum-based theories, *suggested* by methods or insights arising in the discrete causal context. For example, one might consider the possibility of "shape-causal duality" between certain classes of objects arising in shape dynamics and domain theory. Or, one might examine connections between domain theory and twistor theory involving relation space, analogous to the relationship between $\mathcal{R}(\mathbb{R}^{3+1})$ and the twistor space T, mentioned in Section 5.1.

Graph theory; matroid theory; random topology. The historical development of graph theory has been somewhat idiosyncratic, determined by internal factors to a greater extent than the evolution of larger and older fields, such as algebra and number theory. Partly for this reason, graph theory possesses a considerable lexicon of specialized terminology, which is often inconvenient when studying graph-theoretic aspects of mathematical structures that are *not* purely graph-theoretic in their own right. For this reason, I have chosen in this book to dispense with graph-theoretic terminology almost entirely, even though the mathematical objects of principal interest here; namely, directed sets and multidirected sets, are abstractly equivalent to directed graphs and directed multigraphs, respectively. Of course, this choice is largely motivated by the fact that directed sets and multidirected sets are accorded specific "set-like" interpretations in discrete causal theory for physical reasons, and it is only sensible to highlight these interpretations. For example, it is obviously advantageous to refer to an ordered pair of elements representing "causally connected" events as a *relation,* rather than as a *directed edge,* and it is far preferable to refer to the corresponding global construction as a *relation space* rather than as a *line digraph.*

Terminology and viewpoint aside, there are many essentially graph-theoretic *topics* that are of great interest in the discrete causal setting. Of these, perhaps the most fruitful are those with an algebraic flavor, such as the theory of *quivers.* Another closely-related field is *matroid theory,* mentioned briefly in Section 3.7, which involves a notion of *independence* similar to linear independence in linear algebra, and distinct from the notion of independence specified by the *independence convention* (IC) for directed sets and multidirected sets. The potential role of matroid theory in discrete causal theory stems ultimately from the existence of many interesting and physically suggestive matroids arising in graph theory. For example, the matroid-theoretic objects called *gammoids* are relevant in this context, due to their obvious potential for physical interpretations involving initial and terminal conditions. Another related topic of interest is the higher-dimensional theory of *simplicial complexes* and *cell complexes.* Here, I am referring not so much to the traditional uses of these objects in homological algebra and algebraic topology, but to their more recent application in subjects with a *dynamical* flavor. Spin foams and causal dynamical triangulations represent prototypical examples of these applications, but the rapidly developing field of *random topology* is worthy of serious consideration in a more abstract and general setting. This field may be viewed as a higher-dimensional generalization of Erdös and Rényi's theory of *random graph dynamics* [ER59].

At a more specific level, there are a number of important "open" or "active" problems in graph theory whose solutions would be of great interest in discrete causal theory. Perhaps the most obvious example, given recent events, is the *graph isomorphism problem* [BA15], mentioned briefly in Section 1.4. However, the largest source of interesting open problems in this context is probably random graph dynamics, and random topology more generally. This is true for at least two reasons. First, random topology itself is still relatively young and undeveloped as a mathematical field. Second, certain specific types of random topological problems of particular interest in discrete causal theory remain largely untouched to date, simply because there has been no obvious motivation for studying them. By contrast, many of the graph dynamical processes of greatest prominence in the mathematical literature, such as those mentioned in Section 8.5, are relevant only as analogies in discrete causal theory.

Physical applications of category theory. Specific employment of category theory in modeling *physical* structure, as opposed to its general use as a mathematical framework, has already been considered at several points throughout this book. First, a few established attempts to describe fundamental physics in category-theoretic terms, such as Isham's theory of *quantization on a category* [IS05], have been cited for their conceptual contributions to some of the material developed here. More generally, Isham's *topos-theoretic program* for studying fundamental physics contains many of the conceptual ingredients underlying the ideas of *iteration of structure* (IS), *co-relative histories,* and *kinematic schemes,* and deserves much more attention in this context than the brief descriptions given in Sections 3.8 and 6.1. Second, a few related ideas have been introduced for the specific purpose of illustrating how category-theoretic notions can lead in the *wrong direction* in representing physical relationships. Example include the analogy between *relation spaces* and *morphism categories* described in Section 5.5, and the comparison between co-relative histories and morphisms of directed sets in Section 6.4.

Third, general hierarchical notions in discrete causal theory, inspired partly by category-theoretic antecedents, are used unsystematically throughout the book, and are briefly elaborated more formally in Section 8.6. In the same section, I reiterate the fact that this book only scratches the surface of the subject of algebraic structure and hierarchy in discrete causal theory, and careful analysis of category-like notions in this context is an important part of this relatively untouched panorama. I remain convinced that category theory will prove most useful, from a physical perspective, as a *source of structural analogies,* rather than as a system that should be imported wholesale as a basic physical framework, but the theory still has much more to offer. In particular, there is great potential in the examination of connections between discrete causal theory and the theory of categories equipped with additional structures, such as *monoidal categories.* The relatively young mathematical topic of *categorification* and *decategorification* also deserves further examination in the discrete causal context, due to its obvious relationship with iteration of structure.

Noncommutative algebra; noncommutative geometry; quantum groups. Historically, ordinary quantum theory has been one of the principal avenues through which the study of *noncommutative structures* has entered mainstream theoretical physics. For example, noncommuting pairs of operators on Hilbert spaces, such as "position" and "momentum" operators on spaces of "wave functions" over Euclidean spacetime $\mathbb{R}^4$, may be associated with pairs of incompatible observables, such as the position and momentum of a particle. The basic mathematical content of the Heisenberg uncertainty principle is often expressed in these terms. Hence, the importance of noncommutative *algebra* in fundamental physics has been clear for nearly a century. Noncommutative *geometry,* by contrast, is a modern development, requiring a much deeper understanding of the overall relationships between the mathematical fields of algebra and geometry than was available when ordinary quantum theory was first developed. Alexander Grothendieck and his immediate collaborators bridged much of this conceptual gap in the 1960s and 1970s, even though most of their work was carried out in the commutative setting, and without any notion or intention of physical applications. Later, Alain Connes and a few others developed noncommutative geometry proper, and connected it to theoretical physics.

A general method for understanding the properties of a "geometric object" is to study families of functions on it, and this approach long predates "modern physics," at least in an operational sense. In many cases, the most interesting such families form rings or algebras; for example, families of real or complex-valued functions inherit such structure from their target fields. What Grothendieck's school accomplished, at a rough conceptual level, was to elevate the correspondence between certain "spaces" and the corresponding rings or algebras of functions on these spaces, to an *equivalence,* under which algebraic structures may be viewed as geometric, and geometric structures may be viewed as algebraic. Here I am referring, of course, to the theory of *commutative algebraic schemes,* in which commutative rings or algebras are equivalent to the affine schemes defined by their *spectra,*[21] and arbitrary schemes may be built by "gluing together" affine schemes. Such schemes make up the "geometry" part of "algebraic geometry," while the corresponding rings, algebras, sheaves, and related objects, make up the "algebraic" part. A generation later, Connes and several contemporaries vastly extended this picture to cover the case of *noncommutative* algebras. Without going into the details, the results of this development enable one to associate geometric objects called *noncommutative spaces* with such algebras. Simple examples include "fuzzy" spheres and tori. Connes has almost singlehandedly constructed a very interesting and impressive corpus of work linking the central ideas of noncommutative geometry to fundamental physics [CO94, CM07].[22] In particular, Connes has proposed a noncommutative structure for spacetime, of which the familiar four-dimensional pseudo-Riemannian structure of relativity is only one aspect. Connes' ideas have been gradually absorbed by other researchers, and have

[21] More precisely, there is a category-theoretic duality between commutative rings and affine schemes. The "spectra" mentioned here are essentially "spaces of prime ideals;" they are not "physical spectra" in any familiar sense.

[22] One of the many features that distinguishes Connes among his contemporaries is a combination of specific mathematical expertise and deep general awareness of many different scientific fields, including theoretical physics.

been synthesized into more mainstream approaches to fundamental physics, such as string theory.

In the context of discrete causal theory, noncommutative geometry offers a completely different geometric viewpoint regarding *causal path algebras* than that provided by the causal metric hypothesis (CMH). In particular, one may view causal path algebras as abstract algebras which are *very* noncommutative, and may proceed to study the associated noncommutative spaces. Ioannis Raptis and Roman Zapatrin discuss similar ideas in their paper *Algebraic description of spacetime foam* [RA01]. Some of the notions underlying the theory of *deformed special relativity* (DSR) involve noncommutative geometry, and are relevant to discrete causal theory, despite the apparent limitations of DSR in its own right. Noncommutative methods play an important role in canonical quantization of general relativity, as described by Thiemann [TH07], and any detailed comparison of loop quantum gravity with discrete causal theory will necessarily involve such considerations. Certain classes of noncommutative algebras that have enjoyed much recent attention are the so-called *quantum groups,* which may be identified, rather vaguely, as "noncommutative algebras with additional structure." For example, they include special types of "deformed" *Hopf algebras.*[23] As described at the end of Section 5.10, causal path algebras possess significant "additional structure," and this leads to an interesting overlap between the theory of quantum groups and discrete causal theory.

My own intuition regarding noncommutative geometry as a theory of fundamental spacetime structure is that it lacks compelling physical motivation, and is unlikely, in its own right, to represent the deepest properties of the universe. This viewpoint may seem to ignore the natural way in which noncommutative algebras entered the physical picture in the context of ordinary quantum theory, but the function spaces and operator algebras involved in this setting are very likely, in my opinion, to be only emergent. Ultimately, the conceptual and philosophical differences between discrete causal theory and noncommutative geometry stem from different *choices* regarding which aspects of previous theories to isolate, elevate, and generalize. Discrete causal theory chooses causal structure, and adds discreteness to the picture because of the metric recovery theorems. Noncommutative geometry chooses algebraic notions arising from ordinary quantum theory, and adds a vast and sophisticated mathematical technology. Whichever approach proves to be more useful from a physical perspective, the appearance of large and complicated noncommutative algebras in discrete causal theory virtually guarantees an interesting interplay between the two theories.

Gyrogroups; loops, "group-like objects." From a naïve structural viewpoint, the noncommutative algebras discussed above enter the picture when one *discards a familiar property that certain "well-behaved" algebraic objects exhibit;* namely, commutativity. From a physical perspective, this is a necessary step, because certain

[23] From a physical standpoint, the choice of terminology for these objects is perhaps unfortunate, because many of them have nothing to do with quantum theory in any useful sense; they merely share certain loose abstract similarities with algebraic notions of quantization. However, some of these objects do possess useful physical interpretations; for example, in quantum information theory.

physical entities, such as pairs of incompatible observables in ordinary quantum theory, display behavior that cannot be suitably described in commutative terms. It turns out that other familiar algebraic properties, such as *associativity,* are also unsuitable in certain physical settings, some of which are actually quite basic. Perhaps the most striking example arises in the "addition of velocities" in special relativity, which is a nonassociative operation. Mathematically, this corresponds to the fact that *the family of Lorentz boosts is not a subgroup of the proper orthochronous Lorentz group L''*. Examining the situation further, Lorentz boosts form an algebraic object called a *gyrogroup,* which obeys "modified" commutativity and associativity laws, defined in terms of an operator called the *Thomas gyration,* which encodes the *Thomas–Wigner rotation* of inertial frames of reference discussed in Section 8.4. This operator corresponds to the well-known physical phenomena of *Thomas precession,* which is a change in the orientation of a physical object undergoing parallel transport in Minkowski spacetime $\mathbb{R}^{3+1}$.

Gyrogroups are special cases of a more general class of "group-like objects" called *loops,* a rather unfortunate coincidence of terminology in the present setting. In particular, these "loops" have nothing to do with the "loops" that generalize causal cycles in Section 8.7, nor the "loops" in loop quantum gravity. To add to the confusion, some researchers refer to special classes of loops as "group-like," implying that the class as a whole is too general to be considered as such. I take a broader view of "group-like" structure, including loops among a variety of similar types of objects. An obvious reason why such objects are of interest in the discrete causal setting is because Lorentz boosts involve the causal structure of Minkowski spacetime, in contrast to rotations, which involve only spatial structure. Hence, discrete causal analogues of algebraic entities such as the Thomas gyration are worth examining, particularly in the context of emergent symmetry groups. A related topic that plays a role in curved spacetime is *geodesic multiplication,* which also leads to nonassociative algebraic structures.

Partial differential equations; boundary value problems; numerical methods. A huge and sophisticated literature exists on the subject of "Schrödinger-type partial differential equations," often referred to as generalized *heat equations, diffusion equations,* or *wave equations.*[24] The basic conceptual ingredients of Sections 6.7 and 6.8 of this book, involving "flux through a boundary," relationships between local "differential" and global "integral" data, generalized Lagrangians, the superposition principle, and so on, are all familiar to this field. It is virtually certain that additional useful connections between discrete causal theory and the theory of partial differential equations can be identified in both directions. In particular, it is possible that some of the "large and complicated" path sums involved in realistic computations of discrete spacetime interactions may be approximated using existing methods in partial differential equations. Of course, this is more likely to occur in idealized situations

[24]As noted in Section 1.1, the ordinary Schrödinger equation $i\hbar\frac{\partial\psi}{\partial t}=\mathbf{H}\psi$ is "more similar" to the classical heat equation $\frac{\partial\psi}{\partial t}=\alpha\nabla^2\psi$ than to the classical wave equation $\frac{\partial^2\psi}{\partial t^2}=c^2\nabla^2\psi$, because it is *first-order in time.*

where a geometric approximation is readily available; for example, in *sprinkling constructions* such as those discussed in Section 4.5. In any case, such an approach would turn the conventional relationship between exact solutions and approximations on its head, since *discrete* approximation methods are presently used in this field to examine the properties of "exact" continuum-based entities. In the other direction, it is likely that the irregular, or "random," nature of the directed sets involved in discrete causal theory might lead to useful insights involving approximation methods for partial differential equations. Some of the existing techniques in this field involve modifying naïve lattice-based approximation methods to admit less-regular structures. This can be particularly useful in the context of sequential or iterative procedures, where the results of one approximation are fed into the next. These latter ideas fall under the umbrella of *numerical methods for partial differential equations,* which attempt to reduce continuum-based problems to discrete approximations in the most efficient possible way. Particularly relevant in this context are *finite element methods* involving *variational techniques.* Finally, general conceptual notions arising in discrete causal theory, such as relation space methods, might possibly prove useful in the study of partial differential equations on pseudo-Riemannian manifolds, even though the objects involved in continuum-based incarnations of these notions are a priori large and cumbersome.

Representation theory; cohomology theories; Galois theory. The natural "next step" beyond studying the emergence of the *Poincaré group* P, as discussed in Section 8.4, is to examine analogues of the corresponding *representation theory.* Representations of P may be used to classify possible "particle types" in conventional quantum field theory, and this renders the examination of discrete causal analogues of such representations automatically interesting. However, the applicability of Lie representation theory in theoretical physics is vastly broader than this. Another prominent area in which the subject arises is in the context of the *gauge groups* of the standard model of particle theory, and in corresponding facets of related theories that attempt to extend the standard model. The challenge of recovering this type of structure *purely from directed sets* is one of the most obvious hurdles facing ambitious versions of discrete causal theory based on the strong interpretation of the causal metric hypothesis (CMH). Other connections to representation theory arise via the theory of *generalized semicategory algebras,* abstracting the role of *group algebras* in group representation theory. Turning to other algebraic topics, *algebra cohomology theories* such as *Hochschild cohomology* and *cyclic cohomology* are relevant in the study of *causal path algebras* and *phase theory.* Some of the unpublished work culled from the final version of this book involves theorems relating such cohomology theories to corresponding cohomology theories involving more primitive "precursor" objects, such as semicategories.[25] *Relative cohomology* plays a role in this setting, via congruences. *Lie cohomology* and *algebraic* K*-theory* also enter the picture here. Finally, the brief remarks regarding *causal Galois groups* and *Galois connections* in Section 6.1 of this book may be greatly elaborated.

[25] James Madden is responsible for suggesting some of these connections to me, via private communication.

8.10 Final Remarks

Final note to the reader. In writing this book, I have adopted the viewpoint that it is more useful to the reader for me to present as much information about discrete causal theory as possible, and to highlight the various gaps and uncertainties I am aware of in the development of the theory, and my own knowledge of it, than to restrict attention to "safe" information that will avoid the likelihood of unfavorable scrutiny. Setting aside the essential aspects of the theory itself, some of the guesses, speculations, and illustrations included in the book will likely prove to be mistaken or irrelevant in the long term, as in any work of comparable length on a similarly young topic. However, I am not concerned about the future status of such details, because the overall research program described in the book represents an effort to accomplish something inherently risky and difficult; namely, to produce a significant body of new ideas and viewpoints that could at least *conceivably* help to break out of the plodding incrementalism of mainstream academic physics. In this sense, the book is virtually guaranteed to provide a more interesting contribution to the field than certain large and well-funded research projects, which would not make much of a lasting impact even if they were completely successful. Further, it seems reasonably certain, on basic information-theoretic grounds, that discrete causal theory itself will always remain relevant to the future of theoretical physics, regardless of the actual details of fundamental spacetime structure.

A well-known quote of Planck may be paraphrased by the statement that *"theories do not die; theorists do,"* meaning that professionally successful scientists are reluctant to abandon mainstream approaches, even when they have been thoroughly discredited. This is perhaps an overly-harsh generalization, since there are always honest and idealistic individuals who are willing to consider, examine, and embrace new ideas, even if these ideas supplant their own work. However, such individuals seem to be less common than one might wish, perhaps partly because the process of "academic natural selection" favors ruthless advocates of proprietary interests. Hence, it is possible that the results of this book may be most interesting to those not yet fully invested in any previously-existing theory of fundamental physics. These include graduate students, young researchers, and members of fields that *involve* the study of fundamental spacetime structure, without depending exclusively on a particular approach to the subject. An advantage of trying to develop or master ideas that are substantially original, rather than small modifications of existing frameworks, is that it is impossible to avoid encountering conflicting viewpoints. One is thus forced away from the common psychological pitfall of seeking continual reinforcement and isolation from challenge, which unfortunately tempts even those of the highest abilities. From this perspective, established advocates of more conventional approaches to fundamental physics might find it worthwhile to read this book, and others like it, simply because it is stimulating to find something to disagree with. For those of lesser ambition, theories such as "Time Cube" can offer a similar service without requiring as much effort.

Genesis of the book. I began to develop the ideas behind this book as a graduate student in mathematics in 2008–2009, with little knowledge of existing versions of causal theory, besides a superficial awareness of domain theory, acquired during my brief time at Tulane University, when Michael Mislove, Keye Martin, and Johnny Feng were present. No serious researcher would deliberately abstain from reading existing results about a subject of interest, but it is perhaps not such a disadvantage to devote considerable thought to a subject in a state of *accidental* ignorance, provided that one is willing afterwards to read thousands of pages of literature, and to face the disappointment of repeatedly discovering that cherished ideas are either already known, or are wrong. My first interaction with the "outside world" on the subject of discrete causal theory came in June 2010, when I began exchanging emails about causal set theory with a leading researcher in the field. By this time, I knew that causal set theory had more in common with my own ideas than other existing approaches to fundamental spacetime structure, but I also felt certain that significant advantages could be gained by treating a few basic structural questions in a different manner. In January 2011, I communicated to this same individual the basic premises of Chapters 3 and 4 of this book, which explain some of these advantages, but never received a response. I have since corresponded with a number of other causal set theorists about narrower and more specific aspects of the theory.

In August 2012, I outlined many of the ideas presented in this book in the essay [DR15], written for a contest hosted by the Foundational Questions Institute. My motivation for introducing these ideas in a public setting was to elicit feedback, and particularly criticism, which I hoped to use to improve the theory. The sociology of such a contest includes both favorable and unfavorable factors from a scientific viewpoint. The nature of the competition discourages sharp honest criticism by other participants, because entrants are allowed to vote on the submissions as part of the evaluation process. However, the public nature of the proceedings enables knowledgeable non-participants to freely weigh in, and discourages dishonest criticism and lazy enforcement of convention by anonymous turf warriors. In my own case, I received both positive and negative feedback from several dozen prominent physicists, which might have been difficult to obtain without the benefit of such a forum. I was also fortunate enough to receive one of the prizes. In the process, I learned some interesting details about other approaches to fundamental physics, but encountered nothing to significantly alter my basic optimism about the viability and importance of discrete causal theory.

In late 2013, I posted a long paper [DR13] on the arXiv, which contains early versions of most of the principal results in this book. Exceptions include some of the material in Chapter 7 and in this chapter, which are largely new. Chapters 1 and 2 are also new, but are mostly expository and motivational. The original draft of the paper, submitted in early November 2013, contained a number of errors, some of which I discovered myself, and some of which were pointed out to me by other people; most notably, Johnny Feng. Subsequent versions, posted in late November and late December 2013, respectively, corrected some of these mistakes, but a number of additional glitches came to my attention while writing the book. For example, the properties of the abstract element space functor were not stated completely correctly, a few conclusions of secondary importance were missing necessary hypotheses, an "inner product formula" was missing some terms, and some of the definitions appeared in slightly different versions in different parts of the paper.[26] None of these errors affect the basic conclusions of the work in any significant way. I am optimistic that the most important conclusions of the book itself, which is sure to contain errors as well, will hold up in a similar fashion.

References

[BR88] Edward A. Bender and Robert W. Robinson. *The Asymptotic Number of Acyclic Digraphs II*. Journal of Combinatorial Theory, Series B, **44**, pp. 363–369, 1988.

[BD10] Dionigi M. T. Benincasa and Fay Dowker. *Scalar Curvature of a Causal Set*. Physical Review Letters, **104**, 18, 181301, 2010. arXiv preprint: http://arxiv.org/pdf/1001.2725v4.pdf.

[RS03] Carlo Rovelli and Simone Speziale. *Reconcile Planck-scale discreteness and the Lorentz-Fitzgerald contraction*. Physical Review D, **67**, 6, 064019, 2003. arXiv preprint: http://arxiv.org/pdf/gr-qc/0205108v1.pdf.

[SH48] Claude Shannon. *Space-Time Approach to Non-Relativistic Quantum Mechanics*. The Bell System Technical Journal, **27**, 3, pp. 379–423, 1948.

[RO04] Carlo Rovelli. *Quantum Gravity*. Cambridge Monographs on Mathematical Physics. Cambridge University Press, 2004.

[SR99] David Rideout and Rafael Sorkin. *Classical sequential growth dynamics for causal sets*. Physical Review D, **61**, 2, 024002, 2000. arXiv preprint: http://arxiv.org/pdf/gr-qc/9904062v3.pdf.

[ZE64] E. C. Zeeman. *Causality Implies the Lorentz Group*. Journal of Mathematical Physics, **5**, 4, pp. 490–493, 1964.

[MA77] David B. Malament. *The class of continuous timelike curves determines the topology of spacetime*. Journal of Mathematical Physics, **18**, 7, pp. 1399–1404, 1977.

[ER59] Paul Erdös and Alfred Rényi. *On random graphs I*. Publ. Math. Debrecen, **6**, pp. 290–297, 2009.

[BO01] Béla Bollobás. *Random Graphs*. Cambridge Studies in Advanced Mathematics, **73**, Second Edition. Cambridge University Press, 2001.

[SR01] David Rideout and Rafael Sorkin. *Evidence for a continuum limit in causal set dynamics*. Physical Review D, **63**, 10, 104011, 2001. arXiv preprint (updated 2008): http://arxiv.org/pdf/gr-qc/0003117v4.pdf.

[KKSR12] Dmitri Krioukov, Maksim Kitsak, Robert S. Sinkovits, David Rideout, David Meyer, and Marián Boguñá. *Network Cosmology*. Nature Scientific Reports **2**, 793, 2012.

[TH07] Thomas Thiemann. *Modern Canonical Quantum General Relativity*. Cambridge Monographs on Mathematical Physics. Cambridge University Press, 2007.

[CS14A] Marina Cortês and Lee Smolin. *Quantum energetic causal sets*. Physical Review D, **90**, 4, 044035, 2014. arXiv preprint: http://arxiv.org/pdf/1308.2206v2.pdf.

[CS14B] Marina Cortês and Lee Smolin. *Spin foam models as energetic causal sets*. arXiv preprint: http://arxiv.org/pdf/1407.0032.pdf.

[CS14C] Marina Cortês and Lee Smolin. *The universe as a process of unique events*. Physical Review D, **90**, 8, 08007, 2014. arXiv preprint: http://arxiv.org/pdf/1307.6167.pdf.

[26] For the sake of clarity, I plan to post an updated version of the paper [DR13], detailing the necessary corrections, when time permits. However, all the errors mentioned here are already corrected in the book.

[BHS09] Luca Bombelli, Joe Henson, and Rafael Sorkin. *Discreteness without symmetry breaking: a theorem*. Modern Physics Letters A, **24**, 32, pp. 2579–2587, 2009. arXiv preprint: http://arxiv.org/pdf/gr-qc/0605006v1.pdf

[AJJR12] Jan Ambjørn, S. Jordan, J. Jurkiewicz, and R. Loll. *Second- and first-order phase transitions in causal dynamical triangulations*. Physical Review D, **85**, 12, 124044, 2012. arXiv preprint: http://arxiv.org/pdf/1205.1229.pdf.

[AJL05] Jan Ambjørn, J. Jurkiewicz, and R. Loll. *Spectral Dimension of the Universe*. Physical Review Letters, **95**, 171301, 2005. arXiv preprint: https://arxiv.org/pdf/hep-th/0505113v2.pdf.

[DP14] Giacomo Mauro D'Ariano and Paolo Perinotti. *Derivation of the Dirac Equation from Principles of Information Processing*. Physical Review A, **90**, 6, 062106, 2014. arXiv preprint: http://arxiv.org/pdf/1306.1934v1.pdf.

[HA76] S. W. Hawking, A. R. King, and P. J. McCarthy. *A new topology for curved space-time which incorporates the causal, differential, and conformal structures*. Journal of Mathematical Physics, **17**, 2, pp. 174–181, 1976.

[MP06] Keye Martin and Prakash Panangaden. *A Domain of Spacetime Intervals in General Relativity*. Communications in Mathematical Physics, **267**, 3, pp. 563–586, 2006.

[BA15] László Babai. *Graph Isomorphism in Quasipolynomial Time*. arXiv preprint: https://arxiv.org/pdf/1512.03547v2.pdf

[IS05] Christopher Isham. *Quantising on a Category*. Foundations of Physics, **35**, 2, pp. 271–297, 2005. arXiv preprint: http://arxiv.org/pdf/quant-ph/0401175v1.pdf.

[CO94] Alain Connes. *Noncommutative Geometry*. Electronic book: URL: http://www.alainconnes.org/docs/book94bigpdf.pdf.

[CM07] Alain Connes and Matilde Marcolli. *Noncommutative Geometry, Quantum Fields and Motives*. Electronic book: URL: http://www.alainconnes.org/docs/bookwebfinal.pdf.

[RA01] Ioannis Raptis and Roman Zapatrin. *Algebraic description of spacetime foam*. Classical and Quantum Gravity, **18**, pp. 4187–4212, 2001. arXiv preprint: http://arxiv.org/pdf/gr-qc/0102048v2.pdf.

[DR15] Benjamin F. Dribus. *On the Foundational Assumptions of Modern Physics*. Questioning the Foundations, The Frontiers Collection, Springer, pp. 45–60, 2015. Early version: http://www.fqxi.org/community/essay/winners/2012.1#dribus.

[DR13] Benjamin F. Dribus. *On the Axioms of Causal Set Theory*. Preprint. arXiv link: http://arxiv-web3.library.cornell.edu/pdf/1311.2148v3.pdf.

Appendix A
Conventions

A.1 Notational Conventions

Directed sets and multidirected sets. The mathematical objects of principal interest in this book are types of structured sets, called *directed sets* and *multidirected sets,* respectively, both defined in Section 3.6. In the context of discrete spacetime models, they may be viewed as generalizations of causal sets. In most cases, only isomorphism classes of such sets are significant, but most constructions are described in terms of particular representatives of isomorphism classes. Individual structured sets are usually denoted by capital letters such as D or M, and their individual elements by lower-case letters such as x or y. Whenever a more detailed treatment is necessary, a directed set may be represented by an ordered pair $(D, \prec)$, where $\prec$ is a binary relation on D, and a multidirected set may be represented by an ordered quadruple (M, R, i, t), where R is a set of relations, and where i and t are *initial* and *terminal element maps,* assigning to each element of R initial and terminal elements in M, as described in Section 3.6. The *precursor symbol* $\prec$ is used for three distinct but related purposes: first, to represent a binary relation on a set, as mentioned above; second, to represent an individual relation between a pair of elements of a set; and third, to indicate that at least one such relation exists. For a directed set $D = (D, \prec)$, the expression $x \prec y$ indicates the existence of an individual relation between x and y, which means that the ordered pair (x, y) is an *element* of the binary relation $\prec$. For a multidirected set, there may exist multiple relations between a given pair of elements in either or both directions. In this case, the expression $x \prec y$ refers either to a particular relation with initial element x and terminal element y, or indicates that at least one such relation exists.

Other general notational remarks. Categories are denoted by mathscript letters, such as $\mathcal{A}$ for the category of acyclic directed sets, $\mathcal{D}$ for the category of directed sets, and $\mathcal{M}$ for the category of multidirected sets. The mathscript letter $\mathcal{C}$ sometimes denotes a general category, and sometimes denotes the category of causal sets. Functors are assigned notation "suggestively," rather than in any systematic fashion; for example, the *transitive closure functor* from $\mathcal{D}$ to $\mathcal{D}$ is denoted by tr, while the

B.F. Dribus, *Discrete Causal Theory*, DOI 10.1007/978-3-319-50083-6 *relation space functor* from $\mathcal{M}$ to $\mathcal{D}$ is denoted by $\mathcal{R}$. The natural numbers, integers, rationals, reals, complex numbers, quaternions, and octonions are denoted by the usual blackboard-bold symbols $\mathbb{N}$, $\mathbb{Z}$, $\mathbb{Q}$, $\mathbb{R}$, $\mathbb{C}$, $\mathbb{H}$, and $\mathbb{O}$, respectively. The natural numbers $\mathbb{N}$ are taken to include zero. The additive group of integers modulo a positive integer n is denoted by $\mathbb{Z}_n$. The p-adic numbers, for a particular prime p, are denoted by $\mathbb{Q}_p$. Kinematic schemes are denoted by blackboard-bold $\mathbb{S}$, due to their analogies with number systems. Subscripts are added to identify a particular kinematic scheme, if necessary; for example, $\mathbb{S}_{\text{PS}}$ denotes the positive sequential kinematic scheme. The term *space* usually means *structured set,* rather than specifically *topological space.* In particular, the structure involved may be order-theoretic, or more generally, multidirected, although this structure may be used to define natural topologies. For example, *power spaces* are multidirected sets whose elements are subsets of a given space. Relativistic spacetime manifolds are denoted by capital letters near the end of the alphabet, such as W and X. A fairly comprehensive list of notation appears at the beginning of the book.

A.2 Terminology

Correcting "wrong" terminology, e.g., "local finiteness." This book contains a few notable disagreements with standard terminology. In some cases, the standard terminology is actually *wrong,* in the sense that it contradicts the meaning of the terms involved. For example, the condition of *interval finiteness* (IF) is called "local finiteness" in the causal set literature, but it is not a local condition in any reasonable sense, and attempting to use it to quantify and study local behavior leads to serious conceptual and technical problems.

Re-purposing inconvenient terminology, e.g., "directed sets." In other cases, terms whose plain meaning evokes a useful general idea have acquired inconvenient proprietary meanings. For example, the term *directed set* is used in this book to mean simply a set equipped with a family of directed relationships between pairs of its elements, i.e., a set equipped with a binary relation. This is a straightforward and suggestive meaning. However, the same term conventionally signifies a meaning much more specific and complicated; namely, that of a set equipped with a reflexive, transitive binary relation, satisfying a common successor property. The alternative term *filtered set* has this same conventional meaning, but conveys it better, so I reclaim the term *directed set* for more general use.

Avoiding awkward terminology, e.g., graph theory. Sometimes more than one "language" is available for describing a particular type of structure, and one must choose which is more useful to convey the desired meaning. In particular, the language of graph theory *could* be used to describe most of the ideas appearing in this book, but set-theoretic language is generally much better for expressing the associated physical connotations. One *could* argue for using the term "directed graph" in place of "directed set," since the latter term has a different standard meaning than the one used in this book, but it would be awkward to do so without importing the rest of the graph-theoretic lexicon at the same time. This would be a bad idea. In particular, the term "multidirected set" is better than the term "directed multigraph" or "quiver," and the term "relation" is much better than the term "directed edge." The principal reason for using set-theoretic terminology, of course, is that spacetime is naturally viewed from a set-theoretic perspective. In particular, it would be especially awkward to flip back and forth between graph theory and set theory when comparing discrete and continuum-based models of spacetime. Graph-theoretic terminology *is* used in a limited fashion to refer to *figures* appearing in the text for illustrative purposes. In particular, "nodes" and "edges" are used to *represent* elements and relations.

A.3 Formatting and Navigation

Highlighting. Formal introduction of technical terms, headings of subsections, definitions, axioms, lemmas, theorems, and important arguments and principles, appear in **bold font**. Some definitions are embedded in the text, but the most important definitions are formally set off from the text and numbered. *Italic font* is used for the introduction of less-important technical terms, and also for the bodies of definitions, axioms, lemmas, theorems, and quotations, for non-English terms and phrases, and for ordinary English emphasis. Italics are also used to emphasize important technical terms referenced *before or after* their formal bold-font introductions. The latter use of italics is generally limited to once per page or less for a given term; its purpose is primarily to highlight when such terms are *reintroduced.* In particular, the same term is not italicized repeatedly in a given paragraph. Quotation marks are used to indicate phrases which are intended to be understood as informal, imprecise, or metaphorical, as well as for actual quotations, and for verbatim repetitions of previous phrases. On occasion, a given phrase may be *both* "important," and "informal," and may therefore be *both* italicized and placed in quotes. In-line literature quotations are sometimes highlighted in the same way, but the difference should be clear from the context. Quotation marks also used to indicate dubious usages of the enclosed terms, as in the case of "local finiteness" appearing above.

Hyperlinks; acronyms. Clickable hyperlinks are included throughout the text to enable easy navigation. Axioms, conventions, and principles of sufficient importance to demand frequent reference are labeled by short acronyms. The axiom of *transitivity* (TR) and the *causal metric hypothesis* (CMH) are examples. These acronyms serve as hyperlinks to the original introductions or definitions of the corresponding terms. As noted at the beginning of Chapter 1, their function is the opposite of the common usage of acronyms as "labor-saving devices" in technical works, which only serves to exasperate the reader. As in the case of italics, the use of such labels is limited to roughly once per page or less for each term, since their objective is to facilitate navigation. The intent is that the reader should have a "nearby hyperlink" available to access the necessary introductory information about whatever is being discussed.

Numbering. Definitions, lemmas, theorems, equations, and figures are all numbered according to a single system, which encodes the chapter, section, and number of the item involved. For example, Equation (6.9.3) is followed in the numbering system by Figure 6.9.4, which is followed by Definition 6.9.5.

Index of topics. A substantial topical index appears at the end of the book. In the electronic version of the book, this index is a clickable "smart index." It has two levels of hierarchy; for example, the term *abstract element space* has a main index entry, followed by twelve indented subentries. Each main entry and subentry is followed by a clickable list of pages on which the relevant term appears. These lists are rather selective, for obvious reasons; it would be both awkward and useless if the index listed every page on which a common term such as *relation* appears in the book. The list of pages following each main entry usually indicates either "initial" or "principal" usage of the relevant term. For example, the first page listed following the main entry for *abstract element space* is the beginning page of Section 5.2, which is itself titled "Abstract Element Space." The list of pages following each subentry indicates important appearances of the relevant main-entry term *in a context specified by the subentry.* For example, the page listed after the subentry *fails to preserve acyclicity* for the main entry *abstract element space* contains the statement of Theorem 5.2.3, which includes the statement that the abstract element space functor fails to preserve the property of acyclicity for directed sets. The index includes several thousand entries altogether, but is far from exhaustive, and its organization is less systematic in some ways than much of the book itself. Its principal intent is to supply a *useful* searchable resource for important topics in the book, rather than a definitive catalogue of content.

A.4 Figures

Generalized Hasse diagrams. Although many of the constructions in this book apply to general multidirected sets, most of the accompanying figures, such as Figure 1.4.3, depict plane diagrams of acyclic directed sets, called *generalized Hasse diagrams,* which are analogous to Minkowski diagrams in special relativity. In a generalized Hasse diagram, elements of an acyclic directed set are represented by nodes, and relations between pairs of elements are represented by edges between pairs of nodes. The directions of relations are inferred via the "up the page" convention, i.e., by comparing the "vertical coordinates assigned to their initial and terminal nodes." More specifically, if x and y are two elements of an acyclic directed set $(D, \prec)$, then by convention, $x \prec y$ in D if and only if there is an edge between the corresponding nodes in a generalized Hasse diagram for D, such that the vertical coordinate of the node representing x is less than the vertical coordinate of the node representing y. Pairs of elements whose representing nodes have no edge between them are assumed to be unrelated, regardless of the coordinates of these nodes. Heuristically, *"causal influence flows upward along edges."*

Intended generality of illustrations. The qualifier "generalized" in the phrase "generalized Hasse diagrams" is included because "'true" or "strict" Hasse diagrams represent only irreducible relations, while the diagrams appearing in this book also admit reducible relations. *Acyclic* multidirected sets may be represented by generalized Hasse diagrams in an analogous way, but directed sets and multidirected sets involving cycles cannot be so represented. Regardless of the actual properties of the structured sets appearing in a figure, these sets are usually labeled and referenced at the same level of generality to which the corresponding constructions or examples are intended to apply. For example, a directed set appearing in a figure may be labeled M, rather than D, if the figure is *intended to apply to multidirected sets in general.* In particular, the acyclic directed set appearing (twice) in Figure 4.3.1 is labeled M, because the topological notion illustrated in the figure applies to multidirected sets in general. Conversely, if part of the purpose of the figure is to convey the fact that a certain possible behavior of multidirected sets *manifests itself even in the special case of directed sets,* then the set appearing in the figure may be labeled D, rather than M. In particular, the permeability problem, which is generic to multidirected sets, is illustrated for an acyclic directed set D in Figure 5.4.1.

A.5 References

General remarks. Prior to the information age, the practice of including a maximal list of references in a technical work could provide a valuable service to the reader, even if a much smaller cross section would suffice to document its actual content. Today, online topical databases provide a much more appropriate setting for exhaustive lists of sources than do bibliographies in specific technical works, with the obvious exception of actual literature reviews. For example, two useful online sources on causal sets are Sumati Surya's resource page and Luca Bombelli's topical page. In this book, I choose to restrict attention to a moderate number of reliable and well-written references, subject to the requirement of conveying the proper priority of authorship. Many of the authors cited here have produced numerous other books or papers of equal quality and relevance, but I have made an effort to distill the necessary concepts and quotations from a more manageable selection. In particular, I cite only sources whose content I mention specifically.

A few of the cited references fall short in quality and/or clarity, but demand mention due to their priority in treating certain crucial topics. These include articles suffering from certain generic technical problems that I wish to point out, and also book and papers about questionable or dubious alternative approaches to fundamental physics. I cite peer-reviewed, edited, and/or updated versions of references whenever possible, but also include links to arXiv preprints when available. The reader should always be aware of the possibility of differences between the two, and of the transient nature of links in general. I do not include formal citations to quotations or paraphrases obtained through private correspondence, though I always acknowledge such sources in the main body of the text. Regarding both content and priority of authorship, I beg indulgence for any unintended slights due to my own ignorance.

Conventions for references. Citations in the text are indicated by alpha-numeric tags such as [FE48] or [BLMS87]. In the electronic version of the book, each tag serves as a clickable hyperlink to the list of references at the end of the chapter in which it appears. Tags for references having only a single author consist of the first two letters of the author's last name, followed by the last two digits of the year in which the reference was published. For example, the tag [FE48] stands for "Feynman, 1948." Similar tags are used for references having multiple authors, with the exception that the first letter of the last name of each author is used in place of the first two letters of the last name of a single author. For example, the tag [BLMS87] stands for "Bombelli, Lee, Meyer, and Sorkin, 1987." In a few cases, this system is not sufficiently specific. To distinguish multiple papers by the same authors in the same year, a letter is added to the end of the tag; for example, the tag [CS14C] stands for "Cortês and Smolin, 2014, third paper." To distinguish multiple authors in the same year sharing the same first two letters in their last names, additional letters are added as needed; for example, the tag [DRE15] stands for "Dreyer, 2015," in contrast to the earlier tag [DR15] for "Dribus, 2015." Finally, for papers involving many authors, a maximum of four are included in the tag; for example, the tag [KKSR12] is short for Krioukov, Kitsak, Sinkovits, Rideout, Meyer, and Boguñá, 2012.

References are listed independently for each chapter. For example, Feynman's 1948 paper *Space-Time Approach to Non-Relativistic Quantum Mechanics* is cited in Chapters 1, 3, and 6, and the same bibliographic information is repeated in the list of references at the end of each of these chapters. The rationale for this practice is to make the chapters as self-contained as possible for individual downloading. In each chapter, references appear in the order in which they are *first cited.* Only the names of the actual books or articles cited appear in italics. Journal names, the names of books containing articles, etc., appear in normal font. Bold digits in typical journal citations indicate the *volume* of the journal cited; succeeding digits indicate the *number* or *issue,* the *page numbers* (preceded by "pp.") or *article number,* and finally the *year published.*

Index

A

Abstract element space, 280, 281, 462
- as a functor, 159, 283, 331, 516
- as an inverse to relation space, 280
- basic properties of, 283
- construction of, 281, 282
- cycles, 494
- definition of, 281
- fails to preserve acyclicity, 283
- fails to preserve star finiteness, 283
- hidden structure (HS), 205
- over the real numbers, 283
- preservation of information, 285, 287
- versus relation space, 276

Accessibility (A), 423

Action, 32
- additivity of, 346
- as a path functional, 344, 346
- causal set theory, 150, 379
- classical, 32, 341, 344, 465
- discrete causal, 465
- Einstein–Hilbert, 150, 466, 498
- Feynman's postulates, 371
- Hamilton's principle, 344, 472
- of a path increment, 380
- phase map via, 345, 379

Acyclic directed set(s), 144, 156, 157, 160, 172, 206
- accessibility conditions, 425
- behavior under abstract element space functor, 287
- behavior under relation space functor, 277–279
- boundedness conditions, 247
- category of, 157, 175
- causal atomic decomposition, 306
- causal set theory ignores most, 139, 140
- chain directed product algebra over, 335
- coarse relation, 299
- completions of kinematic schemes, 446
- countably infinite, 447
- counting classes of, 173
- definition of, 157
- degeneracy, 192
- finiteness conditions, 258, 259
- generational dynamics, 310, 353, 374, 469
- generational kinematics, 442
- generational refinement types, 404
- graph-theoretic viewpoint, 158
- higher induced relations, 302
- higher-level cycles, 374, 494
- incidence algebra over, 335
- interval finiteness (IF), 212, 252
- interval finiteness versus star finiteness, 217
- isomorphism classes of, 172
- kinematic preschemes of, 422
- kinematic schemes of, 415, 466
- Kleitman–Rothschild pathology, 188, 251
- McKay's enumeration, 189
- nontransitive, 430
- obtained by collapsing cycles, 495
- order refinement principle (ORP), 174
- positive sequential kinematic scheme, 351, 398, 429
- refinements, 174
- relative over the integers, 245, 266
- rigidity, 355, 467
- simplicial power spaces, 303
- skeleton of, 191
- Sorkin-Rideout kinematic scheme, 428
- star finiteness (SF), 215

B.F. Dribus, *Discrete Causal Theory*, DOI 10.1007/978-3-319-50083-6

Stern–Brocot tree, 449
transitive reduction, 191
transitive relation, 181
transitivity (TR), 263
Acyclicity (AC), 8, 22, 33, 47, 55–57, 70, 71, 105, 111, 185, 187, 246, 259, 267, 425, 457
abstract element space fails to preserve, 283
arguments for, 100
as a physical relevance criterion, 72
as an axiom for directed sets, 151, 157, 265, 267, 460
causal set theory, 155
generalization to multidirected sets, 161, 164
implied by transitivity and irreflexivity, 152, 157
meaning of, 71
partial orders, 154
pasts and futures, 170
positive sequential kinematic scheme, 432
refinements, 403
Alexandrov topology, 76, 108, 109, 155, 202
alternative mathematical meaning, 208
as a type of interval topology, 108, 202, 208
definition of, 108
physical significance of, 109
Algebraic geometry, 73, 121, 130, 327
algebraic Lie theory, 480
contribution to discrete causal theory, 369
correspondences, 484
flag varieties and kinematics, 323, 441
Grothendieck, 22, 177
hidden structure (HS), 178, 205, 297, 313
homotopy theory, 318
moduli spaces, 5, 352
morphisms for schemes versus for multidirected sets, 159
noncommutative geometry, 511
origin of category theory, 176
origin of topos theory, 342
relative viewpoint (RV), 22, 25, 57
string theory, 114, 177, 498
tangent sheaf, 345
Zariski topology, 235
Antichain(s), 104, 161
addition of (evolution), 468
analogous to acausal subsets, 163
are causal atoms, 307
as a choice of present, 386
as structural increments, 377
definition of, 161
direct evolution, 413
entropic phase maps, 468
finiteness conditions, 258
generalized inner products, 393
generational dynamics, 310
generational kinematics, 442
generational refinement types, 404
impermeability theorem, 291
local, 310
maximal, 105, 289, 373
nontrivial automorphism groups, 468
of relations, 187, 190, 290, 291, 293, 375, 468
path summation, 374
permeability problem, 105, 163, 275, 289, 373, 421, 432, 462
Schrödinger-type equations, 386
symmetric groups, 467
thickening, 290
versus Cauchy surfaces, 163
Asselmeyer-Maluga, Torsten, 16
exotic smoothness, 16
Atomic topologies, 307
Axiom(s), 101, 126, 132, 263, 274, 425, 456, 461, 464
acyclicity (AC), 265, 267, 460
analysis of (revisited), 457
binary axiom (B), 265, 267, 458
causal metric hypothesis (CMH), 68
conservative, 266, 460
convergence issues, 366
countability (C), 265, 267, 459
domain theory, 508
fatal versus merely unexpected behavior, 131
for encoding discreteness, 71
for encoding local finiteness, 200, 212, 215, 222
generalized measure axiom (M*), 265, 458
interval finiteness (IF), 151, 212, 252, 267, 457
involving relative directed sets, 247, 267
irreducibility (IRR), 267
irreflexivity (IR), 151
local versus global, 131
measure axiom (M), 151, 222, 267
of category theory, 176, 296, 328
of causal set theory, 138, 150, 152, 457
of partially ordered sets, 111

of topology, 203
positive sequential kinematic scheme, 430
prescription versus description, 129
star finiteness (SF), 214, 215, 256, 265, 267, 459
suggested list for directed sets, 265, 458
transitivity (TR), 151, 457

B

Background dependence, 340, 368
evolutionary processes, 341
incidence algebras, 336
non-causal quantum correlations, 422
particle motion, 344
path summation principle (PSP), 365
paths versus histories, 188, 436, 441
Poincaré group, 410, 482
quantization, 502
quantum automaton theory, 505
quantum causal set theory, 144, 202, 235, 243
quantum origin of temporal direction, 368
subclassical structure, 448
Background independence, 28, 340, 367, 369, 398
causality paradoxes, 29
causal metric hypothesis (CMH), 67
contribution to discrete causal theory, 54, 369
dynamics, 464
equal magnitudes, 123, 347, 378
evolutionary processes, 341
functors of kinematic schemes, 449
general relativity, 28, 84, 98
generational dynamics, 310
higher-level cycles, 374
loop quantum gravity, 499
non-causal quantum correlations, 302
path functionals, 436
path summation, 373, 415, 462, 469
path summation principle (PSP), 365
paths versus histories, 189, 414
perfect, 28, 97, 99, 368, 457, 494
phase maps, 466
positive sequential kinematic scheme, 429
prescription versus description, 68, 130
quantum causal metric hypothesis (QCMH), 434
quantum causal set theory, 145
quantum origin of temporal direction, 368, 390, 428
quantum theory, 124
ranking theories via, 30
string theory, 507
universal properties, 29, 130
versus general covariance, 28
Binary relation(s), 204, 375
acyclic, 72, 157
binary tree (hierarchy), 493
causal atomic decomposition, 309
causal dynamical triangulations, 504
causal relation, 39, 95, 180, 190, 461
causal set relation, 138, 151
coarse relation, 298
concatenation relation, 324
congruence relations, 332
definition of, 70
directed product relation, 326
enhanced star model, 297
Finkelstein's causal connection relation, 183
higher induced relations, 302
induced by sprinkling, 224
induced relation on relation space, 276
metric recovery, 97
non-causal quantum correlations, 421
nontransitive, 71, 133, 263, 366
on a directed set, 156
on a linear directed set, 320
on causal path spaces, 330
partial order, 152
power spaces, 496
refinement of, 174, 230, 244, 248, 402, 482
relativistic causal relation, 95, 402, 481
skeletal relation, 191
Stern–Brocot tree, 449
strict partial order, 100
subset relation, 152
temporal, 402
total order, 115
transitive relation, 191
transitivity (TR), 160, 179
types of, 153
versus N-ary relations, 496
versus splice relations, 305
Bombelli, Luca, 110
as founder of causal set theory, 140, 155, 208, 229
Lorentz invariance, 227
metric recovery, 110, 112

C

Category
  2-category, 295
Category/categories, 73
  definition of, 175
  enriched, 493
  higher-level algebras, 333
  kinematic scheme analogy, 147, 490
  monoidal, 380, 490, 510
  morphism category, 295
  non-unital (semicategories), 323, 329, 331
  of acyclic directed sets, 157
  of causal sets, 148, 152
  of directed sets, 156, 482
  of multidirected sets, 159
  of partially ordered sets, 481
  of points and paths, 330
  of relative multidirected sets, 248
  of structured sets, 342
  pseudo-Euclidean, 251, 479, 480
  pseudo-Riemannian, 82, 90
  quantization on, 69, 146, 336, 342
  smooth, 78, 81, 201
  smooth conformal, 89
  topological, 77, 201, 204
Category theory, 23, 175
  analogy involving generational refinement types, 411
  as a source of analogies, 134, 172, 176, 189, 297, 307, 491, 510
  as a structural paradigm, 78, 126, 130, 172, 175, 327
  as an organizing principle, 77
  axioms of, 176
  categorification/decategorification, 491, 493, 510
  colimits, 494
  contribution to discrete causal theory, 369
  conventional meaning of directed set, 156
  co-relative histories, 363
  covariance (disambiguation), 477
  functors, 278
  groups via endomorphisms, 215, 329
  hierarchy, 344, 492
  higher category theory, 295
  history of, 177, 328
  Isham's program, 146, 148, 173, 176, 342, 510
  iteration of structure (IS), 351
  kinematic analogy, 148, 412, 415, 419, 421, 448
  limits, 439, 446, 494
  natural transformations, 328
  passive versus active viewpoint, 90
  path algebras arising in, 335
  phase theory, 466, 474
  physical applications of, 510
  physical limitations of, 176, 297, 311, 418
  proper role in physics, 176
  relative versus co-relative terminology, 248
  structure on relativistic spacetime, 73
  universal properties, 451
  versus discrete causal theory, 173, 328, 490
Cauchy sequences, 116
  completions of kinematic schemes, 446
  order-theoretic completeness, 116
Cauchy surface(s), 102, 161, 169
  causal set analogues of, 290
  definition of, 102
  discrete causal analogues of, 104, 163, 187, 190, 289, 376, 444
  filters information flow, 102, 288
  foliation via, 103, 288, 309
  globally hyperbolic condition, 103
  impermeable, 102, 288
  independent modes of influence, 186
  path integration, 372
  refinements, 230
  separating past and future, 167
  state functions, 384
Causal atom(s), 169, 305
  approximation via, 462
  as generalized elements, 307
  atomic accretion, 245
  convex subset analogy, 306
  hidden structure (HS), 205
  local, 310
Causal atomic decomposition, 305, 307
  definition of, 307
  eliminating cycles via, 495
  foliation example, 309
  generational refinement types, 404
  in generational dynamics, 309
  local, 308
  not transitive, 311
  path summation, 374
Causal atomic resolution, 305, 310
  coarse-graining, 184, 472

conservative top-down scenarios involving, 314
definition of, 311
emergence of Poincaré group, 483, 485
generational dynamics, 444
growth factors of, 312
multiresolution analysis analogy, 312
scale-dependent properties, 312
ways to characterize, 311
Causal boundary, 27
definition of, 171
relation space, 275, 281, 462
shrinking of generations near, 476
Causal concatenation space, 324
relation space as a full subobject of, 325, 393
Causal condition, 95, 99, 138, 153, 156
acyclicity (AC), 264
Alexandrov topology, 108
causal metric hypothesis (CMH), 110
metric recovery, 100
relativistic causal relation, 481
rules out closed causal curves, 96, 99
sprinkling, 223
stability issues, 100
strict partial order, 100
Causal directed product space, 325
Causal dynamical triangulations, 45, 498, 503
as a hybrid theory, 497
dimension reduction, 225, 304, 312, 503
includes spatial structure, 203, 497, 504
locally Lorentzian, 55, 504
path summation, 382, 504
phase transitions, 45, 503
pragmatic nature of, 54, 503
random topology, 509
simulations, 55, 225
small valences, 240
uniform valences, 430
versus directed sets, 504
versus quantum automaton theory, 505
versus simplicial power spaces, 304
versus spin foams, 504
Causal Galois groups, 357, 514
entropic phase maps, 467
higher-level relation functions, 439
Causality conditions, 97, 106, 167
causal condition, 99
chronological condition, 98
globally hyperbolic condition, 102
past and/or future distinguishing conditions, 100
stably causal condition, 100
strongly causal condition, 101
Causal locality, 179, 190, 202, 461
applies to all discrete manifolds, 203
dispenses with action at a distance, 203
represented via star topology, 206
should precede choice of model, 203
useful intuition for multidirected sets, 203
versus spatiotemporal locality, 233, 234
Causal metric hypothesis (CMH), 8, 16, 52, 66, 119, 132, 263, 456, 461
alternatives, 497
background independence, 29
causal locality, 202
causal set theory, 141
causality group, 482
choice of terminology for, 66
classical version of, 68, 69, 71, 77, 84, 110, 131, 132, 168, 200, 207, 208, 215, 256, 259, 263, 403, 415, 457
contribution to discrete causal theory, 369
cycles, 494
direct evolution, 413, 417
directed paths as causal paths under, 321
does not require discreteness, 66, 456
domain theory, 508
entropic gravity, 506
fate of Lie theory, 491
fundamental scale, 199
historical roots, 11, 66
in non-fundamental settings, 66
interpretations of, 11, 67
local conditions more fundamental than metric, 201
motivation via metric recovery, 74, 88, 106, 107, 110
noncommutative geometry, 512
phase maps, 379
philosophical content of, 67
Poincaré semigroup, 483
prescription versus description, 67, 129
quantum version of, 68, 132, 433, 440
recovering algebraic structure, 514
relative viewpoint (RV), 275
relativistic causal structure, 91
Sorkin's version of, 66, 71, 72, 141, 200, 214, 256, 264
statement of classical, 71, 456
statement of general, 10, 67, 456
statement of quantum, 433, 456

strong interpretation of, 11, 28, 36, 37, 67, 98, 99, 146, 150, 165, 178, 185, 203, 225, 237, 416, 448, 456, 464
subclassical structure, 495
technical implementation of, 68, 110
unifies kinematics and dynamics, 36
versus extrinsic ring structure, 332, 346
versus loop quantum gravity, 502
versus shape dynamics, 507
Causal morphism, 96
Causal net(s), 158, 182
Causal path algebra(s), 145, 175, 177, 323, 326, 332, 333
additive subgroups of, 334
chain concatenation algebra, 333
chain directed product algebra, 334
cohomology, 514
concerns regarding extrinsic structure, 332
conventional applications of, 335
evaluation maps, 391
generalized graded structure of, 334
motivation for construction of, 332
noncommutative geometry, 512
path summation, 462
phase maps, 347, 348
subalgebras of, 334
target ring of, 333
versus short-circuit algebras, 336
via generalized semicategories, 327, 328, 332
Causal path semicategories, 330
as precursors to causal path algebras, 332
evaluation maps, 391
phase maps, 346
Causal path set(s), 321
as generalized relation sets, 322
causal path sums, 348
definition of, 321
mapping spaces, 348
path summation, 350, 365
phase maps, 346, 347
Causal path space(s), 157, 162, 302, 305, 318, 323, 334, 462
algebraic structure of, 378
are not power spaces, 496
as generalized semicategories, 323, 329
binary axiom (B), 458
causal concatenation space, 324, 325
causal directed product space, 325
examples of, 305
hidden structure (HS), 179
hierarchy, 490
splice relations and, 305
twistor theory, 506
versus causal path semicategories, 393
via splice products, 323
Causal relation, 72, 95, 96, 180, 263, 461
acyclicity (AC), 457
choice of terminology, 181
definition of, 190
generational kinematics, 444
generational refinements of, 404, 409
kinematic accounts, 407, 431
meaning of, 190
pointed refinements of, 483
refinements of, 244, 310, 403, 406, 411, 441, 451, 478, 479, 481
relation functions, 439
versus induced relation on relation space, 276
Causal set theory, 54, 113, 138, 263
axioms of, 150, 263
causal condition, 100
causal metric hypothesis (CMH), 66, 91, 110, 143, 456, 461
cosmological constant, 43, 60, 367
discreteness, 71
distinctions with theory in this book, 55, 179, 429, 457
dynamics of, 147
eigenstate analogies, 235
Einstein–Hilbert action, 379
elegance of, 142
energetic causal sets, 502
hauptvermutung, 142
history of, 139
inflationary hypothesis, 45
interval finiteness (IF), 66, 76, 77, 151, 461
interval topology, 208
irreflexive formulation of, 151
Kleitman–Rothschild pathology, 167, 188, 214, 251
limitations of, 138, 139, 179
local finiteness, 200, 213, 366
Lorentz invariance (systematic), 14, 111, 112, 233
metric recovery, 16
minimally modified version, 266, 430, 456, 460
Moore's objections, 229
nongravitational matter, 11, 29, 146
partial order formulation of, 152
path algebras in, 335
path summation, 33, 343, 374

permeability problem, 105, 290, 373
phenomenology, 149
Poisson fluctuations, 111
quantum, 144, 335, 356, 367
reducible relations dependent in, 166
relation functions, 32
spatiotemporal locality, 150, 233
sprinkling, 143, 223
star finiteness (SF), 81, 233
time-first, 497
transitions, 24, 58
transitivity (TR), 151
valence fields, 211
variable speed of light, 51
versus loop quantum gravity, 503
versus mainstream theories, 16
Causal set(s), 138
as a discrete background, 144, 342, 368, 422
as classical histories, 341, 342, 373
as order-theoretic models, 138
as relative directed sets over the integers, 178, 244
category of, 152
coarse-graining of, 305
connected, 249
co-relative kinematics, 442
cosmology, 489
curvature of, 150, 466
cutoff, 238
decorated, 146
definition of, 138
finite, 251
included in positive sequential kinematic scheme, 430
infinite, 251
infinite bouquet, 218
Kleitman–Rothschild pathology, 461
labeled, 343, 428
manifold-like, 139, 143
morphisms between pairs of, 152
relation space over, 280
sequential growth of, 147, 310, 343, 353, 361, 405, 415, 427, 436, 463, 478, 488
Sorkin-Rideout kinematic scheme, 398, 427
sprinkled, 142, 143, 186, 208, 213, 218, 220, 223, 243, 252, 315, 399, 430, 439, 459, 479, 486, 503
star finite, 243
star infinite, 234, 243
total labeling of, 149, 174
transitions between pairs of, 147
versus causal nets, 182
Causal structure, 7, 75, 175, 476
algebraic data via, 34
background independence, 367
building blocks of, 69
causal dynamical triangulations, 504
causality paradoxes, 98
causal metric hypothesis (CMH), 10, 52, 67, 111, 237, 416, 456, 461
causal relation, 190
classical, 68, 71, 114, 120, 132, 248, 460
discrete, 76, 141, 266, 463
domains of influence, 167
emergence of Poincaré group, 479
emergent scale, 139
frames of reference, 402
fundamental, 240, 243
general relativity, 22, 91
geometry emergent from, 61
historical prominence of, 140
incrementing, 469
induced, 225
inflationary hypothesis, 45, 469
interaction with fields, 33
interpolative, 167
Isham's program, 176
kinematic schemes, 423
local, 185, 207, 208, 214, 215, 222, 225, 255, 264, 312, 458, 459
Lorentz boosts, 480, 513
Lorentz signature, 82
meaning of, 55, 68, 91, 180
metric recovery, 16, 66, 74, 91, 97, 100, 106, 110, 124, 133, 141, 146, 458
non-causal quantum correlations, 302, 421
noncommutative geometry, 512
particles emergent from, 34
phase theory, 464
prescription versus description, 323
quantum, 68
relativistic causal relation, 96
shape dynamics, 506
spatiotemporal locality, 233
spin foams, 502
sprinkled causal sets, 143
strong interpretation of the causal metric hypothesis, 11, 29
table of binary relations, 153
top-down causation, 313
transitivity (TR), 179, 461
versus extrinsic ring structure, 332, 346
versus logical structure, 259

versus spatial structure, 18, 203, 480, 497
Zeeman's theorem, 481
Causality group, 481
Chain algebras, 334, 348
Chain concatenation algebra, 333, 387
as a quiver algebra, 335
as a standard path algebra, 335
generalized quantum amplitudes, 380
past and future chain functionals, 389
Chain concatenation semicategory, 331
generalized quantum amplitudes, 380
higher-level path functionals, 437
past and future chain functionals, 389
path summation, 375
phase maps, 380, 386
Chain directed product algebra, 334
inner product formula, 393
mathematical properties of, 335
Chain directed product semicategory, 331
Chain functional(s), 380
future, 389, 393
maximal, 381, 393
past, 389, 390, 393
Chain semicategory/categories, 331
as causal path semicategories, 331
definition of, 331
Chain(s), 105, 133, 157, 161
accessibility (A), 423
admitting extension by a relation, 388
algebras of, 333
along a path, 374
analogous to causal curves, 163
annihilation of, 388
as causal paths, 162, 322
as evolutionary pathways, 470
as generalized elements, 305
bounded, 249
Cauchy sequence analogy, 447
collapsing (limits), 494
completions of kinematic schemes, 447
complex, 162
co-relative kinematics as, 440
covariance, 475
cyclic directed sets, 374
definition of, 161
entropic phase maps, 470
equivalence classes of, 446
extending to simplices, 303
factorization of, 380, 387
finite accessibility (FA), 425
finite relative accessibility (FRA), 425
generalized action, 472
generalized semicategories of, 331
higher induced relations between pairs of, 302
higher-level, 408, 413, 414, 418
incrementation of, 377, 387
independence of families of, 165
length of, 162
limits, 430
morphism analogy, 329
of simplices, 304
of transitions, 355
originary, 430, 441
permeating, 289, 385, 432
reducing, 164, 181, 186, 187, 191
relative accessibility (RA), 425
representing Cauchy sequences, 450
representing linear kinematic schemes, 494
unbounded, 221
versus causal curves/paths, 309
wrapping around a cycle, 374, 495
Zorn's lemma, 175
Chain set(s), 161
as causal path sets, 322, 331, 495
chain functionals on, 389
definition of, 161
fractal decomposition, 391
higher induced relations on, 303
Chronological condition, 98, 153
Chronon, 18, 48, 70, 445
choice of terminology, 18
classical, 18
Classical causal metric hypothesis (CCMH), 71, 456
Classical history/histories, 4, 12, 69, 88, 114
acyclic, 22
as a classical universe, 5
as a fixed background, 31
as building blocks of quantum theory, 368
as elements of a kinematic scheme, 132, 207, 274, 373, 435
axioms for, 56, 265, 458
background independence, 368
boundary value problems, 28
bounded, 259
causal metric hypothesis (CMH), 456
classes of, 433
classical causal metric hypothesis (CCMH), 71
cobase families of, 350
configuration spaces of, 69
covariance, 476
decorated, 33

domain theory, 508
entropic phase maps, 467
evolution of, 20, 22, 37, 40, 59, 123, 169, 294, 378, 449, 452, 463, 501
expanding universe, 19, 47
family or families of, 8, 34
finite, 259
generalized quantum amplitudes, 392
initial, 9, 57
in terms of relations, 25
kinematic accounts, 407
kinematic schemes of, 30, 37, 59
lattices as, 235
local entropic traps, 474
meaning of diagrams representing, 18
meaning of elements of, 10
modeled as a directed set, 5, 9, 53, 55, 70
non-causal quantum correlations, 421
nontransitive, 133
particle trajectory in, 34
paths versus histories, 436, 441
phase transitions, 50
quantum information theory, 17
quantum origin of temporal direction, 38, 429
recovery from relation space, 26
relationships between pairs of (co-relative histories), 8, 23, 36, 37, 58, 134, 147, 203, 340, 354, 359, 398, 415
relation space as a, 25
relation space over a, 56, 275
relative viewpoint (RV), 22, 25, 57
rigidity, 20, 471
run-down to geometric state, 46, 47
Schrödinger-type equations, 382
sprinkled causal sets as, 235
star finite, 133
target families of, 350
terminal, 9, 57
uncountable, 459
valences of elements of, 34
versus quantum histories, 501
Classical holism, 313, 460
Classical theory/theories, 5
correspondence principle, 6, 341
determining amplitudes, 341
examples of, 6
general relativity imperfect, 500
quantizing, 499
relationship to quantum theory, 500
subclassical structure, 448
unification with quantum theory, 53, 351, 353
versus quantum theory, 5, 58, 121
Concatenation algebra, 333
as a generalized graded algebra, 334
convolution, 347
definition of, 333
Concatenation product, 323
causal Schrödinger-type equation, 390
definition of, 324
inner product formula, 391
phase maps, 347, 379, 380, 386, 389
Configuration space(s), 4
as a quantum object, 7
conformal superspace, 316
iteration of structure (IS), 350, 462
kinematic scheme as a, 4, 30, 36, 53, 59, 463
Kleitman–Rothschild pathology, 167, 188, 214, 461
non-structured, 436
of acyclic directed sets, 192
of causal sets, 145
of classical histories, 56, 69, 147, 342
of co-relative histories, 8, 369
of directed sets, 158, 192, 259, 340, 342, 352, 364
of evolutionary processes, 341
of finite acyclic directed sets, 189
of partial orders, 192
of particles, 274
of relativistic spacetime manifolds, 354
of spacetime regions, 367
of structured sets, 342
paths in, 318
path summation, 6, 341, 366
quantum origin of temporal direction, 429
Schrödinger-type equations, 381
superposition, 351
Conformal boundary, 275
Conformal field theory, 87, 110
Conformal structure, 75, 76, 84, 175, 202, 236
active versus passive viewpoints, 90
angle preservation, 87
metric recovery, 106, 141, 265
path topology, 109
shape dynamics, 316, 506
Conformal superspace, 316
Connes, Alain, 73
noncommutative geometry, 73, 179, 507
Continuum model, 206
cyclic directed sets, 495
Continuum topology, 209, 215

cyclic directed sets, 495
Co-relative history/histories, 8, 23, 36, 58, 59, 69, 133, 134, 147, 172, 177, 207, 266, 359, 463
as equivalence classes of transitions, 418
as families of morphisms, 362
as relations in a kinematic scheme, 132, 274, 364, 415
background independence, 369
between pairs of causal dynamical triangulations, 504
between pairs of directed sets, 361
between pairs of domains, 508
between pairs of spin foams, 502
category theory, 24, 510
choice of terminology for, 25, 248
classes of, 416, 420, 439
classical stochastic dynamics, 38
composition of, 363, 418, 441
congruence relations, 332
cycles of, 362
definition of, 361
direct evolution, 412
entropic phase maps, 467
entropy, 437
foliation scheme, 444
formalizing evolutionary processes, 354
generational kinematic scheme, 442
hierarchy, 490
higher-level analogues of, 448
higher-level cycles, 374
higher-level relation functions, 379, 439, 474
higher-level structure, 37, 340, 342, 347
image-fixed, 364, 420
impermeability theorem, 294
iteration of structure (IS), 59, 351, 364
McKay's example, 360, 364, 415
morphism analogy, 412, 420
multidirected structure, 24, 37, 264
path summation, 441
permissible, 362
phase maps, 464
phase theory, 464
physical meaning of, 359
positive sequential kinematic scheme, 429
proper, full, originary, 360, 416, 463
quantum causal metric hypothesis (QCMH), 434
random graph dynamics, 488
reducible, 418, 431
relative viewpoint (RV), 23, 57, 178, 275, 354, 435
represented by relations, 203, 464
represented by transitions, 24, 354, 419
Shannon entropy, 41
skipping generations, 417
S-matrix analogy, 435
Sorkin-Rideout kinematic scheme, 148, 427
structural increments, 469
technical properties of, 58
universal kinematic schemes, 451
versus category theory, 329, 363
versus causal paths, 321
Co-relative kinematics, 440
definition of, 440
Correspondence principle, 6, 341
Cosmology, 4, 43, 237
causal dynamical triangulations, 55
causal set theory, 66, 230, 290, 489
conformal cyclic, 87, 90, 221
finiteness conditions, 259
inflationary hypothesis, 43, 44, 60, 356, 367, 469, 489
interval finiteness (IF), 256
multiverse in string theory, 4
phase transition, 50
relativistic, 44, 81, 288
Countability (C), 55, 56, 251
as an axiom for directed sets, 265, 459
as an axiom of causal set theory, 151, 247
Covariance, 20, 409, 434, 475
automorphism groups, 478
category theory, 477
disambiguation, 477
discrete causal, 464
discrete general, 425
discrete general (causal set theory), 478
general (relativistic setting), 477
higher-level, 478
in terms of extraphysical information, 477
kinematic schemes, 467
lower-level, 478
phase theory, 464
Poincaré group, 479

D

Degeneracy, 189
causal dynamical triangulations, 504
involving reducible relations, 192
Kleitman–Rothschild pathology, 189, 192

Diffusion equation, 7, 513
Dirac, Paul, 6
  causal metric hypothesis (CMH), 11
  delta function, 349
  Dirac equation, 505
  path summation, 6, 58, 370
Directed graph(s), 56
  directed sets as, 18, 55, 70, 156, 487, 509
  enumeration of, 406
Directed multigraph(s)
  multidirected sets as, 159, 276, 509
Directed product, 323
  additivity of classical action, 346
  continuum-based analogue, 372
  definition of, 325
  homotopy theory, 346
  inner product formula for, 393
  phase maps, 347
Directed product algebra, 333
  as a generalized graded algebra, 334
  definition of, 334
  versus incidence algebras, 335, 343
  versus monoids of arrow fields, 343
Directed set(s), 66, 72, 95, 100, 105, 114, 131, 156, 159, 172
  abstract element space over a, 281
  accessibility (A), 423
  acyclic, 70, 72, 140, 144, 156, 157, 159, 172, 173
  alternative accessibility conditions, 425
  as a pair, 156
  as classical histories, 5, 9, 18, 53, 55, 68, 69, 132, 263
  as elements of a kinematic scheme, 73, 415
  as generalized causal sets, 138, 139, 156
  automorphism groups of, 467, 481
  axioms for, 68, 132, 265, 457
  cardinality conditions for, 459
  category of, 156
  category theoretic analogy, 176
  causal dynamical triangulations, 304
  causal Galois groups of, 439
  causality group of (same as automorphism group), 482
  causal metric hypothesis (CMH), 416
  causal path spaces as, 323
  causal relation on, 151, 190, 276
  choice of terminology for, 158
  classical causal metric hypothesis (CCMH), 70, 456
  cobase family of, 362
  concatenation algebra over, 333
  configuration spaces of, 158, 259, 342, 436
  conservative axioms for, 266
  containing cycles, 99, 494
  co-relative histories between pairs of, 24, 58, 133, 342, 361, 364
  cyclic, 374, 495
  definition of, 156
  degeneracy, 192
  discrete, 71, 111, 121, 123, 146, 168, 461
  domains of influence, 167
  dynamics of, 105
  emergence of Poincaré group, 479
  emergent geometry, 223
  enhanced star model of, 297
  evolution of, 354, 398, 411, 412, 416, 422
  factorization in, 380
  finiteness conditions, 258
  fundamental scale, 199
  generalized Hasse diagram of, 10, 21, 70
  generational refinement types, 411
  graph-theoretic viewpoint, 18, 55, 70, 509
  hierarchy, 340, 490
  higher-level, 412, 414, 416, 419, 437
  independence of relations in, 170, 187
  inevitability of, 17
  interior of, 287
  interpolative, 217, 508
  interval finiteness (IF), 212
  interval topology, 208
  irreducibility of relations in, 170, 187
  iteration of structure (IS), 353
  kinematic preschemes of, 416
  kinematic schemes of, 398
  lattice of subsets of, 301
  linear, 118, 160, 162, 319, 375, 405
  local finiteness, 33, 212, 263
  local irregularity of, 146, 201, 514
  locally infinite, 219
  local properties of, 108
  Lorentz invariance, 14, 227, 356, 462, 483
  manifold-like, 223
  mathematical conditions on, 161, 164
  McKay's example, 360
  metric recovery, 43, 77, 83, 84
  morphisms between pairs of, 23, 24, 58, 97
  nontransitive, 11, 147
  ordinal number analogy, 122
  path summation over, 341
  paths versus histories, 435

permeability problem, 104, 105, 289
permissible, 410, 422, 431
phase maps, 347
physical interpretation of, 165, 274
Poincaré semigroup of, 483
pointed, 484
quantum causal histories, 184
quantum causal metric hypothesis (QCMH), 433, 456
quantum circuits as, 505
quantum origin of temporal direction, 429
random graph dynamics, 487
realizable, 422
recovery of algebraic data, 514
refined, 403
relation space as a, 57, 274–276
relation space over a, 105, 133, 178
relative over a fixed base, 122, 178, 244, 248, 320, 446, 483
relative viewpoint (RV), 22
reverse-engineered, 225
rigidity, 74, 232, 236, 355, 438, 439, 467
rotational notions, 486
scalar fields, 211
spatiotemporal locality, 462
star finite, 166, 257, 462
star infinite, 223
subclassical structure, 448
subsets of, 156
suggested list of axioms for, 265
symmetry considerations, 356
target families of, 362
terminal, 416
transitions between pairs of, 354, 360
transitive, 220
transitive closure of, 181, 191
tubular diagrams, 507
unconventional meaning in this book, 5, 9, 18, 156
valences of elements of, 33, 503
variations on star finiteness, 216
versus causal dynamical triangulations, 504
versus spin foams, 502
Discreteness, 256, 320
a posteriori (loop quantum gravity), 498
causal metric hypothesis does not require, 456
contribution to discrete causal theory, 369
Lorentz invariance versus spatiotemporal locality, 233
meaning of, 71, 459
measure-theoretic, 71
metric recovery, 456
of fundamental spacetime structure, 140
order-theoretic, 71
rigidity, 319
special relativity, 214
structural subtleties introduced by, 138, 141
topological, 71
via local finiteness, 200
Discrete topology, 207
atomic only in the acyclic case, 307
Domain(s) of influence, 167, 169, 400
as full subobjects, 171
causal path sets, 321
definition of, 169
direct future, 170
direct past, 170
future, 168
maximal past, 170
minimal future, 170
Minkowski spacetime, 402
past, 168
quantum information theory, 505
relativistic prototypes, 95, 168
total, 169, 211, 224, 393
Domain theory, 17, 217
as applied order theory, 507
as nondiscrete causal theory, 54, 91, 167, 206, 267, 460, 482
causal metric hypothesis (CMH), 508
interpolative property, 203
Martin and Panangaden, 167
metric recovery, 11
relative viewpoint (RV), 508
Scott continuity, 204
shape-causal duality, 507, 509
twistor space, 509
Dowker, Fay, 145
causal set actions, 379, 466
diffusion, 150
discrete manifolds, 158, 203
Lorentz invariance, 227
spatiotemporal locality, 145
Dynamical law(s), 366
arises at quantum level, 6
causal Schrödinger-type equation as a, 7, 36, 59, 379, 386, 463
correspondence principle, 7
foliation-type approach to deriving, 103
model-dependence of, 382
of causal set theory, 147

of discrete causal theory, 167
Schrödinger's equation as a, 382
unaltered by gauge transformations, 39
Dynamics, 36, 59
background independence, 28
causal set, 55, 144, 146, 147, 185, 230, 234, 367
classical stochastic, 38, 122, 473
covariance, 39
describes favored behavior, 36, 407, 416
discrete causal, 50, 104, 225, 258, 275, 294, 376, 398
foliation-type approaches to, 57, 103
generational, 295, 309, 312, 316, 353, 357, 374, 442, 443, 468
generational automorphism groups, 472
interference via path summation, 30, 52
iteration of structure (IS), 36
local, 234
network, 45, 185, 276
nonlinear, 318
of directed sets, 105, 440
path integration, 373
permeability problem, 25, 289
phase maps, 379
phase theory, 463
quantum, 134, 275
random graph, 45, 53, 464, 487
Schrödinger-type equations, 381
sequential growth, 24, 38, 59, 144, 146, 147, 152, 155, 174, 188, 230, 310, 328, 353, 405, 415, 425, 427, 436, 463, 478
shape, 15, 54, 288, 316, 497, 506
spin, 122
thermodynamics, 51, 188, 274, 357, 440
toy, 37, 122, 398, 407, 473
unification with kinematics, 36, 416, 464
versus kinematics, 36, 407, 416
via a Lagrangian, 465
via path summation, 415
via phase maps, 416, 463

E
Einstein field equation, 8, 12, 13, 15, 29, 36, 74, 80, 81, 83, 84, 95, 354, 466, 477
Einstein manifold, 83
Einstein metrics, 354
Einstein, Albert, 5
as founder of general relativity, 5, 74, 129, 141
causal metric hypothesis (CMH), 11, 67
cosmological constant, 131
hole argument, 29, 74, 94
influence of Mach, 316
misgivings about continuum, 140
pseudo-Riemannian geometry, 78
scientific philosophy, 127, 130
Einstein–Hilbert action, 150, 466, 498
Element(s), 8, 15
abstract, 282
acyclicity (AC), 72
addition of (evolution), 20, 58, 59, 148, 343, 358, 427, 438
antichains of, 290, 376, 506
as a 0-chain, 162
as objects, 22, 275, 491
concrete, 282
converting morphisms to (limits), 494
demoting objects to, 493
distinguishable, 471
element field, 31, 144
emergence of group elements, 484
extremal, 170, 174, 207
fundamental spacetime, 91, 139, 311
generalized, 157, 159, 172, 184, 205, 305, 313, 458
hidden structure (HS), 178, 205
hierarchy, 490
higher-level, 73
infinitesimal, 116
initial, 18, 26, 144, 158
in mathematics, 73
internal structure of, 448, 494
labeling of, 149, 343, 405
local structure, 200
maximal, 171, 175
maximal cycles as, 374
meaning of (classical history context), 10, 18
meaning of (relativity context), 29
minimal, 171
nonextremal, 71, 153
number of, 51
of a binary tree, 493
of a causal path algebra, 333, 347
of a causal set, 138
of a classical history, 8
of a directed set, 18, 69, 111, 156
of a domain of influence, 167
of a generalized semicategory, 330
of a generation, 404, 417
of a group, 329
of a linear continuum, 117
of a loop set, 495
of a moduli space, 5

of a multidirected set, 158
of a power space, 300
of a pseudo-Riemannian manifold, 74, 94
of a random graph, 487
of a ring, 327, 349, 439
of a topology, 204
of an algebraic scheme, 179
of an automorphism group, 356, 473
of generalized spheres, 123
of simplicial complexes, 496
of the Poincaré group, 410, 480
of the Poincaré semigroup, 483
of the Stern–Brocot tree, 450
of unit magnitude, 378
oxymoronic, 25
physical interpretation of, 274
Poisson sprinkling of, 14
relative viewpoint (RV), 22
representing algebraic varieties, 313
representing causal atoms, 305, 495
representing causal paths, 321
representing classical histories, 37, 132, 274
representing co-relative histories, 8, 274, 415
representing cycles, 495
representing directed sets, 353, 408, 415, 420, 435
representing geometric objects (moduli spaces), 352
representing lines (projective space), 352
representing null lines (twistor theory), 315
representing objects of a category, 184
representing operators, 501
representing pentachora, 504
representing phase maps, 347, 348
representing relations, 25, 56, 210, 274
representing spacetime events, 9, 10, 151
representing spacetime regions, 313
representing transitive degeneracy classes, 192
representing two-element subsets, 317
self-related, 154
size of, 199
sprinkled, 143, 223, 315
terminal, 18, 26, 144, 158
valences of, 33, 112, 185, 211, 430, 467
versus relations, 111
zero element, 119
Ellis, George, 44
classical holism, 313
general relativity, 80, 208
relativistic cosmology, 44
top-down causation, 313
Emergent matter, 201
Enhanced causal morphism, 96
Enhanced star model, 297
Entropic phase map(s), 463, 467
covariance, 475
emergent particles, 474
examples of, 472
extraphysical data, 475
for kinematic schemes, 470
generational dynamics, 468
local physics, 474
Entropy, 50
causal atomic resolution, 472
coarse-graining of phase space, 471
covariance, 475
dependence on local structure, 474
discrete causal analogues of Lagrangian, 379
distinguishable states, 471
entropic gravity, 51
entropic phase maps, 41, 440
expansion of the universe, 47
generational dynamics, 357
generational measures of, 469
instantaneous, 469, 471
Kleitman–Rothschild pathology, 188, 459
local entropic traps, 474
multiplicities of states, 467
of antichains of relations, 295
of thickened spatial sections, 467
relation functions, 437
Shannon, 41, 473
via generational automorphism groups, 469
von Neumann, 473
Entropy function, 472
as a generalized Lagrangian, 472
Boltzmann-type, 472
negative, 472
Shannon-type, 473
Equal magnitudes, 346, 378, 438
covariance, 478
entropic phase maps, 472
phase maps, 466
Evolution, 4
between cobase and target families, 350
causal dynamical triangulations, 55
co-relative histories, 361
covariance, 478
curves (shape dynamics), 506

direct, 412
domain theory, 508
entropic phase maps, 468
generational dynamics, 309
in time, 122
kinematic accounts of, 343, 379, 407, 416, 417, 424, 435, 440
likelihoods, 37, 40
of causal sets, 148, 249
of classical histories, 20, 24, 37, 59, 69
of conformal geometries, 87, 316
of directed sets, 398
of spacetime, 178
of subsets of a directed set, 300
path summation, 59
pathways, 6, 7, 22, 30, 33, 37, 39, 40, 123, 146, 149, 294, 362, 365, 398, 399, 422, 425, 445
processes, 30, 189, 316, 319, 328, 332, 341, 360, 378, 392, 403, 408–410, 415, 422, 435, 444, 451, 452
random graph dynamics, 487
relationships, 8, 9, 22, 59
relative viewpoint (RV), 57
relevance of distant past, 471
sequential process, 4, 20
stages of, 5, 20, 21, 37, 42, 58, 147, 169, 310, 413, 426, 435, 439, 449
transitions, 354
unitary, 505
viewpoints, 398

F

Feng, Johnny, 105, 516
abstract element space, 283
antichains, 105, 162
open intervals, 151
Feynman, Richard, 6, 369
background independence, 368
causal metric hypothesis (CMH), 11
equal magnitudes, 478
Feynman diagrams, 370
ideal measurements, 237
inner product formula, 391, 393
path increments, 372, 376
path summation/integration, 6, 23, 30, 32, 58, 69, 93, 121, 134, 145, 188, 190, 285, 294, 318, 340, 341, 348, 350, 365, 369, 371, 383, 415, 434, 435, 441, 462, 501
phase map, 123, 345, 346, 372, 378, 383, 438, 465
postulates for path summation, 370, 371, 384, 392
propagator, 150
quantum amplitudes, 372
Schrödinger equation, 42, 382, 385
state functions, 382, 390
Field (algebraic), 57, 58
algebraically closed, 121
Archimedean, 116
as a coefficient ring, 335
as a target object, 344, 347, 511
complete, 116
complex numbers, 115, 123, 346
definition of, 115
p-adic, 117
rational numbers, 115
real numbers, 115, 319
totally ordered, 115
Field (physical), 30, 60
arrow (quantization on a category), 184, 336, 343
as nongravitational matter, 11, 28
background dependence, 367
cardinal-valued, 211
contribution to Lagrangian, 465
electromagnetic, 382
element, 31, 144
emergence of, 29, 33
field lines, 329
field theories, 275
gravitational, 29, 383, 502
in dynamical spacetime, 33
inflaton, 44
of Ising spins, 145
on a causal set, 144, 145, 202, 235, 342, 368, 422
on a directed set, 340, 350, 356, 448
on a field, 29
on a multidirected set, 211
on a static background, 31
on a toy history, 31
on Euclidean spacetime, 382
relation, 31, 144
scalar, 31, 150, 185, 211
tetrad, 12
valence, 33, 111, 211, 215, 379
Finite accessibility (FA), 425
Finiteness condition(s), 258
Finite relative accessibility (FRA), 425
Finkelstein, David, 54, 461
anticipated discrete causal theory, 140
arguments against transitivity, 181, 182, 186, 190

causal metric hypothesis (CMH), 11
causal nets, 158, 182, 190
credits Sorkin, 184
locality, 202, 220, 237
Foliation, 57
causal atomic decomposition, 309
causal set analogy, 105
generational dynamics, 310
generational refinement types, 405
leaves of, 103
local generational dynamics, 310
loop quantum gravity, 498
of relativistic spacetime, 103, 231
spacelike, 442
theorem, 447
via Cauchy surfaces, 288, 309
Foliation scheme, 426, 443, 444, 450, 460
covariance, 475
definition of, 444
generational growth in, 446
hierarchy, 490
Fractal decomposition, 391
Fundamental question of phase theory, 465
Fundamental scale, 83, 91, 112, 113, 141, 240
causal structure at, 11, 179
causal versus spatial structure, 203
dimension reduction, 304
discrete manifolds, 497
emergence of Poincaré group, 479
independence at, 185
local behavior, 199
measure axiom (M), 266
particle theory, 474
physics at, 497
Planck scale, 199
quantum information theory, 505
size of, 199, 240, 310, 472, 489
spacetime structure at, 8, 15
via elements and relations, 199
Zeeman's theorem, 481
Fundamental spacetime structure, 10, 16, 52, 53, 72, 83, 87, 106, 124, 340, 366
building blocks of, 198
category theory, 177
causal set theory, 149
discreteness, 140
duality, 275
in terms of relations, 294
line digraphs (relation spaces), 276
noncommutative geometry, 512
nontransitive relations, 183
path summation, 436
quantum theory of, 369
relation space, 462
shape dynamics, 316
Future, 168
abstract (in a kinematic scheme), 435
causal atoms, 245
causal (relativity), 95, 167
chronological, 96
cobordism, 507
definition of, 169
direct, 170
interval topology, 207
minimal, 170
of a spacetime region, 373
shrinking due to cutoff, 238
Future region, 375
continuum-based setting, 383
discrete causal setting, 375, 386
Future state function, 386
continuum-based setting, 382, 383
definition of, 389
discrete causal setting, 386

G

General relativity, 42, 72, 124, 425
ADM formulation, 316
as a classical theory, 5, 353
as a smooth approximation of discrete causal theory, 77
background independence of, 28, 33, 84, 145, 345, 367
binary relations, 153
canonical quantization of, 498
Cauchy surfaces, 288
causality conditions, 97
causality paradoxes, 29
causal structure, 91, 140
conformal structure, 84
contribution to discrete causal theory, 369
curves versus paths in, 93, 318
cycles (closed causal curves), 22, 55, 70, 158, 265, 267, 307, 375, 460, 494
diffeomorphism invariance, 74
dimension of spacetime, 201
domains of influence, 168
domain theory, 110
Einstein–Hilbert action, 466
emergence of, 12
experimental success of, 83
foliation, 309
fundamental scale, 198

governs large-scale phenomena, 5
imperfection of, 500
independence convention (IC), 167
kinematics and dynamics of, 36
locality defined in metric terms, 201
local Lorentz invariance, 13
local regularity, 201
local structure, 201
loop quantum gravity, 42, 126, 498
Mach's influence, 316
metric recovery, 11, 74, 110, 141
metric structure, 10, 78
morphisms, 23
prescription versus description, 129, 168
pseudo-Riemannian geometry, 78
quantizing, 113
residual background structure, 29
Rovelli, 29
transitivity (TR), 180
unification, 8
variable speed of light, 51
versus discrete causal theory, 74, 129, 225, 226
Generalized semicategory algebra(s), 333
versus group algebras, 514
Generalized semicategory/semicategories, 327
abbreviations, 331
as grouplike structures, 329
causal path semicategories, 330
chain, 331
concatenation, 330
definition of, 330
directed product, 330
morphisms of, 346, 347, 386
multiplication via, 327
versus categories, 328
Generational dynamics, 442
entropic phase maps, 468, 474
first-order information, 471
self-replicating structures, 474
versus shape dynamics, 506
Generational kinematics, 442
Generational kinematic scheme, 442
hierarchy, 490
structural increments, 469
Generational property (G), 424
Glaser, Lisa
spatiotemporal locality, 145, 150, 234
Globally hyperbolic condition, 102
foliation, 310
Grothendieck, Alexander, 22
algebraic geometry, 177, 313, 511
causal metric hypothesis (CMH), 11
hidden structure (HS), 205
relative viewpoint (RV), 22, 56, 57, 130, 133, 244, 267, 275, 354, 359, 435, 448, 459, 461, 508
topos theory, 342
Guth, Alan, 44
inflationary hypothesis, 44

H

Hamiltonian methods, 127, 383, 498
Hamiltonian operator, 7, 382
versus Lagrangian, 383
Hamilton's principle, 344, 372, 472
Hawking, Stephen, 10
causal metric hypothesis (CMH), 11
general relativity, 80, 208
metric recovery, 10, 54, 66, 74, 76, 98, 106–108, 138, 140, 141, 508
philosophy is dead, 127
Heat equation, 7, 513
Hereditary property (H), 423
Hidden structure (HS), 119, 205, 313, 461
abstract element space, 205
algebraic geometry, 178, 205
causal atoms, 205
causal path spaces, 179
contribution to discrete causal theory, 369
Grothendieck, 173, 205
kinematic schemes, 179
power spaces, 179
radical possibilities, 313
relation space, 179
star model, 297
statement of principle of, 178
topological, 204
Hierarchy, 4, 489
algebraic, 278, 333, 489
binary trees, 491
category theory, 24, 344, 492
category theory versus discrete causal theory, 328, 510
classical versus quantum, 5, 36, 59, 134
discrete causal, 489, 492
global, 491
higher-level cycles, 494
higher-level paths, 437
horizontal, 287, 490
hyperquantum theory, 122, 448, 491
iteration of structure (IS), 490
levels of structure, 4, 18, 22, 24, 30, 32, 37, 53, 55, 58, 59, 340, 347, 421

limits, 494
local, 491
multidirected sets encoding, 158
non-causal quantum correlations, 422
of finiteness conditions, 258
promotion and demotion, 493
reducing (limits), 390
subclassical, 491
top-down causation, 313
transitions, 355
vertical, 287, 490
Higher induced relations, 302
Hilbert space(s), 58, 122, 124
extrinsic, 34, 343
fields on a directed set, 31
in conventional quantum theory, 6, 7, 134, 274, 505
operators on, 347
origins of noncommutative geometry, 511
presheaf of, 343
quantization on a category, 343
rays representing states, 382, 410
Holism, 67
action principles, 344
causal metric hypothesis (CMH), 67
classical, 264–267, 274, 313
quantum, 470
shape dynamics, 316
top-down causation, 313
treating cycles via, 374
via modified binary axiom, 274, 458, 460
via power spaces, 316, 462
Holistic power spaces, 300, 316, 458
Hyperquantum theory, 122, 448, 491

I

Imaginary unit, 7, 116, 465
Impermeability theorem, 291, 314, 373, 377, 389, 393, 468
statement of, 291
Independence (chains), 165
definition of, 165
versus linear independence, 165, 509
Independence convention (IC), 165, 235, 263, 297, 461, 509
causal locality, 203
contribution to discrete causal theory, 369
definition, 166
encoding actual influence, 167
enforces objectivity, 165
facilitates relation space methods, 190
permeability problem, 290
star finiteness (SF), 215, 257
versus interpolative property, 167
versus transitivity, 181, 186, 220
Inflationary hypothesis, 43, 44, 218
causal set theory, 45
discrete causal alternatives to, 45, 60, 469, 476, 489
horizon problem, 44
phase transitions, 45, 356, 367
Interior of a multidirected set, 171
relation space recoverable from, 285
Interval finiteness (IF), 55, 212, 217, 219, 246, 247, 251, 256, 260, 261, 266, 267
arguments against, 252, 457
as an axiom of causal set theory, 66, 71, 76, 77, 138, 151, 190, 207, 263
generalization to multidirected sets, 164
imposes unjustified restrictions on global structure, 131, 256
incomparability with star finiteness, 217, 219, 243, 255
mislabeled as local finiteness, 200
not a local condition, 133, 212, 217, 253, 254, 263
not natural in discrete causal theory, 214, 230
permits pathological behavior, 213, 243, 255
possible reasons for considering, 220
prescription versus description, 129
prominent in the literature, 212, 252
reasons for discussing, 212
sprinkled causal sets, 213
technical inconvenience of, 214, 368
versus democracy of bases, 248
versus star finiteness, 212
versus topological local finiteness for the interval topology, 208
Interval topology, 207, 214, 252–256
alternative meanings, 207
analogue of Alexandrov topology, 77, 108, 208
and interval finitness, 253
fails to capture or isolate local structure, 208
interval finitness, 253
is atomic, 307
Isham, Christopher, 25, 176, 342, 369
doubts about transitivity, 184
generalized truth values, 366, 377, 439

mentions Raptis, 343
monoids of arrow fields, 184, 343
quantization on a category, 32, 33, 54, 69, 134, 146, 148, 176, 179, 184, 336, 342, 347, 510
topos theory, 25, 172, 176, 267, 329, 342, 490, 507, 510
Iteration of structure (IS), 5, 9, 36, 58, 59, 134, 274, 340, 350, 355, 365, 369, 415, 436
as a general unifying principle, 350, 351, 353
categorification/decategorification, 510
co-relative histories, 364
definition, 353
generalization of the binary axiom, 353
hierarchy, 490
hyperquantum theory, 448
in discrete causal theory, 353
kinematic preschemes, 421
moduli spaces, 352
quantum causal metric hypothesis (QCMH), 433
unifies classical and quantum theory, 353, 462
versus Ellis' hierarchy of complexity, 313
vertical hierarchy, 287
via kinematic schemes, 351
via the relative viewpoint, 354

J

Jungle-desert metaphor, 36, 146, 201

K

Kaluza and Klein, 11, 312
causal metric hypothesis (CMH), 67
matter in causal set theory, 146
matter in discrete causal theory, 201
Kinematic account(s), 39, 189, 398, 407
as co-relative kinematics, 440
as higher-level causal paths, 379, 407
as higher-level chains, 413
as sequences of co-relative histories, 434
causal set theory, 249, 343
complete, 440
equal magnitudes, 478
extraphysical information, 415, 416
organization of, 410
partial, 440
path summation, 415
phases of, 416
versus curves in shape space, 506
versus refinements, 409
via refinements, 407
Kinematic prescheme(s), 59, 259, 351, 416
category-theoretic analogy, 419
class of co-relative histories of, 416
definition of, 416
functors of, 448
irreducible, 452
iteration of structure (IS), 421
number system analogies, 449
object class of, 416
of infinite sets, 447
permeability problem, 421
possible structural deficiencies, 422
represented by multidirected sets, 416
underlying directed set of, 419
underlying multidirected set of, 419
underlying transition structure of, 419
Kinematics, 36, 59, 416
co-relative, 147
describes permissible behavior, 36, 407
discrete causal, 398
generational, 442
kinematic accounts, 407
relativistic, 92
unification with dynamics, 36, 416, 464
versus dynamics, 36, 407, 416
via kinematic schemes, 59
Kinematic scheme(s), 36, 38–40, 59, 73, 172, 207, 259, 266, 324, 415, 422
accessibility (A), 191, 365, 423
as configuration spaces of classical histories, 4, 318, 340, 342, 364
as families of kinematic accounts, 410, 411, 422, 434
as well-behaved kinematic preschemes, 416, 422, 435
category-theoretic analogy, 176, 181, 415, 490, 510
completions of, 147, 446
conceptual pitfalls, 435
congruence relations, 332
covariance, 425, 434, 475
definition of, 423
entropic phase maps, 467
evolution, 350, 415, 435
extraphysical information, 399
foliation scheme, 30, 353, 379, 444
frames of reference, 398
fundamental question of phase theory, 465
generational, 30, 353, 442, 469

generational property, 424
hereditary property, 423
hidden structure (HS), 179
hierarchy, 353, 490
higher-level cycles, 374, 416, 447
higher-level phase maps, 347
hyperquantum theory, 122
iteration of structure (IS), 350, 353
limits of, 447
McKay's example, 361
motivation for path summation, 435
motivation from Isham's program, 147, 343
object class of, 173
of causal dynamical triangulations, 504
of domains, 508
of spin foams, 502
path functionals for, 436
paths in, 407
path summation over, 365, 369, 373, 434, 436, 441
paths versus histories, 435, 442
permeability problem, 294, 421, 432, 469
phase theory, 464
positive sequential, 4, 19, 20, 24, 30, 37, 59, 139, 147, 351, 353, 364, 368, 398, 429
proper, full, originary, 417
quantum causal metric hypothesis (QCMH), 68, 433, 456
quantum origin of time, 368, 390
quantum theory, 133, 340
random graph dynamics, 488
relation functions, 437
relative viewpoint (RV), 178
represented by multidirected sets, 415
representing sequential growth, 20, 463
Sorkin-Rideout, 147, 343, 353, 361, 368, 398, 427
subschemes of, 443
toy example of, 412
underlying multidirected set of, 133, 156, 158, 165, 274, 364, 375, 432
universal, 451
versus non-structured configuration spaces, 436
versus S-matrix-like approaches, 436
Kinematic spaces, 451
definition of, 452
hierarchy, 490
hyperquantum theory, 451
Kleitman–Rothschild pathology, 139, 143, 155, 167, 188
definition of, 188

L

Lagrangian, 8, 32, 344, 465
as an infinitesimal path functional, 345, 346
classical action via, 344, 379, 465
discrete causal, 465
entropy, 379
entropy functions, 472
first-order information, 345, 471
gauge theory, 477
Hamilton's principle, 344, 372
path integration, 371, 372
relation function analogy, 345, 346, 380
Schrödinger equation, 383
versus Hamiltonian, 383
versus Lagrangian density, 344
Lagrangian methods, 127, 344, 383, 465
Linde, Andrei, 44
Local condition(s), 200, 217, 221, 222, 251, 256
formalizing via stars, 216
formalizing via topology, 203
interval finiteness is not a, 152, 212, 214, 252–254, 263, 266
more fundamental than metric notions, 201
on a multidirected set, 200, 297
star finiteness is a, 152, 217, 257
versus transitivity, 190
Local finiteness condition(s), 213, 220, 243, 254, 256, 263
convergence of path sums, 366
discreteness via, 71, 133, 459
encode meaning of discreteness, 200
interval finiteness is not a, 131, 230, 252
need for, 66, 133, 200, 214, 256
star finiteness (SF), 72, 133, 166, 173, 179, 206, 459
transitivity (TR), 220
Local properties/behavior, 197, 199, 206
absolute, 200
absurd under interval finiteness, 255
as key to new physics, 198
causal metric hypothesis (CMH), 68
causally, 215
contextual, 204
examples of, 200
interval topology poor for defining, 108, 208
local finiteness, 133

of algebraic schemes, 121
of directed sets, 108
safe under star finiteness, 257
sprinkled causal sets, 143
topology, 108
Loop quantum gravity, 4, 126, 498
a posteriori discreteness, 113, 125, 498
common classical/quantum language, 353
connections to discrete causal theory, 54, 502
conservative, 498
conventional origin of, 42
co-relative histories, 502
energetic causal sets, 502
foliation, 288
history of, 498
holonomy, 330
kinematic schemes, 502
loops (disambiguation), 513
Lorentz invariance, 356, 472, 503
noncommutative geometry, 512
path spaces, 318
path summation, 382
relation space, 502
spatial structure, 497
spin foams, 502
unproven, 129
versus causal dynamical triangulations, 504
versus causal set theory, 503
versus quantum automaton theory, 505
versus string theory, 498
Lorentz group, 13, 228, 480, 484
boosts, 480, 513
orthochronous, 480
proper orthochronous, 228, 480, 513
rotations, 480
spatiotemporal locality, 234
versus Zeeman's causality group, 78, 481
Lorentz invariance, 13, 223
aesthetic motivations, 15
approximate, 236, 356, 430, 462
eigenstate analogy, 235
group representation theory, 20
in causal set theory, 14, 111, 112, 139, 143, 155, 458
local, 13, 80, 232
of Maxwell's equations, 66
Poincaré group, 479
quantum-level accommodation, 472, 503
systematic, 14, 233, 399
versus small valences, 356, 472
versus spatiotemporal locality, 46, 60, 233
violations of, 13, 46, 60, 226, 367, 463

M

Mach, Ernst, 316
Malament, David, 10
causality conditions, 97, 101
causal metric hypothesis (CMH), 11
causal morphisms, 78
metric recovery, 10, 54, 66, 74, 75, 97, 106–108, 124, 138, 140–142, 508
path topology, 109
Poincaré group, 481
Manifold topology, 76, 78, 108, 109, 202, 224, 233
versus Alexandrov topology, 108, 208
versus path topology, 109
Markopoulou, Fotini, 32
doubts about transitivity, 183, 186
quantum causal histories, 32
relation space, 276
Martin, Keye, 11, 54
domain theory, 167, 206, 508
metric recovery, 110, 143
Matroid theory, 165, 509
gammoids, 509
independence, 165, 509
Maxwell, James Clerk, 6, 66
McKay, Brendan, 147
acyclic directed sets, 189
co-relative histories, 147
example of inequivalent transitions, 360, 412, 415
Metric (distance function), 201
continuum topology, 209
meaning of, 201
on a space of refinements, 484
on Poincaré semigroup, 486
on the Poincaré semigroup, 484
topology, 202
Metric (pseudo-Riemannian), 29, 36, 76, 78, 106, 229
almost FLRW, 81
anti-de Sitter, 81
as a tensor, 12, 29, 79
as extrinsic structure, 114, 124
causal curves, 91
conformal equivalence, 86, 89
conformal structure, 85
constant, 80
de Sitter, 81

definition of, 79
determines local notions in relativity, 201
Einstein, 354
Euclidean, 80, 82
examples of, 80
Friedman-Lemaître-Robertson-Walker (FLRW), 81
metric recovery, 142
Minkowski, 13, 47, 80, 82, 92, 233, 240
pulling back, 81
Schwarzchild, 80
signature of, 82
smooth conformal isometries, 88
versus metric space distance function, 116, 201
Metric recovery, 10, 23, 43, 53, 54, 73–75, 84, 106, 124, 133, 138, 140, 211, 214, 223, 236, 265, 317, 340, 386
background independence analogy, 125
causality conditions, 97
causal set theory, 141, 142
causal set *hauptvermutung*, 142
causal structure, 91
conformal structure, 77, 84
discrete measures, 139
discreteness, 77, 123, 459
domain theory, 143, 508
hidden structure (HS), 313
loop quantum gravity, 502
measure axiom (M), 458
modern improvements on, 109
motivates causal metric hypothesis, 66, 74, 88, 110, 456, 507
natural scale, 121
star finiteness (SF), 257, 462
statement of, 106
topological structure, 76, 107, 201
versus matter recovery, 146
Metric structure, 11, 75, 175
Minkowski spacetime, 82, 125
abstract element space over, 283
as a fixed background, 368
causality group, 481
choice of notation for, 370
covariance, 410
domains of influence, 168
frames of reference in, 451
generalized, 240
gyrogroups, 513
interval in, 92, 242
isomorphic to tangent space, 92
null cone, 85, 95
particle trajectory in, 344
path integration over, 369
Poincaré group, 248, 356, 464, 478, 479
recovery of, 112, 149, 232
relation space over, 280
relative directed set analogy, 251
relativistic causal relation, 401
relativity of simultaneity, 399
spatiotemporal locality, 233
spin foams, 503
sprinkling into, 110, 111, 186, 213, 218, 223, 252, 430, 459
twistor space over, 315
twistor theory, 505
Moduli space(s), 5
algebraic geometry, 352
iteration of structure (IS), 351
of Einstein metrics, 354
Monoid of arrow fields, 184, 336, 343
Morphism category, 295
relation space analogy, 296, 328, 415, 510
Multidirected set(s), 66, 73, 108, 130, 138, 156, 158
abstract element space as a, 280
abstract structure of a kinematic scheme, 147, 149, 274, 294, 351, 353, 369, 491
analogous to small categories, 159, 176
antichains in, 161
as a directed multigraph, 159, 276
as a directed simplicial complex, 304
as a quadruple, 158
as a small quiver, 159
as generalized order-theoretic objects, 172
associated with power spaces, 317
boundary of, 170
cardinality conditions for, 459
category of, 159
causal atomic decomposition of, 308
causal atomic resolution of, 311
causal atoms in, 305
causal directed product space over, 325
causal path algebras over, 326, 335
causal path functionals for, 346
causal path set of, 321
causal paths in, 320, 321
causal path spaces over, 323, 329
causal splice algebra over, 333
causal splice module over, 333
chains in, 161
choice of terminology for, 159
classical holism, 313
continuum topology, 209

cycles in, 374
definition of, 158
directed product algebra over, 333, 334
discrete topology, 207
domains of influence in, 167
enhanced star model as, 297
enhanced star model of, 297
fields on, 211
finiteness conditions on, 259
generalizations of, 496
generalized causal locality, 203
generalized elements of, 305
generalized interpretation of, 274
graph-theoretic viewpoint, 509
hidden structure (HS), 204
hierarchy, 490
higher-level, 148, 415, 419, 437
interior of, 170
interval topology, 207
iteration of structure (IS), 351, 353
local conditions on, 190, 200, 216
local finiteness, 263
local irregularity of, 201
mathematical conditions on, 164
mathematical versus physical properties of, 164
morphisms between pairs of, 159
path summation over, 190, 318, 340, 348, 365, 369, 373, 375, 415, 421, 434, 440, 462
path summation principle (PSP), 365
permeability problem, 275, 289, 374, 421, 462
power space over, 301
quantum causal metric hypothesis (QCMH), 132
relation functions, 345
relation space over, 275, 462
relative over a fixed base, 172, 178, 243, 320, 362, 462
star finite, 215
star infinite, 223
star model of, 179, 205, 210
star topology, 210
subobject of, 158
tangent sheaf, 159
topological local finiteness, 215
topology, 204, 206
types of, 159
Multiverse, 4, 128
kinematic schemes versus string theory, 4

N
Noncommutative geometry, 73, 134, 177, 179, 507, 510
as a field of mathematics, 507
Connes, 511
deformed special relativity, 512
foliation, 288
loop quantum gravity, 501, 512
origins of, 511
path algebras, 335, 512
physical motivation, 512
Null cone, 85, 92, 402
convergence to, 229
domains of influence, 168
relation space, 280
sprinkled causal sets, 224
terminology, 95

O
On-shell, 6
Open interval(s)
in causal sets, 155
in multidirected sets, 164
interval finiteness (IF), 212
interval topology, 207
in the real numbers, 108, 203
Operator algebra(s), 6
as a target object of a phase map, 347
loop quantum gravity, 501
noncommutative geometry, 512
on a Hilbert space, 6, 34
vertex, 333
Order refinement principle (ORP), 174, 461
Order theory, 115
accessibility (A), 427
algebraic flavor of, 327
as a structural paradigm, 126, 130, 172
contribution to discrete causal theory, 369
conventional meaning of directed set in, 156
discreteness, 71
domain theory, 507
not general enough, 114, 172
notions of size, 251
real manifolds, 118
relative directed sets, 245
Scott continuity, 204
Order topology, 108

P
Panangaden, Prakash, 11, 54

domain theory, 167, 206, 508
metric recovery, 110, 143
permeability problem, 290
pre-causal graphs, 212, 276, 329
Particle(s), 30
as nongravitational matter, 11, 28, 31
atomic accretion, 246
background independence, 367
elementary, 10, 199, 356
emergence of, 29, 33–35, 474
entropic phase maps, 474
Feynman's postulates, 371
group theory, 474
Hamilton's principle, 344
in Newtonian physics, 318
in quantum field theory, 35
interaction with spacetime, 345
jungle-desert metaphor, 36
Lagrangian, 465
massless, 92
motion of, 344, 345, 382
on a causal set, 144, 422
on a directed set, 340, 350, 448
pair production, 19
path integration, 370
phase space, 471
Poincaré group, 479, 514
random graph dynamics, 489
trajectories, 22, 32, 34–36, 341, 342, 345, 369, 378, 392, 414, 465, 474
Wheeler's geons, 35
Particle theory, 59
desert hypothesis, 198
emergence of, 481
energy crisis, 61, 199
entropic phase maps, 474
Poincaré group, 410
quantum automaton theory, 505
standard model, 5, 8, 15, 28, 33, 42, 125, 353, 463, 479, 498
Past, 168
abstract (in a kinematic scheme), 435
causal atoms, 245
causal (relativity), 95, 167
chronological, 96
definition of, 169
direct, 170, 219
interval topology, 207
maximal, 170
shrinking due to cutoff, 238
Past and/or future distinguishing conditions, 100
Past region, 375
continuum-based setting, 383
discrete causal setting, 375, 386
Past state function, 386
continuum-based setting, 382, 383
definition of, 389
discrete causal setting, 386
phase theory, 465
Past to future, 38
flow of information, 103, 187, 288, 291, 372
quantum origin of temporal direction, 38, 368, 390
Path integration, 318, 341, 348, 350, 365, 369–371, 373, 376, 377, 383, 385, 463
discrete causal analogues of, 373
equal magnitudes, 378
Feynman diagrams, 370
non-structured configuration space, 436
paths versus histories, 414, 441
permeability, 373
Schrödinger's equation, 381, 382
simplified Lagrangian, 345
Wick rotation, 370
Path summation, 6, 344
as a general approach to quantum theory, 6, 7, 51, 58, 69, 123, 127, 134, 318, 340, 462
background independence, 369, 415
causal dynamical triangulations, 504
causal set theory, 143, 145, 188, 230
cycles, 374, 495
discrete causal theory, 25, 58, 59, 134, 158, 188, 259, 285, 294, 310, 318, 341, 350, 375, 463
equal amplitudes, 377
higher-level analogues, 439
iteration of structure (IS), 350
Kleitman–Rothschild pathology, 251
ordinary quantum theory, 23, 30, 32, 37, 93, 121, 318, 341, 345, 348, 365, 369, 501
over a kinematic scheme, 434, 436, 440, 469
over a multidirected set, 190, 340, 348, 373, 415, 421, 462, 465, 470
path summation principle (PSP), 365, 415
paths versus histories, 414, 435
phase maps, 377
phase theory, 466
quantum causal metric hypothesis (QCMH), 434

quantum measure theory, 343
quantum origin of temporal direction, 368
Schrödinger equation, 381
spin foams, 499, 503
target objects, 377
versus S-matrix-like approaches, 362
Path summation principle (PSP), 365
contribution to discrete causal theory, 369
Path topology, 76, 108, 202
advantages, 109
definition, 109
Penrose, Roger, 45
abstract index notation, 80
conformal cyclic cosmology, 87, 221
doubts about inflation, 45
light cone versus null cone, 95
quantum nonlinearity, 341
spin networks, 502
twistor theory, 54, 82, 179, 267, 275, 280, 315, 505, 507
Permeability problem, 103–105, 161, 163, 190, 235, 275, 285, 288–291, 293, 294, 310, 314, 366, 373–376, 385, 386
hierarchy, 469
impermeability theorem, 291
in causal set theory, 105, 290, 373
kinematic schemes, 421, 432
path integration, 373
relation space solution, 25, 105, 291, 366, 462
Phase map(s), 59, 345, 377, 463
arbitrary, 440
as causal path functionals, 346
as formal sums, 347
as generalized action functionals, 466
as generalized functors, 467, 473
as generalized semicategory morphisms, 347
as path algebra elements, 348
background dependent context, 465
background independent context, 465
causal metric hypothesis (CMH), 464
causal path sums, 348
choice of terminology, 346
covariance, 39, 467, 475, 478
defining state functions, 386
discrete causal theory, 346, 379
emergent particles, 474
entropic, 41, 440, 442, 463, 467, 472, 474
entropic gravity, 506
equal magnitudes, 378, 466
Feynman, 345, 372, 378, 438, 465
for kinematic schemes, 441, 464, 467
fundamental question of phase theory, 465
generated by a relation function, 350, 380, 439, 440, 464
generational automorphism groups, 465
generational dynamics, 468
higher-level, 343, 347, 380, 467, 473
multiplication of (convolution), 347, 348
multiplicative, 346, 437
ordinary quantum theory, 6, 22, 32, 37, 123
path summation, 365
paths versus histories, 414
phase theory, 463
rigidity, 465, 469
second law of thermodynamics, 465
specifies dynamics, 379
support of, 348
target object of, 348, 466
uniqueness of, 416
via classical action, 379
via generational automorphism groups, 469
Phase theory, 463
fundamental question of, 465
Phase transition(s), 45
causal dynamical triangulations, 503
conventional, 45, 50
cyclic to non-cyclic, 312
graph-dynamical, 24, 45, 50, 147, 367, 488
inflationary hypothesis, 356, 367
sharpness of, 51
toy model, 45
Phenomenology, 42, 60, 367
causal set, 55, 147, 149
cosmological constant, 43, 149, 367, 463
discrete causal, 42, 60, 367, 410, 463
emergence of Poincaré group, 60, 410
inflationary hypothesis, 43, 367
Lorentz invariance, 43, 60, 367, 463
nongravitational matter, 150
phase transitions, 367
spatiotemporal locality, 43, 60, 150, 367, 463
statistical deviations, 367
Planck, Max, 7
length, 120, 199
reduced constant, 7, 345, 371, 465

scale, 198, 240, 312, 315, 367, 499, 501, 505
solution of blackbody radiation problem, 124, 131
theories do not die, 515
time, 199
units, 240
Positive sequential kinematic scheme, 427
entropic phase maps, 476
hierarchy, 490
random graph dynamics, 488
Power space(s), 173, 177, 179, 267, 298, 314, 462, 496
alternatives to, 460, 496
as generalizations of multidirected sets, 300
binary axiom (B), 458
causal atomic decomposition, 305, 308
definition of, 300
enhanced star model as, 301
examples of, 300
holistic, 264, 300, 301, 316, 458
induced, 300
path summation, 365
simplicial, 303
top-down causation, 314
twistor space, 315
via higher induced relations, 302
via splice relations, 305
Precursor object(s), 514
causal path semicategories, 332
causal relation to transitive relation, 181
causal Schrödinger-type equation, 390
classical to quantum, 379, 466
cohomology theories, 514
path functionals to quantum amplitudes, 380
path functionals to state functions, 389, 393
Poincaré semigroup, 485
Precursor symbol, 8, 9, 41, 69
representing a transition, 446
Predecessor(s), 153, 169
direct, 170, 215, 257, 390, 391
maximal, 170, 213, 225, 230, 237
Prescription versus description, 67, 129, 132, 460
background independence, 367
causal paths, 323
cycles, 460
finiteness conditions, 259
general relativity, 168
interval finiteness (IF), 219
relative directed sets, 251
transitivity (TR), 167, 181, 189
Pseudo-Euclidean geometry, 82, 251, 479, 480
Pseudo-Riemannian geometry, 78, 82, 232
domain theory, 508
emergence of, 91
ingredients of, 142
lacks natural scale, 88, 124
sprinkling, 218
versus noncommutative geometry, 511
via metric recovery, 146
Pseudo-Riemannian manifold(s), 74, 202, 231
as classical histories, 425
Einstein manifolds, 83
emergence, 77
metric recovery, 77, 236
partial differential equations on, 294, 514
phantom role (loop quantum gravity), 501
relation space over, 280
sprinkling into, 110, 143, 213, 439
structure on, 75
Pseudo-Riemannian metric, 202

Q

Quantization on a category, 69, 134, 146, 184, 336, 342, 510
Quantum amplitude(s), 341, 367, 372, 380, 383, 392
generalized, 59, 294, 350, 380, 381, 389–392, 449
in Feynman's first postulate, 371, 384
interpretation of, 474
via Feynman's path integral, 369
via path summation, 6, 7, 381, 415, 436
via phase maps, 466
Quantum causal metric hypothesis (QCMH), 433, 456
Quantum field theory, 8, 34, 80, 82, 113, 124, 134, 198, 505, 508
background dependent, 125, 368, 501
discrete, 67, 145
external symmetries, 356
Feynman diagrams, 370
particle types, 514
path integration, 42, 318, 341, 345, 370, 382
Poincaré group, 410, 479
quantum amplitudes, 369
Schrödinger-type equations, 381

S-matrix, 22, 37, 435
standard model, 28
Thiemann, 499
topological, 507
Quantum gravity, 11, 14, 16, 53, 198, 213, 219, 225, 226, 340
a posteriori discreteness, 124
causal set theory, 140, 230
curves versus paths, 318
cycles, 460
dimension reduction, 110
discreteness, 71
exotic smoothness, 16
foliation, 288
fundamental scale, 113
generalized truth values, 366, 439
histories as complete objects, 436
Isham's program, 172, 176
loop quantum gravity, 498
low-energy limit, 502
nonlinearity, 341
path summation, 382
quantum automaton theory, 505
simplicial complexes, 304
star finiteness (SF), 243
technical difficulties, 113
unorthodox approaches to, 507
Quantum information theory, 17, 54, 61, 212, 498, 504
domain theory, 507
entropy, 473
no-cloning theorem, 494
number systems, 122
quantum groups, 512
spacetime analogies, 67
Quantum theory, 7, 139
as an improvement on Newtonian physics, 500
background dependent, 341, 505
background independent, 124, 274, 368, 369, 382
basic nature of, 500
causal sets, 251
contribution to discrete causal theory, 369
correspondence principle, 6, 341
cycles, 158
discrete causal, 259, 318, 462
discreteness, 124
higher-level structure, 37, 53
Hilbert space approach, 6, 134, 274, 490, 505, 510
hyperquantum theory, 448
incompatible observables, 235, 513
information content, 34
in relation space, 145
interval finiteness (IF), 256
multidirected structure, 37
non-Archimedean, 117
non-causal quantum correlations, 302, 421
noncommutative geometry, 510
of causal sets, 139, 143, 144, 188
path summation approach, 6, 23, 25, 30, 32, 33, 37, 51, 58, 93, 127, 134, 145, 285, 294, 340, 365, 370, 415, 435, 462, 501, 504
paths versus histories, 415, 435
primacy of, 6
quantum measure theory, 343
role of complex numbers in, 121, 122, 378
Schrödinger equation, 7, 42, 381
superposition, 214, 340
uncertainty principle, 235
unification with classical theory, 53, 126, 351, 353
versus classical theory, 5, 58, 499
via iteration of structure, 351
via kinematic schemes, 59
via positive sequential kinematic scheme, 430

R
Random graph dynamics, 45, 487, 509
phase transitions, 488
Raptis, Ioannis, 145, 181, 461
incidence algebras, 184, 335, 343
locality, 202
noncommutative geometry, 512
quantum causal set theory, 145
transitivity (TR), 183
Refinements, 174, 410
binary axiom (B), 458
covariance, 478
emergence of the Poincaré group, 479
order refinement principle, 174
organization of, 410
Poincaré semigroup, 483
pointed, 482
sprinkled causal sets, 230
total orders, 174
Relation field, 32
induced by valence field, 212
on a causal set, 144, 146

Relation function(s), 8, 32, 42, 59
as generalized Lagrangians, 345, 379, 380, 383, 465
causal Schrö-dinger-type equation, 464
definition of, 346
entropy, 437
entropy (Boltzmann-type), 472
generating a phase map, 350, 379, 380, 386, 388, 389, 437
generational dynamics, 469
higher-level, 439, 474
matrix-valued (quantum information theory), 505
naïve, 472
on binary trees (hierarchy), 493
phase theory, 464
rigidity, 439
toy, 437
via antichains of relations, 469
via generational automorphism groups, 472
Relation(s), 8
addition of (refinement), 402, 405, 411
adjacent, 294
as causal paths, 322
as 1-chains, 162
as directed edges, 156
as elements of relation space, 25, 56, 211
as elements of the star model, 179, 211
as lines in a line digraph, 276
as links, 162, 315
as operators (incidence algebras), 335
as path increments, 377
between a pair of causal paths, 334
between a pair of chains, 302
between a pair of generalized elements, 305
between a pair of pentachora, 504
between a pair of relations, 26, 276, 297
between a pair of simplices, 303
between a pair of subsets of a multidirected set, 298
binary axiom (B), 265, 458
composition of, 329
connecting (concatenation product), 324
dependent, 167, 189
dual role of, 297
extending a chain, 388
fundamental, 199
generalized, 172
hidden structure (HS), 205
hierarchy, 492
higher-level, 420
in a causal set, 138, 232
in a classical history, 9, 21
in a directed set, 18, 23
in a multidirected set, 158
independence convention (IC), 166
independence of, 71, 72, 133, 166, 181, 187, 461
irreducible, 161, 315
local behavior in terms of, 200
meaning of (classical history context), 18
morphism analogy, 176
notation for, 70
of family descent, 9
reciprocal, 160
reducible, 144, 161
reflexive, 153, 154, 164
relation field, 32, 144
relation functions, 32, 346, 464
relation space, 274
representing co-relative histories, 203, 420, 464
representing direct influence, 146
representing functors of kinematic preschemes, 452
representing image-fixed co-relative histories, 420
representing transitions, 420
rigidity, 467
size of, 199
symmetry breaking, 355
timelike tangent vector analogy, 32, 159
viewing elements as (abstract element space), 280
Relation set, 158
as the source of a relation function, 345, 346, 438
at a specific element, 209, 279
causal path set generalization, 322
colored (N-ary relations), 496
continuum model, 209
dual role of, 297
of a multidirected set, 158, 205
over the real numbers, 320
power spaces, 301
relative viewpoint (RV), 275
star finiteness (SF), 215
star model, 179, 210
versus relation space, 162, 179
Relation space, 25, 56, 145, 162, 177, 178, 187, 265–267, 274, 342, 365, 462
analogy with morphism categories, 295, 328, 510
antichains in, 291, 375, 386, 468, 469

as a directed set, 157, 274
as a functor, 159, 278, 331, 421
as a line digraph, 276, 509
basic properties of, 278
better-behaved than abstract element space, 283
Blute, Panangaden, and Straßburger, 276
causal dynamical triangulations, 504
causal path spaces as generalizations of, 323
construction of, 276
definition of, 275
domain theory, 508
duality involving, 275
evaluation maps, 349
generalized semicategories, 331
Harary and Norman, 276
hidden structure (HS), 179
hierarchy, 490
higher induced relations, 302
higher-level, 415
immune to permeability problem, 105, 163, 275, 288, 290, 314, 366, 373–375
impermeability theorem, 291
incidence algebras, 336
induced relation on, 276, 297, 298, 326
irreducible, 278
Kleitman–Rothschild pathology, 252
loop quantum gravity, 502
Markopoulou, 276, 280
over a pseudo-Riemannian manifold, 280
over the real numbers, 280
partial differential equations, 514
path increments, 385
path summation over, 369, 375
permeability problem, 462
phase maps, 464
phase theory, 464
power space generalization, 298
preservation of information, 275, 285
preserves star finiteness, 278
quantum theory in, 373
relative viewpoint (RV), 133, 275
representing a space of co-relative histories, 274
scale data, 111
star topology, 211
twistor space via, 280, 315
twistor theory, 505
useful for formulating spatial notions, 294, 376
useful for studying kinematic schemes, 432
versus abstract element space, 280
versus transitivity, 167, 190
well-suited to discrete context, 280
Relative accessibility (RA), 425
Relative viewpoint (RV), 22, 56, 57, 130, 133, 267, 461
binary axiom (B), 462
classical level, 22
contribution to discrete causal theory, 369
co-relative histories, 23, 36, 57, 133, 178, 360, 435
countability (C), 459
cycles, 374
definition of, 178
domain theory, 508
Grothendieck, 173, 178
hyperquantum theory, 448
in algebraic geometry, 25, 57, 178
in discrete causal theory, 178
initial and terminal conditions, 354
permeability problem, 25
quantum level, 22
regarding causal paths, 332
relation space as an application of, 25, 133, 275
relative multidirected sets, 244
versus relationism, 178
Relativistic causal relation, 95, 96, 100, 110, 211
definition of, 95
on Minkowski spacetime, 224, 401
Poincaré group, 481
refinement of, 230, 244, 402
Rideout, David, 24
discrete general covariance, 425
fields on causal sets, 144
kinematic scheme, 343, 353, 361, 368, 398, 427, 430, 452, 490
permeability problem, 290, 373
random graph dynamics, 488
sequential growth dynamics, 24, 32, 38, 59, 146, 147, 155, 174, 188, 230, 247, 249, 310, 343, 353, 361, 415, 427, 436, 463, 478, 488
Riemann, Bernhard G.F., 139, 179
causal metric hypothesis (CMH), 11
discrete manifolds, 125, 140, 158, 203
Grothendieck-Riemann-Roch theorem, 178
Riemann hypothesis, 449
surfaces, 110
Rovelli, Carlo, 12, 54

background independence, 29
covariance, 478
curves versus paths, 94, 163, 318
diffeomorphism invariance, 74
fields on fields, 29
loop quantum gravity, 498
Lorentz invariance, 503
Mach's principle, 316
scope of general relativity, 129

S

Scale, 74
cardinality too coarse to supply, 123
causal atomic decomposition, 307
causal metric hypothesis (CMH), 110
causal set theory, 110, 142
conformal cyclic cosmology, 87
conformal equivalence, 87
conformal factor, 141
conformal geometry, 87
conformal structure, 84, 85
correspondence principle, 341
currently-accessible, 199
cycles, 375
danger of extrapolation, 130
dependence of properties on, 71
dimension reduction, 304
domain theory, 508
emergence of, 71, 84
extrinsic in pseudo-Riemannian context, 88, 97, 124
FLRW scale factor, 81
for number systems, 122
fundamental, 84, 113, 141, 179, 185, 199, 472, 474, 479, 489, 497, 505
intermediate, 233
measure axiom (M), 458
metric recovery, 133, 214, 223, 456
missing conformal factor, 74, 77, 91
natural in discrete causal context, 77, 88, 91, 110, 121, 250
nonlocality, 235
Planck, 198, 312, 367, 499, 501, 505
rigidity, 355
scope of general relativity, 83
spatial, 199, 311
temporal, 199, 311
ways of assigning, 111
Zeeman's theorem, 481
Schrödinger equation, 7, 121, 341, 381, 382, 386
as a dynamical law, 382
derivation of, 384
versus heat, diffusion, wave equations, 513
via path integration, 382, 383
Schrödinger, Erwin, 7
gruppenpest, 130
Schrödinger-type equation, 7
algebraic precursor to, 390
causal, 7, 9, 32, 36, 39, 59, 190, 275, 379, 382, 386, 390, 391, 463
definition of, 390
derivation of, 390
emergence of Einstein's equation via, 466
fractal decomposition, 392
general, 381, 513
higher-order analogues, 391
phase theory, 463
relation space, 316
toy analogue, 39
Second law of thermodynamics, 50, 463, 467
coarse-graining of phase space, 471
entropic unified theory, 506
generational measures of entropy, 469
phase maps, 440
via counting argument, 471
Sequential growth dynamics, 146, 147, 152, 155, 174, 188, 230, 328, 405, 415, 436, 463, 488
discrete general covariance, 425
iteration of structure (IS), 353
kinematic accounts, 249
quantum measure theory, 343
relative directed sets, 247
Sorkin-Rideout kinematic scheme, 427
versus generational dynamics, 310
Shannon, Claude, 6
entropy, 41, 473
information theory, 6
Shannon entropy, 473
Shape dynamics, 316, 506
conformal structure, 87
extreme holism, 316
favors spatially closed universe, 316
foliation, 288
Mach's philosophy, 316
Simplicial power space(s), 303, 496
causal dynamical triangulations, 503
Skeleton, 191, 432, 452
fails to preserve accessibility, 144, 191
star finiteness in, 216
S-matrix, 22, 37
co-relative histories analogy, 435

versus path summation, 362, 436
Smooth causal curve, 92
Smooth structure, 74, 75, 88, 175
diffeomorphism, 78
inadequate for modeling classical spacetime, 121
path topology, 109
smooth category, 78
underlying pseudo-Riemannian geometry, 79, 83
versus continuous structure, 76
Sorkin, Rafael, 11, 369
as founder of causal set theory, 54, 140, 155, 208, 229, 507
causal metric hypothesis (CMH), 11, 66, 71, 72, 110–112, 122, 139, 142, 200, 214, 256, 264, 458, 508
causal set *hauptvermutung*, 143
cosmological constant, 43, 225
diffusion, 150
discrete general covariance, 425
fields on causal sets, 144
history of causal set theory, 140
kinematic scheme, 353, 361, 368, 398, 427, 430, 452, 490
links and null segments, 315
local finiteness, 200, 459
Lorentz invariance, 227
path summation, 33
paths versus histories, 442
quantum measure theory, 69, 147, 342, 343
sequential growth dynamics, 24, 32, 38, 59, 146, 147, 155, 174, 188, 230, 247, 249, 310, 343, 353, 361, 415, 427, 436, 463, 478, 488
spatiotemporal locality, 60, 145, 150, 202, 233, 315
sprinkling, 143
transitivity (TR), 184
Spatiotemporal locality, 223, 233
approximate, 236, 462
causal set theory, 85, 145, 150
eigenstate analogy, 235
loop quantum gravity, 503
phenomenology, 367, 463
rigidity, 440
top-down causation, 316
twistor theory, 315
versus systematic Lorentz invariance, 233
Spin foams, 37
a posteriori discreteness, 498
as decorated cell complexes, 502
causal locality, 203
classical versus quantum histories, 501
co-relative histories between pairs of, 502
evolutionary processes, 341
kinematic schemes of, 37
Lorentz invariance, 232, 503
path summation, 499
random topology, 509
valences, 240, 503
versus causal dynamical triangulations, 504
versus directed sets, 502
Splice relations, 305
Sprinkled causal set(s), 110, 142, 143, 186, 208, 213, 223, 240, 243
construction of, 223, 224
cosmological constant, 225
generalization to directed sets, 236, 258
in Minkowski spacetime, 224, 231, 252, 256
indispensable for study purposes, 143, 225, 230, 243, 252, 459
induced by pseudo-Riemannian geometry, 224, 229
induced relation on, 224
kinematic schemes, 430
locally infinite behavior of, 114, 213, 218, 219, 223, 225, 229
Lorentz invariance (systematic), 111, 220, 226–228, 232, 242, 243, 479, 483
modified via cutoff, 237, 238
Moore's objections to, 229
not necessary for quantum gravity, 230
not physically realistic, 220, 230
Poisson process, 224
refinements, 230
relation functions, 439
relations nearly null, 315
relativity of simultaneity, 399
spaces of, 226, 228
spatiotemporal locality, 85, 233, 235
versus causal dynamical triangulations, 225
versus spin foams, 232, 503
versus star finite directed sets, 229
Stably causal condition, 100
Standard model, 42, 125, 463
as a quantum theory, 5, 353
background dependence, 28
gauge groups of, 15, 480, 514
governs small-scale phenomena, 5

grand unification, 17
Poincaré group, 479
recovery via discrete causal theory, 53
unification, 8
Standard model of cosmology, 44
Star finiteness (SF), 55, 56, 71, 72, 81, 133, 164, 173, 200, 206, 211, 212, 214, 216, 230, 251, 258, 259, 261, 266, 267, 353, 369, 373, 447
arguments for, 256
as a replacement for interval finiteness, 263
as an axiom for directed sets, 265, 459
contribution to discrete causal theory, 369
definition of, 215
does not impose unjustified restrictions on global structure, 257
foliation scheme, 444
for multidirected sets, 215
generational kinematic scheme, 442
in the skeleton, 216
in the transitive closure, 216
incomparability with interval finiteness, 217, 243, 255
independence convention (IC), 166
infinite intervals, 243
is a local condition, 152, 190, 217, 257
Lorentz invariance, 258
metric recovery, 257, 462
natural in discrete causal theory, 257
not preserved by abstract element space, 283
path summation, 348
positive sequential kinematic scheme, 432
preserved by relation space, 278
spatiotemporal locality, 258
star model, 179
topology, 215
transitivity (TR), 220, 221
variations on, 216
Star model, 206, 307
causal locality, 179
enhanced, 297
generalizations of, 304
versus 2-categories, 297
Star topology, 210, 215, 222, 235, 257, 462
definition of, 210
irreducible, 250
is atomic, 307
scalar fields, 211
star finiteness, 215
subbasis for, 210
variations of, 211
State function, 7
future, 382, 386
generalized, 8
in Schrödinger's equation, 121
past, 8, 382, 384, 386
phase theory, 465
versus wave function, 382
via path summation, 436
String theory, 4
AdS/CFT correspondence, 81, 87, 275
algebraic geometry, 177, 498
Atiyah-Singer index theorem, 178
background dependence of, 28
Calabi–Yau manifolds, 114
conventional origin of, 42
dominant paradigm, 113
efforts to rescue, 42
inflationary hypothesis, 44
landscape, 30, 56, 114
metric recovery, 110
multiverse, 128
noncommutative geometry, 507
path spaces, 318
phenomenological shortcomings, 463
revolutions, 179
tubular diagrams, 507
versus discrete causal theory, 497
versus loop quantum gravity, 498
vertex operator algebras, 333
wrong dimension for spacetime, 114, 500
Strongly causal condition, 101, 208
Successor(s), 153, 169
accessibility (A), 427
direct, 170, 215, 257
minimal, 170, 213, 225, 237
Suggested list (of axioms for directed sets), 68, 157, 264–266, 274, 313, 425, 457, 458, 508
statement of, 265
Superposition, 17, 340
configuration spaces, 351
hierarchy, 490
incidence algebras, 336
kinematic schemes, 434
of spin foams, 498, 502
partial differential equations, 513
quantum causal metric hypothesis (QCMH), 69, 434
Surya, Sumati, 110
causal set phenomenology, 150
metric recovery, 110

permeability problem, 290, 373
topology, 206

T
Thermodynamics, 232, 357
distinguishability, 471
Kleitman–Rothschild pathology, 188
multiplicities of states, 467
phase spaces, 274
quantum theory, 51
second law of, 50, 440, 463
sharp properties, 51
Thiemann, Thomas, 29, 54
category of points and paths, 330
curves versus paths, 94, 163, 318
general covariance, 28
loop quantum gravity, 498
noncommutative geometry, 507, 512
Thomas–Wigner rotation, 480, 483, 485, 486
gyrogroups, 513
Thomas precession, 513
Time Cube, 11, 515
Top-down causation, 302, 313
via holistic power spaces, 316
Topological continuum, 117
Topological hidden structure, 204
Topological local finiteness, 205
Topological quantum field theory, 507
Topological space(s), 204
cohomology theories, 328
paths in, 318, 330
topos theory, 342
wedge sum of, 326
Topological structure, 75, 107, 175
elementary properties, 108
homeomorphisms, 78
metric recovery, 106, 141
of the real numbers, 119
on relativistic spacetime, 76
strongly causal condition, 101
topological category, 77
Topology, 107, 201, 203
Alexandrov, 76, 108, 155, 202
algebraic, 177, 287, 318, 509
atomic, 307
basis for, 203
coarseness of, 204
continuum, 206, 209, 210
definition of, 203
discrete, 204, 206, 207
Euclidean metric, 202, 206, 209
fineness of, 204
general, 203
geometric, 143
global, 131
hidden structure (HS), 204
interval, 77, 202, 204, 206, 207
locally finite, 205
manifold, 76, 202, 208
meaning of discreteness, 71
metric, 202
path, 76, 109, 202
quotient, 209
random, 509
star, 206, 210
subbasis for, 204
topological categories, 77
Zariski, 235
Topos theory, 172, 267, 296, 342, 490, 507, 510
Toy model, 18
causal Schrödinger-type equation, 39
classical history, 18, 31
classical stochastic dynamics, 37
phase theory, 470
phase transition, 45, 47, 49, 50
Transition(s), 354
between pairs of causal sets, 147, 174
between pairs of classical histories, 24, 58
between pairs of directed sets, 354
causal set theory, 24
congruence of, 364
definition of, 354
equivalence classes of, 359, 416
higher-level, 449
interpretation of, 354
McKay's example, 360
phase theory, 470
representing co-relative histories, 354, 361, 413, 418
self-transitions, 447
S-matrix, 22
symmetry considerations, 355, 438
too specific, 356
Transitive closure, 72, 191, 405
as a functor, 278, 283
causal paths, 319
linear directed sets, 320
star finiteness in the, 216
Transitivity (TR), 12, 55, 111, 138, 179
advantages of abstaining from, 96, 146, 185, 190, 296, 368, 430, 457
arguments against, 181, 186

as an axiom of causal set theory, 138, 151, 263, 336
clutters structural picture, 167, 283
definition, multidirected case, 164
degeneracy, 192
Finkelstein's arguments against, 182
generalization to multidirected sets, 164
Isham's doubts about, 184
Kleitman–Rothschild pathology, 155, 188
local finiteness, 220
locally infinite behavior, 214
Markopoulou's doubts about, 183
obscures direct relationships, 180
of the relativistic causal relation, 94, 95
possible ways to rescue, 185
prescription versus description, 129, 132, 181, 189
Raptis' arguments against, 183
Sorkin's current view, 184
versus independent modes of influence, 167, 181, 186
versus star finiteness, 221
via transitive closure functor, 180, 191

Twistor space, 275, 315, 505
via relation space, 280, 505

Twistor theory, 179, 267, 315, 324, 505
as an alternative representation of spacetime structure, 275
correspondences, 484, 506
domain theory, 509
metric signature, 82
null lines in, 315

U

Underlying directed set, 420
definition of, 420
of positive sequential kinematic scheme, 432

Underlying multidirected set, 420
co-relative kinematics, 440
definition of, 420
hierarchy, 491
higher-level cycles, 494
of generational kinematic scheme, 442
of positive sequential kinematic scheme, 430, 432
path summation, 441

Underlying transition structure, 420
definition of, 420
of positive sequential kinematic scheme, 432

Unification, 8, 11, 53, 198, 340
grand, 17
hierarchy, 494
of classical and quantum theory, 53
of kinematics and dynamics, 36, 416, 464
of relativity and quantum theory, 126

V

Valence field, 211
future, 212
past, 212
phase maps, 379

Von Neumann entropy, 473

Von Neumann, John, 175
entropy, 473
set theory, 175

W

Wave equation, 7, 382, 513
Cauchy surfaces, 102

Weyl, Hermann, 90, 179
causal metric hypothesis (CMH), 11, 67
scientific philosophy, 128

Wheeler, John, 11
causal metric hypothesis (CMH), 11, 67
geons, 35
spacetime and matter, 68

Y

Yang–Mills theory, 126

Z

Zeeman, E.C., 78
causality group, 481
causality theorem, 481
motivation for fundamental scale, 481

Zeno's paradoxes, 11
interpolative property, 125, 167

MIX
Papier aus verantwortungsvollen Quellen
Paper from responsible sources
FSC® C105338

If you have any concerns about our products,
you can contact us on
ProductSafety@springernature.com

In case Publisher is established outside the EU,
the EU authorized representative is:
Springer Nature Customer Service Center GmbH
Europaplatz 3, 69115 Heidelberg, Germany

Printed by Libri Plureos GmbH
in Hamburg, Germany